Advances in Intelligent Systems and Computing

Volume 818

Series editor

Janusz Kacprzyk, Polish Academy of Sciences, Warsaw, Poland
e-mail: kacprzyk@ibspan.waw.pl

The series "Advances in Intelligent Systems and Computing" contains publications on theory, applications, and design methods of Intelligent Systems and Intelligent Computing. Virtually all disciplines such as engineering, natural sciences, computer and information science, ICT, economics, business, e-commerce, environment, healthcare, life science are covered. The list of topics spans all the areas of modern intelligent systems and computing such as: computational intelligence, soft computing including neural networks, fuzzy systems, evolutionary computing and the fusion of these paradigms, social intelligence, ambient intelligence, computational neuroscience, artificial life, virtual worlds and society, cognitive science and systems, Perception and Vision, DNA and immune based systems, self-organizing and adaptive systems, e-Learning and teaching, human-centered and human-centric computing, recommender systems, intelligent control, robotics and mechatronics including human-machine teaming, knowledge-based paradigms, learning paradigms, machine ethics, intelligent data analysis, knowledge management, intelligent agents, intelligent decision making and support, intelligent network security, trust management, interactive entertainment, Web intelligence and multimedia.

The publications within "Advances in Intelligent Systems and Computing" are primarily proceedings of important conferences, symposia and congresses. They cover significant recent developments in the field, both of a foundational and applicable character. An important characteristic feature of the series is the short publication time and world-wide distribution. This permits a rapid and broad dissemination of research results.

More information about this series at http://www.springer.com/series/11156

Sebastiano Bagnara · Riccardo Tartaglia
Sara Albolino · Thomas Alexander
Yushi Fujita
Editors

Proceedings of the 20th Congress of the International Ergonomics Association (IEA 2018)

Volume I: Healthcare Ergonomics

Springer

Editors
Sebastiano Bagnara
University of the Republic of San Marino
San Marino, San Marino

Riccardo Tartaglia
Centre for Clinical Risk Management and Patient Safety, Tuscany Region
Florence, Italy

Sara Albolino
Centre for Clinical Risk Management and Patient Safety, Tuscany Region
Florence, Italy

Thomas Alexander
Fraunhofer FKIE
Bonn, Nordrhein-Westfalen
Germany

Yushi Fujita
International Ergonomics Association
Tokyo, Japan

ISSN 2194-5357 ISSN 2194-5365 (electronic)
Advances in Intelligent Systems and Computing
ISBN 978-3-319-96097-5 ISBN 978-3-319-96098-2 (eBook)
https://doi.org/10.1007/978-3-319-96098-2

Library of Congress Control Number: 2018950646

This Springer imprint is published by the registered company Springer Nature Switzerland AG
The registered company address is: Gewerbestrasse 11, 6330 Cham, Switzerland

Preface

The Triennial Congress of the International Ergonomics Association is where and when a large community of scientists and practitioners interested in the fields of ergonomics/human factors meet to exchange research results and good practices, discuss them, raise questions about the state and the future of the community, and about the context where the community lives: the planet. The ergonomics/human factors community is concerned not only about its own conditions and perspectives, but also with those of people at large and the place we all live, as Neville Moray (Tatcher et al. 2018) taught us in a memorable address at the IEA Congress in Toronto more than twenty years, in 1994.

The Proceedings of an IEA Congress describes, then, the actual state of the art of the field of ergonomics/human factors and its context every three years.

In Florence, where the XX IEA Congress is taking place, there have been more than sixteen hundred (1643) abstract proposals from eighty countries from all the five continents. The accepted proposal has been about one thousand (1010), roughly, half from Europe and half from the other continents, being Asia the most numerous, followed by South America, North America, Oceania, and Africa. This Proceedings is indeed a very detailed and complete state of the art of human factors/ergonomics research and practice in about every place in the world.

All the accepted contributions are collected in the Congress Proceedings, distributed in ten volumes along with the themes in which ergonomics/human factors field is traditionally articulated and IEA Technical Committees are named:

I. Healthcare Ergonomics (ISBN 978-3-319-96097-5).
II. Safety and Health and Slips, Trips and Falls (ISBN 978-3-319-96088-3).
III. Musculoskeletal Disorders (ISBN 978-3-319-96082-1).
IV. Organizational Design and Management (ODAM), Professional Affairs, Forensic (ISBN 978-3-319-96079-1).
V. Human Simulation and Virtual Environments, Work with Computing Systems (WWCS), Process control (ISBN 978-3-319-96076-0).

VI. Transport Ergonomics and Human Factors (TEHF), Aerospace Human Factors and Ergonomics (ISBN 978-3-319-96073-9).
VII. Ergonomics in Design, Design for All, Activity Theories for Work Analysis and Design, Affective Design (ISBN 978-3-319-96070-8).
VIII. Ergonomics and Human Factors in Manufacturing, Agriculture, Building and Construction, Sustainable Development and Mining (ISBN 978-3-319-96067-8).
IX. Aging, Gender and Work, Anthropometry, Ergonomics for Children and Educational Environments (ISBN 978-3-319-96064-7).
X. Auditory and Vocal Ergonomics, Visual Ergonomics, Psychophysiology in Ergonomics, Ergonomics in Advanced Imaging (ISBN 978-3-319-96058-6).

Altogether, the contributions make apparent the diversities in culture and in the socioeconomic conditions the authors belong to. The notion of well-being, which the reference value for ergonomics/human factors is not monolithic, instead varies along with the cultural and societal differences each contributor share. Diversity is a necessary condition for a fruitful discussion and exchange of experiences, not to say for creativity, which is the "theme" of the congress.

In an era of profound transformation, called either digital (Zisman & Kenney, 2018) or the second machine age (Bnynjolfsson & McAfee, 2014), when the very notions of work, fatigue, and well-being are changing in depth, ergonomics/human factors need to be creative in order to meet the new, ever-encountered challenges. Not every contribution in the ten volumes of the Proceedings explicitly faces the problem: the need for creativity to be able to confront the new challenges. However, even the more traditional, classical papers are influenced by the new conditions.

The reader of whichever volume enters an atmosphere where there are not many well-established certainties, but instead an abundance of doubts and open questions: again, the conditions for creativity and innovative solutions.

We hope that, notwithstanding the titles of the volumes that mimic the IEA Technical Committees, some of them created about half a century ago, the XX Triennial IEA Congress Proceedings may bring readers into an atmosphere where doubts are more common than certainties, challenge to answer ever-heard questions is continuously present, and creative solutions can be often encountered.

Acknowledgment

A heartfelt thanks to Elena Beleffi, in charge of the organization committee. Her technical and scientific contribution to the organization of the conference was crucial to its success.

References

Brynjolfsson E., A, McAfee A. (2014) The second machine age. New York: Norton.

Tatcher A., Waterson P., Todd A., and Moray N. (2018) State of science: Ergonomics and global issues. Ergonomics, 61 (2), 197–213.

Zisman J., Kenney M. (2018) The next phase in digital revolution: Intelligent tools, platforms, growth, employment. Communications of ACM, 61 (2), 54–63.

Sebastiano Bagnara
Chair of the Scientific Committee, XX IEA Triennial World Congress
Riccardo Tartaglia
Chair XX IEA Triennial World Congress
Sara Albolino
Co-chair XX IEA Triennial World Congress

Organization

Organizing Committee

Riccardo Tartaglia (Chair IEA 2018)	Tuscany Region
Sara Albolino (Co-chair IEA 2018)	Tuscany Region
Giulio Arcangeli	University of Florence
Elena Beleffi	Tuscany Region
Tommaso Bellandi	Tuscany Region
Michele Bellani	Humanfactor[x]
Giuliano Benelli	University of Siena
Lina Bonapace	Macadamian Technologies, Canada
Sergio Bovenga	FNOMCeO
Antonio Chialastri	Alitalia
Vasco Giannotti	Fondazione Sicurezza in Sanità
Nicola Mucci	University of Florence
Enrico Occhipinti	University of Milan
Simone Pozzi	Deep Blue
Stavros Prineas	ErrorMed
Francesco Ranzani	Tuscany Region
Alessandra Rinaldi	University of Florence
Isabella Steffan	Design for all
Fabio Strambi	Etui Advisor for Ergonomics
Michela Tanzini	Tuscany Region
Giulio Toccafondi	Tuscany Region
Antonella Toffetti	CRF, Italy
Francesca Tosi	University of Florence
Andrea Vannucci	Agenzia Regionale di Sanità Toscana
Francesco Venneri	Azienda Sanitaria Centro Firenze

Scientific Committee

Sebastiano Bagnara (President of IEA2018 Scientific Committee)	University of San Marino, San Marino
Thomas Alexander (IEA STPC Chair)	Fraunhofer-FKIE, Germany
Walter Amado	Asociación de Ergonomía Argentina (ADEA), Argentina
Massimo Bergamasco	Scuola Superiore Sant'Anna di Pisa, Italy
Nancy Black	Association of Canadian Ergonomics (ACE), Canada
Guy André Boy	Human Systems Integration Working Group (INCOSE), France
Emilio Cadavid Guzmán	Sociedad Colombiana de Ergonomia (SCE), Colombia
Pascale Carayon	University of Wisconsin-Madison, USA
Daniela Colombini	EPM, Italy
Giovanni Costa	Clinica del Lavoro "L. Devoto," University of Milan, Italy
Teresa Cotrim	Associação Portuguesa de Ergonomia (APERGO), University of Lisbon, Portugal
Marco Depolo	University of Bologna, Italy
Takeshi Ebara	Japan Ergonomics Society (JES)/Nagoya City University Graduate School of Medical Sciences, Japan
Pierre Falzon	CNAM, France
Daniel Gopher	Israel Institute of Technology, Israel
Paulina Hernandez	ULAERGO, Chile/Sud America
Sue Hignett	Loughborough University, Design School, UK
Erik Hollnagel	University of Southern Denmark and Chief Consultant at the Centre for Quality Improvement, Denmark
Sergio Iavicoli	INAIL, Italy
Chiu-Siang Joe Lin	Ergonomics Society of Taiwan (EST), Taiwan
Waldemar Karwowski	University of Central Florida, USA
Peter Lachman	CEO ISQUA, UK
Javier Llaneza Álvarez	Asociación Española de Ergonomia (AEE), Spain
Francisco Octavio Lopez Millán	Sociedad de Ergonomistas de México, Mexico

Contents

Study on the Pause Effects During the Work Day in the Cardiovascular Load in the Line of Production of High Cadence With Heart Rate Assessment

Debora Caroline Dengo(✉), Diana Henning do Amaral, and Patrícia Rossafa Branco

Positivo University, Curitiba, Brazil
debora.dengo@hotmail.com

Abstract. **Objective:** evaluating the effects of the break during the working day in cardiovascular burden on high production line cadence with heart rate assessment. **Method:** The research has conducted in a pharmaceutical industry cosmetic products in a High Production Line cadence filling hydroalcoholic. The sample was composed of 10 volunteers, also divided into control and study groups. Breaks were carried out after 2 h of working hours and duration of 5 min. The control group (CG) accomplished standard labor gymnastic (GL) while the study group (SG) practiced compensatory labor gymnastics, whose exercise was composed with both upper limbs in total neutral position (UL), working only the lower limbs. The heart rate (HR) has measured at rest, before and after break with finger oximeter. The evaluation of cardiovascular burden (CCV) has evaluated using Apud score (1997). **Result:** It has been shown statistically that HR pos pause of SG was effective, $p < 0.05$, as the result of CG was not statistically significant, $p > 0.05$, demonstrating that the GL with total break of the upper limbs is better than GL of complete body purposing to reduce the HR in upper limbs high production line cadence. In relation to CCV, GE received a CCV 463% lower than the GC, although not significant statistically ($p > 0.05$), due to the small number of participants in the sample. **Conclusion:** This study has showed that the break had satisfactory results for a cardiovascular burden in the volunteers.

Keywords: Ergonomics · Cargo cardiovascular · Heart rate
Cargo physical labour · Pause · Occupational therapy · Energy expenditure

1 Introduction

According to the monthly follow-up of the accidental sickness benefits (benefits paid to insured persons who have suffered accidents at work or were affected by occupational diseases), granted under ICD-10 Codes (M00-M99: diseases of the musculoskeletal system and connective tissue), represented by code B-91, indicates that from January to December 2016, benefits were designed for an average of 8,378 workers per month. Totaling 100,528 benefits in a single year. The amount spent only with insured persons

S. Bagnara et al. (Eds.): IEA 2018, AISC 818, pp. 1–8, 2019.
https://doi.org/10.1007/978-3-319-96098-2_1

who suffered occupational accidents or illnesses, paid by INSS (National Social Security Institute), exceeded the 100 million reais mark.

In that same year, the companies paid to the justice of Brazil more than 3 billion reais in labor lawsuits. And for the claimants (employees who filed labor lawsuits), 22 billion reais.

It is assumed that many of these occupational characteristics disorders develop when the duration of recovery time between successive operational activities or periods of work is insufficient (Colombini et al. 2014).

This type of disturbance is common in high-cadence production line. The cadence of work would be a quantitative aspect, referring to the speed of the movements that are repeated in a unit of time. Therefore, high-speed production line is equivalent to a high-speed work. (Ministry of Health: Repetitive strain injuries and work-related musculoskeletal disorders, 2001)

If there are a large number of hours with muscle contractions intense pains may arise, requiring relaxation to restore blood circulation. A rest period should be provided so that the circulation has time to remove metabolic products accumulated within the muscles (ITIRO IDIA 2005).

In all the functions of the human body can be verified the rhythmic exchange between energy expenditures and force replacement, or in a simple way, between work and rest. The pause of work is therefore an indispensable physiological condition in the interest of maintaining production capacity (Grandjean 1998).

It can be defined as a recovery or pause period in which there is a substantial inactivity of one or more myotendine groups previously involved in the execution of work actions (Colombini et al. 2014).

According to sub-item 17.6.3 of NR 17 (Regulatory Norm that aims to establish parameters that allow the adaptation of working conditions to the psychophysiological characteristics of the workers, in order to provide maximum comfort, safety and efficient performance), on the work organization, this subitem says that to prevent psychic overload, static or dynamic muscular neck, shoulders, back, upper limbs and lower limbs industries should provide respite for rest.

In case of repetitive work it is advisable to have a recovery period every 60 min with a ratio of 5 (work): 1 (recovery); it results that the ideal distribution relation of repetitive work is 50 min of repetitive work and 10 min of recovery (Colombini et al. 2014).

The measurement of heart rate has been used by several authors to estimate the physical workload of workers in different types of work (Sullman et al. 2007).

The evaluation of physical workload was the first parameter treated by the physiology of work and remains a central issue for most workers in the world (Fiedler et al. 2008). Heart rate (HR) is an important indicator for assessing the workload, due to the innumerable knowledge acquired in human physiology and the great ease of recording the data (Edholm 1968).

For continuous work over an 8-h work day, HR should not increase continuously, and after the end of work the HR should return to normal resting values after about 15 min. HR also should decrease after recovery (pause). The maximum HR limit should occur when the mean pulse rate reaches 30 bpm (women) and 35 bpm (men) above the resting pulse. In relatively light work the pulse rate rises rapidly and remains at a height corresponding to the intensity of the work and remains constant throughout the duration of the workload. When the work ends, within a few minutes the pulse rate returns to the initial levels (Grandejean 2001).

When heavy work is done, the heart rate rises steadily while it is performed, until work is stopped or the person is in a state of exhaustion that forces them to stop.

The activities related to the type of compensatory gymnastics or pause gymnastics, interrupt the tasks of the work place, being in the middle of the day or peak of fatigue. It prevents postural addictions from activities of daily living (ADLs) and from practical life activities (PBLs). In addition to breaking the monotony of work, it is used to compensate for some overloaded structures (LIMA, 2003, apud Carvalho and Mesquita 2006).

For Couto (1995), during a workday of eight hours the heart rate should not exceed 110 bpm, because above this limit there is indication of fatigue. Souza and Minette (2002) say that as fatigue increases, work rhythm, reasoning and attention are reduced, making the worker less productive and more susceptible to errors, incidents and accidents.

Thus, this study has the objective of evaluating the effects of the pause through the heart rate of the volunteers, calculating the energy expenditure and classifying the physical workload.

2 Methods

The study was carried out in a pharmaceutical company of cosmetic products, in a production line of high cadence with sampling of 10 volunteers, with average age of 34,8 years (±5,65), being 30% of the male sample and divided into control and experimental groups. The control and experimental group had 80% of the sedentary sample, but the experimental group had 20% of the population with systemic arterial hypertension and 60% had a habit of alcoholism once a week. In both groups there is no smoker. The choice of the line was for convenience and indication of the company. The chosen line was the hydroalcoholic packaging sector, which is characterized by the mechanical packaging of the bottles, and manually empty bottles of the cardboard boxes are placed in the jars on the mats, after the mechanical containers are placed: valve, cache, finishing of the packaging, packaging of the product and placing the product ready in cardboard boxes, having several units in a box, then going through 11 points of manual intervention with 70 technical actions per minute, with static postures standing and sitting.

The routine of the volunteers involves a preparatory gymnastics before the work day and a 5-m break with part-time postural gymnastics, within an 8-h workday, with an hour apart for an extra-day lunch. The control group performed work gymnastics and the experimental group had the gymnastics replaced by compensatory gymnastics working only the lower limbs, in order to completely rest the upper limbs.

The physical workload or cardiovascular load was calculated with the HR collected through a pulse oximeter, PalmSAT 2500a, 7.4 oz, with two operating buttons of the NONIN MEDICAL, INC. Battery. The HR values were collected by the factory's physiotherapists, who wrote down a standard table at three times of the day: entrance (before pre-gym), pre-pause and post-pause. The methodology of Apud (1997) was used to calculate the cardiovascular load (CCV) of the work, corresponding to the percentage of HR during the work in relation to the maximum usable HR, through the following equation:

$$CCV = FCT - FCR \times 100/FCM - FCR$$

On what:
CCV = cardiovascular load;
FCT = working heart rate;
FCM = maximum heart rate (220 - age); and
FCR = resting heart rate.

The CCV calculation used the FCR as the FC collected at the entrance of the production line and the FCT was the pre-pause. The HRC should be measured in the same position as the person working (RIO, PIRES, 2001). The FCR of the experimental group was 93.15 bpm (±11.9) and the control group was 88.9 bpm (±6.3). In order to carry out continuous work without health risks, the CCV, according to Apud (1997), should not exceed 40% in a work shift of 8 h, and the CCV was higher than this value, it is calculated the heart rate limit (FCL). According to Minette et al. (2007), the breaks are necessary to avoid the work overload when physical loads are detected.

The table proposed by Apud (1997) was used to classify the physical workload, in which the FCT is taken into account for this classification (Table 1).

Table 1. Classification of physical workload and heart rate

Physical charge of work	FC em BPM
Very slight	<75
Light	75–100
Moderately heavy	100–125
Heavy	125–150
Heavy duty	150–175
Extremely heavy	>175

Source: Apud 1997.

The plant was contacted by SESMT (Specialized Service of Safety Engineering and Occupational Medicine) and then three meetings were scheduled with the company's board of directors. The first one was for presentation of the theme, the second for clarification and delivery of the terms of acceptance of the company and free and informed consent of the employees, and the third for training of physiotherapists who collected the data. The company's physical therapists gave the employees the free and informed consent form, assuring all their rights in relation to the study and with a form on habits of life and personal data, filled by the volunteers themselves. In addition to the meetings held in the factory, two more visits were made, to collect the terms and to collect the material with the data of the heart rate.

The data analysis was performed in a descriptive way, tabulating the data in Microsoft Office Excell® in order to identify and characterize the sample and the cardiovascular responses related to HR values. To determine the homogeneity of the groups, the Shapiro-Wilk test was used. Significance was set at $p < 0.05$.

3 Results and Discussion

The sample consisted of 10 volunteers, with a mean age of 34.8 years (±5.65), with 70% of the female sample divided equally between the control and experimental groups.

According to the normality test of Shapiro-Wilk, the values of the samples are normal, since $p < 0.05$. Regarding the results, it was statistically shown that the post-pause heart rate of the experimental group was effective, since $p < 0.05$, whereas the result of the control group was not statistically significant, since $p > 0.05$, demonstrating that gymnastics labor with total pause of MMSS is better than full-body workout with the objective of decreasing the HR in high-speed production line of MMSS. Regarding CCV, the experimental group had a CCV of 463% lower than the control group, but not statistically significant ($p > 0.05$), due to the small number of samples.

The CCV calculation indicated that on average the CCV of the experimental group was 10% and the control group of 2%, that is, no group reached 40% of the maximum load, indicating that the time of 5 min of pause already implemented in the company is effective for normalizing cardiovascular load.

The study by Carvalho (2012), who also evaluated the CCV and the pause, entitled "Ergonomic conditions of workers in chicken broiler sheds during the heating phase", analyzed the CCV to evaluate the need for pause and concluded that in this population the CCV was 50% and that the worker then needed 25 min rest for every 35 min worked, calculated using the formula Tr = [Ht x (FCT-FCL)] ÷ (FCT-FCR), where Tr of the pause and Ht - working time (in minutes), being used only when the CCV exceeds 40%.

Also, the study of Carmo (2010), "Ergonomic evaluation of the gel application operation in two forest companies", the CCV required for the manual planting operation with gel application was 41.0% in A and 40, 0% in B, concluding that every 60 min of the daily workday the worker should work 58 min and rest for another 2 min. Both studies had CCV above 40%, which according to Apud (1997) indicates physiological overload. These data show us the possible relation of CCV and the pause with regard to cardiovascular overload, indicating that the hourly pause regime is necessary when the CCV is equal to or greater than 40%.

The data of the mean working heart rate presented in Fig. 1, from both groups, allows to classify the workload of this production line as light, since it is within the range of 75 bpm to 100 bpm proposed by Apud (1997), however, classification by the OCRA checklist, described by Colombini et al. (2014) as an instrument for the mapping (identification and estimation) of biomechanical overload risk of the upper limbs, is classified in the violet area, where the exposure indexes are equal to or higher than 22.6, showing high risks. That is, in the general state of the worker there is no cardiovascular overload, but the upper limbs are exposed to repetitive strain injury through the analysis that the industry performed 6 months earlier through the Check List OCRA, in which it was classified as risk).

As the heart rate of the control group and the experimental group decreased after the pause, as shown also in Fig. 1, it indicates that the 5-min pause is effective for reducing energy expenditure, since there is no increase in cardiac output. According to Lemura and Von Duvillard (2006), as the demand for strength increases, intermediate and fast-contracting fibers are recruited by the central nervous system, increasing energy expenditure and provoking fatigue more quickly, evidenced by an increase in cardiac output (HR × systolic volume) to meet the demands of increased blood flow in the active muscles.

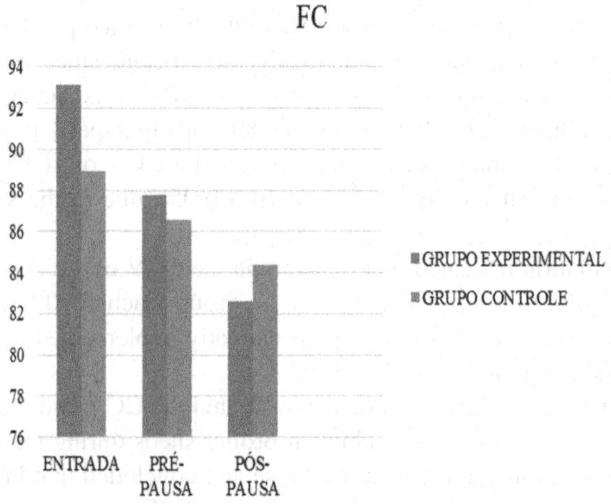

Fig. 1. Heart rate in the control group and in the experimental group, evaluated at the worker's entrance, before the pause and after the pause.

A study recently published by Souza et al. (2015) conducted research to compare HR changes in different work shifts. Regarding the parameters of HR variability, the study found that this population working in other shifts has the autonomic cardiovascular system with less favorable regulation (greater sympathetic modulation and lower parasympathetic modulation) compared to the control group (day shift workers), increasing HR during work, these changes, which are associated with a higher risk of cardiovascular disease. As our study evaluated second shift workers, we can emphasize the importance of breaks for these workers, and the need for further studies in this area for greater understanding.

4 Conclusion

This study demonstrated that the pause presented satisfactory results for the energetic recovery and cardiovascular stabilization of the volunteers, both with the work gymnastics already offered by the company and compensatory work gymnastics, since the volunteers do not have CCV above 40%. However, the experimental group with total rest of upper limbs with compensatory labor gymnastics during the 5 min presented an important reduction, indicating that in the same pause period, it is possible to reduce still more the energetic expenditure, being an alternative favorable to the higher income of the collaborator.

The data provided by the CCV, classification of Apud (1997) and Check List OCRA, showed us that more studies are necessary in this area, showing that the fact that there is no physiological, cardiac overload and less fatigue propensity, the upper limbs can be exposed to injury by repetitiveness in the same way. Another factor to consider is the need for a larger sample due to the expressive values of CCV.

References

Lemura LM, Von Duvillard SP et al (2006) Physiology of clinical exercise: application and physiological principles. Guanabara Koogan, Rio de Janeiro

Colombini D, Occhipinti E, Fanti M (2014) OCRA method: for analysis and risk prevention by repetitive movements. Publisher LTR, São Paulo

Rocha GC (2004) Work, health and ergonomics: relationship between legal and medical aspects. Juruá, Curitiba (2004)

Abrahão J (2009) Introduction to ergonomics: from practice to theory. Blucher, São Paulo

ITIRO IIDA (2005) Ergonomics: project production. 2. ed. rev. e ampl. Blucher, São Paulo

Laville A (1977) Ergonomics. EPU, Ed. of the University of São Paulo, São Paulo

Couto HA (1995) Ergonomics applied to work: the technical manual of the human machine. Ergo Editora, Belo Horizonte

Grandjean E (1998) Manual of ergonomics: adapting the work of man, 4th edn. Medical Arts, Porto Alegre

Nunes AS, Mejia DPM (2013) The importance of the work physiotherapist and his attributions within the companies: bibliographic review (2013)

Enoka RM (2000) Neuromechanical bases of kinesiology, 2nd ed. São Paulo, Manole

Francischetti AC (1990) Sedentary work: a problem for the health of the worker. UNICAMP, Campinas

Statistical Yearbook of Social Security - Secatery of Social Security Policies. Department of the General Regime of Social Security, General Coordination of Statistics, Demography and Actuary (2013). <http://previdencia.gov.br>. Accessed 30 Mar 2015

Regulatory Standard 17: Public Ministry of Labor and Employment. <http://portal.mte.gov.br/legislacao/normas-regulamentadoras-1.htm>. Accessed 29 Mar 2015

Regulatory Rule 36: Public Ministry of Labor and Employment. <http://portal.mte.gov.br/legislacao/normas-regulamentadoras-1.htm>. Accessed 29 Mar 2015

Oliveira AS, Freitas CMS et al (2006) Electromyographic evaluation of muscles of the shoulder girdle and arm during exercises with axial and rotational load. Revista Brasileira de Medicina Esporte, 12(1), São Paulo

Barnes RM (2008) Study of movements and times: design and measurement of work. Blucher, São Paulo
Ministry of Health (2001) Repetitive strain injuries (RSI) and work-related musculoskeletal diseases (DORT): series a. Norms and Technical Manuals, Brasília
Silva FRD, Gonçalves M (2003) Analysis of muscle fatigue by the amplitude of the electromyographic signal. Brazilian Journal of Science and Movement. São Paulo
Carvalho MMG et al (2014) Behavior of the EMG signal of the muscles of the upper limbs before and after workday: ergonomic analysis. XXIV Brazilian Congress of Biomedical Engineering - CBEB 2014

Body Balance Estimation in Standing and Walking Conditions Using Inertial Measurement Units and Galvanic Skin Response

Andris Freivalds(✉) and Shubo Lyu

Penn State University, University Park, PA 16802, USA
axf@psu.edu

Abstract. Falls can be detrimental, especially in the elderly. We focused on balanced and unbalanced conditions while standing and walking. Body accelerations on the sternum, right wrist, waist, left and right ankles were measured, as well as galvanic skin response (GSR). The results showed that the unbalanced conditions were significantly different from balanced conditions. Acceleration ranges increased during unbalanced conditions, along with large peaks, which represent detectable sways and trips. The correlations between accelerations and GSR were well correlated, which is promising for predicting body balance and unbalance leading to falls.

Keywords: Balance · Inertial measurement units · Galvanic skin response

1 Introduction

In 2010, an estimated 524,000,000, or 8% of the world population, were aged 65 and older. By 2050, this number is expected to double because of the trends in fertility declines and improvements in longevity [1]. As people age, their balance and agility tends to decline, and dangerous activities like postural sway and falls may occur. In fact, falls are the second leading cause of accidental deaths in the United States, with 30% of the elderly experiencing a fall annually. Among those experiencing a fall, 20–30% suffer moderate to serious injuries, with consequences ranging from a loss of independence to an increased risk of death. The average hospital cost for a fall injury is over $54,000 with over 800,000 patients hospitalized for hip/head injuries in the US alone. In 2015, these injuries amounted to total costs of over $31 billion [2]. Current research in fall prevention has utilized inertial measurement units (IMUs) which measure inertial properties (Gyro) and acceleration (ACC) as a function balance/stability. However, those measures may not give sufficient time for the person to prevent or recover from the ongoing fall [3]. The proposed project will develop wearable sensors to measure three additional measures: galvanic skin response (GSR), heart rate (HR) and blood pressure (BP). The first may give an indication of a surprise event or potential fear of falling through the arousal of the sympathetic system and concurrent drop in skin resistance, which precedes the actual fall by several hundred

S. Bagnara et al. (Eds.): IEA 2018, AISC 818, pp. 9–12, 2019.
https://doi.org/10.1007/978-3-319-96098-2_2

milliseconds; while the latter two measure the effects of a common problem in the elderly, syncope or loss of blood pressure leading to a potential fall [4] (Fig. 1). The gathered data will be analyzed using a pattern recognition algorithm developed the during research development phase. Once an imminent fall is detected, the user will be notified with a warning alarm and lights to sit down, grab onto something, and/or produce an automated call to emergency services or caregivers.

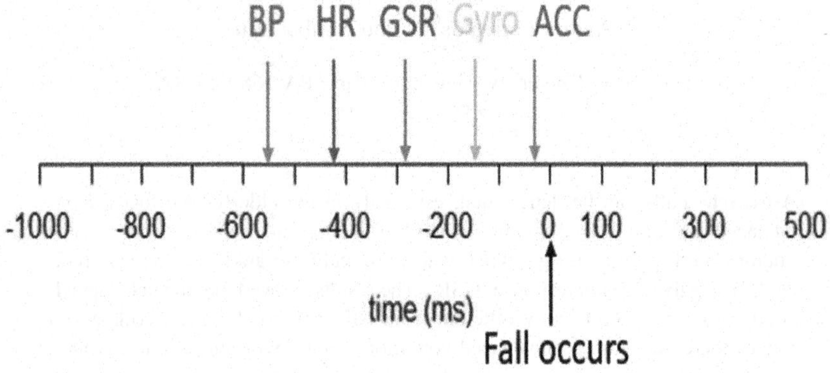

Fig. 1. Projected time frame of events leading to a fall.

2 Pilot Results

Pilot experiments, with Mbient (https://mbientlab.com/) IMUs placed on the subject head, sternum and waist while walking on a balance beam and on balance hemispheres (Fig. 2) showed significant acceleration peaks during instances of major balance

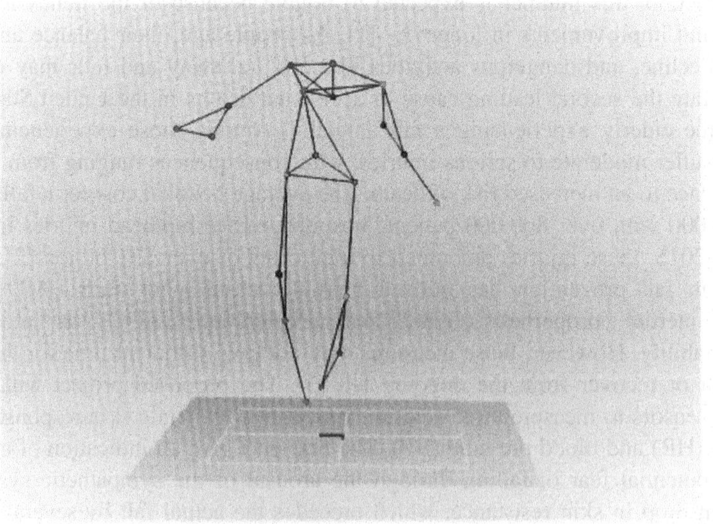

Fig. 2. Motion capture for walking on hemispheres.

deviations (Fig. 3). However, these were immediate (short time period) changes that probably would not provide sufficient warning time to prevent an imminent fall. GSR sensors (presented as conductance in Fig. 4) would increase the time window for more adequate warnings. These will be tested with additional subjects using the above balance procedures as well inducing falls (with the subjects protected with a harness system).

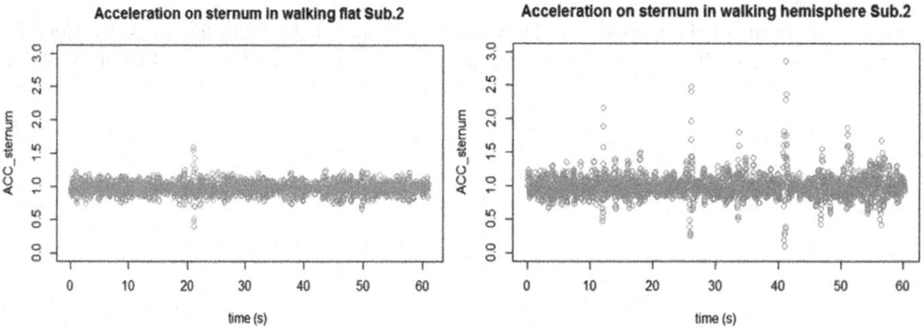

Fig. 3. Acceleration changes, flat surface vs hemispheres.

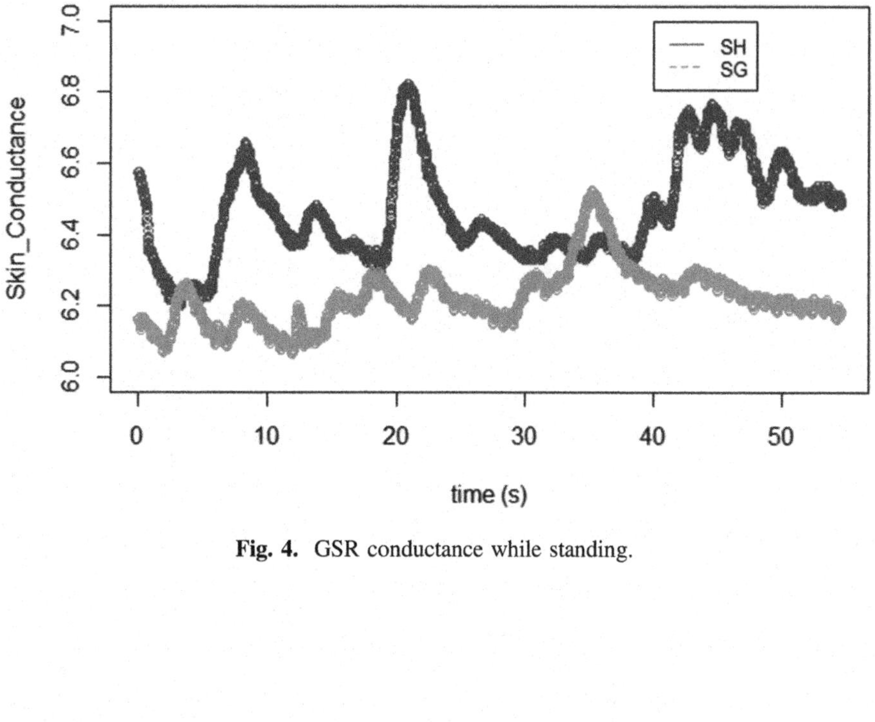

Fig. 4. GSR conductance while standing.

References

1. United Nations, World Population Prospects. http://www.un.org/esa/population/publications/wpp2008/wpp2008_highlights.pdf
2. Costs of Falls Among Older Adults (2016). http://www.cdc.gov/HomeandRecreationalSafety/Falls/fallcost.html. Accessed 02 Nov 2016
3. Amini N, Sarrafzadeh M, Vahdatpour A, Xu W (2011) Accelerometer-based on-body sensor localization for health and medical monitoring applications. Pervasive Mob Comput 7(6):746–760
4. Pirozzi G, Ferro G, Langellotto A, Della-Morte D, Galizia G, Gargiulo G, Cacciatore F, Ungar A, Abete P (2013) Syncope in the elderly: an update. J Clin Gerontol Geriatr 4:69–74

Ergonomics Intervening Cases in Hospitals for Patient Safety Improvement

Chih-Wei Lu[1(✉)], Yan-Teng Lian[1], and Hsun-Hsiang Liao[2]

[1] Department of Industrial System and Engineering Chung, Yuan Christian University, Taoyuan City, Taiwan
chwelu@cycu.edu.tw

[2] Taiwan Joint Commissions on Hospital Accreditation (TJCHA), New Taipei City, Taiwan

Abstract. Due to several medical accidents breaking out, patient safety is getting more emphasis. People have raised attention to ergonomics approach to improve patient safety. This study is to evaluate the situation of ergonomics intervening medication for patient safety improvement cases and provide the advanced human factors knowledge to medication in hospitals and to reduce the risk of human error in hospitals. The cases are collected from Taiwan Joint Commission on Hospital Accreditation (TJCHA). The cases have been reorganized by the domains view of ergonomics (physical ergonomics, cognitive ergonomics and organizational Ergonomics); graded by the level how they approach of ergonomics and quantified to do statistical analysis. The results shows that percentage distribution of ergonomics knowledge using for medication safety promotional, ensuring infection control and improving operations Safety, such as the results shows that percentage distribution of ergonomics knowledge using for medication safety promotional are physical ergonomics (5.41%), cognitive ergonomics (21.62%) and organizational Ergonomics (72.97%). Some recommendations for increasing medication safety have been presented to all levels of medical institutions for reference study.

Keywords: Ergonomics · Patient safety improvement · Hospitals

1 Introduction

1.1 Patient Safety

To maintain health and lives of patients is the mission of the workers in the hospitals. However, there are considerable risks and injuries in the medical treatment itself and medical treatment must be based on the principle of no-harm. Therefore, medical institutions and related people are obliged to reduce medical errors (system or man-made), and to eliminate or even prevent them. According to the book was entitled "To Err is Human: Building a Safer Health System", published in 1999 by the Institute of Medicine (IOM) of the National Institutes of Science, have pointed out many manmade medical events or accidents, occurring at U.S, can be prevented before they happen. However, the medical events have made great impact on the healthcare industry and in terms of money and the number of casualties [1].

S. Bagnara et al. (Eds.): IEA 2018, AISC 818, pp. 13–23, 2019.
https://doi.org/10.1007/978-3-319-96098-2_3

Since 2002, several major medical error events have occurred in Taiwan, hospitals have paid more attention to the patient safety [2]. Patient safety is no longer only focused on hospitals to improve patient safety actions, but also has expanded the sense of experienced by the people about to build and evaluate patient safety culture. In order to reduce human errors in the hospital and the interaction errors between the human and the environment, the relationship between humans & environment and people & objects must be considered. Therefore, it is also linked to field of ergonomics or human factors engineering.

1.2 Ergonomics (or Human Factors)

Based on website of International Ergonomics Association (IEA), the definition of ergonomics are: Ergonomics (or human factors) is the scientific discipline concerned with the accepting interactions between humans and other elements of a system, and the profession that applies theory, principles, data and methods to design to optimize human well-being and overall system performance [3].

To improve the performance of patient safety, the factors of people and fit the job to the people have been considered to establish a human-center and patient-center healthcare environment, and deeply implant the patient safety culture of healthcare people and based on patient center principle to improves the quality of healthcare.

2 Methods

2.1 Source of Samples

Due to limited to research time and research resources, and the study was based on the final reports of the 14 Healthcare Quality Improvement Circle (HQIC) Council. The TJCHA focuses on the goal of the Patient Safety Works in 2014 such as: Goal One: Raising prescription or medicine safety; Goal Two: Implement Infection Control; Goal Three: Improve surgical safety and discuss the current status safety culture of hospital patient, and the relevance of patient safety culture and ergonomics or human factors engineering.

2.2 Process of the Research

The process of the research are: (1) setting goals, (2) collecting case of improvement cases, (3) putting cases into three major ergonomics types (Physical Ergonomics, Cognitive Ergonomics, Organization Ergonomics), (4) setting into the subgroup of the three major group, (4) statistic analysis. And we have combined with the Ergonomics checklist, the main research methods, and using Plato's analysis method, data quantification and statistical analysis, to analyze patient safety improvement cases in Taiwan hospitals.

2.3 Ergonomics Checklist

Ergonomics or Human factors engineering means that human factors are taken into account and in the interaction process between humans and the environment, systems or machinery and equipment so humans and other components can properly cooperate to produce the best performances. According to the definition of International Ergonomics Association (IEA), the main areas include physical ergonomics, cognitive ergonomics, organizational ergonomics, and other related fields which are not completely independent, but complementary to actual job needs.

The key of checklists is to design easily to understand in order to help users overcome the limitations of human short-term memory to make sure the core investigation is thorough and complete. Based on three major classifications, the detail items have been labeled in each category, coded as: A: physical ergonomics, B: cognitive ergonomics, C: organizational ergonomics shown in the Table 1 shows.

Then, detailed project factors are coded. For example, if the improvement suggestion is related to the organization of work in the organizational human factors project, the job organization code is C-1. The analogy is similar to that shown in Table 2.

Table 1. The ergonomics classification and code

Ergonomics classification	Code
Physical ergonomics,	A
Cognitive ergonomics,	B
Organizational ergonomics	C

Table 2. The category of detail items

Ergonomics major fields	Detail Items (sub-item)	Code
Physical ergonomics, cognitive ergonomics,	Operation space	A-1
	Physical demand	A-2
	Work environment	A-3
Physical ergonomics, cognitive ergonomics,	Mental demand	B-1
	Human-computer interaction	B-2
	Work stress	B-3
Physical ergonomics,	Work organization	C-1
	Task design	C-2
	Occupational background	C-3
	Staff characteristics	C-4

In the results report, the improvement suggestions have been rated according to the degree of ergonomics skills. The higher score obtained, based on the degree of relevance to ergonomics, the score is from 0 to 3. A score of 0 indicates that the improvement suggestion is not associated with the human factor principle; a score of 1 indicates that the improvement suggestion is some related to the human factor, but the correlation is low; a score of 2 indicates that the improvement suggestion is middle related to the human factor; the score 3 means the highest correlation.

The scoring processes have done by the ergonomics research term. Based on the improvement suggestion of the results report, the term has decided a best match detail items for each the improvement suggestion. According to the improvement suggestion, the percentage of the matching ergonomics principle that should be setting in the detailed project checklist and scored: If the suggestion does not correspond to any item on the ergonomics checklist, the score is 0; if the improvement suggestion corresponds to 30% or less of the checklist, the score is 1; if the improvement suggestion corresponds to more than 30% to less than 60%, the score is 2; if the improvement suggestion corresponds to more than 60% of the checklist, the score is 3. The actual results have been determined by the research team together with the professional opinion.

2.4 Data Analysis

On the fourteenth HQIC final reports, the improvement suggestions have been written description. First of all, in order to make a large number of cases can be integrated and analyzed, categorized improvement suggestions have been quantified, and the frequency analysis or numerical analysis performed by the statistical software Minitab. The research team has established standard operating procedures to quantify the data:

Step 1. Collecting the improvement cases and records;
Step 2. Rating the improvement suggestions by using the checklist;
Step 3. Quantify the improvement suggestion case.
Step 4. Use statistical methods to find out the relationship between patient safety goals and human factors or ergonomics.

In this study, the statistical analysis has divided into several parts, the three parts are the main points of statistical analysis as follow:

(1). Collect the number of major group the improvement suggestion for the number of major areas and do descriptive statistics. Then, divided the subgroup items in the three major areas, discuss the number of occurrences, and analyze the reasons.
(2). the statistics of the number of detail items for the three major fields are collected, and the results of the collection are analyzed. The Platonic analysis of is used and conducted to discuss the first 80% of the detail items, and the number of fields of each detail item can also be realized.
(3). Based on the scores obtained from the Checklist, the Kruskal-Wallis Test was used to analyze the selected patient safety missions because the data types collected id not meet the normality assumptions. The goal is whether there are significant differences between using the three ergonomics main fields.

3 Results

3.1 HQIC Improvement Suggestion Analysis

In the study, we have collected 100 HQIC improvement suggestions. To match the healthcare quality and patient safety mission, there are 37 (prescription safety improvement), 23 (infection control measures), and 40 (surgical safety improvements).

In the analysis of 100 improvement suggestions in the current stage, according to the main categories of ergonomics, 11 are classified in the physical ergonomics field, 22 are classified in cognitive ergonomics, and 67 are classified in the organization ergonomics, and among them, the physical ergonomics has been accounted for 11%, the cognitive ergonomics for 22%, and the organization ergonomics for 67%.

3.2 Prescription Safety Improvement

Regard to the prescription safety promotion, the number of ergonomics classifications is shown in Table 3. From the improvement suggestion, written in the results report, the most common are the organization of ergonomics, accounting for approximately 73.0%, cognitive ergonomics accounted for approximately 21.6%, the least are the physical ergonomics, accounted for approximately 5.4%.

The detail item distribution, has shown in Fig. 2. The physical ergonomics section, (operation space, physical demand, work environment) the operating space and physical demand were classified one time and physical demand is zero; in cognitive ergonomics section (mental demand, human-computer interaction, work stress), the number of machine interactions was 1 and the work stress was 0; in the organizational ergonomics (work organization, task design, occupational background, staff characteristics, the number of task design categories was 22, the work organization was 5, and the occupational background and staff characteristics were 0.

Table 3. The percentage of ergonomics field for prescription safety improvement

Ergonomics	Prescription safety improvement (N = 37)		
Classification	Physical	Cognitive	Organizational
Number	2	8	27
%	5.41%	21.62%	72.97%

Based on the K-W test, the resulting scores has shown in Table 4, show a significant difference in scores obtained from different ergonomics classifications (p-value = 0.022) (Fig. 1).

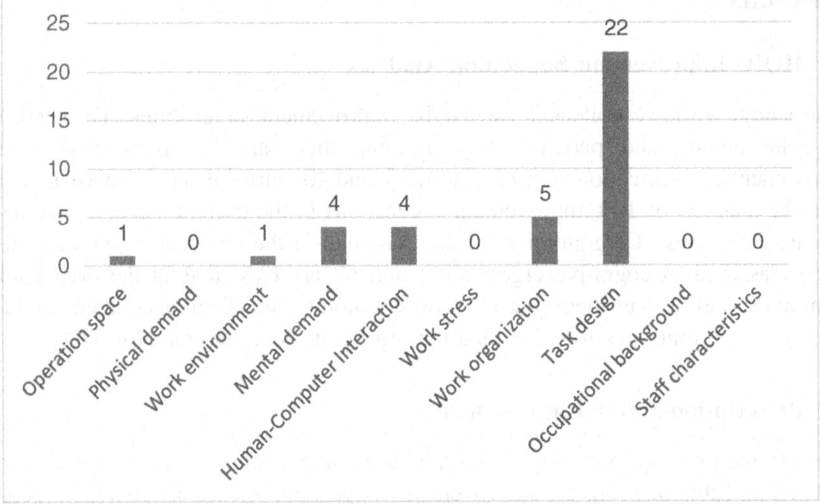

Fig. 1. The detail items distribution of suggestion of prescription safety improvement

Table 4. K-W Test of Prescription safety improvement

Major classification	Sample size	Median	Average grade
Physical	2	2.00	24.0
Cognitive	8	2.00	27.8
Organization	27	1.00	16.0
Summery	37		19.0

H = 7.68; DF = 2 P = 0.022

3.3 Infection Control Implement

About the infection control Implement, the number of ergonomics classifications is shown in Table 4. From the improvement suggestion, written in the results report, the most common are the organization of ergonomics, accounting for approximately 65.2% cognitive ergonomics accounted for approximately 21.7%, the least are the physical ergonomics, accounted for approximately 13.0% (Table 5).

The detail item distribution, has shown in Fig. 2. The physical ergonomics section, (operation space, physical demand, work environment) the work environment and physical demand were classified 0 time; in cognitive ergonomics section (mental demand, human-computer interaction, work stress), the number of machine interactions and the work stress were both 0; in the organizational ergonomics (work organization, task design, occupational background, staff characteristics, the number of occupational background, staff characteristics and work organization was 0, the task or job design was 15, and the occupational background and staff characteristics were 0. Based on the K-W test, the resulting scores has shown in Table 6, show null significant difference in scores obtained from different ergonomics classifications (p-value = 0.186).

Table 5. The percentage of ergonomics field for infection control Implement

Ergonomics	Prescription safety improvement (N = 37)		
Classification	Physical	Cognitive	Organizational
Number	3	5	15
%	13.04%	21.74%	65.22%

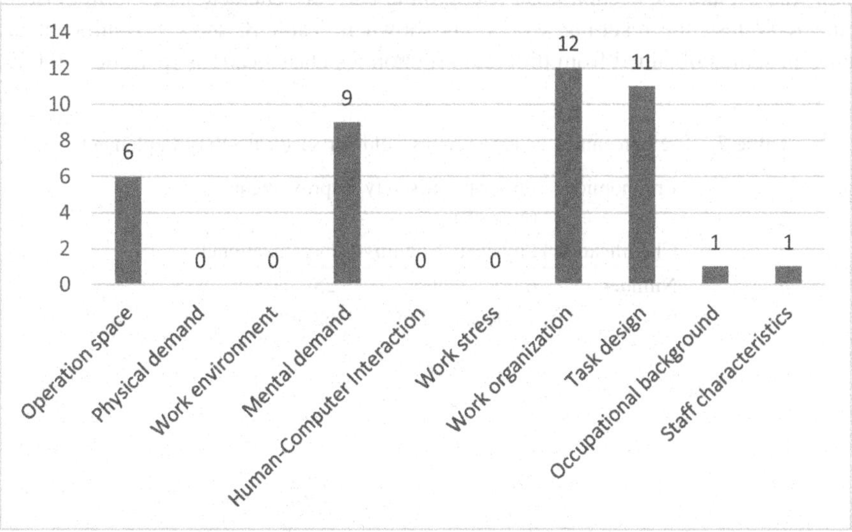

Fig. 2. The detail items distribution of suggestion of infection control Implement

Table 6. The K-W Test of infection control Implement

Major classification	Sample size	Median	Average Grade
Physical	3	1.00	10.0
Cognitive	5	2.00	16.9
Organization	15	1.00	10.8
Summary	23		12.0

H = 3.37; DF = 2 P = 0.186

3.4 Surgical Safety Implement

Regard to the surgical safety Implement, the number of ergonomics classifications is shown in Table 7. From the improvement suggestion, written in the results report, the most common are the organization of ergonomics, accounting for approximately 62.5%, cognitive ergonomics accounted for approximately 22.5%, the least are the physical ergonomics, accounted for approximately 15.0%.

The detail item distribution, has shown in Fig. 3. The physical ergonomics section, (operation space, physical demand, work environment) the operation space were classified 6 times, work environment and physical demand were 0; in cognitive ergonomics section (mental demand, human-computer interaction, work stress), the number of mental demand was 9, the number of machine interactions and the work stress were both 0; in the organizational ergonomics (work organization, task design, occupational background, staff characteristics) the number of work organization was 12, occupational background and staff characteristics both were 1, the task or job design was 11, and the occupational background and staff characteristics were 0. Based on the K-W test, the resulting scores has shown in Table 8, show a significant difference in scores obtained from different ergonomics classifications (p-value = 0.005).

Table 7. The percentage of ergonomics field for surgical safety Implement

Ergonomics	Prescription safety improvement (N = 40)		
Classification	Physical	Cognitive	Organizational
Number	6	9	25
%	15.00%	22.5%	62.50%

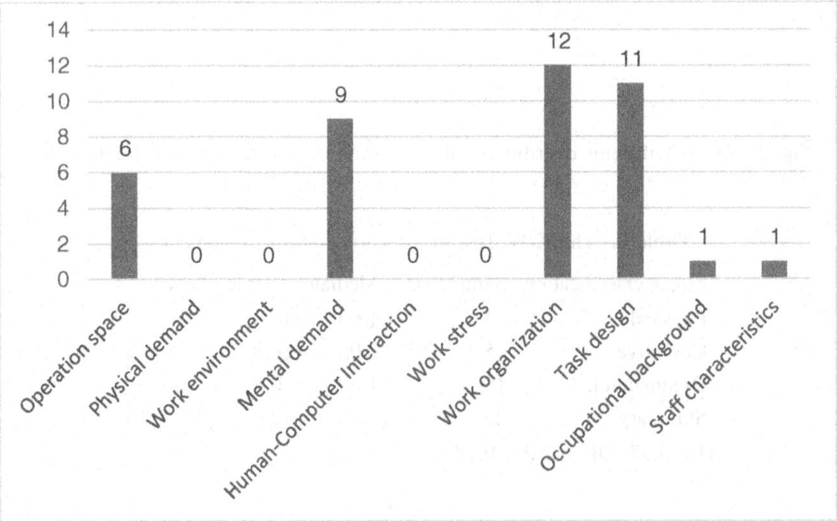

Fig. 3. The detail items distribution of suggestion of surgical safety Implement

Table 8. The K-W Test of surgical safety Implement

Major classification	Sample size	Median	Average Grade
Physical	6	2.00	34.0
Cognitive	9	1.00	22.1
Organization	25	1.00	16.7
Summary	40		20.5

H = 10.80; DF = 2; P = 0.005

3.5 Plato Analysis for Improvement Suggestions

In the section, based on the healthcare quality and patient safety goal and the checklist to classify the HQIC improvement suggestion, we have used the Plato Analysis Method to analyze the detail items of improvement suggestions. The main categories of the three major fields have already been coded are: A, B, and C, which are physical, cognitive, and organization ergonomics. Next, the detail item numbers of each major category are searched, and the number of appearances of each detailed item is investigated, and the percentage of each detailed item is cumulatively increased to nearly 80% from the order of occurrence of the largest number of occurrences to the occurrence of the minimum number of times.

Regard to prescription safety improvement, the percentage of each detailed item is cumulatively increased from C-2 (22, task design), C-1 (5, work organization), B-1 (4, mental demand) to 83.78%, near 80%, so the number of detail item to cumulate near 80% is 3.

About infection control implement, the percentage of each detailed item is cumulatively increased from C-2 (15, task design), B-1 (5, mental demand) to 86.96%, near 80%, so the number of detail item to cumulate near 80% is 2.

About surgical safety improvement, the percentage of each detailed item is cumulatively increased from C-1 (5, work organization) C-2 (11, task or job design), B-1 (9, mental demand)to 80.00%, near 80%, so the number of detail item to cumulate near 80% is 3.

4 Discussion

On the prescription safety improvement, according to the data & statistical analysis results, the main categories of ergonomics are organized to organization ergonomics to appear most frequently, the scores obtained from the KW test. As shown in Table 4, there are significant differences (p-value = 0.022) in the scores obtained from different ergonomics classifications. The organizational ergonomics is compared with physical ergonomics such as the cognitive engineering implementation suggestion, the scores obtained are low, indicating that the organization principle has been widely used by healthcare industrials at all levels, but the technical level was still not as familiar as the other two ergonomics fields due to the project; Plato analysis results are shown a total of C-2 (work design), C-1 (work organization), and B-1 (psychological needs), which are the items with the highest relevance to human factors engineering in the proposed improvement.

About the implementation of infection control, according to the data analysis, the main categories of ergonomics is organization to appear most frequently. The resulting scores, based on the K-W test, as shown in Table 6, show that there are no significant differences in the scores obtained from different ergonomics classifications (p-value = 0.186). The results of Plato analysis has shown C-2 (work design) and B-1 (psychological needs), which are the items with the highest relevance to ergonomics in the proposed improvement.

In terms of surgical safety improvement, according to the data analysis, the mainly classified ergonomics field is organization ergonomics. Based on the KW test, there are significant differences in the scores obtained from different ergonomics classifications (p-value = 0.005). The organization ergonomics is compared with the entity ergonomics such as the cognitive engineering, the scores obtained are lower, indicating that the organization ergonomics principle has been widely used in healthcare workers in hospitals at all levels, but the technical level was still not as familiar as the other two ergonomics fields. Plato analysis results has shown C-1 (work organization), C-2 (job design), and B-1 (psychological needs) which are the items with the highest relevance to ergonomics in the proposed improvement.

5 Results

One of the key goals of ergonomics is to reduce human error and accidents, thereby improving the safety of work and life. From long time, the medical safety has depended on the personal responsibility of healthcare worker, centered on physicians; healthcare operators provide services based on their professional knowledge and technology for patients with different diseases and symptoms. If events or accidents have happened, it is regarded as personal responsibility. But, due to the evolution of the system and times, with the new development of science and technology and the advancement of medical technology, healthcare is no longer a matter of individuals but systematically. Instead, in a complex healthcare system, it is composed of several different categories of "people", including professional and non-professional people.

In the related processes, using the "matter" of various medicines and materials, the complexities formed by the collaboration and cooperation of the hundreds of thousands of beds and thousands of hospitals combined with more than tens of specialist and outpatient clinics.

Based on the management system, the complex system and cooperation to service is "environment". James Reason believes that the medical malpractice occurred because in the course of the accident, the environment were actually some potential failures (system deficiencies). In such work environment, due to the failure of active response, an accident was detonated. So during reviewing the cause of the accident, especially when the active failure was a human factor, human error should not be considered as the only factor, but should focus on the causes of system failure caused by human error.

Due to the improvement suggestions collected in this study, all from hospitals volatile reporting the cases to the TJCHA. There are still many unknown medical errors that have not yet occurred, as Bird has talked the iceberg effect and mentioned: "Serious injuries are only at the tip of the iceberg, the serial thing that should be noticed is the iceberg that is hidden under the surface of the water. It is also the most often overlooked."

Therefore, the ergonomics project done by the checklist analysis can be used to prevent medical malpractice that has not yet occurred which was based on improvement suggestion case studies as a reference. Patient safety is the core value of healthcare quality. In the future, more information needs to be collected to conduct analysis and exploration to create a good healthcare environment.

Acknowledgement. The study term thanks for the help and award from the hospitals and the Taiwan Joint Commission on Hospital Accreditation (TJCHA). Moreover, are also thanks funding from the MOST of Taiwan (Ministry of Science and Technology). The funding codes are 105-2221-E-033-038- and 106-2221-E-033-038-MY3.

References

1. Kohn LT, Corrigan JM, Donaldson MS (eds). To Err is human: building a safer health system. Institute of Medicine (US) Committee on Quality of Health Care in America; National Academies Press (US), Washington (DC). https://www.ncbi.nlm.nih.gov/pubmed/25077248
2. Chihwei L (2014) Special experiences- using ergonomics or human factor engineering in medical care systems. J Healthc Qual 8(6):17–20
3. Definition and Domains of Ergonomics. http://www.iea.cc/whats/index.html
4. Bird FE, Germain GL (1996) Practical loss control leadership, Det Norske Verita. Loganville, GA. LNCS Homepage. http://www.springer.com/lncs. Accessed 21 Nov 2016
5. Ben-Tzion K, Roger B (2010) Macroergonomics and patient safety: the impact of levels on theory measurement, analysis and intervention in patient safety research. Appl Ergon 41 (5):674–681
6. Bridger RS (2008) Human error, accidents, and safety. Introduction to ergonomics, 607–655
7. Gavriel S (2012) Human factors and ergonomics in health care. Handbook of human factors and ergonomics, 1575–1589
8. Blackman HS, Gertman DI, Boring RL (2008) Human error quantification using performance shaping factors in the spar-h method. In: Proceedings of the human factors and ergonomics society annual meeting, vol 52, p 1733
9. Kleiner BM (2006) Macroergonomics: analysis and design of work systems. Appl Ergon 37 (2006):81–89

Estimation Accuracy of Step Length by Acceleration Signals: Comparison Among Three Different Sensor Locations

Tomoya Ueda[1(✉)], Naoto Takayanagi[1], Yoshiyuki Kobayashi[2], Motoki Sudo[1], Hiroyasu Miwa[2], Hiroaki Hobara[2], Satoru Hashizume[2], Kanako Nakajima[2], Yoshifumi Niki[1], and Masaaki Mochimaru[2]

[1] Tokyo Research Laboratories, Kao Corporation, 2-1-3 Bunka, Sumida-ku, Tokyo, Japan
ueda.tomoya@kao.com

[2] Digital Human Research Group, Human Informatics Research Institute, National Institute of Advanced Industrial Science and Technology, Waterfront 3F, 2-3-26, Aomi, Koto-ku, Tokyo, Japan

Abstract. Methods to estimate step length using accelerometers are gaining attention in recent times. However, the influence of the sensor location on the accuracy of step length estimation is still unknown for models fabricated in a uniform manner. Therefore, the purpose of this study was to compare the accuracy of step length estimations among the following three body parts: ankle, pelvis, and wrist. Ten time-normalized acceleration signals from one gait cycle were obtained from 247 healthy adults aged 20 to 77 while walking barefoot at a comfortable, self-selected speed. Linear multiple regression analyses with leave-one-participant-out cross validation technique were used to build the algorithms. The absolute value of mean error for each participant (AME) was computed to compare the accuracies among the body parts. Mean (standard deviation) values of AME for each part were as follows: ankle, 2.66 (2.24) cm; pelvis, 3.09 (2.39) cm; and wrist, 4.05 (3.01) cm. Statistical analyses revealed significant differences for the ankle–wrist and pelvis–wrist estimations. We found that step length can be estimated from the acceleration signal of the ankle or pelvis with almost the same accuracy (approximately 3 cm of average error between participants). Also, estimation of step lengths with the acceleration signals obtained from the wrist needs to be conducted more carefully than those obtained from the ankle or pelvis.

Keywords: Step length estimation · Acceleration signal · Sensor location

1 Introduction

Step length is one of the important indicators associated with the decline of an individual's functional abilities [1, 2]. Traditionally, step length has been measured in the clinical or laboratory setting by using a motion capture system [3] or a sheet-type pressure sensor [4]. Recently, step lengths have also been obtained during normal daily

S. Bagnara et al. (Eds.): IEA 2018, AISC 818, pp. 24–30, 2019.
https://doi.org/10.1007/978-3-319-96098-2_4

activities using wearable sensors [5–8]. However, these studies only investigated the accuracy of step length estimation using the acceleration signals obtained from sensor (s) attached at limited locations of the body. Furthermore, the methodologies and algorithms used to create the estimation models are different among the previous studies. Therefore, it is difficult to understand the extent to which the sensor locations affect the accuracy of step length estimations when the models are built in a uniform manner.

The purpose of this study was to compare the accuracy of step length estimations among the following three body parts: ankle (lateral malleolus), pelvis (anterior superior iliac spine), and wrist (stylion radiale). These locations were decided from the availability of actual wearable products in the market and the literature regarding the activity monitoring technologies [5, 9, 10]. In the present study, measured step length (MSL) during walking was defined as the absolute anterior/posterior length between the right and left heel markers at heel contact events. Therefore, we hypothesized that the acceleration signals obtained from the ankle could estimate the step length most accurately because the ankle is the closest location to the heel marker from which this parameter is calculated.

2 Methods

2.1 Participants

In this study, gait data of 247 healthy adults (114 males and 133 females) aged 20 to 77, which were obtained from the National Institute of Advanced Industrial Science and Technology (AIST) gait database [11], were analyzed. The demographic information of the participants are presented in Table 1. All of the participants were able to walk independently without assistive devices (e.g., canes, crutches, or orthotic devices), had normal or corrected-to-normal vision, had no history of neuromuscular disease, and lived independently in the local community. Those who had trauma or any orthopedic diseases were excluded. The experimental protocol was approved by the local institutional review board, and all the participants gave their written informed consent before participating.

Table 1. Mean (SD) demographic data of the participants.

	Age: years	Height: cm	Weight: kg	Step length: cm
All	50.3(18.8)	162.3(8.5)	59.2(10.4)	66.0(5.9)
Females	49.7(18.3)	157.3(6.1)	53.7(7.8)	65.1(5.9)
Males	50.9(19.4)	168.1(7.0)	65.6(9.3)	66.9(5.8)

2.2 Data Collection

Measurements were performed in a room with a straight 10-m path on which the participants could walk. Three-dimensional (3D) positional data were obtained using reflective markers and a 3D motion capture system (VICON MX, VICON, Oxford,

UK) using a sampling frequency of 200 Hz. A total of 57 infrared reflective markers were attached to the subjects by one of three experts (with more than 10 years of experience) in accordance with the guidelines of the Visual 3D software (C-Motion Inc., Rockville, MD, US). We also recorded ground reaction forces (GRFs) using six force plates (BP400600-2000PT × 4, and BP400600-1000PT × 2, AMTI, Watertown, MA, US) that were sampled at 1 kHz intervals.

The participants were asked to walk barefoot at a comfortable, self-selected speed. Before the walking trials, the positions of the markers were recorded while the participants stood stationary. The participants were then allowed sufficient practice walks to ensure a natural gait. After the practice, ten successful trials in which each participant properly stepped on the force plates were recorded and further analyzed.

2.3 Data Analysis

The raw data were digitally filtered using a fourth-order Butterworth filter with zero lag and cut-off frequencies of 10 Hz for the positional data and 56 Hz for the GRFs. The acceleration signals were represented by a global coordinate system and were calculated using the second derivative method. Gravitational acceleration was added to the vertical component of the calculated acceleration. The acceleration signals were then time-normalized by the gait cycle duration determined from the force plate data and divided into 101 variables ranging from 0 to 100%. Thus, we obtained a dataset of 909 variables from each trial (i.e., 101 time points, 3 body locations, and 3 axes). Lowpass filtering and calculation of the variables were performed using the Visual 3D software (C-Motion, US). MSL during walking was defined as the absolute anterior/posterior length between the right and left heel markers at heel contact events. This parameter was extracted from the gait cycle at the center of a 10-m path.

2.4 Statistics

In this study, a stepwise multiple linear regression analysis, which is one of the fundamental statistical methods, was used to build the models to estimate the step lengths from the acceleration of each body location. The independent variables used to build the models were the following 303 variables: 101 time-normalized acceleration values from 3 planes for each body location. The dependent variables were the step lengths of each trial. The inclusion and exclusion criteria for the stepwise approach were set as follows: inclusion for $p < 0.05$ and exclusion for $p > 0.10$. Further, the leave-one-subject-out cross validation technique was used to build the model and to obtain the estimated step length. Therefore, the estimated step length of each trial or participant was obtained from his/her own acceleration data and the model, which was built from the dataset without his/her own data.

Furthermore, the absolute value of with-in participant mean of errors (AME) between MSL and estimated step length (ESL) was computed by the following equation.

$$\mathrm{AME\,(cm)} = \left|\left(\sum_{1}^{10}\mathrm{MSL} - \sum_{1}^{10}\mathrm{ESL}\right)/10\right| \quad (1)$$

These parameters were used to assess the accuracy of the step length estimations among the body locations. Further, a one-way analysis of variance (ANOVA) for the repeated measure was conducted on AME to assess the main effect of the sensor locations (on the body) on step length estimation. Bonferroni's multiple comparison and point-biserial correlation coefficient (r) were used for post-hoc analysis if a significant main effect was observed. Because of the large number of data (247 participants), the differences in the means were considered statistically significant if the p values were less than 0.05, the partial eta-squared (η^2) values were greater than 0.06, and the r values were greater than 0.30, thus indicating a medium effect size [12]. All of the statistical analyses were performed using the SPSS statistical software package (IBM SPSS Statistics Version 23, SPSS Inc., Chicago, IL, USA).

3 Results

Figure 1 shows the mean (standard deviation) of AME obtained from each body location. The AME at each location was as follows: ankle, 2.66 (2.24) cm; pelvis, 3.09 (2.39) cm; and wrist, 4.05 (3.01) cm. The one-way ANOVA with repeated measure revealed a significant main effect of the sensor location on the accuracy of step length ($F_{(2,492)} = 33.87$, $p < 0.001$, $\eta^2 = 0.12$). Post-hoc analysis revealed that significant differences were noted between ankle–wrist and pelvis–wrist ($p < 0.001$, $r = 0.52$ and $p < 0.001$, $r = 0.35$).

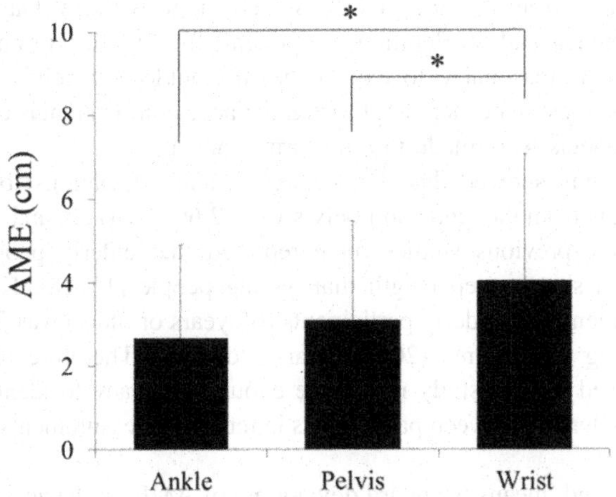

Fig. 1. Mean and standard deviations of AME obtained from each body location. Data are shown as the mean + standard deviation. AME: absolute value of with-in participant mean of errors between measured and estimated step length. *$p < 0.05$, $r > 0.30$

4 Discussion

The objective of this study was to compare the accuracies of step length estimations among different body locations (sensor locations). We hypothesized that the acceleration signals obtained from the ankle could estimate the step length most accurately compared to the pelvis and the wrist because the ankle is the closest location to the heel marker, where this parameter is calculated.

The present study showed that the accuracy of step length estimation from the ankle was significantly higher than that of the wrist. However, contrary to our initial hypothesis, there was no difference in the accuracy of step length estimation between the ankle and pelvis. Further, the accuracy of step length estimation from the pelvis was also significantly higher than that of the wrist. These results suggest that step length can be estimated from the acceleration signals of the ankle or pelvis with almost same accuracy.

Acceleration sensors attached to the lower limbs detect the reaction force received from the ground during heel contact [13]. The ankle is the closest location to the heel marker, which may enable detection of the accurate time of heel contacts from the vertical and anterior/posterior acceleration sensors. Although the pelvis is far from the heel marker compared to the ankle, it is close to the body center of mass (COM). Previous studies reported that COM motion appears to increase in the vertical direction and decrease in the mediolateral direction as walking speed increases [14]. Therefore, this may enable detecting the acceleration component in the direction of progression, in addition to the reaction force at heel contact. However, the wrist is not only far from the heel marker but also from the body COM, which suggests that it may be difficult to detect the heel contacts and acceleration components from vertical or anterior/posterior axis. Therefore, it is reasonable to expect that the ankle and pelvis were the more accurate locations to estimate step length from the acceleration signals compared to the wrist when the models were made in a uniform manner.

The present study showed that the means (standard deviations) of AME on the acceleration signals from the ankle and pelvis were 2.66 (2.24) cm and 3.09 (2.39) cm, respectively. Many previous studies have reported that elderly people walk more slowly and with a shorter step length than young people [15–18]. Kobayashi et al. reported that step length in elderly participants (65 years or older) was 3.86 cm shorter than that in young participants (20–39 years old) [18]. Therefore, the step length estimation proposed by this study may have enough accuracy to identify age-related differences in step length between participants in terms of measurements from the ankle and pelvis.

On the other hand, means (standard deviations) of AME on the acceleration signals from the wrist was 4.05 (3.01) cm. Therefore, estimation of step length with acceleration signals obtained from the wrist needs to be conducted more carefully than those from the ankle or pelvis. Zihajehzadeh et al. [10], who reported the model to estimate the walking speed from wrist acceleration, used principal component analysis to find the direction of the highest acceleration variation in the horizontal plane to improve the estimation accuracy. Applying such analysis may be effective to the step length estimation as well.

There are certain considerations that must be acknowledged when interpreting the results of the current study. First, the participants were required to walk barefoot whereas in most of the previous studies mentioned above, the participants were required to walk with shoes. Second, the algorithm in the present study is applicable to healthy adults, but it may not hold for children or elderly subjects aged 78 years or over. Hence, further studies are necessary to clarify these points.

5 Conclusion

This study compared the accuracy of estimated step length among different body landmarks (ankle, pelvis, and wrist). The results revealed that the ankle and pelvis were more accurate locations to estimate the step length from the acceleration signals compared to the wrist (approximately 3 cm of average error between participants). Further, estimation of step length with the acceleration signals obtained from the wrist needs to be conducted more carefully than that compared with the ankle or pelvis.

References

1. Kim H, Suzuki T, Yoshida H, Shimada H, Yamashiro Y, Sudo M, Niki Y (2013) Are gait parameters related to knee pain, urinary incontinence and a history of falls in community-dwelling elderly women? Nihon Ronen Igakkai Zasshi 50(4):528–535
2. Taniguchi Y, Yoshida H, Fujiwara Y, Motohashi Y, Shinkai S (2012) A prospective study of gait performance and subsequent cognitive decline in a general population of older Japanese. J Gerontol A Biol Sci Med Sci 67(7):796–803
3. Dang HV, Živanović S (2015) Experimental characterisation of walking locomotion on rigid level surfaces using motion capture system. Eng Struct 91:141–154
4. Demura T, Demura S (2010) Relationship among gait parameters while walking with varying loads. J Physiol Anthropol 29(1):29–34
5. Köse A, Cereatti A, Della Croce U (2012) Bilateral step length estimation using a single inertial measurement unit attached to the pelvis. J Neuroeng Rehabil 9:9
6. Shin SH, Park CG (2011) Adaptive step length estimation algorithm using optimal parameters and movement status awareness. Med Eng Phys 33(9):1064–1071
7. Renaudin V, Susi M, Lachapelle G (2012) Step length estimation using handheld inertial sensors. Sensors (Basel) 12(7):8507–8525
8. Pepa L, Verdini F, Spalazzi L (2017) Gait parameter and event estimation using smartphones. Gait Posture 57:217–223
9. Peruzzi A, Della Croce U, Cereatti A (2011) Estimation of stride length in level walking using an inertial measurement unit attached to the foot: a validation of the zero velocity assumption during stance. J Biomech 44(10):1991–1994
10. Zihajehzadeh S, Park EJ (2016) Regression model-based walking speed estimation using wrist-worn inertial sensor. PLoS ONE 11(10):e0165211
11. Kobayashi Y, Hobara H, Mochimaru M (2015) AIST Gait Database. https://www.dh.aist.go.jp/database/gait2015/index.html. Accessed 11 Apr 2018
12. Cohen J (1988) Statistical power analysis for the behavioral sciences, 2nd edn. Lawrence Erlbaum Associates, Hillsdale

13. Collins JJ, Whittle MW (1989) Impulsive forces during walking and their clinical implications. Clin Biomech (Bristol, Avon) 4(3):179–187
14. Orendurff MS, Segal AD, Klute GK, Berge JS, Rohr ES, Kadel NJ (2004) The effect of walking speed on center of mass displacement. J Rehabil Res Dev 41(6A):829–834
15. Lord SR, Lloyd DG, Li SK (1996) Sensori-motor function, gait patterns and falls in community-dwelling women. Age Ageing 25(4):292–299
16. Elble RJ, Thomas SS, Higgins C, Colliver J (1991) Stride-dependent changes in gait of older people. J Neurol 238(1):1–5
17. Oberg T, Karsznia A, Oberg K (1993) Basic gait parameters: reference data for normal subjects, 10–79 years of age. J Rehabil Res Dev 30(2):210–223
18. Kobayashi Y, Hobara H, Heldoorn TA, Kouchi M, Mochimaru M (2016) Age-independent and age-dependent sex differences in gait pattern determined by principal component analysis. Gait Posture 46:11–17

Occupational Exposure to Agrochemicals: A Literature Review

L. E. A. R. Junqueira(✉) and L. Contrera(✉)

Federal University of Mato Grosso do Sul, Campo Grande, Brazil
laura_elis@hotmail.com

Abstract. The use of agrochemicals plays an important role in the control of fungi, insects and weeds in agriculture and livestock, as well as in the control of infectious diseases. The aim of the study was to analyze the scientific production in the period from 2007 to 2017 regarding the health risks of professionals exposed to agrochemicals in the work environment. A search was conducted in the Portal of Periodicals CAPES-MEC by the library of the Federal University of Mato Grosso do Sul, in national and international journals in the Portuguese, English and Spanish languages. We found 72 complete articles that after systematic reading and exclusion of the duplicates we reached the number of 16 articles that were pertinent on the subject studied. This study revealed that there is little research production on the subject matter and that most of them are concentrated in the developing countries, since it is the localities that use these products mainly in agriculture.

Keywords: Agrochemicals · Occupational risks · Human engineering

1 Introduction

In Brazil, it is one of the most adopted measures as part of sustainable and integrated management for vector control in Public Health. The continuous use of these products can cause great damage to the health of exposed workers, being harmful to the function of different organs of the body, including the nervous, endocrine, immune, reproductive, renal, cardiovascular and respiratory systems.

2 Objective

The aim of the study was to analyze the scientific production in the period from 2007 to 2017 regarding the health risks of professionals exposed to agrochemicals in the work environment. Method: literature review, using as integrative revision method. A search was conducted in the Portal of Periodicals CAPES-MEC by the library of the Federal University of Mato Grosso do Sul, with the range of publication from 2007 to 2017 in national and international journals in the Portuguese, English and Spanish languages. We used the Boolean descriptors and operators: Agrochemicals and Occupational Risks and Human Engineering. Only full-text and online articles were used (Table 1).

S. Bagnara et al. (Eds.): IEA 2018, AISC 818, pp. 31–34, 2019.
https://doi.org/10.1007/978-3-319-96098-2_5

Table 1. Distribution of studies, according to the title, research site and year of publication.

TITLE	AUTHORS	RESEARCH SITE	PUBLICATION
Exposure to chemicals and beliefs associated with its use in the Morote River basin, Guanacaste, Costa Rica: A case study [1]	Trejos, V.Y.	Costa Rica	2015
Behavior adopted by family farmers, in the use and pesticide handling: a case study in the Settlement Guapirama, in Campo Novo do Parecis-MT [2]	Sznitowski, A. M. Menegon, N. L.	Brazil	2012
A contribution of the design to agriculture trough an ergonomic evaluation of the 20 liters plastic packings bottles for agro-chemicals [3]	Zerbetto, C. A. A.Gimenez, A. O.; Kague, N. A.	Brazil	2009
Pesticide exposure and its repercussion in the health of sanitary agents in the State of Ceará, Brazil [4]	Lima, E.P. Lopes, S.M.B. Amorim, M.I.M. et al	Brazil	2009
Socio environmental policies and conflicts: the case of land tenure and the monocultives in the caribbean of Costa Rican Caribbean (2006-2012) [5]	Llaguno, J. J. Mora, S. S. Gutiérrez, E. A. Barrios, A. P.	Costa Rica	2014
Spatial distribution of pesticide use in Brazil: a strategy for Health Surveillance [6]	Pignati, W.A. Francco, F.A.N.S. de Lara, S.S.	Brazil	2017
Efficiency of new clothes water-repellent in the protection of the tractor-driver in pesticide spraying in guava orchards with the air-assisted sprayer [7]	Tácio, M.B. Oliveira, M.L. Neto, J.G.M.	Brazil	2008
Major rural accident: the pesticide "rain" case in Lucas do Rio Verde city – MT [8]	Machado, J. M. H. Cabral, J. F.	Brazil	2007
Tobacco cultivation in the south of Brazil: green tobacco sick-ness and other health problems [9]	Riquinho, D.L Hennington, E.A	Brazil	2014
Pesticide use in soybean production in Mato Grosso state, Brazil: A preliminary occupational and environmental risk characterization [10]	Belo, M.S.S.P. Pignati, W. Dores, E.F.G.C. Moreira, J.C. Peres, F	Brazil	2012
Self-perception of hearing disorders, habits, and hearing loss risk factors in farmers [11]	Stadler, S.T. Ribeiro V.V França, D.M.V.R.F	Brazil	2016
Are tick medications pesticides? Implications for health and risk perception for workers in the dairy cattle sector [12]	da Silva, T.P.P . Costa, J.M. Peres, F.	Brazil	2012
Analysis of rural workers' exposure to pesticides [13]	Siqueira, D.F. Moura, R.M. Carneiro, G.E.	Brazil	2013
Evaluation of the auditory system of farm workers exposed to pesticides [14]	Kós, M.I. Miranda, M.F Guimarães, R.M. Meyer, A.	Brazil	2014
Risk perception, attitudes and practices on pesticide use among farmers of a city in Midwestern Brazil [15]	Recena, M. C. P. Caldas, E.D.	Brazil	2008
Auto-perception of auditory and vestibular health in workers exposed to organophosphate [16]	Hoshino, A.C.H. Pacheco-Ferreira, H. Taguchi, C.K. Tomita, S. Miranda,M.F	Brazil	2008

3 Results

We founded 72 complete articles that after systematic reading and exclusion of the duplicates we reached the number of 16 articles that were pertinent on the subject studied. Regarding the themes approached, four categories of study were found: occupational exposure (75.00%); ergonomics (6, 25%); labor rights (6.25%) and knowledge about risks (12.50%) (Fig. 1).

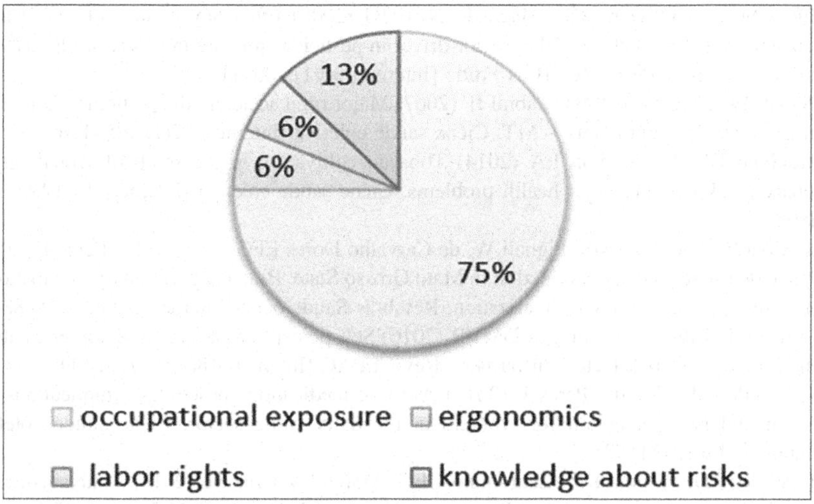

Fig. 1. Categories of study

4 Discussion

Was observed that the studies are incipient in relation to strategies to prevent the exposure of these risks, demonstrating that more studies need to be carried out on the subject, as well as studies related to the consequences of the prolonged use of these agrochemicals in workers' health.

References

1. Trejos YV (2015) Exposure to chemicals and beliefs associated with its use in the Morote river basin, Guanacaste, Costa Rica: a case study. Cienc Trab [Internet] 17(52):54–68
2. Sznitowski AM, Menegon NL (2012) Behavior adopted by family farmers, in the use and pesti-cide handling: a case study in the Settlement Guapirama, in Campo Novo do Parecis-MT. Rev GEPROS [Internet] 2(2):51–64
3. de Almeida Zerbetto CA, Gimenez AO, Kague NA (2009) A contribution of the design to agriculture trough an ergonomic evaluation of the 20 liters plastic packings bottles for agrochemicals. Semin Ciências Agrárias [Internet] 30(2):259–270
4. Lima EP, de Matos Bandeira Lopes S, de Amorim MIM, Araújo LHSA, Neves KRT, Maia ER (2009) Pesticide exposure and its repercussion in the health of sanitary agents in the State of Ceará, Brazil. Ciênc saúde coletiva [Internet]. 14(6):2221–2230
5. Llaguno JJ, Solano SM, Espelata ALG, Alfaro PB, Moraga FM (2014) Socio environmental policies and conflicts: the case of land tenure and the monocultives in the caribbean of Costa Rican Caribbean (2006–2012). Rev Ciencias Soc. 2014(3):81–98
6. Pignati AW, Neri de Souza e Lima FA, de Lara SS, Correa MLM, Barbosa JR, Leão LHC, Pgnati MG (2017) Spatial distribution of pesticide use in Brazil: a strategy for Health Surveillance. Ciênc saúde coletiva [Internet]. 22(10):3281–3291

7. Tácio MB, de Oliveira ML, Machado Neto JG (2008) Efficiency of new clothes water-repellent in the protection of the tractor-driver in pesticide spraying in guava or-chards with the air-assisted sprayer. Rev Bras Frutic [Internet] 30(1):106–111
8. Pignati WA, Machado JMH, Cabral JF (2007) Major rural accident: the pesticide "rain" case in Lucas do Rio Verde city – MT. Ciênc saúde coletiva [Internet] 12(1):105–114
9. Riquinho DL, Hennington ÉA (2014) Tobacco cultivation in the south of Brazil: green tobacco sickness and other health problems. Ciênc saúde coletiva [Internet] 19(12):4797–4808
10. da Silva Peixoto Belo MS, Pignati W, de Carvalho Dores EFG, Moreira JC, Peres F (2012) Pesticide use in soybean production in Mato Grosso State, Brazil: a preliminary occupational and environmental risk characterization. Rev bras Saúde ocup [Internet] 37(125):78–88
11. Stadler ST, Ribeiro VV, França DMVR (2016) Self-perception of hearing disorders, habits, and hearing loss risk factors in farmers. Rev CEFAC [Internet] 18(6):1302–1309
12. Silva TPP, Moreira JC, Peres F (2012) Are tick medications pesticides? Implications for health and risk perception for workers in the dairy cattle sector. Ciênc saúde coletiva [Internet] 17(2):311–325
13. de Siqueira DF, Moura RM, Laurentino GEC, Araújo AJ, Cruz SL (2013) Analysis of rural workers' exposure to pesticides. Rev bras promoç saúde [Internet] 26(2):176–185
14. Kós IM, de Fátima Miranda M, Guimarães RM, Meyer A (2014) Evaluation of the auditory system of farm workers ex-posed to pesticides. Rev CEFAC [Internet], 16(3):941–948
15. Piazza MCR, Caldas ED (2008) Risk perception, attitudes and practices on pesticide use among farmers of a city in Midwestern Brazil. Rev Saúde Pública [Internet] 42(2):294–301
16. Hoshino ACH, Pacheco-Ferreira H, Taguchi CK, Tomita S, de Fátima Miranda M (2009) Auto-perception of auditory and vestibular health in workers exposed to organophosphate. Rev CEFAC [Internet], 11(4):681–687

Relationship Between Working Environment Factors, Burnout Syndrome and Turnover Intentions Among Nurses – A Cross-Sectional Study in Bulgaria

Rumyana Stoyanova(✉)

Medical University-Plovdiv, 15A Vassil Aprilov blvd., 4000 Plovdiv, Bulgaria
rumi_stoqnova@abv.bg

Abstract. The aim of this study was to establish the relationship between working environment factors, burnout syndrome and turnover intentions among Bulgarian nurses. An anonymous survey was carried out among 391 nurses working in outpatient or hospital medical health care in the South Central Region of the Republic of Bulgaria in the period October 2013–January 2014. The questionnaire contained several panels: socio-demographics; personal self-assessment and satisfaction with organizational, socio-economic and psychological working conditions, Maslach Burnout Inventory and question related to turnover intentions. Results were analyzed by descriptive statistics of the data, non-parametric and factor analysis, using the SPSS 17.0. The study revealed that lower level of satisfaction with key areas of the work environment, such as: remuneration (including equity); relationship among staff, working conditions and safety; opportunities for career development and workload leaded to a higher probability of developing burnout syndrome. On the other hand the results found correlation between the number of burnout's dimensions with high score and turnover intentions among respondents ($P < 0.05$). The present study noted that work environment factors influence the development of burnout syndrome and turnover intentions of nurses.

Keywords: Factors of work environment · Burnout · Turnover intention Nurses

1 Introduction

Ageing of the population, which means that more and more people live with severe and chronic diseases, leads to increasing demand for health services. Therefore, the offer of such services needs to increase as well, and that would exacerbate the issues related to the existing shortage of nurses [1–3]. In the same time, there is a risk of growing emigration of well-educated and well-trained nurses due to unfavorable working conditions which will pose a serious problem for health care systems [4].

Employees of the health care systems are exposed to higher mental, physiological and cognitive requirements. The labour of nurses has long been pointed at as among the most stressful in the world and among the few jobs where employees have to cope with

S. Bagnara et al. (Eds.): IEA 2018, AISC 818, pp. 35–43, 2019.
https://doi.org/10.1007/978-3-319-96098-2_6

different situations under stress [5, 6]. High levels of stress are connected with the basic characteristics of their activities: extremely dangerous, fast and intensive. Therefore, these employees have to approach their professional duties with the required attention and responsibility because there is a direct threat or clear and present life threat. They often work in teams, which require good collaboration between team members. Given the continuous 24-h work process, they work in shifts.

Working under potentially dangerous conditions and unexpected situations often lead to emotional exhaustion [7]. Emotional exhaustion is a specific, stress related reaction that is considered as a key component of burnout syndrome [8].

Burnout syndrome is considered one of the newest diseases of modern people trying to build up a career. It is a psychological syndrome which occurs as a response to intensive and prolonged exposure to stress in a work environment.

The term 'burnout syndrome' was coined by Freudenberger in 1974 and was further developed by Maslach [9].

Maslach defined burnout as a state of physiological, emotional and intellectual depletion characterized by chronic fatigue, a feeling of helplessness and hopelessness, development of a negative perception and behavior towards oneself, the job, life and the others [10]. The main factors that cause its development are physical and mental overload, lack of support and adequate remuneration, and poor working conditions.

The significance of burnout for the individual and the work place resides in its connections with consequences like achievement, productivity, absenteeism, turnover of professionals, low satisfaction with the job and impaired health status [11].

The modern theoretical framework of the burnout syndrome integrates individual and situational factors based on the work/personality model focusing on the compatibility between the working person and six key areas of the working environment: work load, control, reward, community, justice and values that bring different perspectives in the interaction between the individuals and the specificities of their working environment [11–16]. These six areas of the organization's life constitute the framework of the organization's characteristics that precede the development of burnout. The higher the chronic incompatibility, the higher is the probability of burnout.

On the basis of extensive research for more than 20 years, Tzenova came to this conclusion: "Burnout syndrome is closely related to job dissatisfaction but the direction of correlation has not been explained, namely: whether it is burnout that makes people feel frustrated with their work or job dissatisfaction that causes the development of burnout syndrome. Or maybe both burnout and job dissatisfaction are caused by other factors such as poor working conditions" [11].

The aim of this study was to establish the relationship between working environment factors, burnout syndrome and turnover intentions among Bulgarian nurses.

2 Materials and Methods

The study was conducted among nurses working in outpatient or hospital medical health care in the South Central Region of the Republic of Bulgaria between October 2013 and January 2014.

The selection of units was made using two-stage cluster sampling. A determined in advance number of health-care establishments were randomly drawn and then, randomly again, the necessary number of units to survey were drawn. The principles of representativity of the sample were observed: random selection and necessary number of units (see Table 1).

Table 1. Distribution of nurses by regions and health care establishments

Regions	Total number of registered nurses (NSI data to 31.12.2013)	%	Number of medical establishments for hospital care, included in the research	Number of medical establishments of outpatient care, included in the research	Total number of sent questionnaires in the selected medical institutions
Plovdiv	3242	53,84	8	3	270
Pazardzhik	885	14,7	2	1	70
Haskovo	832	13,82	2	1	70
Smolyan	479	7,94	1	1	40
Kardzhali	584	9,7	1	1	40
	6022	**100,00**	**14**	**7**	**490**

Primary information was collected through a voluntary and anonymous inquiry among respondents administered at their respective work place. The front page of the survey explained that participation was voluntary and the results would be kept in strict confidence.

For data collection, a self-administered questionnaire and Maslach Burnout Inventory (MBI) were used as the research instruments.

The self-administered questionnaire included questions relating to the demographic characteristics of the respondents (6 closed questions) as well as questions relating to the organisation and management, social and economic factors of work place and turnover intentions of nurses. Social-economic, organisation and management factors were determined through assessment of different signs pertaining to such factors (e.g. salary, work conditions, relations with the management and with peers, etc.). The questionnaire included 30 questions aiming at assessing these signs, of which 6 questions were open-ended in view of clarifying the answer to certain close-ended questions.

Turnover intentions of nurses were researched using a single-component model. This model studies intentions through the positive or negative answer for a question [17]. In our case, the question that surveyed the turnover intentions of the respondents was: “Do you plan to leave your current workplace in the next 12 months?” with possible answers: “yes”, “neither yes, nor no”, and “no”.

The MBI is a self-administered instrument, designed to measure hypothetical aspects of the burnout syndrome. The inventory is composed of 22 statements that evaluate the three dimensions independently of one another, which are: Emotional Exhaustion (9 statements), Depersonalization (5 statements) and Personal Achievement

(8 statements). Each statement is rated on frequency scale that goes from 0 ("never") to 6 ("every day").

The score in each subscale was obtained by means of the sum of the respective values. For example, in the subscale of Emotional Exhaustion (EE), a score equal to or higher than 27 was considered indicative of a high level of exhaustion; the interval 14–26 corresponded to moderate values, and values equal to or lower than 13 would indicate a low level of exhaustion.

On the subscale of Depersonalization (DP) a score equal to or higher than 10 would be a high level; scores between 6 and 9 a moderate level and a score between 0 and 5 a low level of depersonalization.

And in the subscale of Personal Achievement (PA), scores equal to or lower than 40 would be a low level; between 34 and 39 a moderate level, and a score between 0 and 33 a high level of personal accomplishment [10].

Results were analysed by descriptive statistics of the data (mean, standard deviation, number and percentage), and presented in the form of tables, using the SPSS 17.0 program.

To compare the variables the non-parametric Chi Square ($\chi 2$) test was applied, and for this a level of significance of 5% probability ($P < 0.05$) was adopted. To measure the strength of correlations was used contingency coefficient (C). Other statistical analysis method that was used is factor analysis.

3 Results and Discussion

In the course of the study 490 questionnaires were sent of which 435 were returned after a reminder (response rate - 88.76%). The number of validly filled-in questionnaires was 391.

The average age of the respondents was 44,79 ± 11,207. Respondents had an average experience 21,25 ± 11,95 years. The average managerial experience was 1,97 ± 4,62 years. Their socio-demographic characteristics are shown in Table 2.

The results have shown that a significant percentage of nurses have high levels of emotional exhaustion and reduced performance (see Table 3).

Figure 1 shows the distribution of the total levels of burnout in those affected by the syndrome.

Non-parametric and factor analysis were used to assess the factors that affect burnout syndrome.

Table 4 shows all signs where were established a statistically significant relationship with development and level of burnout syndrome.

To reduce and highlights the main determinants that influence development of burnout syndrome was used factor analysis. Some of the studied signs with factor weight < 0.5 were removed from analysis (№ 3 and № 16). The others signs were subjected to rotation using the Varimax method that aims at increasing their factor weights on account of those that were removed. In this way, the studied variable, namely burnout, is explained using the least possible number of factors containing logically interconnected signs. The Kaiser-Meyer-Olkin measure of sampling adequacy

Table 2. Socio-demographic characteristics and work place of the respondents.

		N	%
Marital status	Married	253	64,7
	Not married	65	16,6
	Divorced	35	9,0
	Widow/Widower	18	4,6
	Domestic partnership	20	5,1
	Total	391	100,0
Number of children	No child	88	22,5
	One child	161	41,2
	Two children	135	34,5
	Three and more children	7	1,8
	Total	391	100,0
Educational level	College degree	206	52,7
	Bachelor degree	138	35,3
	Master degree	47	12,0
	Total	391	100,0
Rank	Nurse	337	86,2
	Senior nurse	44	11,3
	Head nurse	10	2,6
	Total	391	100,0
Property's form of health organization	State	56	14,3
	Municipal	150	38,4
	Private	123	31,5
	Mixed	62	15,9
	Total	391	100,0
Activity's area (workplace)	Outpatient care	50	12,8
	Hospital care	341	87,2
	Total	391	100,0

is 0,869 (i.e. >0.5) and Bartlett's Test of sphericity is Sig. = 0.00, which proves that the use of factor analysis is acceptable.

The results from factor analysis have shown that there are four factors that affect the development of burnout syndrome (see Fig. 2).

The first factor explains 35,55% of burnout development and include next signs: № 8 ($r = -0,525$), № 9 ($r = 0,862$), № 10 ($r = 0,816$), № 11 ($r = 0,854$), № 12 ($r = 0,787$), № 13 ($r = -0,712$), № 14 ($r = -0,674$), № 15 ($r = 0,803$).

The second factor explains 15,86% of burnout development and include next signs: № 4 ($r = 0,814$), № 5 ($r = 0,814$), № 6 ($r = 0,671$); № 7 ($r = 0,536$).

The third factor explains 9,32% of burnout development and include next signs: № 1 ($r = -0,720$), № 2 ($r = 0,786$).

The fourth factor explains 7,57% of burnout development and include only sign № 17 ($r = 0,836$).

Table 3. Assessment of burnout syndrome dimensions.

Dimension	Level	N	%	Mean	SD
Emotional exhaustion	High	126	32,2	37,33	7,552
	Moderate	131	33,5	19,76	3,823
	Low	134	34,3	7,52	4,129
	Total	391	100,0	21,23	13,343
Depersonalization	High	32	8,2	13,78	3,066
	Moderate	52	13,3	6,87	0,971
	Low	307	78,5	1,0	1,469
	Total	391	100,0	2,83	4,149
Personal achievement	High	145	37,1	24,81	6,633
	Moderate	98	25,1	36,64	1,645
	Low	148	37,8	44,35	2,852
	Total	391	100,0	35,17	9,612

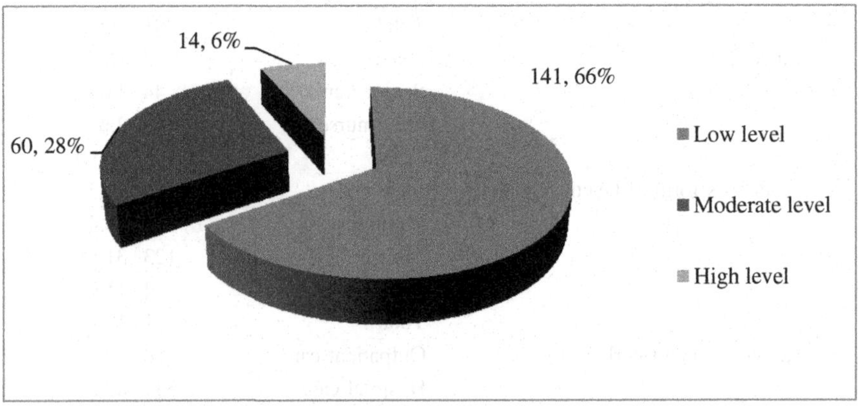

Fig. 1. Levels of burnout

Results of the analysis have proved that workplace, organizational and some demographic factors contribute to the development of burnout among nurses. On the other hand were found significant relationships between nurses' turnover intention and their level of burnout syndrome (see Table 5).

Development of burnout syndrome leads to increased turnover intentions of nurses (see Fig. 3).

Table 4. Relationship between some demographics, organizational and economic signs and burnout.

N	Signs		Non-parametric analysis	
			χ^2	P<
1.	Age	Level of burnout syndrome	44,479	0,001
2.	Level of education		17,547	0,01
3.	Region		117,786	0,001
4.	Relations with the management		39,078	0,001
5.	Relations with peers		28,881	0,01
6.	Safety		49,850	0,001
7.	Prestige of the profession		27,432	0,01
8.	Career opportunities		47,166	0,001
9.	Payment method		30,147	0,001
10.	Distribution of the salary fund		34,355	0,001
11.	Salary/Duties ratio		35,468	0,001
12.	Remuneration compared to that of other professions		20,894	0,05
13.	Social evaluation on the scale "Poor/Rich"		64,134	0,001
14.	Social evaluation on the scale "I deprive myself of everything/nothing"		79,042	0,001
15.	General satisfaction with the remuneration		51,816	0,001
16.	General satisfaction with the profession.		48,835	0,001
17.	Shift work		31,414	0,01

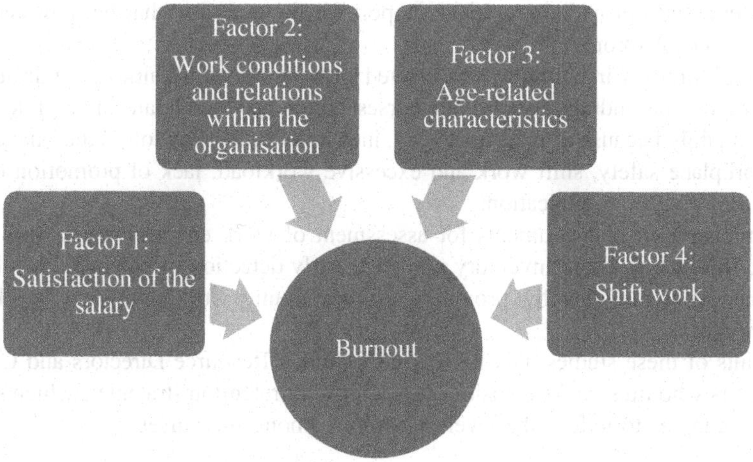

Fig. 2. Factors affecting burnout

Table 5. Relationship between burnout and turnover intention among nurses

Variables		Non-parametric analysis		
		χ^2	P<	Contingency coefficient
Level of burnout	Turnover intentions	39,680	0,001	C = 0,304

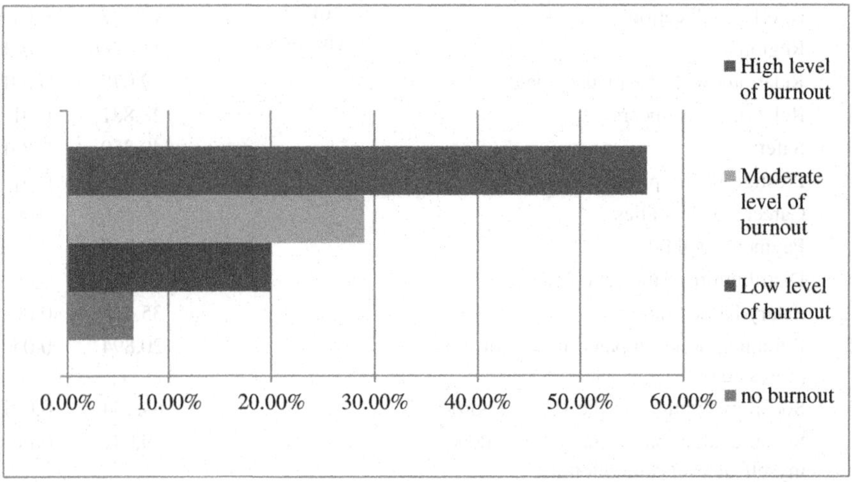

Fig. 3. Nurses' turnover intention according to the level of burnout

4 Conclusions

The current study provided empirical support for the assertion that burnout is linked with turnover intention.

Nurses' turnover intention was associated with their level of burnout syndrome. On the other hand the findings showed that nurses experienced moderate and high levels of burnout, mainly because of inadequate pay, inequality of distribution of the salary fund, poor workplace safety, shift work and excessive workload, lack of promotion opportunities and poor communication.

Administration of instruments for assessment of work environment factors along with the Maslach Burnout Inventory may offer early detection of potential, developing or actual problems, thereby providing an opportunity for appropriate workplace interventions.

Results of these studies may be useful to Human Resource Directors and Clinical Supervisors who must make decisions regarding staff retention strategies, which should include strategies to reduce the level of burnout among the nurses.

References

1. Buchan J, Parkin T, Sochalski J (2003) International nurse mobility: trends and policy implications. Report funded by World Health Organisation, International Council of Nurses and Royal College of Nursing
2. Simon M, Hasselhorn HM, Kuemmerling A, Van der Hejden B (July 2005) Nurses' work home interference in Europe. In: Hasselhorn HM, Muller BH, Tackenberg P (eds) NEXT Scientific Report, pp 57–59
3. Stoyanova R (2016) The labor market for health professionals in Bulgaria. Knowledge 14 (2):684–689
4. Stoyanova R (2016) Why Bulgarian nurses want to emigrate? JSHER 1(1):56–64
5. Bakker AB, Heuven E (2006) Emotional dissonance, burnout, and in-role performance among nurses and police officers. Int J Stress Manage 13(4):423
6. Schaufeli WB, Enzmann D (1998) The burnout companion to study and research: a critical analysis. Taylor & Francis, London. https://books.google.co.in/books?id=cL88XbNVv8QC&printsec=frontcover&hl=bg&source=gbs_ge_summary_r&cad=0#v=onepage&q&f=false. Accessed 15 Jan 2018
7. Maslach C, Schaufeli WB, Leiter MP (2001) Job burnout. Ann Rev Psychol 52:397–422
8. Maslach C (1982) Burnout: The cost of caring. Prentice Hall, Englewood Cliffs, NJ
9. Freudenberger HJ (1974) Staff Burnout. J Soc Issues 30(1):159–165
10. Maslach C, Jackson SE (1981) The measurement of experienced burnout. J Organ Behav 2 (2):99–113
11. Tzenova B (2005) Personality correlates of burnout syndrome. National congress of psychology, vol. 3, pp 316–322. Sofi-R, Sofia. (in Bulgarian)
12. Lee RT, Ashforth BE (1996) A meta-analytic examination of the correlates of the three dimensions of job burnout. J Appl Psychol 81(2):123
13. Leiter MP, Maslach C (2000) Preventing burnout and building engagement: a complete program for organizational renewal. Jossey-Bass, San Francisco
14. Maslach C (1998) A multidimensional theory of burnout. In: Cooper CL (ed) Theories of organizational stress. Oxford University Press, Oxford, UK, pp 68–85
15. Schaufeli WB, Bakker AB (2004) Job demands, job resources, and their relationship with burnout and engagement: a multi-sample study. J Organ Behav 25(3):293–315
16. Tzenova B (1993) Phenomenon of burnout. Bul J Psychol 4:35–56 (in Bulgarian)
17. Neshev P (2010) Social attitudes and racial prejudices. notifications of the union of scientists, vol. 1, p 145. Varna (in Bulgarian)

Ergonomic Design of a Tube Head Unit (THU) for Radiographers in Digital X-Ray Environment

Guk-Ho Gil(✉), Sungwoo Sul, Yun-Su Kim, Eunmee Shin, Miyoung Lee, and Seolynn Park

Samsung Electronics,
B33, Seongchon-gil, Seocho-gu, 06765 Seoul, South Korea
gil.gukho@gmail.com

Abstract. The ergonomic design of a tube head unit (THU) in a digital radiography (DR) system has been developed in order to reduce work-related musculoskeletal disorders and improve working efficiency operated by radiologic technologists (RT). In this paper, the five ergonomic design guidelines have been studied for a new type of the THU design considering: (1) the height of a table based on knuckle height of 5th percentile female RTs to maintain appropriate posture of table top operation, (2) viewing distance and eye-touchscreen coordination to reduce the shoulder and neck musculoskeletal fatigue and to define the elements of user interface, (3) the type and angle of a hand grip of the THU to operate with less effort, (4) the thickness of handle for strong hand grip, and (5) the range of the THU movement by one or two hand operation. The new type of THU has been designed to minimize the size and weight including a circular type handle with controllers to feel comfortable with soft handling and natural posture without physical overexertion of wrists and shoulders caused by repeated THU movements. The User Interface for the new THU provides consistent user experiences for both landscape and portrait mode to maintain same screen layouts when it rotates. It is expected that the conceptualized THU design could provide RTs with less musculoskeletal disorders and higher working efficiency.

Keywords: Tube head unit · Ergonomic design · User experience

1 Introduction

Since the first digital imaging system was introduced in 1980 [1], digital radiography is widely used in all radiographic applications due to several advantages including a fully digitalized picture archiving and communication system (PACS), with images stored digitally and available anytime electronically [2]. The DR system consists of a tube head unit (THU) including ceiling suspension, a digital detector, a patient bed, and a workstation. The THU allows users to easily operate the system. The THU's most important task is to manipulate beam size of x-ray as narrow as it could be by moving the THU and control collimators in order to reduce x-radiation overexposure.

S. Bagnara et al. (Eds.): IEA 2018, AISC 818, pp. 44–49, 2019.
https://doi.org/10.1007/978-3-319-96098-2_7

In order to locate THU on the affected area and maintain appropriate source to image distance (SID), RTs frequently involve manual handling of patients and materials including carrying cassettes on the hip using one side of body, lifting patients by one radiologic technician (RT), bending from their waist with straight legs, and so on [3]. For this reason, RTs reported 59% of work-related musculoskeletal disorders (WRMSD) while other professions reported only 34%. Among these WRMSDs, 64% of RTs were attributed to occupational overexertion caused by poor posture, faulty body mechanics, decreased flexibility and poor physical fitness [4]. The most affected areas of RTs' WRMSDs were trunk (40.2%) and back (35.5%) related disorders followed by shoulder and neck pain [4].

The new type of THU provides touch interaction and multi-directional control grip while conventional type of THUs only provides physical user interface with uni- or bi-directional control grip. Although Weigner et al. [5] provides ergonomics design guidelines for medical hand tools to enhance human motor capabilities, to control and help individuals perform tasks, and a visual display to present dynamic information to the user, the control unit including touch display, physical user interface and control hand grip together was not fully considered for its ergonomic guidelines.

Thus, the purpose of this research is to provide the ergonomic design guidelines for a THU in a DR system to reduce work-related musculoskeletal disorders and improve working efficiency operated by RTs. It also provides the application of the ergonomic design guideline to the new type of THU and its corresponding user interface design.

2 Methods

This section will present five different ergonomic design guidelines that can be applied to the ergonomic THU of the DR system, especially considering location of THU and handle design with controllers.

2.1 Location of THU

Operating the THU is similar to the manual materials handling (MMH) which includes a wide variety of activities such as loading and unloading boxes, removing materials from a conveyer belt, stacking items in a warehouse, etc. [6]. With respect to the ceiling suspended THU, RTs require manual handling of the THU with one or two hands to move and rotate it in 5 different axis, although it could move automatically into a designated position. In order to locate a patient to get an accurate image according to a doctor's order, RTs first need to adjust the height of the table and locate a patient maintaining appropriate SID between a patient and radiation source.

First, for this reason, a table should be located near knuckle height of RTs and adjustable for RTs of different heights and for different patient postures. In Fig. 1(a) shows the height of a table based on knuckle height of 5th percentile female RTs to maintain appropriate posture of table top operation.

Second, with respect to the static strength values from 5th percentile female and 95th percentile male workers considered in this research, pull and push movements are the strongest influenced by the angle of the elbow at 150° and 180° and up and down movements are the strongest at 76° and 116° [6]. In Fig. 1(b) shows the example of viewing distance influenced by the angle of 120°. In order to manipulate the touch screen on the THU, however, the viewing distance should be considered based on the range of viewing distance from 33 cm to 72 cm in Fig. 1(c) [7]. Moreover the max eye rotation up and down should be considered in order to manipulate the touch screen with less muscle and eye fatigue. In Fig. 1(c) shows the location of the THU based on view angle and distance.

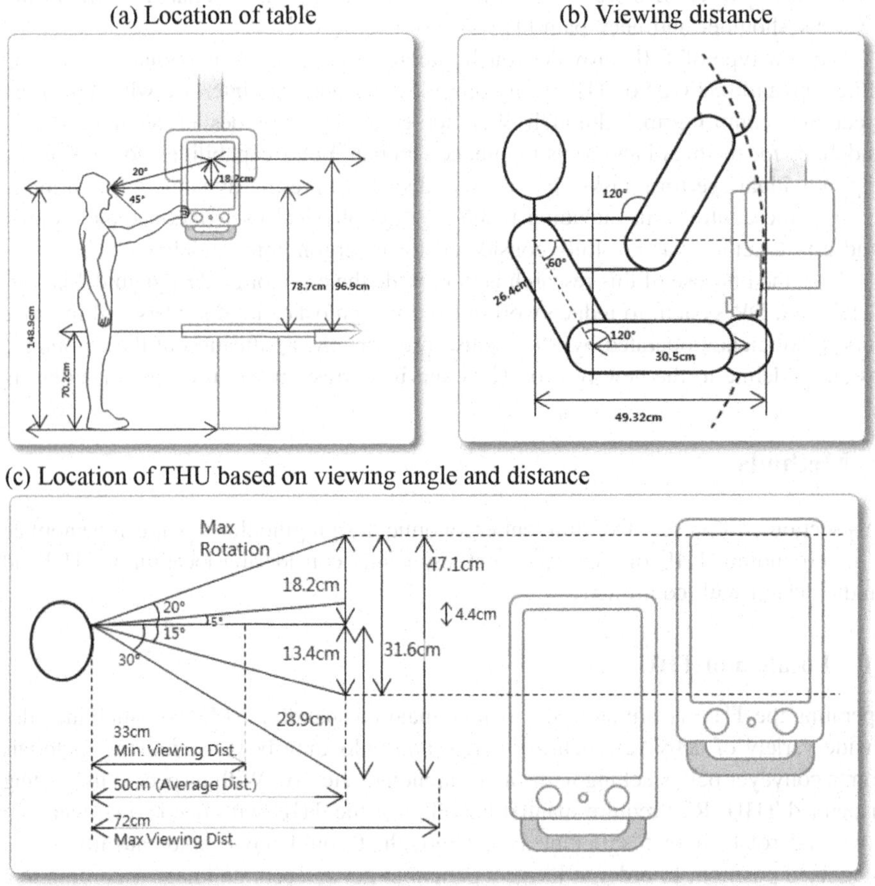

Fig. 1. Ergonomic analysis of THU.

2.2 Handle Design

In order to design an ergonomic THU, hand grip and moving range should be considered. **Third**, the horizontal line of handle should be aligned with 12° grip line to maintain greatest comfort and least stress [7]. Figure 2(a) shows the example of the inclined hand grip for one handed operation. When RTs use inclined hand grip, an elbow and shoulder of RTs could be in neutral body posture in order to reduce stresses placed on the wrist and shoulder.

Fourth, the diameter of hand grip should be in optimal range from 3.2 cm to 3.8 cm that can feel secure for one handed grip [7]. Figure 2(b) shows the example of optimal range of hand grip. It also shows the location of controllers if the handle is capable of operating the THU with one hand and thumb.

Fifth, while one handed operation have longer moving range than two handed operation, two handed operation can use less strength than one handed operation. Thus, the THU should use one of the operational methods between less stress and higher efficiency. In Fig. 2(c) shows the THU moving range by one handed or two handed operation.

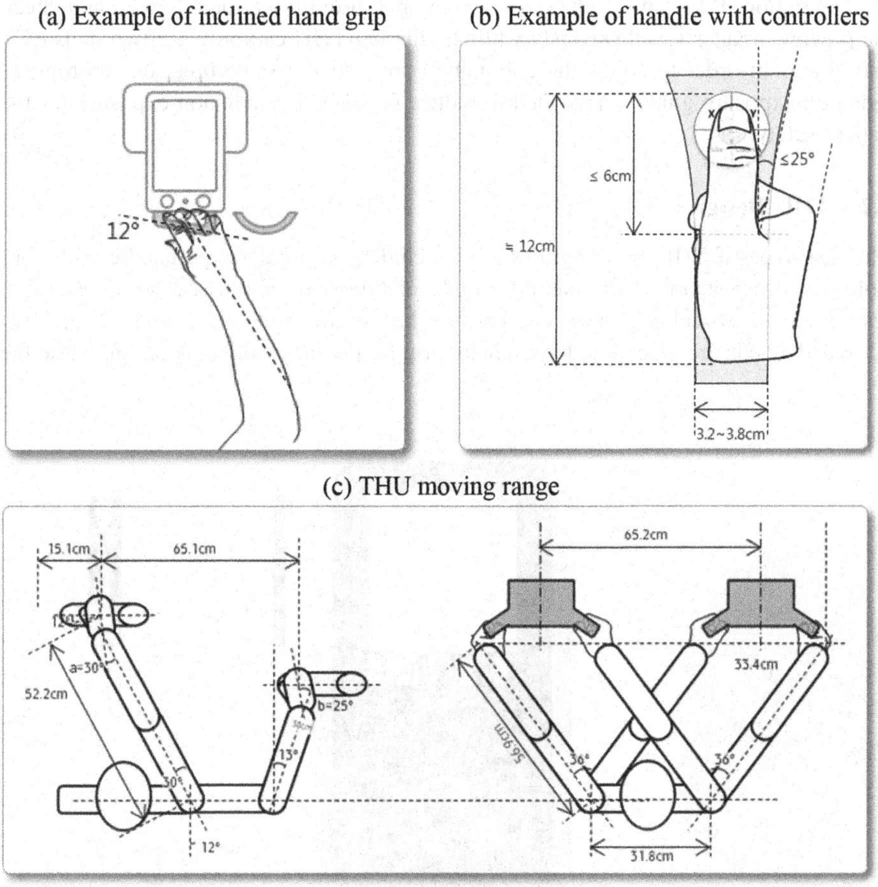

Fig. 2. Ergonomic analysis of handle and hand grip in THU.

3 Results

3.1 Visual Display Design

The User Interface for the new THU provides consistent user experience for both landscape and portrait mode to maintain same screen layouts regardless of THU rotations. In other words, the same functions and components are provided at the same position in both vertical and horizontal modes reducing cognitive workload and human errors caused by layout changes during the THU rotations. For instance, when the RTs adjust the angle and distance with precise control, RTs can easily notice that the value is displayed instantly on the same location of the screen. This functionality increases the work efficiency and reduce the pre-processing time. Figure 3(a) shows the same layouts when THU is rotated.

In addition, RTs can set up a collimation area through the touch screen and check the preview images simultaneously while traditional THU can only use two dials of X and Y axis in order to adjust the collimation area. At the same time, the appropriate radiation exposure guide is provided to reduce unnecessary radiation exposure for the patient safety.

3.2 THU Design

The new type of THU has been designed including a circular type handle with controllers that can maintain the natural posture of the hand and wrist at all times. With respect to the vertical grip in the circular type handle, the buttons to move the THU on X, Y and Z axis are placed under the hand grip on the index finger position, while the

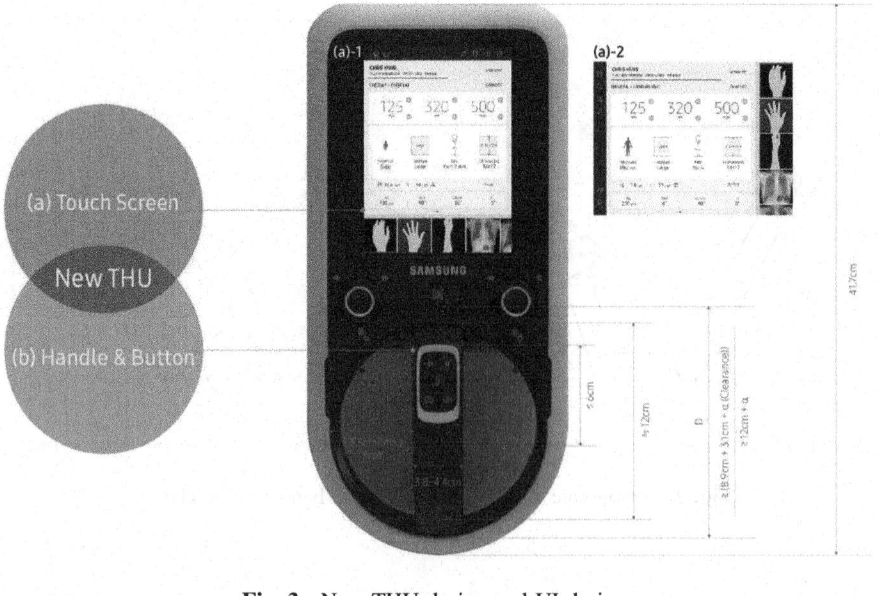

Fig. 3. New THU design and UI design

buttons for the vertical rotation are arranged in the place of the thumb-position. With this type of controllers, the RTs can adjust the THU with less muscle strength by one hand. Figure 3(b) shows the circular type handle design with controllers on the front side of the grip.

Moreover, the size and weight of the THU is minimized compared to the existing THU for RTs to feel comfortable with soft handling functionality with less physical overexertion of wrists and shoulders caused by repeated THU movements.

4 Discussion and Conclusion

Based on ergonomic design guidelines, the compact THU can provide the touch interaction and multi-directional control grip to reduce work-related musculoskeletal disorders and improve working efficiency operated by RTs. With respect to the visual display design, the THU can provide consistent user experience for both landscape and portrait user interfaces to reduce cognitive workload and human errors caused by layout changes during the THU rotation. Since a medical device considering usability can help RTs easy to learn and use, a usability test for the new THU should be conducted by RTs and patients in controlled environment first. However, the ergonomic design guidelines will help medical device designers create user interfaces that can be less prone to use error.

References

1. Ovitt TW, Christenson PC, Fisher HD 3rd, Frost MM, Nudelman S, Roehrig H, Seeley G (1980) Intravenous angiography using digital video subtraction: X-ray imaging system. Am J Roentgenol 135(6):1141–1144
2. Korner M, Weber CH, Wirth S, Pfeifer KJ, Reiser MF, Treitl M (2007) Advances in digital radiography: physical principles and system overview. Radiographics 27(3):675–686
3. Sharan D, Mohandoss M, Ranganathan R, Jose J, Rajkumar S (2014) Work related musculoskeletal disorders among radiologists and radiographers. In. Nordic Ergonomics Society Annual Conference in Human Factors in Organizational Design and Management, vol 46
4. Kling MY, West N (2015) Radiology and safe patient handling. Am J Safe Patient Handling Mobility 4:148–153
5. Weinger MB, Wiklund ME, Gardner-Bonneau DJ (eds) (2010) Handbook of human factors in medical device design. CRC Press
6. Sanders MS, McCormick EJ (1992) Human factors in engineering and design, 7th edn. McGRAW-HILL Book Company
7. Tilley AR (2002) The measure of man and woman: human factors in design. Wiley, New York

Ergonomics/Human Factors in Healthcare: A Vision for the Future

Alexandra Lang[1,2] and Sue Hignett[1,2(✉)]

[1] Chair of Special Interest Group in Healthcare Ergonomics, Chartered Institute of Ergonomics and Human Factors, MindTech, University of Nottingham, Nottingham, UK
alexandra.lang@nottingham.ac.uk
[2] Chair of Professional Affairs Board, Chartered Institute of Ergonomics and Human Factors, Loughborough Design School, Loughborough University, Loughborough, UK
S.M.Hignett@lboro.ac.uk

Abstract. How could and should Human Factors/Ergonomics (HFE) be implemented in healthcare? The Chartered Institute of Ergonomics & Human Factors (CIEHF, UK) has developed a vision for the future to complement the strategic views expressed by the Care Quality Commission, NHS Education for Scotland (NES), Health Education England (HEE), Medicines and Healthcare products Regulatory Agency (MHRA) and others. The initiatives delivered in the National Health Service (NHS, UK) have succeeded in sparking an interest in HFE. Whilst there have been successes in introducing local improvements using risk management, quality improvement and patient safety methods, the participative nature of HFE supports collaboration by all stakeholders ensuring that a system-wide approach is taken. This offers a new beginning for a common understanding of HFE in healthcare, accessible to all healthcare service stakeholders and provides a vision for education, capacity building and implementation. This paper will describe the consultation process and give examples of '*what good looks like*'; for example making the best use of human capabilities (physical, cognitive, psychological and social characteristics); mitigating for human limitations; and utilizing people in ways that maximize system safety and minimize risk.

Keywords: Healthcare · Vision · Integration

1 Introduction

Human Factors/Ergonomics (HFE) input in any industrial sector often follows a major incident or change in legislation. In the 1990s, the removal of Crown Immunity from prosecution under the Health and Safety Act 1974 meant that the NHS had to comply with safety legislation as hospitals and other care locations were considered to be places of work. From the 1980s to 2000s, HFE input focused on occupational health [1], building design [2], and systems approaches to embed HFE as part of the health care organizational culture [3].

S. Bagnara et al. (Eds.): IEA 2018, AISC 818, pp. 50–57, 2019.
https://doi.org/10.1007/978-3-319-96098-2_8

In 2004 the National Patient Safety Agency (NPSA) recommended the use of HFE as part of the 'Seven Steps to Patient Safety' [4]. There has been much good progress in the last 14 years including a National Concordat [5] bringing together 16 organisations. Health and social care has already started to benefit from HFE approaches and there will be many opportunities in the future. However, although the need for HFE in healthcare has been recognised the development and growth have been slow. We suggest that, in the UK, this is partly due to a misconception about what HFE is. The driving influences for 'Human Factors' in the UK have been from academic discipline of organisational (or industrial) psychology [6] and the industrial aviation sector [7]. There was a lack of engagement with the UK HFE professional body (Chartered Institute of Ergonomics & Human Factors, CIEHF) which, in our opinion, contributed to a dysfunctional separation (or lack of integrated safety) for the human elements into occupational health for staff, and patient safety [8]. This response is similar to that for occupational musculoskeletal problems where, for many years guidance from clinical professional bodies focused on technique (behavior) training. However, after 20 years, there was strong research evidence that '*interventions predominantly based on technique training had no impact on working practices or injury rates*' [9] and the focus moved to a Systems approach [3].

Catchpole [10] commented that '*while entirely legitimate and increasingly well evidenced,* [this behavioural safety approach] *is limited... Frequently espoused by well-meaning clinicians and aviators, rather than academically qualified HF professionals, it has led to misunderstandings about the range of approaches, knowledge, science and techniques that can be applied from the field of HF to address patient safety and quality of care problems*'. Russ et al. [11, 12] suggested that these influences have contributed to the tendency to blame '*the failures of people as the underlying cause of adverse events or broken healthcare delivery processes, a stance that is contrary to human factors science and counterproductive for advancing patient safety*' as '*little attempt is made to explore and address the underlying systemic causes that lead to errors.*'

This paper describes the process to develop a professional White Paper as a vision for future HFE integration in health and social care in the UK. The White Paper is aimed at health and social care service managers and Human Factors champions in care settings, as well as anyone interested in ensuring health and social care is the best it can be.

2 Developing the White Paper

The need for a White Paper to address the misconceptions has previously been discussed with respect to quality improvement [13]. In 2016, a workshop was held to explore how HFE knowledge and experience in defence could both inform and possibly be transferred to healthcare [14]. It was felt that there was a better 'fit' between defence and healthcare rather than the usual comparison between aviation and healthcare. Both defence and healthcare have complex and potentially chaotic working environments; activities take place 24 h a day, 365 days a year; and there are professional 'silos' (e.g. army, navy, air force). The workshop resulted in a 1, 5, 10 and 20 year principles (framework for a route map) including:

1. Service level professional collaboration agreements with partner organisations
2. Clear message and aim with plan for way forward
3. Examples as case studies (and/or safety cases)
4. Competency matrix
5. HFE applied in investigations (local incident reports through to Coroner's Court)
6. HFE and Quality Improvement linked
7. HFE included in Care Quality Commission inspections
8. HFE mandatory in processes for audit, procurement etc.
9. HFE as uniform approach across all Trusts, sectors (primary, secondary, mental health, ambulance, community & home care)
10. HFE competency matrix – clear, embedded and audited
11. HFE capacity to deliver improved safety.

It was proposed that, by 2036 HFE should be mandatory in healthcare processes for audit, procurement and so forth with a uniform approach across all healthcare providers as in defence [15]. The role of HFE providers (SQEP; Suitably qualified/experienced person) will be clearly defined based on CIEHF professional competencies to provide assurance in the same way as other professional regulators (e.g. General Medical Council, Nursing and Midwifery Council, General Pharmaceutical Council).

2.1 White Paper Project Proposal

A Project Initiation Document was submitted to the Chartered Institute of Ergonomics & Human Factors (CIEHF) Council in 2017 with a proposal to develop the CIEHF vision for Human Factors/Ergonomics (HFE) in Healthcare as the UK authoritative guide to help readers understand how HFE could and should be used. This included:

- HFE best practice in processes including audit, investigation and procurement from e.g. local pharmacy services and medical devices to national inter-agency information communication systems across all service providers (public and private), and sectors (primary, secondary, mental health, ambulance, community, social and home care).
- Strategies for HFE Implementation (e.g. integration with quality improvement) including a sector specific competency matrix.
- HFE capacity to deliver improved safety as a resilient system for safety culture and work load.
- Education including undergraduate curricula for Pharmacy, Academy of Medical Royal Colleges and postgraduate training.
- Competencies and experience needed to fill the various roles required for effective HFE implementation.

The project was announced at the CIEHF annual conference 2017 to explore interest and start the development/engagement process as part of a healthcare ergonomics symposium.

2.2 Workshop to Develop Key Messages

The purpose of the workshop was to agree key messages and bring together source material. A wide range of participants included CIEHF members representing most grades (Associate, Graduate, Registered, Fellow); Chairs of two CIEHF Special Interest Groups (Healthcare and Pharmaceutical), current and previous representatives of the Healthcare Safety Investigation Branch [16] and National Patient Safety Agency [17] and input from NHS Education Scotland [18], Clinical Human Factors Group (advocate) [19].

The possible scope of the White Paper was discussed in detail including:

- HFE best practice in processes and design
- All service providers (public and private), and sectors (primary, secondary, mental health, ambulance, community, social and home care)
- Boundaries, e.g. exclude pharmaceutical manufacturing pipeline
- What HFE is and is not…?
- Strategies for Implementation including competency and experience the various HFE roles.

3 The White Paper

3.1 Vision Statement

The draft vision statement is that '*through ongoing collaboration, co-creation and discovery with health and social care providers, professionally qualified Human Factors Specialists will contribute towards developing and embedding sustainable system-level improvements*'. It is proposed that this will be achieved by:

- Building on existing good work
- Broadening the scope of Human Factors understanding
- Guiding the understanding of shared aims and offerings from partner organisations
- Promoting the integration of Human Factors to optimise human (patient and staff) wellbeing and overall system performance
- Raising awareness of the discipline as an accredited, professional career
- Ensuring and maintaining the standard of Human Factors practice through demonstration of competence and experience
- Encouraging the contribution of professional (qualified) Ergonomists & Human Factors Specialists via consultation and employment
- Championing an accessible, user-focused approach.

3.2 What Is Human Factors

This section provides the IEA definition [20] and dispels myths about differences between Human Factors and Ergonomics. 'Systems' are introduced as a fundamental HFE concept with examples for micro and macro systems.

Ideas for integrating HFE into healthcare systems include ensuring HFE practice is design-led and proactive; working with existing risk management programmes (including quality improvement); and being creative in developing inclusive solutions so that products and services will be accessible to, and usable by, as many people as reasonably possible.

3.3 Increasing Human Factors Competency and Capacity

HFE awareness already exists for a range of professionals in healthcare including psychologists, occupational therapists, physiotherapists, medical device designers and engineers. So, part of the vision is to increase Human Factors competence with, for example, inclusion of HFE content in clinical curricula. The World Health Organization [21] developed a patient safety curriculum, but little is known about how providers ensure learners develop safety competencies. A set of 12 tips has been proposed for incorporating HFE principles and methods in clinical curricula to enhance the effectiveness of safety and improvement work in frontline healthcare practice [22]. The 12 tips include the systems framework, HFE tools and competency, and ideas for implementation.

Professional behaviour in HFE is guided by the CIEHF Code of Professional Conduct [23] and professional competencies [24] in 5 domains with 6 levels of competency (Fig. 1).

1. **Unaware**: No knowledge or understanding of this competency.
2. **Aware**: Knowledge or an understanding of basic techniques and concepts particular competency for a particular competency.
3. **Novice**: Limited experience gained in a classroom and/or as a trainee on-the-job for a particular competency
4. **Intermediate**: Can successfully complete tasks independently in a particular competency; can demonstrate the appropriate use of different techniques and methods in the application of Human Factors research or consultation.
5. **Advanced**: Can perform actions associated with this competency without assistance; can bring together disparate theories and techniques or the application of novel solutions to complex problems
6. **Expert**: Recognised authority in this area; can provide guidance, troubleshoot and answer questions related to this area of expertise with consistent excellence in applying this competency across multiple projects and/or organisations.

Fig. 1. Proficiency levels for professional competencies

Ideas for increasing capacity include HFE content in clinical curricula, input from qualified HFE professionals; recommending that every health and social care organisation has an identified Human Factors advisor at a senior level. The goal is to develop sufficient HFE capability to deliver a resilient system that encompasses safety culture and acceptable workloads for all healthcare providers.

3.4 Understanding How to Use Human Factors

Three areas are covered for understanding how to use HFE. These are (1) using HFE in investigations; (2) thinking about systems; and (3) thinking about design. For investigations, where human error may seem at first to be the cause of an incident, an HFE approach takes a wider view to encompass contributing factors such as poor product design. For example an interim report by HSIB [25] has identified systems issues including different processes for portable and installed systems (1–2 step tasks) and lack of visual indicators (feedback) for both flow and residual volume.

The Systems of Systems in healthcare have been described as '*nested and overlapping systems; 'a bed in a hospital is a system, the patient monitoring equipment is a sibling system, the two together plus the patient's room comprise another system, ...; whereas the radiology or scanning equipment, the drugs dispensary, the beds, the ambulances are all systems, but together can be seen as a system of systems when looking at maintenance and replacement regimes'* [26].

HFE is relevant at all stages (upstream and downstream) in the life-cycle of equipment, environment and services; from early stages of planning and design, right through to implementation and evaluation, and re-design. Healthcare is starting to recognise the benefits of HFE where the systems view of safety and an inclusive, human-centred design process can be applied in nearly all work situations.

3.5 Implementing Human Factors

Health and Social Care organisations face challenges in embedding and sustaining improvement initiatives. Successful implementation often relies on individuals having sufficient knowledge, time and resources to effectively play their part.

The vision for Human Factors implementation is:

- Human Factors good practice is common across all health and social care processes including audit, new and redesigned services, investigation and procurement.
- Human Factors good practice informs all areas from local pharmacy services and medical devices to national inter-agency information communication systems across all service providers (public and private), and sectors (primary, secondary, mental health, ambulance, community, intermediate, social and home care).
- Proven strategies and frameworks exist and are used for Human Factors implementation, such as integration with Quality Improvement, including a sector-specific competency matrix to which all practitioners adhere.
- International standards for Ergonomics and Human Factors are embedded in healthcare design and systems for planning, acquisition and safety.
- The underlying culture is a learning culture, keen to drive continuous improvement in human performance and wellbeing.

3.6 Case Study Example: Design for Patients

Birth in water gained momentum in the early 1990s however the design of birthing pools at that time had not been designed with users in mind. Mothers found it difficult to get into the birthing pool and almost impossible to get out of in an emergency.

Midwives had to adopt poor posture when monitoring and examining mothers in the pool. User needs were identified and incorporated into a new design which included steps and rails to assist entry and exit, shaped edges to support the mother, knee room to allow the midwife easier access for monitoring and a seat that allowed the mother to be evacuated rapidly in an emergency. The work revolutionised the design of birthing pools with resultant improvements in safety and wellbeing of the mother, baby and midwife. Support staff carrying out maintenance, cleaning and infection control also benefited from the improved design [27].

4 Conclusion

The White Paper is going through the final consultation stages (May 2018). Through ongoing collaboration, co-creation and discovery involving clinicians, HFE experts and other professionals within and beyond healthcare, this approach will contribute towards developing, and implementing sustainable system level improvements.

References

1. Straker LM (1990) Work-associated back problems: collaborative solutions. Occup Med 40:75–79
2. Hilliar P (1981) The DHSS ergonomics data bank and the design of spaces in hospitals. Appl Ergon 12(4):209–216
3. Hignett S (2001) Embedding ergonomics in hospital culture: top-down and bottom-up strategies. Appl Ergon 32:61–69
4. http://www.nrls.npsa.nhs.uk/resources/collections/seven-steps-to-patient-safety/?entryid45=59787. Accessed 12 Apr 2018
5. https://www.england.nhs.uk/wp-content/uploads/2013/11/nqb-hum-fact-concord.pdf. Accessed 12 Apr 2018
6. Flin R, Maran N (2004) Identifying and training non-technical skills in acute medicine. Qual Saf Health Care 13(Suppl I):i80–i84
7. Flin R, Martin L, Goeters K, Hoermann J, Amalberti R, Valot C, Nijhuis H (2003) Development of the NOTECHS (Non-Technical Skills) system for assessing pilots' CRM skills. Hum Factors Aerosp Saf 3:95–117
8. Hignett S, Carayon P, Buckle P, Catchpole K (2013) State of science: human factors and ergonomics in healthcare. Ergonomics 56:1491–1503
9. Hignett S (2003) Intervention strategies to reduce musculoskeletal injuries associated with handling patients: a systematic review. Occup Environ Med 60(9):e6
10. Catchpole K (2013) Spreading human factors expertise in healthcare: untangling the knots in people and systems. BMJ Qual Saf 22:793–797
11. Russ AL, Fairbanks RJ, Karsh B-T, Militello LG, Saleem JJ, Wears RL (2013) The science of human factors: separating fact from fiction. BMJ Qual Saf 22:802–808
12. Russ AL, Militello LG, Saleem JJ et al (2013) Response to separating fact from opinion: a response to 'the science of human factors: separating fact from fiction'. BMJ Qual Saf 22:964–966

13. Hignett S, Jones E, Miller D, Wolf L, Modi C, Shahzad MW, Banerjee J, Buckle P, Catchpole K (2015) Human factors and ergonomics and quality improvement science: integrating approaches for safety in healthcare. BMJ Qual Saf 24:250–254
14. Hignett S, Tutton W, Tatlock K (2017) Human Factors Integration (HFI) in UK Healthcare Route map for 1 year, 5 years, 10 years and 20 years. In Charles R, Wilkinson J (eds) Contemporary ergonomics 2017; Proceedings of the annual conference of the chartered institute of ergonomics and human factors. Taylor & Francis, London
15. MOD (2015) JSP 912. Human Factors Integration for Defence Systems. Part 1: Directive. https://www.gov.uk/government/uploads/system/uploads/attachment_data/file/483176/20150717-JSP_912_Part1_DRU_version_Final-U.pdf. Accessed 12 Apr 2018
16. https://www.hsib.org.uk/. Accessed 10 Apr 2018
17. http://www.npsa.nhs.uk/. Accessed 10 Apr 2018
18. https://learn.nes.nhs.scot/800/patient-safety-zone/human-factors. Accessed 10 Apr 2018
19. https://chfg.org/. Accessed 10 Apr 2018
20. IEA (2001) International Ergonomics Association. Core competencies. www.iea.cc/project/PSE%20Full%20Version%20of%20Core%20Competencies%20in%20Ergonomics%20Units%20Elements%20and%20Performance%20Criteria%20October%202001.pdf. Accessed 11 Apr 2018
21. WHO (2011) World Health Organisation The Multi-Professional Patient Safety Curriculum Guide. http://www.who.int/patientsafety/education/curriculum/en/. Accessed 15 June 2017
22. Vosper H, Bowie P, Hignett S (2017) Twelve tips for embedding Human Factors and Ergonomics principles in healthcare educational curricula and programmes. Med Teacher. https://www.tandfonline.com/doi/abs/10.1080/0142159X.2017.1387240
23. CIEHF Charter Documents, Professional Code of Conduct, 20 p. https://www.ergonomics.org.uk/Public/About_Us/CIEHF_Documents/Public/About_Us/CIEHF_Documents.aspx?hkey=8df03a4a-ab8a-482d-8a50-99c4a052f0c7. Accessed 11 Apr 2018
24. CIEHF Professional Competencies. https://www.ergonomics.org.uk/Public/Membership/Registered_Member/Public/Membership/Registered_Member.aspx?hkey=32fc9cb9-6d12-45fd-a3cb-8063e4c256f4. Accessed 11 Apr 2018
25. https://www.hsib.org.uk/investigations-cases/design-and-safe-use-portable-oxygen-systems/interim-bulletin/. Accessed 13 Apr 2018
26. Wilson JR (2014) Fundamentals of systems ergonomics/human factors. Appl Ergon 45:5–13
27. Case Study 18: Improving birthing pool design. The Human Connection. CIEHF 2015. https://www.ergonomics.org.uk/Public/Resources/Publications/Human_Connection/Public/Resources/Publications/Case_Studies.aspx?hkey=6cef60ed-99a9-498e-b2b7-edd072ff31dc. Accessed 13 Apr 2018

A Comparison of Front and Rear Wheel Shock Magnitudes for Manual Tilt-in-Space Wheelchairs with and Without Suspensions

Molly Hischke and Raoul F. Reiser(✉)

Colorado State University, Fort Collins, CO 80523, USA
raoul.reiser@colostate.edu

Abstract. When encountering an obstacle, both the front and rear wheels encounter a shock that may be unhealthy to the wheelchair user. Furthermore, depending on the obstacle there may be two peaks at each wheel - one when the wheel first encounters the obstacle and one when it leaves the obstacle. Limited research reports on these multiple peak accelerations at each wheel when investigating shock and vibration exposures in wheelchair users. There is also limited information comparing the front wheel impact to those at the rear wheel. One of the few studies available suggests the front wheel incurs greater shock than the rear wheel. However, this study had the wheelchair mounted on a treadmill for one of their obstacles. Although a treadmill controls for speed, the participants were not operating the wheelchair as they would during daily use. Using an attendant propelled tilt-in-space wheelchair, the present study investigated the initial impact at the front wheel versus the rear wheel, and the final impact at the front wheel versus the rear wheel. The obstacles included a door threshold, 2 cm descent, and 2 cm ascent. Front and rear wheel un-weighted and frequency- weighted (per the ISO 2631-1 standards) peak accelerations were significantly higher depending on the obstacle the wheelchair was traversing (door threshold, 2 cm descent or 2 cm ascent).

Keywords: Shock · Whole-body vibrations · Acceleration

1 Introduction

The U.S. Census estimates 3.6 million people use wheelchairs ranging from manual to motorized chairs (Brault 2010). In addition to their primary physical disability, wheelchair users may develop secondary health conditions. For example, many wheelchair users have a higher prevalence of back pain, pelvic pain, and neck pain than the general population (Jensen 2013; Cooper 1995; Boninger 2003). Low back pain, neck pain, muscle fatigue, and spinal discomfort has been associated with shock and whole-body vibration exposure (Bovenzi 1996; Milosavljevic et al. 2011; Ebe and Griffin 2000; Mansfield et al. 2014; Bovenzi et al. 2015; Zimmermann et al. 1993; VanSickle et al. 2001; Requejo et al. 2008; Maeda et al. 2003). Because of the association between pain and whole-body vibration exposure, the International Organization for Standardization (ISO) established methods (ISO 2631-1) to quantify shock and whole-body vibration exposure (International Organization for Standardization 1997).

S. Bagnara et al. (Eds.): IEA 2018, AISC 818, pp. 58–70, 2019.
https://doi.org/10.1007/978-3-319-96098-2_9

Research has reported wheelchair users experience shock and vibrations exceeding the ISO 2631-1 standards (VanSickle et al. 2001; Wolf et al. 2004, 2007; Garcia-Mendenz et al. 2013). Front and rear wheel suspension systems have been incorporated into the wheelchair design to help reduce shock and vibration exposure (Cooper et al. 2003; Requejo et al. 2008, 2009; Kwarciak et al. 2008). Limited research has reported shock and vibration exposure for manual tilt-in-space wheelchair users, a type wheelchair potentially used by those with cerebral palsy, muscular dystrophy, or a spinal cord injury (Physical and Mobility Impairments: Information and News 2015). Furthermore, prior to the QuadshoX LLC (Fort Collins, CO) spring-damper suspension design, manual tilt-in-space wheelchair users did not have a suspension system available for use.

When evaluating the impact on peak accelerations with the use of suspension systems, previous research has primarily isolated the acceleration at the rear wheel or reported the highest peak acceleration (Cooper et al. 2003; Requejo et al. 2008, 2009; Kwarciak et al. 2008). When evaluating front wheel suspension, one group included a peak acceleration at the front wheels and a peak acceleration at the rear wheels, and reported the largest accelerations occurred at the front wheels (Gregg and Derrick 1998). The researchers used a treadmill with a 2 mm bump and an over ground ramp with a 3 cm drop at the end which resulted in subjects impacting the obstacles between 1.53 and 2 m/s (Gregg and Derrick 1998). The speeds at which subjects were impacting the obstacles might not be generalizable to all wheelchair users, and may not accurately represent how wheelchair users would interact with an obstacle. Using an attendant propelled tilt-in-space wheelchair, the present study investigated the differences between the initial impact at the front and rear wheels, and the differences between the final impact at the front and rear wheels. Because tilt-in-space wheelchairs are typically attendant propelled, the investigators hypothesized large magnitude accelerations would occur equally at the front and rear wheels in both initial and final impacts of an obstacle.

2 Methods

2.1 Participants

Ten non-wheelchair users (7 men, 3 women: age = 22.1 ± 3.36 yrs., height = 1.75 ± 0.067 m, and mass = 73.9 ± 8.87 kg (mean ± SD)) voluntarily participated in this institutional review board-approved study after providing written informed consent. Participants were healthy (pain and injury free at the time of data collection) 18 year olds or older recruited from the student population of the investigator's institution.

2.2 Instrumentation

A single Quickie IRIS ® Tilt-in-Space manual wheelchair (17.69 kg, aluminum frame, maximum user weight capacity: 136 kg, 1.19 m L × 1.02 m W × 1.37 m H, front caster wheels: 0.20 m) manufactured by Sunrise Medical (Phoenix, AZ), was used by all subjects and for all trials. The tilt system was adjusted to a posterior tilt of 15° for all conditions. For the rigid trials, the wheelchair was kept as manufactured. For the suspension trials, the wheelchair was fitted with a QuadshoX, LLC (Fort Collins, CO) suspension kit (Fig. 1). The suspension system was adjusted per manufacturer instructions for each subject. Since rear wheel diameter varies, two solid Primo wheels of different diameter (Cheng Shin Rubber, Yuanlin, Taiwan), were also studied: a 0.381-m (small) diameter wheel and a 0.508-m (large) diameter wheel. There was no suspension on the front caster wheels.

A tri-axial accelerometer (Model339A31, PCB Piezotronics, Depew, NY) was mounted at the rear of the seat pan of the wheelchair (Fig. 1). A custom data-collection program was written in LabVIEW software (National Instruments, Austin, TX) to interface with the data-acquisition assistant (DAQ, National Instruments, Austin, TX) sampling at 2000 Hz. Raw voltages were amplified with a gain of 10 by the power supply (Model 480B21, PCB Piezoelectronics, Depew, NY). The equipment (DAQ and power supply) was kept in a small pack fastened to the wheelchair, and connected to a hand held laptop computer.

Fig. 1. The accelerometer and mount (circled) was securely attached to the rear seat pan of the wheelchair. Also visible is the QuadshoX suspension system on the right side of the wheelchair as well as the small diameter wheel. The suspension kit has moment arm brackets (green in image) that attach at the rear axle and carriage shaft, and the spring/damper unit attaches to the carriage shaft as well as moment arm bracket. The suspension kit replaces the manufactured axle bracket on Quickie IRIS® (Sunrise Medical, Phoenix, AZ) tilt-in-space wheelchair. (Color figure online)

2.3 Trials

The surfaces were designed by the investigators to represent common obstacles encountered by wheelchair users. An indoor course was made out of three sheets of plywood (laying over carpet/concrete, 1.22 m wide × 2.44 m long × 0.02 m high). The first and last sheets of plywood remained in place, and the center sheet was interchanged between two surfaces with four different obstacles (Fig. 2, other obstacles discussed in Hischke and Reiser, in press). The peak accelerations associated with each wheel sets impacts from the obstacles analyzed in the present study include:

Fig. 2. (A) Surface consisting of a door threshold (B) Close up of the 2 cm descent/ascent (descent if traversing from left to right, and ascent if traversing from right to left)

1. Door threshold (1.27 cm high × 91.44 cm long (M-D Building Products, Inc., Oklahoma City, OK))
2. 2 cm descent made by overlapping one sheet of plywood over another
3. 2 cm ascent made by overlapping one sheet of plywood over another.

2.4 Data Collection

All rigid trials were collected first followed by the suspended trials on a different day, separated by ~2 weeks. Subjects were pushed by the same trained investigator over all three obstacles. A minimum of three acceptable trials were recorded for each condition performed in the following order: rigid chair/small wheel (RS), rigid chair/large wheel (RL), suspended chair/small wheel (SS), suspended chair/large wheel (SL). Acceptable trials had to be within 0.2 s of each other and the average of the other conditions as determined by hand timing. During each trial subjects were instructed to stay as relaxed as possible, not reacting any more than necessary to the obstacle. The investigator pushed subjects as consistently as possible when traversing the obstacle.

2.5 Data Processing and Analysis

The accelerations were processed with a custom MATLAB code (version 8.4, The MathWorks, Inc; Natick, MA). Each channel was first zero-meaned to remove any potential DC offset and then converted from volts to m/s^2 by factoring out the gain and incorporating manufacturer supplied conversions. While each trial was hand timed as described above in an attempt to keep time spent over the obstacle in each condition the same, post-hoc analysis using the peak accelerations showed the time spent on obstacles was significantly different. Therefore, the trials used for statistical analysis were selected based on comparable time spent over the obstacle between the rigid and suspended conditions.

The ISO 2631-1 states vibrations should be evaluated separately in each direction. When assessing the health effects of vibration at the seat, the vibration evaluated should be the highest acceleration. If the accelerations are comparable in two or more directions, they can be combined to evaluate (International Organization for Standardization (ISO) 1997). Therefore, anterior/posterior and vertical directions were selected and combined for analysis.

Acceleration data was frequency weighted (FW), using a MATLAB algorithm adapted from Kwarciak et al. 2008, according to the standard vibration evaluation methods and parameters as stated in ISO 2631-1. The ISO 2631-1 recommends different frequency weighting for assessing vibrations' effects on health, comfort, and perception. The weightings put the most emphasis on the frequencies from ~2–12 Hz and the frequencies below and above this range gradually receive less weighting. Each direction is first weighted in isolation before being combined. The resultant FW acceleration equation used with x being the anterior/posterior and z being the vertical direction was:

$$a_w = \left(1.4a_{wx}^2 + a_{wz}^2\right)^{\frac{1}{2}} \tag{1}$$

where 1.4 and 1 are the x and y multiplying factors, respectively, as defined by ISO 2631-1.

2.6 Statistical Treatment

Outliers were identified using box plots, and extreme outliers (greater than three box lengths) were removed. After removal of outliers, the data was determined to be normally distributed by assessing the ratios between skewness and its standard error, and kurtosis and its standard error.

The data was analyzed using a two by two (rigid/suspended by front/rear wheel) repeated measures analysis of variance (ANOVA) to compare the mean differences between groups in un-weighted and weighted peak accelerations. For the door threshold, the initial front wheel impact (F1) and initial rear wheel impact (R1) were analyzed for differences (Fig. 3). Similarly for the door threshold, the final front wheel impact (F2) and final rear wheel impact (R2) were analyzed for differences (Fig. 3). The variables analyzed over the 2 cm descent (D) and ascent (A) were weighted and un-weighted peak accelerations at the front (F1) and rear (R1) wheels (F1_D, R1_D, F1_A, R1_A), since only one peak consistently existed within the acceleration data (Fig. 4). Small and large diameter rear wheel trials were analyzed separately. All analyses were performed using IBM SPSS Statistics software Version 24.0 (IBM Corp., Armonk, NY, USA), with a significance level of $p < 0.05$. Main effects of front and rear wheel are reported in the present study (whereas the main effects of suspension are further expanded upon in Hischke and Reiser, in press).

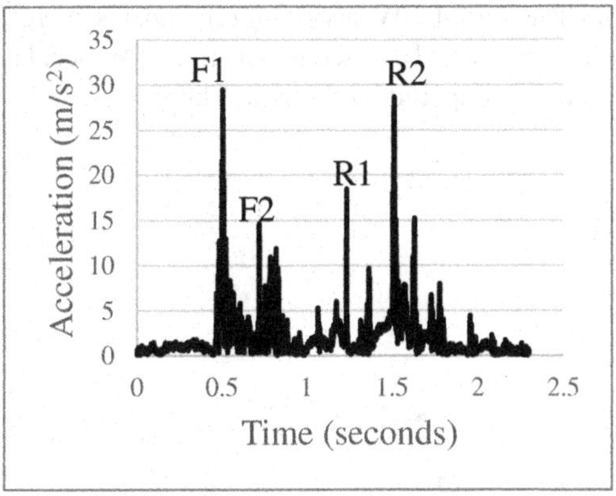

Fig. 3. Raw un-weighted resultant acceleration profile of the wheelchair traversing the door threshold. Peak F1 = when the front caster wheel first hits the door threshold, Peak F2 = when the front caster wheel leaves the door threshold, Peak R1 = when the rear wheel first hits the door threshold, and Peak R2 = when the rear wheel leaves the door threshold.

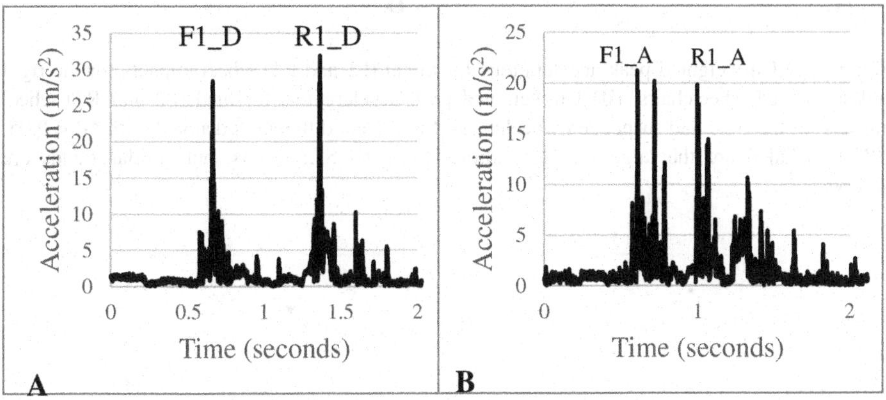

Fig. 4. **(A)** Raw un-weighted resultant acceleration profile of the wheelchair over the 2 cm descent. Peak F1_D = front caster wheel dropping off, and Peak R1_D = rear wheel dropping off. **(B)** Raw resultant acceleration profile of the wheelchair over the 2 cm ascent. Peak F1_A = front caster wheel ascending the obstacle, and Peak R1_A = rear wheel ascending the obstacle.

3 Results

3.1 Door Threshold Peak Accelerations

There were significant differences in un-weighted peak accelerations between the impacts of the front wheel (F1 and F2) versus the rear wheel (R1 and R2) for all conditions (RS, SS, RL, SL) ($p \leq 0.003$) (Fig. 5). Similarly, there were significant differences in FW peak accelerations between the impacts of the front wheel (FW_F1 and FW_F2) versus the rear wheel (FW_R1 and FW_R2)) for all conditions (p ≤ 0.008) (Fig. 6). The initial FW acceleration impacts were greater at the front wheel (compared to the rear wheel), whereas the final FW acceleration impacts were smaller at the front wheel (compared to the rear wheel).

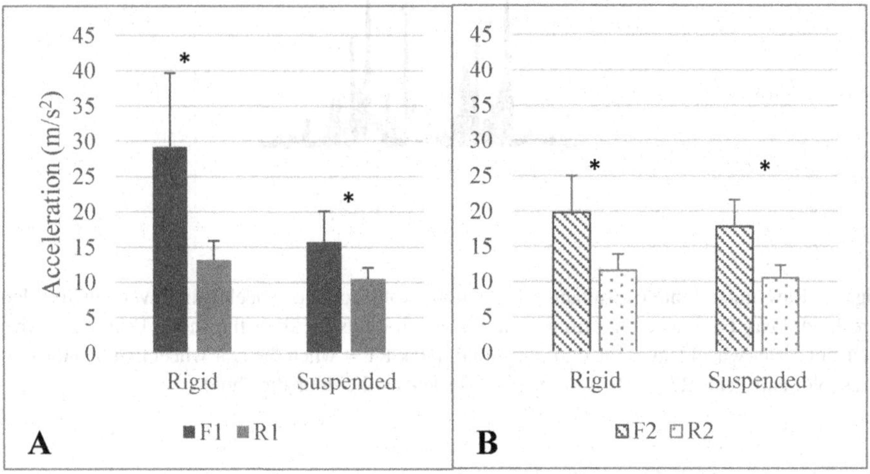

Fig. 5. **(A)** Un-weighted peak accelerations for initial (F1 and R1) wheel impacts for the rigid and suspended wheelchairs. **(B)** Un-weighted peak accelerations for final (F2 and R2) wheel impacts for the rigid and suspended wheelchairs. Significant differences denoted with *P < 0.05, **Not included are the large wheel condition (RL and SL) results, but similar trends are observed.

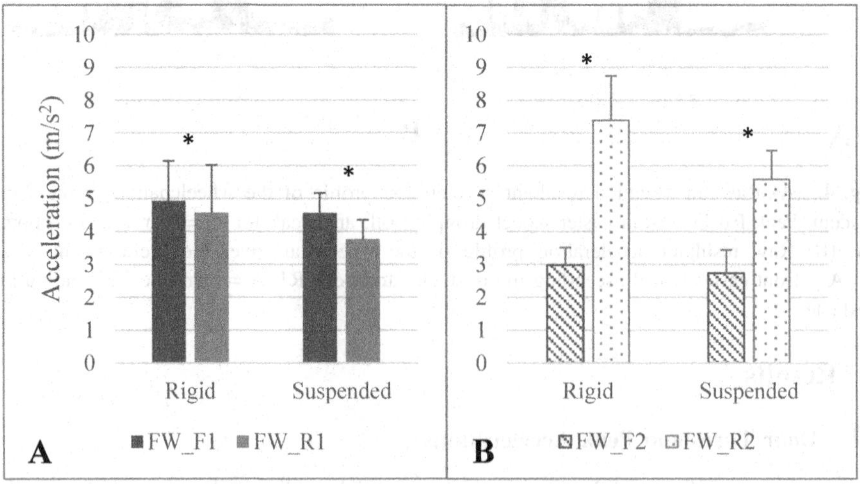

Fig. 6. **(A)** Frequency weighted (FW) peak accelerations for initial (FW_F1 and FW_ R1) wheel impacts for the rigid and suspended wheelchairs. **(B)** Frequency-weighted (FW) peak accelerations for final (FW_F2 and FW_R2) wheel impacts for the rigid and suspended wheelchairs. Significant differences denoted with *P < 0.05, **Not included are the large wheel condition (RL and SL) results, but similar trends are observed.

3.2 Decent and Ascent Peak Accelerations

There were no significant differences between front and rear wheel un-weighted peak accelerations during the 2 cm descent ($p \geq 0.103$) for all conditions (Fig. 7). However, during the 2 cm ascent, the un-weighted peak accelerations at the front wheel were significantly higher than the accelerations at the rear wheel ($p < 0.0001$) (Fig. 7). There were significant differences between front and rear wheel frequency weighted peak accelerations (FW_F1_D and FW_R1_D) during the 2 cm descent and ascent ($p \leq 0.0001$) (Fig. 8). During the 2 cm descent, the rear wheel frequency weighted accelerations were higher than the front wheel accelerations. Conversely, during the 2 cm ascent (FW_F1_A and FW_R1_A), the front wheel (FW_F1_A) frequency weighted accelerations were higher than the rear wheel accelerations (FW_

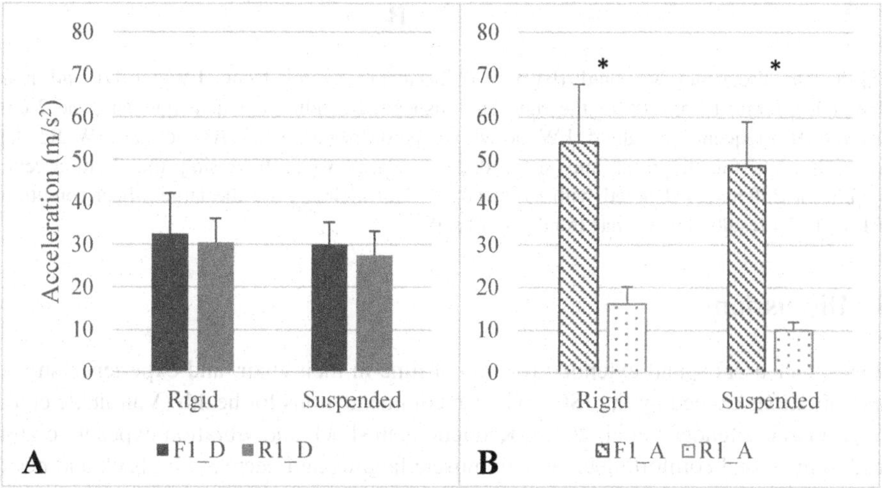

Fig. 7. **(A)** Un-weighted peak accelerations for front (F1_D) and rear (R1_D) wheel impacts for the rigid and suspended conditions when traversing the 2 cm descent. **(B)** Un-weighted peak accelerations for front (F1_A) and rear (R1_A) wheel impacts for the rigid and suspended conditions when traversing the 2 cm ascent. Significant differences denoted with *$P < 0.05$, **Not included are the large wheel condition (RL and SL) results, but similar trends are observed.

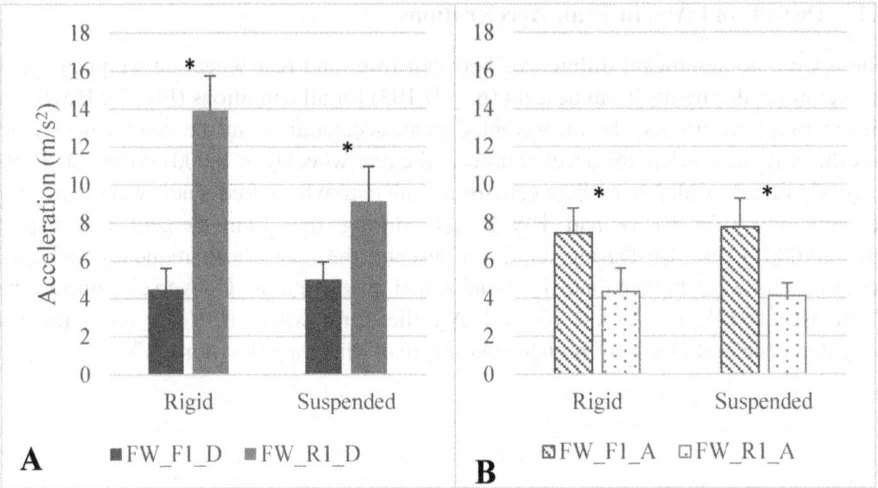

Fig. 8. **(A)** Frequency-weighted (FW) peak accelerations for front (FW_F1_D) and rear (FW_R1_D) wheel impacts for the rigid and suspended conditions when traversing the 2 cm descent. **(B)** Frequency-weighted (FW) accelerations for front (FW_F1_A) and rear (FW_R1_A) wheel impacts for the rigid and suspended conditions when traversing the 2 cm ascent. Significant differences denoted with *$P < 0.05$, **Not included are the large wheel condition (RL and SL) results, but similar trends are observed.

4 Discussion

Wheelchair users spend extended periods of time in their chair, and experience shock and vibrations exceeding the ISO 2631-1 recommendations for health (VanSickle et al. 2001; Garcia-Mendenz et al. 2013). Reductions in shock and vibration exposure could lead to increased comfort, decreases in muscle fatigue, and decreases in back and neck pain, although, there is no conclusive evidence to the percent reduction needed for a decrease in health risks. Therefore, it is recommended to reduce shock and vibration exposure as much as possible. Previous literature has concluded the greatest magnitude shocks occur at the front caster wheel of a wheelchair (Gregg and Derrick 1998). In the Gregg and Derrick (1998) study, the investigators were evaluating the effectiveness of front caster wheel suspension whereas the present experiment was part of the evaluation of rear wheel suspension expanded on in more detail in Hischke and Reiser, in press. The subjects in the Gregg and Derrick (1998) experiment used a manual self-propelled wheelchair while traversing a ramp with a 3 cm drop at the end. Additionally, controlling the speed with a treadmill, the subjects traversed repeated 2 mm bumps. The speeds at which subjects were impacting the obstacles might not be generalizable to all wheelchair users (i.e. tilt-in-space wheelchair users). The present study investigated the differences in initial front and rear wheel impact accelerations (un-weighted and frequency-weighted per ISO 2631-1 standards), and final front and rear wheel impact accelerations for attendant propelled tilt-in-space wheelchair users.

4.1 Un-weighted Peak Accelerations

When traversing the door threshold and 2 cm ascent, the front wheel continuously had significantly greater un-weighted peak accelerations (ranging from 29–80% higher than the rear wheel). The analysis for the door threshold only included comparing the initial impacts of the front and rear wheels, and comparing the final impacts of the front and rear wheels. For the 2 cm descent, there were no significant differences in un-weighted peak accelerations between the front and rear wheel. The present findings were similar to the results from Gregg and Derrick (1998) where they found greater peak accelerations at the front caster wheels in comparison to the rear wheels. Three key differences exist between the present study and Gregg and Derrick (1998) including: (1) different obstacles, (2) the subjects in the present study were pushed by a trained investigator, and (3) different speeds at which subjects were impacting the obstacles. Gregg and Derrick (1998) used a treadmill with 2 mm bumps and a ramp with a 3 cm drop at the end. In contrast, the present study used a door threshold, 2 cm descent, and 2 cm ascent, and subjects were pushed by a trained investigator over the obstacles. Gregg and Derrick (1998) had subjects impacting the obstacles at 1.53–2 m/s. In the present study, the subjects were pushed over the obstacles approximately less than 1 m/s. The differing speed and investigator pushing subjects over the obstacles could explain the smaller peak accelerations at the front wheel compared to those reported in Gregg and Derrick (1998).

Overall, the un-weighted peak accelerations were greater at the front wheel, but during the 2 cm descent there were no differences in the magnitudes of the accelerations. Previous research investigating the impact of front and rear wheel suspension only collecting one peak acceleration (regardless of front or rear wheel) found no differences between the rigid and rear wheel suspended chairs (Cooper et al. 2003). When evaluating rear wheel suspension, some researchers have isolated the shock at the rear wheel to measure the impact of suspension eliminating the need to identify which wheel a peak acceleration was occurring (Kwarciak 2003; Requejo et al. 2008, 2009). As demonstrated by the present study, the front caster wheels may not always experience the highest magnitude accelerations indicating a need for acceleration reductions at both wheels to improve rider comfort and potentially reduce health risks associated with shock exposure. Although initial impacts of the wheels were not compared to final impacts of the wheels, it appears the initial impact is greater than the final impact when two distinct peaks exist for the front wheel and two distinct peaks exist for the rear wheel.

4.2 Frequency – Weighted Peak Accelerations

The ISO 2631-1 frequency-weightings put more emphasis on the frequencies most harmful to the human body; therefore, it is proposed that they are more useful in concluding the injury risk associated experiencing high magnitude shocks when traversing an obstacle (Kwarciak 2003). Thus, peak accelerations were also analyzed by the ISO 2631-1 standards in the present study, and were not included in the work completed by Gregg and Derrick (1998). Traversing the door threshold, 2 cm descent, and 2 cm ascent, frequency-weighted peak accelerations were significantly different between the front and rear wheel impacts. When the front and rear wheel were leaving the door threshold and during the 2 cm descent, the FW peak accelerations were 40–70% higher at the rear wheel. Conversely, during the initial wheel impact on the door threshold and the 2 cm ascent, the FW peak accelerations were 17–60% higher for the front wheels. High magnitude FW peak accelerations occurred equally at the front and rear wheels depending on the type of obstacle. Shock and vibration exposure has been correlated with low back pain, neck pain, muscle fatigue, early degeneration of the lumbar spine, herniated lumbar discs, and spinal discomfort and pain (Bovenzi 1996; Milosavljevic et al. 2011; Ebe and Griffin 2000; Mansfield et al. 2014; Bovenzi et al. 2015; Zimmermann et al. 1993; VanSickle et al. 2001; Requejo et al. 2008; Maeda et al. 2003). With high magnitude FW peak accelerations occurring at both the front and rear wheels equally, future research on evaluating the effectiveness of front and rear wheel suspension might be more comprehensive if ISO 2631-1 frequency-weighted accelerations and multiple peak accelerations are included in the analysis.

Limitations in the present study include recruiting subjects who did not use tilt-in-space wheelchairs and a small sample size, however, other researchers have used non-wheelchairs users and a similar sample size when evaluating suspension systems. Based on the p-values of our main effects, adding only a few more subjects would be unlikely to change the results. It is not desirable to expose wheelchair users to additional shock and vibration in a research study when it is not clear if wheelchair users respond differently to shock and vibration than able-bodied subjects. Other researchers have proposed the ISO 2631-1 standards do not represent the shock and vibration exposure for individuals who use wheelchairs as they have different mass distributions and trunk control than the individuals used deriving the ISO 2631-1 standards (Requejo et al. 2008). However, no alternative approaches have been established when measuring the shock and vibration exposure in this population.

5 Conclusions

To our knowledge, limited research has investigated shock and vibration exposure when using tilt-in-space wheelchairs as most of the previous literature focuses on self-propelled manual wheelchairs or motorized power wheelchairs. Un - weighted peak accelerations were almost always higher at the front wheel, but ISO 2631-1 higher frequency - weighted peak accelerations occurred equally at the front and rear wheel. Therefore, because large accelerations occur at both the front and rear wheel, both front and rear wheel accelerations should be considered when evaluating the effectiveness of wheelchair suspension.

References

Brault M (2010) Americans with disabilities. U.S. Census Bureau. http://www.census.gov/prod/2012pubs/p70-131.pdf. Accessed 10 Dec 2016

Physical and Mobility Impairments: Information and News. Disabled World. https://www.disabled-world.com/disability/types/mobility/. Accessed 3 Mar 2015

Zimmermann CL, Cook TM, Goel VK (1993) Effects of seated posture on erector spinae EMG activity during whole body vibration. Ergonomics 36:667–675

Bovenzi M (1996) Low back pain disorders and exposure to whole-body vibration in the workplace. Semin Perinatol 20:38–53

Ebe K, Griffin MJ (2000) Qualitative models of seat discomfort including static and dynamic factors. Ergonomics 43:771–790

VanSickle DP, Cooper RA, Boninger ML, DiGiovine CP (2001) Analysis of vibrations induced during wheelchair propulsion. J Rehabil Res Dev 3:409–421

Maeda S, Futatsuka M, Yonesaki J, Ikeda M (2003) Relationship between questionnaire survey results of vibration complaints of wheelchair users and vibration transmissibility of manual wheelchair. Environ Health Prev Med 8:82–89

Requejo PS, Kerdanyan G, Minkel J et al (2008) Effect of rear suspension and speed on seat forces and head accelerations experienced by manual wheelchair riders with spinal cord injury. J Rehabil Res Dev 45:985–996

Milosavljevic S, Bagheri N, Vasiljev RM et al (2011) Does daily exposure to whole-body vibration and mechanical shock relate to the prevalence of low back and neck pain in a rural workforce. Ann Occup Hyg 56:10–17

Mansfield NJ, Mackrill J, Rimell AN, MacMull SJ (2014) Combined effects of long-term sitting and whole-body vibration on discomfort onset for vehicle occupants. ISRN Autom Eng 2014:1–8

Bovenzi M, Schust M, Menzel G et al (2015) A cohort study of sciatic pain and measures of internal spinal load in professional drivers. Ergonomics 58:1088–1102

International Organization for Standardization (ISO) (1997) Mechanical vibration and shock-Evaluation of human exposure to whole-body vibration- Part 1: General requirements. ISO, Geneva

Wolf EJ, Cooper RA, DiGiovine CP et al (2004) Using the absorbed power method to evaluate effectiveness of vibration absorption of selected seat cushions during manual wheelchair propulsion. Med Eng Phys 26:799–806

Wolf EJ, Cooper RA, Pearlman J et al (2007) Longitudinal assessment of vibrations during manual and power wheelchair driving over select sidewalk surfaces. J Rehabil Res Dev 44:573–580

Garcia-Mendenz Y, Pearlman JL, Boninger ML, Cooper RA (2013) Health risks of vibration exposure to wheelchair users in the community. J Spinal Cord Med 36:365–375

Dicianna BE, Margaria E, Arva J, et al (2008) RESNA Position on the application of tilt, recline, and elevating legrests for wheelchairs. Assistive Technol 21:1–19

Kwarciak AM, Cooper RA, Fitzgerald SG (2008) Curb descent testing of suspension manual wheelchairs. J Rehabil Res Dev 45:73–84

Requejo PS, Maneekobkunwong S, McNitt-Gray J et al (2009) Influence of hand-rim wheelchairs with rear suspension on seat forces and head accelerations during curb descent landings. J Rehabil Med 41:459–466

Wheelchair Suspension/Shock Absorption (2009). http://atwiki.assistivetech.net/index.php/Wheelchair_suspension/shock_absorption. Accessed 18 May 2017

Instruction Manual: QX-1L (2017). http://quadshox.com/instruction-manual-qx-1l/. Accessed 9 Aug 2017

Cooper RA, Wolf E, Fitzgerald SG et al (2003) Seat and footrest shocks and vibrations in manual wheelchairs with and without suspension. Phys. Med Rehabil 84:96–102

Kwarciak AM (2003) Performance analysis of suspension manual wheelchairs. Master's thesis, University of Pittsburgh, Pittsburgh, Pennsylvania

Gregg MT, Derrick TR (1998, June) Wheelchair vibrations using shock-absorbing front castor forks. Frog Legs Inc. http://cdn.shopify.com/s/files/1/0229/9999/files/Frog_Legs_study.pdf?6260. Accessed 15 Feb 2017

Cooper RA (1995) Rehabilitation engineering applied to mobility and manipulation. Taylor & Francis Group, LLC, New York

Frank AO, De Souza LH (2017) Clinical features of children and adults with a muscular dystrophy using powered indoor/outdoor wheelchairs: disease features, comorbidities and complications of disability. Disabil Rehabil 40:1–7

Hischke M, Reiser RF (in press) Effect of rear wheel suspension on tilt-in-space wheelchair shock and vibration attenuation. Phys Med Rehabil

Comparison of Triage Models of Suspected ACS Patients: A Case Study of the Far Eastern Memorial Hospital

Ray F. Lin[1(✉)], Chieh Lee[1], and Kuang-Chau Tsai[2]

[1] Department of Industrial Engineering and Management, Yuan Ze University, 135 Yuan-Tung Road, Chungli, Taoyuan 32003, Taiwan
juifeng@saturn.yzu.edu.tw
[2] Department of Emergency, Far Eastern Memorial Hospital, 21 Sec. 2, Nanya S. Rd., New Taipei City 220, Taiwan

Abstract. This study aimed at evaluating four existing models, comprising he Zarich's model [1], the flowchart model [2, 3], and the Heart Broken Index (HBI) model [4], for triaging potential acute coronary syndrome (ACS) patients who presented at the emergency department. The 793 clinical cases, randomly selected from 7,962 clinical cases that applied the HBI in the ED of the Far Eastern Memorial Hospital in Taiwan, were used for the model testing. The results showed that although the chest-pain and HBI models had high sensitivity (both 99.24%), they had very low specificity (3.93% and 4.08%), whereas the Zarch's and flowchart models had relatively higher specificity (14.98% and 17.25%), but they had lower sensitivity (96.97% and 93.18%). To increase specificity and maintain high sensitivity while triaging suspected ACS patients, future research can focus on using systematic methods to develop more effective ACS triage models.

Keywords: Acute coronary syndrome · Emergency department
Decision making · Triage

1 Introduction

At the emergency department (ED), triaging potential acute coronary syndrome (ACS) patients to admission of a chest pain unit (CPU) or observation unit (OU) is critical to reduce mortality and morbidity of ACS patients. Chest pain [5] has been accepted as a critical symptom to rapidly triage possible ACS patients for any fast and efficient protocol-driven diagnostic testing. However, the use of chest pain alone is inadequate and would result in false-positive test results and unnecessary downstream procedures [2, 6–8].

Besides chest pain, there were studies that used additional criteria in the triage stage. To make the triage more effective, several models were developed using additional criteria. Zarich et al. [1] expanded the criteria to any male patient over the age of 35 and female over the age of 40 presenting with any non-traumatic chest pain. Sánchez et al. [2] and López et al. [3] derived and validated a five-step triage non-ACS flowchart to rule out ACS patient. The HBI model triaged possible ACS patients based

S. Bagnara et al. (Eds.): IEA 2018, AISC 818, pp. 71–74, 2019.
https://doi.org/10.1007/978-3-319-96098-2_10

on chest pain (with age ≥ 30), epigastric pain, cold sweating, and dyspnea; the suspected ACS patients were those who had the symptom of chest pain or any two of the other three symptoms. However, the predictive performance of the abovementioned triage models has not been tested and compared for Taiwanese.

2 Research Objective

This study aimed at comparing the predictive performance of the four abovementioned ACS triage models, comprising the Zarich's model [1], the flowchart model [2, 3], and the HBI model [4], using clinical data from a Taiwanese hospital.

3 Method

3.1 Testing Data

The testing data were clinical cases that applied the HBI in the ED of the Far Eastern Memorial Hospital that was a 1000-bed regional teaching hospital located in Northern Taiwan having a PCI center. In total, 793 clinical cases were randomly selected from 7,962 clinical cases in a period from June 2012 to December 2014. To collect data, ED nurses went to the medical record room to take pre-randomly-selected paper-based medical records and then typed relevant information into an Excel file. For each case, gender, age, relevant medical history (heart disease, high blood pressure, diabetes mellitus, kidney disease), presenting symptoms at the ED, and the final diagnosis of the patient made by physicians were recorded.

3.2 Testing Models

Four triage models, comprising the chest-pain model, the Zarich's model [1], the ACS flowchart model [2, 3], and the HBI model [4], were tested to predict the two states (ACS patient who should be admitted to an OU or CPU and non-ACS patient) of all the 793 clinical cases. The chest-pain model triaged possible ACS patients simply using chest pain. The Zarich's model triaged patients based on chest pain, with an additional age criterion: male patients over the age of 35 years or female patients over the age of 40 years. The flowchart model ruled out ACS patients without chest pain, or with chest pain but without 1) age of 40 years or younger, 2) diabetes, 3) coronary artery disease (not previously known), and 4) non-retrosternal pain. The HBI model triaged possible ACS patients based on chest pain (with an age of 30 years or older), epigastric pain, cold sweating, and dyspnea; suspected ACS patients were those who had the symptom of chest pain or any two of the other three symptoms.

4 Results

The predictive performance of the four triage models is shown in Fig. 1. As shown in the figure, the chest-pain and HBI models had high sensitivity (both 99.24%). However, they had very low specificity (3.93% and 4.08%). In the other hand, the Zarch's model and flowchart model had relatively higher specificity (14.98% and 17.25%), but they had lower sensitivity (96.97% and 93.18%).

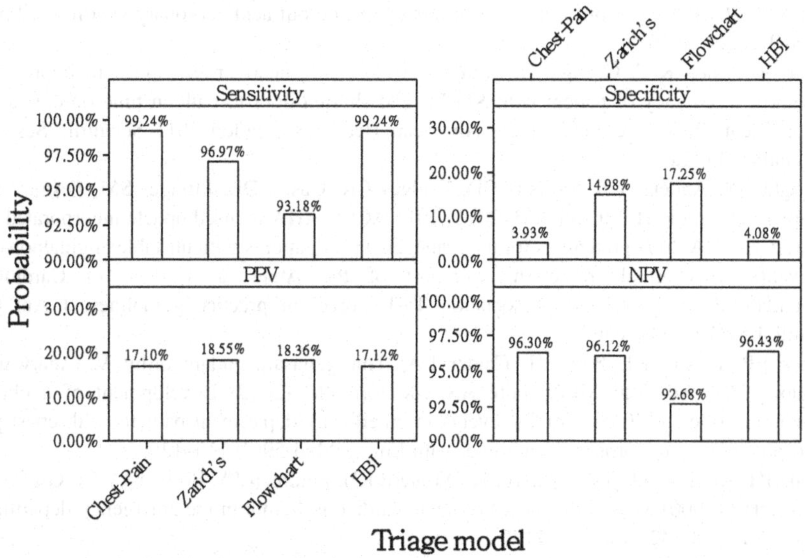

Fig. 1. A figure caption is always placed below the illustration. Short captions are centered, while long ones are justified. The macro button chooses the correct format automatically.

5 Discussion

This preliminary study shows that the four existing ACS triage models could not satisfy the need of EDs, especially when they have limited ED resources, but a lot of visits. Although the chest-pain and HBI models had high sensitivity, they would result in false-positive test results and unnecessary downstream procedures. This is reasonable because a missed diagnosis of ACS may lead to further ischemic events and a potentially preventable death or disability. Hence, hospitals would like to apply this low-threshold method to triage possible ACS patients. However, with unnecessary electrocardiography inspections and cardiac biomarker measurements these procedures expose patients to ionizing radiation and increase costs and workload of ED physicians and nurses. To increase specificity and maintain high sensitivity while triaging suspected ACS patients, future research can focus on using systematic methods to develop more effective ACS triage models.

References

1. Zarich SW, Sachdeva R, Fishman R, Werdmann MJ, Parniawski M, Bernstein L, Dilella M (2004) Effectiveness of a multidisciplinary quality improvement initiative in reducing door-to-balloon times in primary angioplasty. J Interv Cardiol 17(4):191–195
2. Sánchez M, López B, Bragulat E, Gómez-Angelats E, Jiménez S, Ortega M, Coll-Vinent B, Alonso JR, Queralt C, Miró Ò (2007) Triage flowchart to rule out acute coronary syndrome. Am J Emerg Med 25(8):865–872
3. López B, Sánchez M, Bragulat E, Jiménez S, Coll-Vinent B, Ortega M, Gómez-Angelats E, Miró Ò (2010) Validation of a triage flowchart to rule out acute coronary syndrome. Emerg Med J 28:841–846
4. Hsu J-C, Chen K-C, Cheng I-N, Li A-H (2011) Using heart broken index to improve the diagnostic accuracy of patient with STEMI and shorten door-to-balloon time on emergency department. Paper presented at the American Heart Association 2011 Scientific Sessions, Orlando, Florida
5. Wright RS, Anderson JL, Adams CD, Bridges CR, Casey DE, Ettinger SM, Fesmire FM, Ganiats TG, Jneid H, Lincoff AM (2011) 2011 ACCF/AHA focused update incorporated into the ACC/AHA 2007 guidelines for the management of patients with unstable angina/non–ST-elevation myocardial infarction: a report of the American College of Cardiology Foundation/American Heart Association Task Force on practice guidelines. J Am Coll Cardiol 57(19):e215–e367
6. Hess EP, Brison RJ, Perry JJ, Calder LA, Thiruganasambandamoorthy V, Agarwal D, Sadosty AT, Silvilotti MLA, Jaffe AS, Montori VM (2012) Development of a clinical prediction rule for 30-day cardiac events in emergency department patients with chest pain and possible acute coronary syndrome. Ann Emerg Med 59(2):115–125
7. Pope JH, Aufderheide TP, Ruthazer R, Woolard RH, Feldman JA, Beshansky JR, Griffith JL, Selker HP (2000) Missed diagnoses of acute cardiac ischemia in the emergency department. N Engl J Med 342(16):1163–1170
8. Rosenfeld AG, Knight EP, Steffen A, Burke L, Daya M, DeVon HA (2015) Symptom clusters in patients presenting to the emergency department with possible acute coronary syndrome differ by sex, age, and discharge diagnosis. Heart Lung J Acute Crit Care 44(5):368–375

Comparative Analysis of Subjective Workload in Laparoscopy and Open Surgery Using NASA-TLX

Giovanni Miranda[1(✉)], Mario Casmiro[2,3], Giorgio Cavassi[1], Riccardo Naspetti[4], Egidio Miranda[5], Riccardo Sacchetti[6], Emanuele Dabizzi[7], and Rosario Tranchino[8]

[1] Ergonomics Unit, Faentia Consulting, Faenza, RA, Italy
gmiranda@faentia-consulting.com
[2] Research & Development Unit, Faentia Consulting, Faenza, RA, Italy
[3] Unit of Neurology, Ospedale per gli Infermi, AUSL della Romagna, Faenza, RA, Italy
[4] Digestive Endoscopy, Azienda Ospedaliero - Universitaria Careggi, Florence, Italy
[5] Unit of General Surgery, Ospedale del Casentino, Bibbiena, AR, Italy
[6] Oncologic Surgery, Azienda Ospedaliero - Universitaria Careggi, Florence, Italy
[7] Digestive Endoscopy, Policlinico, Modena, Italy
[8] Unit of General Surgery, Ospedale per gli Infermi, AUSL della Romagna, Faenza, RA, Italy

Abstract. Background: The mental resources required by laparoscopy (LS) and open (OS) surgery may be different; if this hypothesis is correct, the analysis of subjective total workload (STW) could allow to identify the causes of such differences and reduce the risk of error.
Objective: We tested the hypothesis that STW is different between LS and OS.
Methods: The NASA-TLX questionnaire was self-administered by trained physicians at the end of each procedure; STW was calculated using NASA-TLX software.
Results: Fourteen surgeons performed 66 LS and 48 OS procedures. The OS group showed a higher STW. Sub-item analysis showed higher temporal demand and frustration values in the OS group. In both groups STW was not normally distributed, showing a high (HWS) and a low (LWS) STW subgroup; the HWS within the LS group exhibited a higher mental and physical demand.
Conclusions: NASA-TLX is a valuable tool for assessing STW in the surgical setting. Higher STW was observed in the OS group, possibly related to a longer duration of such procedures and a greater experience of the "open surgeons". These results should be viewed with caution because of potentially confounding variables; larger studies will be required to identify STW determinants among different surgical groups.

Keywords: Ergonomics · Laparoscopy · NASA-TLX · Mental workload

S. Bagnara et al. (Eds.): IEA 2018, AISC 818, pp. 75–84, 2019.
https://doi.org/10.1007/978-3-319-96098-2_11

1 Introduction

Surgeons operating with the laparoscopic technique are at risk of developing several physical injuries, among which have been reported joint diseases (shoulder and wrist) and diseases of lumbosacral spine. These injuries may be related to the following causal factors: the postures (static and unfavourable) maintained for prolonged periods [1, 2], the type of movements required for upper limbs (often outside the range set by International Standards used in the design of workstations), the fixed position of the "input ports" of the instruments, the use of devices burdened with a "user-gap" (namely, designed without taking into account the possibility, for example, that different surgeons have different hand sizes) and the greater effort (up to six times for hand grip) required by the surgeon because of the very nature of the operation and of the tools which are used [2–5]. The frequency of these pathologies may also depend on the specific role of the surgeon during the operation.

Moreover, laparoscopic surgeons could also be subject to a higher "cognitive load" compared to open surgery, depending on several factors: they do not have a direct visual contact with the instruments and the operating field, they do not have a direct feedback from arms and hands, they must "blend" what they observe on the display with mechanical feedback (in some instances, for example in case of ERCP, the task of mixing the information is made even more difficult by the need to use multiple monitors and the relative positions of these monitors), and the fact of working on two-dimensional images of the operative field [2], the so-called scaling effect [6]. Therefore, it could be hypothesized that the share of "mental resources" that the laparoscopic surgeon must put in place could be higher than that required by open surgeons.

The National Aeronautics and Space Administration-Task Load Index (NASA-TLX) is a multidimensional tool used for assessing subjective workload. The tool is based on a weighted average of the score obtained in six subscales (mental, physical, temporal demand, performance, effort and frustration). The scores of the six subscales are combined and weighted according to the subjective importance assigned by the subject in relation to the specific task assigned, rather than to the relevance assigned a priori. NASA-TLX has been validated by several authors [7–10] in different experimental conditions (flight simulation, simulation control surveillance, laboratory tasks), showing lesser inter-observer variability with respect to other one-dimensional methods for the estimation of workload.

In this study we tested the hypothesis that subjective workload determined by means of NASA-TLX is higher in surgeons performing laparoscopic interventions compared with those performing open surgery.

2 Materials and Methods

Surgeons recruited patients at five Italian surgical centres (Digestive Endoscopy and Oncologic Surgery of the Azienda Ospedaliero-Universitaria Careggi, Firenze; Unit of General Surgery, Ospedale per gli Infermi, Faenza, Ravenna; Unit of General Surgery, Casentino Hospital, Bibbiena, Arezzo; Digestive Endoscopy, Policlinico, Modena).

The NASA-TLX questionnaire was completed by trained physicians at the end of the surgical procedure. The NASA-TLX software was used to calculate the subjective total workload (STW) and single items' values from each paper questionnaire. The resulting value provided by the NASA-TLX is not an absolute value. For this reason we compared the mean values between the two groups, laparoscopic and open surgery.

3 Statistical Analysis

Statistical analysis was performed using SPSS for Windows (version 13.0, SPSS Inc.) and Epi Info (version 3.5.1). Comparisons between the two groups were performed with a parametric test for inequality of population means (ANOVA). The Bartlett's Test for Inequality of Population Variances was used when variances were not homogeneous (ANOVA potentially inappropriate): in these cases the Mann-Whitney/Wilcoxon two-sample test (Kruskal-Wallis test for two groups) was used instead.

4 Results

In this study, 66 procedures were performed in the laparoscopy (LS) group and 48 in the open surgery (OS) group. The general features of the two groups of surgeons who completed the questionnaires are described in Table 1.

Table 1. general features of the surgeons performing laparoscopic and open surgery procedures

	Laparoscopy	Open surgery	p**
Gender (male/female ratio)	61/5	48/0	–
Age (years)*	40.8 ± 9.1	50.8 ± 9.8	0.000
Experience (years of activity)*	9.5 ± 6.9	20.2 ± 11.1	0.000
Duration of surgical procedure (minutes)*	98.9 ± 87.4	120.8 ± 47.6	0.001
Surgical specialization (%)			
- general surgery	80.3	95.8	–
- digestive tract surgery	19.7	0	–
- urology	0	4.2	–
Role of the operator (%)			
- 1st operator	81.8	87.5	–
- 2nd operator	16.7	10.4	–
- 3rd operator	1.5	2.1	–

*Values are expressed as mean ± standard deviation. **Kruskal-Wallis test for two groups

The male:female ratio was comparable in both groups and showed a clear-cut preponderance of male sex. The surgeons in the OS group were significantly older. The surgeons' "experience" was defined in terms of years of surgical activity for each surgeon and was significantly greater in the OS group. The duration of the surgical operations was significantly longer in the OS group (119.8 ± 48.9 vs 98.9 ± 87.4 min; p = 0.0026). The vast majority of surgeons had a specialty in general surgery (laparoscopy = 80.3%, open surgery = 95.3%); compared to digestive tract surgery (laparoscopy = 19.7%, open surgery = 0%) and urology (laparoscopy = 0%, open surgery = 4.7%). The questionnaire was completed by the 1st operator of the intervention in more than 80% of cases in both groups, by the 2nd operator in less than 20% of cases, and by the 3rd operator in 1–2% of cases.

4.1 Type of Surgical Intervention

Surgical procedures included in the LS group were mainly represented by video-laparo-cholecystectomy (63.6%), followed by appendectomy (10.6%) and abdominoplasty (7.6%). Hemicolectomy, Nissen fundoplication, adrenalectomy, anterior rectal resection, peptic ulcer repair, fifth hepatic segment resection accounted for less than 3 cases each.

In the OS group the series included: hemicolectomy (23.3%), gastrectomy (18.6%), thyroidectomy (11.6%) and abdominoplasty (9.3%); appendectomy, exploratory laparotomy, transvescical adenomectomy, cholecystectomy, ileocecal resection, anterior rectal resection and nephrectomy accounted for less than 3 cases each.

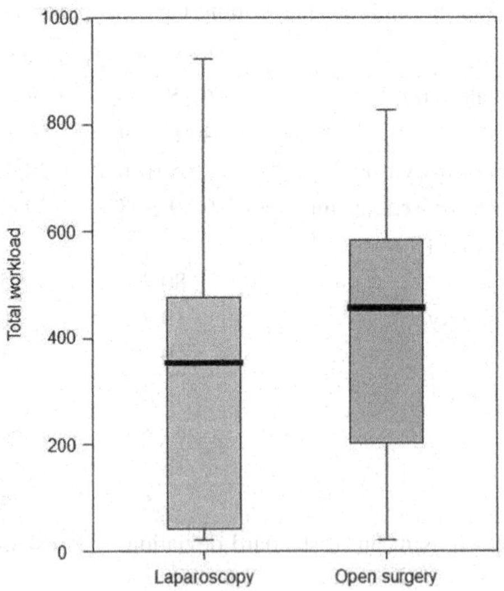

Fig. 1. Boxplots showing the distribution of the values of NASA-TLX total workload in laparoscopy (left) and open surgery (right). The lower, upper and middle lines of boxes represent respectively the 25th percentile, 75th percentile and median values. The top and bottom whiskers represent the 95th and 5th percentiles, respectively.

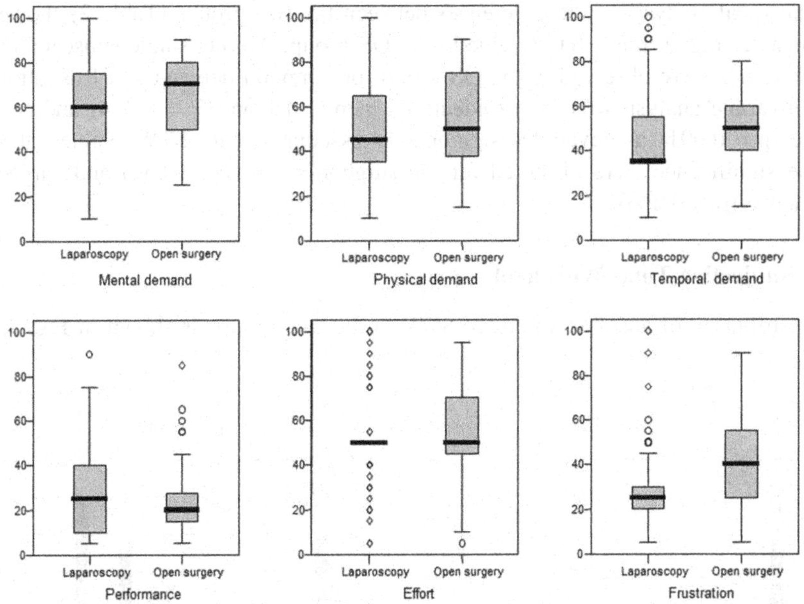

Fig. 2. Boxplots show the distribution of NASA-TLX subscales values for LS (left) and OS (right). Lower, upper and middle lines of boxes represent respectively the 25th percentile, 75th percentile and median values. The top and bottom whiskers represent the 95th and 5th percentiles, respectively; open circles and diamonds represent outliers.

The boxplots in Figs. 1 and 2 show, respectively, the distribution of the values of STW and of single subscales (mental, physical and temporal demands, performance, effort and frustration) for both groups.

Table 2. NASA-TLX: comparison of STW and single items values between LS and OS groups

	Laparoscopy	Open Surgery	p**
Total workload*	316.8 ± 265.5	392.4 ± 232.4	0.045
Single items*			
- Mental demand	59.6 ± 24.9	65.4 ± 18.9	0.231
- Physical demand	51.4 ± 22.4	53.9 ± 20.6	0.433
- Temporal demand	45.8 ± 21.6	51.6 ± 16.9	0.026
- Performance	30.5 ± 22.7	24.5 ± 16.0	0.351
- Effort	52.2 ± 20.7	53.9 ± 18.8	0.250
- Frustration	29.5 ± 20.8	44.3 ± 21.9	0.000

*Values are expressed as mean ± standard deviation. **Mann-Whitney/Wilcoxon two-sample test (Kruskal-Wallis test for two groups).

Statistical analysis of the differences between the two groups (Table 2) showed a significantly higher mean STW value in the OS group. Among single subscale items, higher values were observed in the OS group for temporal demand and frustration.

Univariate analysis of variance identified mental demand ($p = 0.004$) and performance ($p = 0.001$) as covariates significantly associated with STW; values of borderline significance were observed for the surgeon's age ($p = 0.052$) and surgeon's experience ($p = 0.066$).

4.2 Subjective Total Workload

The distribution of absolute values of STW in the two groups is shown in Fig. 3.

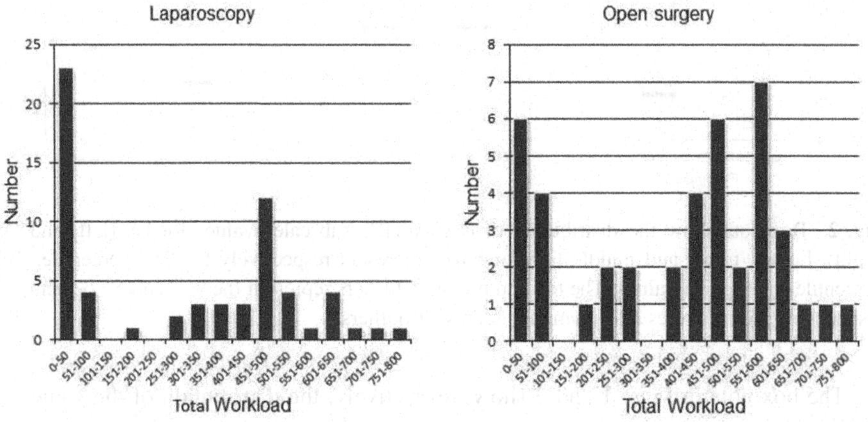

Fig. 3. Distribution of total workload in the laparoscopy (left) and open surgery (right) groups.

In both groups the distribution of STW was bimodal; setting an arbitrary cut-off value at 100, we identified a "low workload subgroup" (LWS: values $\leq$ 100) and a "high workload subgroup" (HWS: values > 100). In the LS group, LWS accounted for 42.9% of cases and a second, broader peak for HWS was observed around 500. In the OS group LWS accounted for 23.3% of cases and the second peak was observed around 500–600.

4.3 Subgroup Analysis

Table 3 shows the results of the comparison of the features of the two subgroups (LWS and HWS). No significant difference was observed with regards to the demographic features, experience of the surgeons, and duration of surgical interventions. Among single item values of NASA-TLX, mental and physical demand were significantly higher in the LS HWS subgroup.

Table 3. Comparison of the general features of the surgeons performing laparoscopic and open surgery procedures, subdivided by total workload

	Laparoscopy			Open surgery		
	LWS	HWS	p**	LWS	HWS	p**
Gender (male/female ratio)	25/2	36/3	–	11/0	37/0	–
Age (years)*	38.9 ± 8.5	42.0 ± 9.3	0.256	52.1 ± 9.5	50.4 ± 10.0	0.730
Experience (years of activity)*	8.8 ± 6.0	10.0 ± 7.5	0.652	23.3 ± 10.4	19.3 ± 11.2	0.205
Duration of surgical procedure (minutes)*	85.7 ± 49.1	108.1 ± 105.9	0.901	120.5 ± 40.8	120.9 ± 49.9	0.674
Surgical specialization (%)						
general surgery	81.5	79.5	–	90.9	97.3	–
digestive tract surgery	18.5	20.5	–	0	0	–
urology	0	0	–	9.1	2.7	–
Role of the operator filling in the questionnaire (%)						
1st operator	85.2	79.5	–	90.9	86.5	–
2nd operator	11.1	20.5	–	0	13.5	–
3rd operator	3.7	0	–	9.1	0	–
Values of single subscales*						
Mental demand	49.1 ± 27.8	66.9 ± 20.1	0.003	61.4 ± 16.9	66.6 ± 19.5	0.401
Physical demand	44.8 ± 21.7	55.9 ± 21.9	0.037	49.5 ± 20.1	55.1 ± 20.9	0.336
Temporal demand	41.3 ± 21.9	48.8 ± 21.0	0.108	50.5 ± 14.4	51.9 ± 17.7	0.711
Performance	25.5 ± 16.0	33.8 ± 25.9	0.454	23.6 ± 15.2	24.7 ± 16.5	0.651
Effort	47.6 ± 19.8	55.4 ± 20.9	0.123	48.6 ± 19.5	55.5 ± 18.6	0.299
Frustration	30.6 ± 19.1	28.8 ± 22.1	0.382	47.3 ± 17.7	43.4 ± 23.1	0.489

LWS: low (≤ 100) workload subgroup; HWS: high (>100) workload subgroup. *Values are expressed as mean ± standard deviation. **Kruskal-Wallis test for two groups

5 Discussion

Supporting and enhancing a surgeon's cognitive capabilities has been indicated as one of the main goals of both training and engineering interventions in laparoscopy [11]. Simplifying the terms of the problem, given a defined amount of mental resources available for a task, the surgeon will be required to allocate them between a primary task (the execution of the surgical procedure via instruments and visual feedback) and a secondary task (the process of decision making throughout the intervention aimed at determining the better strategy to be adopted, taking account of unforeseen variations that can occur during surgery). Therefore, if at a certain moment of the intervention an increased amount of resources will be required for the primary task, fewer will be available for the secondary one; this could increase the risk of error if a strategy re-modulation is required during the surgical intervention. This hypothesis supports the need for relevant cognitive measures in evaluation protocols of surgical skills and performance in laparoscopy [11].

The problem of a surgeons' mental workload has been addressed in several papers. Youssef et al. [12] used NASA-TLX to assess mental workload with regards to a surgeon's standing position during laparoscopic cholecystectomy performed on a virtual reality simulator. In this study they reported a significant association between the side-standing position and high physical demand, effort, and frustration, whereas the two-handed technique in the side-standing position required more effort compared to the one-handed technique. NASA-TLX was also utilized by Yurko et al. [13] to study 28 novices during simulator training on a complex laparoscopic task (Nissen fundoplication in animal model). In this study they showed that the NASA-TLX scores declined during training and significantly correlated with performance scores and that higher workload scores correlated with an increased number of inadvertent injures. The authors concluded that NASA-TLX may help identify individuals that experience higher workload during skill transfer to the clinical environment and are therefore more prone to errors. In a porcine Nissen fundoplication model, robotic assistance was shown to significantly improve suturing performance while decreasing novice's workload [14]. In an experimental setting comparing localizing anatomical structures in either three-dimensional (3D) or two-dimensional (2D) views, 3D viewing produced lower values of mental workload (specifically within the mental demand component) [15].

Vocal versions of the NASA-TLX are available and are acceptable alternatives to standard written formats, provided that caution be used when comparing individual subscales between studies using different administration modalities [16]. In addition, the mental workload of anaesthetists during routine surgical procedures (measured using NASA-TLX) showed a correlation between response times and self-reported mental load, physical load and frustration [17]. NASA-TLX was also identified [18] as the most reliable and valid questionnaire to measure workload of nurses in the Intensive Care Unit setting. The NASA-TLX has been proposed by Young et al. [19] to assess the subjective workload in perianesthesia nursing, with the aim to support decision making in conditions of high pressure with potentially life threatening implications. Finally, some authors [20] have also proposed a different test for the evaluation of cognitive tasks (cognitive task load index or CTLX) which is considered to be psychometrically pure and allows summing the scale scores.

Our study was designed as a pilot study aimed at testing the NASA-TLX as a tool for assessing subjective workload in the surgical setting and at generating hypotheses about factors determining the total workload for possible ergonomic interventions. Contrary to our initial hypothesis, we found higher subjective workload for the open surgery; this result could correlate with longer duration of surgical procedures in this group and with longer experience of "open surgeons"; the determining factors could be represented by the amount of mental and perceptual activity required, a more demanding task, and how successful or satisfied the surgeon feels in accomplishing the goal. However, the results of our study should be viewed with caution because of several potentially confounding variables, some of which include the relatively small sample size, the absence of a matching for the type of intervention and for the learning curve in laparoscopy. Larger studies, controlling for the main variables such as the type and duration of the intervention and the experience of the surgeons, will be required to identify the distribution of subjective workload among different surgical groups and its determinants in order to introduce improvement actions.

In conclusion, NASA-TLX is a valuable tool for assessing the STW in the surgical setting as well as in experimental conditions and surgical training. The mental workload of the surgeon may correlate with the duration of surgery and with the surgeon's experience, however because of the several potentially confounding variables involved, larger studies will need to be performed to confirm such conclusions.

Acknowledgements. Preliminary findings of this study were presented at the 14th Italian Congress of Neuroepidemiology, held in Milan, Italy, November 21–22, 2014. The authors thank Luca Vignatelli, MD, for his valuable suggestions in the reading of the data and the following surgeons who collaborated in the collection of cases: Andrea Valeri, MD, Andrea Rinnovati, MD, Bernardo Boffi, MD, Chiara Linari, MD, Silvia Nesi, MD, Alessandra Vegni, MD, Lapo Bencini, MD, Luis Jose Sanchez, MD, Marco Bernini, MD, Marco Farsi, MD, Massimo Calistri, MD, Nicola Antonacci, MD, and Silvia Aldrovandi, MD.

References

1. Toffola ED, Rodigari A, Di Natali G, Ferrari S, Mazzacane B (2009) Postura ed affaticamento dei chirurghi in sala operatoria [Abstract in English]. G Ital Med Lav Ergon 31 (4):414–448
2. Berguer R (1999) Ergonomics & Laparoscopic Surgery. In: Kavic MS, Levinson CJ, Wetter PA (eds) Prevention and management of laparoendoscopic surgical complications. Society of Laparoendoscopic Surgeons, Miami, pp 8–11
3. Berguer R (1999) Surgery and ergonomics. Arch Surg 134(9):1011–1016
4. Berguer R, Hreljac A (2004) The relationship between hand size and difficulty using surgical instruments: a survey of 726 laparoscopic surgeons. Surg Endosc 18(3):508–512
5. Supe AN, Kulkarni GV, Supe PA (2010) Ergonomics in laparoscopic surgery. J Minim Access Surg 6(2):31–36
6. Hodgson AJ, Person JG, Salcudean SE, Nagy AG (1999) The effects of physical constraints in laparoscopic surgery. Med Image Anal 3(3):275–283
7. Hart SG, Staveland LE (1988) Development of NASA-TLX (task load index): results of empirical and theoretical research. In: Hancock PA, Meshkati N (eds) Human mental workload. North Holland, Amsterdam, pp 1–46
8. Reid GB, Nygren TE (1988) The subjective workload assessment technique: a scaling procedure for measuring mental workload. In: Hancock PA, Meshkati N (eds) Human mental workload. North Holland, Amsterdam, pp 185–218
9. Vidulich MA, Tsang PS (1985) Assessing subjective workload assessment: a comparison of SWAT and the NASA-bipolar methods. In: Proceedings of the human factors society twenty-ninth annual meeting. Human Factors Society, Santa Monica (CA), pp 71–75
10. Vidulich MA, Tsang PS (1986) Techniques of subjective workload assessment: a comparison of SWAT and the NASA-bipolar methods. Ergonomics 29(11):1385–1398
11. Carswell CM, Clarke D, Seales WB (2005) Assessing mental workload during laparoscopic surgery. Surg Innov 12(1):80–90

12. Youssef Y, Lee G, Godinez C, Sutton E, Klein RV, George IM, Seagull FJ, Park A (2011) Laparoscopic cholecystectomy poses physical injury risk to surgeons: analysis of hand technique and standing position. Surg Endosc 25(7):2168–2174
13. Yurko YY, Scerbo MW, Prabhu AS, Acker CE, Stefanidis D (2010) Higher mental workload is associated with poorer laparoscopic performance as measured by the NASA-TLX tool. Simul Healthc 5(5):267–271
14. Stefanidis D, Wang F, Korndorffer JR Jr, Dunne JB, Scott DJ (2010) Robotic assistance improves intracorporeal suturing performance and safety in the operating room while decreasing operator workload. Surg Endosc 24(2):377–382
15. Foo JL, Martinez-Escobar M, Juhnke B, Cassidy K, Hisley K, Lobe T, Winer E (2013) Evaluating mental workload of two-dimensional and three-dimensional visualization for anatomical structure localization. J Laparoendosc Adv Surg Tech A 23(1):65–70
16. Carswell CM, Lio CH, Grant R, Klein MI, Clarke D, Seales WB, Strup S (2010) Hands-free administration of subjective workload scales: acceptability in a surgical training environment. Appl Ergon 42(1):138–145
17. Byrne AJ, Oliver M, Bodger O, Barnett WA, Williams D, Jones H, Murphy A (2010) Novel method of measuring the mental workload of anaesthetists during clinical practice. Br J Anaest 105(6):767–771
18. Hoonakker P, Carayon P, Gurses A, Brown R, McGuire K, Khunlertkit A, Walker JM (2011) Measuring workload of ICU nurses with a questionnaire survey: the NASA task load index (TLX). IIE Trans Healthc Syst Eng 1(2):131–143
19. Young G, Zavelina L, Hooper V (2008) Assessment of workload using NASA task load index in perianesthesia nursing. J Perianesth Nurs 23(2):102–110
20. Bridger RS, Brasher K (2011) Cognitive task demands, self control demands and the mental well-being of office workers. Ergonomics 54(9):830–839

Handover: An Experimentation in the Territory Hospital Transition

Mario D'Amico[1(✉)], Maria Maddalena Freddi[1], Amedeo Baldi[2], and Chiara Lorenzini[2]

[1] U.O.S. Clinical Risk North Area ASL Tuscany North West, Tuscany, Italy
mario.damico@uslnordovest.toscana.it

[2] U.F. Lunigiana Community Health Activities ASL Tuscany North West, Tuscany, Italy

Abstract. Handover is considered one of the most delicate moments of medical activity as it requires the utmost attention, in fact there has been an increase in adverse events that have occurred with regard to taking care of the patient. The purpose of the drafting and implementation of the good practice is to start a standardization process to ensure the flow of a minimum set of information necessary for the professionals who are called to assist the patient.

The study that we have set ourselves in this experimentation is to verify the correct passage of information from a hospital ward to an intermediate territorial structure. The aim is to draft a standard passage document containing all the necessary clinical information so that local health workers can manage the patient in an appropriate manner.

As we shared the document necessary for a correct transition, the expected results are that for each patient who accesses the intermediate care places, the document containing the clinical information necessary for a patient's arrival arrives in the accepting structure, correct allocation and care management

Keywords: Internal medicine · Intermediate care

1 Introduction

Handover is considered one of the most delicate moments of medical activity as it requires the utmost attention, in fact there has been an increase in adverse events that have occurred with regard to taking care of the patient, for lacking communication between operators and misunderstandings due to incomplete or inadequate information about the patient's clinical condition at the level of an acceptable functional structure. The purpose of the drafting and implementation of the good practice is to start a standardization process to ensure the flow of a minimum set of information necessary for the professionals who are called to assist the patient, to do it correctly and above all safe with regard to diagnostic and therapeutic decisions.

S. Bagnara et al. (Eds.): IEA 2018, AISC 818, pp. 85–88, 2019.
https://doi.org/10.1007/978-3-319-96098-2_12

2 State of the Art

The Tuscany Region has drafted a Patient Safety Practice that deals with the passage of information. The information sheet of the Good Practice specifies that the two units of passage of the patient, the issuing and the receiving, must agree on what are the minimum information that must reach the medical staff of the receiving unit so that the patient is gestioti, from the point of assistance, adequately and above all in safety.

3 Objective and Methods

The study that we have set ourselves in this experimentation is to verify the correct passage of information from a hospital ward to an intermediate territorial structure. The aim is to draft a standard passage document containing all the necessary clinical information so that local health workers can manage the patient in an appropriate manner.

In the Lunigiana Zone of the ASL Toscana northwest, two functional structures were identified, one representing the issuing department and the other the receiving department. The two wards of Internal Medicine of the Hospitals of Fivizzano and Pontremoli (respectively 23 and 25 beds) have been identified as the issuing department, while the receiving units have identified the territorial structures in which the intermediate care beds are allocated (6 beds at the BIC Hospital of Fivizzano, 4 beds at the Cabrini Institute in Pontremoli, 4 beds at the Sanatrix in Aulla and 2 beds at the RSA La Fontana d'Oro in Mommio). In a preliminary meeting, a paper tool was shared which identified the minimum clinical information necessary for the recipients to assist the patient who was discharged from the two medicines and transferred, for clinical or social conditions, to the treatment beds intermediate.

The evaluation items were:

(1) Reporting form
(2) Registry
(3) Reference family
(4) Relevant assistance aspects
(5) Evaluation scales
(6) Main pathology
(7) Concurrent diseases
(8) Pharmacological therapy
(9) Signaling of pressure ulcers
(10) Venous access signaling
(11) Pain
(12) State feed

In July–December 2016, all the health records of patients admitted to the intermediate care beds that came from the Departments of Medicine were checked to see if they were present and if the transition form had been completed correctly (Fig. 1).

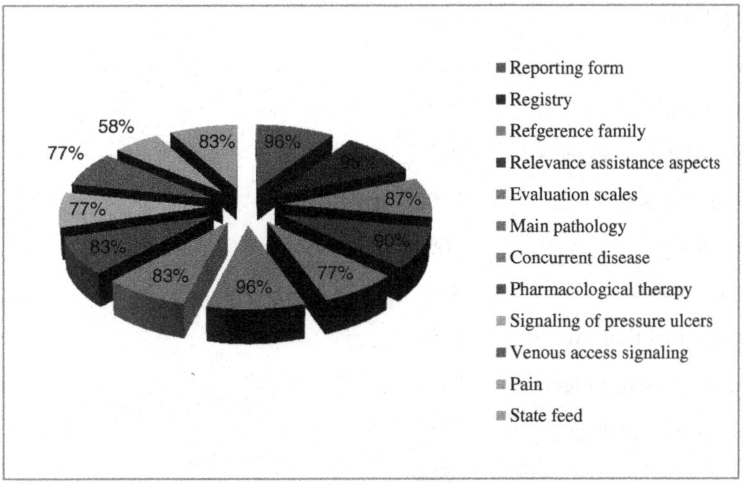

Fig. 1. Evaluation items

4 Results and Discussion

During the period indicated, the clinical risk of the Territorial Territory of Massa Carrara was received, the communication of the names of the patients who are sent from the issuing department to the receiving department, within 5 days from the arrival of the patient in the receiving department.

At the time of sending the patient to the receiving department, the issuing department had to fill in the form attached to the Project and send it to the receiving department at the same time as the patient arrived.

The clinical staff of the Clinical Risk has provided, every two weeks, to carry out check-ups in the receiving departments to verify the presence and correct compilation of the card in question.

51 patients were examined including 25 males and 26 females.

The origin was: 27 from the Department of Medicine of the Hospital of Pontremoli and 24 from that of the Hospital of Fivizzano. The destination of these patients was as follows: 24 at the RSA Cabrini, 25 at the BIC of the Fivizzano Hospital, 1 at the RSA of Mommio and 1 at the RSA Sanatrix. Average age was 78.7 years (77.8 males 79.5 females).

The data are referred to the presence of the item compiled correctly are the following

Reporting form	96%
Registry	96%
Reference family	87%
Relevant assistance aspects	90%
Evaluation scales	77%
Main pathology	96%
Concurrent diseases	83%
Pharmacological therapy	83%
Signaling of pressure ulcers	77%
Venous access signaling	90%
Pain	58%
State feed	83%

5 Conclusion and Perspectives

The detection of the data obtained suggests that there has been a good adhesion by the personnel concerned to the use of the agreed instrument for the transmission of information between the two units; the transmitter and the receiving one.

It has also been reported that in some cases the timely arrival of the model duly completed to the receiving unit has failed and, in a single case, a reminder has been necessary for the documentation to reach its destination.

There is also a lack of information regarding the detection of pain. This practice was highlighted by regional regulatory acts and by the approval of the information sheet of a practice for patient safety; however, it should be noted that not all operators have reached that degree of sensitivity that allows it to trace this essential aspect for proper care of the patient.

This suggests that it is appropriate to provide training opportunities for health personnel who implement the application of the good practice itself.

References

1. Factsheet Safety practice Handover Regione Toscana (2016)
2. Poletti P (2012) Handover: Passing Key Care Deliveries Care 4/2012 24–32
3. Jeffcott SA, Evans SM, Cameron PA et al (2009) Improving measurement in clinical handover. Qual Saf Health Care 18:272–276
4. Hesselink G, Zegers M, Vernooij-Dassen M, Barach P, Kalkman C, Flink M, Öhlen G, Olsson M, Bergenbrant S, Orrego C, Suñol R, Toccafondi G, Venneri F, Dudzik-Urbaniak E, Kutryba B, Schoonhoven L, Wollersheim H, European HANDOVER Research Collaborative (2014) Improving patient discharge and reducing hospital readmissions by using intervention mapping. BMC Health Serv Res 14:389

Ergonomics Systems Mapping for Professional Responder Inter-operability in Chemical, Biological, Radiological and Nuclear Events

Graham Hancox[1], Sue Hignett[1(✉)], Hilary Pillin[2], Spyros Kintzios[3], Jyri Silmäri[4], and C. L. Paul Thomas[5]

[1] Design School, Loughborough University, Loughborough, UK
s.m.hignett@lboro.ac.uk
[2] HRP Professional Services Ltd., Witney, UK
[3] R&D Project Management Department, Hellenic Navy, Athens, Greece
[4] Rescue Services, South-Savo Regional Fire Services, Mikkeli, Finland
[5] Centre for Analytical Science, Department of Chemistry, Loughborough University, Loughborough, UK

Abstract. A European consensus was developed as a concept of operations (CONOPS) for cross-border, multi-professional chemical, biological, radiological and nuclear (CBRN) responses. AcciMaps were co-designed with professional responders from military, fire, ambulance, and police services in UK, Finland and Greece. Data were collected using document analysis from both open and restricted sources to extract task and operator information, and through interviews with senior staff representatives (Gold or Silver Command level). The data were represented on the Accimaps as a high level Socio-Technical Systems (STS) map of CBRN response using the themes of communication, planning, action, and reflection. Despite differences between service sectors and in terminology, a macro systems level consensus was achieved for the command structures (Gold, Silver and Bronze), and Hot Zone responders (Specialist Blue Light Responders and Blue Light Responders). The detailed tasks and technologies have been analysed using Hierarchical Task Analysis (HTA) to represent both complex response scenarios (macro) and detailed technologies (micro interfaces) for detection, diagnosis and decontamination. The outputs from these two systems mapping tools (Accimaps and HTAs) are being used in two field trials/exercises.

Keywords: Ergonomics · CBRN · NATO · AcciMap · Sociotechnical systems

1 Introduction

Chemical, Biological, Radiological and Nuclear (CBRN) and terrorist events will lead to emergency services having to respond in environments which are dangerous, complex, fast paced, high-stakes, unpredictable and substantially novel [1, 2]. CBRN incidents in particular can cause a great deal of psychological stress due to the level of uncertainty both before and after an event has occurred, resulting not knowing where and when they might take place, and when responding through difficulties in

S. Bagnara et al. (Eds.): IEA 2018, AISC 818, pp. 89–96, 2019.
https://doi.org/10.1007/978-3-319-96098-2_13

identifying the substances involved and how best to deal with them [3]. This can be especially true for emergency service responders (Fire, Police and Ambulance) whose actions and decisions within an event can directly save lives [4]. Yet, a lack of knowledge, skills or awareness in CBRN incidents may lead to such professionals being reticent to engage in their duties in the event of such an incident [5].

The scale of mass casualty incidents (MCIs) such as CBRN events often require the cooperation of many varied organisations both on the ground and at the higher levels of command [6, 7]. This multi-agency interoperability can bring with it many issues when the different emergency services are required to work together, possibly having unfamiliar working practices, technology, communication styles and goals [8–11]. For a very large scale event there may even be a need for cross border assistance, increasing the likelihood of such issues of interoperability occurring [7, 12].

For these reasons having a better understanding of the emergency services' procedures, both their own organisation (intra-team) and those of other emergency service responders (inter-team) who they will be required to work alongside, can be advantageous [2]. High level guidelines for response to CBRN incidents is provided by the North Atlantic Treaty Organisation (NATO) as a concept of operations (CONOPS) covering information gathering, assessment and dissemination, scene management, saving/protecting lives, and specialist support [13]. A previous model of CBRN response by Healey et al. [14] chronologically mapped an event for prevention, preparedness, alerting/early response, and remediation. This model allowed for response phases and events to be established but also had no detailed information about tasks that should occur, lines of communication needing to be established and where technologies might fit into the system. Having access to technologies, such as those which can quickly detect and identify substances or track patients and evidence in this chaotic environment can aid in situational awareness, thus reducing the psychological distress placed on emergency responders [15–18].

Systems mapping methods that have previously been used to visualise a Social Technical System (STS) in other domains include AcciMaps [19] and Hierarchical Task Analysis (HTA) [20, 21]. This paper describes the development of a CBRN AcciMap as a platform for development and evaluation of technology. An example of an HTA for a CBRN diagnostic technology is used to show how the 2 systems mapping tools can be used in conjunction to visualize macro and micro systems.

2 Method

2.1 Accimaps

Participants were recruited through purposive sampling requiring them to have experience of operating at Silver or Gold levels of command and having previously worked at Bronze level, giving them knowledge of all levels of the STS, and ensuring they could provide relevant CONOPS documents.

An iterative approach was used for the development of the AcciMaps with empirical data taken from both document analysis and interviews. Both open source (e.g. NATO [13]), and restricted (if access was approved) documents were read and the information from these was used for visualization of the multiple tasks and responsibility levels on a single map. These initial data were subsequently expanded with interviews to describe specific task activities from first blue light responder arriving on scene upwards through the command chain.

The Accimap was reviewed with a participant from each stakeholder service (fire, ambulance, police, and military) to discuss differences between written procedures (work as imagined) and operational activities (work as done). The individual service AcciMaps were compared to look for similarities and a higher abstraction level consensus AcciMap was developed. The EU final AcciMap was validated by all participants as meaningful for their country.

2.2 Hierarchical Task Analysis (HTA)

HTAs were created on both a macro level (mapping of the Field Trial/Exercise; FTX) and micro level (mapping of prototype technology). The macro HTAs were compiled through interviews with the lead organizer of each FTX; with information from these interviews mapped in a macro systems HTA for each FTX. These were reviewed and revised through an iterative process until the lead organisers confirmed them as an accurate depiction of the tasks in the FTX.

The micro level HTAs were created for individual technologies which were at least Technology Readiness Level (TRL) 6 and above in development indicating that it is at least sufficiently developed to be demonstrated in a relevant environment. Data were collected from document analysis (e.g. instruction manuals and user guides) and interviews with the technology developers to map out mental models of how the technology should work ('work as imagined'). The HTAs were then reviewed and validated by the developers as accurate representations.

3 Results

3.1 Accimaps

A final harmonised (consensus) Accimap was created (Fig. 1) with cross-cutting themes of Communication, Planning, Action, and Reflection. Although the structure of command in an incident was similar across the EU the terminology used sometimes varied with Silver command and Bronze Command terms being described as 'Tactical', 'Incident', and 'Operational'. These levels always reported to a Gold Command (Strategic) and managed (were supported by) Specialist Blue Light Responders (S-BLR) and initial Blue Light Responders (BLR).

Working from the base of the hierarchy, the BLRs would often be first on scene as part of the conventional response to everyday emergency calls. Their role in CBRN is limited by their level of equipment and training but it was agreed across all countries that at a minimum they would recognise the scene as a possible CBRN event and pass information to the control rooms to initiate dispatch of specialists (S-BLR) with greater CBRN response capabilities.

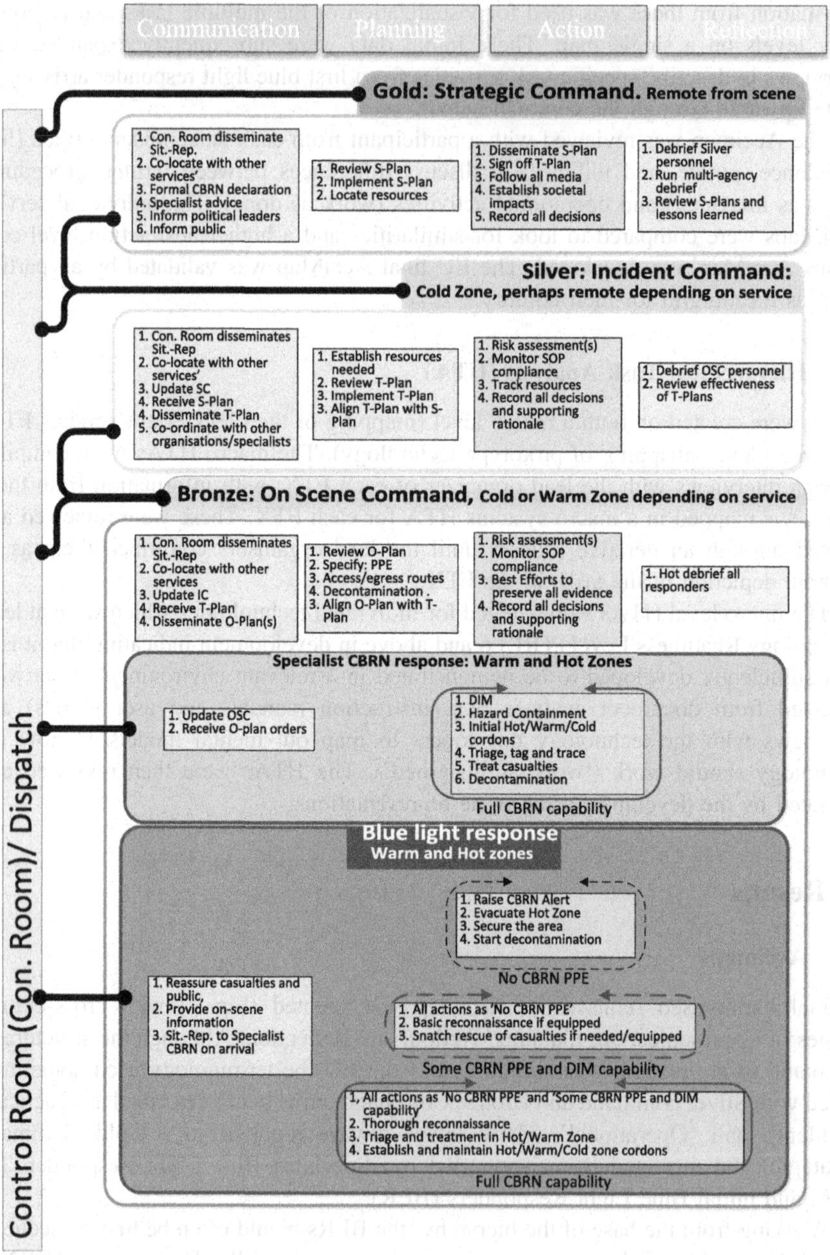

Fig. 1. EU wide Accimap (previously published in Hancox et al. [22])

The Bronze/On-Scene Command is responsible for ensuring all resources (equipment and personnel) are optimally used by applying the Incident Command Tactical Plan (T-Plan), and Standard Operating Procedures (SOPs). They are also responsible for managing the BLR and S-BLR, and apply the Operational Plan (O-Plan).

Silver/Incident Command manages the T-Plan to track and monitor the resources needed in the CBRN response. They are typically, but not always (depending on service and country), located at a distance from the CBRN scene and have a broader perspective of the incident which allows them to offer advice to the Bronze/On-scene Command. They act a point of contact for Gold and Bronze commands, controlling the information flowing up and down the STS, and reducing the likelihood of information overload.

Gold/Strategic Command implements the Strategic Plan (S-Plan) based on policies, legal frame-works and protocols. This requires managing resources on a regional, national and international level with a more 'outward facing' perspective, and acting as a point of contact for Government representatives and public messages (via the media). They also consider (plan for) limiting the after effects of the incident, so make decisions based to plan for returning to readiness ('business as usual').

Control Room/Dispatch is represented in Fig. 1 as a communication spine, transmitting information up and down the hierarchy. This is most frequently achieved with the use of dedicated radio and wireless channels across all levels; communication in the STS mostly takes place only between adjacent levels. A dedicated Major Incident Control Room (Con.Room) is usually set up in later stages of the incident for each agency.

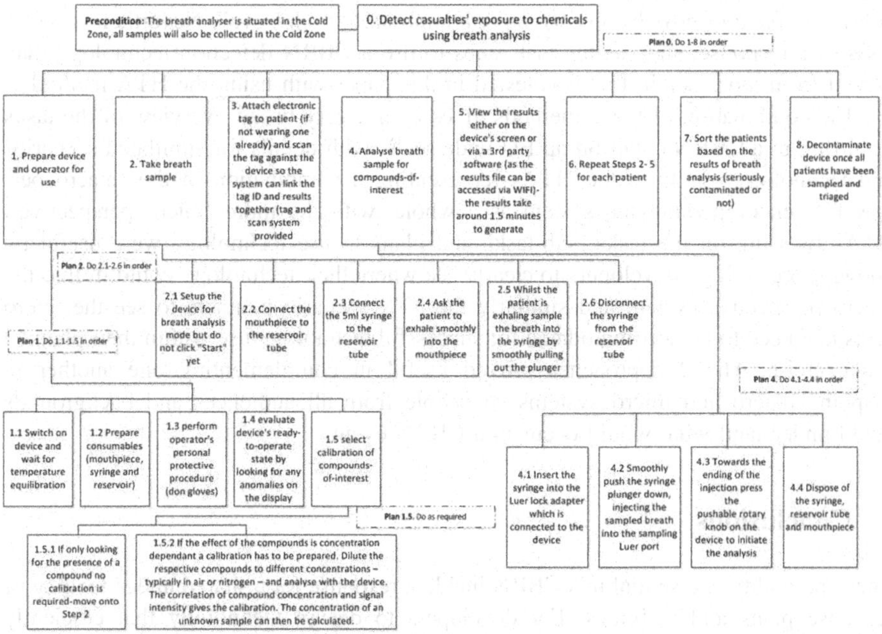

Fig. 2. Breath analysis hierarchical task analysis

3.2 Hierarchical Task Analysis

HTA was used in 2 ways, firstly to expand the Accimaps and map the events (as goals, tasks, sub-tasks and plans) for two FTXs to support testing of prototype technologies in a simulated CBRN event; this was the 'macro' element to look at the system as a whole. Secondly HTAs were created for individual technologies to represent the 'micro' element of the system - how each technology would be used within the system.

Figure 2 shows an example of a technology HTA to map the series of tasks, sub-tasks and plans for a prototype breath analyser. The 8 subtasks give a clear and concise process from preparing the device for use through to decontamination.

4 Discussion

A new consensus of an EU harmonised civilian CBRN systems (CONOPS) was achieved through cross mapping the planned responses to CBRN events in a number of EU countries (Greece, Finland, Czech Republic and UK). This shows that there are at commonalities when dealing with CBRN incidents across services and borders which support interoperability. Technology developers who may be less familiar with CBRN response procedures, may also benefit from Fig. 1 with a greater understanding of the systems to support their technology integration.

The Accimap model approach taken in this paper may be sufficiently detailed for mapping EU wide CBRN responses as more specific inter-operating procedures may not be achievable as commented by Mendonça et al. [11], "*extreme events occur infrequently, and no two are exactly the same. A comprehensive set of procedures to cover the space of possible events may be impossible to achieve*". Nevertheless, some parts of a response, such as the task steps to use a CBRN detection technology may benefit from more detail; this was tested in this paper with using the HTA method..

The combination of the 2 methods gives both a very broad overview of the tasks occurring in a CBRN event on an EU wide scale, with details for simulated scenarios and technology use from the HTAs representing a combination of the 'macro perspective' encompassing the system as a whole, with the more 'micro perspective'. HTAs focusing on the tasks, sub-tasks and plans to use technology were useful by allowing technology developers to clearly see where their technology could fit into the macro perspective system and similarly the FTX scenario planners to see the micro tasks that need to be accommodated for successful technology use within their planned trial/exercise. The 2 approaches proved useful in complementing one another in mapping macro and micro systems so people from all audiences and backgrounds could understand what would occur in a CBRN event.

5 Conclusions

Inter-operability is essential in a CBRN incident response for a shared understanding of response plans and activities. For developers to design technology that efficiently works within systems of response requires an understanding of the planned responses.

An adapted Accimap methodology proved useful in mapping an EU wide CBRN response in conjunction with more specific HTAs to show the numerous specific actions at both macro and micro-perspectives.

References

1. James K (2011) The organizational science of disaster/terrorism prevention and response: theory-building toward the future of the field. J Organ Behav 32(7):1013–1032
2. Power N (2017) Extreme teams: towards a greater understanding of multi-agency teamwork during major emergencies and disasters. Am Psychol. http://eprints.lancs.ac.uk/88335/1/POWER_AP_PRE_PRINT.pdf. Accessed 13 Apr 2018
3. Palmer I (2004) The psychological dimension of chemical, biological, radiological and nuclear (CBRN) terrorism. J R Army Med Corps 150(1):3–9
4. Alexander DA, Klein S (2003) Biochemical terrorism: too awful to contemplate, too serious to ignore: subjective literature review. Br J Psychiatry 183(6):491–497
5. Kako M, Hammad K, Mitani S, Arbon P (2018) Existing approaches to chemical, biological, radiological, and nuclear (CBRN) education and training for health professionals: findings from an integrative literature review. Prehospital Disaster Med 33(2):182–190
6. Wilkinson D, Waruszynski B, Mazurik L, Szymczak AM, Redmond E, Lichacz F (2010) Medical preparedness for chemical, biological, radiological, nuclear, and explosives (CBRNE) events: gaps and recommendations. Radiat Protect Dosimetry 142(1):8–11
7. Centre for strategy and Evaluation Services (2011) Ex-post evaluation of PASR activities in the field of security and interim evaluation of FP7 security research. https://ec.europa.eu/home-affairs/sites/homeaffairs/files/e-library/documents/policies/security/pdf/interim_evaluation_of_fp7_security_ex_post_pasr_final_report_en.pdf. Accessed 13 Apr 2018
8. Waring S, Alison L, Carter G, Barrett-Pink C, Humann M, Swan L, Zilinsky T (2018) Information sharing in interteam responses to disaster. J Occup Organ Psychol. https://onlinelibrary.wiley.com/doi/full/10.1111/joop.12217. Accessed 13 Apr 2018
9. Power N, Alison L (2017) Offence or defence? Approach and avoid goals in the multi-agency emergency response to a simulated terrorism attack. J Occup Organ Psychol 90(1):51–76
10. House A, Power N, Alison L (2014) A systematic review of the potential hurdles of interoperability to the emergency services in major incidents: recommendations for solutions and alternatives. Cognit Technol Work 16(3):319–335
11. Mendonça D, Jefferson T, Harrald J (2007) Collaborative adhocracies and mix-and-match technologies in emergency management. Commun ACM 50(3):44–49
12. Home Office. CONTEST: the United Kingdom's strategy for countering terrorism. https://www.gov.uk/government/uploads/system/uploads/attachment_data/file/97994/contest-summary.pdf. Accessed 13 Apr 2018
13. NATO (2011) Project on minimum standards and non-binding guidelines for first responders regarding planning, training, procedure and equipment for chemical, biological, radiological and nuclear (CBRN) incidents. http://www.nato.int/nato_static_fl2014/assets/pdf/pdf_2016_08/20160802_140801-cep-first-responders-CBRN-eng.pdf. Accessed 13 Apr 2018
14. Healy MJ, Weston K, Romilly M, Arbuthnot K (2009) A model to support CBRN defence. Defense Secur Anal 25(2):119–135
15. Bolic M, Borisenko A, Seguin P (2012) Automating evidence collection at the crime scene using RFID technology for CBRN events. Forensic Sci Policy Manage Int J 3(1):3–11

16. Stewart KA (2018) NPS, international special forces groups, NATO collaborate to counter CBRN threats. https://calhoun.nps.edu/bitstream/handle/10945/41344/Naval%20Postgraduate%20School%20%20NPS%2c%20International%20Special%20Forces%20Groups%2c%20NATO%20C.pdf?sequence=1&isAllowed=y. Accessed 13 Apr 2018
17. Sferopoulos R (2009) A review of chemical warfare agent (CWA) detector technologies and commercial-off-the-shelf items. http://www.dtic.mil/dtic/tr/fulltext/u2/a502856.pdf. Accessed 13 Apr 2018
18. Pacsial-Ong EJ, Aguilar ZP (2013) Chemical warfare agent detection: a review of current trends and future perspective. Front Biosci 5(1):516–543
19. Salmon PM, Cornelissen M, Trotter MJ (2012) Systems-based accident analysis methods: a comparison of Accimap, HFACS, and STAMP. Saf Sci 50(4):158–170
20. Annett J, Duncan KD (1967) Task analysis and training design. Occup Psychol 41(1):211–221
21. Stanton NA (2006) Hierarchical task analysis: developments, applications, and extensions. Appl Ergon 37(1):55–79
22. Hancox G, Hignett S, Pillin H, Kintzios S, Silmäri J, Thomas PCL (2018) Systems mapping for technology development in CBRN response. Int J Emerg Serv. https://www.emeraldinsight.com/eprint/ZBSH8EQ76IFFWCDCSINB/full. Accessed 13 Apr 2018

Creative 'Tips' to Integrate Human Factors/Ergonomics Principles and Methods with Patient Safety and Quality Improvement Clinical Education

Helen Vosper[1,2,3], Sue Hignett[1,2,3](✉), and Paul Bowie[1,2,3]

[1] Robert Gordon University, Aberdeen, UK
h.vosper@rgu.ac.uk
[2] Loughborough Design School, Loughborough University, Loughborough, UK
S.M.Hignett@lboro.ac.uk
[3] Medical Directorate, NHS Education for Scotland, Glasgow, UK

Abstract. The goal of these 12 tips is to enhance the effectiveness of safety and improvement work in frontline healthcare practice by providing a framework for integration of Human Factors and Ergonomics (HFE) theory and approaches within undergraduate curricula, postgraduate training and healthcare improvement programs. This paper offers both support and challenges to healthcare educators when planning the inclusion of HFE principles within existing curricula. The 12 tips include the systems framework (Tip 1, 3), HFE tools and competency (Tips 2, 7), misunderstandings (Tips 4, 5), and ideas for implementation (Tip 6, 8, 9, 10, 11, 12). They will support the goal of enhancing the performance of care systems (productivity, safety, efficiency, quality) and the wellbeing of all the people (patient outcomes, staff presenteeism).

Keywords: Education · Competency · Accreditation

1 Introduction

There is an increasing focus on safety in healthcare education but progress is slow with little direction for teaching provided by professional, statutory and regulatory bodies. The World Health Organisation [1] developed a patient safety curriculum, but little is known about integrating Human Factors and Ergonomics (HFE) concepts.

This paper outlines 12 tips [2] to provide a preliminary platform for healthcare educators to explore how to integrate key HFE principles within existing curricula. The tips include the systems framework (Tip 1, 3), HFE tools and competency (Tips 2, 7), misunderstandings (Tips 4, 5), and ideas for implementation (Tip 6, 8, 9, 10, 11, 12) with the knowledge sequence indicated in Fig. 1. We believe that these tips will support the goal of enhancing the performance of care systems (productivity, safety, efficiency, quality) and the wellbeing of all the people (patient outcomes, staff presenteeism).

S. Bagnara et al. (Eds.): IEA 2018, AISC 818, pp. 97–101, 2019.
https://doi.org/10.1007/978-3-319-96098-2_14

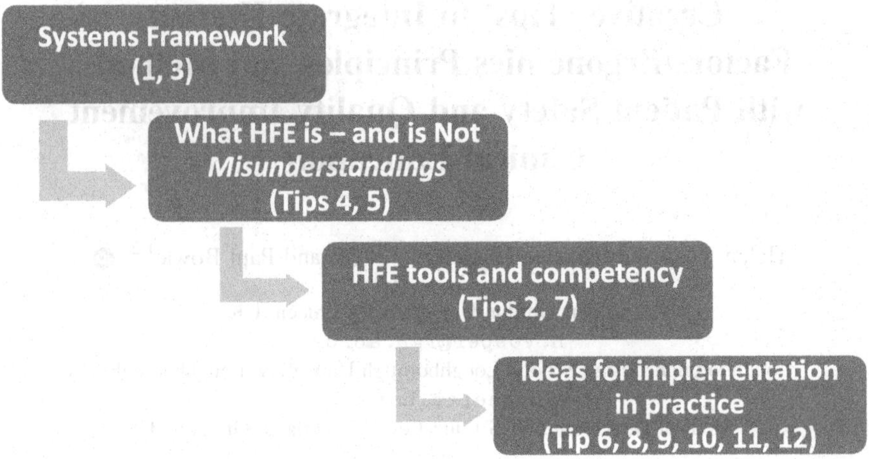

Fig. 1. Four general categories for 12 Tips

2 12 Tips

2.1 Tip 1: Jointly Optimise Systems Performance and Human Wellbeing

The HFE core concept is to jointly optimise systems performance and the wellbeing of people. Understanding and applying the systems approach must be the starting point for embedding HFE; it is a fundamental HFE concept and underpins all the other tips. Systems are defined as "*a set of inter-related or coupled activities or entities (hardware, software, buildings, spaces, communities and people) with a joint purpose*". There needs to be an increased focus on Safety-II to optimise overall systems performance and human wellbeing.

- Safety-I is an error-reductionist approach which seeks to identify and rectify the root cause(s) of 'errors' and often focuses on people (e.g. technique training) rather than wider systems when trying to understand and resolve issues
- Safety-II recognises that systems, despite their inherent imperfections, operate safely most of the time (through resilience adjustments, Tip 9).

2.2 Tip 2: Teaching Faculty

Teaching faculty must be competent to deliver theory and practice (knowledge and skills). A minimum HFE competency (see Tip 7) is needed for appropriate application of HFE tools (including systems modelling and task analysis) and interpretation of results.

2.3 Tip 3: Consider Adopting Human-Centred Organisation Principles

Practice what you preach: consider adopting human-centred organisation principles.

- Address the hidden curriculum/culture where the safety values and attitudes are more influenced by implicit learning from in-practice behaviours than from taught curricula.
- In other sectors, international standards are used to ensure products and services are of appropriate quality.

2.4 Tip 4: Recognise What HFE Is…

Recognise what HFE is… A specific way of thinking and doing which needs to be fully embraced if benefits are to be realised. Healthcare systems range from micro systems (single tasks/tools) through meso systems (teams), up to complex (macro) systems. HFE interventions may focus on optimisation of a micro-system, but there will *always* be clear mapping of the relationship of the micro with the macro system. Incorporate HFE from project inception across micro, meso and macro systems.

2.5 Tip 5: …And Recognise What HFE Is Not

HFE is not behaviour-based training and rarely identifies a single "root cause" after a systems analysis. An HFE intervention would not focus on requiring people to adapt behaviours (non-technical/specialist skills) to accommodate poorly designed systems of work and/or technology.

2.6 Tip 6: HFE and QI Synergies

Don't throw the baby out with the bath water: Recognise that HFE and QI can offer synergies. Explore a combined HFE and QI approach for real-life problems to inform more meaningful design and evaluation of improvement interventions.

- HFE focusses more on wellbeing and performance
- QI focusses more on process issues; the processes are mostly delivered by people but the people are not the focus of the improvement.

2.7 Tip 7: Three Levels of HFE Competencies

Curriculum design and content should be driven by learning outcomes to develop appropriate HFE competencies. Signpost HFE competency to support professional practice within a code of conduct (IEA) and additional education.

2.8 Tip 8: Participatory Approach

Use the participatory approach central to HFE to strengthen your specific curriculum or programme of training. Provide space to actively promote recognition of mismatches; this requires partnership with staff/students *('experts' about the hidden curriculum).*

2.9 Tip 9: Learning from Errors

Recognise that to err is not just human, but is highly desirable as part of a learning strategy to develop transferable skills in building resilient systems. It is not possible to prevent all errors (normal part of work and learning) in complex sociotechnical systems. A zero-error approach to patient safety is the least effective approach. Need opportunities to consider safety in day-to-day routines and the systems resilience to absorb inevitable errors and deliver outcomes. Exploring factors that reduce/prevent risk requires educational-based opportunities to make errors and follow the trajectory to the natural end.

2.10 Tip 10: Build on What Is Already There

Build on what is already there. Many curricula may have 'human factors' teaching, even if it is focussed on 'non-technical skills' *(see Tip 5)* or 'patient safety' training. Review whether content includes the HFE fundamental principles in collaboration with HFE experts from IEA Federated Societies.

2.11 Tip 11: Inter-Professional Education

Take an Inter-Professional Education perspective to curriculum design and content. Systems optimisation can only happen if all relevant stakeholders are engaged. Consider 4 groups for HFE interventions:

- Systems actors: healthcare staff, patients (service users), carers etc.
- Systems experts: including HFE professionals
- Systems decision makers: e.g. managers, with power to effect change
- Systems influencers: political bodies, regulators etc.

2.12 Tip 12: HFE Capacity and Capability

Build HFE capacity and capability creatively. The following ideas can be considered:

- Input from qualified and regulated HFE professionals to ensure both credibility and that professional standards are adhered
- Build collaborations with other healthcare disciplines to enrich the expert pool
- Train healthcare staff with postgraduate academic training to be HFE "champions" with responsibility for supporting others in HFE practice and educational provision including developing "train the trainer" activities to support basic competency with fundamental concepts and approaches that can be applied in frontline care
- Develop HFE expertise by setting objectives within existing reward and recognition frameworks for current healthcare teaching and learning, and at postgraduate level through CPD activity.

3 Conclusion

An important opportunity exists for current undergraduate curricula, postgraduate training and healthcare safety and improvement programs to be considerably strengthened by the integration of HFE theory and methods. While current guidance for patient safety teaching appears to recognize this, it is apparent that the potential impact is compromised by conflation of HFE and a limited and potentially misleading focus on "factors of the human." The WHO multiprofessional patient safety curriculum [1] (for example) is one of the few resources available to healthcare faculty for designing and delivering curricula that support the development of safety competence. While this document contains a great deal of excellent guidance, its practical application is perhaps undermined by an apparently conflicted understanding of HFE.

References

1. WHO World Health Organisation (2011) The multi-professional patient safety curriculum guide. http://www.who.int/patientsafety/education/curriculum/en/. Accessed 15 June 2017
2. Vosper H, Bowie P, Hignett S (2017) Twelve tips for embedding Human Factors and Ergonomics principles in healthcare educational curricula and programmes. Med Teach. https://doi.org/10.1080/0142159X.2017.1387240

Applying Human Factors Methods to Explore 'Work as Imagined' and 'Work as Done' in the Emergency Departments Response to Chemical, Biological, Radiological, and Nuclear Events

Saydia Razak(✉), Sue Hignett, Jo Barnes, and Graham Hancox

Loughborough University, Loughborough, UK
s.razak@lboro.ac.uk

Abstract. The Emergency Department (ED) is a complex, hectic, and high-pressured environment. Chemical, Biological, Radiological, and Nuclear (CBRN) events are multi-faceted emergencies and present numerous challenges to ED staff (first receivers) with large scale trauma, consequently requiring a combination of complex responses.

Human Factors and Ergonomics (HF/E) methods such as Hierarchical Task Analysis (HTA) have been used in healthcare research. However, HF/E methods and theory have not been combined to understand how the ED responds to CBRN events.

This study aimed to compare Work as Imagined (WAI) and Work as Done (WAD) in the ED CBRN response in a UK based hospital. WAI was established by carrying out document analyses on a CBRN plan and WAD by exploring first receivers response to CBRN scenario cards. The responses were converted to HTAs and compared.

The WAI HTAs showed 4–8 phases of general organizational responsibilities during a CBRN event. WAD HTAs placed emphasis on diagnosing and treating presenting conditions. A comparison of WAI and WAD HTAs highlighted common actions and tasks. This study has identified three key differences between WAI and WAD in the ED CBRN response: (1) documentation of the CBRN event (2) treating the patient and (3) diagnosing the presenting complaint.

Findings from this study provide an evidence base which can be used to inform future clinical policy and practice in providing safe and high quality care during CBRN events in the ED.

Keywords: Human Factors and Ergonomics · CBRN events
Emergency Department

S. Bagnara et al. (Eds.): IEA 2018, AISC 818, pp. 102–110, 2019.
https://doi.org/10.1007/978-3-319-96098-2_15

1 Introduction

A CBRN event is "the exposure (or risk of exposure) of a large number of individuals to hazardous Chemical, Biological, Radiological (and Nuclear) materials" [1]. The UK the Civil Contingencies Act (2004) places statutory duties on category one organizations' such as Emergency Departments (ED) to plan, prepare, and respond to CBRN events effectively [2].

CBRN events present multifaceted demands on the ED [3, 4] which is already a complex, hectic, high-pressured, and often short-staffed environment [5, 6]. CBRN events are rare [7] resulting in first receivers being unfamiliar with the clinical assessment, containment, and treatment unique to patients who have been exposed to CBRN materials. This means that plans are used as a reference point and are implemented into policy and procedures to respond to CBRN events. Additionally, it is legislative practice in the UK to have well–practiced emergency plans [8].

Human Factors and Ergonomics (HF/E) "is the scientific discipline concerned with the understanding of interactions among humans and other elements of a system" [9]. The use of HF/E is advocated as a means of improving healthcare quality and safety [10].

Hierarchical Task Analysis (HTA) is a central method of analysis in HF/E. HTA describes a task as a higher level goal with a hierarchy of superordinate and subordinate tasks. At each level of the subtasks, a plan directs the sequence and possible variance of task steps [11].

Responding to a CBRN event involves clinical work, based on guidelines and policies which represent Work as Imagined (WAI). This is what designers, managers, regulators, and authorities believe happens, or should happen in the workplace. Compliance to policies and guidelines suggests that Work as Done (WAD), what actually happens in the workplace, are similar or identical [12, 13].

This study combined the HF/E method (HTA) and theory (WAI vs. WAD) to better understand the ED response to CBRN events.

2 Method

2.1 Design

An exploratory qualitative design was used to ensure a data driven understanding of CBRN plans, with a continuous and thorough method of representation (HTA). Qualitative methods have been suggested to be effective to find out why people make choices and carry out tasks in certain ways in healthcare [14].

2.2 Pilot Study

A publically available CBRN plan was downloaded from the Internet, and analyzed as an HTA as a pilot exercise to prepare for data collection and trial of various software for HTAs (e.g. Human Factors Risk Manager, Microsoft PowerPoint, and Microsoft Visio). Microsoft Visio was predominantly used, because it provided the flexibility of gradually building up the HTA and an easy flow of the HTA over numerous pages. The scenario cards were piloted on three participants (1) SR (Emergency Department Practitioner), (2) an ex ED nurse, and (3) a first receiver in the ED.

2.3 Work as Imagined

WAI was established from a document analysis of a CBRN plan from a NHS hospital Trust in the UK. The Trust employs 15,000 staff; serves 1 million residents and treated 237,000 ED patients during 2016–2017. Document analysis was used as it requires data to be thoroughly examined and interpreted to elicit meaning, understanding, and develop empirical knowledge [15, 16]. The HTA was produced using Stanton [17] HTA guidance which encourages iterative verification of the analysis with subject matter experts.

2.4 Work as Done

During the WAD stage, CBRN based scenario cards representing chemical (Sarin), Biological (Severe Acute Respiratory Syndrome), and Radiological (Acute Radiation Syndrome) situations were presented to 29 first receivers. Scenario cards were used to create a hypothetical CBRN situation, scenario cards have effectively been used whilst testing incident command systems in hospital based disaster simulation exercises [18].

The inclusion criteria ensured that participants were 18 years and over, and had been employed in the ED for a minimum of three months on a substantial contract; this was to ensure that all participants had the opportunity to attend an induction training. This resulted in a sample of 15 females and 14 males, aged 21–60 years. The length of employment in the ED ranged from 5 months to 17 years. Purposive sampling was used to identify and select individuals knowledgeable about or experienced with a phenomenon [19].

Scenario cards were presented during a range of shifts. First receivers were given time to read the scenario card, and then asked to talk through their actions in response to the scenario. Field notes were used to record the data and converted to HTAs.

Stratified purposive was used to validate the findings from the scenario card presentations with subsequent interviews to compare, contrast, and identify similarities and differences in the phenomenon of interest [20]. The validation interview began by giving the first receiver the scenario card as a memory aid. The HTA was explained and the participants were asked “do you think this diagram is a true representation of what you would do in a CBRN event?” They were then given the opportunity to discuss and make amendments to the HTA.

2.5 Work as Imagined vs. Work as Done

The comparison of WAI (action cards) and WAD (scenario cards) HTAs was a two-staged analysis. The first stage consisted of highlighting the similarities and differences between WAI HTAs and corresponding scenario card WAD HTAs. The second stage involved a deeper analysis of understanding similarities and differences in tasks between WAI and WAD.

2.6 Ethical Approvals, Good Practice, and Recruitment

Ethical approval was given by the Loughborough University sub-committee (C17-22) and NHS Health Research Authority (HRA) ethical approval (Integrated Research Application (IRAS) (219968). Trust specific Research and Development (R&D) department approval was also given. Data collection methods anonymized responses and participants' details in accordance with the Data Protection Act (1998) [21]. First receivers were recruited by placing posters in staff only areas in the ED and by SR attending staff handovers.

3 Results

Both types of HTAs were verified by the hospital CBRN lead and were later reviewed by Ergonomists as a service evaluation of the CBRN plan.

3.1 Work as Imagined

Two types of HTAs were used to represent WAI. The first HTA visualized the general organizational responsibilities of the hospital during a CBRN event (Fig. 1), which were to (1) understand roles and responsibilities (2) take notification of casualties (3) establish command and control (4) activate CBRN plan (5) manage scene (6) decontaminate (7) initiate recovery and (8) debrief.

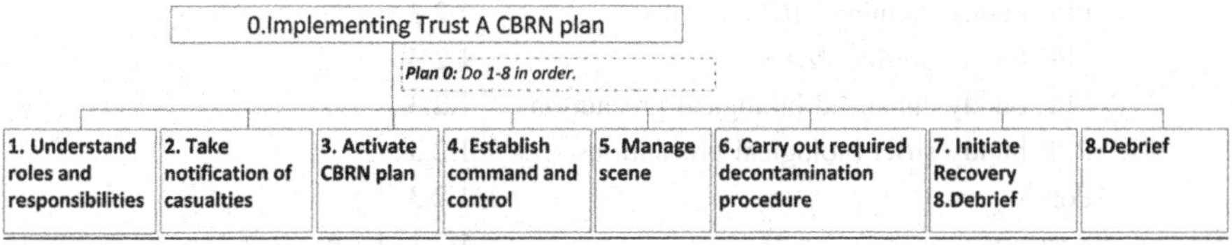

Fig. 1. Superordinate tasks of Trust A general organisational responsibilities during a CBRN event

The analyses of the general organizational responsibilities highlighted a limitation within the plan in superordinate task 6 (carry out required decontamination procedure) with repetition and a crossover of information. As the focus of this study was the ED, the first superordinate task of understanding the roles and responsibilities required from the ED during a CBRN are further described in Fig. 2 as 7 subtasks. The 7 subtasks are achieved by first receivers as shown in Table 1.

The second WAI HTA (n = 17) represented actions required from first receivers during a CBRN event. Action card superordinate tasks were to (1) prepare to respond to CBRN incident (2) respond to CBRN incident (3) initiate recovery from CBRN incident and (4) document CBRN incident as shown in Fig. 3. It was identified that there was ambiguous guidance on how to document the event and what to do with the

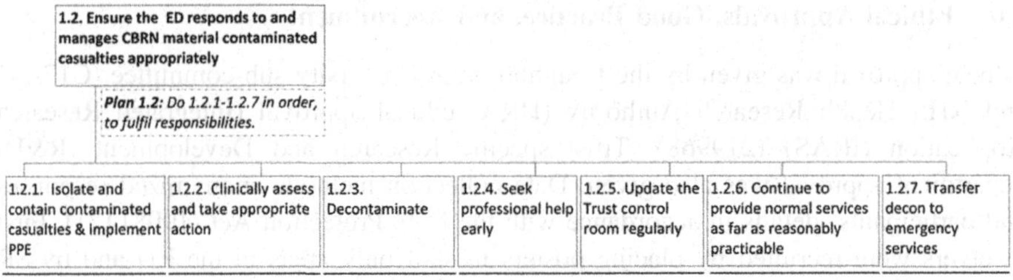

Fig. 2. Subordinate tasks of Trust A ED responsibilities during a CBRN event

Table 1. Trust A WAI action cards and corresponding first receiver responsibilities

Action card	Responsibilities
Receptionist	1.2.1
Nurse in charge	1.2.2, 1.2.4, 1.2.5, 1.2.6
Doctor in charge	1.2.2, 1.2.4, 1.2.5, 1.2.6
Decontamination nurse (team leader)	1.2.3, 1.2.4, 1.2.7
Triage disrobing nurse (dry decontamination)	1.2.3
Triage disrobing nurse (wet decontamination)	1.2.3
Healthcare assistant	1.2.3
Timing board nurse	1.2.2
Exit nurses	1.2.3
Assessment doctor	1.2.2
PPE buddy donning PRPS*	1.2.3
PPE buddy doffing PRPS	1.2.3
PPE buddy enhanced biological precautions	1.2.3
PPE buddy strict biological precautions	1.2.3
Porter	1.2.3
Security	1.2.1, 1.2.6
Medical physics	1.2.2

documentation after the event. This was discussed with the CBRN lead as part of the constructive evaluation of the action cards.

3.2 Work as Done

The key actions first receivers described for the CBRN response were grouped as: containing the contaminant, communicating with first receivers and the command team, diagnosing the presenting condition, escalating to seniors, investigating symptoms and contaminant, implementing PPE, preventing cross contamination, protecting colleagues and the work environment, and treating the patient.

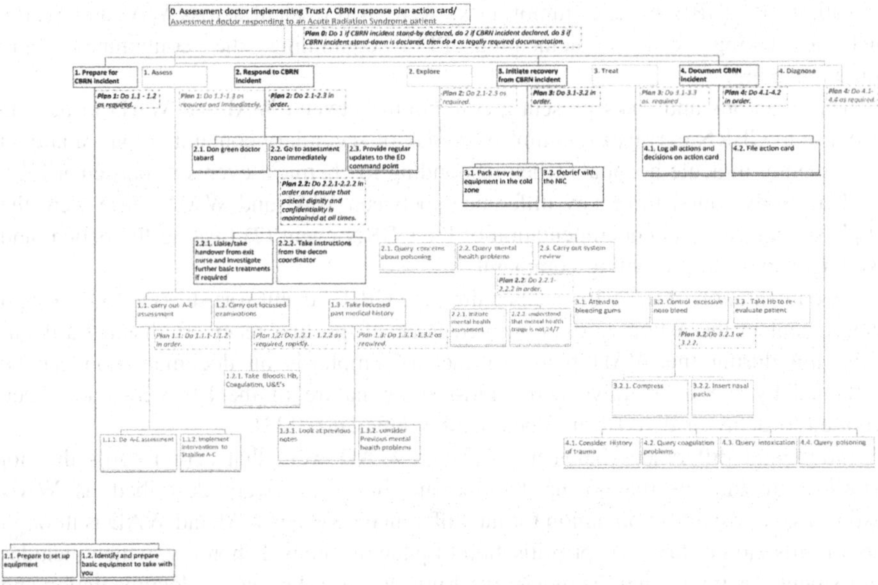

Fig. 3. Trust A WAI (assessment doctor action card) vs. WAD (ED registrar responding to ARS patient)

3.3 Work as Imagined vs. Work as Done

The comparison of WAI and WAD HTAs highlighted common actions such as isolating the patient, escalating the patient presentation to a senior first receiver, and activating the CBRN plan. Further analyses revealed common tasks across WAI HTAS, for example the decontamination team leader would be required to lead decontamination, implement PPE, contain the contaminant, and communicate with first receivers and emergency services. Such tasks were also evident for the WAD HTAs, in which first receivers would take lead of the situation, implement PPE, contain the contaminant, and communicate. Emerging from the results, differences between WAI and WAD were evident in the importance placed on (1) documentation of the CBRN event (2) treating the patient and (3) diagnosing the presenting complaint, as shown in Fig. 3.

4 Discussion

This study has taken a rigorous, and systematic approach using the HF/E method (HTA) and theory (WAI vs. WAD) to better understand the ED response to CBRN events. This study found both commonalities and differences between WAI and WAD in the ED CBRN response.

Common actions emerged from both the action card HTAs and scenario card HTAs, confirming an alignment between WAI and WAD. This alignment was evidenced through actions such as isolation, escalation to senior first receivers, and activating the CBRN plan. Common tasks were also evident across WAI and WAD, such as taking lead, implementing PPE, containing the contaminant, and communicating.

These actions and tasks present a continuum between WAI and WAD in the ED response to CBRN events, to minimise the risk of secondary contamination - a known phenomenon in acute hospitals when responding to chemical events in particular [22].

This study found three key differences between WAI and WAD, these were the emphasis placed on (1) documentation of the CBRN event (2) treating the patient and (3) diagnosing the presenting complaint.

Documentation of the actions taken during a CBRN event was stressed to be a legal requirement in the action cards; however, none of the first receivers discussed documentation during the WAD phase. A lack of emphasis on documentation can be explained by the interruptive and multi-tasking nature of the ED which has been reported to delay or divert from documenting efficiently [23].

Additional differences between WAI and WAD were that action cards did not prioritise treating or diagnosing the patient, but these were described as WAD responses. A possible explanation for the difference between WAI and WAD is through the prioritisation of tasks vs. prioritisation of patients needs dichotomy, which includes assessment, treatment, and diagnosis. WAI prioritised tasks such as documentation and effective decontamination, whereas first receivers (WAD), prioritised patients needs. Although priority is given to minimise the number of deaths, it is advised to prioritise decontamination procedures to reduce casualties exposure to CBRN materials prior to clinical treatment [24]. Therefore, the importance of tasks such as documentation and decontamination should be clearly prioritised in future CBRN plans while taking into consideration the complexities of an interruptive ED environment.

4.1 Limitations

Although a realistic representation of WAD was achieved by carrying out the scenario card presentations in the ED, CBRN events could not ethically be anticipated or created due to their life endangering nature. This was addressed by using realistic scenario cards based on Health Protection Agency CBRN clinical guidance [25] and reviewed by a Hazardous Area Response Team specialist prior to data collection.

4.2 Conclusion

This study has analyzed the ED response to CBRN events using HTA. It was found that the implementation of WAI and WAD had both an alignment and misalignment between policy and practice. The alignment existed in tasks set out by CBRN plans (e.g. implementing PPE and taking lead); the misalignment was a result of the prioritization of tasks (documentation) within the CBRN plan and the prioritization of the patients needs (treatment and diagnosis) by first receivers. Future work should focus on the differences in task priorities between WAI and WAD to enhance clinical practice in the ED response to CBRN events.

Acknowledgements. The authors would like to thank Professor Tim Coates, Aaron Vogel, Elizabeth Cadman- Moore, Lisa Mclelland, and the first receivers for their enthusiasm and support throughout this study.

References

1. Chilcott R, Wyke S (2016) CBRN incidents. In: Sellwood C, Wapling A (eds) Health emergency preparedness and response. CABI, Oxfordshire, pp 166–180
2. NHS England. https://www.england.nhs.uk/ourwork/eprr/
3. Luther M, Lenson S, Reed K (2006) Issues associated in chemical, biological and radiological emergency department response preparedness. Australas Emerg Nurs J 9:79–84. https://doi.org/10.1016/j.aenj.2006.03.007
4. Koenig KL (2003) Strip and shower: the duck and cover for the 21st century. Ann Emerg Med 42:391–394. https://doi.org/10.1067/mem.2003.381
5. Chartier LB, Cheng AHY, Stang AS, Vaillancourt S (2017) Quality improvement primer part 1: preparing for a quality improvement project in the emergency department. CJEM 0:1–8. https://doi.org/10.1017/cem.2017.361
6. Basu S, Qayyum H, Mason S (2016) Occupational stress in the ED: a systematic literature review. Emerg Med Online. https://doi.org/10.1136/emermed-2016-205827
7. Boyd A, Chambers N, French S, Shaw D, King R, Whitehead A (2014) Emergency planning and management in health care: priority research topics. Heal Syst 3:83–92. https://doi.org/10.1057/hs.2013.15
8. Cabinet Office. https://www.gov.uk/guidance/preparation-and-planning-for-emergencies-responsibilities-of-responder-agencies-and-others
9. IEA (2017) International Ergonomics Assocation. http://www.iea.cc/whats/
10. National Quality Board. https://www.england.nhs.uk/wp-content/uploads/2013/11/NQB-13-04-02.pdf
11. Shepherd A (2001) Hierarchical task analysis. Taylor and Francis, London
12. Chuang S, Hollnagel E (2017) Challenges in implementing resilient healthcare. In: Braithwaite J, Wears RL, Hollnagel E (eds) Resilient healthcare: Reconciling Work-as-Imaged and Work-as-Done. Ashgate, Surrey, pp 72–84
13. Saurin TA, Rosso CB, Lacey C (2017) Towards a resilient healthcare. In: Braithwaite J, Wears RL, Hollnagel E (eds) Resilient healthcare: Reconciling Work-as-Imaged and Work-as-Done. Ashgate, Surrey, pp 30–42
14. Hignett S, Wilson JR (2004) The role for qualitative methodology in ergonomics: a case study to explore theoretical issues. Theor Issues Ergon Sci 5:473–493. https://doi.org/10.1080/14639220412331303382
15. Bowen GA (2009) Document analysis as a qualitative research method. Qual Res J 9:27–40. https://doi.org/10.3316/qrj0902027
16. Corbin J, Strauss A (2008) Basics of qualitative research: techniques and procedures for developing grounded theory. Sage, California
17. Stanton N (2006) Hierarchical task analysis: developments, applications, and extensions. Appl Ergon 37:55–79. https://doi.org/10.1016/j.apergo.2005.06.003
18. Thomas TL, Hsu EB, Kim HK, Colli S, Arana G, Green GB (2005) The incident command system in disasters: evaluation methods for a hospital-based exercise. Prehosp Disaster Med 20:14–23. https://doi.org/10.1017/S1049023X00002090
19. Creswell JW, Plano-Clark VL (2011) Designing and conducting mixed methods research. Sage, London

20. Palinkas LA, Horwitz SM, Green CA, Horwitz SM, Green CA, Wisdom JP, Duan N, Hoagwood K (2015) Purposeful sampling for qualitative data collection and analysis in mixed method implementation research. Adm Policy Ment Heal Ment Heal Serv Res 42:533–544. https://doi.org/10.1007/s10488-013-0528-y
21. Legislation UK. https://www.legislation.gov.uk/ukpga/1998/29/contents
22. Larson TC, Orr MF, Auf der Heide E, Wu J, Mukhopadhyay S, Kevin-Horton D (2016) Threat of secondary chemical contamination of emergency departments and personnel: an uncommon but recurrent problem. Disaster Med Public Health Prep 10:199–202. https://doi.org/10.1017/dmp.2015.127
23. Werner NE, Holden RJ (2015) Interruptions in the wild: development of a sociotechnical systems model of interruptions in the emergency department through a systematic review. Appl Ergon 51:244–254. https://doi.org/10.1016/j.apergo.2015.05.010
24. North Atlantic Treaty Organisation. https://www.nato.int/ https://www.nato.int/nato_static_fl2014/assets/pdf/pdf_2016_08/20160802_140801-cep-first-responders-CBRN-eng.pdf
25. Gov UK. https://www.gov.uk/government/publications/chemical-biological-radiological-and-nuclear-incidents-recognise-and-respond

The Co-design Process of a Decision Support Tool for Airway Management

Raphaela Schnittker[1,2(✉)], Stuart Marshall[2], Tim Horberry[1], and Kristie L. Young[1]

[1] Monash University Accident Research Centre, Monash University, Clayton, VIC, Australia
raphaela.schnittker@monash.edu

[2] Department of Anaesthesia and Perioperative Medicine, Central Clinical School, Monash University, Melbourne, VIC, Australia

Abstract. The aim of this study is to design a decision support tool for challenging airway management events in anaesthesia. Major complications in airway management occur infrequently, but have a high risk of causing patient harm. Airway management takes place in complex sociotechnical environments that require anaesthesia teams to make decisions 'on the fly', often under time pressure and uncertainty. Contemporary decision support tools for airway management are too complex and do not involve anaesthesia team members in the design process. This study reports a co-design process to design an airway equipment trolley in conjunction with anaesthetists and anaesthetic nurses. It is part of a decision-centred design research program and is based on previously performed cognitive task analysis methods such as observations, Critical Decision Method interviews and focus groups. The present paper will discuss the co-design process including the elicitation of design requirements, the prototype development and the evaluation using case scenarios.

Keywords: Human factors · Decision making · Airway management

1 Introduction

Airway management is fundamental to anaesthesia and involves the support of breathing functions of an anaesthetised patient. Major airway complications are rare, but potentially life-threatening as they can result in hypoxia, brain injury and even death. A nationwide audit project in the United Kingdom (NAP4) found that major airway complications associated with death, brain damage, emergency surgical airways and unexpected Intensive Care stay occur in one per 22,000 (0.005%) cases of general anaesthetics (Cook et al. 2011).

A feared airway management crisis is 'can't intubate, can't oxygenate' (CICO), where a patient cannot be intubated (placement of a tube into the trachea) nor oxygenated with any rescue airway technique such as masks and tubes; requiring a surgical airway through the front of the neck to provide a passage for oxygen as the last resort (Marshall and Mehra 2014). The incidence of CICO is extremely rare, ranging from 1 in 10,000 to 1 in 50,000 of routine anaesthetics (Australian and New Zealand College

S. Bagnara et al. (Eds.): IEA 2018, AISC 818, pp. 111–120, 2019.
https://doi.org/10.1007/978-3-319-96098-2_16

of Anaesthetists 2012). Nevertheless, despite this low incident rate CICO's still account for roughly 25% of anaesthesia-related fatalities (Cook and Macdougall-Davis 2012).

The low occurrence of CICO reflects that anaesthesia teams are highly skilled in handling evolving airway complications (Larsson and Holmström 2013). However, evidence suggests that anaesthetic safety would benefit from further improvement in a range of Human Factors. Human Factors here specifically refer to the environmental, technical, organisational, psychological and physiological aspects of a system that affect human performance (Flin et al. 2013). Human factors negatively affecting performance were amongst others associated with team work (Bromiley 2009; Cook and Macdougall-Davis 2012), (delayed) decision-making (Cook et al. 2011; Watterson et al. 2014), organisational management (The Royal College of Anaesthetists 2011) and inadequate design of decision support tools (Chrimes 2016; Marshall 2013; Schnittker et al. 2017).

1.1 Decision Support Fir Airway Management

Anaesthesia is a highly protocolized discipline inheriting many guidelines and algorithms that aim to assist anaesthesia teams with the complex (cognitive and technical) work of airway management (Berkow 2004; Heidegger et al. 2005). Cognitive aids in the form of decision trees and symbolic charts have also been developed to assist teams with airway management during emergencies (Chrimes 2016; Heard et al. 2009). The shortcoming with these decision aids is that they were designed from the 'top down' based on how work is imagined to be done in an optimal way, but without basing it on an understanding of how decisions are actually made in the real world (Marshall 2013). In fact, not considering the cognitive demands of practitioners is a common challenge when designing for healthcare including the way they make decisions in naturalistic environments (Lintern and Motavalli 2018).

Anaesthesia team members are extremely skilled clinicians who have undergone at last 5 years of training after obtaining their medical degree before becoming independent practitioners (Australian and New Zealand College of Anaesthetists 2012). Due to the complex sociotechnical nature of anaesthesia, anaesthesia teams have to perform and make decisions under time-pressure, high stakes, uncertainty, multiple goals and organizational norms. Under these circumstances, anaesthesia team members majorly make decisions by employing a recognition-primed process that links cues and action in a prototypical fashion, instead of using a process of option comparison (Schnittker et al. 2017).

The present study is the last element in a broader research framework that addresses the disparity between decision support design and decision-making processes of anaesthesia teams. This will be done by designing a decision support tool for airway management with a Human Factors approach; specifically a decision-centred design process (Klein et al. 1997). The goal of the present study was to design and evaluate the chosen decision support tool in collaboration with anaesthetists and anaesthetic nurses.

1.2 Cognitive Task Analysis for Knowledge Elicitation

As part of the decision-centred design process, three methods from Cognitive Task Analysis were initially conducted to elicit knowledge from subject-matter experts: Critical Decision Method interviews, focus groups and field observations (Schnittker et al. 2016). The first aim was to explore how anaesthesia teams, anaesthetists and anaesthetic nurses, make decisions in challenging airway management situations. The second aim was to identify potential decision support tools that could support the most difficult decisions from a Human Factors perspective. Critical Decision Method interviews (Klein et al. 1989) were initially conducted to elicit knowledge from anaesthesia team members about challenging airway management cases they experienced in their past. Sixteen interviews with anaesthetists and nurses revealed that many challenging decisions were made throughout the operative period, most of them were related to planning and securing the patient's airway during anaesthesia induction (Schnittker et al. 2017).

A more in-depth analysis of the interviews revealed that the location and storage of airway equipment was the main contributor to successful airway management. On the other hand, non-standardization of airway equipment and cognitive aids distracting from the actual work were contributors to unsuccessful airway management (Schnittker et al. 2018). Equipment that was not stored in a unique location or visible to anaesthesia teams were more easily overlooked or forgotten to be prepared. Consequently, leverage points for decision support tools was the standardization of airway equipment, involvement of other operating theatre team members, and a seamless integration with the natural work flow of airway management.

1.3 Selection of Decision Support Tool for Airway Management

Triangulation of the three knowledge elicitation methods identified five potential decision support concepts: a standardized difficult airway trolley, a standardized organised airway trolley for routine airway equipment, an advanced pulse oximeter, a symbolic chart as a cognitive aid and training specifically focused on cue recognition and transitioning between airway techniques. The five design concepts were then evaluated on a conceptual level with a small sample of anaesthetists by using an adaptation of a standardized approach called FACES (Shapell and Wiegmann 2010). Participants were provided with a summary of the decision support concepts and asked to rate those on a scale from 1 to 5 along five criteria: perceived effectiveness, feasibility, acceptance, environmental fit, reliability and costs. All design concepts received high ratings (above 3.3). The highest rankings received were for training (4.4) and an organized airway equipment trolley for nurses (4.3). The airway equipment trolley for anaesthetic nurses was chosen for further research here because (1) it has not been designed previously and (2) is relatively low cost, whereas simulation training is already part of the curriculum, expensive and designed by experienced clinicians. This will paper will describe the design process and method that will be used for the evaluation.

2 Method

2.1 Cognitive Walkthrough and Co-design Process

To make sure practitioners are involved at every stage of the decision-centred design process, the design of the decision support tool prototype will be performed using a co-design method which was adapted from the cognitive walk-through method (Wharton et al. 1994). Participants will (1) provide input for the design and (2) probe on how they would solve a prominent airway challenge using the prototype. The present study will discuss the process and outcomes of this co-design process as an example of how to apply decision-centred design in health care.

2.2 Participants and Research Design

A total of 16 participants will be recruited from an urban tertiary hospital in Melbourne, Australia. Half of the sample will be anaesthetic nurses and the other half (junior) anaesthetists. No particular level of experience is required to participate in the study.

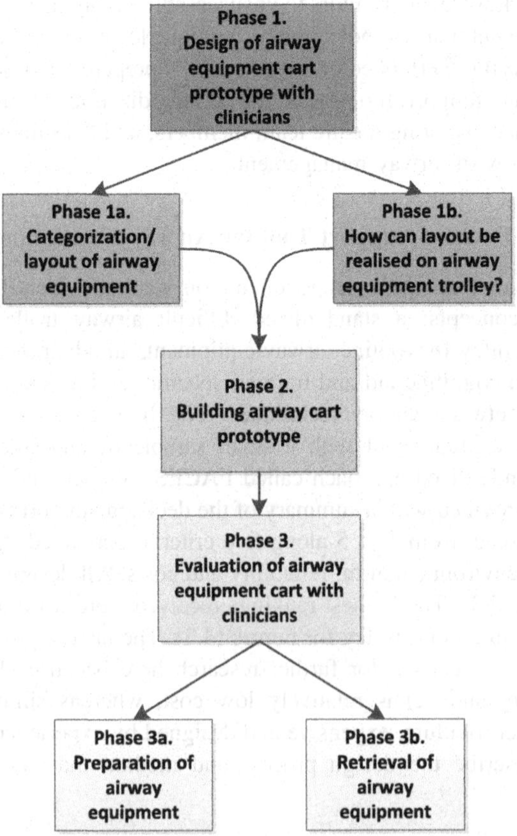

Fig. 1. Co-design process of the present study

The study consists of three phases: (1) design of airway equipment cart with clinicians, (2) building the airway cart prototype and (3) evaluation of the airway equipment cart with clinicians. The process of the study is outlined in Fig. 1.

The study employs a between-subjects design which means that participants are either allocated to the design or evaluation phase. The allocation will depend on when participants will be recruited, with those who are recruited first participating in the design phase. See Table 1 for the allocation of anaesthetists and anaesthetic nurses in both phases.

Table 1. Allocation of participants in the design and evaluation phase

Participants	Design	Evaluation
Nurses	4	4
Anaesthetists	4	4
Totals (N = 16)	**8**	**8**

2.3 Procedure

Ethics approval has been obtained from the hospital's Human Research Ethics Committees and Nursing Advisory Committees. The study is taking place at the hospital of the respective participant in an isolated room either at the hospital's anaesthetic department or patient simulation facilities. A recruitment email with the Participant Information and Consent Form is sent out via the Anaesthetic department to invite clinicians to participate in the study. Furthermore, flyers are distributed in the hospital's surgical suites. Upon arrival, the participants will have to sign the consent form and fill in a short questionnaire about their experience in airway management. Depending on the participants' allocation, the study will then follow the process illustrated in Fig. 1.

Phase 1. Design of Airway Equipment Cart. As indicated in Fig. 1, this study will start with the design phase. Participants allocated to the design phase will be asked to categorize standard airway equipment routinely used for airway management. This does also include airway equipment they need to have available immediately but may not actually have to use (i.e. back up equipment). It does not include advanced airway equipment to manage difficult airways that is not routinely available in the operating theatre such as video laryngoscopes and fiberoptic intubations. Advanced airway equipment is stored in a separate 'difficult airway trolley' which has to be shared across operating theatres. The difficult airway trolleys have been designed by local airway management special interest groups. This study does not interfere with this but is regarded as a complement by focusing on the standardization of routine airway equipment in the operating theatre.

Categorization of Airway Equipment. The airway equipment will be randomly laid out in front of the participant on a table in the room where the study will take place. The participant will then be asked to categorise the airway equipment according to their common flow of work and what they perceive as useful. The participant is free to not use all the pieces of equipment laid out in front of them. After the participant has finished grouping the airway equipment, a photo of the airway equipment will be taken for later analysis. Next, the participant will be asked how to the categorised airway equipment could be transferred onto an airway trolley. The participant will be prompted to be creative and think about different airway trolley designs that will incorporate the categorisation of the airway equipment but consider space constraints. Table 2 lists questions that will be used to prompt the participant. Pens and paper will be available for the participant to draw and adjust sketches. Also, this part of the session will be audio-recorded for later analysis. The participant will be further prompted until they are satisfied with the design.

Table 2. Probes for the categorization of airway equipment and trolley design

Content	
Grouping and inclusion of airway equipment	• How would you group the airway equipment together? What is most helpful to you? • Where should the airway equipment be located and in which order?
Design	
Space/Environmental fit	• How can we get the most out of the space? How can we best organise the airway trolley you are currently using to realise this layout? • Where can we put the airway equipment to support your work flow optimally and better than currently? The layout can be totally unique, and you can iterate it as you go along. • What would help your decision-making (in preparation, transitioning between techniques), effectiveness and efficiency?

Phase 2. Building The Airway Cart Prototype. After each participant session, the photographs will be saved and the audio-recordings taken of the participant's comments on the design will be transcribed. Eventually, a prototype aggregating all participants' responses will be developed. A prototype will then be built by the researchers with simple materials (i.e. cardboard, pouches, stickers, etc.). The aim is to incorporate the main design concepts gathered from this study into a simple mock-up prototype.

Phase 3. Evaluation of Airway Cart Prototype. Participants allocated to the evaluation phase will complete a scenario with both the new airway cart design and the currently used airway cart/approach (within-subjects design). Two different airway emergency scenarios will be created that will require the participants to perform the key activities related to the key decisions: preparation, grasping and passing airway equipment to assist transitioning between techniques and suggesting/prompting certain airway equipment when the current technique does not work. Participants are randomly allocated to (1) order of using new airway cart or currently used airway cart and (2) the scenario they will complete with each of the two carts (see Fig. 2).

Airway Management Activities and Scenarios. As indicated in Fig. 1, participants will evaluate the airway equipment cart by completing two airway management activities with both airway carts (preparation and retrieval of airway equipment, phase 3a and 3b). In order to prevent carryover effects, four patient scenarios will be created. Two of them will require the set-up and preparation of the airway cart for a routine adult surgery, and two will involve a difficult airway management scenario where transitions between airway techniques are required to manage the airway successfully. The scenarios covering similar activities will be unique with differing clinical contexts, but will require a similar set up and retrieval of equipment to make a comparison possible.

Counterbalance Scheme. The interaction with both airway carts will be counterbalanced by alternating the type of airway cart the participants started with (see Table 3). Participants will always start with the preparation scenario and finish with the retrieval scenario in order to adhere to the natural progression of airway management activities as occurring in reality. Thereby, all four scenarios are completed with both airway carts four times, twice with each group of participants (anaesthetic nurses and anaesthetists).

Table 3. Counterbalancing order of airway cart and airway management scenario.

Participant	Occupation	Scenario 1	Scenario 2	Scenario 3	Scenario 4
1	Anaesthetist	P1*	R1*	P2*	R2*
2	Anaesthetist	P2	R2	P1	R1
3	Anaesthetist	P1	R1	P2	R2
4	Anaesthetist	P2	R2	P1	R1
5	Nurse	P1	R1	P2	R2
6	Nurse	P2	R2	P1	R1
7	Nurse	P1	R1	P2	R2
8	Nurse	P2	R2	P1	R1

**Note.* P = Preparation activity, R = Retrieval activity, 1 = new airway cart, 2 = existing airway cart

2.4 Apparatus – Airway Equipment

The list of standard airway equipment recommended for safe administration of anaesthesia has been obtained from the *Australian and New Zealand College for Anaesthetists* (ANZCA 2012). The list ranges from syringes to inflate cuffs of endotracheal tubes to appropriate sizes of face masks, tubes and other artificial airways. For the evaluation phase, the specific sizes of equipment required for the adult patient scenarios were informed by an experienced anaesthetist. To maintain face validity and a valid recall of sizes, real airway equipment will be used for the study.

2.5 Measures

In the design phase, photographs will be taken of the categorised equipment to inform the layout of the airway equipment trolley. The design phase will also be audio-recorded to capture design suggestions by participants which will be incorporated in the design of the prototype. In the evaluation phase, participants will be video-recorded. While participants complete the preparation scenario, completeness of preparation and time for preparation will be measured. During the retrieval scenario, time for suggesting and retrieving equipment and completeness (or omissions) of suggesting particular airway equipment, and time and completeness of identifying missing airway equipment will be measured. After having both scenarios completed with one airway cart, participants will fill in the System Usability Scale; for each airway cart separately (Bangor et al. 2008; Brooke 1996). After having both scenarios completed with both airway carts, participants will fill in a short generic evaluation questionnaire.

3 Analysis

Due to the small sample size and observational nature of this pilot study, the analysis will be descriptive. Percentages will be calculated for the completeness of preparation and retrieval of airway equipment. Time for preparation, retrieval and identification of missing airway equipment will be presented in seconds. The SUS scores will be interpreted in relation to percentile ranking.

4 Limitations

This study will use a small sample size and therefore findings cannot be generalized without caution. While traditional design and usability studies consider five participants as a sufficient 'magic number', more recent research has found that much larger numbers are required (Schmettow et al. 2013). However, the goal of this study was to conduct an efficient pilot study only. The study would require replication with a larger number of participants if the new design proves to be beneficial. Furthermore, designing a prototype should ideally consist of a complete design iteration, bringing the designed prototype back to the participants for further feedback and improvement before evaluation. Due to time and resource limitations this was not possible for this pilot study. However, as the prototype design is based on a full decision-centred design process and Human Factors approach we believe we have mitigated this limitation. If this study will be conducted on a larger scale with additional iterations, more randomization in relation to the order of scenarios and airway carts would increase validity of the experimental set up. This study only counterbalanced the order of airway carts and scenarios. Finally, evaluating both airway carts between and not within subjects would further improve the internal validity of the study.

5 Future Research

If this study reveals superiority of the new airway cart, a larger scale study will be planned addressing the limitations outlined in the previous section. The airway cart will be taken back to participants for another iteration, followed by the design of a fully functional prototype. If the larger scale evaluation study proves to be beneficial, the airway cart will be validated in a clinical setting. This research acknowledges the relevance of considering decision requirements of healthcare practitioners in order to improve the design of healthcare environments.

References

Australian and New Zealand College of Anaesthetists. (2012) ANZCA Handbook for Training and Accreditation. Melbourne, Australia. http://www.anzca.edu.au/documents/training-accreditation-handbook

Berkow LC (2004) Strategies for airway management. Best Practice Res Clin Anaesthesiol 18 (4):531–548. https://doi.org/10.1016/j.bpa.2004.05.006

Bromiley M (2009) Would you speak up if the consultant got it wrong?…and would you listen if someone said you'd got it wrong? J Perioperative Pract 19(10):326–330

Chrimes N (2016) The vortex: a universal "high-acuity implementation tool" for emergency airway management. Br J Anaesth, aew175. http://doi.org/10.1093/bja/aew175

Cook TM, Macdougall-Davis SR (2012) Complications and failure of airway management. Br J Anaesth 109(SUPPL1):i68–i85. https://doi.org/10.1093/bja/aes393

Cook TM, Woodall N, Frerk C (2011) Major complications of airway management in the UK: results of the Fourth National Audit Project of the Royal College of Anaesthetists and the Difficult Airway Society. Part 1: anaesthesia. Br J Anaesth 106(5):617–631. https://doi.org/10.1093/bja/aer058

Flin R, Fioratou E, Frerk C, Trotter C, Cook TM (2013) Human factors in the development of complications of airway management: preliminary evaluation of an interview tool. Anaesthesia 68:817–825

Heard AMB, Green RJ, Eakins P (2009) The formulation and introduction of a "can"t intubate, can't ventilate' algorithm into clinical practice. Anaesthesia 64(6):601–608. https://doi.org/10.1111/j.1365-2044.2009.05888.x

Heidegger T, Gerig HJ, Henderson JJ (2005) Strategies and algorithms for management of the difficult airway. Best Pract Res Clin Anaesthesiol 19(4):661–674. https://doi.org/10.1016/j.bpa.2005.07.001

Klein G, Calderwood R, Macgregor D (1989) Critical decision method for eliciting knowledge. IEEE Trans Syst Man Cybern 19(3):462–472

Klein G, Kaempf GL, Wolf S, Thorsden M, Miller T (1997) Applying decision requirements to user-centered design. Int J Hum Comput Stud 46(1):1–15. https://doi.org/10.1006/ijhc.1996.0080

Larsson J, Holmström IK (2013) How excellent anaesthetists perform in the operating theatre: a qualitative study on non-technical skills. Br J Anaesth 110(1):115–121. https://doi.org/10.1093/bja/aes359

Lintern G, Motavalli A (2018) Healthcare information systems: the cognitive challenge. BMC Med Inform Decis Mak 18(1):1–10. https://doi.org/10.1186/s12911-018-0584-z

Marshall S (2013) The use of cognitive aids during emergencies in anesthesia: A review of the literature. Anesth Analg 117(5):1162–1171. https://doi.org/10.1213/ANE.0b013e31829c397b

Marshall SD, Mehra R (2014) The effects of a displayed cognitive aid on non-technical skills in a simulated "can"t intubate, can't oxygenate' crisis. Anaesthesia 69(7):669–677. https://doi.org/10.1111/anae.12601

Schmettow M, Vos W, Schraagen JM (2013) With how many users should you test a medical infusion pump? Sampling strategies for usability tests on high-risk systems. J Biomed Inform 46(4):626–641. https://doi.org/10.1016/j.jbi.2013.04.007

Schnittker R, Marshall S, Horberry T, Young K (2018) Human factors enablers and barriers for successful airway management – an in-depth interview study. Anaesthesia, 1–10. http://doi.org/10.1111/anae.14302

Schnittker R, Marshall S, Horberry T, Young K, Lintern G (2017) Exploring decision pathways in challenging airway management episodes. J Cogn Eng Decis Making 11(4):353–370. https://doi.org/10.1177/1555343417716843

Schnittker R, Marshall S, Horberry T, Young K, Lintern G (2016) Examination of anesthetic practitioners' decisions for the design of a cognitive tool for airway management. In: Proceedings of the Human Factors and Ergonomics Society 60th Annual Meeting, Washington D.C., pp 1763–1767

Shapell S, Wiegmann DA (2010) Integrating human factors into system safety. In: O'Connor JV, Cohn PE (ed) Human performance enhancement in high risk environments: insights, developments and future directions in military research. Praeger Security International, Santa Barbara, pp 189–209. http://doi.org/10.1037/0003-066X.55.11.1196

The Royal College of Anaesthetists and The Difficult Airway Society (2011) Major complications of airway management in the United Kingdom. https://www.rcoa.ac.uk/system/files/CSQ-NAP4-Full.pdf

Watterson L, Rehak A, Heard A, Marshall S (2014) Transition from supraglottic to infraglottic rescue in the 'can't intubate can't oxygenate' (CICO) scenario. http://www.anzca.edu.au/documents/report-from-the-anzca-airway-management-working-gr.pdf

Wharton C, Rieman J, Lewis C, Polson P (1994) The cognitive walkthrough method: a practitioner's guide. Usability Inspection. https://doi.org/10.1108/09685220910944731

Towards More Interactive Stress-Related Self-monitoring Tools to Improve Quality of Life

Corinna A. Christmann, Gregor Zolynski, Alexandra Hoffmann(✉), and Gabriele Bleser

AG wearHEALTH, Technische Universität Kaiserslautern, 67663 Kaiserslautern, Germany
{christmann, zolynski, hoffmann, bleser}@cs.uni-kl.de

Abstract. Self-monitoring with diaries is one way to identify stress causing events and the respective personal reactions. Considering the broad distribution of smartphones over the past decade, an interactive stress diary application (app) was developed. Diary entries are linked to changes in the appearance of an avatar to support regular usage behavior through vicarious reinforcement.

To investigate the effectiveness of this interactive feature on actual user behavior, 55 young adults randomly received one of two versions of the self-monitoring app, one with vicarious reinforcement (experimental group) and one with no changes in the avatar (control group). After a four week test interval, participants were asked for feedback. Moreover, participants filled out standardized psychometric questionnaires measuring the subjective stress level, occurrence of daily hassles, quality of sleep, and physical symptoms.

Diary entries were correlated with the scores of the respective standardized psychometric questionnaires, indicating convergent validity of the diary categories. A significant increase of missing diary entries over time was found for the control group only. In line with this finding, participants of the experimental group stated that watching the avatar's change over time was fun. These results are a first step towards more interactive stress-related self-monitoring tools to improve quality of life.

Keywords: Avatar · Vicarious reinforcement · Diary · Smartphone App

1 Introduction

Stress-related illness is a major cause for long-term sick leave and is assumed to even increase in future. Self-monitoring with diaries is one way to identify stress causing events and to understand the personal reactions to these events. In the present contribution we present an interactive smartphone app for stress-related self-monitoring.

1.1 Self-monitoring in Health Apps

In order to change maladaptive health behaviors, self-regulatory skills are required [1]. The self-regulation process can be split up into the following three distinct stages: self-

S. Bagnara et al. (Eds.): IEA 2018, AISC 818, pp. 121–130, 2019.
https://doi.org/10.1007/978-3-319-96098-2_17

monitoring, self-evaluation, and self-reinforcement [2]. Self-monitoring involves deliberately paying attention to one's own behavior [1] and is assumed to precede the following stages. This is why self-monitoring has been identified as an important behavior change technique for intervention programs [3]. Three properties are required for effective self-monitoring: truthfulness, consistency, and temporal proximity [4]. Especially the last two requirements can be supported through digital diaries within health applications for smartphones, as mobile technology allows making entries anywhere and anytime. Following this idea, diaries have already been implemented, for example, in the context of exercise and nutrition [5, 6], pain [7], emotion [8], and stress management [9].

While exercise and nutrition represent observable behavior, care must be taken for measurements in the stress management context. Psychological constructs like subjective stress level, quality of sleep, or daily hassles [10] are not only dependent on objective factors such as the amount of arousal or the number of stressful events during the day, but also on the person's appraisal of his or her own condition and the interpretation of events [11]. In order to prove the convergent validity of the stress-related entries within a smartphone health diary, established questionnaires for the assessment of subjective stress level (Perceived Stress Scale, PSS, [12]), daily hassles (Daily Stress Inventory, DSI, [13, 14]), physical symptoms (Cohen–Hoberman Inventory of Physical Symptoms, CHIPS, [12]) as well as sleep quality (Pittsburgh Sleep Quality Index, PSQI, [15]) were used in the present study.

Besides the validity of diary categories, another critical factor for the effectiveness of mobile health diaries is the regularity of self-monitoring and the corresponding entries. Therefore, a crucial question of this contribution is how regular app usage can be supported by interactive app design.

1.2 Vicarious Reinforcement in Health Apps

According to social cognitive theory, behaviors can be learned by observing models [16]. There are several factors that influence whether the new behavior will be adopted or not. One of them is vicarious reinforcement: the observation and interpretation of rewards and punishments experienced by others. If another person is rewarded or punished for a certain behavior, this could also happen to the observer, who might then change his/her own behavior dependent on the consequences [17].

Digital self-representations, so-called avatars, are one way to improve the user experience of interactive systems through gameful design [18], and additionally allow providing vicarious feedback to the user [19]. This can be achieved by linking the avatar's condition and appearance to the user's behavior, e.g., by linking the appearance of an avatar to food decisions [20], to the user's exposure to cell phone RF emissions [21] or to the daily step count [22].

The usage of avatars in health apps, however, depends on app genre: Whereas about 50% of sport and fitness apps provide an avatar for the user [23], a current review of 63 apps identified not a single stress management that made use of avatars [24].

Randomized, controlled evaluation studies on the effectiveness of vicarious reinforcement through an avatar are also rare. One study demonstrated the motivational power of vicarious reinforcement by using an avatar presented in a virtual mirror via a head-mounted display (HMD). The avatar changed its abdominal girth depending on the extent of the user's training on a treadmill [25]. In the context of health apps, Byrne and colleagues demonstrated that adolescents, who received positive or negative feedback through a virtual pet as response to a photo of their breakfast, were more likely to consume breakfast compared to a control condition without pet and a control condition with positive feedback only [26].

In line with these results, we conducted a randomized controlled trial on the effectiveness of the inclusion of an avatar that provides vicarious reinforcement based on the user's diary entries in the app with regard to regular app usage.

2 App Design

The main functionality of the app is a diary of eight categories that are supposed to be rated each day. The first four categories consist of different health behaviors: duration of sleep (in hours), caffeine intake (in 100 mg units), consumption of fruits and vegetables (number of different portions whereby one portion is defined as 80 g, [27]), and the duration of moderate aerobic training activities (in minutes). Moreover, the overall emotional state can be indicated on a nominal scale with different smileys representing "happy", "neutral", "sad", and "angry". The subjective stress level is rated on a scale from 0 (not at all) to 5 (extreme). The last two scales represent the occurrence of positive and negative events during the day. These events are rated on a scale from 0 (none) to 3 (many). For each entry, the color of the answer changes depending on whether the behavior was in line with official health recommendations or not (see Table 1, Fig. 1, and [28] for details). Beside this traffic lights feedback, a short description of each health behavior and its link to health and stress is provided when tapping on the symbol of each category.

Table 1. Diary categories with entry evaluation and descriptions of the avatar's changing appearance depending on the diary entries.

Diary categories	Green evaluation	Yellow evaluation	Red evaluation	Change of avatar's appearance
Sleep in hours	7–9	6,10	≤ 5, ≥ 11	Position of head, dark circles under the eyes
Caffeine intake in 100 mg units	0–3	4–5	≥ 6	Trembling of the whole avatar
Portions of fruits and vegetables	≥ 5	3–4	0–2	Color saturation of plumage
Sport in minutes	≥ 30	10–20	0	Girth of abdomen
Emotional state	happy	neutral	sad, angry	Emotional expression
Stress level	0–1	2–3	4–5	Tidiness of plumage
Positive events	many	some	none, few	Position of the ears
Negative events	none, few	some	many	Position of the ears

Fig. 1. Left: diary dialog with traffic lights feedback based on latest health recommendations. Right: avatar and diary entries.

A birdlike, mythical creature called Elwetritsch, which is typical for the region where this app was developed and tested (Palatinate in Germany), was chosen as the user's avatar. It is displayed above the diary entries (see Fig. 1 right). The appearance of the Elwetritsch is dependent on the user's entries. Missing entries and entries that do not follow the above-mentioned health recommendations result in a degradation of the avatar's physical condition (see Table 1 and Fig. 2 for details). Although the changes of the avatar are not explicitly explained within the app, the user learns to understand them in the process of using the app.

3 User Study

3.1 Participants

In accordance with the declaration of Helsinki, all participants (N = 55) gave written consent after the nature of the experiment was explained to them. They were randomly assigned to one of the two groups (Experimental Group "EG" with vicarious reinforcement, Control Group "CG" without changes in the avatar), completed the pretest and installed the diary app on their own android device (n(EG) = 29, n(KG) = 26).

Fig. 2. Changing appearance of avatar depending on diary entries

Eight participants (n(EG) = 3, n(KG) = 5) were not able to complete the posttest, resulting in a final sample size of N = 47 (n(EG) = 26, n(KG) = 21).

Both groups did not differ with respect to age (EG: M = 23.38 years, SD = 3.01 years, KG: M = 23.67 years, SD = 3.47 years, t(45) = 0.30, p = .77) and distribution among sexes (EG(m/f) = 16/10, KG(m/f) = 8/10, $\chi^2(1) = 2.56$, p = .11). All participants were students (82.98%) or employed (17.02%). This distribution was comparable for both groups ($\chi^2(1) = 0.20$, p = .650). Comparable results concerning age, sex and profession were also found for the pretest sample (N = 55).

All participants used their smartphones at least several times a day. 42% of the sample did not have prior experiences with health apps. There were no group differences concerning the prior use of health apps ($\chi^2(1) = 0.38$, p = .537).

3.2 Procedure

At the beginning, the nature of the experiment was explained to every participant. The principal investigator, however, did not reveal that there were two versions of the app and did not point out that there might be changes in the avatar within the app. This information was given to each participant after having completed the whole experiment.

Participants filled out the following questionnaires: a smartphone usage checklist (frequency of smartphone usage, regularly used functions, prior experience with health apps), the Perceived Stress Scale (PSS) [12] measuring the subjective stress level, the Daily Stress Inventory (DSI) [13, 14] measuring daily hassles, the Cohen–Hoberman Inventory of Physical Symptoms (CHIPS) [12] for the assessment of stress-related physical symptoms, and the Pittsburgh Sleep Quality Index (PSQI) [15].

After having installed the diary app on their own device, the participants received an instruction of the basic app functions. Then, they had to complete their first diary entry and sent the data in the presence of the principal investigator to ensure that every participant had understood the procedure.

Participants were instructed to keep the diary for four weeks and to return for the posttest not later than two weeks after having completed the four weeks. After having used the diary for three weeks, all participants received a reminder for the posttest.

During the posttest, participants were once again asked to send the diary data via the app. Afterward they completed the same stress-related questionnaires as in the pretest (PSS, DSI, CHIPS, PSQI, and the health behavior questionnaire).

At the end, the principal investigator fully explained the nature of the experiment to each participant, who received 15€ and an overview of his/her diary entries from all four weeks.

3.3 Results

As the assumption of normal distribution was rejected for the diary data and questionnaire ratings as revealed by the Shapiro-Wilk test (all ps $\leq$ .20), non-parametrical tests were used for the data analysis, namely: Friedman test instead of ANOVA with repeated measures, Wilcoxon signed-rank test for two related samples as well as Mann-Whitney U test for two independent samples, and Spearman correlations. Bonferroni correction method was used to control for effects of multiple testing.

The number of missing diary entries per week was calculated for each participant. The Friedman test (α = .025 due to Bonferroni correction) revealed significant changes over time for the control group ($\chi^2(3) = 10.31$, p = .016), whereas no changes could be found for the experimental group ($\chi^2(3) = 0.80$, p = .850). The increase over time in the control group was largest during the third (Z = −2.21, p = .027) and fourth week (Z = −2.12, p = .034) compared to the first week (see Fig. 3 for details). However, the groups did not differ in the number of missing diary entries at any point in time as revealed by Mann-Whitney U-tests (all ps $\geq$ .05).

There were significant correlations between the subjective stress level, negative and positive events, and sleep diary entries with the posttest questionnaires (see Table 2 for an overview).

4 Discussion

The pattern of results underpins the convergent validity of the diary categories. The subjective stress level diary entries were not only found to be associated with the posttest scores from the PSS, but also with the daily hassles scale DSI and the physical symptoms checklist CHIPS. Subjective stress level is known to be increased by daily hassles [10] and daily stress was frequently shown to be accompanied by subsequent health problems [29].

Although positive as well as negative events were associated with the PSS score (negative correlation for the positive events), only the negative events were associated with the DSI scale. This pattern of results underlines the validity of the separation of

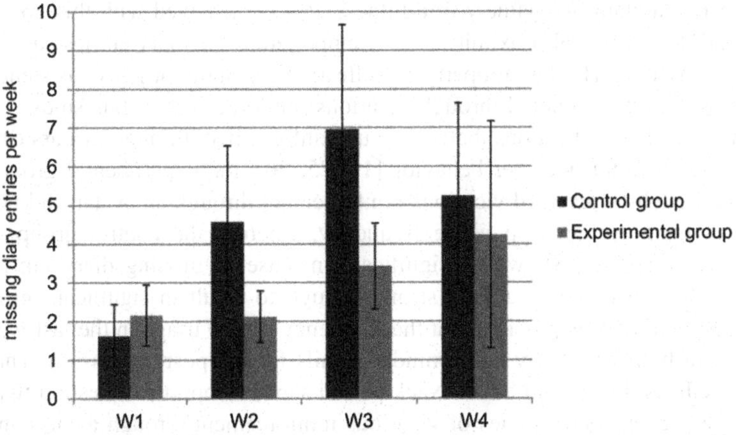

Fig. 3. Mean and standard error for all missing diary entries per week (W1 = first week, W2 = second week et cetera) for both groups.

Table 2. Spearman correlations r between diary categories and posttest questionnaires

Diary category	Posttest questionnaire	r	p
Subjective stress level	PSS	.538	<.001
Subjective stress level	CHIPS	.496	<.001
Subjective stress level	DSI	.574	<.001
Negative events	DSI	.357	.014
Negative events	PSS	.401	.005
Positive events	PSS	−.394	.006
Sleep red entries	PSQI	.528	<.001
Sleep yellow entries	PSQI	.299	.041
Sleep green entries	PSQI	−.440	.002

negative and positive events into two different scales: Daily hassles and social uplifts are both known to influence the subjective stress level, social uplifts in a reducing way [29], whereas daily hassles tend to increase the subjective stress level [30]. They are, however, only modestly related [10]. Therefore, positive events were not expected to be correlated negatively with the DSI score.

For the sleep entries it was not appropriate to use the number of sleep hours for the correlation with the PSQI, as too little and too much sleep are both harmful to a person's well-being [31] (see also Table 1). Therefore, the number of red (not recommended), yellow (borderline) and green (recommended) entries was calculated separately. Red and yellow entries were positively associated with the PSQI score, whereas green entries were negatively associated, underpinning the validity of the sleep category.

To summarize, the overall pattern of results supports the validity of the diary categories subjective stress level, negative and positive events and sleep. Provided that the users do not deceive themselves, the first property for effective self-monitoring according to Bandura [4], namely truthfulness, can be achieved with this application. The second one, temporal proximity is also supported as the diary entries can be made anywhere, anytime. The last property for effective self-monitoring, consistency [17], was supposed to be promoted through vicarious reinforcement in this study.

In line with first indications that vicarious reinforcement through avatars can be an effective way to influence user behavior [19, 25, 26], the experimental group in the present study, which received vicarious reinforcement through an avatar, did not show changes in their average app usage frequency, whereas the control group without vicarious reinforcement showed a significant increase of missing diary entries over time. The effect was, however, not strong enough to result in significant group differences. Moreover, it is possible that the frequency of app usage in the last week was influenced in both groups by the reminder emails for the posttest session. Therefore, long-term effects that extend a four week period stay an important question that needs to be addressed in future studies of vicarious reinforcement through avatars in health apps.

As some members of the experimental group stated that watching the avatar's change over time was fun, we conclude that vicarious reinforcement through the avatar is a first step towards more interactive stress management tools, but it is probably not enough to keep the user entertained over a long usage period. Moreover, participants from both groups asked for a broader set of stress management methods besides self-monitoring (see [32] for taxonomy). Therefore, based on the here presented stress diary and avatar, an extensive stress management app which links established stress management methods to a complete gamification framework was developed [33] and will be further evaluated in future studies.

Acknowledgements. Special thanks are due to Ann-Kathrin Beck, Nesibe Elibol and Stefanie Scülfort for supporting the data collection. The junior research group wearHEALTH is funded by the Federal Ministry of Education and Research (Bundesministerium für Bildung und Forschung, BMBF, reference number: 16SV7115).

References

1. Kanfer FH, Gaelick-Buys L (1991) Self-management methods. In: Kanfer FH, Goldstein AP (eds) Helping people change. A textbook of methods, pp 305–360. Allyn and Bacon, Boston
2. Kanfer FH (1970) Self-regulation: research, issues and speculations. In: Neuringer C, Michael JL (eds) Behavior modification in clinical psychology. Appleton-Century-Crofts, New York, pp 178–220
3. Abraham C, Michie S (2008) A taxonomy of behavior change techniques used in interventions. Health Psychol 27:379–387 Official Journal of the Division of Health Psychology, American Psychological Association
4. Bandura A (1998) Health promotion from the perspective of social cognitive theory. Psychol Health 13:623–649
5. Breton ER, Fuemmeler BF, Abroms LC (2011) Weight loss-there is an app for that! But does it adhere to evidence-informed practices? Transl Behav Med 1:523–529
6. Sharp DB, Allman-Farinelli M (2014) Feasibility and validity of mobile phones to assess dietary intake. Nutrition 30:1257–1266

7. Hundert AS, Huguet A, McGrath PJ, Stinson JN, Wheaton M (2014) Commercially available mobile phone headache diary apps: a systematic review. JMIR mHealth uHealth 2: e36
8. Mattson DC (2016) Usability evaluation of the digital anger thermometer app. Health Inf J 23:234–245
9. Smedberg Å, Sandmark H (2012) Design of a mobile phone app prototype for reflections on perceived stress. In: eTELEMED 2012: the fourth international conference on eHEALTH, telemedecine, and social medicine, pp 243–248. IARIA
10. Kanner AD, Coyne JC, Schaefer C, Lazarus RS (1981) Comparison of two modes of stress measurement: daily hassles and uplifts versus major life events. J Behav Med 4:1–39
11. Folkman S, Lazarus RS, Dunkel-Schetter C, DeLongis A, Gruen RJ (1986) Dynamics of a stressful encounter. Cognitive appraisal, coping, and encounter outcomes. J Pers Soc Psychol 50:992–1003
12. Cohen S, Hoberman HM (1983) Positive events and social supports as buffers of life change stress. J Appl Soc Pyschol 13:99–125
13. Brantley PJ, Waggoner CD, Jones GN, Rappaport NB (1987) A daily stress inventory: development, reliability, and validity. J Behav Med 10:61–74
14. Traue HC, Hrabal V, Kosarz P (2000) Alltagsbelastungsfragebogen (ABF): Zur inneren Konsistenz, Validierung und Stressdiagnostik mit dem deutschsprachigen Daily Stress Inventory. Verhaltenstherapie und Verhaltensmedizin 21:15–38
15. Buysse DJ, Reynolds CF III, Monk TH, Berman SR, Kupfer DJ (1989) The Pittsburgh Sleep Quality Index: a new instrument for psychiatric practice and research. Psychiatry Res 28:193–213
16. Bandura A (2001) Social cognitive theory of mass communication. Media Psychol 3:265–299
17. Bandura A, Ross D, Ross SA (1963) Vicarious reinforcement and imitative learning. J Abnorm Soc Psychol 67:601–607
18. Seaborn K, Fels DI (2015) Gamification in theory and action: a survey. Int J Hum Comput Stud 74:14–31
19. Parks P, Cruz R, Ahn SJ (2014) Don't hurt my avatar: the use and potential of digital self-representation in risk communication. Int J Robots Educ Art 4:10–20
20. Hswen Y, Murti V, Vormawor AA, Bhattacharjee R, Naslund JA (2013) Virtual avatars, gaming, and social media: designing a mobile health app to help children choose healthier food options. J Mob Technol Med 2:8–14
21. Burigat S, Chittaro L (2014) Designing a mobile persuasive application to encourage reduction of users' exposure to cell phone RF emissions. In: Hutchison D, Kanade T, Kittler J, Kleinberg JM, Kobsa A, Mattern F, Mitchell JC, Naor M, Nierstrasz O, Pandu Rangan C et al (eds) Persuasive technology, vol 8462, pp 13–24. Springer International Publishing, Cham
22. Lin JJ, Mamykina L, Lindtner S, Delajoux G, Strub HB (2006) Fish'n'Steps: encouraging physical activity with an interactive computer game. In: Dourish P (ed) Proceedings of the 8th international conference on ubiquitous computing, vol. 4206, pp 261–278. Springer, Berlin, Heidelberg
23. Lister C, West JH, Cannon B, Sax T, Brodegard D (2014) Just a fad? Gamification in health and fitness apps. JMIR Serious Games 2:e9
24. Hoffmann A, Christmann CA, Bleser G (2017) Gamification in stress management apps. A critical app review. JMIR Serious Games 5:e13
25. Fox J, Bailenson J, Binney J (2009) Virtual experiences, physical behaviors. the effect of presence on imitation of an eating avatar. Pres Teleoper Virtual Environ. 18:294–303

26. Byrne S, Gay G, Pollack JP, Gonzales A, Retelny D, Lee T, Wansink B (2012) Caring for mobile phone-based virtual pets can influence youth eating behaviors. J Child Media 6:83–99
27. Agudo A (2018) Measuring intake of fruit and vegetables. http://www.who.int/dietphysicalactivity/publications/f&v_intake_measurement.pdf. Accessed 17 Apr 2018
28. Christmann CA, Zolynski G, Hoffmann A, Bleser G (2017) Effective visualization of long term health data to support behavior change. In: Duffy VG. (ed) Proceedings of the Digital human modeling. Applications in health, safety, ergonomics, and risk management: ergonomics and design: 8th international conference, DHM 2017, held as part of HCI International 2017, Vancouver, BC, Canada, 9–14 July 2017, pp 237–247. Springer, Cham
29. DeLongis A, Folkman S, Lazarus RS (1998) The impact of daily stress on health and mood: psychological and social resources as mediators. J Pers Soc Psychol 54(3):486–495
30. Lu L (1991) Daily hassles and mental health: a longitudinal study. Br J Psychol 82:441–447
31. Hirshkowitz M, Whiton K, Albert SM, Alessi C, Bruni O, DonCarlos L, Hazen N, Herman J, Katz ES, Kheirandish-Gozal L, Neubauer DN, O'Donnell AE, Ohayon M, Peever J, Rawding R, Sachdeva RC, Setters B, Vitiello MV, Ware JC, Adams Hillard PJ (2015) National Sleep Foundation's sleep time duration recommendations: methodology and results summary. Sleep. Health 1(1):40–43
32. Christmann CA, Hoffmann A, Zolynski G, Bleser G (2017) Stress management apps with regard to emotion-focused coping and behavior change techniques: a content analysis. JMIR Mhealth Uhealth 5(2):e22
33. Christmann CA, Hoffmann A, Zolynski G, Bleser G (2018) Stress-Mentor: linking gamification and behavior change theory in a stress management application, In: Stephanidis C (ed) HCII posters 2018, CCIS, vol 851. Springer, Cham

Nursing Home Manager Role in Managing Aging Workers: A Qualitative Research in Nursing Homes

Massenti Denise[1], Angela Carta[2](✉), and Livia Cadei[3]

[1] RSA "Pietro Beretta" Gardone Val Trompia, Gardone Val Trompia, BS, Italy
[2] Department of Medical and Surgical Specialities, Radiological Sciences and Public Health, University of Brescia, Brescia, Italy
angela.carta@unibs.it
[3] Faculty of Psychology, Catholic University of Sacred Heart, Brescia, Italy
livia.cadei@unicatt.it

Abstract. Nursing home manager (NHM) has now a great challenge: managing an aging staff ensuring higher level of quality assistance and enforcing health and safety regulations.

Aim of this research was to analyze NHM awareness of professional role in managing aging workers in order to think up solutions to improve working condition and quality of assistance.

On the basis of preliminary literature search, an ethnographic quality approach was applied. This study was based on a retrospective data collection of administrative data and on the analysis of responses to semi structured interviews on these topics. A comprehensive description was made and results were discussed with an expert occupational physician. Key points emerging are: (a) NHM were usually involved in individual job task description, but their awareness of responsibilities in health and safety management of nursing staff was not clear; (b) NHM were not usually consulted by their occupational physician (OP); (c) work modifications indicated by OP were often considered by NHM generic and not contextualized; (d) organization consequences of work restrictions were not shared among OP, Chief of staff and NHM; (e) when work restrictions were preceded by OP's case discussion with relevant parties, there was a better management of nursing staff. On the bases of these results, main suggestions to improve management will include: new approaches for nursing staff risk assessment, better definition of NHM role and duty, specific pathways of learning for NHM and for registered nurses, involvement of registered nurses in nursing assistants' management, OP case discussion with relevant parties, improvement in scientific documentation for fitness to work process, implementation of procedure for work restriction management.

Keywords: Aging workers · Nursing home managers · Work restriction

M. Denise—Nursing Home Manager.
A. Carta—MD.
L. Cadei—Professor of Pedagogy.

S. Bagnara et al. (Eds.): IEA 2018, AISC 818, pp. 131–138, 2019.
https://doi.org/10.1007/978-3-319-96098-2_18

1 Introduction

Nowadays in Italy, as in many other countries, a nursing home manager (NHM) has a great challenge: managing an aging staff facing an increasing clinical complex blend of residents' disorders, ensuring higher level of quality assistance and enforcing health and safety regulations.

As defined by Local authority nursing homes (NH) are authorized public or private facilities to guarantee health and social interventions aimed at improving the levels of autonomy of residents, to promote well-being, to prevent and treat chronic diseases and their exacerbation.

In Brescia province there are 86 Nursing Homes for a population of about 1,250,000 resident population with an aging index (defined as N > 65 years old/100,000 resident people) of 18.5 in 2011 [3].

All persons aged $\geq$ 65 are eligible for the service under conditions of partial or total non self-sufficiency or affected by Amyiotrophic Laterl Sclerosis; in some structures patients in Vegetative State can also be accepted [1].

Admission to the facilities is carried out by a territorial team called Multi-Dimensional Assistive Continuity Unit with the role of welcoming any request for action by non-self-sufficient citizens or their families and supporting them. In decoding the real needs, proposing the most suitable solutions among those available. This team guarantees its intervention also for the management of the entry list of the NH of the Val Trompia, defining the overall picture of the elderly and monitoring the situations during the waiting for a vacant bed.

People who access the NH are significantly compromised in autonomy, carriers of severe comorbidity: 61% have a complete dependence on walking, 73% have a severe degree of comorbidity (from 5 to 7 pathologies) and 86% has psychiatric-behavioral pathology. The S.OS.IA classification is the system for assessing the fragility of the residents. It is used in Lombardy for the nursing home, with a diversification for severity levels: from class 1 (more serious) to class 8 (less severe). The classes determine the daily quota of the regional health fund provided to the nursing home, as well as the assistance to be guaranteed to the residents. The remaining costs are to be paid by the elderly person, by the family members or by the Municipality of residence [16].

Over the years there has been a progressive aging of people accepted in nursing home and a worsening of their functional, cognitive and clinical conditions. The need to manage problems related to functional and cognitive disability, prevention and treatment of acute events must be combined with more attention to the relational aspects of each resident and family [7].

At the same time we are observing the aging of the working population, with a consequent reduction in work capacity [8]. In particular, there is a progressive aging in the health sector of National Health System employees, with a quota for workers over 55, which is still below 25%.

The main risk factors for healthcare workers are:

- Ergonomic factors: lifting and handling of patients; uncomfortable or painful postures;

- Psychosocial factors: high work rates or excessive workload; psychological requests; threats and physical violence; shift work; home-work reconciliation;
- Biological factors: risk of exposure to biological fluids;
- Accident risk: accidental falls, needle and cutting injuries.

The aging of health workers resulting from the increase of retirement age will give rise in the next 5–10 years to a situation where a significant proportion of workers (probably 15–20%) will not be able to carry out their duties or will meet strong difficulties, worsening their health and quality of care, and risking dismissal for non-fitness or sick leave [6].

The safety of employees is one of the activities that the NHM must take on as the person in charge for protection of workers' health. It is a fundamental element that can influence the quality of care (safety of the assisted persons) and the organization as a whole.

NHM must create the organizational conditions: this means "organize", a word used as the synthesis of management applied to the context and to the actual situation. The classic management functions are: planning, organization, guidance, control.

The issue of personnel management with job limitations requires the implementation of shared strategies and the ability of the NHM to create the organizational conditions that benefit the operator, the assisted and the organization itself [15].

2 Methods

On the basis of preliminary literature search, an ethnographic quality approach was applied.

Qualitative research responds to a wide variety of questions related to the interest of nursing care for current and potential human responses to health problems. The purpose of qualitative research is to describe, explore and explain phenomena, the questions often are: "what is this?" Or "what's going on here?" And they mainly concern processes rather than outcomes [4, 5, 18].

Ethnographic method is one of the most well-known approaches in anthropology and has also been used for some time in nursing sciences. Thanks to the studies of Leininger, nurses are realizing the importance, in daily care, of understanding cultures, especially today that we live and work more and more in multicultural realities.

Leininger's theories of care are at present finding important repercussions and applications not only in practical field, but also in education and research [4, 5].

This method allows nurses to gain insights into the health system of different cultures, as well as to deepen the different values, customs and habits in providing nursing care. All this is done so that nursing personnel can provide attentive assistance as social actors according to social values (emic) and professional values (etic). Therefore an "Ethno-care" (ethno-assistance) takes into account the actions and meanings of different social and cultural backgrounds [11].

The main objective is to develop the understanding of the theme for how it exists and for how individuals conceive it. This means describing and interpreting, in each NH, the specific modes of behavior, in order to understand the management of workers' health and safety. NHM can provide a better organization and quality of care to the assisted trying to understand habits, customs, social rules and how these variables affect the management of staff.

The sampling is "propositive" or "reasoned choice" represented by the NHMs of 8 Assisted Health Residences (RSA) and given that not always the figure of the NHM deals with the topic studied, a "network sampling" was adopted.

This is a retrospective study to investigate the research topic, a semi-structured interview was used. Table 1 reported the guide questions and the related research objectives. Some institutional documents of each RSA have been acquired and examined (Charter of Services, main duties of the NHM, total number of employees, age and employee length of service, number of employees with restricted fitness, suitability judgments with limitation to the task).

Table 1. Guide questions and research objectives of semi-structure questionnaire

	Questions	Research objectives
1	Are there any staff work restriction? If so, what kind of work restrictions and how many?	Existence of work restrictions type, size of the "problem"
2	Do the work restrictions seem to fit the work environment?	Discrepancy between judgments expressed and work environment
3	Does the OP consult the NHM in order to acquire information on the organizational impact deriving from work restrictions?	What collaboration among the various figures?
4	Is there a shared procedure on the management of work restrictions?	Sharing in teams the management of staff work restriction
5	Have you found it difficult to interpret work restrictions? If so what was the difficulty?	Meaning attributed to the work restriction and misunderstanding

The collected data were processed coding them through specific topics and precise knowledge objectives. Condensing the collected material we tried to give a complete, holistic, deep and rich description of the problem.

The results were then discussed with an expert in the field and the main demographic variables were elaborated, trying to represent them analytically with respect to the general results.

3 Results

The 8 NHs of Val Trompia have 651 beds.

There are 401 employees, socio-demographic characteristics are described in Table 2.

Mean age was 45 (range 21–67) and mean length of employment in each facility was 11 years (range 0–34);

Age distribution show an aging workforce in all the examined NHs examined; as better described in Fig. 1 workers aged over 40 represent 72% of all workers.

Table 2. Characteristics of the study population.

	N	%
Age		
<45	183	45,6
45–55	145	36,2
<55	73	18,2
Gender		
Male	35	8,7
Female	366	91,3
Length of employment		
<10	169	44,5
10–20	137	36,2
>20	73	19,3
Work Restrictions		
All	108	26,9
Manual handling	85	78,7
Manual handling and shiftwork	10	9,3
Shiftwork	7	6,4
Other	6	5,6

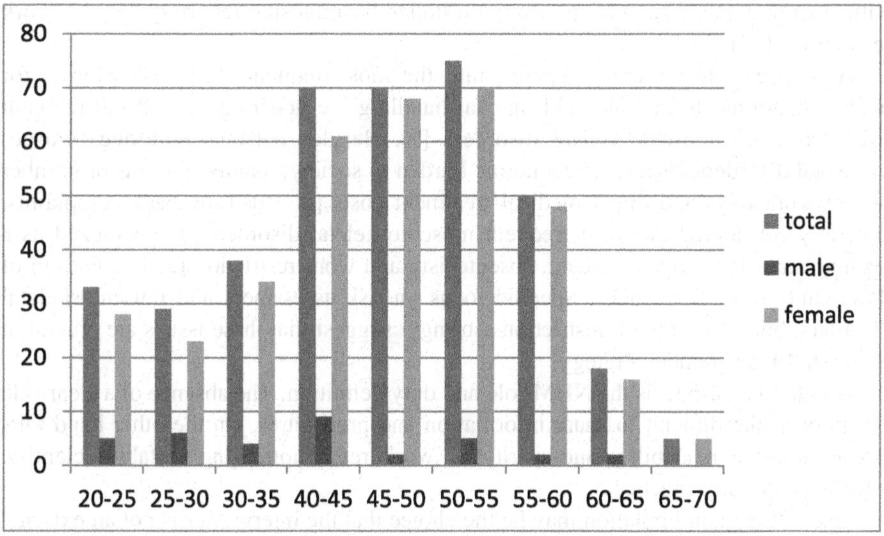

Fig. 1. Frequency distribution by age

The staff with work restrictions is present in all examined NHs. Employees with work restrictions are 26.9% and the most frequent work restriction is "manual handling" (more than 88%), followed by limitation for night- shift. About 9% of the staff have multiple work restrictions.

NHMs are present in 7 out of 8 NHs but just 3 of them have a Master degree in nursing management. There is not always a card of the main tasks of the NHM and the organizational role is different from structure to structure. Few facilities have a formal description of NHM role and duty; the majority of the NHMs is involved in the drafting of individualized work plans. Occupational physician (OP) are external professionals in all NHs examined.

Key points emerging from the interviews were: (a) almost all NHMs were involved in individual job task description, but NHM's awareness of responsibilities in health and safety management of nursing staff was a problem; (b) the management of staff with limited fitness for work was critical; NHMs has met difficulties in carrying out this task; (c) they were not usually consulted by OP; (d) work modifications indicated by OP were considered by of NHM often generic and not contextualized; (e) organization consequences of OP's work restrictions were not shared among OP, Chief of staff and nursing managers with negative consequences on nursing staff management; (f) when Ops' fitness for work judgment was preceded by case discussion with relevant parties, management of work restrictions carried to specific work task descriptions and a better management of nursing staff.

4 Discussion

Data on aging working staff and limited job fitness in Val Trompia area are consistent with emerging data from a recent study on Italian hospital staff reporting 23.28% work restrictions [13].

As expected in an aging nursing staff, the most frequent work restriction is for manual handling; indeed age and manual handling are generally considered relevant risk factors for musculoskeletal disorders [9]. Besides personal suffering, musculoskeletal disorders dispose an economic burden to society, mainly in terms of number of lost work days and direct medical treatment costs [2, 10]. Low back complaints, generally considered the most frequent musculoskeletal disorders, is recognized as a leading cause for early retirement, absenteeism and work restriction [2, 14]. Design of these study did not include a specific focus on risk assessment and musculoskeletal disorders, but data on work restrictions strongly suggest that these issues are crucial in NHs' staff management strategy.

Another key aspect is the NHM role and duty definition. The absence of a clear role definition make difficult to share information and procedures. On the other hand OPs would improve readability and clarity of work restriction using suitable scientific references in daily practice.

One of the main limitation may be the choice that the interviewer is not an external but a member of the system itself. The research topic emerged from the direct experience of the researcher, which led him to ask questions and develop interest in the subject.

In the past anthropology claimed that researchers should detach the examined subjects in order to prevent their personal points of view from guiding the research process. It was feared that personal involvement would make the topic more stressful, that researchers would lose objectivity and objective results. Today these objections are largely resized but "a great effort is needed to understand what comes from you and what is not. I think this can be done successfully, but I think a lot of experience and a lot of hard work is needed to get to this awareness " [12, 17]. However, qualitative methods usually do not require "de-personalizing" those who are being studied or denying the researcher's influence on the behavior of the research participants.

In carrying out the research, new emerging problems have been highlighted. The production of data, in addition to the use of semi-structured interview, required the use of other strategies to have sufficient insight into the problem. Furthermore, the material given by Institutional documents allowed to verify the veracity of what was expressed in the interviews [12, 17].

5 Conclusion

Fitness for work with work restrictions in an aging nursing staff is a challenge. These study highlights the need to: (a) a better definition of NHM role and duty; (b) implementation of shared procedures for the management of work restrictions; (c) a clear definition of the actors involved in health and safety of NHs' staff; (d) unambiguous and shared work restrictions formulated by OP. In order to improve the system.

NHM would be the leader trying to create conditions of a better collaboration with OP with the help of all figures involved in staff management. The specific training of NHM should be carried out to all nurses to raise awareness of their role in health and safety of both nursing staff and residents.

Finally, it is very important to experiment and implement new methods of risk assessment that are more specific to the work context and which, from a participatory perspective, can foster a culture of safety, an improvement of working conditions for aging workers and, ultimately, an improvement also in the quality of assistance.

References

1. ATS Brescia. https://www.ats-brescia.it/bin/index.php. Accessed 31 Mar 2018
2. Bergstrom G, Hagberg J, Busch H, Jensen I, Bjorklund C (2014) Prediction and sickness absenteeism, disability pension and sickness presenteeism among employees with back pain. J Occup Rehabil 24:278–286. https://doi.org/10.1007/s10926-013-9454-9
3. Camera di Commercio. Brescia in cifre 2017. http://www.bs.camcom.it/files/Studi/Approfondimenti_tematici_2017/brescia-in-cifre-2017-def.pdf. Accessed 31 Mar 2018
4. Cadei L (2012) Legami e significati nell'esperienza della sofferenza. Un'indagine qualitativa. in L. Pati (a cura di), Sofferenza e riprogettazione esistenziale. Il contributo dell'educazione, La Scuola, Brescia, pp 25–66. ISBN 978-88-350-3071–3
5. Cadei L (2005) La ricerca e il sapere per l'educazione, I.S.U., Milano, p 273. ISBN 10: 8883113853, ISBN-13: 978-88-831-1385-7
6. D'Errico A (2015) Invecchiamento e lavoro in sanità. Dati e Prospettive. https://www.ciip-consulta.it/images/MilanoExpo/invecchiamento/Angelo_DErrico.pdf
7. Guerrini G, Ramponi JP, Scarcella C, Trabucchi M (2014) Manuale di igiene ed organizzazione sanitaria delle residenze sanitarie assistenziali. Maggioli Editore, 1st edn

8. Guardini I, Deroma L, Salmaso D, Palese A (2011) Invecchiamento della popolazione infermieristica di due ospedali del Friuli Venezia Giulia: applicazione di un modello matematico deterministico. G Ital Med Lav Erg, 55–62
9. Hoy D, Bain C, Williams G, March L, Brooks P, Blyth P et al (2012) A systematic review of the global prevalence of low back pain. Arthritis Res Ter 64:2028–2037. https://doi.org/10.1002/art.34347
10. Krismer M, van Tudler M (2007) Low back pain (nonspecific). Best Pract Res Clin Rehum 21:77–91
11. Leininger M (1985) Transcultural care diversity and universality: a theory of nursing. Nurse Healt Care 6(4):208–212
12. Morse M, Richards L (2009) Fare ricerca qualitativa. In: Graffina FG Fare Ricerca Qualitativa. FrancoAngeli, Milano
13. Maricchio R, Ferraresi A, Bonamici F, Bertelli A, Passarini L, Bagnasco A, Sasso L (2013) Invecchiamento dei professionisti sanitari e fenomeno delle inidoneità al lavoro: studio osservazionale. L'infermiere 50:e9–e16. http://www.ipasvi.it/ecm/rivista-linfermiere/rivista-linfermiere-page-32-articolo-373.htm. Accessed 18 Apr 2018
14. Pattani S, Costantinovici N, Williams S (2001) Who retires early from NHS because of ill health and does it cost? A national cross sectional study. BMJ 322:208–209
15. Pennini A, Barbieri G (2017) Le Responsabilità del coordinatore delle professioni sanitarie. II Edizione Mc Graw Hill Education
16. Podavitte F, Scarcella C, Mattana E, Trabucchi M (2008) Uno strumento per gestire razionalmente le liste d'accesso alle RSA. I Luoghi della Cura 3:16–20
17. Polit F, Beck C (2014) Fondamenti di Ricerca Infermieristica. McGraw Hill, Milano
18. Robb MC, Mosci D, Chiari P (1999) EBN Identificazione dei disegni di ricerca. Evid Based Nurs 4:1–3

An Ergonomic User Interface Design for a New Extremity MRI Focusing on the Patient Chair

Eui S. Jung[1(✉)], Kimin Ban[1], Jinyoung Kim[1], Jiwon Ahn[1], Sangkyun Na[1], Jinho Yim[2], and Kyungjin Oh[2]

[1] School of Industrial and Management Engineering, Korea University, 145 Anam, Seongbuk, Seoul 02841, South Korea
ejung@korea.ac.kr

[2] Samsung Electronics, 129 Samsungro, Yeongtong, Suwon 16677, South Korea

Abstract. Ergonomic design guidelines for the new development of an extremity MRI have been developed to minimize patient's postural discomfort on different scanning types with a specific focus given to the patient chair and leg supporter. The research started with a known zero gravity position as an optimal body posture and did a market survey on various industrial chairs. Based on the anthropometric characteristics of the populations being considered, the comfortable ranges of the dimensions and angles were defined for four scanning types: knee, ankle, elbow and wrist scannings. In order to validate these guidelines, a simple mockup was made and tested for a group of participants. The test was to find out 3-dimensional comfortable postures of 14 participants with respect to the scanning type and scanning duration, which subsequently yielded design dimensions and adjustable ranges of angles for MRI bore and exterior, chair and leg supporter.

Keywords: Extremity MRI · Chair design guidelines
Patient postural discomfort

1 Introduction

The interaction between people and technologies affects people to have active user experiences. It is important to focus on providing a good user experience in designing products and services [1]. It has been shown that understanding human factors when designing medical devices is expanding to convey a good user experience [2].

Aging global population has led to an increase in the number of diseases associated with arthritis and spinal diseases, and a steady increase in the utilization of related medical services [3]. In particular, the general public, health organizations, and even the scientific community have enthused magnetic resonance video devices [4]. In order to provide a guide for the progress and treatment of musculoskeletal diseases, Magnetic Resonance Imaging (MRI) is widely used [5]. However, according to Meléndez and McCrank [6], 30 to 40 percent of patients reported negative experiences associated with anxiety during existing MRI scans.

In order to reduce the anxiety and inconvenience of patients, an extremity magnetic resonance imaging (xMRI) was developed to provide partial magnetic resonance

S. Bagnara et al. (Eds.): IEA 2018, AISC 818, pp. 139–147, 2019.
https://doi.org/10.1007/978-3-319-96098-2_19

imaging of the arms and legs [7]. This xMRI has low noise levels in the equipment, can provide patients with an easygoing attitude, and is effective in terms of user satisfaction [8]. In fact, the system is designed so as to be able to photograph a patient without fear of a clandestine room phobia, and provides the patient with an easy seated experience [9]. The xMRI is known to be cost-effective reducing overall construction, installation and maintenance costs [5].

However, in the case of xMRI, only a small number of local joint parts can be examined and it takes a longer process to have the image by providing a smaller visual field (FOV) than existing MRI [7]. Also, standard seating postures during the scannings of the local joint parts, which must consider human anthropometric characteristics, have not yet been properly defined. The patient's comfortable seating posture is still challenged. Therefore, it is necessary to determine the optimum standard posture for each scanning type.

The aim of this study was first to suggest a set of optimal standard seating postures for each scanning type to guarantee patient's comfort, and also to suggest the design guidelines for the chair and supporter that properly support patients' postures. As the zero gravity posture from NASA has been known to represent a comfortable neutral body posture considering various anthropomorphic characteristics, the study initially adapted the posture and subsequently the other industrial seating postures in order to provide the optimal standard postures [10].

2 Standard Seating Postures for xMRI

2.1 Body Size in Percentiles and Anthropometric Variables

xMRI standard seating postures require consideration of wide ranges of anthropometric structures of the human body. The body sizes of Koreans and Americans were chosen as the standard groups. The representative percentiles of each standard population considered were 5, 25, 50, 75, and 95 percentiles for both male and female populations.

The angle between torso and thigh was assumed to be at approximately 128° as applied to the sitting position from neutral body posture that NASA suggested [10]. Based on this neutral body posture, human anthropometric variables associated with the xMRI standard postures were selected: seating height, toe to eye seating height, toe to neck seating height, toe to neck seating height, toe to thigh seating height, toe to knee seating height, toe to shoulder seating height, toe to hamstring seating height, elbow height of the seating posture, buttocks to hamstring width of seating postures, buttocks to knee width of seating postures, and buttocks width of seating postures and knees. The foot and shoulder widths are additionally selected because they indirectly influence the sitting posture.

2.2 Selection of Chair Design Variables and Mapping Between Anthropometric and Design Variables

Design variables of the other industrial chairs were reviewed in this study. Chair design variables were divided into three categories according to their purposes: existing xMRI, resting, and performing tasks. As depicted in Table 1, the seatpan height was calibrated as 500 mm, which is 16 mm higher than the traditional xMRI design. Seatpan depth took the range of 200–350 mm. Although headrest height had not previously been included in existing xMRI design variables, it was added as having the adjustable range of 745–862 mm, taking into account the range of 300 mm to 930 mm found in other industrial chairs. Armrest height made it possible to support arm and shoulder by adding 49 mm to the existing xMRI design.

Anthropomorphic variables mentioned above are mapped to the chair design variables as illustrated in Fig. 1. The headrest height is correlated to the seating height, and the toe to neck seating height. The seatback height is relevant to the seating height, the toe to shoulder seating height, and the toe to neck seating height. Armrest is associated with the elbow height of the seating posture. Seatpan height is related to the toe to knee seating height, the toe to hamstring seating height, and the toe to thigh seating height. The leg supports are mapped to the toe to knee and the toe to hamstring seating height.

2.3 Adjustable Ranges of Human Joint Angles Suggested for the Chair

The adjustable ranges of human joint angles for xMRI chair were derived from the research results on the comfortable angle of the joint, shown in Fig. 2. Figure 2 shows the adjustable ranges of the five joints: ankle, hip, knee, elbow, and shoulder joints. The angle of ankle was set at 110, Hip was 125–135, knee 120–150, Elbow 110, and Shoulder 65°. Ranges of joint angles are depending on anthropomorphic variables of xMRI.

2.4 Adjustable Angles of the Chair Depending on Scanning Types

Different design angles depending on the four scanning types were derived for xMRI scanning based on the standard postures mentioned above. For four scanning types, knee, ankle, and elbow angles were set to have the same ranges of 115–125, 125–135, and 105–115°, respectively. But hip angle took the different range as being 130–140 degrees for knee scanning, 120–130 degrees for ankle scanning, and 150–160 degrees for elbow and wrist scannings. Since the upper leg must take the flatter posture for knee scanning than that for ankle scanning, a higher angle range was assumed than the ankle scanning. The angles for elbow and wrist scannings were set to be identical due to the similarity of scanning postures. Because the ranges of shoulder and arm angles must be horizontal, the angle of hip was set to be higher, as shown in Fig. 3.

3 A Validation Study on the Standard Posture with a Mock-up

3.1 Methods

To see the adequacy of the proposed postures, a mock-up chair and supporter was made and tested by evaluating postural comfort on each scanning type. For eleven different

Table 1. Design variables drawn from existing xMRIs and the other industrial chairs (in mm)

Chair part	Variables	XMRI		Massage chair	Car seat	Comfort chair	Office chair	Train seat	
		This study	Traditional	Brams comeback P-1010	SAE D segment	Folding deck chair	FURSYS ITIS2	KTX 1st seat	KTX 2nd seat
Seatpan	Height	500	495	550	500	400	380–460	370	370
	Depth	200–350	220	500	520	550	440	430	430
	Width	458	560	770	530	520	492	470	430
Seatback	Height	961	945	1130	820	750	598	800	800
	Width	620	550	770	480	660	398	650	530
Headrest	Height	745–862	–	300	640	930	675	555	555
Armrest	Height	249	200	300	–	220	220–300	260	260
	Depth	333	265	460	330	520	332	370	370
	Width	548	550	450	500	530	478	470	430

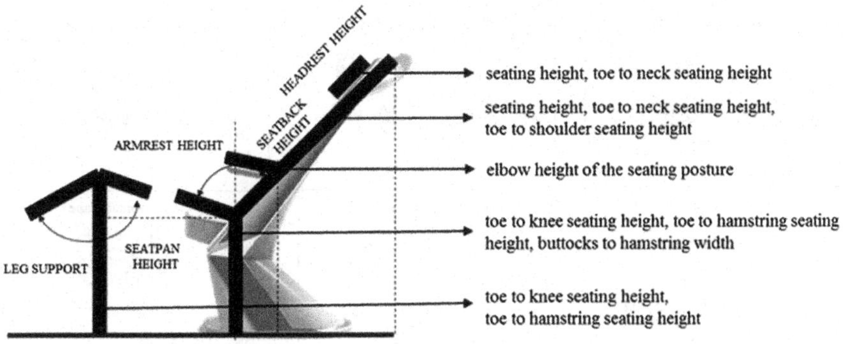

Fig. 1. Mapping between anthropometric variables and chair design variables

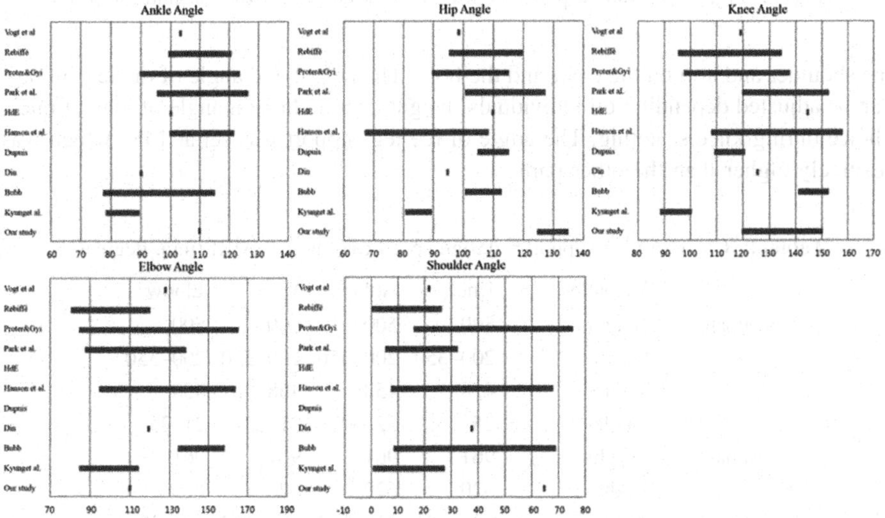

Fig. 2. Adjustable ranges of joint angles suggested from existing research

local body parts (neck, shoulders, upper back, upper arms, mid back, lower arms, lumbar, buttocks, wrist, thighs, and legs) and four scanning types (ankle, knee, elbow, and wrist), the subjective ratings of local body part discomfort for four scanning types were measured during a 30 min' session. Fourteen participants who were in their 20 s and 30 s volunteered. The local body part discomfort was assessed by using a 7-point unipolar Likert scale.

Table 2 shows the design specifications of the chair and supporter built for the test in terms of height, width, length and angles. The lengths of the seatpan, seatback, headrest, leg support (front), and leg support (rear) were 200–350 mm, 961 mm, 234 mm, and 400 mm, respectively, regardless of different scanning types. The angle of the seatpan was set be 32–39°, allowing the ankle to be bent during scanning. The seatback angle for the wrist and the elbow scannings was to be at 151–152°, supporting

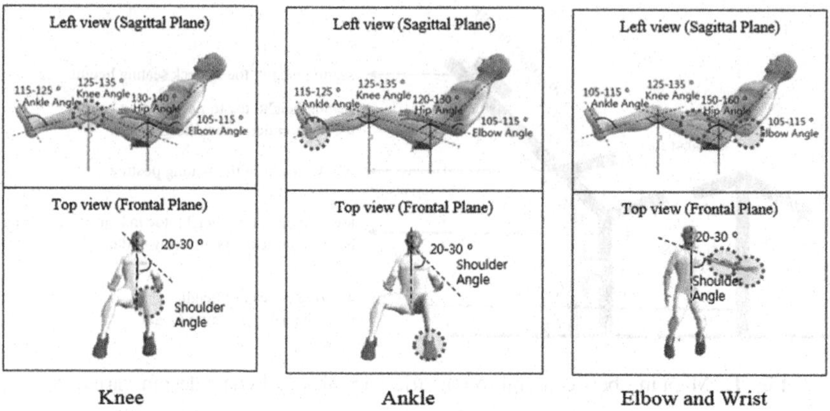

Fig. 3. Proposed standard postures in body joint ankles for four scanning types

the shoulder and arm on the chair and the bore. The height and angle of the leg support can be adjusted depending on individuals' height and the flexion angle of legs of one's choice during knee scanning, The angle of the leg support was set at 148° which was relatively higher than the other parts.

Table 2. Chair and leg supporter design specifications for xMRI (mm, degrees)

	Variables	Knee	Ankle	Wrist	Elbow
Seatpan	Height	500	500	500	500
	Depth	200–350	200–350	200–350	200–350
	Width	458	458	458	458
	Angle	25	32–39	21–23	21–23
Seatback	Height	961	961	961	961
	Width	620	620	620	620
	Angle	130–131	110–113	151–152	151–152
Headrest	Length	234	234	234	234
	Height	745–862	745–862	745–862	745–862
Leg support	Length (Front)	275	275	275	275
	Length (Rear)	400	400	400	400
	Angle	148	135	130	130

3.2 Results

In this study, there was an interaction between scanning duration and body parts ($p < .001$). There was no significant difference between the different scanning types and the body parts ($p > .001$). The discomfort level of the wrist and elbow scannings which have the same standard posture was lower than 2 in 7-point scale except the upper arm. This explains that the upper limb scanning was perceived to be quite comfortable. The highest discomfort was observed in the thigh being 2.9 during knee scanning and next in the leg being 2.8 during ankle scanning. In general, all the discomfort levels of the body parts were under 3 which indicates no significant or noticeable discomfort during 30 min' scanning.

The final design guidelines for the xMRI were determined from the results obtained from the user evaluation test in the following: headrest height of 745 ± 117 mm, seatback height of 961 mm, 410 ± 30 mm for seatpan height, 200 + 150 mm for seatpan width, 524 ± 44 mm for leg support. 110–152 degrees for the adjustable angle between seatback and seatpan, and 21–39 degrees for the adjustable seatpan angle (Figs. 4 and 5).

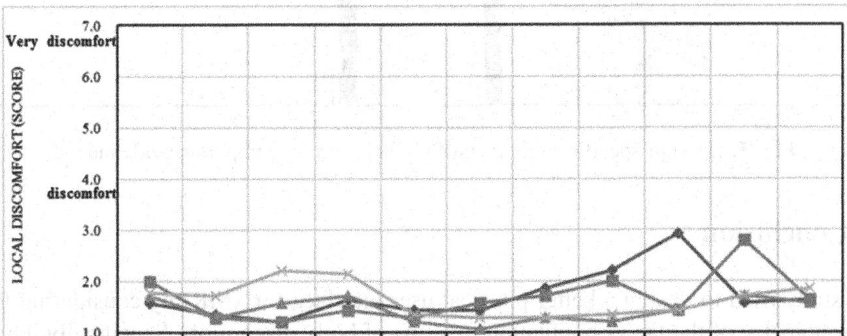

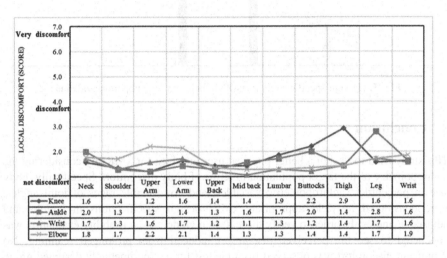

	Neck	Shoulder	Upper Arm	Lower Arm	Upper Back	Mid back	Lumbar	Buttocks	Thigh	Leg	Wrist
Knee	1.6	1.4	1.2	1.6	1.4	1.4	1.9	2.2	2.9	1.6	1.6
Ankle	2.0	1.3	1.2	1.4	1.3	1.6	1.7	2.0	1.4	2.8	1.6
Wrist	1.7	1.3	1.6	1.7	1.2	1.1	1.3	1.2	1.4	1.7	1.6
Elbow	1.8	1.7	2.2	2.1	1.4	1.3	1.3	1.4	1.4	1.7	1.9

Fig. 4. The level of body part discomfort for four different scanning types during 30 min' session

It was shown that the standard posture of xMRI during the 30-min scanning was not uncomfortable from the result of usability evaluation. However, a higher discomfort level from the upper arm during the wrist scanning was observed because of the unavoidable gap between the core and chair. An extra accessory that connects bore and chair must be provided for better comfort. Although the bore was designed to be slightly angled at 7° to accommodate the comfortable angle of lower leg and the positioning of the foot, a relatively higher discomfort occurred because the coil size of the bore is limited due to hardware specifications for knee and ankle scannings. In order to further mitigate discomfort, it would be helpful to provide a larger bore by approximately 20 mm.

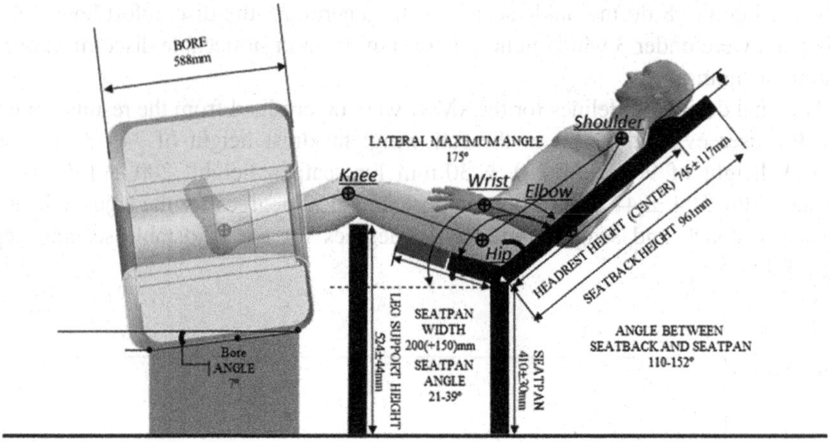

Fig. 5. Design specifications for xMRI chair and leg supporter guidelines

4 Conclusion

This study tried to design a better physical user interface for xMRI by considering the anthropometry of the target population in terms of body discomfort. Specifically, body size in percentiles, anthropometric variables, and comfortable rages of joint angles, were considered to yield appropriate xMRI design specifications in lengths and adjustable angles. In order to verify these design guidelines, a usability evaluation was conducted to see the perceived discomfort induced from prolonged scanning. No significant discomfort was observed from the test for a experimentally designed group of participants and necessary modifications were finally made following the results of the test. Findings of the study will help the designer ergonomically develop a xMRI in the future.

References

1. Mival O, Benyon D (2015) User Experience (UX) design for medical personnel and patients. In: Requirements engineering for digital health, pp 117–131. Springer, Cham
2. Privitera MB, Murray DL (September 2009) Applied ergonomics: determining user needs in medical device design. In: Annual international conference of the IEEE engineering in medicine and biology society, EMBC 2009, pp 5606–5608. IEEE
3. Schiff MH, Hobbs KF, Gensler T, Keenan GF (2007) A retrospective analysis of low-field strength magnetic resonance imaging and the management of patients with rheumatoid arthritis. Curr Med Res Opin 23(5):961–968
4. Steinberg EP (1986) The status of MRI in 1986: rates of adoption in the United States and worldwide. Am J Roentgenol 147(3):453–455
5. Chung M, Dahabreh IJ, Hadar N, Ratichek SJ, Gaylor JM, Trikalinos TA, Lau J (2011) Emerging MRI technologies for imaging musculoskeletal disorders under loading stress

6. Meléndez JC, McCrank E (1993) Anxiety-related reactions associated with magnetic resonance imaging examinations. JAMA 270(6):745–747
7. Lindegaard HM, Vallø J, Hørslev-Petersen K, Junker P, Østergaard M (2006) Low-cost, low-field dedicated extremity magnetic resonance imaging in early rheumatoid arthritis: a 1-year follow-up study. Ann Rheum Dis 65(9):1208–1212
8. Extremity-only MRI (2013) Health Services Research (HSR). Health Care Knowledge Centre (KCE), Brussels, Belgian
9. Sharp JT (2006) Magnetic resonance imaging in rheumatologic practice: low field or standard. J Rheumatol 33(10):1925–1927
10. Griffin BN, Lewin JL, Louviere AJ (1978) The influence of zero-G and acceleration on the human factors of spacecraft design. JSC-14581, NASA-JSC

A Cross-Cultural User Evaluation of the Prototype Extremity MRI

Kimin Ban[1], Eui S. Jung[1(✉)], Kibum Park[1], Dasol Kwon[1], Jinsang Park[1], Jinho Yim[2], and Kyungjin Oh[2]

[1] School of Industrial and Management Engineering, Korea University, 145 Anam Seongbuk, Seoul 02841, South Korea
ejung@korea.ac.kr

[2] Samsung Electronics, 129 Samsungro, Yeongtong, Suwon 16677, South Korea

Abstract. The proposed ergonomic design guidelines for the new extremity MRI led to the development of a prototype MRI. A cross-cultural user evaluation was conducted to guarantee the adequacy of the prototype for both Korean and Caucasian populations from 5th percentile female to over 95th percentile male potential patients. First, a scenario of MRI usage was defined from the chair setting, patient seating, scanning, and the egress of a patient. The user evaluation was done to measure participants' local and whole body discomfort for the combinations of scanning types and durations, specifically focusing on the ingress and egress of the participant and the whole duration of scanning. The statistical analyses revealed that no significant change in discomfort was observed up to 40 min of scanning duration. Since no electrical adjustments can be implemented due to magnetic interferences, it is noted that existing mechanical adjustments caused local discomfort to certain extreme population of patients in specific body postures. Thus, a set of modifications of the design was suggested in terms of supplementary supporters.

Keywords: Extremity MRI · Cross-cultural user evaluation · Anthropometry

1 Introduction

As medical device market is developing, the number of users is increasing. So, value of user experience on medical devices becomes an important topic. In order to provide a good user experience, it is necessary to apply user-centered design as well as effective and safe design when developing interactive medical products infrastructures and services [1]. Therefore, delivering a good user experience is needed to enhance user satisfaction when developing medical devices [2].

The healthcare field is increasingly interested in Magnetic Resonance (MR) imaging devices, which enable 90% high diagnostic accuracy and early diagnosis to facilitate disease treatment [3, 4]. Existing Magnetic Resonance Imaging (MRI) requires high constructing and operating costs [5]. To improve existing MRIs, Extremity MRI (xMRI) has been developed since 1990s. xMRI is equipped with a low electric field of less than 1.5 T to provide a magnetic resonance image for partial scan such as arms and legs in small areas while reducing installing and operating costs [6, 7].

S. Bagnara et al. (Eds.): IEA 2018, AISC 818, pp. 148–156, 2019.
https://doi.org/10.1007/978-3-319-96098-2_20

Especially, unlike traditional MRI, xMRI is expected to improve patient usability by diagnosing through scanning only the local body parts.

However, in the case of xMRI that scans local body parts, a precise examination requires a long duration, which causes higher discomfort during scanning with limited movements. This higher discomfort occurring during scanning becomes a major factor that degrades user experience. Callaghan and Trapp [8] found that medical devices that must be used continuously for an extended period of time are also uncomfortable with some pressure. Therefore, it is necessary to asses discomfort and evaluate its usability based on standard sitting postures over time.

The purpose of this study is to verify the usability of the xMRI prototype derived from the standard postures and the design guidelines for xMRI as presented in Jung et al. [9]. Therefore, we conducted two usability evaluations based on using patients' xMRI scenarios. First, we focused on patients' discomfort occurring during the sitting and later standing. The second experiment was performed to see the level of discomfort on four scanning types of knee, ankle, wrist and elbow. Specifically, the change in patient's discomfort was investigated over time.

2 Methods

2.1 Participants

In this experiment, 30 participants aged between 20s and 30s were involved in the experiment. With an outlier excepted, the remaining 29 participants were consisted of 14 Koreans and 15 Caucasians. And there were 7 males and 7 females among Koreans, and 8 males and 7 females among Caucasians. We classified the groups into below 5 to 25%, 25% to 50%, 50% to 75%, and 75% to over 95% according to their heights, and the redundant range of the classification was integrated into ten participant groups, as illustrated in Fig. 1. The first group included participants with their heights below 1500 mm and the last group did participants with their heights over 1900 mm. Those groups are defined as extreme groups (M = 172.43, SD = 10.98).

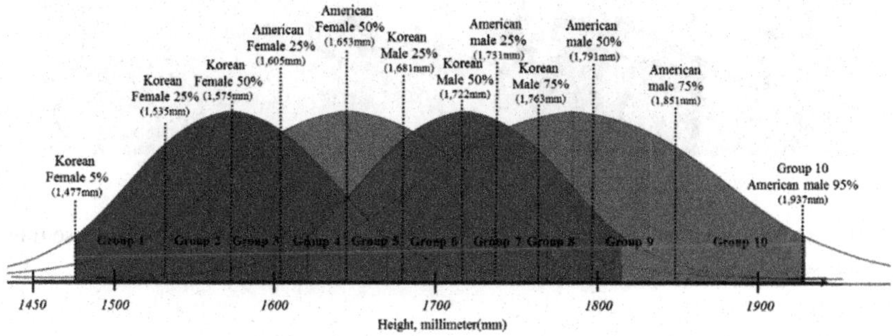

Fig. 1. Classification of participants in ten groups

2.2 Experimental Setup

The experiment was conducted in the space of 3000 mm by 3300 mm. The xMRI mock-up was equipped with the chair, the leg supporter, and the bore which adopted the design specifications based on anthropometric data [9]. In this experiment, adjustable angles were considered as they constitutes the standard postures for four scanning types. Figures 2(a) and (b) shows the arrangement of chair and supporter for the lower limb scanning and the upper limb scanning, respectively.

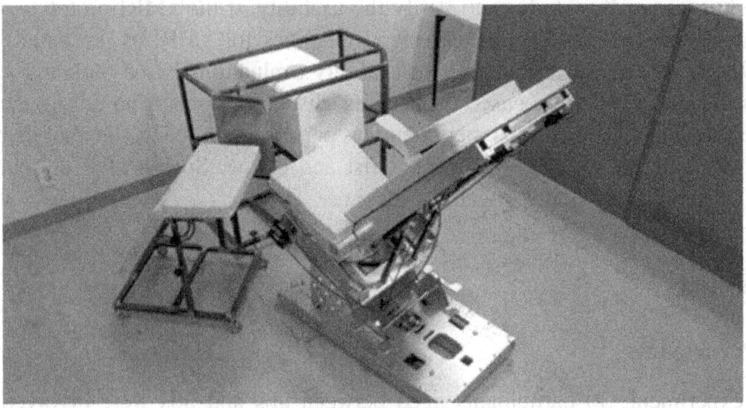

(a)

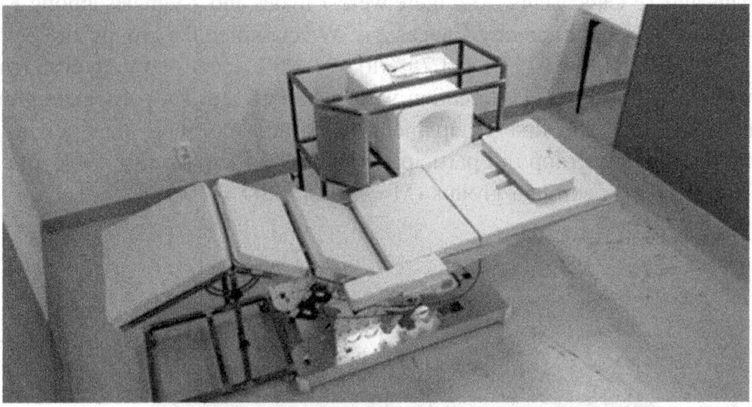

(b)

Fig. 2. (a) xMRI prototype setting for lower limb scanning, (b) xMRI prototype setting for upper limb scanning

In addition, xMRI chair and leg supporter were designed to adjust its reclining, tilting, height, and depth, considering anthropometrically sensitive variables. Also, the cushion used in xMRI chair and leg supporter were made with a 2-layered cushion in varying hardness and a separate sponge was included in the experiment to fix the bore area during scanning.

2.3 Experimental Design

Two experiments were conducted based on the xMRI scenario. The first experiment concerns the use of a 470 mm footboard because there was an excessive discomfort observed by an extreme group of participants that even the lowest seatpan height which was limited by the bore height was felt to be too high, when they were sitting and later taking off from the chair. In the second experiment, evaluations of perceived discomfort occurring during scanning were made for four scanning types (knee, ankle, wrist, and elbow), which are representative of the most scanned local regions in the previous study [5, 6]. Participant group according to the percentile of height. Scanning lasted for 40 min and perceived discomfort were measured at the onset and every ten minutes: the onset, after 10, 20, 30, and 40 min.

2.4 Experimental Procedure

The experimenter first informed the participants of the experiment and its purpose. After the informed consent forms are obtained, he or she measured the participants' height and weight for classifying groups. The participants were then given a training session to get familiar with the experiment.

Based on the use scenario, as listed in Fig. 3, this experiment was conducted in 9 steps. In Step 1, the experimenter positioned the chair at a default setting and the participant was asked to decide if he or she uses the foot board for the first task of sitting. In Step 2, the participant was seated in the chair and an evaluation rating was made for the degree of whole body discomfort during sitting by using a 7-point unipolar scale. The purpose of this step is to find the difference of the whole body discomfort among ten participant groups for the use of footboard. In Steps 3 to 6, the chair and the leg supporter were positioned and adjusted in relation to the bore to guarantee the best comfort for each scanning types: knee, ankle, wrist, and elbow.

Step 1. Setting the chair at default setting
Step 2. Participant sitting and discomfort rating
Step 3. Positioning of the chair for scanning
Step 4. Adjusting the chair for standard posture
Step 5. Adjusting the supporter for standard posture
Step 6. Apply the accessories
Step 7. Begin scanning and discomfort rating at intervals
Step 8. Participant taking off the chair and d. r.
Step 9. Setting the chair back at default setting

Fig. 3. Steps of scanning and discomfort ratings based on the use scenario

The whole body discomfort and local body discomforts (neck, lumbar, leg, thigh, mid back, lower arm, buttocks, upper back, upper arm, wrist, and shoulder) were evaluated. All the ratings were made while participants tried to maintain the standard postures. Accessories were then used to assist if necessary in Step 7. The ratings of whole and local body discomforts over time during the scanning were taken using the same scale from the participant for all scanning types. In Step 8, the participant was asked to leave the chair and again to rate his level of discomfort. In the last step, the experimenter set the chair and supporter back to default setting eight for next scanning.

3 Results

In this study, the usability of xMRI was evaluated through two types of tasks. First, the whole body discomfort of sitting and standing was statistically significant depending on the use of foot board ($F_{(1,19)} = 9.18, p = .02$). Second, the difference in whole body discomfort among scanning types was statistically significant over time ($F_{(4,199)} = 18.88, p < .001$). There was also a statistically significant difference in whole body discomfort according to the participant group ($F_{(9,199)} = 9.68, p < .001$). However, there was only the interaction between the participant group and scanning types ($F_{(27,199)} = 2.58, p < .001$), as summarized in Table 1.

Table 1. Statistical analyses of whole body discomfort observed

Task	Independent variable	DF	F	p-value
Sitting/standing	Participant groups	9	1.65	=.25
	Use of footboard	1	9.18	=.02
Scanning	Time	4	18.88	<.001
	Participant groups	9	9.68	<.001
	Scanning types	3	0.6	=.62
	P. groups * S. types	30	1.15	<.001

3.1 Sitting and Standing

In the first task, there was a statistically significant difference of whole body discomfort (p = .02) in the use of foot board. The whole body discomfort was lower when the 470 mm foot board was used (M = 1.87, SD = 0.71).

In particular, the 470 mm footboard used in this experiment showed the level of discomfort as being 3 for the case of extreme short women and 1 for the extreme tall men as depicted in Fig. 6. However, the whole body discomfort according to the user group was not statistically significant (p = 0.246).

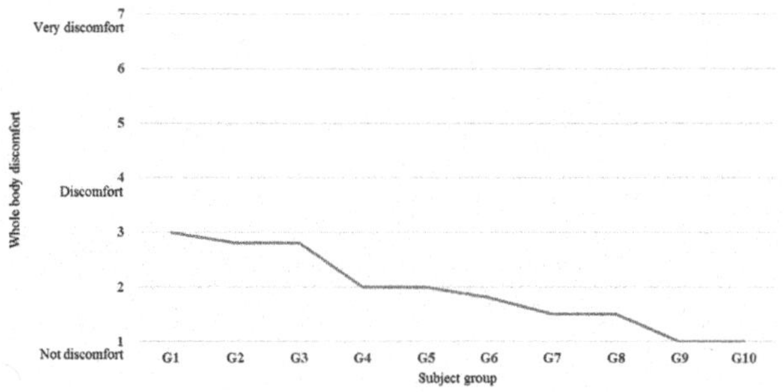

Fig. 6. Whole body discomfort according to participant groups (Participant groups are formed according to the body percentile range, G1 means extremely short, and G12 means extreme tall)

3.2 Scanning

In the second task, the participants expressed that discomfort became higher as the scanning progressed. Figure 7 shows the increase in the level of discomfort score (M = 2.58, SD = 0.13). Specifically, the discomfort observed during the ankle scanning was higher than that of other scanning types. As depicted in Fig. 8, the discomfort during ankle scanning was shown to increase rapidly after 30 min. However, the discomfort at 40 min did not increase as drastically as 30 min. Also, when the scanning continued up to 30 min, the highest discomfort was observed in the area of neck and lumbar. Leg and thigh discomfort were the next local body discomfort, as shown in Fig. 9.

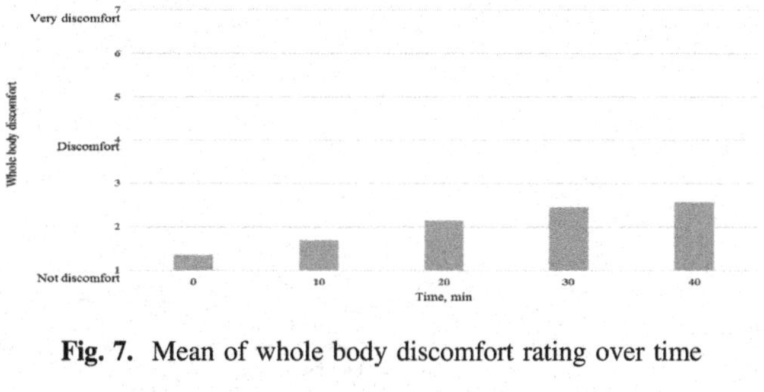

Fig. 7. Mean of whole body discomfort rating over time

In the case of whole body discomfort according to the participant group and the scanning type, extreme short women showed higher discomfort during knee scanning, while extreme tall men showed higher discomfort during ankle and wrist scanning, as shown in Fig. 10.

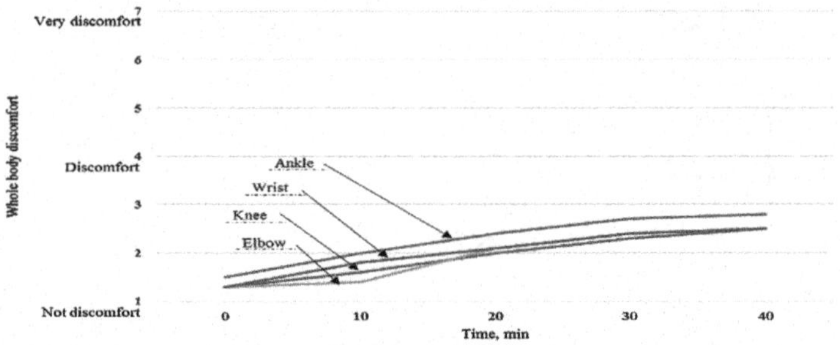

Fig. 8. Mean of whole body discomfort rating according to the scanning type

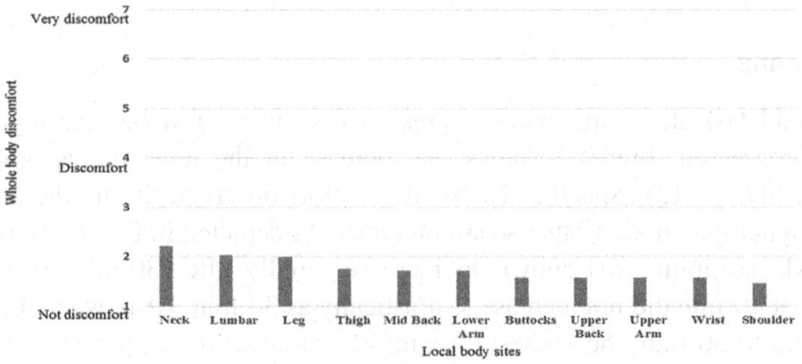

Fig. 9. Local body discomfort rating after 30-min scanning

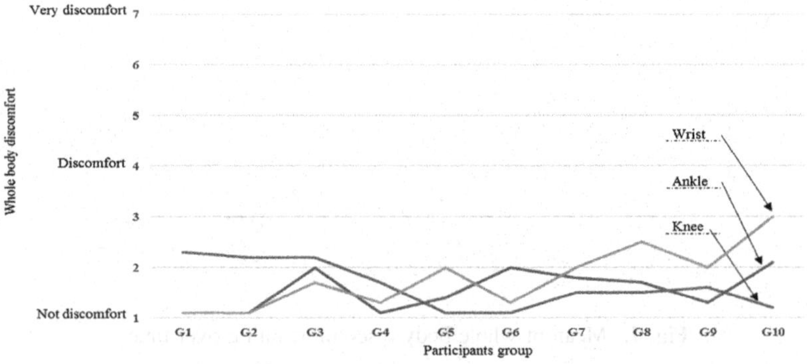

Fig. 10. Interaction effects on whole body discomfort

4 Discussion

470 mm of height for the foot board was appropriate according to the maximum acceptance range of 475 mm of chair height suggested in the Standard American Guidelines. It is supported by our result that the discomfort ratings were lower than 3 points when using the footboard. However, the extremely short women group under

1500 mm of height indicated that they had higher discomfort than others. This was caused by the current limit of bore height and a better experience can be achieved if the bore height can be adjusted.

As expected, the whole body discomfort tended to increase over time. The discomfort rating, however, did not exceed 3 points in all scanning types, which meant the prototype xMRI did not incur excessive discomfort. It should be noted, however, that the discomfort ratings for ankle scanning were higher than those for the other types, which implies the extremely tall group over 1900 mm of height felt quite uncomfortable on the ankle within the bore. In terms of local body part discomfort, the discomfort of neck and lumbar are relatively higher than those of other parts. In order to reduce this discomfort, it is necessary to consider further the shape of lumbar contour or headrest in the future design of xMRI. In addition, if the scanning time can be reduced down to under 30 min, a significant reduction in the discomfort will be achieved.

The extremely short group showed a higher discomfort level especially in the thigh for knee scanning. It is related to their length of leg because the range of angles between two legs is narrower than that of larger users. In the case of ankle and wrist scannings, the extremely tall group showed a higher discomfort on the legs and upper arms. Those discomfort over time had the highest scores due to gap between the bore and the chair and the bore. An accessory that links two parts must be provided to mitigate the discomfort of extreme users in the future design of xMRI.

5 Conclusion

This study evaluated the usability of the prototype xMRI from the patients' standard postures specified in the design specifications from [9]. It is shown that there is no excessive discomfort observed from the xMRI prototype in terms of the whole body and local body discomfort evaluations. There was, however, some level of discomfort occurred for extreme user groups and for certain scanning types. A further modifications are being made to reduce such potential discomfort. It is believed that this study will help designers and developers of medical devices reflect human characteristics on their products and deliver a better user experience.

References

1. Martin JL, Clark DJ, Morgan SP, Crowe JA, Murphy E (2012) A user-centred approach to requirements elicitation in medical device development: a case study from an industry perspective. Appl Ergon 43(1):184–190
2. Mival O, Benyon D (2015) User experience (UX) design for medical personnel and patients. In: Requirements Engineering for Digital Health, pp 117–131. Springer
3. Steinberg EP (1986) The status of MRI in 1986: rates of adoption in the United States and worldwide. Am J Roentgenol 147(3):453–455
4. Brydie A, Raby N (2003) Early MRI in the management of clinical scaphoid fracture. Br J Radiol 76(905):296–300

5. Roemer FW, Lynch JA, Niu J, Zhang Y, Crema MD, Tolstykh I, El-Khoury GY, Felson DT, Lewis CE, Nevitt MC, Guermazi A (2010) A comparison of dedicated 1.0 T extremity MRI vs large-bore 1.5 T MRI for semiquantitative whole organ assessment of osteoarthritis: the MOST study. Osteoarthritis Cartilage 18(2):168–174
6. Health Services Research (HSR) (2013) Extremity-only MRI. Health Care Knowledge Centre (KCE), Brussels
7. Brooks S, Cicuttini FM, Lim S, Taylor D, Stuckey SL, Wluka AE (2005) Cost effectiveness of adding magnetic resonance imaging to the usual management of suspected scaphoid fractures. Br J Sports Med 39(2):75–79
8. Callaghan S, Trapp M (1998) Evaluating two dressings for the prevention of nasal bridge pressure sores. Prof Nurse (London, England) 13(6):361–364
9. Jung ES, Ban KM, Kim JY, Na SK, Ahn JW, Yim JH (2018) An ergonomic user interface design for a new extremity MRI focusing on the patient chair. In: International ergonomics association 20th congress

Constructing Quality of Care in Neurology: Anticipation Strategies of a Team of Nurses

Nicolás Canales Bravo(✉), Adelaide Nascimento, and Pierre Falzon

Ergonomics Laboratory,
Research Center on Work on Development (CRTD) – Conservatoire National des Arts et Métiers, 41 rue Gay Lussac, 75005 Paris, France
nicolas.canalesbravo@lecnam.net

Abstract. Neurological patients are characterized by multiple handicaps and by physical, behavioral, and psychological symptoms that are diverse and fluctuating. They need specialized, multidisciplinary and constant health care, a large amount of hygiene care and technical care, but also relational care. The objective of this communication is to present the anticipatory operative strategies used by a team of nurses working in a neurology hospitalization sector and their representations of healthcare quality. Healthcare quality is studied from the constructive perspective of activity ergonomics: it considers quality as a permanent process on-going in the work situation, articulating the effective conditions provided by the organization for the realization of work-as-imagined and the individual and collective resources available by the agents who deal with the actual context. The methodology used is based on an exploratory field study using the clinical method of work analysis in ergonomics. The results show that quality is the result of arbitrations and individual and collective adjustments built in context, which incorporate formal and informal standards of care quality according to the content of care, the type of patient and the current state of the latter. The strategies identified allow the agents to reconcile in more or less satisfactory ways the different existing standards.

Keywords: Healthcare quality · Quality of work · System performance
Work strategies

1 Introduction

The improvement of the quality of healthcare is today an important issue for neurology in the hospital, especially in France, where it is estimated that the medical activity in neurology has increased by almost 15% since the year 1998 [1]. The patients involved have multiple disabilities, in addition to various and fluctuating physical, cognitive, psychic and behavioural symptoms [2]. They benefit of hyperspecialized and multidisciplinary support throughout life.

The literature shows that the quality of healthcare in neurology is strongly related to the ability of professionals to achieve coordinated and constant management during hospitalization [3]. For example, delays in the distribution of drugs to patients with Parkinson's disease lead to "blockages" and are a frequent cause of prolonged stay in

S. Bagnara et al. (Eds.): IEA 2018, AISC 818, pp. 157–165, 2019.
https://doi.org/10.1007/978-3-319-96098-2_21

hospitalization [4]. Knowing the conditions that promote the realization of quality care in an organized and safe way throughout the stay in the hospital is therefore a major challenge.

Many disciplines have addressed the quality of hospital care in the last twenty years (nursing, public health, psychology, ergonomics). However, many of these studies adopt a "normative" approach to quality [5]. In such a view, the quality of care is a stable object that can be defined beforehand: it is the "degree of conformity of the standards of care given in advance to health professionals [6] (p. 257)". The standard corresponds here to the level to be attained. Deviating from it or transgressing it decreases quality. The standards are developed from representations of what would be a "standard care situation" and/or an "average patient", regardless of the actual situation in context.

In contrast, ergonomics based on the analysis of real activity have developed in recent years a constructive approach to quality [7, 8]. Quality production is considered as a dynamic process implemented by the operators taking into account the actual situation, combining in context the resources provided by the organization to carry out the planned work—formal rules, procedures, physical conditions—and the individual and collective resources available to the actors—competencies, autonomy, health—[9]. This perspective differs from the normative view approach in different aspects [10]. First, it establishes a radical distinction between the ideal quality defined by the organization and the real quality of care in healthcare. There is an irreducible gap between them, as every health care situation is singular. Secondly, the differences to the norm should not necessarily be seen as factors of degradation of quality, but as contributions to the preservation of quality (by example, not respecting a procedure in an emergency situation makes it possible to achieve a better quality of care because the operator has more time to act). Finally, quality is a "moving target" [5], always unstable, which depends on both the human and technical conditions available at a particular time, and the social and cultural context in which practices are carried out.

This communication is part of this perspective, and aims to present how a team of nurses working in a neurology department manages the constraints and resources available to build the quality of care. In particular, it is interested in the anticipatory strategies implemented by these agents to carry out the care, taking into account their own vision of quality.

2 Situation and Methods

The study responds to a joint request from the quality manager and the doctor in charge of the Neurology Department of a Parisian hospital. This hospital is planning an architectural and organizational change with a fusion of services, including the pooling of nursing teams between different sectors of the department. The study focused on the quality of care. In this communication we present the analyses carried out in one of the two sectors studied.

2.1 Work Context

The study was conducted over six months in two areas of the Department of Neurology hosting sixteen patients in hospitalization. The reference pathologies are Parkinson's disease and behavioral disorders. The first is a multisystemic, chronic and progressive neurodegenerative pathology. Its clinical signs are heterogeneous and its diagnosis is based on a triad of symptoms: akinesia, rest tremors and hypertonia (stiffness Excessive muscles). The second refers to a group of neurobehavioural symptoms including compulsions, apathy, disinhibition, altered emotional or food behavior. The team consists of 13 nurses who alternate in the morning and afternoon every 15 days, and who switch their area (Parkinson's disease/neurological disorders) every month. The average age is 31 years, the average seniority in the service is 3 years (seven years to one month).

Nurses are responsible for the preparation and administration of treatment and technical care in accordance with medical requirements. The day is organized according to a sequence of tasks: oral communications with the preceding team, organisation/preparation of medical prescriptions, first tour of the service, second tour, oral communications with the successor team. Nurses are in charge of the safe preparation, organization and distribution of the medical prescriptions at the scheduled times. During the day, nurses must organize their actions in coordination with those of other professionals who work in the sector: doctors, caregivers, para-medical staff, etc.

2.2 Data Collection Methods, Analysis Modes

The data collection was done in two stages. In the first, 25 h of open observation and 4 systematic observation programs (pen-and-paper data collection) were conducted during 4 different working days (2 mornings, 1 afternoon and 1 weekend) with several nurses. The organizational strategies of nurses have been identified. Then, 4 nurses who were observed in the first stage participated in semi-structured interviews, in order to achieve a fine understanding of the invisible aspects of the activity (operative strategies, logical of action, quality criteria, etc.).

Observation data were treated qualitatively and quantitatively according to the following categories: displacement, care actions and collective activity. Verbal data from interviews were recorded and analyzed through a manual content analysis. The themes coded for analysis are: quality of care, anticipation strategies and determinants of quality of care.

3 Results

3.1 Demands and Constraints Related to the Characteristics of the Patient and the Working Day

The results show that the demands related to the management of the patients differ depending on the type of pathology and the condition of the patient during the visit (a visit last between 1 to 21 min). On the one hand, Parkinson's patients - whose mobility fluctuates over the course of the day, ranging from involuntary tremors to total body stiffness - need a large number of oral treatments administered every hour. The administration must be at specific times in order to avoid "blocking", a situation in which the person can no longer move, being literally blocked. On the other hand, patients with behavioral disorders are more physically independent, but regularly exhibit psychological decompensations that result in sudden changes in behavior (aggressions, attempts of fugue, weeping). Caregivers need to spend time managing their seizures, or to constantly monitor them to avoid attempts to run away or aggressive behaviors. All these elements characterize a work situation with strong temporal constraints and very limited leeway, forcing caregivers not to spend enough time with patients.

On the other hand, the systematization of actions during the tour for distribution of care shows that these demands and restrictions vary widely according to the day and the type of working day (see Table 1). Thus, on Tuesdays and weekends, the duration of the distribution of care tour exceeds the time prescribed by the organization (1 h). A large number of external and internal exams are performed on Tuesdays (tour time of 99 min, 9 patients being taken care of, 8 min on average for each visit). The weekend (tour time of 128 min, 8 patients being taken care of, 13 min on average for each patient), half of the staff and half of the patients are present, and the nurses must work with the caregivers. On the other days, the tour lasts approximately one hour, both in the morning and in the afternoon, and every visit to a patient lasts an average of 4 min. The time spent with the patient varies little over the week, between 71% and 80% of the duration of the tour of care distribution, except in the afternoon, where it does not exceed 48%. The care actions are mainly technical and are regularly associated with relational care. Basic care (care of hygiene and comfort) occupies an important place on weekends when nurses work in pairs with caregivers.

Table 1. Day of work observed, duration of visit, characterization of care actions and collective activity during the tour of distributions of care (4 nurses observed).

	Nurse1	**Nurse 2**	**Nurse 3**	**Nurse 4**
	Morning 1	Afternoon	Morning 2	Weekend
Tour time	99 min	60 min	59 min	128 min
Number of patients being taken care of	9	7	9 (2)	8
Average length of visit	8 min	4,14 min	4,8 min	13 min
Maximum duration of the visit	16 min	13 min	10 min	21 min
Interruptions	1	1	1	0
Patient Time	70 min	29 min	44 min	104 min
% time with the patient	71%	48%	74%	80%
Technical Care	78%	93%	79%	25%
Basic Care	0%	6,90%	0%	72%
Relational Care	51%	21%	23%	45%
Time off patient	29 min	31 min	15 min	24 min
% off patient time	**29%**	**52%**	**25%**	**20%**
% Group Activity	16%	15%	20%	55%

3.2 Reconcile and Prioritize Formal and Informal Criteria to Build Care Quality

In order to carry out the programmed care, nurses do not apply the prescriptions made by the doctors in a passive way: they interpret and adapt them according to the context and the condition of each patient, taking into account formal and informal criteria for each type of pathology. These differences are highlighted by the systematic analysis and categorization of the content of the interviews (see Table 2). The formal criteria are related to the technical and normative aspects foreseen by the organization concerning the conformity to the medical prescription (cure dimension). On the contrary, the informal criteria include the specific initiatives that each nurse must implement in their care or education approach according to their leeway (care dimension). For nurses, the joint application of these criteria during the care situation can *"stabilize or improve the health status of patients, taking into account their specificities as individuals and as patients"* (Nurse 1, 30 years, 4 years of professional experience, 4 years of seniority in the service).

Table 2. Criteria that characterize the quality care for each nurse

	Nurse 1	Nurse 2	Nurse 3	Nurse 4
Formal criteria (cure dimension)				
Respect of the prescription	X	X	X	X
Access to the necessary equipment	X	X		X
Respect of the dose (proper dosage)	X	X	X	X
Respect to the route of administration	X	X	X	X
Administration to the right patient	X	X	X	
Administration at the scheduled time	X	X	X	X
Adapt the prescription to the state and needs of the patient	X	X		X
Informal criteria (care dimension)				
Verify that the long-term treatment effects are those that were expected		X		X
Explain the care to the patient	X	X	X	X
Understand the rationale for the prescription	X			
Position the patient correctly	X	X		X
Being present with the patient	X	X	X	X
Provide psychological support	X	X	X	X
Preventing pain	X	X	X	

When asked about quality care, nurses are almost unanimous in terms of formal criteria, but do not consider the same informal criteria. The latter mainly refer to the knowledge developed by nurses from experience and which allows them to implement rules of action in order to produce quality. For example, to put an infusion on a patient, a nurse explains *"that it is important to know how to do a gentle handling, how to manage the patient's movement"* (Nurse 4, 32 years, 5 years professional experience, 3 years seniority in the service).

Even if the nurses manage to achieve the technical care programmed on the day, they do not always manage to reconcile the different quality criteria. They explain that the high number of prescriptions, the lack of time and the requirements of each patient lead to lead to difficult arbitrations and to sacrifices regarding their own quality standards. For example, distributing prescriptions to a Parkinsonian patient while an emergency occurs for another patient (psychological decompensation, attempt to run away, blockage). or having to take care of a patient unable to receive care (due to a blockade, to physical pain, to confusion disorders, to delusion of persecution). In such cases, nurses are forced to give up what they see as the essence of their work: spending time with the patients, respond to their needs, monitor their health state in a suitable way, etc.

3.3 Strategies of Anticipation Based on the Context for Construct the Quality of Care

To reconcile the quality criteria defined by the organization and by themselves and avoid difficult trade-offs, nurses are implementing different strategies of anticipation over the working day. These strategies are common to experienced nurses and have a double purpose. On the one hand, it helps them to develop an organization of care that respects the time provided for the needs of patients, and on the other hand to build up some leeway allowing them to deal with potential unforeseen events, while maintaining their own health.

Different strategies of anticipation were identified. The first is essentially to ensure adaptation of care to each patient, in connection with the personal work organization. The second seeks to preserve the collective activity of work, considering the overall organization of the system. Most of these strategies are coordinated during the stage of preparation and care planning.

Individual Strategies of Anticipation for the Construction of Quality

Individual strategies vary according to the type of patient and include the priority given to some requirements and/or changes of the order of the visits during the tour of duty. For example, in order to administer at the same time a large amount of care to patients with Parkinson's disease, a nurse will prioritize distribution of Parkinsonian treatment, in view of the risk of blocking of these patients. Afterwards, nurses can focus on other tasks of care and have a margin of extra time to spend with patients at the end of the tour. A strategy frequently used with behavior disorder patients is to reassure them by explaining in advance the course and the purpose of the scheduled care, in order to avoid a refusal or a psychological decompensation. A nurse explains: *"you have to reassure them and take the time to make a perfusion (...), it would be more logical to do it fast, but if the patient is distressed and you're going fast, you're sure he'll make a crisis. You have to make it slow to go fast*". (Nurse 2, 31 years old, 7 years of professional experience, 3 years of seniority in the service).

Collective Strategies of Anticipation for the Construction of the Quality
Collective strategies are also directed towards the patient and involve colleagues (nurses or other caregivers). A common example is the case of patients severely handicapped or placed in isolation. Nurses use to coordinate with the assistants to perform together the hygiene and technical care. This strategy facilitates a heavy handling and brings more comfort to the patient who will not be mobilized twice in the

Table 3. Individual and collective strategies of anticipation for constructing care quality

Situations	Strategies	Goals
Individual strategies		
Large amount of treatments for the Parkinson's patients	Administer all treatments rapidly to these patients at the beginning of the tour of care and then come and see them more quietly	Respecting the hour of prescription of the oral treatments to avoid the blockade of the patients
Many blood samples collection to achieve (fasting patients)	Do blood collection before starting the tour of care, even at the cost of beginning work earlier	Avoiding samples collection to interfere with the fixed requirements and avoiding breaking the fast
Anxious patient, or potentially painful care	Take the time to explain to the patient the care beforehand and its realization with anticipation	Informing the patient and avoiding a refusal
Collective strategies		
'Difficult' patients (who attempt to run away or easily have crises)	Agree with the concerned caregiver to not wake up the patient and postpone the care at the end of the tour (if the patient has no prescription at a set schedule)	Avoiding interruptions during the priority care and fugues of the patients, in order to have more time to take care of patients in need
Co-presence with doctors during their visits	Orally ask doctors if they intend to change the prescription before they mention it on the computer	Anticipating in last-minute changes to medical prescriptions in order to save time.
Threat of refusal of treatment in relation to a psychiatric symptom (delusions of persecution, for example)	At the time of planning, the nurse informs the colleagues	Making sure that the patient accepts the treatment at the time of the visit
Patients who have a severe physical handicap or are in isolation	Coordinate with caregivers in order to perform jointly the technical care and hygiene care	Facilitating patient handling and comfort
Patients who need assistance to swallow their medication	Coordinate with caregivers so that the treatment is given with breakfast.	Avoiding complications related to a bad ingestion of the drug; having more time for the rest of the care

same day. Another strategy is to coordinate with a caregiver in order not to wake up a patient with behavior disorder before completing the priority care actions (for example, Parkinsonian treatment). This allows interruptions to be avoided during the tour and provides more room for manoeuvre to spend time later with patients with behavioral disorders (Table 3).

4 Conclusion

The results show that quality is the result of arbitrations and of individual and collective adjustments built in context, which allow agents to reconcile individual, collective and organizational criteria in unstable singular situations. However, the ability to manage the various constraints and resources within a care situation depends on a 'work of organization' [11] with "oneself" but also with "colleagues" and "patients". In the case of neurology, this work should help supporting a particular type of patient, with symptoms that affect all aspects of the body (physical, mental, cognitive, etc.), as well as handicaps varying in nature and in degrees.

In view of the variability of care situations in neurology, the knowledge of the terrain, the ability to anticipate the unexpected and the vagaries of support are central to meet quality objectives. The forward-looking strategies observed during the study contribute to providing spaces of action for carrying out all planned treatments, while articulating in a reasoned way the formal and the informal.

Finally, the results show that the implementation of these strategies depends not only on the individual and collective resources that can be mobilized to respond to a particular situation of care: the role of work organisation is here fundamental: it must allow one to implement these resources in an effective manner for the benefit of the patient. Contributing to work organisation design is an essential issue for ergonomics.

References

1. Cnamts Paris (2017) Les personnes en affection de longue durée en France. Points repère 6 (27)
2. Bradley D, Daroff R, Fenichel G, Jankovic J (2004) Neurology in clinical practice, 4th edn. Butterworth Heinemann, Oxford
3. Cordesse V (2013) Analyse du parcours de santé au cours des maladies neurologiques handicapantes et évolutives. Revue Neurologique 169(6):476–484
4. Cesaro P (2011) Maladie de Parkinson: les enjeux du traitement. La lettre du pharmacologue 25(3):85–88
5. Vincent C, Amalberti R (2016) Safer healthcare, 1st edn. Springer, Berlin
6. Morel M (2012) Qualité des soins. In: Formarier M, Jovic L (eds) Les concepts en sciences infirmières. ARSI, Toulouse, pp 256–260
7. Falzon P, Nascimento A, Gaudart C, Piney C, Dujarier MA, Germe JF (2012) Performance-based management and quality of work: an empirical assessment. Work 41(Suppl 1):3855–3860
8. Nascimento A, Cuvelier L, Mollo V, Dicioccio A, Falzon P (2014) Constructing safety. In: Falzon P (ed) Constructive ergonomics. CRC Press, London, pp 95–109

9. Falzon P, Dicioccio A, Mollo V, Nascimento A (2014) Qualité reglée, qualité gérée. In: Lhuilier D (ed) Qualité du travail, qualité au travail. Octares, Toulouse, pp 27–37
10. Canales Bravo N (2017) La construction de la qualité des soins: le cas des infirmières dans un service d'hospitalisation en neurologie. Master thesis on ergonomics. Cnam, Paris
11. de Terssac G (2011) Théorie du travail d'organisation. In: Maggi B (ed) Interpréter l'agir: un défi théorique. Presses Universitaires de France, Paris, pp 97–121

A User-Centered Virtual Reality Game System for Elders with Balance Problem

Chen-Wen Yen[1], Ping-Chia Li[2], Tzu-Ying Yu[2], Sheng-Shiung Chen[3], Jui-Kun Chang[4], and Shih-Chen Fan[2(✉)]

[1] National Sun Yat-sen University, Kaohsiung 824, Taiwan
[2] I-Shou University, Kaohsiung 824, Taiwan
Maggiefan15@isu.edu.tw
[3] E-Da Hospital, Kaohsiung 824, Taiwan
[4] Kaohsiung Chang Gung Memorial Hospital, Kaohsiung 824, Taiwan

Abstract. Balance problem is one of the leading causes of fall in elderly. Fall in elderly results in fracture, head trauma, death or other serious consequences. Virtual reality training is a novel and potentially useful rehabilitation method. In the virtual environment, elderly receive multimodal sensory feedback and engage in task-oriented activities safely. However, existed customized systems are often expensive and were not designed on the clinical needs of elderly or therapists. There has been limited research focusing on the user friendliness of VR balance training systems. The purpose of this study is to develop a user-friendly VR balance training system for elderly, christened VirReB system.

The hypothesis of this study is the usability of the VirGReb system based on the user-centered design will be significantly better than an off-the-shelf system. In the first year of this project, user needs were identified by focus groups and ethnography study. 30 therapists joined the focus group discussion. User needs were transformed into system requirements. Results showed that elders with stroke or cerebral impairments are most often received balance therapy. Therapists thought that conventional treatment are not fun and lack of variety. Wii it based VR therapy can be fun and provide various treatment in one system. However, levels of difficulty and the game content must fit elder's ability and their life style. We designed three games and wish to provide better user experiences for elders and therapists.

Keywords: Balance · Virtual reality · User interface · Rehabilitation

1 Introduction

Balance problem has a detrimental effect on an elder's functional ability and increases their risk of falling. According to National Health Institute, US, more than one third of elders age ≥ 65 fell down due to impaired balance (Health 2015). Repeated falls and instability are very common precipitators of nursing home. Existed studies highlight that healthy older adults have larger center of pressure displacements (CoP) and sway velocity in one or two feet quiet stances under different conditions (e.g., eyes opened/closed; stable/unstable surface) compared with young adults (Abrahamova and Hlavacka 2008). Several studies used Wii games and Wii-balance board to treat

S. Bagnara et al. (Eds.): IEA 2018, AISC 818, pp. 166–169, 2019.
https://doi.org/10.1007/978-3-319-96098-2_22

balance problems of elders (Afridi et al. 2018; Morrison et al. 2018; Phillips et al. 2018). In the virtual environment, patients receive multimodal sensory feedback and engage in task-oriented activities safely. VR treatments allow for three dimensional fantasy interaction, and can create a fun and sociable environment for users. However, existed systems often neglect the needs of clinical therapists. Limited research focused on the system requirements of a motor rehabilitation VR systems from therapists' perspective. The purpose of this study is to develop a user-friendly VR balance training system for elderly, christened VirReB system. At the first stage of this project, user needs were identified by focus groups and ethnography study.

2 Methods

2.1 Participants

Participants were all registered therapists recruited from the E-Da hospital and two affiliation clinics. The composition of the group members can be seen in Table 1, which details the therapists' gender, age, years of clinical treatment and relevant VR treatment experiences.

Table 1. Participant characteristics

	Gender (n)	Age (years) (mean, SD)	Years of Practice (years) (mean, SD)	Clinical usage of Wii or Kinect Gaming (n)
Female	19	32 (13.4)	8.3 (12.1)	66.7%
Male	13	38.5 (4.35)	12.5 (10.6)	100%

2.2 Data Collection

32 occupational or physical therapists were randomly divided into 8 focus groups. Semi-opening questions regarding clinical reasoning were discussed. Sample questions were: What diagnosis do patients with balance problems have? What kind of balance problems do patients have? What therapeutic theory and techniques do you often apply to patients? How do you evaluate patient's balance? What are the advantages and disadvantages of the therapeutic tools you currently use?

Grounded theory methodology and MAXQDA 11.0 software were used to collect and analyzed verbatim. 3 researchers coded the verbatim independently to ensure triangulation. Initial themes were reviewed by an independent qualitative expert.

3 Results

3.1 Qualitative and Quantitative Data

Question 1: What diagnosis do patients with balance problems have?
Answers: Stroke, Cerebellum Dystrophy and Traumatic Brain Injury are the leading causes of balance problems.

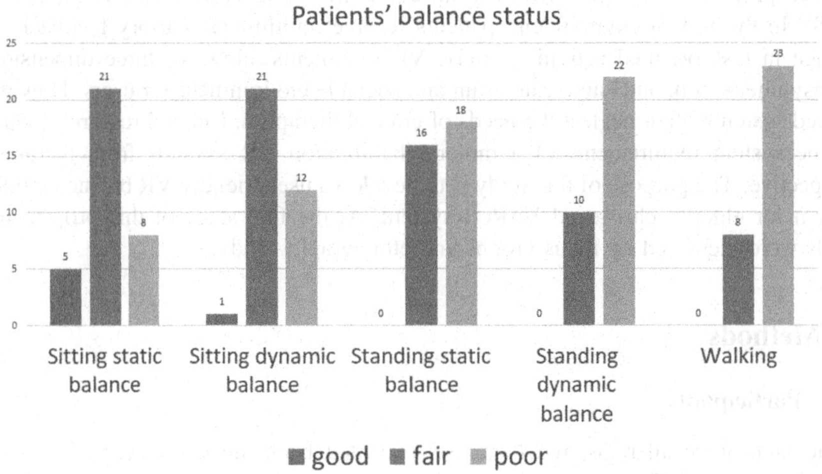

Fig. 1. Patient's balance status

Question 2: What kind of balance problems do patients have?
Answers: Most patients have good static or dynamic sitting balance, but needs help on static or dynamic standing balance (Fig. 1.). Patients usually cannot climb up stairs or walk independently.
Question 3: How do you evaluate patient's balance?
Answers: Berg Balance Scale is most often used by therapists. Berg Balance Scale is a 14 item tools. Every item consisting of a five-point ordinal scale ranging from 0 to 4, with 0 indicating the lowest level of function and 4 the highest level of function. Items include sitting to standing, standing unsupported, sitting unsupported, standing to sitting, transfers, standing with eyes closed, standing with feet together, reaching forward with outstretched arm, retrieving object from floor, turning to look behind, turning 360 degrees, placing alternate foot on stool, standing with one foot in front, standing on one foot.
Question 4: What skills do you first target at?
Answers: Therapists usually first target at patients' trunk control skills, followed by Weight shifting skills. Patients trained to shift weight to weaker side first, then shift weight to stronger side.
Question 5: What are the advantages and disadvantages of the therapeutic tools you currently use?
Answers: stepping box, bean bag, therapeutic ball, tilting board are most used in clinical setting. The advantages of existed tools include easily and quickly set-up, applicable to various patients, with various difficulty levels. The disadvantages of existed tools include unsafe, few motivating feedbacks, and not applicable at home environment. Patients need to be supervised by therapist all the time.

4 Discussion

Based on the data above, we found that providing different level of difficulty and various treatment programs are the key points of a good treatment tools. VR treatment provides audio, visual or haptive feedback which is the advantages of VR treatment. Motor learning theory is the major clinical theory. In the learning theory, faded feedback and summary feedback are effective for patients to re-learn balance skills. At the beginning of learning, Knowledge of performance is helpful but at the later stage of learning, Knowledge of Results is better.

We used AMTI AccuGait Optimized force plate and Unity software to develop our system. We built a game to evaluate patients balance. In the game, patient stands on the force plate and try to shift weight to control the avatar on the screen. The avatar need to dodge obstacle to receive points of the game. The velocity of center of pressure (CoP) is the main parameter to justify the result of balance evaluation.

5 Conclusion

We designed three games and wish to provide better user experiences for elders and therapists.

References

Abrahamova D, Hlavacka F (2008) Age-related changes of human balance during quiet stance. Physiol Res 57(6):957–964

Afridi A, Malik AN, Ali S, Amjad I (2018) Effect of balance training in older adults using Wii fit plus. J Pak Med Assoc 68(3):480–483

Health NS (2015) Balance problems. http://nihseniorhealth.gov/balanceproblems/aboutbalanceproblems/01.html. Accessed 2015

Morrison S, Simmons R, Colberg SR, Parson HK, Vinik AI (2018) Supervised balance training and Wii fit-based exercises lower falls risk in older adults with type 2 diabetes. J Am Med Dir Assoc 19(2):185.e7–185.e13

Phillips JS, Fitzgerald J, Phillis D, Underwood A, Nunney I, Bath A (2018) Vestibular rehabilitation using video gaming in adults with dizziness: a pilot study. J Laryngol Otol 132 (3):202–206

Effect of Gender, Age, Air-Conditioning and Thermal Experience on the Perceptions of Inhaled Air

Yuxin Wu[1,2](✉), Hong Liu[1,2], Baizhan Li[1,2], Yong Cheng[1,2], and Deyu Kong[1,2]

[1] Joint International Research Laboratory of Green Buildings and Built Environments (Ministry of Education), Chongqing University, Chongqing 400045, China
wuyuxin1988@cqu.edu.cn

[2] National Centre for International Research of Low-Carbon and Green Buildings (Ministry of Science and Technology), Chongqing University, Chongqing 400045, China

Abstract. The comfort range of inhaled air temperature is meaningful to the design of air conditioning parameters at breathing zone. In summer, eight college students and eight middle-aged people were recruited to conduct a thermal comfort study in the natural and air-conditioning environment respectively. The subjects were exposed to different inhaled air temperatures from 18 °C to 34 °C at an interval of 2 °C. The study found that the perceived air quality and thermal pleasure of warm inhaled air is better in the surrounding of natural environment than that of air-conditioning environment. The neutral temperature of inhaled air is 28 °C and 26 °C, respectively. The thermal sensation vote has no significant difference between middle-aged and young people, while the thermal pleasure, air freshness and perceived air quality of middle-aged people are better than that of the young people. When the temperature of the inhaled air is 2 °C higher than the ambient temperature, the SBS symptoms are significantly increased. Therefore, the comfort range of temperature of heated air in winter is worth to be further studied.

Keywords: Breathing zone · Inhaled air · Thermal environment
Thermal comfort · Air quality

1 Introduction

With the development of air-conditioning (AC) system and the increasing concern of energy crisis, increasing numbers of non-centralized AC system are applied in the buildings [1], including the personalized conditioning [2] and other air supply system focused on the breathing zone [3]. One main function of these systems was that they delivered the conditioned air directly to the activity or breathing area of human body. However, most of these previous studies were focused on the evaluation of the non-uniform thermal environment. The effect of inhaled air on human comfort in these non-uniform thermal environment was not systematic studied yet.

S. Bagnara et al. (Eds.): IEA 2018, AISC 818, pp. 170–179, 2019.
https://doi.org/10.1007/978-3-319-96098-2_23

Inappropriate temperature of inhaled air would deteriorate physiological and psychological well-being of occupants, including thermal pleasure vote (TPV), perceived air quality (PAQ), and even some symptoms of sickness. Thermal pleasure based on the conception of alliesthesia [4], which is only occurred in the dynamic environment: A peripheral thermal stimulus that offsets or counters a thermoregulatory load-error will be pleasantly perceived and vice versa, a stimulus that exacerbates thermoregulatory load-error will make people feel unpleasant. The PAQ was worse when the inhaled air enthalpy increased [5–7], and warm discomfort had negative effect on PAQ, either [8, 9]. Whereas, whether the PAQ is only effected by inhaled air· enthalpy or also effected by overall thermal environment/sensation is still unclear.

For a long time, the neutral temperature of a large group of people was believed to be independent of age, body size, and gender in a moderate uniform thermal environment [10]. However, variety of studies have shown thermal comfort differences exit, especially in the non-neutral thermal environment [11, 12], due to the difference of body conditions (e.g. heath, body mass and fate rate), behavioural habits (e.g. clothing style), thermal expectation [13], these factors highly affect the ability of thermal adaption. Overall, the females were more dissatisfied in the cool environment [14], and the aged people prefer the warm environment [15].

Adaptive thermal comfort theory indicates that occupants are more thermal neutrally in naturally-ventilated (NV) buildings than in air-conditioned (AC) buildings [16]. One reason is that occupants' thermal sensations are highly effected by thermal experience [17].

All these would limit the application of the local air conditioning system. Thus, the study focused on the perception of inhaled air, and the effect of gender, age and thermal experience difference on that.

2 Methods

2.1 Experimental Platform

Experimental platform combined of climate chamber A and Room B. In climate chamber A, the air temperature (T_a) was adjustable within the range of −5 °C to 40 °C, the accuracy is ±0.30 °C. Relative humidity (RH) was controlled between 10%–90% with the accuracy of ±5%. The air in climate chamber A could be induced into the thermal insulation box (refer to the box), as shown on Fig. 1. The box and air supply duct were trapped by thermal insulation cottons (30 mm-thick), while a silver paper covering on their surface to prevent from smell.

The Room B was equipped with air-conditioning unit, which can run both naturally-ventilated (NV) and air-conditioned (AC) models. The subjects were in Room B, evaluating air in the box. A portable thermal environment sensor was put into the box to monitor Ta and RH of the air in the box, with the accuracy of ±0.50 °C and ±5% respectively.

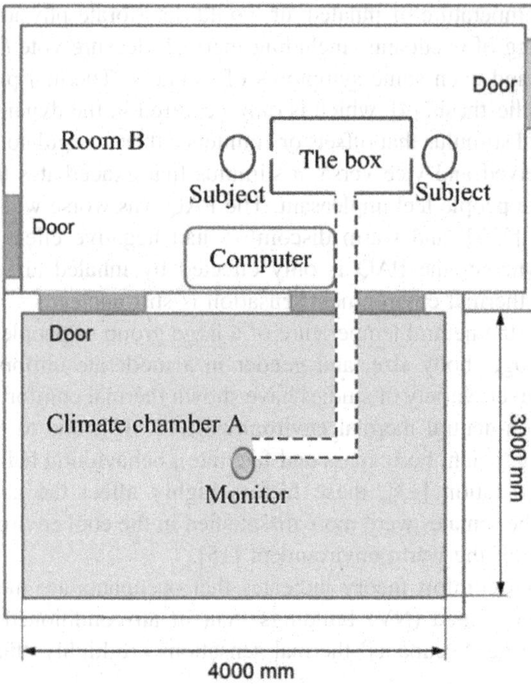

Fig. 1. Diagram of the experimental platform.

2.2 Experimental Conditions and Procedure

The experiments were conducted from July 10th to August 15th in Chongqing, China, where is renowned for its hot summer. In a typical year, there are average 25 days with outside temperature exceeding 35 °C [18]. During the experiment, the outdoor and indoor air temperatures were all above 30 °C. Figure 2 shows how subjects perceived the air in the box during the experiment.

Fig. 2. Photo of the experiment about subjects perceiving the air in the box

Before the experiment, the subjects were asked to stay in Room B for 30 min to eliminate the previous thermal experience effect and fill in their personal information. In the meantime, they were taught by instructors to be familiar with the experiments' requirements and questionnaires. There are two conditions in Room B, i.e. the NV and AC models. In the NV model, no AC was used and the average indoor temperature was about 32 °C. While in the AC model, the average indoor temperature was about 26 °C. The air temperatures in climate chamber A were set to be from 18 to 34 °C at an interval of 2 °C. The RH was set to be 55%. Less than 10 min was needed to increase/decrease 2 °C in the climate chamber A by using the air conditioning control system. After the thermal conditions in Chamber A were stable, the conditioned air was supplied to the box where placed in Room B. The subjects were seated in Room B during the whole time in the experiment, and they were allowed to read or use mobile phones, but any discussion related to the experiment was forbidden. They were asked to breathe the air in the box for 1–3 min and filled in the questionnaire immediately for 9 conditions at 15 min's interval. Thus, about 150 min was needed per experiment, as Fig. 3 shows.

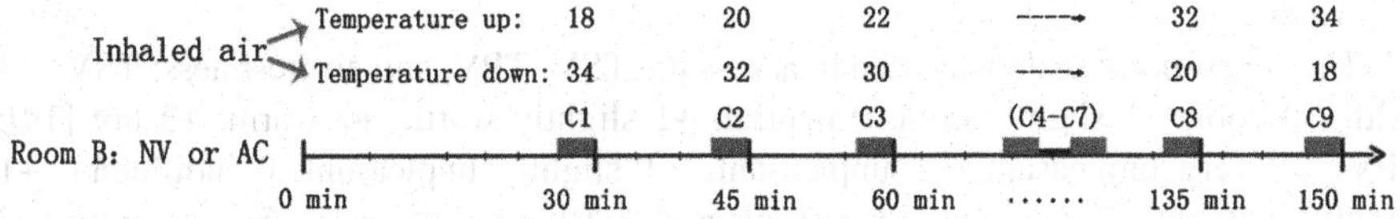

Fig. 3. Experiment procedure

To avoid the effect of previous condition, the subjects were randomly divided into two groups with two different temperature changing modes: one was increasing from 18 to 34 °C, and the other was decreasing from 34 to 18 °C. Due to heat loss (gain) through the duct/box, the actual air temperature in the box may be different from that in climate chamber A, so the actual air temperature was used to analyze in the paper.

2.3 Subjects

16 gender-balanced young and mid-aged subjects participated in the experiments. All of the mid-aged subjects are local people living in the hot-humidity climate zone for their whole life. Some of young subjects came from the cold climate zone and have lived in this hot climate zone for the first year. Their information is listed in Table 1, as we can see, the mid-aged persons spent less time in air-conditioning space per day, and their weights and body fat rates were slightly higher than the young ones. There were no significant gender difference on AC time. The height and weight of female were less than male. The average body fat rate of female (25.0%) was much higher than male (13.7%), which is the most significant difference.

Table 1. Information of Subjects

Group setting		Age	AC* (h/d)	Height (cm)	Weight (kg)	Body fat rate (%)
Age dif.	Young	22 ± 1	17.0 ± 4.4	166 ± 6	56.8 ± 8.7	18.0 ± 4.8
	Mid-aged	50 ± 5	8.4 ± 5.9	161 ± 8	59.8 ± 10.0	21.3 ± 9.4
Gender dif.	Male	36 ± 15	12.1 ± 6.2	170 ± 3	66.0 ± 5.5	13.7 ± 2.2
	Female	38 ± 15	13.3 ± 7.6	158 ± 5	51.7 ± 6.3	25.0 ± 6.4
All		37 ± 14	12.7 ± 6.7	163 ± 7	58.4 ± 9.3	19.7 ± 7.5

Note: AC* means the self-report time (h) stay in air-conditioning space per day.

2.4 Questionnaires

During the experiments the subjects were asked to evaluate the air supply on the scale of During the experiments, the subjects were asked to evaluate the inhaled air on a scale representing the comfort using questionnaires that covered the thermal sensation vote (TSV), thermal pleasure vote (TPV), and PAQ. The overall perception was also evaluated.

The seven-point scale was used to assess the TSV, TPV, and air freshness: TSV: −3 cold, −2 cool, −1 slightly cool, 0 neutral, +1 slightly warm, +2 warm, +3 hot [19]; TPV: −3 very unpleasant, −2 unpleasant, −1 slightly unpleasant, 0 indifferent, +1 slightly pleasant, +2 pleasant, +3 very pleasant [20]; air freshness: −3 very stuffy, −2 stuffy, −1 slightly stuffy, 0 indifferent, +1 slightly fresh, +2 fresh, +3 very fresh. The subjects recorded the perceived air quality within the range −1 (completely unacceptable) to +1 (completely acceptable). The scale was divided in the middle by points +0.1 (just acceptable) and −0.1 (just unacceptable).

3 Results

3.1 Overall Perceptions

Overall perceptions in NV and AC models on surrounding environment (without inhaled air in the box) were showed in Fig. 4. Generally, the thermal environment was warm in NV mode and neutral in AC mode. The TPV was higher in AC mode with the mean value about 1.0, and lower in NV mode with the mean value about −0.3. The PAQ was also slightly higher in the AC mode due to lower temperature and humidity.

3.2 Effect of Air-Conditioning

Subjects felt warmer about the inhaled air with higher temperature in AC than NV, because they were less adapt to high air temperature in AC neutral environment. The quality of in-haled air was also evaluated poorer in AC than NV. The subjects were less pleasant to in-haled air in AC than NV environment, as less "thermoregulatory load-error" was offset in AC environment (Fig. 5).

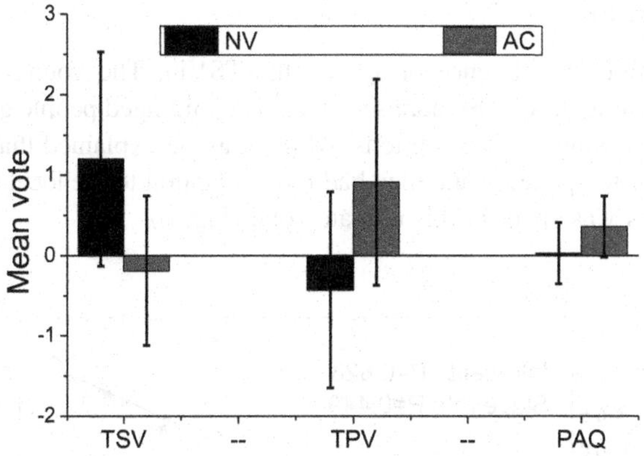

Fig. 4. Overall perceptions in NV and AC modes about surrounding environment

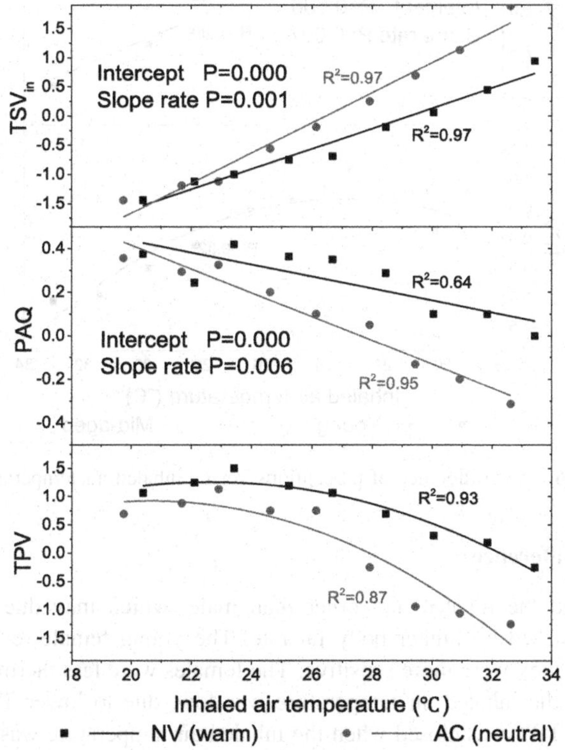

Fig. 5. Effect of air-conditioning on perceptions versus inhaled air temperature

3.3 Age Difference

There is no significant difference of age on the TSVin. The young female reported poorer PAQ, meaning they were more sensitive. The mid-aged people got more thermal pleasure than the young in all conditions, which may be explained that they were less sensitive to the air temperature and they had more adaption to the local climate, as most time of their lives was spent in this climate zone (Fig. 6).

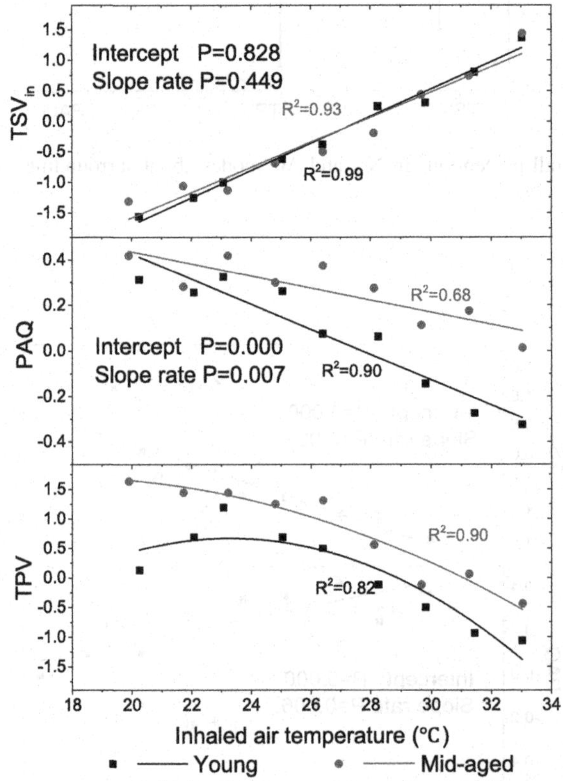

Fig. 6. Age difference of perceptions versus inhaled air temperature

3.4 Gender Difference

Female perceived the inhaled air cooler than male, which may due to their lower metabolic rate caused by higher body fat rate. The young female reported the PAQ poorer, meaning they were more sensitive. The females were less thermal pleased than the males when the inhaled air temperature was low, due to lower TSV and poorer PAQ. While the TPV was closed when the inhaled air temperature was high, it can be explained by that the unpleasant caused by poorer PAQ was offset by lower TSV in warm conditions (Figs. 7 and 8).

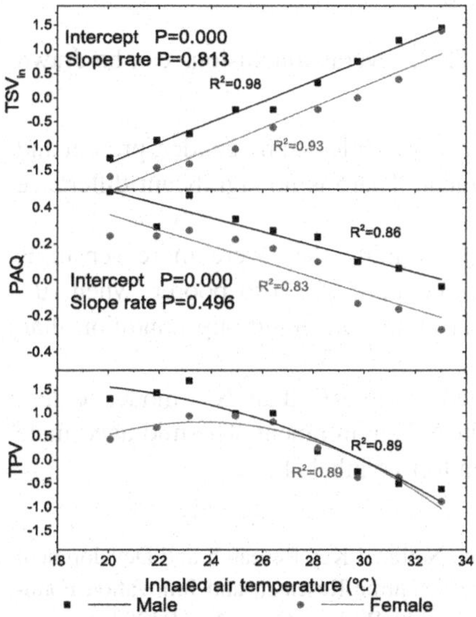

Fig. 7. Gender difference of perceptions versus inhaled air temperature

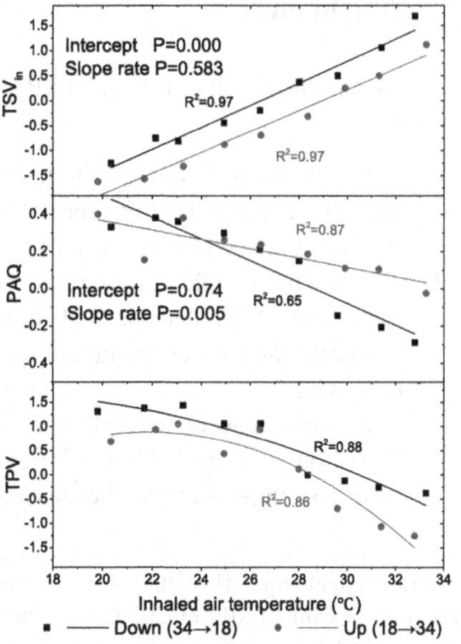

Fig. 8. Effect of thermal experience on perceptions versus inhaled air temperature

3.5 Effect of Thermal Experience

The inhaled air was evaluated to be warmer in temperature-up condition than temperature-down condition, indicating the effect of thermal experience or the subjects tend to vote for a different scale when feeling a change [17]. The PAQ of inhaled air A was evaluated poorer in temperature up condition than down condition. The subjects were less pleased to inhaled air in temperature-up condition than temperature-down condition either, due to thermal experience and the effect on evaluation exerted by previous state.

4 Discussion and Limitation

Obviously, the TSV, PAQ, and TPV on inhaled air were affected by the background environment, thermal experience, gender and age in different degrees. Generally, the subjects were more sensitive to the uncomfortable inhaled air, when they had an experience of being in a more comfortable thermal environment before. Also, the young female were sensitive.

Only sixteen subjects were involved in the experiment and they were divided into two groups depending on the gender or age, meaning the sample size of the experiment is small, which leads to a conclusion not that valid. For the same reason, the quantify comparisons were not conducted, which need to be further studied.

5 Conclusions

Considering the results of the TSV, PAQ, and TPV, several conclusions can be drawn as following:

(1) Subjects' thermal sensation vote of inhaled air was affected by gender, present and previous background on thermal environment. There is no significant difference on TSV between different ages.
(2) The young female reported poorer PAQ, meaning they were more sensitive. The PAQ of inhaled air with different temperature was also poorer, when surrounding thermal environment was neutral or in temperature-up condition than otherwise.
(3) The subjects were less pleasant to inhaled air in AC than NV mode, as less "thermoregulatory load-error" was offset in AC environment. The mid-aged male were less unpleasant to the thermal discomfort of inhaled air.

Acknowledgement. This study was supported by the National Key Research and Development Program of China (2017YFC0702700), the Graduate Scientific Research and Innovation Foundation of Chongqing, China (No. CYB18000), and the 111 Project (Grant No. B13041).

References

1. Zhang H, Arens E, Zhai YC (2015) A review of the corrective power of personal comfort systems in non-neutral ambient environments. Build Environ 91:15–41
2. Vesely M, Zeiler W (2014) Personalized conditioning and its impact on thermal comfort and energy performance—a review. Renew Sustain Energy Rev 34:401–408
3. Wu Y, Liu H, Li B, Cheng Y, Tan D, Fang Z (2017) Thermal comfort criteria for personal air supply in aircraft cabins in winter. Build Environ 125:373–382
4. Parkinson T, de Dear R (2015) Thermal pleasure in built environments: physiology of alliesthesia. Build Res Inf 43(3):288–301
5. Fang L, Clausen G, Fanger PO (1998) Impact of temperature and humidity on the perception of indoor air quality. Indoor Air 8(2):80–90
6. Toftum J, Jorgensen AS, Fanger PO (1998) Upper limits of air humidity for preventing warm respiratory discomfort. Energy Build 28(1):15–23
7. Fang L, Wyon DP, Clausen G, Fanger PO (2004) Impact of indoor air temperature and humidity in an office on perceived air quality, SBS symptoms and performance. Indoor Air 14(Suppl 7):74–81
8. Lan L, Wargocki P, Wyon DP, Lian Z (2011) Effects of thermal discomfort in an office on perceived air quality, SBS symptoms, physiological responses, and human performance. Indoor Air 21(5):376–390
9. Brager G, Zhang H, Arens E (2015) Evolving opportunities for providing thermal comfort. Build Res Inf 43(3):274–287
10. Fanger PO (1970) Thermal comfort: analysis and applications in environmental engineering. Danish Technical Press, Copenhagen
11. Karjalainen S (2012) Thermal comfort and gender: a literature review. Indoor Air 22(2):96–109

12. Mishra AK, Ramgopal M (2013) Field studies on human thermal comfort—an overview. Build Environ 64:94–106
13. Bischof W, Brasche S, Kruppa B, Bullinger M Do building-related complaints reflect expectations?
14. Choi J, Aziz A, Loftness V (2010) Investigation on the impacts of different genders and ages on satisfaction with thermal environments in office buildings. Build Environ 45(6): 1529–1535
15. Hong L, Yuxin W, Heng Z, Xiuyuan D (2015) A field study on elderly people's adaptive thermal comfort evaluation in naturally ventilated residential buildings in summer. J HV&AC 6(45):50–58
16. de Dear RJ, Brager GS (2002) Thermal comfort in naturally ventilated buildings: revisions to ASHRAE Standard 55. Energy Build 34(6):549–561
17. Schweiker M, Fuchs X, Becker S, Shukuya M, Dovjak M, Hawighorst M, Kolarik J (2016) Challenging the assumptions for thermal sensation scales. Build Res Inf 45(5):572–589
18. Li B, Yu W, Liu M, Li N (2011) Climatic strategies of indoor thermal environment for residential buildings in Yangtze River Region, China. Indoor Built Environ 20(1):101–111
19. ASHRAE (2013) ASHRAE standard 55-2013: thermal environmental conditions for human occupancy. ASHRAE, Atlanta
20. Parkinson T, de Dear R, Candido C (2016) Thermal pleasure in built environments: alliesthesia in different thermoregulatory zones. Build Res Inf 44(1):20–33

Effects of an Industrial Logic Implemented in Service Relation: The Case of Drivers of Ambulances of a Brazilian University Hospital

Daniele Pimentel Maciel, Ruri Giannini(✉), Fabiana Raulino da Silva, José Dib Júnior, Laerte Idal Sznelwar, and Cláudio Marcelo Brunoro

Escola Politécnica, University of São Paulo, São Paulo, Brazil
ruri.giannini@gmail.com, laertesz@usp.br

Abstract. The transportation of patients by ambulance is part of the services offered by a well-known Brazilian University Hospital. Ambulance service is extremely important as it connects different institutes involved in the global patient care. However, the increase in the complaints by nurses of irritability and aggressiveness of drivers after a strategic change in the organizational structure of the hospital attracted the attention of the research team. Based on ergonomic work analysis (EWA), the data were obtained through observation and interview with managers, leaders and drivers of the transportation sector of the hospital. The main results found were: (1) the key measurement of drivers' work is time. However, time is a measure which is not manageable by drivers—as it doesn't consider all variabilities that occur during their work. Moreover, this measurement also doesn't consider the real content of their work, (2) prescribed work of drivers only considers the task of driving an ambulance, whilst their real work includes taking care of the patient as well, (3) drivers are not recognized as health professionals despite working with nurses and physicians in the care of patients.

Keywords: Ergonomics · Service operation · Health care

1 Introduction

In the current economic scenario, the main job generator is the service sector. The relevance of the service sector to the overall economy has increased lately. Such growth is related to the changes in the post-industrial society when industrial and manufacturing companies get services together as a way to aggregate value to the manufactured product and there are service activities in all other sectors of economy.

However, service is an activity which is not so easy to be characterized. In a traditional service model definition, the production and consumption of the service occur simultaneously. Moreover, in the contemporary world, the service activity has various characteristics, such as managerial practices, ways of working and relationship with clients and general public (Salerno 2001).

S. Bagnara et al. (Eds.): IEA 2018, AISC 818, pp. 180–191, 2019.
https://doi.org/10.1007/978-3-319-96098-2_24

Among service activities, the health sector might be considered a typical service activity, because it includes a direct and simultaneous relation between the patient and the health professional. However, a hospital performs other activities which are not typical service activities but are involved in the care to the patient, such as laundry, hospitality, alimentation, maintenance and transportation.

To better understand the characteristics of service activities, Zarifian (2001) proposes the use of the word "service" in singular form. In this sense, service production can be understood as a process which transform the existing conditions of the recipient (client or user) to meet his/her expectations. Therefore, the service should act on the use or on life conditions of the recipient (Zarifian 2001).

Zarifian (2001) also highlights the difference between the industrial logic and the service logic. In the industrial logic, the traditional management models from Taylorism-Fordism prevail, based on the concept of mass production. On the other hand, the service logic emerged from the French public system, where the provision of services should follow a mission or a goal which guides its results (Zarifian 2011).

In recent years, researchers argue that the service sector went through a process known as "industrialization" of service. It means the application of the conceptual and methodological model from the Taylorism-Fordism (standardization, mass production, work segmentation, segregation of planning and executing activities), which were typically related to the industrial logic. Thus, the productivity logic based on quantity and time were applied to the service relations.

The evaluation of services based on an industrial logic may generate obstructions and distortions. Since remote times the need of evaluating the work exists, so a lot of discussion has emerged around the choice of the best indicator to evaluate it. Work measurement, in its turn, is strongly controversial, as work is a subjective experience which should not be measured (Dejours 2008).

Moreover, the application of the industrial logic in the health sector can be contradictory to the concept of care. It is implicit to the care the idea of equity or fairness, that is, the need of treating different people or different needs in different ways. The concept of care has its origins in the ethical discussion related to caring for others; its definition also encompasses concern for others, attention and an attitude towards the other considering their needs (Hirata and Guimarães 2012; Molinier 2012). That is why the industrial logic of time management applied to the care is contradictory, as it is opposed to the real of the care.

Another aspect that must be analyzed is the evaluation of quality. Typically, the quality of a service is determined by the part of the process which involves an interaction between the work and the client. In this way, the word "quality" contains the subjective perception of the client built during the production of the provision of the service, and it has no direct relation with the efficiency of the provision of the service. Some quality models are beginning to consider the concept of service, meaning meeting the needs of the client, throughout all department of the company.

2 Object of the Study and Objectives

The object of this study was the activity of ambulance drivers located on an University Hospital in São Paulo, Brazil (we thank Eduardo Costa Sá for promoting this opportunity). The hospital was founded in 1944 and provides public health services, in addition to be a teaching hospital with the goal of educate health professionals. It is an autarchy of the Government of São Paulo, linked to the State Secretary of Health for administrative purposes and to the public School of Medicine for academic purposes. Autarchy in the public administration means an autonomous entity, decentralized from the public administration and with its own resources, but supervised by the State. The activities of the hospital are funded by the Brazilian national health system and its main differential from other public hospitals are technological innovation and technical quality of the professionals.

The main products of the hospital are education (graduation, medical residency, internship programs, specialization course for non-physicians), research (post-graduation, research programs linked to other national and international universities) and health (medical appointment, urgency and emergency assistance, hospitalization, surgery, complementary imaging examination, pharmaceutical and nutritional assistance and motor rehabilitation). Patients and their companions are, thus, the main clients of the hospital, together with education institutions, students and researchers.

The hospital is composed by different institutes which offer specialized medical service to the public. Each institute has a decentralized management subordinated to the central administration. The transportation of patients by ambulance between institutes is part of the health services offered by this hospital. Ambulance service is extremely important as it connects different specialized institutes involved in the global patient care. Patients receive general care during transportation and that is why ambulance service is relevant in the final value perceived by the patient.

Until the year of 2012, the management of the transportation department was also decentralized, and each institute would have an own team of drivers and would be responsible to organize and distribute their work. In 2012, a strategic change impacted the organization of the work of ambulance drivers, who started to work in a centralized way, serving all institutes.

The goal of this change, which occurred at the whole hospital, was the optimization of resources available and revenue increase. In the case of ambulance drivers, some institutes have greater demand of patient transportation than others, and reallocation of driver used to be necessary in the decentralized management. However, it would impact on salary politics and on budget for fuel and maintenance of ambulances which used to be managed by each institute.

With the centralization of the service, all drivers were allocated in a single garage and started to have two instances of leadership: one who is responsible to receive all demands for transportation, get all necessary information about the patient and define priorities in terms of transport and another one who allocates one driver for each demand. Previously, nurses would contact the drivers directly to know about their availability. The criteria for a driver to occupy this second instance of leadership is not clear and they are usually drivers who have constraints to develop the driving activity or to carry weight.

The increase in the number of work accidents involving drivers and in the complaints by nurses of irritability and aggressiveness of drivers after that strategic change attracted the attention of the research team. With this in view, the objective of this study is to understand whether and how the organizational changes and issues have affected the work of ambulance drivers and what kind of impact they have generated for the drivers besides accidents, irritability and aggressiveness, through the analysis of drivers' new work situation.

3 Theoretical Reference

In the 70's, Wild (1977) differentiated manufacturing from services by the positioning of customers in the production system: customer would be an output of a manufacturing system and an input of a service production system. This author also characterized service as a system where the customer or a belonging of the customer is transformed (Wild 1977). However, since those years authors would already affirm that it is difficult to characterize a company as purely manufacturing or service driven because offered products are usually a combination of goods and service (Sasser et al. 1978; Corrêa and Corrêa 2005).

Sasser et al. (1978), Haynes (1990), Gianesi and Corrêa (1994), Cuatrecasas (2002) and Apte and Goh (2004) discussed the main characteristics of service and highlighted those ones which differentiated service production from manufacturing. First, service is a product with no physical properties which is experienced by customers and therefore difficult to be measured. The evaluation of a service taken into consideration intangible aspects while the evaluation of a manufactured good is based on physical attributes.

Further, customers are considered an input of a service system and they initiate the production of a service by requesting them. Equally important, service is perishable and highly variable, which means service varies according to each customer and is hard to be standardized. Subsequently the evaluation of a service will depend on the expectations and needs of each customer, while the quality of manufactured products is usually measured by tangible aspects. For those reasons, every decision made in a service system should take into consideration the level of contact between the customer and the production process (Chase 1978).

Even with all this in mind, the transfer of concepts originally applied to manufacturing operations to service systems has been widely studied during the past decades, focusing the increase of productivity, cost reduction, timing decrease, waste elimination. For a large number of authors, service operations would benefit from all the advantages of the productivity and quality programs developed for manufacturing systems (Allway and Corbett 2002; Bowen and Youngdahl, 1998; Ahlstrom 2004; Sanchez and Pérez 2004).

The application of the industrial logic in service operations can be contradictory to the characteristics of service, specially intangibility and variability. The industrial logic of time management, waste elimination and productivity increase won't fit the real work of service providers, as they face the subjective perception of customers and are evaluated through the expectation of each customer.

4 Methodology

The object of an ergonomic work analysis (EWA) is the activity of work. Ergonomists consider important to deeply understand the distinction between activities and tasks. A task is prescribed by the employer for the employee to perform, based on predetermined conditions that are usually far away from the real conditions of work.

On the other side, an activity is the strategy which subjects create to adapt his/her task according to the real condition of work. In other words, an activity is what the employee really does to perform a task. The EWA analyzes, therefore, the strategies used by the employee to manage the distance between the prescribed work and the real work (Guérin et al. 2001).

Tasks are attributed to every work as a prescription to guide employees to reach expected results. Ergonomics study how activities are created to overcome constraints and limitations of the prescribed work and to transform the work through adaptations that could make work situations more compatible to abilities and limits of the human being (Abrahão et al. 2009; Guérin et al. 2001; Wisner 1987).

The methodology proposed by the activity-centered ergonomics, is structured in several linked phases, as an iterative process. It's not a linear process because the confrontation with the real work might lead to a return to previous phases in order to a deeper understanding of the situation of work. Therefore, EWA is an interactive and participative approach, which is appropriate to the comprehension work and to conceive solutions (Abrahão et al. 2009).

Based on the EWA, the data were obtained through observation and interview with managers, leaders and drivers of the transportation sector of the hospital. The data collection took two months, totalizing 40 h of observation of the activities of ambulance drivers. The results demonstrated common elements identified both in the workers' discourse and in the observations.

5 Collection of Data and Results

According to Guérin et al. (2001) and Wisner (1987), an EWA should begin by analyzing the demand (request), further to build the problem, which must be explicitly expressed. The demand (request) of this study surpassed musculoskeletal complaints and reached the questioning of nurses and other coworkers of the ambulance drivers about their irritability and aggressiveness when they were called to work. As a response, managers of the hospital planned to evaluate drivers by a psychiatrist. Thereafter the problem of the study contains different variables of the activity of the ambulance drivers, which could be related to the questions pointed by nurses.

Initially, the researchers studied the organization to better understand the specificities of the hospital, such as its history and the relative positioning in the health system; as well as, its organizational structure, hiring politics, epidemiological data on the driver population, indicators of quantity of patient transportations performed by the drivers, legislative requirements for the job. The investigation of the technical procedures was performed by initial interviews with workers to an overall comprehension of the service's processes and their relations with other departments of the institution.

According to the demand, the work situation which was studied was the transportation of patients by ambulance from one institute to another. Such work situation was chosen because this kind of transportation represents the majority of the work performed by the drivers and requires the interaction between drivers and patients (other kind of transportation would be organ for transplants, biological material, office supplies, documents and managers of the hospital). When transporting a patient from one institute to another, drivers and patients are always accompanied by a nurse (and by a physician when the condition of the patient is severe).

The researchers performed a free observation of the activities for two days and built a logbook of what the drivers did, with the objective of understanding the specific constraints of the work situation. Furthermore, individual and semi-structured interviews were conducted and recorded so that the researchers could compare them with the observations and the logbook.

Until the year of 2012, the management of the transportation department was decentralized, and each institute had an own team of drivers. In a decentralized organization, the process allowed nurses to contact the drivers directly to know about their availability and to schedule patient transportation, which means drivers had autonomy to divide and organize their own work for their institute. As in 2012, their service has been centralized, serving all institutes, with the goal of optimization of resources available. Some institutes have greater demand of patient transportation than others, and reallocation of drivers seemed to be necessary for the hospital managers.

With centralization, all drivers were allocated in a single garage and started to have two instances of leadership: one who is responsible to receive all requests for transportation, get all necessary information about the patient and define priorities in terms of transport and another one who allocates one driver for each demand. The leader of the second instance is chosen among the team of drivers, but the criteria for a driver to occupy the leadership role is not clear. They are usually drivers who have constraints to develop the driving activity or to carry heavy weight, which means the leader is not necessarily a senior or more experienced driver or one who was promoted to the job.

The Fig. 1 below represents the main tasks performed by drivers and leaders.

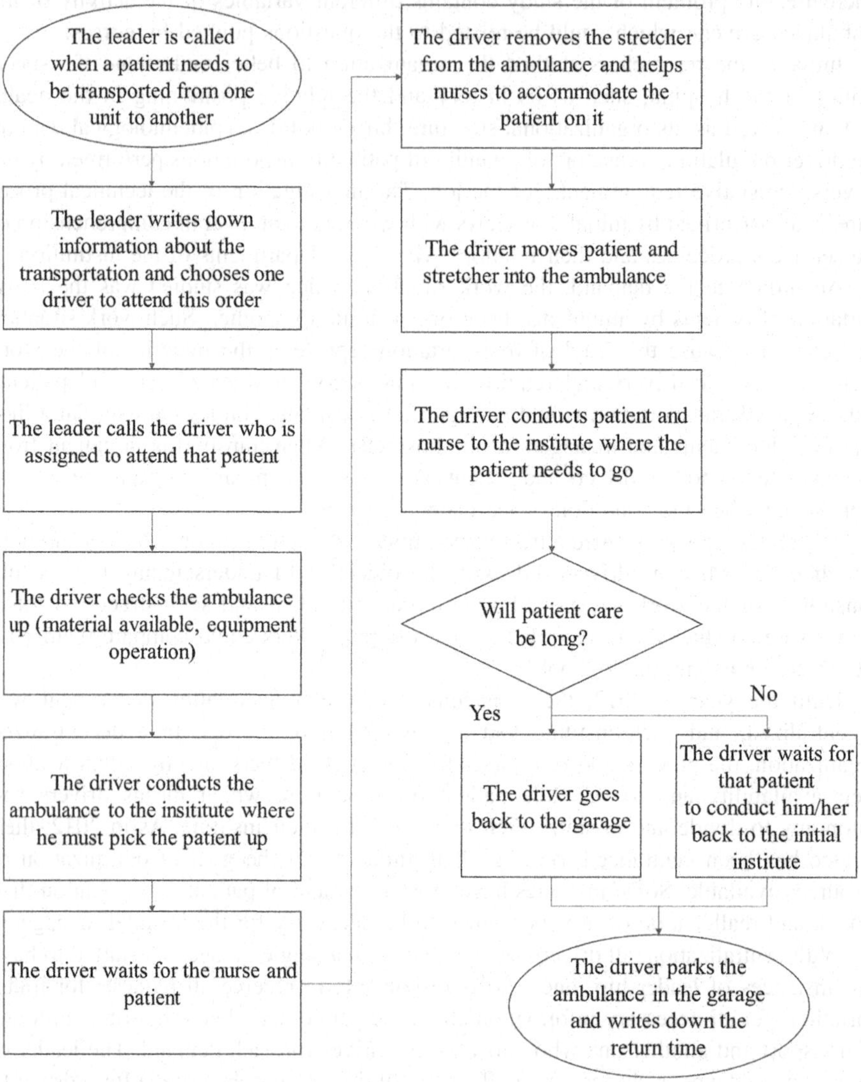

Fig. 1. Flowchart with main tasks performed by drivers and leaders after the centralization of transportation department (created by the authors)

As presented in the flowchart, the nurses don't contact directly the drivers when they have a patient who needs to be transported. The leader receives all orders from the nursing team and assigns drivers for them. When drivers receive a call, they have five minutes to get to the institute where the patient is. In that route, they usually face variabilities such as intense traffic jam, road obstructions, traffic accidents, climate variations which impact on vehicle traffic.

Drivers have autonomy to choose the best route to get to a place and they learn by experience the best ones according to time of the day and location. When they get to the institute to pick a patient up, they need to park the ambulance. Some institutes have parking space, others don't. In that cases, drivers need to look for an allowed space to park or they park the ambulance in a no parking space. After parking the ambulance, they wait for the patient. If patient and nurse are delayed, they call the leader to solve the issue.

When the patient arrives, the drivers remove the stretcher from the ambulance to accommodate the patient on it. Different situations can occur during this task:

1. Patient is on a wheel chair and can reach the stretcher by him/herself or with low supervision;
2. Patient is bedridden and nurse team and drivers need to accommodate him/her on the stretcher;
3. Patient is bedridden in a specific stretcher and can't be accommodate on the stretcher of the ambulance. Drivers should be aware of these situations before leaving the garage so that they can leave their stretcher there. When they are not warned in advance, they leave their stretcher in the institute and need to go back late to catch it;
4. Patient in severe conditions can't be accommodated on the stretcher near the ambulance. In these situations, nurses take the stretcher inside the institute to accommodate the patient and drivers wait in the ambulance.

Drivers then request the documents of the orders to write down departure point, destination, departure time, mileage and waiting time, and moves patient and stretcher into the ambulance. They usually perform this task with no assistance, but there are situations (obese patients, stretchers with different adjustments etc.) when the nursing team, security team of the institute or any other one must assist them to steady the stretcher. Finally, they drive the patient to the specified destination.

When they arrive in such destination, they expect to have a nursing team waiting the patient with another stretcher to carry him/her inside the institute. Thus, the prescribed task would be passing the patient from the ambulance stretcher to the institute stretcher and decide whether to wait or go back to the garage. However, uncontemplated situations may occur:

1. Nurses take the patient inside the institute using the ambulance stretcher and drivers must wait them return the stretcher back, as the ambulance isn't functional without the stretcher;
2. There are delays in the liberation or authorization of patients;
3. Drivers must stay alone with patients while nurses will take another stretcher, wheel chair or any documents inside the institute;
4. Other people rather than the driver and nurse need to assist the removal of the patient from the ambulance stretcher (in adverse situations as obese patients or stretchers with different adjustments).

When driving the patient to the destination and when going back to the garage, drivers face the same variabilities which impact on traffic and on the time to perform tasks. Lastly, they go back to the garage and fill in a logbook with information about the whole task, including adverse situations that may have occurred, and wait for another call in the waiting room, inside the ambulance or at the cafeteria.

The prescribed tasks of drivers are divided into three parts: driving the ambulance from the garage to the departure location, driving the patient from one institute for another one and driving the ambulance back to the garage. Based on these three parts leaders follow up the work of drivers, based on time indicators, which goal is five minutes. It is important to emphasize that this target was established by the organization during the centralization of the work of drivers, who in their turn were not involved in such definition.

Other drivers and pedestrians can impact on the activities of the ambulance drivers; however, they can't use the ambulance siren inside the hospital complex (only emergency lights are allowed) to notify that a patient is being transported. For this reason, the drivers routinely have to choose between getting stuck in traffic or breaking traffic laws. Breaking traffic laws, by its turn, will generate traffic tickets and verbal advertences from their leaders. Still considering the disrespect to the traffic laws, many institutes don't offer an adequate parking space for ambulances. In those cases, drivers have to look for an allowed space to park or to park the ambulance in a forbidden space, where they are susceptible to traffic tickets.

After the observation and analysis of logbooks, we learned that adverse situations related to traffic are not the only situations, which impact on time indicators. On the contrary, two situations are pointed to strongly jeopardize time: (1) when nurses delay to bring the patient who will be transported to the ambulance and (2) when the patient is taken inside the institute with the ambulance stretcher and drivers need to wait nurses to return them back.

To sum up, we organize the main results found in three subjects:

1. **Work evaluation:** the key measurement of drivers' work is time. However, time is a measure which is not manageable by drivers—as it doesn't consider all variabilities and emerging events that occur during their activities, including traffic jam, climate conditions and delays of other professionals involved in the job. Moreover, drivers were not responsible to define the targets of time indicators, so they don't necessarily agree that five minutes is time enough to complete each part of the transportation (driving the ambulance from the garage to the departure location, driving the patient from one institute for another one and driving the ambulance back to the garage);
2. **Real content of the work:** besides the lack of control of time indicators, this measurement also doesn't consider the real content of their work. Prescribed work of drivers only considers the task of driving an ambulance, whilst their real work includes taking care of the patient as well. Drivers are also responsible to accommodate patients into the stretchers, giving special attention for adverse situations, as obese patients, bedridden patients etc. Moreover, they usually are the only companion for patients while nurses are dealing with bureaucratic matters inside the institutes this requires an activity of care;

3. **Feeling of exclusion:** drivers are not recognized as health professionals despite working with nurses and physicians and taking the care of patients. In hospital services, there are workers who have direct contact with patients and are considered part of the health team, while others don't have this contact but provide many important services which are necessary to the production of a hospital service. There are still those workers who have direct and constant contact with patients and their families, but, because they aren't responsible for the clinical care, this contact with patients is little or no recognized at all. These is the case of ambulance drivers and such lack of recognition as being part of the health team conflicts with the meaning they attribute to their work.

6 Discussion and Conclusion

The change in the organizational structure, which involved the work of drivers prioritized an industrial logic, jeopardizing the service relation (Hubault 2015, 2017) between drivers and patients. Within this industrial logic, ambulance drivers are supposed to deliver one specific part of the whole service—driving ambulances—in the minimum time as possible. The key measurement for their work is time, despite traffic jam, climate conditions or any other variability that may occur during the transportation of a patient.

The analysis of the results also demonstrated the invisible work of drivers, who perform several activities, which are not recognized by the other actors involved in the process, mainly the managers. Among this invisible work (or real work) is the activity of taking care of patients, which is not taken into consideration in their prescribed work. The lack of recognition of their real work, the lack of recognition of drivers being health professionals and part of the health team generates feelings of devaluation and conflicts.

The strategic change in the organization of the work of the ambulance drivers which centralized them in a single management follow an industrial logic based on Taylorism-Fordism, which prioritizes the concentration of the production (Zarifian 2001). The centralization of the service had the goal of optimizing the resources of the transportation department, but it didn't take into consideration elements of the actual situation and drivers' activities, which have impact on the production of service (such as knowledge about the specialization of each institute or the personal relationship with coworkers and patients).

The centralization of work brought a centralization of the power of decision and the reduction of the autonomy of drivers, by the creation of the job title "leader". The lost of autonomy by workers is an example of the application of an industrial logic into the service activities (Zarifian 2001). The creation of the leader role also increased one intermediary hierarchical level which jeopardizes the communication.

There is a lot of controversial around the evaluation of work. For Dejours (2008), it is not possible to measure the work because it is a subjective experience of the individual. For Dejours (2008), when we face the impossibility of measuring the work, we start to evaluate the time of the work. In this sense we understand erroneously that by evaluating the time of work we can evaluate the work.

The logic of measuring the work to evaluate it was applied in the restructuring of the transportation department. This method of management of time follows the industrial logic and is an example of the industrial production of service discussed by Zarifian (2001). However, this method is proved to be inadequate, as the measurement of time of work explains the duration of the effort and despises the intensity, the quality or the content of the work (Dejours 2008). In addition, this kind of approach doesn't consider the immaterial values of servicing, such as confidence, cooperation and pertinence (Hubault 2015, 2017; Du Tertre 2017).

The choice of a management of time testifies the application of an industrial logic in the health sector, specifically in the care. In this study, choosing to measure the time to evaluate the work of ambulance drivers evidences that the organization can not recognize the role of drivers in the care to the patient.

Besides confirming the application of an industrial logic to a service activity, the indicator of time is not manageable by the drivers. Drivers are not completely able to manage the time of their work because they face a lot of variables as traffic jam, climate conditions, internal logistics for patient transportation, bureaucracy to admit patients, which they can't control.

Moreover, the indicator of time doesn't take into consideration the real work of ambulance drivers and the recognition oh their role in the care to patients. Sznelwar (2015) advocates that all professionals who are somehow involved with patients are also involved with care. That means that a meaningful part of the work of the ambulance drivers is participating on the care to the patients who are transported from an institute to another one.

The importance of the care to the patients was identified in the analysis of the activity of ambulance drivers. Those workers are directly involved in care activities when they carry stretchers, when they stay together with patients waiting for nurses or when they develop compassion for patients.

According to Sznelwar (2015), the work of care can only be performed where there is room for compassion. The subject compassion was clear in the verbalizations of the drivers: "I am so sorry when a kid is very ill, I pray for him/her", "This is an adrenaline pumping job and I don't want anything bad to happen inside my car". Therefore, when time management is used as the only way to evaluate the work of ambulance drivers we can't recognize their role as health professionals.

This lack of recognition induces the appearance of irritability and aggressiveness of the ambulance drivers, which are described by their coworkers. By not recognizing their real work the organization can't recognize the meaning of work for them. The reductionist stance to health professionals was identified by Sznelwar (2015) in other studies and proves the deterioration of work procedures caused by the lack of recognition of the relevance of care for those professionals.

References

Abrahão J et al (2009) Introdução à Ergonomia: da prática à teoria. Blucher, São Paulo

Ahlstrom P (2004) Lean service operations: translating lean production principles to service operation. Int J Serv Technol Manag 5(5/6):545–564

Allway M, Corbett S (2002) Shifting to Lean Service: stealing a page from manufactures' playbooks. J Organ Excellence 21(2):45–54

Apte UM, Goh CH (2004) Applying lean manufacturing principles to information intensive services. Int J Serv Technol Manag 5(5/6):488–506

Bowen DE, Youngdahl WE (1998) Lean Service: in defense of a production-line approach. Int J Serv Ind Manag Bradfort 9(3):207

Chase RB (1978) Where does the customer fit in a service operation. Harvard Bus Rev 56 (6):137–142

Corrêa HL, Corrêa CA (2005) Administração de Produção e Operações. Manufatura e serviços: uma abordagem estratégica. Atlas, São Paulo

Cuatrecasas LC (2002) Design of a rapid response and high efficiency service by lean production principles: methodology and evaluation of variability of performance. Int J Prod Econ 80:169–183

Dejours C (2008) Avaliação Do Trabalho Submetida à Prova Do Real: Crítica Aos Fundamentos Da Avaliação. [s.l.] Blucher

Du Tertre C (2017) L'économie de la fonctionnalité et de la coopération, A paraitre. https://hal.archives-ouvertes.fr/hal-01471420. Accessed 22 May 2018

Falzon P (2007) Ergonomia. Blucher, São Paulo

Gianesi IG, Corrêa HL (1994) Administração Estratégica de Serviços: operações para a satisfação do cliente. Atlas, São Paulo

Guérin F, Laville A, Daniellou F, Duraffourg J, Kerguelen A (2001) Compreender o trabalho para transformá-lo: a prática da ergonomia. Blucher, São Paulo

Haynes RM (1990) Service typologies: a transaction modelling approach. Int J Serv Ind Manag 1 (1):15–26

Hirata H, Guimarães NA (2012) Introdução. In: Hirata & Guimarães (org) Cuidado e Cuidadoras - As várias faces do trabalho do care, São Paulo, Atlas, pp 1–11

Hubault F (2015) Le bien être un enjeu sensible pour le management dans l'économie du service. http://www.atemis-lir.fr/wp-content/uploads/2017/02/Le-Bien-Etre-un-enjeu-sensible-pour-le-management-dans-leconomie-du-service-Hubault-2015.-1.pdf. Accessed 22 May 2018

Hubault F (2017) Corps, activité, espace – nouvelles interpellations de l'économie dématérialisée. In: Hubault F. (coord.), 2017. Les espaces du travail ; enjeux savoirs, pratiques – actes du séminaire Paris1. Editions Octarès, Toulouse, pp 3–12. http://www.atemis-lir.fr/wp-content/uploads/2017/08/Corps-activite-espace-Hubault-2017.doc. Accessed 22 May 2018

Molinier P (2012) Ética e trabalho do Care in: Introdução in HIRATA & GUIMARÃES (org) Cuidado e Cuidadoras – As várias faces do trabalho do care, São Paulo, Atlas, pp 29–43

Salerno MS (2001) Relação de serviço: produção e avaliação. SENAI, São Paulo

Sznelwar (2015) In: Quando trabalhar é ser protagonista e o protagonismo no trabalho. Blucher, São Paulo

Sasser WE, Olsen RP, Wyckoff DD (1978) Management of service operations. Allyn and Bacon, Boston

Sanchez AM, Pérez MP (2004) The use of lean indicators for operations management in services. Int J Serv Technol Manag 5(5/6):465–478

Wild R (1977) Concepts for operations management. Interscience, Chichester

Wisner A (1987) Por dentro do trabalho. Ergonomia: método e técnica. FTB: Oboré, São Paulo

Zarifian P (2001) Mutação dos sistemas produtivos e competências profissionais: a produção industrial de serviço. In: Salerno MS (ed) Relação de serviço: produção e avaliação. SENAI, São Paulo

The Impact of Telehealth Video-Conferencing Services on Work Systems in New Zealand: Perceptions of Expert Stakeholders

Nicola Green(✉), David Tappin, and Tim Bentley

School of Management, Massey University, Palmerston North, New Zealand
n.j.green@massey.ac.nz

Abstract. Telehealth, the provision of health care services at a distance, is one way to address increasing problems of resource scarcity and equity of access to healthcare. However, the literature suggests there are difficulties with embedding telehealth into routine care and that in the complex system of healthcare consideration of the multiple system components would perhaps aid understanding. Using video-conferencing for the delivery of services is one approach to telehealth and the most commonly used in New Zealand. In this study, a sociotechnical systems approach was used to explore the perspectives of an expert stakeholder group regarding the current characteristics of telehealth video-conferencing services and the impact of these services on work systems. Twenty semi-structured interviews were recorded, transcribed and thematically analysed. Preliminary analysis suggest that key themes include initial experiences of using the technology, changes in the way of working, support to provide and receive telehealth, collaboration, leadership and funding models. The themes are interrelated across the work system components and system mismatches are emerging which may be significant in explaining the lack or slower than expected realization of telehealth goals.

Keywords: Telehealth · Video-conferencing · Sociotechnical systems theory

1 Introduction

Advances in information and communication technologies (ICT) are changing the way health care services are provided. Concurrently, healthcare systems around the world face increasing pressures from a growing and aging population, inequity of access and finite financial and human resources. Telehealth, the delivery of health care services at a distance using information and communication technologies [1] is one solution to problems such as accessibility, quality, professional resource scarcity and cost [2, 3]. Delivery of health care at a distance may be asynchronous, for example, remote reading and reporting of radiology images or synchronous such as real time video-conferencing services.

In New Zealand (NZ) telehealth, particularly video-conferencing, is increasing [4]. This reflects a global increase in telehealth interest [1, 5]. International findings suggest that telehealth can be effective and cost-effective for health outcomes [6, 7]; reduce travel and associated costs [8, 9] and improve access to healthcare for isolated

S. Bagnara et al. (Eds.): IEA 2018, AISC 818, pp. 192–197, 2019.
https://doi.org/10.1007/978-3-319-96098-2_25

communities [10, 11]. However, research indicates that the diffusion of telehealth into routine care remains problematic [12] with many examples of telehealth programs that fail past the pilot stage [13, 14]. The literature suggests that the barriers to success include societal factors (e.g. law); organizational factors (e.g. cost, change management), consumer factors (e.g. technology literacy), provider factors (e.g. resistance to change) and technological factors (e.g. bandwidth, usability) [15, 16]. Moreover, researchers have suggested that for telehealth to be implemented effectively and sustainably it is important to understand how the various components work together as a complex, whole system [14, 17] and that a systems approach is needed to consider the multiple, interrelated aspects of telehealth [3, 13]. Despite much potential contribution from a human factors/ergonomics approach [18, 19] relatively little attention has been given to telehealth in the published ergonomics literature.

Using a sociotechnical system as a theoretical lens this study explores, from the perspective of an expert stakeholder group in NZ, the current characteristics of telehealth video-conferencing services and the impact of these services on work systems.

2 Methods

2.1 Study Design

The study follows a qualitative approach using semi-structured interviews with all twenty members of the NZ Telehealth Forum leadership group as at 8 August 2017. An ethics review was conducted by peers and a low risk ethics notification made to the Massey University's Human Ethics Committee.

2.2 Participants

The role of the NZ Telehealth Forum is to provide advice to the government's Ministry of Health on telehealth. The area of expertise or role of the twenty leadership group members interviewed is shown in Table 1. The participants were located throughout NZ, with 9 of the 20 District Health Boards represented. Each participant was interviewed between the end of October 2017 and February 2018 either in-person (n = 6), by video-conference (n = 10), by telephone (n = 3) and one which was a combination of video and then telephone.

2.3 Data Analysis

Nineteen interviews were audio recorded. The interviews were transcribed and stored in NVivo (QSR International). One participant did not consent to audio recording of the interview. In this case notes were made during the interview, typed up and sent to the participant to amend or elaborate on and then included with the transcript documents. The data were analyzed and coded using thematic analysis [20] and the Framework Method [21] in NVivo.

Table 1. Area of expertise/role of participants

Area of expertise	Number
Clinical – doctor	4
Telehealth programme manager/coordinator	4
Telehealth consultant	1
Technical	3
Consumer panel	1
Professional body/regulator/industry group	3
Governmental	1
Administration	1
Research	1
Clinical governance	1

3 Results

The average interview length was 58 min with a range of 33–98 min. Most of the telehealth video-conferencing in NZ is occurring within the secondary health care system with provision of services from large hospitals to smaller satellite hospitals. There is very limited activity in the primary care sector, i.e. family doctor or general practitioner care, or directly to people at home or on their mobile device. A wide range of disciplines use video-conferencing including pediatrics, renal, dermatology, emergency medicine, speech language therapy and oncology. There is no comprehensive system of recording the use video-conferencing, however the estimated percentage of consultations remain low.

At the time of writing, the data analysis is not complete. However, some broad, key themes of telehealth video-conferencing services' influence on the work system are emerging (noted in italics below). These can be described in terms of the components of the work system. i.e. technology, tasks, people, organization and the external environment [22].

From a technology perspective, telehealth video-conferencing requires that the equipment used needs to be reliable and user friendly to provide an *excellent first experience* for both provider and receivers. This impacts on the interest of users and the continued use of the tools to provide services.

Provision of healthcare using video-conferencing necessitates a *change in the way of working*, changes in the way that clinicians practice and use of an alternative model of care, for example using technology to provide healthcare from a different location rather than an in-person service. This has an impact on the tasks performed for all people involved in the service delivery, including clinicians, support staff, patients and their families. Adjustments typically need to be made to the way the consultations are booked and organized, where people are physically located, how the technology is used and the change to an on-screen relationship.

People in the work system need *support to provide and receive telehealth.* This includes technology support, logistical support and support from the wider telehealth community. Utilizing video-conferencing appears to have a positive impact on people in teams through increased *collaboration* and professional development. Additionally, clinical and professional support is reported to be facilitated by easier and more frequent face to face contact with peers. For example, a rural diabetes nurse specialist can feel less isolated, seek advice more easily and feel more part of the specialty team based in another location by using video-conferencing.

Organizational *leadership* appears important for the development of telehealth video-conferencing services, this may be in the form of strategy, champions and dedicated role creation (e.g. telehealth program managers). Using telehealth video-conferencing as an alternative model of care challenges the existing *funding models* particularly in cross-district interactions and in the primary care sector. For example, general practices are often funded by the Ministry of Health through patients being registered to a 'bricks and mortar' practice, will this funding be affected if the consultation is provided with video-conferencing and not in-person?

4 Discussion and Conclusion

In addition to the key themes of the impacts of telehealth described above, there are also themes emerging on interactions and inter-relationships between system components. For example, technical reliability and usability interacts with the first telehealth video-conference experience of a provider user, which is modified by the support available and this in turn may have an effect on the attitudes of clinicians and their adoption of new ways of working. Similarly, the use of video-conferencing of the provision of a service appears to enhance collaboration and clinical support which may impact on the quality of outcomes both for the patient (e.g. health outcomes) and the clinicians (e.g. job satisfaction and professional development).

Concurrently, there are potential system mismatches [23] that have come to light through the thematic analysis. For example, it is considered by many that the current funding models for the delivery of services do not satisfactorily fit with video consultations. If clinicians use video-conferencing for a consultation instead of travelling in-person to that location it is unclear which service would pay for their time, the provider end or the receiver end? Other potential system mismatches include those of national and/or organizational strategy and clinical attitudes, for example, in NZ the national health strategy includes a goal of the provision of health care closer to home with telehealth a component of this. This may not be realized if clinicians are reluctant to change the way they work [24].

Considering the impact of telehealth video-conferencing services from a socio-technical systems lens helps to identify the multiple components and interactions of sub-systems that potentially influence the success or failure of such technology innovations in the healthcare environment [25]. Further analysis of the data will consider the work processes and outcomes in the system, using the SEIPS 2.0 model as a theoretical framework [26].

This study is limited in that it explores the perceptions of an expert group involved in the promotion of telehealth in NZ. This may mean that the results are skewed towards the positive impacts of telehealth video-conferencing services. A subsequent phase of the study will explore the perception of those directly involved in the providing and receiving of THVCs, i.e. patients and their families and health professionals.

Acknowledgements. The participation of the NZ Telehealth Forum members is greatly appreciated. Nicola Green acknowledges the support of Massey University, by means of the Massey University Conference Presentation Grant.

References

1. World Health Organization (2010) Telemedicine: opportunities and developments in Member States: report on the second global survey on eHealth. Geneva. http://www.who.int/goe/publications/goe_telemedicine_2010.pdf
2. Bradford NK, Caffery LJ, Smith AC (2016) Telehealth services in rural and remote Australia: a systematic review of models of care and factors influencing success and sustainability. Rural Remote Health 16(4):3808 (online)
3. van Dyk L (2014) A review of telehealth service implementation frameworks. Int J Environ Res Public Health 11(2):1279–1298
4. New Zealand Telehealth Forum (2014) New Zealand Telehealth Stocktake: District Health Boards. www.telehealth.co.nz
5. World Health Organization (2016) Global diffusion of eHealth: making universal health coverage achievable. Report of the third global survey on eHealth. Geneva. http://www.who.int/goe/publications/global_diffusion/en/
6. Akiyama M, Yoo BK (2016) A systematic review of the economic evaluation of telemedicine in Japan. J Prev Med Public Health 49(4):183–196
7. Wade VA, Karnon J, Elshaug AG, Hiller JE (2010) A systematic review of economic analyses of telehealth services using real time video communication. BMC Health Serv Res 10:233
8. Müller KI, Alstadhaug KB, Bekkelund SI (2016) Acceptability, feasibility, and cost of telemedicine for nonacute headaches: a randomized study comparing video and traditional consultations. J Med Internet Res 18(5):e140
9. Wootton R, Bahaadinbeigy K, Hailey D (2011) Estimating travel reduction associated with the use of telemedicine by patients and healthcare professionals: Proposal for quantitative synthesis in a systematic review. BMC Health Serv Res 11:185
10. Birns J, Roots A, Bhalla A (2013) Role of telemedicine in the management of acute ischemic stroke. Clin Pract 10(2):189–200
11. Moffatt JJ, Eley DS (2010) The reported benefits of telehealth for rural Australians. Aust Health Rev 34(3):276–281
12. Zanaboni P, Wootton R (2012) Adoption of telemedicine: from pilot stage to routine delivery. BMC Med Inform Decis Mak 12(1):1
13. Eason K, Waterson P, Davda P (2014) The sociotechnical challenge of integrating telehealth and telecare into health and social care for the elderly. In: Healthcare administration: concepts, methodologies, tools, and applications, vol 3, pp 1177–1189
14. Hendy J, Chrysanthaki T, Barlow J, Knapp M, Rogers A, Sanders C, Bower P, Bowen R, Fitzpatrick R, Bardsley M, Newman S (2012) An organisational analysis of the implementation of telecare and telehealth: the whole systems demonstrator. BMC Health Serv Res 12(1):403

15. Brewster L, Mountain G, Wessels B, Kelly C, Hawley M (2014) Factors affecting front line staff acceptance of telehealth technologies: a mixed-method systematic review. J Adv Nurs 70(1):21–33
16. Kruse CS, Karem P, Shifflett K, Vegi L, Ravi K, Brooks M (2016) Evaluating barriers to adopting Telemedicine worldwide: a systematic review. J Telemed Telecare 0(0):1–9
17. McLean S, Sheikh A, Cresswell K, Nurmatov U, Mukherjee M, Hemmi A, Pagliari C (2013) The impact of telehealthcare on the quality and safety of care: a systematic overview. PLoS ONE 8(8):e71238
18. Demiris G, Charness N, Krupinski E, Ben-Arieh D, Washington K, Wu J, Farberow B (2010) The role of human factors in telehealth. Telemed J E-health Off J Am Telemed Assoc 16(4):446–453
19. Hignett S, Carayon P, Buckley P, Catchpole K (2013) State of science: human factors and ergonomics in healthcare. Ergonomics 56(10):1491–1503
20. Bazeley P (2009) Analysing qualitative data: more than 'identifying themes'. Malays J Qual Res 2:6–22
21. Gale NK, Heath G, Cameron E, Rashid S, Redwood S (2013) Using the framework method for the analysis of qualitative data in multi-disciplinary health research. BMC Med Res Methodol 13:117
22. Carayon P, Schoofs Hundt A, Karsh BT, Gurses AP, Alvarado CJ, Smith M, Brennan PF (2006) Work system design for patient safety: the SEIPS model. Qual Saf Health Care 15 (suppl. 1):i50–i58
23. Ammenwerth E, Iller C, Mahler C (2006) IT-adoption and the interaction of task, technology and individuals: a fit framework and a case study. BMC Med Inform Decis Mak 6:3
24. Greenhalgh T, Shaw S, Wherton J, Vijayaraghavan S, Morris J, Bhattacharya S, Hanson P, Campbell-Richards D, Ramoutar S Hodkinson I (2018) Real-world implementation of video outpatient consultations at macro, meso, and micro levels: mixed-method study. J Med Internet Res 20(4):e150
25. Cresswell K, Sheikh A (2013) Organizational issues in the implementation and adoption of health information technology innovations: an interpretative review. Int J Med Inform 82(5): e73–e86
26. Holden RJ, Carayon P, Gurses AP, Hoonakker P, Hundt AS, Ozok AA, Rivera-Rodriguez AJ (2013) SEIPS 2.0: a human factors framework for studying and improving the work of healthcare professionals and patients. Ergonomics 56(11):1669–1686

Tailored Patient Experiences

A Research Through Design Study

Bob Groeneveld[1(✉)], Marijke Melles[1], Stephan Vehmeijer[2], Nina Mathijssen[2], Lisanne van Dijk[1], and Richard Goossens[1]

[1] Faculty of Industrial Design Engineering, Delft University of Technology, Landbergstraat 15, 2628 CE Delft, The Netherlands
b.s.groeneveld@tudelft.nl
[2] Orthopaedics Group, Reinier de Graaf Hospital, Reinier de Graafweg 5, 2625 AD Delft, The Netherlands

Abstract. To achieve optimal patient-centered care for people undergoing a Total Hip Arthroplasty (THA), communication should ideally be tailored. In previous studies, three clusters of patients or patient 'roles' were identified based on communication preferences and clinical and psychological characteristics as a starting point for tailored communication in orthopedics. The current study aims to formulate initial guidelines for the design of tailored communication and information provision based on these roles. Two design cases were each evaluated as storyboards with twelve patients (three, seven, and two patients of each role, respectively). Generic and functionality-specific preferences were indicated by participants for both design proposals. Similarities in feedback per role provided the basis for generating an initial set of role-specific guidelines, that can be used to design tailored information and communication solutions.

Keywords: Communication design · Healthcare · Patient experience

1 Introduction

People undergoing a Total Hip Arthroplasty (THA) find communication with healthcare professionals and information provision important [1, 2]. To design effective communication support, a holistic, user-centered approach is essential [3]. However, most healthcare products or services that support information provision and communication are designed as one-size-fits-all solutions. Tailored solutions can contribute to patient-centered care [4], this way enhancing patient engagement and quality of care [5, 6].

As a starting point for developing tailored solutions in hip surgery, previous studies determined clusters of patients who are similar in their needs regarding communication and in their psychological and clinical characteristics. Quantitative data was gathered in a survey (n = 191) and included socio-demographic, psychological, and surgery-related characteristics, as well as communication preferences [7]. This survey resulted in three subgroups or 'roles' of THA patients: The 'optimistic' role, the 'managing' role, and the 'modest' role. A subset of survey participants was included in subsequent generative research [8]; in four sessions, 19 patients in total constructed and shared

S. Bagnara et al. (Eds.): IEA 2018, AISC 818, pp. 198–207, 2019.
https://doi.org/10.1007/978-3-319-96098-2_26

their experiences in the past and hopes for the future [9]. The resulting qualitative insights were aggregated into personal values and design leads for each of these roles.

These insights are expected to be useful for designing products or services that are tailored to each sub group. Next step is to investigate how these insights will benefit the design process. Furthermore, differences between individual patients still have to be done justice to in practice [6, 10]. Finally, the resulting design should be acceptable for the healthcare professional as well.

This study investigates how patient roles can be embedded in tailored products or services to support information provision and communication. Two design proposals are evaluated in a narrative way with patients from each patient role. The insights gathered from these evaluations are summarized as design recommendations or guidelines for each patient role and design case. In addition, we reflect on the differences and similarities between the guidelines for each design case.

2 Methods

In this study, the process of designing and doing user research (using prototypes, gathering data through interviews and observation) is combined in a so-called Research through Design setup [11]. Specifically, knowledge is gathered through creation of design proposals and evaluating these proposals with stakeholders. Two interventions were studied: The first design is called BiConnect, which is an information application that supports the communication between patient and physician during consultations. It also aims to help in managing patient expectations of the period after surgery. The second design is a rehabilitation device called BioCoach. This product-service system supports outpatients by providing feedback on rehabilitation exercises at home. Exercise data generated by the BioCoach can also be used to support meetings with e.g. a physiotherapist. These interventions were developed on a generic embodied level in earlier research. An impression of these designs is provided in Fig. 1.

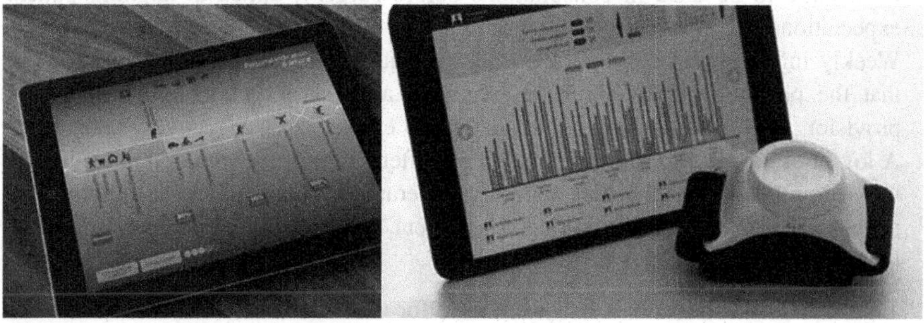

Fig. 1. Impression of BiConnect application (left), BioCoach application and leg band (right).

Further elaboration of both designs is done in an iterative process, where insights from user testing and evaluation are used to improve preliminary designs [12]. Prototype testing is key in this process. In the early stages of the design process such prototypes can be storyboards depicting interaction or use; many insights about patient preferences can be gathered through narrative evaluation of such storyboards [13]. This study focuses on storyboard evaluation of both design proposals.

2.1 Participants

Storyboards of both designs were evaluated with twelve THA patients, with multiple patients from each 'role'. Which 'role' a patient belonged to was determined using a survey developed in earlier research. Participants were selected deliberately to have as much variation in roles as possible. Three patients were classified as having an 'optimistic' role, seven patients as having a 'managing' role, and two patients as having a 'modest' role. Eleven of these patients participated in the survey and generative research described above [7, 8]; one additional patient was recruited as part of one author's research project (LvD). This participant was in the seventh week of recovery, whereas all other participants had their surgery at least six months ago. Five participants were male, average participant age was estimated at around 75 years.

2.2 Storyboards

The two designs were elaborated as storyboards depicting interaction and possible functionalities of the two design proposals.

BiConnect Proposal. This design was adopted in a paper version as a booklet for patients, that informs the patients and can be used to keep track of his or her rehabilitation experiences. This booklet is adjusted based on survey responses to suit his/her preferences. (These adaptations are not yet specified in this scenario.) The following functionalities were incorporated into the booklet and evaluated:

1. A timeline to be used in consultations before surgery, aimed at aligning patient expectations.
2. Weekly information in the booklet (possibly augmented with online information) that the patient receives about his recovery after surgery, to align information provision with needs arising over time and to emphasize that recovery takes time.
3. A log book to fill in during the first weeks after surgery, and the option to discuss this with a healthcare provider (e.g. physiotherapist). This is aimed at monitoring a patient's progress, and reassuring the patient that rehabilitation takes time and there's often no need to worry about this.

Figure 2 shows the visuals that were used for this scenario.

Fig. 2. Storyboard for BiConnect proposal: A patient receiving a booklet (top left); using a timeline in consultation with a healthcare provider aimed to align expectations and answer questions (top right); examining information and filling in a log book in the first weeks after surgery (bottom left); and discussing log book insights with a healthcare provider (e.g. physiotherapist, bottom right)

The BioCoach. The storyboard developed for the second product focused on four different functionalities that the BioCoach fulfils:

1. Tracking and mapping exercise activity, aimed at promoting an optimal amount of exercises over time and to record activity progress.
2. A dial on the product to indicate pain during or after exercises, including feedback (e.g. 'is this pain level normal'); this function could support pain management.
3. Opportunity for digital communication with peers or caregivers, aimed at discussing personal situation, progression and needs, and at providing motivation during rehabilitation.
4. Motivational feedback based on exercise patterns, to reinforce training behaviour and provide reassurance.

Figure 3 shows the visuals that were used for this storyboard.

Both storyboards included several visuals to introduce context and the general idea behind both designs. A general 'THA patient journey' was also shown, in which these scenarios were contextualized. This overview served as an articulation of the researchers' assumptions and knowledge of the THA process, and as a starting point for conversation with the participant.

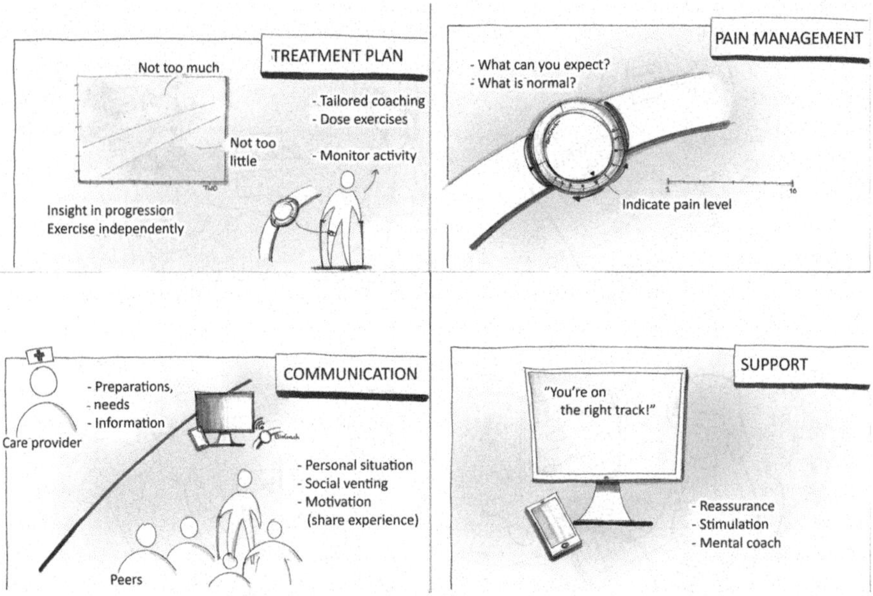

Fig. 3. Storyboard for the BioCoach study: Tracking and mapping exercise activity (top left), a dial on the product to indicate pain during or after exercises, including feedback (top right), opportunity for digital communication peers or caregivers (bottom left), motivational feedback based on exercise patterns (bottom right; text replaced for legibility)

2.3 Procedure

Each scenario was introduced and shown to the participants, with a short explanation per slide; the BiConnect scenario was always shown first as this product was introduced earlier in the patient journey. In the BiConnect case, the entire scenario was shown first and then questions were asked to the participant. In the BioCoach case, several questions were asked after presenting each individual functionality. Questions included: "What is your overall impression of this product proposal/function? What do you think of [specific functionality] in this proposal? How would you like to use [specific function] in this proposal? How could we further adapt this proposal to your needs and preferences?" Participants could freely discuss any comments on the scenarios and other associations they had that were relevant to the presented materials. Figure 4 presents an impression of the evaluations.

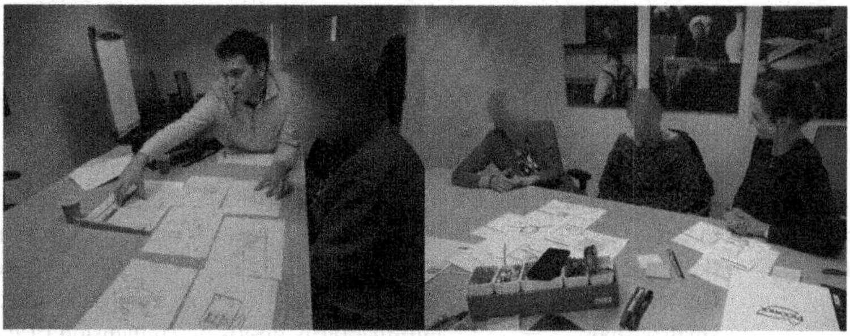

Fig. 4. An impression of the scenario evaluations. On the right, a participant was joined by her neighbour who travelled together with her to the session.

2.4 Data Analysis

A general inductive approach was used to analyse the data. For the BiConnect or patient booklet scenario, answers of each patient were digitalized in a table. Participant answers were grouped based on the three roles and specific functions or generic preferences for e.g. communication style throughout the booklet. These aggregations were summarized into preliminary guidelines, and similarities and differences were compared between the roles. For the BioCoach proposal, notes were digitalized and structured based on the four functionalities proposed (see also Fig. 4) as well as generic preferences and other comments. Detailed responses can be requested at the corresponding author.

3 Results

Overall, the conversations were experienced as vivid and rich. Participants seemed to enjoy the opportunity to provide feedback on the design proposals, and in most cases would like to use either (part of) one or both designs in future care. Tables 1 and 2 outline design guidelines based on user feedback for each product, functionality and role.

Table 1. Generic and function-specific guidelines for each patient role for the BiConnect proposal, based on participant feedback.

Product aspect or function	Optimistic role (n = 3)	Managing role (n = 7)	Modest role (n = 2)
Generic style or tone of communication in booklet	Use positive, but strict tone (e.g. 'You have to …, or else …'; 'keep going!'); Complement text with small visuals	Use upbeat, positive tone; emphasize positive stories; use cheerful visual style; emphasize that information is up-to-date	Include simple, straightforward information and humorous elements; Emphasize affective dimension of care & patient experience
Function 1: Timeline	Include recovery scheme for comparison: 'Am I on track'	Emphasize that recovery takes time; Avoid potentially irrelevant information	(Apply adaptations based on generic guidelines above)
Function 2: Weekly information	Encourage patient to ask questions when needed	Include pain management information (e.g. on medication)	Include stories of comparable patients
Function 3: Log book	Facilitate that patient can see his/her progression over several weeks	Include checklists for e.g. arranging transfer to home; Include open fields to write down experiences; Facilitate that patient can see his/her progression over several weeks	Use short questions/answers (e.g. more box-ticking or indications on scale)

3.1 Overarching Guidelines for Tailored Communication Tools

Generic and function-specific preferences were indicated by participants for both, leading to guidelines for communication tools tailored to the patient roles; managing, optimist and modest. Several similarities were observed between the feedback of the three patient groups on both design proposals. These overarching guidelines are described below.

The 'Optimistic' Role. For the 'optimistic' role, participants would like products or services to feel *positive, but realistic or at times strict* in terms of information provision or interaction. Also, participants liked to gain *insight into their rehabilitation process*, and if possible the *reassurance that they were 'on track'*. In the design cases, if patients have the insight that they are doing well, they may simply require little further support.

The 'Managing' Role. A *friendly or upbeat interaction* was preferred, but it also seemed that participants wanted *accurate and up-to-date or trustworthy information*. In this managing group, there seemed to be a slightly higher need to *be clearly told that rehabilitation takes time*; insight in the treatment plan might help to meet this need. Furthermore, for both design proposals patients indicated that they would like to have *specific information when they need it*; for instance when they experience high pain, have no spouse at home, or when they have specific questions for a care provider. It seems that these participants want to have *initiative* to determine which information they acquire, and when they do so.

Table 2. Generic and function-specific guidelines for each patient role for the BioCoach proposal, based on participant feedback. Text (in brackets) indicates low preference for the given functionality.

BioCoach aspect or function	Optimistic role (n = 3)	Managing role (n = 7)	Modest role (n = 2)
Information level	Provide right amount of information, no overload; realistic view on recovery	Provide friendly formulated, sufficient information to be well-prepared	Provide clear, accessible information: Simple language and guidance
Interaction qualities	Realistic, practical, positive	Controlled, trustworthy, friendly	Simple, consistent, guiding, empathic
Function 1: Tracking and mapping exercise activity	Create insight in progression: Motivate or slow down patient	Create possibility to exercise independently of others; give insight into treatment plan; also motivating or slowing down (similar to role 1)	Give insight in progression, positive feedback
Function 2: Pain dial on product	(Give pain information when needed)	Provide information on medication. (Prepare for pain experience, showing what is 'normal')	Give advice about medication use. Prepare patient for pain experience
Function 3: Opportunity for digital communication with peers or caregivers	(Facilitate peer contact)	Facilitate contact with one specific care provider when needed	Provide option to digitize advice of care provider, to see it again. (Peer experience sharing)
Function 4: Motivational feedback based on exercise patterns	(Use function 1 for reassurance and support; positive comparison with others may help. May also help with acceptance.)	Give confirmation and reassurance. Show positive stories to make rehabilitation more pleasant. (Send messages to limit uncertainty)	Provide support, security, limit feelings of anxiety. Show a face or positive visual icons during contact

The 'Modest' Role. Finally, participants with characteristics from the 'modest' role were also similar in their preferences for both design proposals. They preferred *accessible information*. The *need to be taken seriously* (proposal 1) and the *need for guidance in rehabilitation* (proposal 2) were also seen as similar, reflecting perhaps a need to be able to *rely on care providers and to have them close-by.*

4 Discussion

This study aimed to formulate initial guidelines for the design of tailored communication and information provision solutions in total hip arthroplasty (THA) rehabilitation based on three patient roles. Two design cases were evaluated as story-boards with twelve patients that, based on earlier research, could be grouped into three different 'roles' (three, seven, and two patients of each role, respectively). Generic and function-specific preferences were indicated by participants for both design proposals; This lead to a preliminary set of generic guidelines for the development of communication products and services that fit the preferences of the different patient roles.

Several similarities can be observed between the feedback of the three patient groups and earlier research [7]. For the 'optimistic' role, findings seem to align with low reported feelings of anxiety and perhaps little need for coping in general, as they already seem quite satisfied with care. Patients in the 'managing' role appear to have high communication abilities and needs, and they experience more pain and higher anxiety; this seems to be in line with the initiative and insights they seem to desire in both design proposals. Finally, participants with characteristics from the 'modest' role stressed the need for accessible information which can be related to relatively lower education levels. Both the need to be taken seriously (proposal 1) and the need for guidance in rehabilitation (proposal 2) can be seen as reflections of their relatively high anxiety, and the related higher need for emotional support by healthcare providers. This might also explain why they preferred the pain management function in the BioCoach to prepare them for the experienced pain after surgery, and why they would like to receive positive feedback in this design (functions 1 and 4) as well.

4.1 Limitations and Further Work

Whereas it was possible to formulate design guidelines for the three patient roles find similarities between the two proposals for each role, this study has several limitations. First, sample sizes of individual roles were small, especially for the 'modest' role with only two participants interviewed. This was a formative user evaluation in which insights collected are part of a design process so there is no formal requirement for sample sizes, but it is suggested to involve around five participants from a homogenous group in such studies [13] (p. 91). Furthermore, it is uncertain whether these guidelines lead to design proposals that patients from each role actually prefer. This will be evaluated in future research through user research with working prototypes.

4.2 Conclusion

In healthcare, one-size-fits-all communication is not necessary. But the wheel can't be reinvented for every single patient as well. This study suggests design guidelines for three different THA patient roles, to adjust information products to differences between patients meaningfully. This will contribute to tailoring communication in healthcare, which should be beneficial for patients and the healthcare system alike.

Acknowledgements. This work was supported by the Netherlands Organisation for Scientific Research (NWO) and Zimmer Biomet Inc [grant number 314-99-118].

References

1. Zwijnenberg NC, Damman OC, Spreeuwenberg P et al (2011) Different patient subgroup, different ranking? Which quality indicators do patients find important when choosing a hospital for hip- or knee arthroplasty? BMC Health Serv Res 11:299. https://doi.org/10.1186/1472-6963-11-299
2. De Boer D, Delnoij D, Rademakers J (2010) Do patient experiences on priority aspects of health care predict their global rating of quality of care? A study in five patient groups. Heal Expect 13:285–297. https://doi.org/10.1111/j.1369-7625.2010.00591.x
3. Buckle P, Clarkson PJ, Coleman R et al (2006) Patient safety, systems design and ergonomics. Appl Ergon 37:491–500. https://doi.org/10.1016/j.apergo.2006.04.016
4. Goodwin C (2016) Person-centered care: a definition and essential elements. J Am Geriatr Soc 64:15–18. https://doi.org/10.1111/jgs.13866
5. Randström KB, Asplund K, Svedlund M (2013) "I have to be patient"-a longitudinal case study of an older man's rehabilitation experience after hip replacement surgery. J Nurs Educ Pract 3:160–169
6. Cott CA (2004) Client-centered rehabilitation: client perspectives. Disabil Rehabil 26:1411–1422. https://doi.org/10.1080/09638280400000237
7. Dekkers T, Melles M, Mathijssen NMC, de Ridder H (n.d.) Connecting preferences to satisfaction: three profiles of total joint replacement surgery patients
8. Groeneveld BS, Melles M, Mathijssen NMC et al (n.d.) Exploring patient experiences in Total Lower Joint Arthroplasty through generative research
9. Visser FS, Stappers PJ, van der Lugt R, Sanders EB-N (2005) Contextmapping: experiences from practice. CoDesign 1:119–149. https://doi.org/10.1080/15710880500135987
10. Wolf JA, Niederhauser V, Marshburn D, Lavela SL (2014) Defining patient experience. Patient Exp J 1:7–19
11. Stappers P, Giaccardi E (2017) Research through design. In: Soegaard M, Dam R (eds) The encyclopedia of human-computer interaction, 2nd edn. Interaction Design Foundation, Aarhus
12. Roozenburg NFM, Eekels J (1998) Productontwerpen: Structuur en methoden [Product Design: Fundamentals and methods], 2nd edn. Lemma, The Hague
13. Wiklund ME, Kendler J, Strochlic AY (2011) Usability testing of medical devices, 2nd edn. CRC Press, Boca Raton

Analysis of Bio-signal Data of Stroke Patients and Normal Elderly People for Real-Time Monitoring

Damee Kim[1,2], Seunghee Hong[1,2], Iqram Hussain[1,2,3], Young Seo[1], and Se Jin Park[1,2,3](✉)

[1] Korea Research Institute of Standards and Science, Daejeon, South Korea
[2] Electronics Telecommunication Research Institute, Daejeon, South Korea
[3] University of Science and Technology, Daejeon, South Korea

Abstract. We have recently studied the rapidly increasing stroke in the elderly. Stroke focuses on extracting meaningful variables for early diagnosis because early diagnosis has a strong influence on the survival probability. Therefore, we proceeded as follows. We measured vital signs and motion data from 80 stroke patients and 50 normal elderly. This study is part of a study to compare the data patterns of the elderly people by measuring daily life data, motion data, body pressure, EEG (electroencephalogram), ECG (Electrocardiogram), EMG (electromyography), GSR (galvanic skin reflex) data of stroke patients. We experimented with scenarios (walking, moving objects, sitting, etc.) to get natural daily data from stroke patients.

We found that the data of the stroke patients and the normal elderly group were clearly differentiated by the R-R interval parameter of the ECG data and the brain wave data of the frontal and temporal lobe among the EEG data ($p < 0.05$). These features are analyzed to develop algorithms that can detect strokes early, compared with the conventional NIHSS questionnaire to determine stroke patients or the way physicians diagnose. In addition, the bio-signal data is extracted from the experiment, and a judgment model is established by taking the data of the participant's 10-year health examination together. This data includes various screening data such as height, smoking, exercise, triglyceride, LDL-cholesterol, and HDL-cholesterol.

In addition to analyzing these vital signs and analysis data, we are analyzing the cohort data of 2.5 million health checkup patients in the stroke patients group to improve the accuracy of the diagnostic algorithm by extracting the factors influencing the stroke.

The purpose of our research is to detect stroke in advance using big data and bio-signal analysis technology, and contribute to human health promotion. The data we are measuring is the data that elderly people often live in daily life. In this experiment, data were measured with professional measurement equipment, but items measurable in wearable device were selected for future service commercialization. Because, in the future, the patient must be informed about his or her health condition before going to the hospital. Therefore, if we introduce the early detection algorithm of stroke, we think that many people will be able to detect the stroke early and save many lives without going to the hospital.

S. Bagnara et al. (Eds.): IEA 2018, AISC 818, pp. 208–213, 2019.
https://doi.org/10.1007/978-3-319-96098-2_27

Keywords: Internet of Things · Elderly healthcare · Brain stroke
Real-time monitoring

1 Introduction

The Internet of Things (IoT) performs a significant role in the development of smart home, smart vehicles, which offers cloud connectivity, vehicle-to-vehicle interaction, smartphone integration, safety, security, and e-healthcare services. Recent development trends show that healthcare manufacturers are already paying attention to develop IoT healthcare system that could integrate health status and work safety. Several researches are going on in order to develop smart health monitoring system in daily regular activities such as driving, smart home etc. [1, 2]. Aging originates from increasing longevity, and results in deteriorating fertility [3]. Population aging is taking place in nearly all the countries of the world. As age increases, older people become more conservative in performance. Age-related decline in cognitive function hampers safety and quality of life for an elder. The aged population in the developed world is increasing. Among of all health complexity during driving, stroke is the top one. Stroke is the sudden collapse of brain cells due to lack of oxygen, caused by blockage of blood flow to the brain or breakdown of blood vessels. Stroke is the second top reason of death above the age of 60 years, and its proportion is rising [4]. Many health abnormality happens after stroke. Postural disorders is observed as one of the most common disabilities after stroke.

The purpose of our research is to detect stroke in advance using big data and bio-signal analysis technology, and provide emergency assistance. The data we are measuring is the data that elderly people often live in daily life. In this experiment, data were measured with professional measurement equipment, but items measurable in wearable device were selected for future service commercialization. Because, in the future, the patient must be informed about his or her health condition before going to the hospital. Therefore, if we introduce the early detection algorithm of stroke, we think that many people will be able to detect the stroke early and save many lives without going to the hospital.

2 Model and Methodology

A sensor-based integrated health monitoring system has been proposed in order to measure physiological measurement of elderly stroke patient and normal elderly adults. Elderly drivers' health monitoring system is governed by Hyper-connected self-machine learning engine (Fig. 1). EEG, EMG, ECG, PPG, GSR, Accelerometer, Foot pressure will monitor corresponding physiological signals of elderly adults in daily activities. Face tracking and eye tracking camera are employed in this system in order to detect abnormality in appearance due to stroke or other heart diseases in home and driving condition. All described sensors are able to measure basic physiological parameters for predicting health status of an elderly adults.

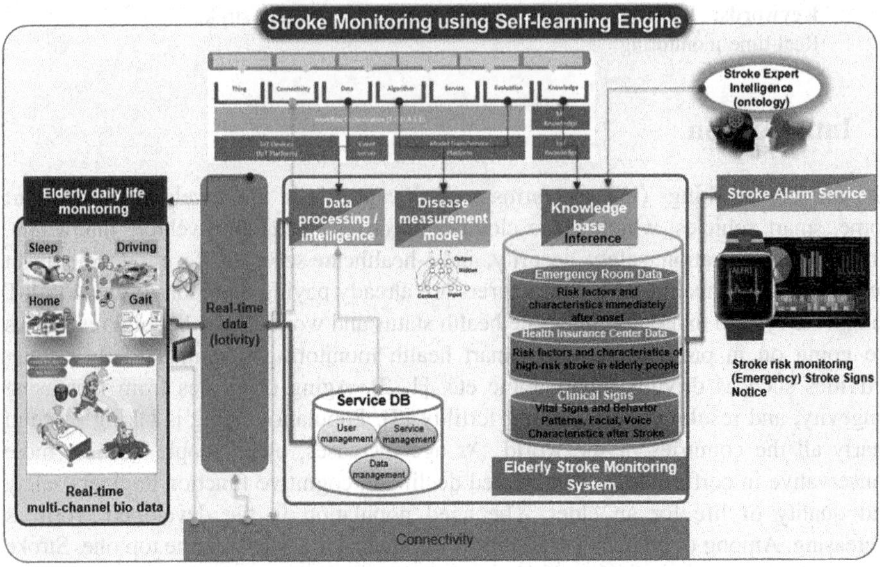

Fig. 1. Model of self-learning engine based elderly stroke monitoring system.

Real-time heath monitoring is most important for detecting stroke onset. Brain Stroke symptoms are active for very small period of time. Most of stroke symptoms automatically eliminates within few moments after stroke. Real-time monitoring system solves this problem. Smart IoT devices and wearable cloths monitor physiological

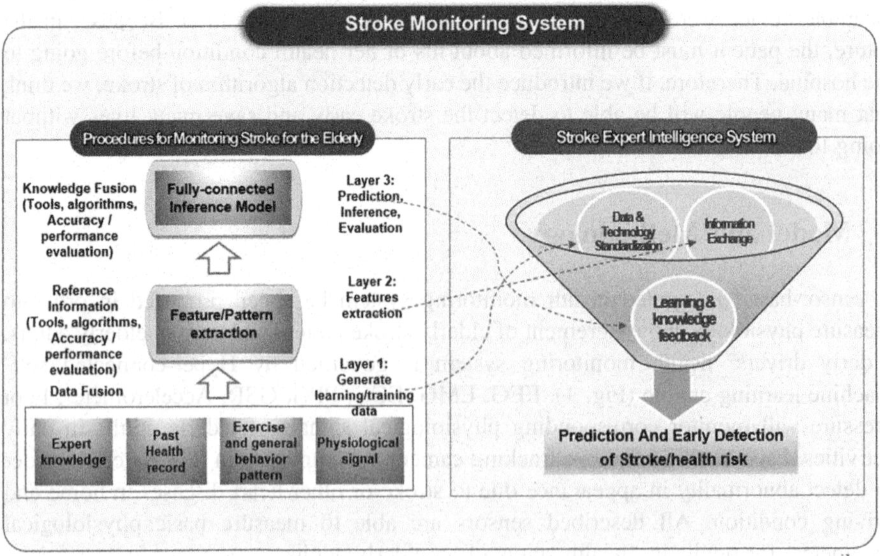

Fig. 2. Stroke monitoring procedures for real-time health monitoring.

parameters in order to understand physiological condition of elderly adults. ECG, Body pressure, Heart beat etc. data will feed to local control system using Bluetooth or infrared wireless network. In next step, local control system will feed physiological data to cloud for further analysis, comparison and response. Stroke monitoring procedures are described in Fig. 2.

3 Framework of Elderly Stroke Monitoring System

Framework of biosensor based elderly health monitoring system is presented in Fig. 3. Real-time data also train-up big data and capture ECG/EMG/heart rate/face pattern that boost up Self-learning engine. For real-time IoT sensor data, MapR Streams is used for scalable data collection, Spark streaming is used for data processing. Processed data is stored using MapR-DB(HBase). In cloud, drivers' health record, normal physiological data have to be stored first as reference data. Real-time ECG/EMG/heart rate will be compared with reference normal data in order to find out health abnormality during driving. Complete system will feed elderly people's physiological data to cloud engine for comparison of real-time data and already stored reference data in order to detect stroke onset of elderly people.

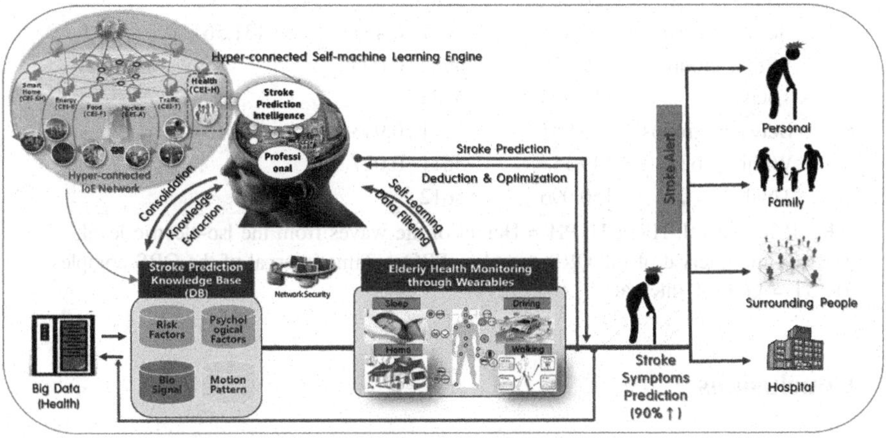

Fig. 3. Framework of elderly stroke monitoring system.

4 Result and Discussion

As shown in Table 1, ECG signal has been analyzed and key features (RRI, PRG, QT, QTC, ST segments) has been extracted for both stroke patients and normal control subjects. Then statistical test, ANOVA (Analysis of Variance) has been conducted. Outcome of ANOVA test has been presented in Fig. 4. Heart rate also showed significant difference between elderly normal adult and elderly stroke patient. System will also predict variation of physiological signal pattern in order to generate an alarm and deliver messages to emergency services, family of the victim, people around the victim, and healthcare professionals in order to ensure the timely medical assistance. Each sensor stroke prediction result contributes own probability in large set of IoT sensor network. The more sensor in health monitoring system, the more reliability of monitoring system.

Table 1. ANOVA test of ECG features between stroke patients and normal control subjects.

ANOVA test: 100 Subjects (50 Stroke patients and 50 non-stroke elderly persons)		Sum of squares	df	Mean square	F	Significance
RBI	Between groups	4.548	1	4.548	195.552	0
	Within groups	139.047	5979	0.023		
	Total	143.594	5980			
PRQ	Between groups	4.222	1	4.222	6.238	0.013
	Within groups	3644.577	5385	0.677		
	Total	3648.799	5386			
QT	Between groups	1.507	1	1.507	346.656	0
	Within groups	24.703	5682	0.004		
	Total	26.21	5683			
QTC	Between groups	0.451	1	0.451	131.564	0
	Within groups	19.472	5682	0.003		
	Total	19.923	5683			
ST	Between groups	0.943	1	0.943	35.464	0
	Within groups	149.153	5611	0.027		
	Total	150.096	5612			

*RRI: R-R interval, HR, RH, PH = Height of the waves from the Iso-electric level, QRS = Time interval of the QRS complex, PRQ = Time interval of the QRS complex, QT, ST = Q-T/ST interval.

5 Conclusions

This paper provides information about framework of Real-time health monitoring such as, stroke detection system using IoT sensors for elderly adults. To evaluate the sensitivity of developed elderly adults' stroke detection system, experimental tests were performed. In addition, ECG/EEG, heart rate sensor can monitor and detect abnormality when brain stroke onset happens. In Overall, ECG, heart rate, Accelerometer, Foot pressure, PPG data can detect abnormal health status of elderly adults during regular activities. In future study would consider a range of bio-sensors and techniques such as image processing of facial condition using face tracking and eye tracking, thermal imaging of elderly adults in order to detect stroke onset during regular daily activities.

References

1. Park SJ, Hong S, Kim D, Seo Y et al (2018) Development of a real-time stroke detection system for elderly drivers using quad-chamber air cushion and IoT devices. SAE Technical Paper 2018-01-0046. https://doi.org/10.4271/2018-01-0046
2. Park SJ, Subramaniyam M, Hong S, Kim D, Yu J (2017) Conceptual design of the elderly healthcare services in-vehicle using IoT. SAE Technical paper (No. 2017-01-1647)
3. Park SJ, Subramaniyam M, Kim SE, Hong SH, Lee JH, Jo CM (2017) Older driver's physiological response under risky driving conditions–overtaking, unprotected left turn. In: Duffy V (ed) Advances in applied digital human modeling and simulation, AISC, vol 481. Springer, Heidelberg, pp 107–114. https://doi.org/10.1007/978-3-319-41627-4_11
4. Park SJ, Min SN, Lee H, Subramaniyam M (2015) A driving simulator study: elderly and younger driver's physiological, visual and driving behavior on intersection. In: IEA 2015, Melbourne, Australia

Prevention of Onycholysis During Cancer Treatment Using an Active Local Cooling Device: Comparison of Three Different Cooling Strategies

Muriel De Boeck(✉), Jochen Vleugels, Katrien Verlaet, Joren Van Loon, Laurens Van Glabbeek, Sam Smedts, Sarah Serneels, Laure Herweyers, and Guido De Bruyne

Department Product Development, University of Antwerp, Ambtmanstraat 1, 2000 Antwerp, Belgium
{muriel.deboeck, jochen.vleugels, guido.debruyne}@uantwerpen.be

Abstract. Nail changes are a common side effect of systemic chemotherapy. Onycholysis is a severe form of nail toxicity in which the nail detaches from the nail bed. Cryotherapy can be effective in preventing chemotherapy induced nail changes as it enables cold-induced vasoconstriction (CIVC), or reduction of blood flow, and therefore limits the transport of chemotherapeutic agents towards the nail beds. Unfortunately, CIVC is inevitably followed up by cold-induced vasodilation (CIVD), causing the blood flow to increase again which reduces the effectiveness of the treatment. Moreover, the local use of extremely low temperatures induce pain and additional distress during chemo treatment. The objective of this article is to examine the usefulness of an active local cooling device for controlling blood flow in the fingertips and reducing CIVD, while limiting pain and discomfort. Three different cooling strategies are evaluated to compare their effectiveness in reducing CIVD. It is hypothesized that pulsating cooling reduces CIVD as it may limit body heat storage during the treatment.

Keywords: Onycholysis · Cold-induced vasodilation · Cryotherapy

1 Introduction

Nail toxicity is a common adverse side effect of chemotherapeutic agents [1]. Chemotherapy induced onycholysis, a severe form of nail toxicity, is characterized by partial or complete detachment of the nail from the nail bed. This medical condition is nearly exclusively associated with chemotherapy chemicals taxane and anthracycline [2], of which taxane is probably the most common anticancer drug implicated in causing nail toxicity [3]. The incidence of nail toxicity induced by cytotoxic agents is reported with up to 44% of the patients [4].

Changes in the nail unit can be associated with pain and functional impairment. Onycholysis may consequently further reduce the quality of life of cancer patients. It is

S. Bagnara et al. (Eds.): IEA 2018, AISC 818, pp. 214–221, 2019.
https://doi.org/10.1007/978-3-319-96098-2_28

therefore important to understand how onycholysis can be prevented during cancer treatment [5].

Nail matrix cells are continuously dividing cells and are the commonest to be affected during chemo treatment [6]. An acute toxic damage of the matrix, as it may occur during chemotherapy, causes a decreased cell survival which can result in nail plate detachment [4]. Cryotherapy, the application of low temperatures as a medical treatment, has been found to decrease the effects of onycholysis and other chemotherapy-induced complications [7]. Localized cryotherapy invokes cold-induced vasoconstriction (CIVC), which causes a decreased blood flow and therefore limits the transport of chemotherapeutic agents towards the nail beds. About 10 to 15 min after the initiation of cold exposure of the hand, the vessels start to dilate again as a protection mechanism of the body to prevent cold-induced tissue damages [8, 9]. This effect is called cold-induced vasodilation (CIVD) and will disrupt the effectiveness of the cryotherapeutic treatment [10]. Furthermore, during prolonged cold exposure, a so-called hunting reaction will occur which is characterized by alternating periods of vasodilation and vasoconstriction [11].

Currently, cryotherapy for the prevention of onycholysis is applied using passive cooling. Examples hereof are gel-filled frozen gloves and socks which are usually refrigerated at −20 °C to −30 °C and worn for 90 min [12]. Since the cooling is applied passively, the temperature inside the gloves and socks does not remain constant and is only low during the first 15 to 20 min [13]. Hence, they are usually replaced by new ones after 45 min. These frozen gloves and socks significantly reduce nail toxicity [14, 15], with Scotté et al. reporting a drop in overall occurrence of nail toxicity from 11% to 51% [12]. However, during such treatments using passive cooling, temperature is not controlled which can lead to the occurrence of CIVD due to persistent local cooling. This method of treatment is also painful, which means that therapy adherence is limited [5]. Research on the understanding of CIVD has been done by Tyler et al. by comparing the thermal responses of the finger to 0 °C and 8 °C water immersion [10]. They concluded that 8 °C may be more suitable when looking to optimise the CIVD fluctuations while minimising participant discomfort. Further, Sawada et al. [16] have studied the occurrence of CIVD at different ambient room temperature conditions. Their results suggest that CIVD reactivity weakens after repeated cooling of the finger in a cooler environment where the body core temperature is liable to decrease. Within this study, core body temperature should not be affected. However, cooling of extremities such as hands may reduce overall body heat loss, such that heat is stored when metabolic heat production is constant. It is therefore hypothesized that CIVD occurs as a reaction to excess heat storage [13].

As opposed to passive cooling, an active cooling system is capable of maintaining a low temperature over longer periods of time, with higher repeatability [13]. Also, active cooling allows continuous control of skin temperature such that the level of cooling can be adjusted during the cooling period. Furthermore, instead of cooling the entire hand, it can be sufficient to cool the area around the nail bed to limit the reduction of heat loss due to extremity cooling [13, 17, 18]. This way, it may be possible to design an effective and less painful method to prevent nail toxicity.

In this study, it is hypothesized that active pulsed cooling prevents heat storage due to extremity cooling and will as such also reduce CIVD, while inducing less pain for patients during the treatment. Three different cooling strategies are evaluated to compare their effectiveness in reducing CIVD.

2 Methods and Materials

2.1 Participants

Twelve healthy subjects – six women and six men, aged between 18 and 24 years – took part in this study. People who suffer from cold intolerance, perniosis, Raynaud syndrome and other cardiovascular diseases were excluded from participating. Each subject was asked not to drink any alcohol the night before the test and ensure a good and regular night's rest.

Before the start of the test, each participant received a document explaining the course of the study. By signing an informed consent document, the participant declared they understand the possible discomforts of participating in the study and that their participation is voluntary.

2.2 Materials and Set-Up

A prototype of an active local cooling system was developed that allows the cooling of the palmar side of the distal phalanx of digitus III (middle finger). Cooling was induced by two Peltier elements and controlled by a microcontroller. The microcontroller was connected to a computer to monitor the data.

The cooling system included three temperature sensors (NTC-thermistors). The first thermistor was placed on the Peltier elements for controlling its temperature. The second thermistor was placed on the nail bed to quantify finger temperature. The third temperature sensor – covered in the fingertip of a latex glove for hygienic reasons – was placed under the armpit as indicator for body temperature. The complete test setup is shown in Fig. 1.

2.3 Protocol

The tests were conducted during three consecutive weeks. Each subject was tested three times in total, with one test per week. The examination room was kept at the same temperature. Participants were asked to wear similar clothes during each test: a long pair of trousers, a T-shirt and a thin sweater. They were not allowed to eat right before or during the test. Each person retained the right to discontinue his or her participation in the study at any time. Before being tested, each person had to acclimatise for half an hour at a room temperature of 21 °C while being seated. Subsequently, the test person was asked to place the middle finger of their dominant hand onto the Peltier elements, whereafter one of the three different cooling strategies was initiated.

2.4 Cooling Strategies

Three cooling strategies (independent variables) were evaluated in this research. In all three strategies, each test started with a period of 10 min where the Peltier elements

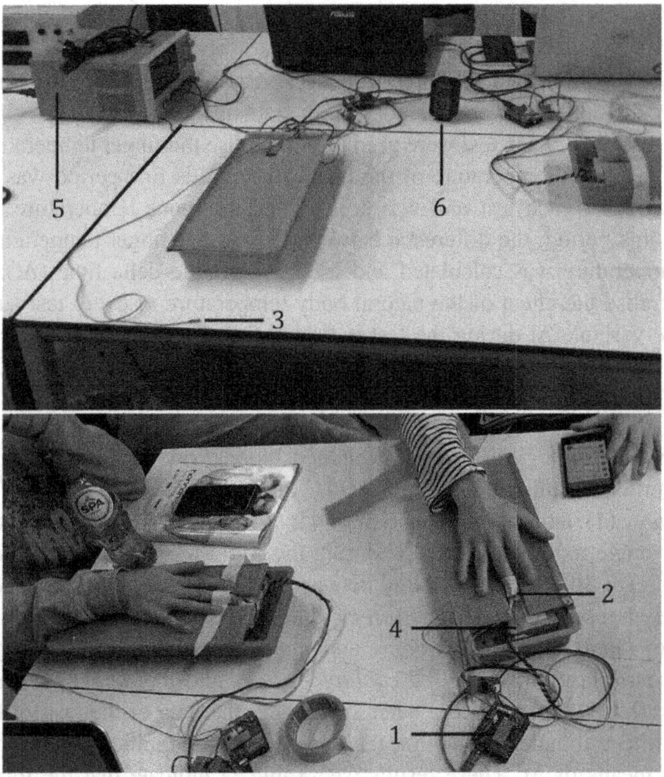

Fig. 1. Test setup consisting of: (1) a microcontroller, (2) a thermistor for the nail bed, (3) a thermistor for the armpit, (4) the active local cooling system within a cooling bath, comprising two Peltier elements + a controlling thermistor, (5) the power supply and (6) a digital room thermometer.

were controlled at a temperature of 20 °C as a reference for statistical analysis. Afterwards, the test continued with one of three strategies: (1) a *linear cooling strategy*, in which the Peltier elements were controlled at 2 °C during a period of 60 min, (2) a *pulsed cooling strategy*, in which two-minute periods of cooling at 2 °C alternated with two-minute periods of 20 °C and (3) a *delayed pulsed cooling strategy*, in which the Peltier elements were controlled at 2 °C during the first 12.5 min, after which strategy 2 was employed for the remaining 47.5 min. The dependent variable is blood flow at the nail bed, quantified as the temperature measured on the dorsal side of the finger. The cooling effectiveness is quantified as the reduction in blood flow during the cooling strategy as compared to the blood flow 10 min prior to the test.

CIVD in the fingertips occurs on average 12.5 min after the start of local cooling. Therefore, it is hypothesised that starting with a cooling period of 12.5 min would induce CIVC, while continuing with pulsed cooling would prevent CIVD.

3 Results

Twelve subjects participated the three test cases, resulting in 36 experiments. Due to technical errors, four tests were excluded from statistical analysis.

The first 10 min of each test were used to standardise the finger temperature of each participant. The mean temperature of the finger during this first period was calculated for each person for each test and was seen as the reference temperature. For every second after this period, the difference between the current finger temperature and the reference temperature was calculated and used as variable delta time (Δt). This was done to neutralise the effect of the natural body temperature of every test subject. The course of the variable Δt during the test indicated the effectiveness of the cooling.

3.1 Effectiveness of Cooling

All three cooling strategies allowed to reduce blood flow at the nail bed, with an average finger temperature reduction of 4.0 °C during the 60 minute-trial of the *linear cooling strategy* (1), an average reduction of 4.2 °C during the *pulsed cooling strategy* (2) and an average reduction of 4.5 °C during the *delayed pulsed cooling strategy* (3).

No significant difference was found between the means of Δt of the *linear cooling strategy* and the *pulsed cooling strategy* during the 60 min ($p = 0.666$); neither between the *linear cooling strategy* and the *delayed pulsed cooling strategy* ($p = 0.524$); nor between the *pulsed cooling strategy* and the *delayed pulsed cooling strategy* ($p = 0.751$). Furthermore, when using only the data of the last thirty minutes and the last fifteen minutes of the test, there were also no significant differences found between the means of Δt. These results (see Table 1) indicate that the three cooling strategies have the same effect on the finger temperature and thus on the occurring amount of CIVC.

Table 1. Repeated measures ANOVA - comparing means of Δt-values

60 min	N	p	μ	μ
Strategy LC-PC	9	0.666	$\mu_1 = -3.951$	$\mu_2 = -4.158$
Strategy LC-DPC	10	0.524	$\mu_1 = -4.043$	$\mu_3 = -4.547$
Strategy PC-DPC	7	0.751	$\mu_2 = -4.212$	$\mu_3 = -3.919$
Last 30 min				
Strategy LC-PC	9	0.791	$\mu_1 = -4.816$	$\mu_2 = -4.525$
Strategy LC-DPC	10	0.582	$\mu_1 = -4.896$	$\mu_3 = -5.547$
Strategy PC-DPC	7	0.641	$\mu_2 = -4.795$	$\mu_3 = -4.189$
Last 15 min				
Strategy LC-PC	9	0.718	$\mu_1 = -5.064$	$\mu_2 = -4.708$
Strategy LC-DPC	10	0.805	$\mu_1 = -5.233$	$\mu_3 = -5.550$
Strategy PC-DPC	7	0.583	$\mu_2 = -4.749$	$\mu_3 = -3.915$

3.2 Amount of CIVD

The average standard deviation of Δt-values is a measure for the fluctuation of the finger temperature and thus indicates the blood flow and the amount of CIVD. The means of the standard deviations were compared each time between two strategies through a repeated measures ANOVA. The results are summarized in Table 2.

Table 2. Repeated measures ANOVA – comparing means of standard deviations of Δt-values

60 min	N	p	μ	μ
Strategy LC-PC	10	0.604	$\mu_1 =$ 1.740	$\mu_2 =$ 1.560
Strategy LC-DPC	10	0.297	$\mu_1 =$ 1.675	$\mu_3 =$ 2.098
Strategy PC-DPC	8	0.380	$\mu_2 =$ 1.373	$\mu_3 =$ 1.701
Last 30 min				
Strategy LC-PC	8	0.027	$\mu_1 =$ 0.491	$\mu_2 =$ 0.202
Strategy LC-DPC	8	0.411	$\mu_1 =$ 0.498	$\mu_3 =$ 0.395
Strategy PC-DPC	7	0.065	$\mu_2 =$ 0.185	$\mu_3 =$ 0.379
Last 15 min				
Strategy LC-PC	8	0.029	$\mu_1 =$ 0.413	$\mu_2 =$ 0.161
Strategy LC-DPC	8	0.015	$\mu_1 =$ 0.321	$\mu_3 =$ 0.188
Strategy PC-DPC	6	0.025	$\mu_2 =$ 0.150	$\mu_3 =$ 0.185

No significant difference ($p > 0.05$) was found between the means of the standard deviations of the three different cooling strategies when calculated with data of the complete sixty minutes. However, there is a significant difference ($p = 0.027$) between the means of the standard deviations when comparing the *linear cooling strategy* with the *pulsed cooling strategy*, with data of the last 30 min of the test. Moreover, using data of the last fifteen minutes of the test, there is a significant difference between all of the cooling strategies ($p < 0.05$). This difference is significant when comparing *linear cooling strategy* ($\mu_1 = 0.413$) with the *pulsed cooling strategy* ($\mu_2 = 0.161$), when comparing the *linear cooling strategy* ($\mu_1 = 0.321$) with the *delayed pulsed cooling strategy* ($\mu_3 = 0.188$) and when comparing the *pulsed cooling strategy* ($\mu_2 = 0.150$) with the *delayed pulsed cooling strategy* ($\mu_3 = 0.185$).

These results indicate that the finger temperature fluctuates significantly less when the *pulsed cooling strategy* or the *delayed pulsed cooling strategy* is applied, compared to the *linear cooling strategy*. Moreover, the *pulsed cooling strategy* gives significantly less fluctuation than the *delayed pulsed cooling strategy*. The results cannot confirm the hypothesis that pulsed cooling results in a reduction of heat storage and thus in a reduction of CIVD. The result do suggest that pulsed cooling may enhance cooling stability.

4 Conclusion

Decreased blood flow may reduce nail toxicity at cancer patients undergoing chemotherapy treatment. The aim of this study is to investigate the effectiveness of an active local cooling device for controlling blood flow in the fingertips and reducing

CIVD, while limiting pain and discomfort. It is hypothesized that pulsed cooling improves control of blood flow at the nail bed as it may limit heat storage during the treatment. Three different cooling strategies are evaluated to compare effectiveness in reducing CIVD: (1) a *linear cooling strategy* in which the cooling was controlled at a temperature of 2 °C for 60 min, (2) a *pulsed cooling strategy* in which two-minute periods of cooling at 2 °C alternate with two-minutes periods of 20 °C and (3) a *delayed pulsed cooling strategy* in which the cooling was controlled at 2 °C during the first 12.5 min, after which strategy 2 was employed for the remaining 47.5 min. Local blood flow at the nailbed was assessed with the use of skin temperature (°C) measurements on the dorsal side of the cooled finger. The cooling effectiveness was quantified as the reduction in blood flow during the cooling strategy as compared to the blood flow 10 min prior to the test, defined as the reference period. It is hypothesised that starting with a cooling period of 12.5 min would induce CIVC, while continuing with pulsed cooling would prevent CIVD.

The results indicate that all three tested cooling strategies allowed reducing blood flow at the nail bed, with an average finger temperature reduction of 4.0 °C during the complete 60 minute-trial of the *linear cooling strategy* (1), an average reduction of 4.2 °C during the *pulsed cooling strategy* (2) and an average reduction of 4.5 °C during the *delayed pulsed cooling strategy* (3). However, no significant differences ($P > 0.05$) were found with respect to the average cooling effectiveness of the three different cooling strategies. These results indicate that the three cooling strategies have a similar effect on the finger temperature and thus on the occurrence of CIVC.

Further, the average standard deviation of cooling effectiveness was measured to indicate the fluctuation of the finger temperature and thus the amount of CIVD. These results show that the finger temperature fluctuates significantly less when the *pulsed cooling strategy* or the *delayed pulsed cooling strategy* is applied, compared to the *linear cooling strategy*. Moreover, the *pulsed cooling strategy* gives significantly less fluctuation than the *delayed pulsed cooling strategy*, indicating that the inclusion of the first cooling period of 12.5 min before starting the pulsed cooling shows no positive effect on the control of the blood flow. In conclusion, pulsed cooling may enhance controlling finger temperature. As such, pulsed cooling may help in designing a less painful therapy for preventing nail toxicity.

References

1. Roh MR, Cho JY, Lew W (2007) Docetaxel-induced onycholysis: the role of subungual hemorrhage and suppuration. Yonsei Med J 48(1):124–126
2. Hussain S, Anderson DN, Salvatti ME, Adamson B, McManus M, Braverman AS (2000) Onycholysis as a complication of systemic chemotherapy: report of five cases associated with prolonged weekly paclitaxel therapy and review of the literature. Cancer 88(10):2367–2371
3. Gilbar P, Hain A, Peereboom VM (2009) Nail toxicity induced by cancer chemotherapy. J Oncol Pharm Pract 15(3):143–155
4. Minisini AM et al (2003) Taxane-induced nail changes: incidence, clinical presentation and outcome. Ann Oncol 14(2):333–337

5. Robert C et al (2015) Nail toxicities induced by systemic anticancer treatments. Lancet Oncol 16(4):e181–e189
6. Shanmugam Reddy PK, Shyam Prasad AL, Sumathy TK, Reddy RV (2017) Nail changes in patients undergoing cancer chemotherapy. Int J Res Dermatol 3(1):49
7. Kadakia KC, Rozell SA, Butala AA, Loprinzi CL (2014) Supportive cryotherapy: a review from head to toe. J Pain Symptom Manag 47(6):1100–1115
8. D'Haene M, Youssef A, De Bruyne G, Aerts J-M (2015) Modelling and controlling blood flow by active cooling of the fingers to prevent nail toxicity [master thesis]. KULeuven
9. Ulrich RS (2001) Effects of healthcare environmental design on medical outcomes. In: Second international conference on design and health, June, pp 49–59
10. Tyler CJ, Reeve T, Cheung SS (2015) Cold-induced vasodilation during single digit immersion in 0 °C and 8 °C water in men and women. PLoS ONE 10(4):1–13
11. Daanen HAM (2003) Finger cold-induced vasodilation: a review. Eur J Appl Physiol 89(5):411–426
12. Scotté F et al (2005) Multicenter study of a frozen glove to prevent docetaxel-induced onycholysis and cutaneous toxicity of the hand. J Clin Oncol 23(19):4424–4429
13. Steckel J et al (2013) A research platform using active local cooling directed at minimizing the blood flow in human fingers. In: Proceedings of ICTs for improving patients rehabilitation research technology, pp 81–84
14. Scotté F et al (2008) Matched case-control phase 2 study to evaluate the use of a frozen sock to prevent docetaxel-induced onycholysis and cutaneous toxicity of the foot. Cancer 112(7):1625–1631
15. Can G, Aydiner A, Cavdar I (2012) Taxane-induced nail changes: predictors and efficacy of the use of frozen gloves and socks in the prevention of nail toxicity. Eur J Oncol Nurs 16(3):270–275
16. Sawada S, Araki S, Yokoyama K (2000) Changes in cold-induced vasodilatation, pain and cold sensation in fingers caused by repeated finger cooling in a cool environment. Ind Health 38:79–86
17. Bladt L et al (2015) Cold-induced vasoconstriction for preventing onycholysis during cancer treatment. Extrem Physiol Med 4(1):A60
18. D'Haene M, Youssef A, De Bruyne G, Aerts J-M (2015) Modelling and controlling blood flow by active cooling of the fingers to prevent nail toxicity

Evaluating Innovations for the Physical Environment in Home Care – A Workplace for One, A Home for the Other

Johanna Persson[1(✉)], Gudbjörg Erlingsdottir[1], Lotta Löfqvist[2], and Gerd Johansson[1]

[1] Ergonomics and Aerosol Technology, Lund University, Lund, Sweden
johanna.persson@design.lth.se
[2] Occupational and Environmental Medicine, Lund University Hospital, Lund, Sweden

Abstract. To a higher degree, older adults will live to old age in their own homes with the assistance of home care services. One effect of this is an increased number of people working in the home environment. This paper presents the results from a research project that studied the physical environment in Swedish home care. Innovations to support the home care situation were developed, with the aim of contributing to patient safety and improving working conditions, without diminishing the homelike atmosphere. These innovations were evaluated by both the care recipients and home care staff. The number of generated ideas shows the potential for improving the way in which home care is performed. One example is piece of storage furniture that addresses many of the identified problem areas. This was well received by the staff and residents. One challenge with the type of products presented is that they are neither direct work tools nor aids for the elderly, which means that in today's organization in Sweden, there is no system for financing them.

Keywords: Home care · Physical environment · Work environment

1 Introduction

The population of Sweden passed the 10 million mark in January 2017, with approximately one fifth of the population over 65 years of age [1]. Many of these older adults will live to an old age in their own homes with the assistance of both homemaker services and health care services, referred to here collectively as "home care". This will also lead to an increased number of people working in the home environment. A high level of patient safety and acceptable working conditions are central for quality home care, but this is challenging to achieve without introducing a hospital-like feeling in the home environment [2]. There is thus room for innovating and designing new products that can be used in home care to improve the physical environment and benefit both the caring staff and care recipients.

This paper presents results from a research project, which aim was to study the physical environment in home care, identify needs and propose products for improvement. The goal was to generate both useful and attractive solutions, and in so

S. Bagnara et al. (Eds.): IEA 2018, AISC 818, pp. 222–231, 2019.
https://doi.org/10.1007/978-3-319-96098-2_29

doing, improving the caring situation, the work environment, as well as maintaining a homelike atmosphere. Results are presented from the third and final phase of the research project in which the generated product ideas are demonstrated and evaluated. The results from the first phase were presented at the 2015 IEA Conference [3], and from the second phase at the 2016 HEPS Conference [4].

1.1 Related Work

The sense of "being at home" and a homelike atmosphere that represents one's identity, integrity and way of living are highly valued by older adults [5]. Despite this, research is scarce on how the home environment is affected by the introduction and deployment of home care. Studies of the determinants for a homelike atmosphere in nursing homes can be found [6, 7]. These studies discuss how common and private spaces can be combined to make the residents feel as comfortable as possible, or how the nursing homes' organization and management strategies influence the sense of feeling at home. When care is provided in a private home instead, the prerequisites are different and the conclusions about common and private spaces do not apply. The majority of the studies on private home settings focus on helping the person to become more independent and less in need of caring staff by introducing welfare technology, such as bidet toilet seats that enable better self-care [8, 9].

Working in home care is related to many risks where hazardous work postures are one major reason for work-related injuries [2, 10]. Other risks are exposure to infectious agents, slips and falls, and physical and/or verbal violence. The prevalence of long-term sick leave, frequency of absenteeism and staff changes are all high amongst home care workers, and pose a future challenge to dealing with the demographic development of an ageing population [8, 11]. Minimizing risks and making this environment less hazardous and more attractive for the work force is often related to introducing more products and aids in the home to support the staff in the caring situation. This in turn has an effect on the homelike atmosphere. Consequently, efforts to combine home care workers' occupational situation with a homelike atmosphere and care recipient independency are challenging.

2 Method

The research project has gone through three phases of which this paper presents the third and final phase. In the first phase, *exploration*, we performed interviews and observations in the home care situation to gain an understanding of the problem areas and needs related to the physical space of the home care environment. This resulted in a list of areas where potential improvements could be made [3]. In the second phase, *innovation*, thorough descriptions of the selected problem areas were presented to Master and Bachelor level students studying Industrial Design, Product Design and Product Development Engineering. They then generated product ideas [4]. From the initial list of areas, the students approached a problem either as part of a course or as a Bachelor or Master level degree project. In the final phase, *evaluation*, the generated product ideas have been evaluated with potential users.

The methods for conducting the student projects differed depending on the background and scope of work, but all projects have followed an iterative design process (Fig. 1) focused on the users' needs, tasks and contexts of use [12]. The first step in such a process is to generate knowledge about the basic problem area by exploring who the users are, what their tasks involve and which contexts the tasks are performed in. This is followed by the innovative phase where ideas are generated and prototypes built. These are then evaluated with the potential users and the outcomes were fed into a new round in the iterative process.

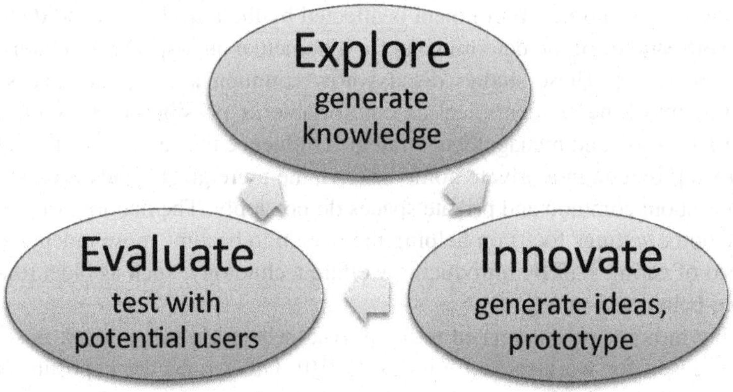

Fig. 1. The research project and all generated product ideas have been developed in an iterative design process.

Working in this way means that the design process is constantly informed through formative evaluation methods. One useful method to generate feedback in a design process is to conduct focus group interviews including sketches or physical prototypes [13]. Focus groups with older adults and home care staff were used at several points in the research project to gain feedback on the product ideas at different stages. Other useful formative evaluation methods that the students used were Pugh matrices, interviews with single users, and user testing in which potential users performed typical tasks with the prototype to understand what needed to be changed in the next iteration.

A summative evaluation was performed at the end of the research project by the researchers to gain a more comprehensive assessment of the products generated. It consisted of two parts: one focused on the staff, and one on the care recipients and their relatives. To obtain the staff perspective, we organized an exhibition and invited the entire elderly care staff in one municipality. During the exhibition the visitors could see, test and discuss prototypes of the product ideas. The participants were also asked to fill in a questionnaire regarding their opinions about the product ideas and how they perceived them as possible work tools as well as how the product ideas could be further developed. The care recipients' opinions were collected through eight semi-structured interviews performed in the home environment.

Two product ideas – a piece of storage furniture and a leg support – were prototyped in versions that could be tested in practice. Two exemplars of the storage furniture were built and placed in two care recipients' homes. The recipients from these two homes were included in a more thorough interview session about the use of the furniture in addition to the interview. The leg support was used in a nursing home and ergonomic measurements of work postures were taken[1]. An overview of the evaluation methods is found in Table 1.

Table 1. Overview of evaluation methods.

	Evaluation method
Formative evaluation	Primarily focus group interviews, but different student projects also used supplementary methods such as Pugh matrices, interviews and user testing
Summative evaluation	An exhibition and a questionnaire to capture the staff perspective
	Semi-structured interviews in the home to capture care perspectives of recipients and relatives
	Usage in two homes of one product prototype: a piece of storage furniture. Evaluated through interviews
	Usage in a nursing home of one product prototype: a leg support. Evaluated through ergonomics measurements

3 Results

A total of 48 product ideas were generated. 47 of these were student projects on different levels, ranging from smaller course projects to Bachelor and Master degree projects. The leg support idea came directly from a nurse working in elderly care. The product ideas addressed the following problem areas: lighting, work posture, medicine handling, packaging, storage of care-related material and surfaces to work on. Examples of product ideas are described below.

Lighting. A flexible and portable lighting solution that has the directional light source needed when giving an injection, for example. It is also designed to look like a piece of jewelry rather than a work tool (Fig. 2). It can either be carried by the staff or be placed in the patient's home and blend in nicely.

Work Posture. A foldable stool that can be used in the home for sitting near the care recipient instead of kneeling on the floor. When not in used, it can easily be folded and placed out of sight (Fig. 3). The concept also included a cover in which the stool could fit when folded and not in use. This was designed such that it could be used as a regular side table by the resident.

[1] The results from these measurements are presented in a poster at the IEA 2018 Conference.

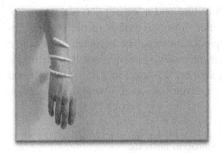

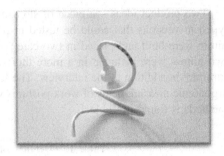

Fig. 2. Flexible and portable lighting designed as a piece of jewelry.

Fig. 3. Foldable stool for better work posture near the care recipient.

Medicine Handling. Different versions of aids used to remove pills from their packaging. This problem area is explicitly focused on hand ergonomics since this process can be very hard on one's thumbs. It was not directly within the scope of the research project, but since it was considered to be a problematic area, it was included in order to find new approaches for solving it. Examples of various products ideas, some to be used in combination with a medicine dispenser, are illustrated in Fig. 4.

Packaging. Disposable material packaging with a more aesthetic design was also addressed. There is much disposable material connected to home care, and different packages with this material are often lying around in the home. Providing more aesthetic packaging is one way to integrate this into the home environment and help preserve a homelike atmosphere. One example is the packaging for disposable gloves that is found in all homes with home care. The example shown in Fig. 5 is designed to

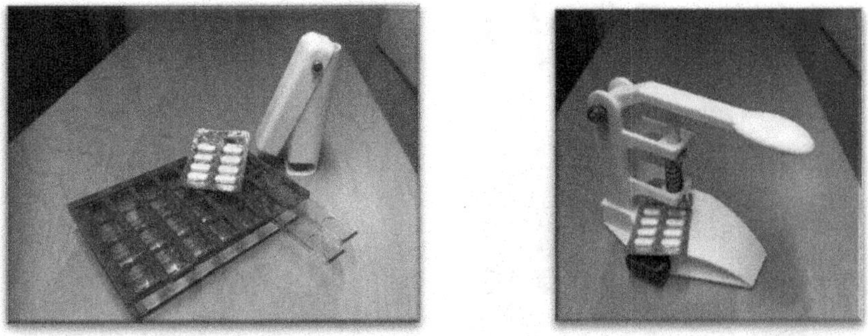

Fig. 4. Ergonomic aids for removing pills from blister packs.

Fig. 5. Aesthetic packaging for disposable material.

be aesthetic, sustainable and to offer usage affordances. The black ribbon encourages the user to close the lid and thereby reduce the risk of contamination.

A few of the most promising ideas were developed into functioning, full-scale models. One example is a *portable tray* with legs (Fig. 6, left) that offers a hygienic working surface and includes storage for a small amount of material. Another example is a *leg support* for dressing leg wounds (Fig. 6, right). The nurse who came up with this design thought that existing leg supports were too heavy to carry or too complex to set up. This leg support is easy to use and transport.

A third example is *a piece of storage furniture* that was developed to store all the material and provide hygienic working surfaces. The amount of material laying around in a home can be immense, and can be both annoying and stigmatic for the care recipient and relatives. A problem for staff is that the material is kept in different places in different home and that it is difficult to quickly inventory. An example of how the home environment may look is shown in the left part of Fig. 7. The piece of storage furniture that was developed is modular and can be combined to suit the needs of each specific care recipient. The standard combination with three modules is shown in the middle part of Fig. 7. It has a large drawer at the bottom, a cabinet in the middle and a cabinet on top with a surface that folds out and can be used to work on. This surface is

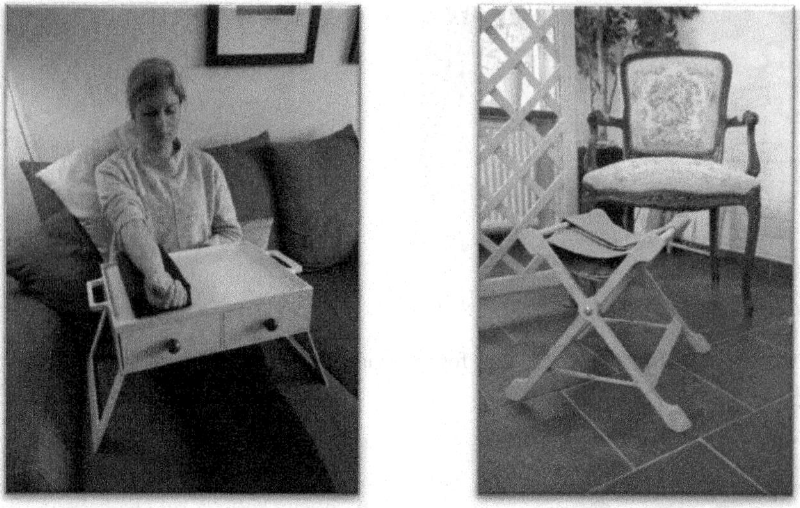

Fig. 6. The picture on the left shows a tray with legs, offering a portable and hygienic working surface including storage for material. The picture on the right shows a leg support designed to be easy to use and transport.

covered by a material that can easily be disinfected. Lighting was also attached, based on feedback from the staff, in order to offer good lighting conditions, as well as to be able to turn the light on during the night without disturbing the care recipient. There is a detachable tray on top that can be used to carry the material. The storage furniture is designed to blend in like a regular piece of furniture into the home and not signal "hospital" in its appearance (Fig. 7, right).

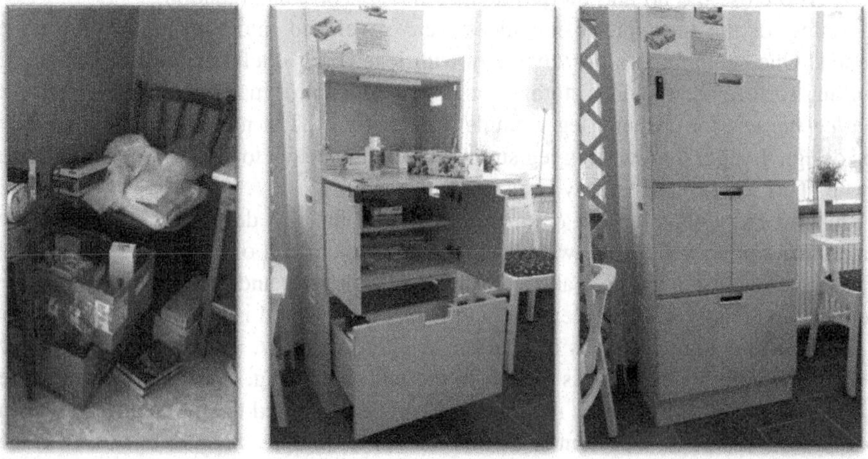

Fig. 7. Example of how the home environment may look with all care-related material lying around (left) and the storage furniture, designed to store all material and offer an ergonomic and hygienic working surface (middle and right).

3.1 Exhibition for Staff

Thirty visitors came to the exhibition, representing home health care nurses, homemaker service aides and nursing home staff. The results presented here are based on comments and observations gathered during the exhibition as well as the questionnaire that was handed out and answered by 10 of the visitors. Figure 8 shows a photo from the exhibition.

Fig. 8. Exhibition for staff.

The overall impression was that the product ideas were interesting and could be useful in the home care situation. A few ideas gained more attention than others did, such as the piece of storage furniture and the leg support. Both homemaker service aides and home health care nurses considered these helpful. The tray with legs (Fig. 6, left), on the other hand, was only considered useful by the home health care staff. The different views of the two groups reflect the different tasks they perform and expresses the nurses' need to bring material with them to the care recipient's side when carrying out some of their tasks.

3.2 Interviews with Care Recipients and Relatives

Interviews with eight residents were conducted to capture their experience of how the home environment was affected by home care. All of them received home health care and some of them also received homemaker services. Two of the interviews involved a relative, and the two prototypes of the storage furniture were in use in two of the homes.

The overall experience is that the home is definitely affected by the home care situation and most residents have taken action to hide the material or at least to collect it in one place. This could mean that part of a closet is dedicated to this material, or a special piece of furniture or storage boxes have been purchased to store the material. In cases where the care recipient suffered from a more severe illness, there was a desire not to be reminded of it all the time, both for the recipient and other family members.

While many pointed out the importance of order and not having the material out in the open where everyone can see it, the ability to get it out of sight is affected by the size of the dwelling and the resident's financial situation. The majority of the interviewees accepted that the home was affected by home care and viewed it as an unattractive but necessary part of the situation. They could even considered adapting more if it would help the staff in their work. The residents and the staff reached an agreeable solution for providing care in the home through collaboration.

When discussing the economic side of this, three alternative models where presented: one in which the care recipient had to purchase the equipment, one where they could rent it from the municipality, and one where they could borrow equipment from the municipality at no charge. The willingness to buy additional equipment or furniture depended on the care recipient's finances, where some had done this already. But it also depended on how long the care would be needed. From an egalitarian perspective, the ability to borrow equipment from the municipality is preferable.

The storage furniture used in two homes was very well received, and the aesthetic aspect was not the only reason for being positive. The main advantage seemed to be that it was easier to obtain an overview of and easy access to the material.

4 Discussion and Conclusion

The number of generated product ideas shows the potential for changing and improving the way in which home care is performed. The care environment does not need to be differentiated based on work environment <u>or</u> home environment. These perspectives can be combined and products that have a positive effect for both parties are possible. The storage furniture is one such example that handles several of the problem areas in one product: it helps organize the care-related material in one place, it offers hygienic and ergonomic working surfaces with good lighting, and it nicely blends into the home.

We are convinced that a user-centered design process in collaboration with care-givers and caretakers has been central to achieving the wide array of products ideas addressing both residents and staff. Further product development and more thorough evaluation is required, however, to generate more knowledge on how the home care environment can be improved. It is definitely a challenging context for which to design. The products need to be easy to use and easy to transport if they are to be handled by the staff. Products that are too complex will not be used even if it means that employees expose themselves to the risk of work related injuries. In addition, the homes and the people living there vary immensely in how they manage their home care situation. While some are very involved in every step together with the staff, others are more passive recipients of care and do not engage in the adjoining activities, such as ordering and organizing material and medicine.

Some of the product ideas generated in this project are neither direct work tools nor aids for the elderly; they fall somewhere in between. This means that there is no system for financing them. Take the piece of storage furniture, for example. Do the staff need it as a work tool to keep things organized and to carry out their work tasks hygienically? Or is it a piece of furniture that the residents need to store their care products? We saw in our interviews that relatively few of the elderly people are willing to invest in new furniture, the only function of which is to support their home care (even though there were examples of some people doing so). Subsequently, the proposed solution is that the municipality owns the products and lends them out to the care recipient as long as they are needed. More knowledge about the impact of home care on the home environment is desirable. Meanwhile, the fundamental approach should be to preserve the homelike atmosphere as far a possible when introducing home care.

References

1. Statistics Sweden (2018) Population statistics
2. National Institute for Occupational Safety and Health (2010) NIOSH Hazard Review: Occupational Hazards in Home Healthcare, D.o.H.a.H. Services, Editor
3. Johansson G et al (2015) Working and living with home care—a workplace for one, a home for the other. In: Lindgaard G, Moore D (eds) 19th Triennial congress of the international ergonomics association. International Ergonomics Association, Melbourne, Australia
4. Persson J et al (2016) Innovations to support people in the physical environment in homecare —a workplace for one, a home for the other. In: HEPS Healthcare and Society, Toulouse, France
5. Gillsjo C, Schwartz-Barcott D, von Post I (2011) Home: the place the older adult cannot imagine living without. BMC Geriatr 11:10
6. de Veer AJ, Kerkstra A (2001) Feeling at home in nursing homes. J Adv Nurs 35(3):427–434
7. Vihma S (2013) Homelike design in care residences for elderly people. In: Hujala A, Rissanen S, Vihma S (eds) Designing wellbeing in elderly care homes. School of Arts, Design and Architecture, Aalto University, Espoo
8. Rechel B et al (2013) Ageing in the European Union. The Lancet 381(9874):1312–1322
9. Hjalmarson J, Lundberg S (2015) Work postures when assisting people at the toilet. Ergon Des 23(2):16–22
10. Hignett S, Edmunds Otter M, Keen C (2016) Safety risks associated with physical interactions between patients and caregivers during treatment and care delivery in home care settings: a systematic review. Int J Nurs Stud 59:1–14
11. Prime Minister's Office Sweden (2013) Future challenges for Sweden—final report of the Commission on the Future of Sweden
12. Kirk D, McClelland I, Fulton Suri J (2015) Involving people in design research. In: Wilson JR, Sharples S (eds) Evaluation of human work. CRC Press, Boca Raton
13. Sharples S, Cobb S (2015) Methods for collecting and observing participant responses. In: Wilson JR, Sharples S (eds) Evaluation of human work. CRC Press, Boca Raton

The Impact of Ventricular Assist Device Therapy on Patients' Quality of Life – A Review

Christiane Kugler(✉)

Faculty of Medicine, Institute of Nursing Science, Albert-Ludwigs-University of Freiburg, Elsässer Str. 2-o, 79106 Freiburg, Germany
christiane.kugler@uniklinik-freiburg.de

Abstract. With prolonged durability of ventricular assist device (VAD) support, as an established treatment strategy for patients with end-stage heart disease, a paradigm shift to focus on psychosocial outcomes becomes necessary. VAD implant procedures can improve heart failure (HF) symptoms and some aspects of quality of life (QoL). Within this article, constructs with importance to the individuals' QoL trajectories while being on long-term VAD support will be addressed. These constructs include physical functioning, professional employment, recreational activities and travel, emotional adjustment, and sexuality for those on durable VAD support. Clinicians should monitor these aspects regularly and should encourage ongoing discussions with the patient and caregiver for treatment options in this regard.

Heart failure (HF) represents a chronic condition, which can be characterized by deteriorating trajectories by those affected [1]. With worldwide growing numbers of HF patients, ventricular assist device (VAD) support has evolved as an alternative treatment strategy for those, which have reached an end-stage within their heart disease therapy [2]. Thus, VAD support holds potential to turn a deteriorating illness trajectory into prolonged survival and improved quality of life (QoL) [2].

QoL can be defined as an 'individual's perception of the impact of the disease and treatment on his/her life' [3]. However, QoL represents a dynamic and multidimensional construct within the framework of VAD therapy [4]. Trajectories of QoL can be influenced in the physical, cognitive, social, emotional, and spiritual domains and should be important to the individual [5]. VAD therapy and dependency on pump function requires a change in and adjustment to self-care needs on a daily basis and may affect an individuals' perception of his/her QoL. Within this article, constructs with importance to the individuals' QoL trajectories while being on long-term VAD support will be addressed.

1 Quality of Life and Physical Functioning

Heart Failure is a muscle wasting disease. Although VADs improve cardiac output and peripheral circulation, exercise intolerance persists in this population [6]. A prospective intervention study in VAD patients showed that age-, and gender-adjusted workload as

S. Bagnara et al. (Eds.): IEA 2018, AISC 818, pp. 232–237, 2019.
https://doi.org/10.1007/978-3-319-96098-2_30

predicted was 42.5% in a VAD cohort at six weeks post-implant and did increase as a result of a supervised reconditioning home ergometry training program to 55.0% during an 18 months period of time [7]. The authors were able to show that improvements in muscle function positively correlated with increases in QoL perceptions in the physical domain [7]. A recent meta-analysis of exercise rehabilitation in VAD patients revealed, that compared to usual care, exercise rehabilitation significantly improved exercise capacity outlined as peak maximal oxygen consumption (VO2max), and six-minute walk test differences [6]. As part of this meta-analysis, six studies assessed QoL, and of these six, four studies reported exercise rehabilitation being beneficial, with no difference observed in two studies [6]. Subsequently, patients after VAD implantation should be referred to cardiac rehabilitation programs in an attempt to improve their physical functioning and QoL.

2 Quality of Life and Professional Employment

Professional integration per se represents a major aspect of an active social life, which can be endangered or even disrupted by end-stage heart disease and subsequent VAD implantation. Despite marked improvements in both, QoL and lifetime expectancy following VAD implantation, only small case studies identified professional employment in patients on VAD support as influencing QoL. Furthermore, data from these case series represented small sample sizes ranging between 10 and 54 patients, short follow-up periods post implant, and almost no specification on how employment status had been measured [8–12]. Employment rates reported for VAD patients have a large variance ranging between 4–60% [8–12]. In the studies available to date, no information on part-time versus full-time employment rates have been given.

In a larger multicenter study, led by Brouwers et al. [11] and a Dutch group, health status and QoL were assessed at 3–4 weeks after implantation, and at 3, 6, and 12 months follow up in 54 VAD patients. Herein, the prevalence rate for employment was 38%, and employment was predictive for better outcomes in terms of QoL. Noteworthy, the authors also pointed out that being employed before VAD implantation was associated with higher employment rates and better QoL scores post-implant. In the grounded theory study by Sandau et al. [13], patient-reported reasoning for professional employment following VAD implantation has been highlighted by the category "being productive" and the associated quote "it does give me a reason to get up in the morning." Potential occupational restrictions, which may include, but are not necessarily exhaustive to infectious and other device-related precautions, need further systematic exploration.

3 Quality of Life and Recreational Activities Including Travel

Being able to participate in recreational activities and to travel are important components of QoL when considering the fact that the majority of VAD implants are performed for long-term bridging or destination therapy. In a recent prospective comparison study [14], the authors compared forty-four patients on long-term VAD support with patients after heart transplantation, HF, and healthy control subjects up to twelve months post-surgery. Herein, VAD patients had the lowest step count and lowest physical activity duration per day, resulting in the lowest active and total energy expenditures (kcal/day), and longest sedentary time per day, in comparison to their counterparts. However, QoL perceptions between the four groups did not differ [14]. Prospective trajectories of physical activity revealed that VAD patients became more active over time [14].

In the literature, there is a scarcity of studies investigating and advising clinicians with regard to travel and travel-related restrictions in VAD patients. Coyle et al. [15] assessed safety and feasibility of national and international travel for destination therapy patients and reported five incidents in fifteen VAD patients, which had traveled 41 long-distance trips by airplane (17), car (23), and cruise (1). Complications occurred included driveline trauma, delay of re-entry into the country of residence (US), missed flight, red heart alarm from bearing wear, and dehydration. All patients returned home safely for routine follow-up. Thus, the authors conclude that long-distance travel is possible, however requires careful planning for safety reasons [15]. Recommendations for air travel, specifically for VAD patients are given by Hammadah et al. [16] in a recent review. VAD patients in a stable clinical status are considered safe to fly. As VAD flow is sensitive to blood volume, the authors recommend (1) maintenance of adequate hydration and (2) avoidance of caffeinated beverages during air travel, (3) carriage of additional sets of fully charged batteries, and (4) copies of medical records, and (5) the VAD card. In addition, VAD patients and their traveling caregivers are advised to instruct the airlines and airline security prior to the trip about their conditions. No interference between VADs and security gates or airplane devices have been reported so far [16]. Nevertheless, patients and clinicians are highly recommended to carefully plan air travel, and to have emergency contact data from their home VAD center and the one in closest proximity to their travel destination. All necessary VAD equipment, medications, and medical records should be transported in the hand luggage.

4 Quality of Life and Emotional Adjustment

Beside the physical recovery, patients' QoL post-implant highly depends on the emotional recovery and adjustment process. Tigges-Limmer et al. [17] describe this process as a challenge due to 'the around-the-clock visibility of life dependency by a pump'. Post-operatively, mental QoL improvements [7, 18, 19], and decreases in symptoms of anxiety and depression have been reported [7, 20]. This phenomenon might be explained, at least in part, by the immediate post-operative reduction of distressing symptoms of severe dyspnea and an increase in exercise capacity, which may diminish ultimate feelings of fear of death.

However, during the long-term adjustment to the device, patients face new challenges including coping with device alarms, the risk of device malfunction, and long-term complications including device-related infections, gastro-intestinal and cerebral bleeding, and subsequent stroke [2, 21]. These stressors can cause an increase of symptoms of anxiety and depression, and post-traumatic stress [22], and other psychiatric comorbidities [23, 24]. Body image assimilation is a clinically commonly observed, but understudied phenomenon in this population [4, 17], which needs further exploration. Nonetheless, the QoL improvements following VAD implantation, careful psychosocial assessment pre-implant, and regular post-implant screening/intervention is necessary for VAD populations and for their caregivers [4, 17, 20, 22]. Professionals with psychosocial and psychotherapeutic skills and a focus on family therapy should become regular members of VAD teams.

5 Quality of Life and Sexuality

Sexual functioning in VAD patients can be impacted by the underlying disease and can be associated with medical therapeutic agents pre- and post-implant. Four studies assessed VAD patients' sexual functioning while being on long-term VAD support [4, 9, 25, 26]. From a methodological perspective, evidence available is limited due to small sample sizes, reports lacking a response rate, diversity in device technologies, and data being primarily driven from non-standardized surveys, limiting generalizability and comparability to build on an evidence-base. In addition, Sandau et al. [13] performed a grounded theory based conceptual definition of QoL for patients on VAD support and identified intimacy as a relevant concept out of the patients' perspective. Interestingly, they placed this category into the social domain of QoL and might have missed that sexuality can be impacted by physical performance, medical agents, and emotional stressors post-implant.

The most recent SALVADOR-study assessed sexual activities in VAD patients and for the first time, considered partners' perspectives with respect to perceptions on illness-related changes in the quality of sexual activity, and its impact on QoL, anxiety, and depression using standardized patient-reported outcome (PRO) instruments. Seventy-two VAD patients and 48 partners participated [4]. Patients and partners expressed interest in sexual activity post-implant, however disturbances occurred due to the device, and were related to battery pockets, the driveline, and changed self-image associated with device-parts [4]. Most importantly, disturbances in sexual activity were independently associated with higher depression rates in patients, and lower mental QoL in partners. Considering these findings, a focus shift and expansion of clinical counseling seems necessary on illness-related changes in the quality of sexual activity including VAD patients and their partners. However, according to a recent review on self-care learning in VAD patients, Kato et al. [27] point out that sexual counseling should also include the use of sexual performance enhancing agents and birth control, if relevant.

6 Conclusions

In conclusion, VAD long-term support necessitates a paradigm and focus shift on QoL and psychosocial elements of care while considering both patients and partners. Available data suggest promising results in terms of QoL improvements for several aspects; however, the majority of these data report exclusively on the short-term recovery period and associated trajectories. Thus, future studies should expand investigations on the long-term psychosocial outcome. Clinicians should monitor psychosocial aspects of QoL routinely and regularly in order to offer timely and tailored interventions for VAD patients and caregivers to support their adjustment to a 'life with quality' while being on the device.

References

1. Roger VL (2013) Epidemiology of heart failure. Circ Res 113:646–659. https://doi.org/10.1161/CIRCRESAHA.113.300268
2. De By TMMH, Mohacsi P, Gahl B et al (2017) The European Registry for Patients with Mechanical Circulatory Support (EUROMACS) of the European Association for Cardio-Thoracic Surgery (EACTS): second report. Eur J Cardiothorac Surg. https://doi.org/10.1093/ejcts/ezx320
3. MacIver J, Ross HJ (2012) Quality of life and left ventricular assist device support. Circulation 126:866–874
4. Kugler C, Meng M, Rehn E et al (2018) Sexual activity in patients with left ventricular assist device and their partners: impact of the device on quality of life, anxiety, and depression. Eur J Cardiothorac Surg 53:799–806
5. Sandau KE, Hoglund BA, Weaver CE, Boisjolie C, Feldman D (2014) A conceptual definition of quality of life with left ventricular assist device: results from a qualitative study. Heart Lung 43:32–40. https://doi.org/10.1016/j.hrtlng.2013.09.004
6. Grosman-Rimon L, Lalonde SD, Sieh N et al (2018) Exercise rehabilitation in ventricular assist device recipients: a meta-analysis of effects on physiological and clinical outcomes. Heart Fail Rev. https://doi.org/10.1007/s10741-018-9695-y
7. Kugler C, Malehsa D, Schrader E et al (2012) Multi-modal intervention in management of left ventricular assist device outpatients—dietary counseling, controlled exercise and psychosocial support. Eur J Cardiothorac Surg 42:1026–1032
8. Morales DL, Argenziano M, Oz MC (2000) Outpatient left ventricular assist device support: a safe and economical therapeutic option for heart failure. Prog Cardiovasc Dis 43:55–66
9. Samuels LE, Holmes EC, Petrucci R (2004) Psychosocial and sexual concerns of patients with implantable left ventricular assist devices: a pilot study. J Thorac Cardiovasc Surg 27:1432–1435
10. Brouwers C, De Jonge N, Kaliskan K et al (2014) Predictors of changes in health status between and within patients 12 months post left ventricular assist device implantation. Eur J Heart Fail 16:566–573
11. Brouwers C, Denollet J, Caliskan K et al (2015) Psychological distress in patients with a left ventricular assist device and their partners: an exploratory study. Eur J Cardiovasc Nurs 14:53–62
12. Overgaard D, Grufstedt Kjeldgaard H, Egerod I (2012) Life in transition: a qualitative study of the illness experience and vocational adjustment of patients with left ventricular assist device. J Cardiovasc Nurs. https://doi.org/10.1097/JCN.0b013e318227f119
13. Sandau KE, Hoglund BA, Weaver CE et al (2014) A conceptual definition of quality of life with a left ventricular assist device: results from a qualitative study. Heart Lung 43(1):32–40. https://doi.org/10.1016/j.hrtlng.2013.09.004

14. Jakovljevic DG, McDiarmid A, Hallsworth K et al (2014) Effect of left ventricular assist device implantation and heart transplantation on habitual physical activity and quality of life. Am J Cardiol 114:88–93. https://doi.org/10.1016/j.amjcard.2014.04.008
15. Coyle LA, Martin MM, Kurien S et al (2008) Destination therapy: safety and feasibility of national and international travel. ASAIO J 54:172–176. https://doi.org/10.1097/MAT.0b013e318167316d
16. Hammadah M, Kindya BR, Allard-Ratick MP et al (2017) Navigating air travel and cardiovascular concerns: is the sky the limit? Clin Cardiol 40:660–666. https://doi.org/10.1002/clc.22741
17. Tigges-Limmer K, Brocks Y, Winkler Y et al (2017) Psychosocial aspects in the diagnostics and therapy of VAD patients. Z Herz- Thorax- Gefaesschirurgie. https://doi.org/10.1007/s00398-017-0171-0
18. Grady KL, Wissman S, Naftel DC et al (2016) Age and gender differences and factors related to change in health-related quality of life from before to 6 months after left ventricular assist device implantation: findings from Interagency Registry for Mechanically Assisted Circulatory Support. J Heart Lung Transplant 35:777–788. https://doi.org/10.1016/j.healun.2016.01.1222
19. Cowger JA, Naka Y, Aarondon KD et al (2018) Quality of life and functional capacity outcomes in the MOMENTUM 3 trial at six months: a call for new metrics for left ventricular assist device patients. J Heart Lung Transplant 37:15–24. https://doi.org/10.1016/j.healun.2017.10.019
20. Bidwell JT, Lyons KS, Mudd JO et al (2017) Quality of life, depression, and anxiety in ventricular assist device therapy: longitudinal outcomes for patients and family caregivers. J Cardiovasc Nurs 32:455–463
21. Magnussen C, Bernhardt AM, Ojeda FM et al (2018) Gender differences and outcomes in left ventricular assist device support: the European registry for patients with mechanical circulatory support. J Heart Lung Transplant 37:61–70. https://doi.org/10.1016/j.healun.2017.06.016
22. Bunzel B, Laederach-Hofmann K, Wieselthaler G, Roethy W, Drees G (2005) Posttraumatic stress disorder after implantation of a mechanical assist device followed by heart transplantation: evaluation of patients and partners. Transplant Proc 37:1365–1368
23. Baba A, Hirata G, Yokoyama F et al (2006) Psychiatric problems of heart transplant candidates with left ventricular assist devices. J Artif Organs 9:203–208
24. Eshelman AK, Mason S, Nemeh H, Williams C (2009) LVAD destination therapy: applying what we know about psychiatric evaluation and management from cardiac failure and transplant. Heart Fail Rev 14:21–28
25. Hasin T, Jaarsma T, Murninkas D (2014) Sexual function in patients supported with left ventricular assist device and with heart transplant. ESC Heart Fail. https://doi.org/10.1002/ehf2.12014
26. Merle P, Maxhera B, Albert A et al (2015) Sexual concerns of patients with implantable left ventricular assist devices. Artif Organs 2015(39):664–669. https://doi.org/10.1111/aor.12535
27. Kato N, Jaarsma T, Ben Gal T (2014) Learning self-care after left ventricular assist device implantation. Curr Heart Fail Rep 11:290–298. https://doi.org/10.1007/s11897-014-0201-0

Ergonomics in Hospital: Prevention of Interruptions and Safety in Drug Administration

Gabriele Frangioni(✉), Klaus P. Biermann, Barbara Caposciutti, Salvatore De Masi, Mario Di Pede, Silvia Prunecchi, and Angela Savelli

Meyer Children's University Hospital, Viale Pieraccini, 24, 50139 Florence, Italy
gabriele.frangioni@meyer.it

Abstract. The lack of signal elements between Healthcare Workers and users can generate errors caused by interruptions during drug administration depending on human, technical or environmental factors that can become a source of stress and adverse events.

Aim of the project is the identification and application of tools for the prevention of interruption that promote the safety of drug administration.

The project is organized in several steps: process mapping; testing; operators training; implementation; monitoring.

The testing observed a 44% reduction of interruption. The introduction of information sheets for families in hospital rooms has contributed to reduce by 28% the interruptions.

The establishment of a "No Interruption Group" had allowed an appropriate training and an organizational and management support to healthcare workers involved.

The active involvement of physicians and nurses revealed critical issues, such as behavioral, organizational and structural ones, and the subsequent improvement actions.

Keywords: Interruptions · Drug Administration Safety · Healthcare design

1 Introduction

1.1 International Background

The combination of multitasking and interruption is a latent source of clinical failure [1–3]. Simulation and analysis of the major negative events in aeronautics [4] and nuclear plants [5] have identified the interruptions as a potential risk of error in task completing.

The interruptions produce quantitative and qualitative effects on performance, due to a mental overload developed in order to avoid errors or because of the effort to recover the mistake. The transition from one task to another is a crucial moment in the error handling [6].

S. Bagnara et al. (Eds.): IEA 2018, AISC 818, pp. 238–247, 2019.
https://doi.org/10.1007/978-3-319-96098-2_31

Research has shown that the interruptions of nurses significantly increase the rate and severity of errors in drug administration [1]. The uncontrolled and untrained interruptions of use in clinical practice is an expensive and dangerous strategy. The need to develop clinical processes that reduce them is a priority [2].

Data in the literature provide discontinuation rates for operators with average values of 6.7 times/h with similar data between physicians, nurses, specialists and residents [3].

The lack of signal elements between operators and users can generate errors caused by interruptions during drug administration depending on human, technical or environmental factors that can become a source of stress and adverse events.

2 Materials and Methods

The method used at the Meyer Children's University Hospital involved simultaneously phases of research and periods of field observations for the purpose to identify and understand the pharmacological path and the related criticalities.

The identification of elements for the prevention of errors related to interruptions required the evaluation of work organization and the mapping of the therapeutic paths in different steps. The first step (2013–2016) included wards of Pediatric Surgery, Pediatric Neurology, Pediatrics, Neonatal Surgery, Neurosurgery, Pediatric Cancer and Hematopoietic Stem Cell Transplantation, Urology, Day Hospital and Day Surgery. The second one (2017–2018) involved wards of PICU, Emergency Department, NICU, Units of Cystic fibrosis, Nephrology and Dialysis and Outpatient Clinic (Allergy, Diabetology, Rheumatology). At last, the third step (2019) will regard wards of Infant e Juvenile Neuro-psychiatry, Operating Rooms and Outpatient clinic (areas not involved in the former phases).

The qualitative investigation was carried out through observations, interviews, documentation collection and analysis related to internal procedures.

Field observations were made with the shadowing method and achieved as "observer observed".

The observations for the assessment of signaling vests for Healthcare Workers (HCWs) was supported by a questionnaire for evaluating the effectiveness in terms of interruption prevention, relevant on: presence of interruptions, their sources and frequency, vest perception in terms of protection and comfort.

Structured interviews with open questions, together with conversations with physicians, nurses, social health worker and health technicians, allowed the acquisition of an adequate understanding of the context and of the processes.

Surveys administered to nurses measured the comfort of the vests, the presence, the number and the origin of interruption.

The analysis was carried out through an ergonomic approach to the problem by adapting the SEIPS Model of Work System and Patient Safety with the observation of physical, cognitive and organizational processes.

The ergonomic assessment of the elements of the process was integrated with the SEIPS model. The focus of assessment was the operator, involved directly or indirectly in the care process, during the execution of the task in relation to the tools and technologies used in the physical environment and the existing organizational conditions. The model identifies five key components (people, tasks, tools and technologies, physical environment, organizational conditions) that interact and influence each other, creating effects on work organization and clinical processes in order to produce outcomes for patients, workers and the organization with respect to the quality and safety of care.

The SEIPS model was applied at various stages of research, placing the patient and the user at the center of the process, through systemic analysis of all interface elements in relation to the tasks conducted [7, 8].

The complexity of the hospital system should make appropriate the use of the SEIPS model for the systemic management of all components and the understanding of physical, cognitive and organizational aspect, in order to identify critical issues and appropriate improvement actions. The interaction between all system components, with continuous feedback of users and operators, is a fundamental aspect for the implementation and enhancement of various components in terms of health, safety and workplace wellbeing.

3 Calculation

3.1 Field Study

The organization of the clinical care process at Meyer Children's University Hospital has an impact on the organization of the spaces.

The hospital has 179 beds for Ordinary Hospitalization (OH) and 72 for the Day Hospital (DH). The admissions per year (data obtained from Hospital Discharge Form for the year 2017) is 8652 in OH and 24.151 in DH.

Defining "No Interruption Zone" and HCWs Ergonomic Intervention Planning. The analysis carried out in the considered wards highlighted common characteristics in the process of drug management during the stage of prescribing, preparation and administration.

The identification of "No Interruption Zones" in wards involves the re-evaluation and re-organization of the existing spaces.

Observations, interviews and signaling, collected during the fieldwork, underlined the criticalities related to the process, particularly to the drug adverse events due to interruptions. The work analysis allowed us to design and develop specific tools tailored to the HCWs needs in terms of comfort, safety, health and work well-being. The significant presence of interruptions during prescribing, preparation and administration of drugs requires the application and integration of two signal elements:

- Static element: a model for identification of spaces where the critical activity takes place (prescription and preparation stages);
- Dynamic element: a wearable device for HCWs (for all phases, where necessary).

Communication. The introduction of new tools requires an appropriate communication to users and HCWs. The HCWs should be trained in the object, purpose, potential problems, correct application and potential benefits of the proposed solution. Training represents a moment of sharing and collaboration for future improvement actions.

With regard to the users, it is necessary to provide information through specific tools such as posters, sheets or table tents that introduce and focus the attention in a simple, clear and repeated way (poster for patient rooms). The moment of the hospital admission is essential for a first users' information, aimed at reducing confusion and future resistances.

Management and Training. Following to data obtained, a multidisciplinary "No Interruption Group" (NIG) has been constituted; this group has been composed of:

1. NIG Management: components of Nursing management and Clinical Risk management;
2. NIG Training operators: one physician and one nurse per ward involved in the first phase (2016) and in the second phase of implementation (2017);
3. Other NIG Training operators will be activated for the wards in further implementation.

NIG Training referent assured a training "on the job" to colleagues of their own ward.

4 Results

4.1 Process Mapping and Space Analysis

The output of the qualitative analysis was the "process mapping", obtained by the study of HCW's work organization through observations, conversations and interviews. The main criticalities in the pharmacological path were identified in every ward together with the most appropriate space to properly carry out the prescription, preparation and administration drug phases in terms of patient safety, staff well-being and the entire therapeutic process.

The work started in the "pilot" departments of Pediatric Neurology and Neurosurgery with the objective to define the main parameters and apply them to the other departments subsequently.

The space analysis allowed each ward to realize a "Thematic Plan of Pharmacological Path" represented by unique colours designating the different areas for the various stages. The choice of different colours was dictated by the need to clarify and make visible for operators, in a single visual framework, all environments where a complex process happens.

Thematic Plans: Pharmacological Path. For each of the stages of prescription, preparation and administration, a reference colour was assigned in order to make the space unambiguous and immediately identifiable to HCWs and users, by limiting cognitive effort and reporting intrinsically obtained data. The colours are identified as follows (Fig. 1):

- Prescription stage: yellow
- Preparation stage: red
- Administration stage: pale blue, violet (only Neurosurgery)

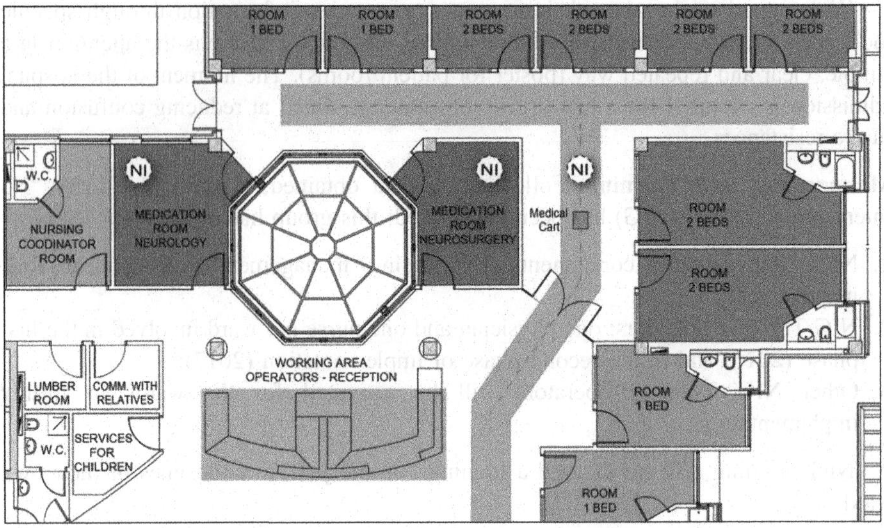

Fig. 1. Pharmacological path plan wards of neurology e neurosurgery (Color figure online)

Double colour usage for administration is due to the simultaneous presence in the same area of the "Pediatric Neurology" and "Neurosurgery" wards; generically the pale blue is used to indicate the administration spaces.

Three types of areas have been identified for mixed or special use. They have been highlighted with different colours:

- Prescription and administration stage: green
- Meeting areas: cream-gray
- Different Specialties areas: magenta

In some departments, both prescription and administration take place at the bedside, making it necessary to identify them with a green colour. The meeting areas identify HCW's rooms, which are used also for handover or other meetings. Special areas are specific to each ward and include the relevant specialties (e.g. Emergency room to the Emergency Department, Burn Center to the Pediatric surgery).

The highlight of the stages of the pharmacological paths was made possible to identify those areas that, for the peculiarities of the activities carried out, can be only temporarily free from interruptions.

The areas identified as "No Interruption" are indicated with a symbol formed of two black letters (NI) within a dashed black circumference on a white field [9].

4.2 Tools for Prevention of Interruptions and Related Errors

The analysis allowed also the definition of interventions on spaces and HCWs that can represent a deterrent to interruptions and not a barrier to communication aimed at ensuring patient care by the health personnel.

The integrated solution of identification signs for spaces and HCWs, properly implemented and verified, represents an important prevention system.

Interventions on Healthcare Workers. The need to identify the HCWs during the execution of a critical task resulted in the design of signalling vests of different colour depending on the process involved:

- Yellow: drug therapy
- Red: transfusions

On the yellow the front and back report the warning "Therapy in progress do not disturb" (Fig. 2), the red ones report the warning "Transfusion in progress do not disturb".

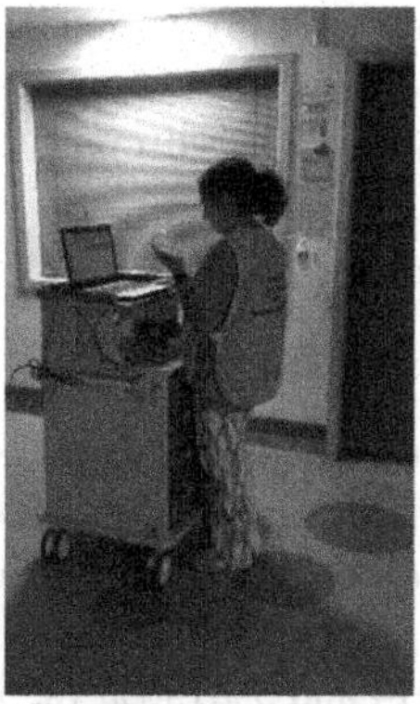

Fig. 2. Nurse signaling vest (Color figure online)

The introduction of vests requires the need of the application of information sheets for parents and patients within the patient's room indicating: the meaning of vests, the available HCWs to ask to in case of need.

The presence of patients of different nationalities has requested translation into foreign languages (English, Spanish, French, Chinese, Arabic, Romanian and Albanian) [9].

Testing of yellow vests (Table 1) in November 2014, performed at Medical Pediatrics A, had a good integration with the present mobile stations, carrying out a function of signaling and safety for HCWs; the 56% of nurses who wore the vest reported they had been interrupting during the administration process.

However, the lack of adequate training of nurses with respect to interruptions and signaling elements resulted in a lack of awareness with respect to their purpose and meaning. Ongoing training increased the collaboration and acceptance by HCWs.

The introduction of information sheets into patient's rooms increased parental awareness, contributing to a reduction of 28% of nurses interrupted in their work during the several weeks of observation.

Table 1. Monitoring results

Wards	Number and percentage on 55 questionnaires	
	n interruptions	%
Pediatrics A		
Testing (Nov 2014)	31	56
Follow up (Jul 2015)	36	65
Post intervention (Sep 2016)	28	51
Pediatrics B		
Baseline (Nov 2014)	52	95
Post-intervention (Sep 2016)	35	64
Pediatric Surgery		
Post-intervention (Oct 2016)	25	45
Neuroscience		
Post-intervention (Nov–Dec 2016)	42	76
Urology		
Post-intervention (Dec 2016)	30	55

The follow up made in July 2015 evaluated an increase of interruptions of 9% (56% vs. 65%) respect to the survey of November 2014. The outcome confirms the need of an appropriate training for nurses before the introduction of the intervention. After training course, we evaluated a reduction of interruption in September 2016 of 14% (65% vs. 51%) respect to July 2014 and 5% (56% vs. 51%) respect to November 2014.

In Medical Pediatrics B, at a baseline performed in November 2017 without signaling elements, 95% of nurses who wore the vest reported they had been interrupted during the administration process. After training course and introduction of signaling elements, we evaluated in September 2016 a reduction of interruption of 31% (95% vs. 64%) respect to November 2014.

After training course and introduction of signaling elements in Pediatric Surgery, Neuroscience and Urology, we observed between September and December 2016 that, respectively, 45%, 76%, 55% of nurses who wore the vest reported they had been interrupted during the administration process.

In test phase, family members have been involved for evaluate a usability of information sheets and the meaning of vests. The results have been positive.

Spaces Interventions. Between August 2015 and December 2016 in Pediatric Cancer and Hematopoietic Stem Cell Transplantation have been realised interventions in terms of prevention to interruptions:

- Management of the access to wards through doors to open with code;
- Acceptation point for Oncologic patient;
- Re-organization of Antiblastic Drugs Unit.

In all wards of interventions, the space of preparation and prescription of drugs had been signaling with information sheets.

Education and Training. The training of wards referents was important to guarantee correct information. The presence of physician and nurses in NIG Training and Wards Training Courses allowed a clear and a constructive conversation about behaviors, internal organization and instruments used. This produced a process of re-evaluation and individuation of appropriate improvement actions inside and between the single wards. Moreover, common questions regarding management items emerged in several wards.

The training on the job has involved 233 HCWs, in the first, and 117 in the second step. This formative method permitted the HCWs to feel themselves part of this project and not component within a top down approach.

The participation of physicians and nurses in the trainings "on the job", which included also social health worker and health technicians, allowed a clear and constructive compare on human behaviors, internal organizations and used tools.

This approach has allowed a revaluation of respective processes and individuation of opportune improvement actions, both within the same and other ward. Other items, critically widespread in several wards, have been treated at management level.

Pediatric Cartoons. Prevention of interruption needed a proper communication to parents and patient regarding the instrument implemented, information sheets and instructions for HCWs.

In test phase, family members have been involved for evaluate a usability of information sheets and the meaning of vests. The results have been positive.

To give a proper communication to the families about instruments for prevention of interruption, we are developed Cartoons for educating parents to best practice for patient safety, with the title "Double check your therapy" (Fig. 3). This and other three Cartoons, entitled "In terms of safety we are a great team", are available from October 2017 in Hospital wards and internet video, reachable with QR code. Furthermore, the World Health Organization for distribution in Europe has adopted those Cartoons.

Fig. 3. Frame to cartoons, title: "Double check your therapy"

5 Discussion

The analysis carried out in the departments of Meyer Children's University Hospital highlights some organizational, cognitive and physical criticalities in drug management processes during prescription, preparation and administration stages. There is a need for introducing operators' supporting systems, which will reduce the cognitive load during repetitive and mnemonic tasks, and allow greater support and patient care customization.

The high specialization of wards requires systems that are customizable.

Prevention of interruptions needs a systematic and multidisciplinary analysis to environmental contest of the work activity and the active participation of the HCWs in all phases of analysis and development.

The analysis and test made, allowed to identify some elements that should be evaluate in term of interruptions:

- Interruptions are prevented with identification systems for HCWs and spaces that should be opportunely integrated;
- Introduction to signalling elements should be a deterrent for the interruption and not a communicative barrier, to guarantee patient assistance on behalf of healthcare staff;
- Interruptions in healthcare are difficult to eliminate but they can be reduced to acceptable levels; we do not want to realise specific spaces like airplane cockpits but organize a proper system to answer to the interruptions;
- Interruptions are prevented throughout three progressive levels of intervention in terms of realization and resources:

1. Behavioral level: First of all, we should change our wrong attitudes;
2. Organizational level: Rethink our activities and procedures;
3. Structural level: this is the most complex level because requires time and resources but, knowing the improvement objective, allows us to plan future implementation.

Communication of the instruments introduced is essential for their acceptation by HCWs, patients and families that should be well trained and informed [9].

6 Conclusions

The interruptions are a blind and underestimated problem in healthcare. We work in a complex system made of many interactions with persons, organization system, technologies and environment. "The interruptions" are a breaking of interaction during the action that can cause an error or a delay in a task conducted.

Patient safety is one of a primary objective in the healthcare system that needs multidisciplinary and systematic approach by the engagement of all HCWs involved in the process. The ergonomic methods can improve physical, cognitive and organizational aspects and enhance the patient safety and healthcare workers well-being.

The interruptions are prevented with an identification system for HCWs and spaces that should be opportunely integrated in the work activities and the communication of its meaning to users and healthcare staff.

The patient safety should necessarily include the operator safety if we want to reach a healthcare system, efficient in terms of outcomes and risk reduction.

This project, started as academic study, tested and implemented in all wards and Day Hospital of a Children's Hospital, has become a referent point for a regional non-interruption project and has showed a height flexibility and transferability of methods applied. It can be reproduced in all Healthcare facilities with simple, clear and low cost instruments because, to reduce the interruptions, we should first improve our personal behaviors and only afterwards, we can act on organization and environment.

Prevention of interruption: Support the Healthcare Workers, Involve the Families in Patient Safety, Give the Right Care to the Child! [10].

References

1. Coiera E (2000) When conversation is better than computation. J Am Med Inform Assoc 7:277–286
2. Westbrook JI, Woods A, Rob M et al (2010) Association of interruptions with increased risk and severity of medication administration errors. Arch Intern Med 170:683–690
3. Westbrook JI, Coiera E, Dunsmuir WT, Brown BM, Kelk N, Paoloni R, Tran C (2010) The impact of interruptions on clinical task completion. Qual Saf Health Care 19:284–289
4. Latorella K (1999) Investigating interruptions. Implications for flightdeck performance. NASA Langley Research Center 1, Hampton, VA
5. Chou C, Funk K (1993) Cockpit task management errors in a simulated flight operation. In: 7th International symposium on aviation psychology, pp 965–969
6. Eyrolle H, Cellier JM (2000) The effects of interruptions in work activity: field and laboratory results. Appl Ergon 31:537–543
7. Carayon P, Schoofs Hundt A, Alvarado CJ, Springman S, Borgsdorf A, Jenkins L (2005) Implementing a systems engineering intervention for improving safety in outpatient surgeries. Adv Patient Saf 3:305–321
8. Carayon P, Schoofs Hundt A, Karsh B-T, Gurses AP, Alvarado CJ, Smith M, Flatley Brennan P (2006) Work system design for patient safety: the SEIPS model. Qual Saf Health Care 15(Suppl I):50–58
9. Frangioni G, Savelli A, Biermann KP et al (2016) Prevenire le interruzioni per assicurare la terapia farmacologica/Prevent interruptions to ensure drug administration. Rivista italiana di Ergonomia, Special Issue 1/2016, 111–116
10. Frangioni G, Biermann KP, Savelli A (2018) Prevention of interruptions and safety in drug administration. In: 2018 quality and safety in children's health conference. www.childrenshospitals.org. Accessed 25 May 2018

Team Adaptation to Complex Clinical Situations: The Case of VTE Prophylaxis in Hospitalized Patients

Megan E. Salwei[1,2(✉)], Pascale Carayon[1,2], Ann Schoofs Hundt[2], Peter Kleinschmidt[3], Peter Hoonakker[2], Brian W. Patterson[2,3], and Douglas Wiegmann[1,2]

[1] Department of Industrial and Systems Engineering, University of Wisconsin-Madison, Madison, WI, USA
msalwei@wisc.edu

[2] Center for Quality and Productivity Improvement, University of Wisconsin-Madison, Madison, WI, USA

[3] School of Medicine and Public Health, University of Wisconsin-Madison, Madison, WI, USA

Abstract. Intensive care units (ICUs) are complex environments, which rely on teams in order to coordinate patient care. Venous thromboembolism (VTE), which includes deep vein thrombosis (DVT) and pulmonary embolism (PE), is a major concern for ICU patients, who are frequently immobile. VTE prophylaxis (prevention) occurs throughout different stages of a patient's stay in the ICU, which range in levels of complexity. The objective of this study is to use social network analysis to understand team adaptation in response to different levels of complexity in the VTE prophylaxis process. The more complex stages of VTE prophylaxis involve more people, more team activities, more team interactions, and more two-way communication compared to the less complex stages. Social network analysis can be used to understand team adaptation to these different levels of complexity in a patient's ICU care.

Keywords: Team adaptation · Social network analysis · Complexity
Patient safety

1 Introduction

Intensive care units (ICUs) in hospitals are complex environments, in which multidisciplinary teams need to coordinate their activities in order to provide safe, effective, and efficient care. Venous thromboembolism (VTE), made up of deep vein thrombosis (DVT) and pulmonary embolism (PE), is a major concern for hospitalized patients, particularly for critically ill patients who are frequently immobile [1]. VTE accounts for approximately 10% of in-hospital patient deaths and is one of the most common complications for ICU patients [1, 2]. The American College of Physicians recommends that all patients be assessed for risk of VTE and administered VTE prophylaxis when appropriate [2]; however, the implementation of such VTE prophylaxis guidelines continues to be a problem [3].

S. Bagnara et al. (Eds.): IEA 2018, AISC 818, pp. 248–254, 2019.
https://doi.org/10.1007/978-3-319-96098-2_32

The VTE prophylaxis process involves multiple team members throughout a patient's stay in the ICU and can be characterized as occurring in the following five stages: admission to the ICU, interruption of VTE prophylaxis during the ICU stay, re-initiation of VTE prophylaxis during the ICU stay, initiation of VTE prophylaxis during the ICU stay (when it was not started at admission to the ICU), and re-evaluation at the time of transfer into the ICU from another hospital unit. The stages of admission to and transfer into the ICU are less complex than the other three stages as there is a clear trigger for considering VTE prophylaxis. In the admission stage, physicians generally use a computerized admission order set that includes consideration for VTE prophylaxis. At the time of transfer into the ICU, physicians conduct medication reconciliation, which includes review of medications for VTE prophylaxis. The stages of interruption, re-initiation and initiation rely on the memory and vigilance of all team members (i.e. physicians, pharmacists, nurses) to recognize the need to interrupt, re-initiate or initiate VTE prophylaxis based on a patient's changing clinical status and care plan.

Teams in complex and uncertain environments such as ICUs rely on efficient and effective interactions and communication in order to meet the needs of the situation [4]. These teams need to adapt their organization. This includes determining who to involve, how they communicate and share information, and how they coordinate their activities based on the level of complexity of the patient's clinical status.

Social network analysis (SNA) is a method that can be used to describe and measure team structure and interactions in situations with varying levels of complexity. For instance, SNA has been used to observe and define network structures for command and control operations within military and emergency services [5–7], to assess naval team readiness [8], to measure team knowledge and its influence on performance [9], and to study information sharing throughout a community [10]. McCurdie et al. [11] used SNA to measure interruptions within the ICU. They observed sixteen unique ICU roles (e.g., administrative officer, pharmacist, bedside nurse) for approximately three hours each, recording each interruption, the role of the interrupter, and the reason for the interruption. From the observational data, they created role networks to understand interruptions from multiple clinicians' perspectives, rather than focusing on one role, in order to develop interventions to reduce interruptions. They found ICU staff were interrupted 7 times per hour on average, with a disproportionate number of interruptions between the bedside nurse and the charge nurse. Based on these results, they developed targeted systems-based interventions to reduce unnecessary interruptions and burden for ICU staff.

Barth et al. [12] used SNA to measure team adaptation in response to changes in complexity in surgical procedures. They observed different phases of 40 pediatric cardiac surgeries that had different levels of clinical complexity. They found higher density and reciprocity during more complex surgical procedures, indicating more frequent discussion and two-way communication. Team communication patterns were more decentralized during complex procedures, showing that teams adapt to changing complexity. In this study, team adaptation to complexity was observed through the use of SNA measures. In our study, we use SNA measures to assess an ICU team's adaptation to changing levels of complexity in the VTE prophylaxis process that occurs during a patient's stay in the ICU.

2 Methods

2.1 Sample and Setting

This study was conducted as a part of a larger study with the aim of developing design requirements for clinical decision support that supports VTE prophylaxis (https://cqpi.wisc.edu/vte-and-health-it/). Data were collected in an academic hospital's ICU. A total of 5 attending physicians were interviewed. The study was approved by all associated institutional review boards.

2.2 Data Collection

The 5 semi-structured interviews were conducted for a total of 3 h and 55 min. Interviews were audio-recorded, transcribed and coded using Dedoose®, a qualitative data analysis software. Interview excerpts were extracted from the transcripts. Using the work system model [13, 14], we identified information in the excerpts about the role of the interviewee, the VTE prophylaxis activity involved, the tool(s)/technology(ies) being used, and other organizational information relevant to the VTE prophylaxis process [15]. The information was then transferred into the diagramming software, Lucidchart®, in which we created 5 role networks for each stage of VTE prophylaxis [3]. Figure 1 shows an example of a role network for the stage of VTE prophylaxis

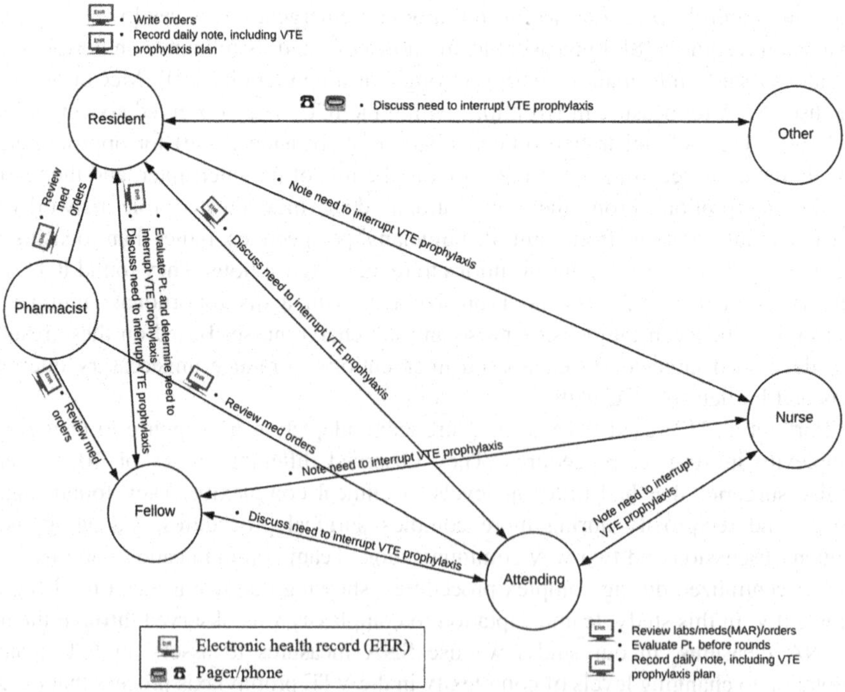

Fig. 1. Role network representing the interruption stage of VTE prophylaxis in the ICU

interruption in the ICU. Data from each role network (5 total) were entered into an Excel spreadsheet, with each row representing data for a single role network.

2.3 Social Network Measures

We used 4 network-level SNA measures to assess each role network: number of roles, number of team activities, number of interactions, and reciprocity. The number of roles is calculated by counting each role represented (e.g., attending, resident, nurse) on the role network. The number of team activities is the sum of activities performed jointly by two roles on a single role network. The number of interactions is the sum of one-way and two-way interactions between roles on a network. Reciprocity is measured on a scale from 0 (low) to 1 (high) and represents the ratio of two-way communications within the network. It is calculated by dividing the number of two-way interactions in a network by the total number of one-way and two-way interactions in that network [16]. A high score of reciprocity means that the network includes a large proportion of two-way interactions as compared to other interactions in the network. We also calculated the team activity EHR use for each VTE prophylaxis stage.

In addition to these network-level measures, we used a role-level measure, degree centrality, which is calculated by dividing the number of connections a role has compared to the total connections possible in the network. Degree centrality scores vary from 0 to 1 with a score of 1 meaning that the role is connected to all other roles possible, and therefore, is a central role in the network.

We organized the data by the complexity (low or high) of the VTE prophylaxis stages: admission and transfer are considered low-complexity stages; interruption, re-initiation, and initiation are considered high-complexity stages (Table 1).

Table 1. Social network analysis measures for VTE prophylaxis stages

SNA measures		Low complexity		High complexity		
		Admission	Transfer	Interruption	Re-initiation	Initiation
Team-level measures	Number of roles	4	4	6	6	6
	Number of team activities	7	7	11	11	12
	Percent of EHR use for team activities	*6/7 (86%)*	*5/7 (71%)*	*5/11 (45%)*	*5/11 (45%)*	*5/12 (42%)*
	Number of interactions	6	6	10	10	11
	Reciprocity	0.5	0.5	0.7	0.4	0.45
Role-level measures	Resident centrality	1	1	1	1	1
	Fellow centrality	1	1	0.8	0.8	1
	Attending centrality	1	1	0.8	0.8	0.8
	Pharmacist centrality	1	1	0.6	0.6	0.6
	Nurse centrality	0	0	0.6	0.6	0.6
	Other centrality	0	0	0.2[a]	0.2[b]	0.4[c]

[a]Proceduralist, [b]Neurosurgery physician assistant, [c]Neurosurgeon

3 Results

As shown in Table 1, the role networks for the less complex stages of admission and transfer were smaller (4 roles versus 6) and involved fewer team activities (7 activities versus 11–12) compared to the more complex VTE prophylaxis stages of interruption, re-initiation, and initiation. The EHR was used more frequently in team activities in low-complexity stages (86% during admission; 71% during transfer). EHR use for team activities decreased in the high-complexity stages of interruption, re-initiation, and initiation, ranging from 42% to 45%. The number of interactions increased from the low-complexity stages (6 interactions) to the high-complexity stages (10–11 interactions). The type of interactions (i.e., one-way versus two-way communication) changed from low to high complexity, as reflected in the reciprocity scores (see Table 1). The reciprocity scores were 0.5 during the low-complexity stages of admission and transfer, and 0.4 and 0.45 in the high-complexity stages of re-initiation and initiation. The highest score of reciprocity, 0.7, occurred during the high-complexity stage of interruption.

The centrality scores varied from low to high complexity stages. The resident, fellow, and attending centrality scores ranged from 0.8 to 1 in the low and high complexity stages. The pharmacist centrality decreased from the low-complexity stages of admission and transfer (1) to the high-complexity stages of interruption, re-initiation, and initiation (0.6). The nurse and other (i.e., proceduralist, neurosurgery physician assistant, or neurosurgeon) centrality increased from 0 in the low-complexity stages,to 0.2–0.6 in the high-complexity stages.

4 Discussion

In this study of VTE prophylaxis for ICU patients, we expanded upon previous research by Barth et al. [12] by assessing the adaptation of an ICU team to situations with differing complexity. We used SNA measures as a quantitative means to study team adaptation across the five stages of VTE prophylaxis, which occur during a patient's stay in the ICU. The number of roles, number of team activities, percent of EHR use for team activities, number of interactions, reciprocity, and centrality were calculated and compared between stages of low complexity (admission and transfer) and stages of high complexity (interruption, re-initiation, and initiation).

We found that in the more complex VTE prophylaxis stages of interruption, re-initiation, and initiation, the number of roles involved in the team was higher than the less complex stages of admission and transfer. More people (i.e., nurse and neuro-surgeon) were involved to adequately assess the patient status and ensure appropriate care; this was in response to the need to communicate changes and care plan about the patient (e.g., upcoming surgical procedure). For example, a nurse may notice a change in the patient's clinical status, and then communicate the change to the pharmacist, resident, and/or fellow who may decide to modify the VTE prophylaxis plan for the patient. This is also demonstrated by a change in centrality of the various roles involved in VTE prophylaxis from low to high complexity stages. As the process complexity increases, the nurse becomes a more central role involved in VTE prophylaxis.

The number of activities performed by the team was higher during complex stages of VTE prophylaxis. Not only are there more people involved in complex stages, but the people who are involved more frequently work together to complete the required activities of VTE prophylaxis. The EHR was often used for team activities in less complex stages of VTE prophylaxis, unlike the more complex stages, where the EHR was less frequently used. The EHR could be better utilized for team activities in the more complex stages of VTE prophylaxis in order to support team activities and the VTE prophylaxis process. Rather than relying on the memory and vigilance of clinicians to monitor the patient's clinical status, the EHR could be designed to serve as a trigger to interrupt, re-initiate, or initiate VTE prophylaxis based on a patient's status and care plan (e.g., surgical procedure).

Similar to Barth et al. [12], we found that complex stages such as interruption elicit more two-way communications between team members, which results in increased reciprocity. When the patient care plan relating to VTE prophylaxis becomes uncertain (e.g., due to a change in patient status or scheduling of a new procedure), the complexity in the process increases. In our study, the critical care team participated in more discussion and two-way communication when the stages of VTE prophylaxis were more complex; this resulted in an increase in reciprocity and an increase in the total number of interactions between team members. In order to respond to uncertainty regarding the patient's VTE prophylaxis plan, the care team adapted (increased) their communication to manage the complexity of the situation.

Previous studies have used SNA as a way to quantitatively evaluate team structure [5–9, 11]. Our study expands upon this work by showing that SNA can be used to identify team adaptation in a process with different stages and varying levels of complexity. According to Burke et al. [4], team members adapt and adjust their actions according to changing situational requirements. Our study as well as the work of Barth et al. [12] show that SNA can be used to measure team adaptation in response to change in complexity.

One limitation of this study is that the data are from one unit in a single hospital, and therefore, results may not apply to other situations and contexts. Future research should include data from additional hospitals and units to gain an understanding on the relationship between complexity and team adaptation in different contexts.

5 Conclusion

Social network analysis is a method that can be used to study team communication and adaptation. In our study, the ICU team adapted throughout the stages of VTE prophylaxis with different levels of complexity. Team adaptation was observed through changes in the numbers of roles, activities, interactions, and two-way communication in the team. Teams adapt to situations with varying levels of complexity, which can be observed through the use of social network analysis.

Acknowledgments. This project was supported by Grant Number R01HS022086 from the Agency for Healthcare Research and Quality. The content is solely the responsibility of the authors and does not necessarily represent the official views of the Agency for Healthcare Research and Quality. The project was also partially supported by the Clinical and Translational Science Award (CTSA) program, through the NIH National Center for Advancing Translational Sciences (NCATS), Grant UL1TR000427. The content is solely the responsibility of the authors and does not necessarily represent the official views of the NIH.

References

1. Wood KE (2011) Major pulmonary embolism. Crit Care Clin 27:885–906
2. Qaseem A, Chou R, Humphrey LL, Starkey M, Shekelle P (2011) Venous thromboembolism prophylaxis in hospitalized patients: a clinical practice guideline from the American College of Physicians. Ann Intern Med 155:625–632
3. Hundt AS, Carayon P, Yang Y, Stamm J, Agrawal V, Kleinschmidt P, Hoonakker P (2017) Role network analysis of team interactions and individual activities: application to VTE prophylaxis. In: Proceedings of the human factors and ergonomics society annual meeting. SAGE Publications, Los Angeles, pp 896–900
4. Burke CS, Stagl KC, Salas E, Pierce L, Kendall D (2006) Understanding team adaptation: a conceptual analysis and model. J Appl Psychol 91:1189–1207
5. Houghton RJ, Baber C, McMaster R, Stanton NA, Salmon P, Stewart R, Walker G (2006) Command and control in emergency services operations: a social network analysis. Ergonomics 49:1204–1225
6. Houghton RJ, Baber C, Stanton NA, Jenkins DP, Revell K (2015) Combining network analysis with cognitive work analysis: insights into social organisational and cooperation analysis. Ergonomics 58:434–449
7. Baber C, Stanton N, Atkinson J, McMaster R, Houghton RJ (2013) Using social network analysis and agent-based modelling to explore information flow using common operational pictures for maritime search and rescue operations. Ergonomics 56:889–905
8. Schraagen JM, Post W (2014) Characterizing naval team readiness through social network analysis. In: Proceedings of the human factors and ergonomics society annual meeting. SAGE Publications, Los Angeles, pp 325–329
9. Espinosa JA, Clark MA (2014) Team knowledge representation: a network perspective. Hum Factors 56:333–348
10. Euerby A, Burns CM (2014) Improving social connection through a communities-of-practice-inspired cognitive work analysis approach. Hum Factors 56:361–383
11. McCurdie T, Sanderson P, Aitken LM (2018) Applying social network analysis to the examination of interruptions in healthcare. Appl Ergon 67:50–60
12. Barth S, Schraagen JM, Schmettow M (2015) Network measures for characterising team adaptation processes. Ergonomics 58:1287–1302
13. Smith M, Carayon-Sainfort P (1989) A balance theory of job design for stress reduction. Int J Ind Ergon 4:67–79
14. Carayon P (2009) The balance theory and the work system model… Twenty years later. Int J Hum Comput Interact 25:313–327
15. Carayon P, Hundt AS, Karsh B, Gurses AP, Alvarado C, Smith M, Brennan PF (2006) Work system design for patient safety: the SEIPS model. BMJ Qual Saf 15:i50–i58
16. Valente TW (2010) Social networks and health: models, methods, and applications. Oxford University Press, Oxford

Perception of Working Conditions and Health by Prison Officers of a Male Prison from Brazil

Fabíola Reinert(✉), Lizandra Garcia Lupi Vergara, and Leila Amaral Gontijo

Universidade Federal de Santa Catarina, Caixa Postal 476, Florianópolis, SC, Brazil
fabiola.reinert@gmail.com

Abstract. This paper presents a case study done with the prison officers of Male Prison from Florianopolis, Santa Catarina, Brazil. Starting from the hypothesis that workers' health is largely influenced by its working condition and that the work in prisons contributes to the incidence of health problems, the real perception of prison workers about their own work and health was investigated through content analysis, using semi-structured interviews. The study emphasizes that the type of work and its content increase the level of stress of the workers analyzed, and the lack of security generates fear and distrust, revealing problems in the organization and the work process. However, a deeper analysis of the work condition is fundamental to understand how the worker deals with the characteristics of the work, the strategies that he finds to solve the problems and how much the work context influences his health.

Keywords: Work conditions · Occupational diseases · Prison officers

1 Introduction

Several authors documented the effects of working conditions on workers' health [1–3], and described the association between a harsh working environment and a wide range diseases, especially mental illnesses [4–6]. In prison staff were observed high levels of risk factors, especially in those in direct contact with detainees, working in an environment characterized by a high level of psychological demands [7, 8]. According to the International Labour Organization [9], the profession is the 2nd most dangerous in the world, as it covers dangerous and unhealthy at the same time.

Dejours [10] states that the conditions and organization of work are closely related to the most diverse diseases. Work can be a source of suffering and illness, and it is necessary to understand how workers maintain their mental balance, even when subjected to destructive working conditions. The work environment in prison units is unlike any other, with the possible exception of the existing environment in psychiatric institutions and other confinement facilities [11]. The work in prison units transforms the workers, being permeated by the phenomenon of violence [12].

The precariousness of the prison units in Santa Catarina - Brazil was studied [13], relating overcrowding, precarious infrastructure and deficits in human resources with health problems such as HIV, tuberculosis, respiratory diseases, dermatoses and mental

S. Bagnara et al. (Eds.): IEA 2018, AISC 818, pp. 255–262, 2019.
https://doi.org/10.1007/978-3-319-96098-2_33

disorders. In the analysis of the working conditions of the social worker in the Women's Prison from Florianópolis, Santa Catarina - Brazil [14], the lack of resources, hygiene and ventilation in the prison was perceived, besides the overload and high level of stress in the worker.

A study on prison health in Rio de Janeiro [15], the threat to the personal integrity of the penitentiary agent appears as inherent in work, as well as emotional tension, psychosomatic manifestations and stress. Research carried out with prison officers from Rio de Janeiro [12], from Salvador [16] and from São Paulo [17] identified biological risks of tuberculosis and hepatitis contamination, due to the poor job. A high risk of violence was also identified [17].

Studies on the negative influences of work in penitentiaries in Rio Grande do Sul and Rio Grande do Norte [18], from the point of view of the prison staff themselves, showed constant fear and dissatisfaction. Also the prison staff at the André Teixeira Lima Psychiatric Custody and Treatment Hospital are exposed to situations of high psychological demands at work, with 83.3% fulfilling criteria for the presence of Common Mental Disorders (CMD) [19], and a research with the teachers who work in Brazilian penitentiary complexes [20], states that 12.5% present mental disorders, considered as indicators of evidence of mental distress.

In this context, starting from the hypothesis that workers' health is largely influenced by its working condition and that the work in prisons contributes to the incidence of health problems, this case study investigated the real perception of prison officers of Male Prison from Florianopolis, Brazil, about their work and health, using the content analysis technique, with guided interview.

2 Method

This research is exploratory, aiming to detect, understand and interpret the phenomenon investigated [21]. Exploratory research allows the researcher to deepen his analyzes within the limits of a specific reality [22].

After the bibliographical survey to construct the problematic, a qualitative research was started, through a case study, which is characterized as a type of research whose object is a unit that is analyzed in depth [22]. Thus, the perception of the work and health of the prison officers of the Male Prison from Florianópolis/SC - Brazil, was investigated through the content analysis method, using a semi-structured interview. The sample was homogeneous (between men and women) and selected by convenience, with the subjects who were on duty at that time.

According to Bardin [23], content analysis, as a method, becomes a set of communication analysis techniques that uses systematic procedures and objectives to describe message content.

The method was applied inside the Male Prison of Florianópolis/SC, in a room granted by the prison's director, during one morning. A simplified process was done, with 15 interviews for comparative analysis. The questions asked were about what they thought of their work, how they considered their relationship with co-workers and inmates, health problems they had, and medications they took. The interview was done individually and recorded for later transcription.

For content analysis three steps were performed: pre-analysis, material exploration and treatment and interpretation of the results. The pre-analysis phase aims to systematize the ideas to elaborate a precise scheme of work development. In this stage the choice of the material, formulation of the hypotheses and the objectives, and elaboration of indicators of the final interpretation are made. In the material exploration phase, the codification, categorization and quantification of the information was done, from techniques such as transcription and reading of the interviews, construction of the synopses, and descriptive analysis through typologies and categorical analysis [23].

The last phase was the treatment of the results, in a quantitative way for validation of the collected data, without excluding the qualitative interpretation. The data obtained are presented in item 3.

The Ethics Committee of the Federal University of Santa Catarina, Brazil, approved this study (inscription 38234114.0.0000.0115).

3 Results

Ten male officers and five female officers were interviewed (Table 1). It is important to note that the prison has 38 male and 20 female officers, which makes the sample homogeneous. The prison officers interviewed were between 28 and 61 years old, with a wide range of age, so that the age group was divided by 33.3% between 25 and 35 years, another 33.3% between 35 and 45 years, 20% between 45 and 55 years and 13.3% between 55 and 65 years ($\sigma = 1.06$).

Table 1. Description of the personal and socio-demographic characteristics (n = 15).

Personal and socio-demographic characteristics			
Independent variables	n (%)	Independent variables	n (%)
Gender		Marital status	
Male	10 (66.7)	Single	6 (40.0)
Female	5 (33.3)	Married	8 (60.0)
Age		Education	
25–35 years	5 (33.3)	High school	10 (66.7)
35–45 years	5 (33.3)	Higher education	5 (33.3)
45–55 years	3 (20.0)		
55–65 years	2 (13.3)		
Service time		Working hours	
4–6 years	5 (33.3)	48 h/w (24 × 72)	12 (80.0)
7–9 years	6 (40.0)	>8 h/d	3 (20.0)
10–12 years	2 (13.3)		
>20 years	2 (13.3)		

The working hours adopted by 80% of the officers interviewed is 48 h/weekly, called by them “24 × 72”, where they work 24 h in a row and pause 72 h. The other 20% worked every day, up to 10 h a day, because they were officers reassigned to the administrative area. The time in the company was also of great variation, between 4 and 41 years, and a division by service time was made, resulting in 33.3% working from 4 to 6 years in the institution, 40% from 7 to 9 years, 13.3% from 10 to 12 years and another 13.3% from more than 20 years ($\sigma = 1.03$).

From the recording of the 15 interviews, it was possible to categorize and organize the information obtained (Table 2), which showed that 66.7% consider that their work affects their psychological state, and 13.3% say that they keep remembering what they saw in jail in their spare time. 40% think the job is dangerous and 13.3% are afraid of detainees attack. In addition, 26.7% commented to have sleeping disorder, 40% commented to have problems with stress, and 66.6% said they often take drugs.

Table 2. Categorized responses of the interviews (n = 15).

Categorized responses of the interviews			
Independent variables	n (%)	Independent variables	n (%)
Work perception		Safety perception	
Remembers at home what saw in prison	2 (13.3)	Feels fear every day	2 (13.3)
		Finds it dangerous	6 (40.0)
Forgets about the prison when leave	1 (6.7)	Thinks there's always someone chasing	1 (6.7)
Enjoys it	1 (6.7)	Fear of inmates attack	2 (13.3)
Conforms to it	2 (13.3)		
Mentally affected	10 (66.7)		
Became suspicious	3 (20.0)		
Stressful	3 (20.0)		
Health problems		Drugs for	
Sleeping disorders	4 (26.7)	Headaches	3 (20)
Tuberculosis	2 (13.3)	Anxiety	2 (10)
Allergic rhinitis	1 (6.7)	Breathing	2 (10)
Irritability	1 (6.7)	Sleeping	3 (20)
Stress	6 (40.0)	High blood pressure	2 (10)
Spain pain	2 (13.3)		
Cancer	1 (6.7)		
High blood pressure	1 (6.7)		
Migraine	1 (6.7)		

Note: the number of citations may exceed the number of observations due to multiple responses.

Comparing the perception about the work with the gender of the interviewed officers, there are not many differences. Both genders believe the job affects their mental state, however, women seem more condescending with the service because they are government employee, which is a stable job in Brazil, although they are more suspicious (Table 3).

Table 3. Comparison between gender and work perception.

	Male n (%)	Female n (%)	Total n (%)
Remembers at home what saw in prison	1 (50.0)	1 (50.0)	2 (100.0)
Forgets about the prison when leave	1 (100.0)	0 (0.0)	1 (100.0)
Enjoys it	1 (100.0)	0 (0.0)	1 (100.0)
Conforms to it	0 (0.0)	2 (100.0)	2 (100.0)
Mentally affected	5 (50.0)	5 (50.0)	10 (100.0)
Became suspicious of everything	1 (33.3)	2 (66.7)	3 (100.0)
Stressful	2 (66.7)	1 (33.3)	3 (100.0)

Comparing the age with the safety perception (Table 4), it can be seen that the older age groups (45–55 years, 55–65 years) did not mention safety issues, being a problem only to the younger ones. The service time also influenced the prison officers perception about safety (Table 5), it is noticed that the youngest in the company (4–6 years) see more security problems than the other employees.

Table 4. Comparison between age group and safety perception.

	25–35 years n (%)	35–45 years n (%)	45–55 years n (%)	55–65 years n (%)	Total n (%)
Feels fear every day	0 (0.0)	2 (100.0)	0 (0.0)	0 (0.0)	2 (100.0)
Finds it dangerous	3 (50.0)	3 (50.0)	0 (0.0)	0 (0.0)	6 (100.0)
Thinks there's always someone chasing	0 (0.0)	1(100.0)	0 (0.0)	0 (0.0)	1 (100.0)
Fear of inmates attack	1 (50.0)	1 (50.0)	0 (0.0)	0 (0.0)	2 (100.0)

Table 5. Comparison between service time and safety perception.

	4–6 years n (%)	7–9 years n (%)	10–12 years n (%)	>20 years n (%)	Total n (%)
Feels fear every day	2 (100.0)	0 (0.0)	0 (0.0)	0 (0.0)	2 (100.0)
Finds it dangerous	4 (66.7)	1 (16.7)	1 (16.7)	0 (0.0)	6 (100.0)
Thinks there's always someone chasing	1 (100.0)	0 (0.0)	0 (0.0)	0 (0.0)	1 (100.0)
Fear of inmates attack	2 (100.0)	0 (0.0)	0 (0.0)	0 (0.0)	2 (100.0)

Comparing the health problems reported with the age of the interviewed officers, it can be seen that sleep disturbance is constant in all age groups. However, stress and irritability problems seem to be more common in younger ones, between 25 and 45 years (Table 6). These data shows that health problems are not related to the advancement of age, but to the content of the work.

Table 6. Comparison between age group and health problems.

	25–35 years n (%)	35–45 years n (%)	45–55 years n (%)	55–65 years n (%)	Total n (%)
Sleeping disorders	1 (25.0)	1 (25.0)	1 (25.0)	1 (25.0)	4 (100.0)
Tuberculosis	1 (50.0)	0 (0.0)	1 (50.0)	0 (0.0)	2 (100.0)
Allergic rhinitis	1 (100.0)	0 (0.0)	0 (0.0)	0 (0.0)	1 (100.0)
Irritability	1 (100.0)	0 (0.0)	0 (0.0)	0 (0.0)	1 (100.0)
Stress	2 (33.3)	2 (33.3)	1 (16.7)	1 (16.7)	6 (100.0)
Spain pain	0 (0.0)	1 (50.0)	1 (50.0)	0 (0.0)	1 (100.0)
Cancer	0 (0.0)	0 (0.0)	1 (100.0)	0 (0.0)	1 (100.0)
High blood pressure	0 (0.0)	0 (0.0)	1 (100.0)	0 (0.0)	1 (100.0)
Migraine	0 (0.0)	0 (0.0)	1 (100.0)	0 (0.0)	1 (100.0)

The field approach carried out shows that the content of the work affects the psychological state of the prison officers, being characterized by a high level of psychological demands [7, 8], and contribute to increase their stress level, for the high tension involved in this type of work [24]. The lack of security structure and equipment causes the agents to become insecure and deal with fear daily, a fact also identified in other studies [18, 25]. The work situation causes the agents to have more contact with crimes than other types of professions, perceiving a constant threat of danger [25], which makes them more suspicious, as this extract from one of the interviews shows:

> "You realize that anyone is subject to anything, and that these things always happen … hence you are always more suspicious, more protective, because I have 2 children, right" (E.13).

However, over the years, this insecurity seems to disappear, as seen in Table 5, possibly by habit with work:

> "If there is danger, we become more accustomed to it" (E.4).

The lack of hygiene and ventilation, besides the high humidity inside the prison are harmful to the health, being able to provoke diverse diseases, like tuberculosis and allergic rhinitis, also identified in penitentiaries of Rio de Janeiro [12], Salvador [16] and São Paulo [17]. About the sleeping disorders, some interviews extracts show the level of the problem:

> "I had no sleep at night anymore, I had nightmares, I stayed completely disturbed after I started working here" (E.8).

"Some years ago I needed a psychiatrist, I had not slept for three days" (E.7).

Sleeping disturbances, irritability problems and even high blood pressure and spinal pain may be related to the work content and stress level of the worker, also described as some of the effects of stress on prison staff in the other studies [26, 27].

4 Conclusion

The results presented through the analysis of the work and health perception by the prison officers show that the type of work and its content increase the level of stress of the workers analyzed, 40% of whom present stress problems. The lack of security (40% find work dangerous) generates fear (13.3%) and distrust (20%), revealing problems in the organization and work process.

With content analysis, it was possible to verify the officers' perception about their work and health, however, a deeper analysis of the work condition is fundamental to understand how the work context influences their health. This case study does not analyze deeply the aspects of work in prison systems, however, it is expected that this study has evidenced the relevance of the interrelationship between work and the health of the worker, allowing the future analysis of those aspects of work that carry risks of illness.

Acknowledgments. To the Federal University of Santa Catarina, to CAPES - Coordination for the Improvement of Higher Education Personnel, to DEAP/SC - Prison Administration Department of Santa Catarina and to the Male Prison of Florianópolis/SC - Brazil.

References

1. Cheng Y, Kawachi I, Coakley E, Schwarts J, Colditz G (2000) Association between psychosocial work characteristics and health functioning in American women: prospective study. BMJ 320:1432–1436
2. Ishizaki M, Kawakami N, Honda R, Nakagawa H, Morikawa Y, Yamada Y (2006) The Japan work stress and health cohort study group: psychosocial word characteristics and sickness absence in Japanese Employees. Int Arch Occup Environ Health 7:640–646
3. Cassito M, Fattorini E, Giliolo R, Rengo C (2003) Raising awareness to psychological harassment at work. Protecting workers' health series. World Health Organization, Milano, pp 16–23
4. Stansfeld S, Candy B (2006) Psychological work environment and mental health—a meta-analysis review. Scand J Work Environ Health 32:443–462
5. Babazono A, Mino Y, Nagano J, Tsuda T, Araki T (2005) A prospective study on the influences of workplace stress on mental health. J Occup Health 47:490–495
6. Higashiguchi K, Nakagawa H, Morikawa Y, Ishizaki M, Miura K, Naruse Y, Kido T (2002) The association between job demand, control and depression in workplaces in Japan. J Occup Health 44:427–428
7. Johnson S, Cooper C, Cartwright S, Donald I, Taylor P, Millet C (2005) The experience of work relates stress across occupations. J Managerial Psychol 20:1–2

8. Ghaddar A, Mateo I, Sanchez P (2008) Occupational stress and mental health among officers: a cross-sectional study. J Occup Health 50:92–98
9. ILO Homepage. http://www.ilo.org/declaration/lang–en/index.htm. Accessed 30 Mar 2017
10. Dejours C (1990) Travail, usure mentale: essai de psychopathologie du travail. Bayard, Paris
11. Bourbonnais R, Jauvin N, Dussalt J, Vézina M (2007) Psychosocial work environment, interpersonal violence at work and mental health among correctional officers. Int J Law Psychiatry 30:355–368
12. Vasconcelos A (2000) A saúde sob custódia: um estudo sobre agentes de segurança penitenciária no Rio de Janeiro. Master Thesis, Escola Nacional de Saúde, Rio de Janeiro
13. Damas FB (2012) Assistência e condições de saúde nas prisões de Santa Catarina, Brasil. Revista de Saúde Pública de Santa Catarina 3(5):6–22
14. Reinert R, Merino ECD, Gontijo LA (2014) Análise das condições de trabalho do assistente social no Presídio Femino de Florianópolis/SC. Ação Ergonomica 9(2):97–106
15. Diuana V, Lhuilier D, Sánchez AR, Amado G, Araújo L, Duarte A (2008) Saúde em prisões: representações e práticas dos agentes de segurança penitenciária no Rio de Janeiro, Brasil. Cadernos de Saúde Pública 8(24):1887–1896
16. Fernandes R, Silvany Neto A, Sena G, Leal A, Carneiro C, Costa F (2002) Trabalho e cárcere: um estudo com agentes penitenciários da Região Metropolitana de Salvador, Brasil. Cadernos de Saúde Pública 18:807–816
17. Rumin C (2006) Sofrimento e vigilância prisional: o trabalho e a atenção em saúde mental. Psicologia: ciência e profissão 26(4):570–581
18. Santos M (2010) Agente penitenciário: trabalho no cárcere. Master Thesis, Universidade Federal do Rio Grande do Norte, Natal
19. Santos D, Dias J, Pereira M, Moreira T, Barros D, Serafim A (2010) A Prevalência de transtornos mentais comuns em agentes penitenciários. Revista Brasileira de Medicina do Trabalho 8(1):33–38
20. Gomes SM (2009) Sofrimento mental e satisfação no trabalho em professores de unidades prisionais em Porto Velho. Master Thesis, Universidade de Brasília, Brasília
21. Gil AC (1996) Como elaborar projetos de pesquisa, 3rd edn. Atlas, São Paulo
22. Triviños ANS (2006) Introdução à pesquisa em ciências sociais: a pesquisa qualitatica em educação. Atlas, São Paulo
23. Bardin L (2013) L'analyse de Contenu. Presses Universitaires de Frances, Paris
24. Schaufeli WB, Peeters MCW (2000) Job stress and burnout among correctional officers: a literature review. Int J Stress Manag 7:19–48
25. Armstrong GS, Griffin M (2004) Does the job matter? Comparing correlates of stress among treatment and correctional staff in prisons. J Crim Justice 32:577–592
26. Anson RH, Johnson B, Anson NW (1997) Magnitude and source of general and occupation-specific stress among police and correctional officers. J Offender Rehabil 25:103–113
27. De Carlo DT, Gruenfeld DH (1989) Stress in the American workplace: alternatives for the working wounded. PA7 LRP Publications, Fort Washington

Ergonomic Evaluation Tools Associated with Biomechanical Risk Factors in Work Activities: Review of Literature

María José González Carvajal(✉), Fernanda Maradei García, and Clara Isabel López Gualdrón

Universidad Industrial de Santander, Bucaramanga, Colombia
MARIAJO.G.DI@gmail.com, {mafermar,clalogu}@uis.edu.co

Abstract. **Objective:** Identify the most used evaluation tools in work situations that involve biomechanical and postural load. Through the literature review, it seeks to answer the following questions, which is the influence of the work or occupational ergonomics in the industry? What are the ergonomic evaluation tools most used in labor situation similar to surgical support? and What is the relation between working conditions and the appearance of symptoms associated with pain and occupational diseases?. **Method:** Were defined keywords related to main areas: Ergonomics, productivity and risk factors, with which it was subsequently made a literature review on these databases: PubMed, Springer, and ScienceDirect. Likewise, it is included in the review, documents related to OMS, OIT, Fasecolda, among others. The documents founded were classified taking into account the inclusion criteria like language (Spanish, English, and French) and the relation with interest topics. Finally, it has been made the analysis of content with Nvivo Software. **Results:** It has been identified that the most recurrent evaluation tools for studies, related to labor activities that involve biomechanical and postural load, were LEST, REBA and VAS methods. Likewise, it was identified the relation between biomechanical risk factors with labors activities that involve load lifting and handling. **Conclusions:** Similar Investigations on the surgical support in companies were reviewed, show that exist a tendency toward the appearance and development of MSDs of occupational origin, especially injuries at the lumbar level due to biomechanical demand of these activities.

Keywords: Ergonomic evaluation · Productivity · Risk factors
Load lifting · Postural load

1 Introduction

The intervention of the labor ergonomics in the industry is based on eliminating the exposure of the worker, to different risk factors implicit in the labor situation [1] and based on the findings, propose solutions to the work transformation. This intervention has a clear purpose of avoiding consequences like absenteeism [2, 3], Labor accidents and incidents, physical and mental exhaustion, occupational diseases [4] and musculoskeletal disorders,

S. Bagnara et al. (Eds.): IEA 2018, AISC 818, pp. 263–271, 2019.
https://doi.org/10.1007/978-3-319-96098-2_34

MSDs mainly [5]. The last ones are the most frequent occupational health problems with 92% of prevalence [6].

In the literature review is noted the absence of optimal working conditions inside a labor situation, influencing in health and productive performance of the worker [7, 8], this has been determined as the origin of occupational symptomatology and diseases, therefore absenteeism and subsequent mobility [9]. Studies related to interventions that have been realized in different labor situations shows like the first action, the implementation of a method and observational and ergonomic evaluation tools [10], for the purpose of characterizing the labor situation until determinate the own critical aspects [11]. Thus, the application of ergonomic evaluation methods to evaluate each factor that makes up a work situation [12, 13], facilitating the detection and characterization of risk factors to which the worker is exposed and based on the findings, define an action plan to transform o become better the workstation or work situation. The above seeks to mitigate the health impact and the productive performance of the worker, as some studies show [1, 14].

According to the statements and developments of some authors in different studies about, identification of risk factors in occupational origin and subsequent intervention by researchers from GEPS research group, It was identified as critic observation of the situation made in real work at the company because of that risk factor were associated with a biomechanical and postural load that is experienced diary. Both factors were defined as key aspects of literature review due to the identification of observation critic's made to the real work situation, wherein it was observed two of four work areas showed exposition related to above risk factors. The work areas observed correspond to administration, quality control, production and surgical support, the last two were estimated as critical, therefore it was pretended to define the most convenient evaluation method to do a detailed ergonomic evaluation in the study area.

Unless worldwide, the professional as surgical instrumentalist makes labors mainly with the support of the surgeon inside the operating theater also, support as trainers of a new generation of professionals, researchers, and commercial administrator in this area, However, it wasn't found studies related to the performance of the identified activities by way of a critic observational study, made on the company, because of that the business increase the search of related risk factors to this and similar labor activities. The consequence of lack investigations around this specific scenario made necessary the implementation of proposal studies about solutions of risk factors and the way how an instrumentalist should do the work in optimal conditions without affecting the health of the professional and the indicators of productivity inside the organization, therefore, it was pretended to make a literature review that allows defining ergonomic evaluation methods.

2 Materials and Methods

2.1 Methodology

The methodology that was developed to identify the relevant literature, started with keywords definition and main areas for the search, Ergonomics, productivity, and risk factors. To each word was assigned a related terminology to encompass a higher quantity of documents to analyzed them inside the interesting context. Taking into consideration the study areas, some databases were selected, between then: PubMed, Springer, and ScienceDirect, where the searches were done. Also, some websites were consulted related to the occupational areas like OMS, OIT, FASECOLDA, Colombian Ministry of Health or work, and others entities relevant to the revision.

The literature search used terms associated with each key areas from where it was obtained some documents being classified by specific inclusion criteria as language (Spanish, English, and French) and quality, this criterion was the most value for the selection and analysis of literature content, therefore, the articles had to be related to application of ergonomics assessment methods and evidence from industrial cases with intervention in problematic labors germane to postural and load risk factors. A revision of titles and abstracts of the literature review was the first step to the detection of interest papers related to the inclusion criteria. Subsequent, it was reviewed the discussion and results of the articles previously selected to corroborate the pertinence and discard papers without relevant thematic. Finally, the systematic review of literature and contents was done with the Nvivo Software where the articles were classified per the relationship with one of three main themes.

2.2 Keywords Definition for Literature Review

The keywords Ergonomics, productivity and, risk factors were defined as main areas to the search, each one was assigned a series of words related between then with the purpose of expanding the useful terminology for the literature review. As results of the keywords definition to the search content, it was generated a list of 18 terms classified in three groups according to initial keywords, which is shown in the Table 1.

Table 1. Matrix of keywords used in the search of scientific literature.

A. ERGONOMICS	B. PRODUCTIVITY	C. RISK FACTORS
"Work ergonomics"	Industry	"Biomechanical load"
"Ergonomic evaluation"	"Work activities"	"Musculoskeletal disorder"
"Ergonomic intervention"	Workflow	Mental load
"Work station"	"Productive cycle"	Lifting
Painful	"Tasks assigned"	"Postural behavior"
Work transformation	"work demand"	"cumulative loading"

The search was carried out with a combination of each key areas of the form A + B + C, then with classified articles, it tried to answer the revision questions like, which is the influence of the work or occupational ergonomics in the industry? What are the ergonomic evaluation tools most used in labor situation similar to surgical support? and What is the relation between working conditions and the appearance of symptoms associated with pain and occupational diseases?

2.3 Content Analysis and Results

The bibliographic content analysis was developed with the Nvivo software and the information was classified and coded according to the interest areas and themes. This analysis allowed identified the most frequent words for which a word cleaning was done, taking into consideration only words related to interest themes. As results were obtained a words cloud which was taken as important input to identify relevant fragments in the reviewed investigations and then to be able to answer the revision questions for a methodological route characterization applied in industrial sectors that involve biomechanical and postural load. 40 articles were selected that met the inclusion and quality criteria, these articles were published between the years 2005 and 2018 and corresponds to ergonomics, occupational, public health and productivity journals. Below are the results of keywords frequency and a findings synthesis related to investigation questions.

2.4 Terminology Frequency Associated with Interest Themes

The content analysis in the Nvivo software allowed the identification of frequent words in analyzed documents. In the search, it was found that the most frequent words are observation, ergonomic methods and, disorder. Once the articles to be analyzed were consolidated, the cloud of words for the study was formed (Fig. 1).

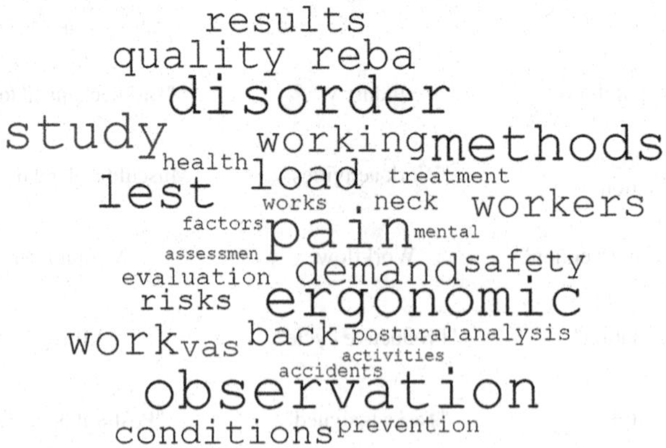

Fig. 1. Cloud of words with the highest co-occurrence, identified in the documents analyzed and coded in the Nvivo software.

3 Summary of the Findings Identified in Relation to the Review Questions Raised

In this section, the findings that were identified as pertinent are exposed, in order to respond to the review questions asked. Each of the questions was posted in order to identify consensus in the literature related to ergonomic interventions in different industries, evaluation methods implemented and results. Likewise, we sought to identify studies related to work activities, specifically those related to risk factors derived from the biomechanical and postural load.

(a) Which is the influence of the work or occupational ergonomics in the industry? The ergonomic intervention in the industry was grounded in the identification of different risk factors which the workers are exposed in the labor situation [1]. The intervention is given in order to avoid the multiple consequences that can trigger risk factors, such as labor absenteeism [2, 3], repetitive disabilities, work incidents and accidents, physical and mental exhaustion, low work performance, occupational diseases [4] performance deficiency and MSDs [5], This last ones are the most frequent problems originate from work, until 92% of prevalence [6].

In fact, the literature exposes cases of study in different work situations wherein the ergonomic has intervened and identified labor risk factors that impact on work performance and workers health lowering the productivity. Elements like the height of the workstation [12], lifting and handling of the load higher than indicated [21], inadequate postures (Natarén and Elío 2004), inadequate postures, repetitive movements, handling tools and equipment [1], among others. These investigations have favored the development of tools and ergonomics evaluation methods used to diagnose work situations and identify relevant aspects that characterize them as critical or painful. The productivity is related to the efficiency and the quality in terms of worker results [15], which depends on the interrelation between health, wellness, satisfaction, labor motivation, psychosocial aspects [10] and availability of the necessary elements for assigned tasks [16, 17].

Meanwhile, Escorpizo proposes a conceptual model about labor productivity respect absenteeism rates, based on three fundamental pillars that are evaluated from the ergonomics: worker metal status, risks that affect the worker, the capabilities, wish of work, and no labor factors in terms of private life that influence psychosocial relationship [18], this model is intended to determine the factors that influence in productivity and which is the relationship between all factors [19, 20].

(b) What are the ergonomic evaluation tools most used in labor situation similar to surgical support? According to the consensus of the authors identified in the literature review [21, 22], it was established that exist different methods to evaluate in a general way a determined work situation and the hardship level. Therefore, it was defined tools and specific evaluation methods which are used then to a global evaluation [23] or in parallel to support the results and conclude about the detailed situation. The observational method, according to some authors [21], is applied to a work situation allowing to identify in real time the activities that are carried out under certain conditions, which are better perceived under this method and, it is ver used to characterize the labor situation intervened and, then define factors that determine the degree of hardship to allow the inclusion of another ergonomic evaluation method in the specific study [24].

Some studies related to different industries wherein the critic factor is the biomechanical load risk. These studies have shown the implementation of LEST methods through by which work conditions are known, determined in terms of the general labor situation. Two important risk factors to consider are load handling and manipulation because can lead to considering a work situation as critical. Likewise, were identified some tools to evaluate specific aspects, as the NIOSH equation, that is used to specifically assess different labor activities that involve load lifting. The exposition to biomechanical load risk has immediate consequences like inadequate postures and the worker try to adapt to the situation responding to the tasks in charge, however, this situation leads to symptomatology development associated with the appearance and growth of MSDs mainly in the back, being one of the biggest causes of absenteeism and labor mobility according to identified reports [6]. In fact, this situation increase costs related to the affected worker by labor risk factors decreasing the enterprise productivity, affecting the economic indicators [9], which has been recognized as dependent on the workers quality of life.

Also, was found that the postural load risk exposition, not only is related to the biomechanical risk even in administrative work situations without implying load manipulation also has evidence of symptomatology and sickness associated with the adapted postural behavior. In both situations is shown the implementation of evaluation methods like the case of REBA method, which has been recognized as a useful tool and pertinent to evaluate the worker postural behavior such as in sitting or standing position [25]. REBA method respect Rula method covers the entire human body, so if the studio needs a load evaluation or a complete postural behavior, the REBA method is the indicated (Kee and Karwowski 2007). Then, it was identified other methods implementation like self-reports in order to know the pain perception and discomfort in different parts of the worker body. In this case, the method related is VAS self-report as an ergonomic evaluation method in which the worker manifests the pain intensity level that experiences [26].

(c) What is the relation between working conditions and the appearance of symptoms associated with pain and occupational diseases? Sundry risk factors related to MSDs are generated by work linked to different biomechanical load aspects, strength application, maintained or prolonged inadequate postures, load lifting, and manipulation, among others, agree with this affirmation, studies carried out by Muñoz (2012) mentioning the strong relationship between the labor risk factors with muscular fatigue manifested in the worker body that affects the labor performance.

With relation to the consequences triggered by the labor activities according to identified Kumar findings reported on the literature that the 50% of worker population is found in a constant exposition to labor risks [25], unchaining lesions and musculoskeletal disorders DSMs [27], being the biggest cause of work absenteeism recorded in case studies [28]. It's pertinent to mention that the DSMs may be due to cumulative trauma lesions and are conditions given by muscle injuries, nervous, tendons, articulations and support members that involve the development of labor activities [29]. Otherwise, according to the European poll about the labor conditions (EwCS 2015), The DSMs for labor cause are the health most reported cases in Europe and USA hereby this pathology affect millions of workers and implies health high costs and low production for the company. In the cases, studies analyzed according to established consensus in the literature is found that the labor risk is identified as a common factor in a different work situation that involves load lifting and manipulation, repetitive movements, among others [30]. In consequence, it's shown the prevalence of lumbago, physical fatigue, and herniated discs, among others professionals sickness that the workers may be developed in determined workstations [31, 32].

4 Discussion

The main purpose of this literature review study was characterized ergonomic evaluation tools that have been used in different industrial sectors. The review focused on analyze studies related to evaluations performed in work situations wherein the biomechanical and postural risk factor is evident. It was necessary to analyze studies made in similar situations to instrumentation or surgical support because it wasn't possible found related studies to this labor activity in a company. This work corresponds to the enlistment of instrumental or surgical material, that involves load lifting and manipulation with inadequate postural behavior. These aspects were taken as a reference to analyze the search and to the rest related interest terminology. Findings in literature affirm the importance of ergonomic evaluation tools and methods in the industry, favoring the identification of risk factors with the final purpose of determinate the hardship level in a work situation. Diverse studies conclude about the need for a full occupational ergonomic evaluation using more than one tool or method to cover in a detailed way of task aspects, workstation or equipment that may be generating risks for operators. It's evident that different ergonomic evaluation tools and methods exist wherein each one allows to deepen in the interest area, ether postural load, pain perception, biomechanical load manipulated, the work conditions, or psychosocial aspects, among others. Likewise, it's considered the inclusion of physics conditions measurement in labor space as sonometer or luxometer. The sonometer measures the sound intensity which is catalog as a determinant factor in the focused or sustained attention process that the worker should have in the work development to avoid mistakes, being an important risk factor from the mental load until generating symptoms related to hearing damage if it exceeds 85 decibels in a production environment. The glucometer is another tool to consider factors in workstation like the illumination in the space, being a determining variable in the quality of the work.

In summary, to perform a full evaluation of a work situation is considered some characteristics related to the specific space or a workstation wherein is developed the labor activities, the mental load implicated in determination labor activities among the work situation, the biomechanical load that is experienced by the worker and the manipulation of the tools, the postural behavior in each evaluated labor situation and finally the pain perception that the patient manifest in relation with labor activities. The LEST, REBA and VAS methods, are efficient tools to the risk factors identification associated with the evaluated work situation, therefore, due to the technological development, currently is taken into account with a direct measurement that allows evaluating in a second phase the work situation in an objective way, which constitutes a complement to the required observational evaluations to the diagnosis in the early stages of the ergonomic studies.

5 Conclusions

It is important that the evaluations are carried out from different areas, in this form may be determined the risk factors in a physical, psychosocial, biomechanical, mental and postural. Likewise, the hardship level in a work situation determines the need and the type of intervention that should be applied, whereby, the ergonomic assessment methods and tools that best fit must be identified depending on the specific situation. Due to the evaluation that it wants to perform in an industrial environment specifically to the work situation in surgical support of instrumental enlistment or surgical material. The use of LEST, REBA, and VAS methods depends on the specific method to takes into consideration due to the work situations and also shown as complete and, easy to apply methods, without intervention in the workflow or the same organization.

References

1. Pérez S, Méndez J, Jiménez A, Pérez S, Méndez J, Jiménez A, Ramos M, Aguilera V (2014) Análisis y optimización de estaciones de trabajo, con enfoque ergonómico para el aumento de la productividad y disminución de riesgos laborales. Ciencias de la Ingenieria y Tecnologia, pp 176–187
2. Ccollana-salazar Y (2015) Rotación del personal, absentismo laboral y productividad de los trabajadores, pp 50–59
3. Guadalupe J, Estrada S, Cristóbal IJ, Pupo G, Yadira II (2009) Clima y cultura organizacional: dos componentes esenciales en la productividad laboral. Rev Cubana de Inf en Ciencias de la Salud (ACIMED) 20(4):67–75
4. Díaz C (2010) Actividad laboral y carga mental del trabajo. Comport Organ 12(36):281–292
5. Arbeláez G, Velásquez A, Tamayo C (2011) Principales patologías osteomusculares relacionadas con el riesgo ergonómico derivado de las actividades laborales administrativas. Rev CES Salud Publica 2(2145–9932):196–203
6. Tolosa-Guzman I (2015) Riesgos biomecánicos asociados al desorden músculo esquelético en pacientes del régimen contributivo que consultan a un centro ambulatorio en Madrid. Rev Ciencia y Salud 13(1):25–38
7. Karlqvist L, Wigaeus E, Hagberg M, Hagman M (2002) Self-reported working conditions of VDU operators and associations with musculoskeletal symptoms: a cross-sectional study focussing on gender differences. Int J Ind Ergon 30:277–294
8. Zare M, Bodin J, Cercier E, Brunet R, Roquelaure Y (2015) Evaluation of ergonomic approach and musculoskeletal disorders in two different organizations in a truck assembly plant. Int J Ind Ergon 50:34–42
9. Caraballo Y (2013) Epidemiología de los trastornos musculo -esqueleticos de origen ocupacional. Temas De Epidemiologia Y Salud Publica, pp 745–764
10. Vernaza P, Sierra C (2005) Dolor músculo-esquelético y su asociación con factores de riesgo ergonómicos, en trabajadores administrativos. Rev de Salud Pública 7(3):317–326
11. Valle E, Manero R (2008) Evaluación integral del nivel de riesgo músculo esquelético en diferentes actividades laborales. Salud de los Trabajadores 16(1):17–28

12. Lite AS, García MG, Ángel M (2007) Métodos de evaluación y herramientas aplicadas al diseño y optimización ergonómica de puestos de trabajo, pp 239–250
13. Escalante M (2009) Evaluación ergonómica de Puestos de Trabajo. In: Seventh LACCEI Latin American and Caribbean Conference for Engineering and Technology (LACCEI'2009), pp 1–7
14. Alonso ML, Dolores M, Aires M, González EM (2011) Análisis de los riesgos musculoesqueléticos asociados a los trabajos de ferrallas. Buenas prácticas Musculoskeletal risks analysis related to steel reinforcement works. Good practices. Ingenieria de Construccion 26:284–298
15. Cequea M, Rodríguez-Monroy C, Bottini MN (2011) Productividad humana. análisis factorial de sus dimensiones y factores, no. February 2014, pp 1–10
16. Lucas RAI, Epstein Y, Kjellstrom T (2017) Excessive occupational heat exposure: a significant ergonomic challenge and health risk for current and future workers, pp 1–15
17. Lan L, Lian Z, Pan L (2010) The effects of air temperature on office workers' well-being, workload and productivity-evaluated with subjective ratings. Appl Ergon 42(1):29–36
18. Escorpizo R (2008) Understanding work productivity and its application to work-related musculoskeletal disorders. Int J Ind Ergon 38:291–297
19. Cequea MM, Monroy CR, Angel M, Bottini N (2014) La productividad desde una perspectiva humana: dimensiones y factores. 7(2): 549–584
20. Arenas-Ortiz L, Cantú Gómez Ó (2013) Factores de riesgo de trastornos músculo-esqueléticos crónicos laborales. Med Interna de Mexico 29(4):370–379
21. Molen HFVD, Mol E, Kuijer PPFM, Frings-dresen MHW (2007) The evaluation of smaller plasterboards on productivity, work demands and workload in construction workers. Appl Ergon 38:681–686
22. Molen HFVD, Kunst M, Kuijer PPFM, Frings-dresen MHW (2011) Evaluation of the effect of a paver' s trolley on productivity, task demands, workload and local discomfort. Int J Ind Ergon 41(1):59–63
23. Molina A (2010) Ergonomia. Espadelada, vol. uno, pp. 3–93
24. Ruíz YR, Brito SV, Martínez RM, No IC, Habana L (2010) ERIN: Un método observacional para evaluar la exposición a factores de riesgo de desórdenes músculo-esqueléticos. Convencion científica de ingeniería y arquitectura, no. 11901
25. Lopez B, González E, Colunga C, Lopez E (2014) Evaluación de sobrecarga postural en trabajadores: revisión de la literatura. Ciencia and Trabajo, pp 111–115
26. Ibáñez RM, Manzanares A (2005) Escalas de valoración del dolor. vol. LXVIII, no. 527, pp 527–530
27. Sirit Y, Amortegui M (2007) Síntomas Músculo Esqueléticos en Trabajadores de una Empresa de construcción civil. Salud de los Trabajadores 15(2):89–98
28. Hurley K, Marshall J, Hogan K, Wells R (2012) A comparison of productivity and physical demands during parcel delivery using a standard and a prototype electric courier truck. Int J Ind Ergon 42(4):384–391
29. Venicio J, Valerio G, Herrera EY, Torres LB, Damian WC (2016) Agroindustrial Science 6: 199–212
30. Morales KL, Acosta G (2017) Diez años de ergonomía en el Banco de la República : De la mano en pro de la salud y la productividad
31. Rodriguez-Ruíz Y, Pérez-Mergarejo E (2014) Procedimiento ergonómico para la prevención de enfermedades en el contexto ocupacional/ergonomic procedure for the prevention of occupational disease. Rev Cubana de Salud Publica 40(2):279–285
32. Caraballo Y (2013) Epidemiología de los trastornos músculo- esqueléticos de origen ocupacional. Temas de epidemiología y salud pública 2:745–764

Development of an Education Scheme for Improving Perioperative Nurses' Competence in Ergonomics

Tamminen-Peter Leena[1(✉)] and Nygren Kimmo[2]

[1] Ergosolutions BC, Turku, Finland
letampe@gmail.com
[2] Satakunta Hospital District, Pori, Finland

Abstract. Perioperative nurses have several high-risk tasks and musculoskeletal disorders are severe problems. Preventive measures are mostly insufficient. The '*Ergonomic patient handling card®*' (later *Card*)—education scheme was introduced in Finland in 2009 with an aim of improving both caregivers' and patient safety. Experiences of this multicomponent programme from both the care units and home care are good; the physical workload and sickness absences due to MSDs are decreasing. As the perioperative nurses' work in Operating Rooms (later *ORs*) varies very much from the work in care units, a special education scheme for them was needed to be developed, aimed at defining the competencies, skills, and knowledge levels required to perform perioperative work safely; ensuring compliance with legislative requirements; and improving both, safety and quality in *ORs*.

The most stressful tasks in *ORs* in Finland were defined by a focus group and by risk assessments in the acute hospital. It produced similar results as the AORN guidance of seven high-risk patient handling tasks in *ORs*. The basic *Card* was a good base for developing a new teaching scheme. The new content has been tested by two pilot-courses. The evaluation results were positive and resulted in minor content changes. The new *Card*—education scheme for perioperative nurses consists of the same four parts as the basic *Card*. The practical training lasts 14 h whereof 7 h is trained in the ORs. Safe working methods and safe usage of appropriated equipment are emphasized in training.

Keywords: Perioperative nurse · Competence · Patient handling

1 Introduction

Perioperative nurses have several high-risk tasks and musculoskeletal disorders are severe problems. The annual prevalence of back pain among Dutch Operating Rooms later *ORs* personnel was 58%; it is a considerably higher number than the one of any other working population. The pain of other body parts, like neck, shoulder, legs and feet are also common [7]. Similar results have been reported also from other parts of the world [3, 4]. Preventive measures are mostly insufficient. AORN Journal published seven guidance statement articles: Safe patient handling and movement in the perioperative

S. Bagnara et al. (Eds.): IEA 2018, AISC 818, pp. 272–278, 2019.
https://doi.org/10.1007/978-3-319-96098-2_35

setting. These articles describe specific ergonomic solutions for high-risk patient handling tasks in the perioperative clinical setting. [6, 13–17].

In Finland, the 'Ergonomic patient handling card®'—education scheme was introduced in 2009 aimed at improving both caregivers' and patient safety. The ***Card***—training is widespread in Finland and experiences of this multi-component education scheme from the care units and home care are good; the physical workload and sickness absences due to MSDs are decreasing [12]. The perioperative nurses' work in ***ORs*** varies very much from the nurses' work in care units. A special education scheme for them was needed to be developed.

Professional competencies in polytechnics require that nurses have a good basic knowledge of rehabilitative nursing and that they can work in an ergonomic way [8]. The curriculum of practical nurses' emphasises the need to observe occupational and patient safety [8]. The legislative requirements to ensure that qualified students are competent to perform their tasks safely are mostly not fulfilled.

2 Objectives

The aim was to develop a perioperative nurses' *Card*—education scheme which was to incorporate compliance with legislative requirements, define the competencies, skills and know-how required for safe patient handling in the *ORs,* to improve both patient safety and the quality of care. Furthermore, the scheme was to enable caregivers to improve their competencies in patient handling and increase the effectiveness of training safe, fluent working practices.

3 Development of an Education Scheme for Perioperative Nurses

3.1 The Stressful Tasks in *ORs* in Finland

A focus group from different *ORs* was convened to define the most stressful tasks in *ORs* in 2011–2012 [9]. The static work exists in many tasks, prolonged standing, holding the same position, tissue retraction, and working with lead vests, horizontal transfers of patients, pushing and pulling. The focus group produced quite similar results as the AORN guidance to seven high-risk patient handling tasks in *ORs* [6, 13–17].

During 2014–2015, the Satakunta Hospital District carried out the development project "Sataplus—ergonomic criteria and good practices" [11]. The aim of the project was to improve ergonomics, occupational and patient safety when handling plus-size patients. The *ORs* took part in the project and the risk assessment revealed the following high–risk patient handling tasks: horizontal patient transfers, assisting patients into operation positions, turning patient from supine to prone or vice versa and, correction of the patient's position in the recovery room.

3.2 Planning the Content of the Education Scheme

The *Card*—education scheme (Fig. 1) constituted a good starting base and provided the needed study material [10]. Based on the earlier assessment and focus group's work two ergonomics experts were considering what was to be taught additionally to the basic *Card*—training and what to be discarded, what could be taught during E-learning and what by practical training. They agreed upon that as done in the basic *Card*—training, theoretical matters, laws and the body awareness exercises could be published for online study. In the practical training, one must focus on the horizontal transfers, turning patient from supine to prone and vice versa and repositioning. For the practical exam 2 the same criteria as in basic *Card*—training could be used to ensure compliance with legislative requirements.

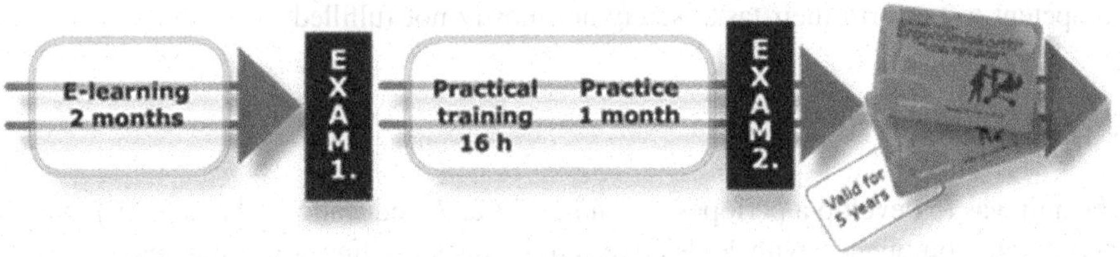

Fig. 1. The *Card*—scheme consists of four parts: E-learning, practical training, application of skills and revision and demonstration of practical skills

3.3 Pilot Courses

The content has been tested by two pilot-courses. The first course was kept as a part of the Sataplus-project and all the participating nurses (n = 9) came from different *ORs* of the same hospital. E-learning was as extensive as in basic *Card*—training but the content was modified for the *ORs*' needs. Practical training was 8 h, 4 h in an *OR* and 4 h in the training room. One month later a 2 h rehearsal took place followed by the exam.

In the second-pilot-course participants (n = 9) came from three hospitals situated in different parts of Finland which gave a good base to exchange experiences. E-learning was kept unchanged but the practical training period was increased from 8 to 12 h. Two physiotherapy based ergonomic experts kept both pilot courses.

3.4 Evaluation of Pilot Courses

The evaluation was carried out with a questionnaire with open and scaled questions. The evaluation results of E-learning are presented in Table 1, the participants appreciated module 1 and 3 the most. In module 1 they assess physical workload and risks at their work and in module 3 they study biomechanical principals and how to use different assisting devices. Despite its importance in the static work, body awareness exercises did not find appreciation. The extent of E-learning surprised the participants but they considered it nevertheless beneficial. Practical training was most liked and participants wished to have more time of practical training especially in the *ORs*. Although the practical training time was increased in the second pilot from 8 to 12 h, again still more time was wished, especially in the *ORs*.

Table 1. Evaluation of E-learning tasks in the pilots 1 and 2.

	Scale 1–5 Mean values	
	1 pilot	2 pilot
1 Ergonomics preventing musculoskeletal disorders (1 = not beneficial - 5 = very useful)		
1.1 Study about physical workload in the nursing work	3.8	3.9
1.2. Assessment of physical load and risks from the pictures and videos	4.0	4.1
2. Body awareness exercises and analyses own movement		
2.1 Study of the natural movements of a human being and about body awareness	3.0	3.1
2.2. Exercises and keeping a diary of own exercises, analyze the results and make a summary	2.2	2.6
3. Biomechanical principals and assistive devices		
3.1. Study about the biomechanics of movement and assistive devices	3.6	3.7
3.2. Task: Usage principals of different assistive devices	4.0	3.9
4. Laws and acts		
4.1 Study the content of occupational safety at work	3.6	2.8
4.2 Discussion about a case based on the laws and acts	3.6	3.0

4 The Perioperative *Card*—Education Scheme

Based on the evaluation results and oral feedback the new *Card*—scheme for perioperative nurses consists of the same four parts: (1) E-learning (2) Practical training of evidence-based principals (3) Application of skills at one's own workplace and (4) Revision and demonstration of practical skills.

4.1 E-Learning

The online platform comprises the theoretical fundamentals: electronic study material, exercises, tests and a discussion forum. Although being quite extensive it was decided to keep it as in the pilot courses i.e. four modules must be completed within two months:

1. Studying the epidemiology of nurses' back problems, different lifting and transferring techniques, the causes of musculoskeletal disorders, ergonomics of the work environment, analysis of risk factors in typical work tasks in *ORs*.
2. Exercises to improve body awareness, keeping a diary about one's own body experiences to become more aware of one's tactile senses.

3. Studying biomechanical principles, becoming acquainted with assistive devices, hoists, analysing the biomechanical principles to apply them in patient handling.
4. Reading the acts related to patient handling, discussing cases with fellow students in order to become familiar with occupational safety responsibilities and obligations.

4.2 Practical Training

The practical training part is prolonged to 14 h whereof 7 h in an ***OR*** as wished. Training emphases on safe working methods and safe usage of appropriated equipment in the risk tasks. Participants are training in the training room: horizontal transfers, turning patient from supine to prone and visa versa, repositioning on the narrow table and the principles of normal human movement in order to move optimally when involved in patient-handling. Participants practice also, how to apply safe, ergonomic handling principles in various handling situations such as: getting up from a lying or sitting position, turning and moving in bed, from sitting to standing. The practical training in the *ORs* concentrates on the different operation positions and how to assist the patient in the correct positions on the operation table.

4.3 Exam

The exam takes place after one month's application time at students' own workplaces. Before the exam, students have the opportunity to rehearse for a few hours. During the exam, two transfers are performed by a team of 2 or 3 students, one manually and one by a hoist. The activities are filmed and two qualified *Card*—trainers evaluate the transfers. The course has been accredited for 3 credit points.

5 Discussion

While the patient safety procedures are emphasised in most *ORs*, less attention is paid to ergonomics and other occupational safety matters. In the United States and Australia nurse's associations [1, 2], have implemented a proactive, multifaceted plan to support the safe patient handling and the prevention of MSDs. As neither the Finnish perioperative nurses association has given any ergonomic guidelines, nor do the professional education of perioperative nurses in Finland fulfil the legal requirements, the *Card*—education scheme improves the ergonomic competence in the *ORs* and prevent MSDs.

The evaluation of the pilot courses showed that participants were mostly satisfied with the *Card*—scheme content hence no bigger adjustments were needed. As wished, the length of practical training was increased. The body awareness exercises remain in force in spite of not being appreciated. The challenge is how to tell the students how crucial the tactile senses are to keep the body in an optimal position in any work but especially in the static work as improved proprioceptive sensation helps to work in a well-balanced position.

There are many ways to improve the ergonomics in the *ORs*. This educational approach is not only teaching how to work ergonomically and safety; it implies the basic know-how of the risks assessment, ergonomic principles in the prevention of musculoskeletal strain and disorders. Two pilot courses are a good starting point, but the whole teaching scheme will be further developed once the training has resulted in more feedback. Training is to be evaluated after every course with a sensitive eye. Surgical work is developing fast and ergonomic experts will have the challenge to keep up, it will be good to have a couple of instructors, a perioperative nurse and an ergonomic expert to compare notes.

References

1. AORN Standards for Perioperative Nursing in Australia, 14th edn. Accessed 11 Apr 2018
2. American Nurses Association (ANA). Handle with care fact sheet. http://www.nursingworld.org/MainMenuCategories/ANAMarketplace/Factsheets-and-Toolkits/FactSheet.htm. Accessed 11 Apr 2018
3. Bos E, Krol B, Star L, Groothoff J (2007) Risk factors and musculoskeletal complaints in non-specialized nurse, IC nurse, operation room nurse, and X-ray technologists. Int Arch Occup Environ Health 80:198–206
4. Choobineh A, Movahed M, Tabatabaie A, Kumashiro M (2010) Perceived demands and musculoskeletal disorders in operating room nurses of Shiraz City hospitals. Ind Health 48:74–84
5. Finnish National Agency for Education. Sosiaali- ja terveysalan perustutkinto 79/011/2014 http://www.oph.fi/download/177323_Maarays_11_011_2016_YTO_16_6_2016.pdf. Accessed 16 Apr 2018
6. Hughes NL, Nelson A, Matz MW, Lloyd J (2011) AORN Ergonomic Tool 4: solutions for prolonged standing in perioperative settings. AORN J 93(6):767–774
7. Meijsen P, Knibbe H (2007) Work-related musculoskeletal disorders of perioperative personnel in the Netherlands. AORN J 86(2):193–208
8. Finnish National Agency for Education. Sosiaali-ja terveysalan perustutkinto 79/011/2014 http://www.oph.fi/download/177323_Maarays_11_011_2016_YTO_16_6_2016.pdf. Accessed 16 Apr 2018
9. Tamminen-Peter L, Peltonen P, Lagus R, Hämäläinen K et al (2012) Ergonomialla kevennystä leikkaussalityöhön. Pinsetti 2:31–32
10. Tamminen-Peter L, Fagerström V (2014) Did the Finnish Ergonomic patient handling card®-scheme evoke changes in vocational education and workplaces? In: The 5th international conference on applied human factors and ergonomics AHFE 2014 proceedings, Krakow, Poland
11. Tamminen-Peter L, Nygren K, Moilanen A (2016) Ergonomic criteria and good practices for bariatric patients' care. In: Proceedings of NES-conference 2016, Kuopio
12. Tamminen-Peter L, Sormunen E (2016) The ergonomic patient handling card®-education scheme. In: HEPS—conference proceedings, Toulouse, France
13. Waters T, Baptiste A, Short M, Plante-Mallon L, Nelson A (2011) AORN ergonomic tool 1: lateral transfer of a patient from a stretcher to an OR bed. AORN J 93(3):334–339
14. Waters T, Baptiste A, Short M, Plante-Mallon L, Nelson A (2011) AORN ergonomic tool 6: lifting and carrying supplies and equipment in the perioperative setting. AORN J 94(2):173–179
15. Waters T, Lloyd JD, Hernandez E, Nelson A (2011) AORN ergonomic tool 7: pushing, pulling, and moving equipment on wheels. AORN J 94(3):254–260

16. Waters T, Short M, Lloyd J, Baptiste A, Butler L, Petersen C, Nelson A (2011) AORN ergonomic tool 2: positioning and repositioning the supine patient on the OR bed. AORN J 93(4):445–449
17. Waters T, Spera P, Petersen C, Nelson A, Hernandez E, Applegarth S (2011) AORN ergonomic tool 3: lifting and holding the patient's legs, arms, and head while prepping. AORN J 93(5):589–592

Impact of Innovative Clothing Design on Caregivers' Workload

Karlien Van Cauwelaert[1,2(✉)], Veerle Hermans[3,4], Kristien Selis[2], and Liesbeth Daenen[5,6]

[1] University College Odisee, Brussels, Belgium
[2] Department of Ergonomics, Group IDEWE (External Service for Prevention and Protection at Work), Brussels, Belgium
karlien.vancauwelaert@idewe.be
[3] Group IDEWE, Louvain, Belgium
[4] Department of Experimental and Applied Psychology, Work and Organisational Psychology (WOPS), Faculty of Psychology and Education Sciences, Vrije Universiteit Brussel, Brussels, Belgium
[5] Knowledge, Information and Research Center, Group IDEWE, Louvain, Belgium
[6] Department of Rehabilitation Sciences and Physiotherapy, Human Physiology and Anatomy (KIMA), Faculty of Physical Education and Physiotherapy, Vrije Universiteit Brussel, Brussels, Belgium

Abstract. *Background*: Prevalence of musculoskeletal disorders in nurses is the highest for the back, followed by the shoulders and the neck. 'Dressing and undressing of patients' is experienced as a physically-demanding task by caregivers. But also for the patient: it is often uncomfortable and sometimes even painful. Due to many variations in patterns and fastening systems, the existing custom-clothing is not in accordance with the ergonomic rules. *Objective*: This study aimed at (1) investigating the impact of innovative custom-clothing on caregivers' physical workload (i.e. percentage of harmful postures at the neck, back and shoulders and duration of exposure to the physically-demanding task) and (2) examining its usability and comfort for caregivers as well as patients. *Methodology*: Eight caregivers and one healthy, elderly person (as patient) were included in the study. An innovative custom-clothing design with magnetic buttons was used to test the study hypothesis. Caregivers were asked to dress the patient (i.e. a pair of trousers, undershirt and shirt) according to 3 conditions: (1) with traditional clothing (control group), (2) with custom-clothing using press buttons (press button group) and (3) with innovative custom-clothing using magnetic buttons (magnetic button group). Duration of the dressing task was evaluated and postures of neck, back and shoulders were measured using Tea CAPTIV instrument. A structured questionnaire was used to evaluate its usability and feasibility. *Results*: Dressing time was significantly reduced in the magnetic and press button group compared to the control group (respectively $p = .003$ and $p = .013$). For postures of the back (i.e. rotation and forward flexion), neck (i.e. flexion) and shoulders (i.e. flexion), no significant differences were found between control, press button and

V. Hermans—Department Head of Ergonomics.

S. Bagnara et al. (Eds.): IEA 2018, AISC 818, pp. 279–284, 2019.
https://doi.org/10.1007/978-3-319-96098-2_36

magnetic button group ($p > .05$). The caregivers agreed that the innovative custom-clothing using magnetic buttons was easier and smoother in use.

Keywords: Innovative clothing design · Caregivers' workload
Usability

1 Introduction

Prevalence of musculoskeletal disorders in nurses is the highest for the back, followed by the shoulders and the neck [1]. As a consequence of the ageing population, there is a noticeably shift towards growing care burden in the elderly care and the rehabilitation centers [1]. Efficient and high-quality health care (for the patients as well as for the caregivers) is of great importance and remains a challenge. Therefore, it is important that caregivers can work in a comfortable manner as this increases patient comfort as well [2].

Back pain is mostly associated with the handling of heavy loads during transfers of patients. But many other stressful trunk postures are related to nursing work [3–5]. The main tasks responsible for 'unhealthy' postures (i.e. postures with trunk inclination of >60°) were 'bed making' (21%), 'basic care' (16%) and 'clearing up/cleaning' (16%) [4]. Only 33% of the exposure time to the abovementioned care tasks takes place in a harmless or healthy posture [6]. Dressing and undressing of care-dependent patients is experienced as uncomfortable, and sometimes painful, by the patient and as physically loaded by caregivers [3, 7]. Due to many variations in patterns and fastening systems, the existing custom-clothing for patients is not in accordance with the ergonomic rules [3, 8, 9]. In addition, this clothing is often stigmatizing and old-fashioned and, not corresponding to patient's choice [9].

Innovative custom-clothing that provides a satisfactory answer to the disadvantages of existing custom-clothing (i.e. reducing workload and exposure for caregivers and enhancing comfort and decreasing painful, manipulative care for patients) is needed [3, 8, 9]. The innovative custom-clothing design of Attractive2 Wear (A2W) [10] attempts to fit to these requirements. It redesigns the existing custom-clothing with respect to patient's dignity and psychosocial wellbeing. It uses a magnetic closure, visual etiquettes and an uniform redesign pattern for each piece of clothing. These adaptations may result in better care quality aspects such as less manipulations, reduced discomfort and pain, more gain in time, higher user-friendly care with respect to patient's dignity and choice.

2 Objective

This study aimed at (1) investigating the impact of innovative custom-clothing on caregivers' physical workload (i.e. percentage of harmful postures at the neck, back and shoulders and duration of exposure to the physically-demanding task) and (2) examining its usability and comfort for caregivers as well as patients. It is hypothesized that dressing time, transfers and percentage of harmful postures of neck, back and shoulders in caregivers will reduce using innovative custom-clothing with magnetic buttons compared to traditional clothing and custom-clothing with press buttons.

3 Methodology

3.1 Study Population

Eight caregivers (four trainee caregivers, two nursing lectors of the University College Leuven-Limburg and two ergonomists of the Belgian occupational service for well-being at work IDEWE) were included in the study. All participants had knowledge of basic patient care and followed a training session in using the innovative custom-clothing. A healthy, elderly person (79 years) was selected as patient. She was asked to participate in a passive way (i.e. not actively cooperating during the dressing task).

3.2 Study Procedure and Materials

An innovative custom-clothing design with magnetic buttons was used to test the study hypothesis. Caregivers were asked to dress the patient (i.e. a pair of trousers, undershirt and shirt) according to 3 conditions: (1) with traditional clothing (control group), (2) with custom-clothing using press buttons (press button group) and (3) with innovative custom-clothing using magnetic buttons (magnetic button group). Duration of the dressing task was evaluated and postures of neck, back and shoulders were measured using Tea CAPTIV instrument [11]. A structured questionnaire was used to evaluate usability and feasibility of the innovative clothing design. E.g. "Did the dressing task go smoothly?", "Did it take more time or require fine actions or visual control?", "Did you experience (dis)comfort or pain (i.e. back, neck, shoulder, knee, hip, wrist) during the dressing task?".

One-way ANOVA analyses by means of the SPSS 23.0 software package (SPSS Inc; Chicago, IL) were used to test the study hypotheses.

4 Results

4.1 Impact on Postures of Neck, Back and Shoulders

The mean total duration time that a caregiver stand in an unhealthy position (i.e. for neck flexion > 10°, neck rotation > 15°, back flexion > 30°, back rotation > 15°, shoulder forward flexion > 60°) was measured and compared in the 3 conditions.

The results showed no significant differences in total duration time in neck posture (i.e. flexion and rotation), back posture (i.e. rotation and forward flexion) and shoulder posture (i.e. forward flexion) between the control group, the magnetic button and press button group ($p > .05$).

4.2 Impact on Dressing Time and Number of Transfers

Figure 1 shows the mean duration of the dressing task in the different groups: 544 s (S. D. 76) in the control group, 438 s (S.D. 68) in the pressure button group and 417 s (S. D. 55) in the magnetic button group. Dressing time was significantly reduced in the magnetic and press button group compared to the control group (respectively $p = .003$ and $p = .013$). No significant differences in dressing time were found between the magnetic and press button group ($p > .05$).

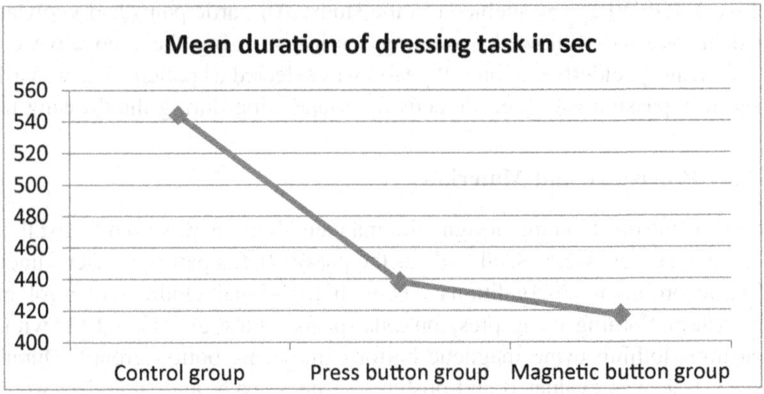

Fig. 1. Mean duration of the dressing task in the different groups

In the control group, four transfers (turn sideways) were needed to dress the patient. In contrast, only one transfer was needed to dress the patient in the press button as well as magnetic button group.

4.3 Impact on Comfort and Feasibility

All participants agreed that the used innovative custom-clothing concept with magnetic buttons was easier and smoother in use than traditional clothing. No discomfort or pain was experienced during the performance of the dressing task. Also for the patient, feeling of discomfort was lower when using the innovative custom-clothing with magnetic buttons compared to using traditional clothing.

5 Discussion and Conclusions

The study results revealed that innovative custom-clothing for patients can reduce dressing time, number of transfers and hence, shorten caregivers' time exposed to harmful postures while dressing a care-dependent patient. A small study population has its limitations. The study was carried out in a lab situation with a healthy patient. Further research is warranted in order to evaluate the impact of innovative custom-clothing in a larger group of patients and caregivers, and conducted in a clinical setting.

Comparing results between different studies is quite difficult because mostly other measurement devices are used. The specialized measurement device used in this study, TEA Captiv, is only used in a few studies. A large study report by Knibbe et al. [8] showed more favorable working postures of back and shoulders when using custom-clothing with push buttons compared to traditional clothing. However, the last study used the Ovako Working Posture Analysing System (OWAS). This system categorize the subjective positioning of postures in certain categories and counts the number of healthy and unhealthy postures instead of total duration of (un)healthy position. The Captiv system measures both (number and duration of (un)healthy postures and is more exact and accurate then the OWAS.

Suggestions were given to optimize the innovative custom-clothing design with magnetic buttons, in order to further improve the usability of the custom-clothing and prove the added value on the custom-clothing with pressure buttons. Extra attention must be paid to good visual instructions and sufficient training for caregivers.

Literature shows that innovative custom-clothing is insufficiently known and used in the care of care-dependent patients [8]. These study results are also of important clinical relevance in the health sector, i.e. in home care or disabled facilities, with in Belgium more than 79.000 care-seeking patients. If future research can strengthen the study results about dressing time, workload and usability, it offers opportunities to expand to other kind of caregivers such as physiotherapists, occupational therapists and related-caregivers. In 2015, 290.000 nurses and nurse assistants, 32.000 physiotherapists and 9400 occupational therapists are working in the health care sector in Belgium.

However, innovative clothing is only one aspect within the "comfort care" or care of high care-dependent patients. Also other aspects of "comfort care" on caregiver's workload need to be taken into account such as adapted bed layout (sliding sheet system), customized incontinence material, comfortable positioning in sitting and lying position and the use of patient hoists for transfers. More in-depth studies in this domain are needed.

References

1. Davis KG, Kotowski SE (2015) Prevalence of Musculoskeletal disorders for nurses in hospitals, long-term care facilities, and home health care: a comprehensive review. Hum Factors 57(5):754–792
2. Owen BD (2000) Preventing injuries using an ergonomic approach. Assoc Perioper Regist Nurses 72(6):1031–1036
3. Nevala N, Holopainen J, Kinnunen O, Hänninen O (2003) Reducing the physical work load and strain of personal helpers through clothing redesign. Appl Ergon 34(6):557–563
4. Freitag S, Ellegast R, Dulon M (2007) Quantitative measurement of stressful trunk postures in nursing professions. Ann Occup Hyg 51(4):385–395
5. Knibbe JJ, Knibbe NE (2012) Static load in the nursing profession; the silent killer? Work 41:5637–5638 LOCOmotion, Research in Health Care, Brinkerpad 29, 6721 WJ Bennekom, The Netherlands
6. Brinkhoff A, Knibbe NE (2003) The ErgoStat Program Pilot study of an ergonomic intervention to reduce static loads for caregivers. Prof Saf 48:32–39

7. Zwakhalen SMG, Koopmans RTCM, Geels PJEM, Berger MPF, Hamers JPH (2009) The prevalence of pain in nursing home residents with dementia measured using an observational pain scale. Pain 13(1):89–93
8. Knibbe JJ, Knibbe NE (2005) Ergonomische aspecten van aangepaste kledij. LOCOmotion Bennekom. Onderzoek in opdracht van het ministerie van Sociale Zaken en Werkgelegenheid
9. Iltanen S, Topo P (2007) Ethical implications of design practices: the case of industrially manufactured patient clothing in Finland. In: Proceedings of design inquiries: 2nd Nordic design research conference, Konstfack 27–30 May 2007
10. Clothing design: Attractive2Wear (A2W). http://attractive2wear.com
11. Captiv Tea, France. http://teaergo.com/wp/

Application of Wearable Device to Develop Visual Load Intelligence Monitoring and Evaluation Technology

Hsin-Chieh Wu[1(✉)], Mao-Lun Chiang[1], Wei-Hsien Hong[2], and Hsi-An Kou[1]

[1] Chaoyang University of Technology, Wufeng District, Taichung City 41349, Taiwan
hcwu1@cyut.edu.tw
[2] China Medical University, Taichung City 40402, Taiwan

Abstract. The visual load investigated in this study is the load imposed on the eyes when working on visual tasks such as computers, mobile phones, or tablets. Based on the gradual maturity of wearable technology, this study applies wearable technology as a device for continuous monitoring of visual loads over a long period of time. The main purpose of this study is to develop automated assessment techniques for visual loads. The developed technology in this study can calculate the watching screen time and rest time of the labor. The proposed technology can be used to automatically monitor the visual load of workers throughout the day to prevent the occurrence of cumulative occupational diseases.

Keywords: Visual load · Wearable device · Feature identification

1 Introduction

The visual load studied in this study is the load imposed on the eyes when working on visual tasks such as computers, mobile phones, or tablets. Prolonged watching on the screen may result in visual discomfort. Symptoms include sore eyes, swollen eyes, photophobia, blurred vision, dizziness, and headache. Most assessments of visual load first collect eye fatigue-related indicators before work. These indicators may be critical fusion frequency (CFF), subjective assessment of visual fatigue, and so on. At the end of work or after working for a period of time, these indicators of eye fatigue are collected again, and finally expert statistics and evaluation results are provided. These data collection and load assessment processes require a considerable amount of expert manpower cost. Based on the gradual maturity of wearable technology, this study applied wearable technology as a device for continuous monitoring of visual loads over a long period of time. The main purpose of this study is to develop automated assessment techniques for evaluating visual loads. The proposed technology is expected to automatically monitor the visual load of workers throughout the day to prevent the occurrence of cumulative occupational diseases.

S. Bagnara et al. (Eds.): IEA 2018, AISC 818, pp. 285–288, 2019.
https://doi.org/10.1007/978-3-319-96098-2_37

2 Methods

Many literatures indicate that the longer the worker keep watching on the screen, the more severe his/her eyes become. Further, the recommendations of work/rest schedule for watching a screen are listed in Table 1. Based on the above-mentioned literatures, this study used the continuous watching time as the main indicator of the magnitude of visual load; and considered 30-min watching time as the action limit.

Table 1. The work/rest schedule recommended for watching a screen

Code	Recommendations	References
1	When employers engage their laborers in precision work, they should shorten their working hours and give them a minimum of 15-min rest for continuous 2-h work	Ministry of Labor [1]
2	When watching the tablet continuously for 40 min, the worker must rest for 10 min.	Wu [2]
3	When watching mobile screen continuously for 30 min, the person must rest for 5–10 min	Tsai [3]

The study used a head-mounted camera (Fig. 1) to record what the worker seen, and upload the collected video to the cloud for the following data analysis and risk assessment. Data analysis is the use of image recognition methods to judge the working hours and rest time of the worker viewing a screen, and then determine whether the worker's watching time is over 30-min.

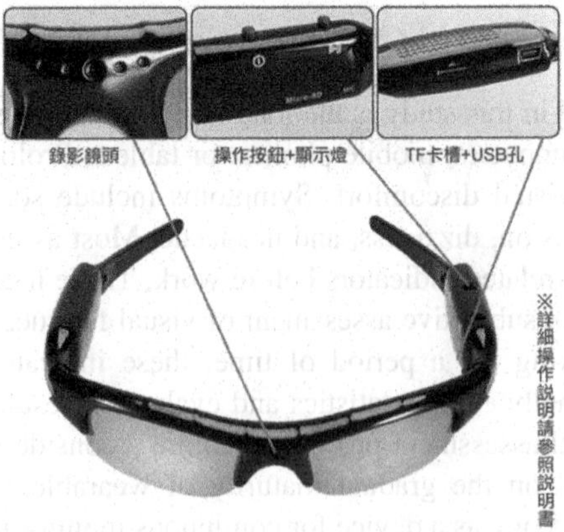

Fig. 1. The multi-functional sunglasses type driving recorder was used to record what the worker watch in the visual work.

After a variety of method tests and comparisons, we finally decided to put a logo in the upper left corner of the screen. If there is a designated logo in the image, it means that the worker is watching the screen. Using the feature identification method SIFT (Scale-invariant feature transform), the program can correctly identifies the logo if there is a designated logo in the image. SIFT is an algorithm in computer vision to detect and describe local features in images. The algorithm uses features points and related scale and orientation descriptions in a picture to get features and feature matching. In order to highlight features, we use the Logo of the Taichung City Government. This Logo has the characteristics of low complexity, high saturation, and large color difference.

3 Results and Discussion

Using the SIFT method, the picture was read and the feature point positions and vectors are calculated. The feature points of the picture are taken out, and the feature points are compared by using the Flann Match method and the similar feature points are taken out. Finally, several similar feature points were listed (Fig. 2). Six videos recording the screen tasks with a resolution of 640 × 480 were produced for testing this method. And one photo was taken every second from the video files. The success rate of the feature identification method was above 91%.

Fig. 2. The SIFT algorithm could successfully identify the logo on the upper left corner of the screen.

4 Conclusion

The developed technology in this study can calculate the watching time and rest time of the labor. Future research can be extended to automatic identification of screen size, screen types, and light source intensity. Enterprises can also use this technology to understand the level of work load in the company's visual work area, and to adjust the distribution of manpower so that human resources can be used more effectively.

Acknowledgements. The authors thank the Ministry of Science and Technology, R.O.C (Grant No. MOST104-2221-E-324-016-MY3), for financially supporting this research.

References

1. The Ministry of Labor, Standards for Precise Operational Workers' Functional Protection Facilities, Ministry of Labor's labor statute inquiry system. https://laws.mol.gov.tw/FLAW/FLAWDAT0202.aspx?lsid=FL015024. Accessed 29 May 2018
2. Wu HC (2014) Study on the occurrence and recovery method of visual fatigue using handheld smart devices, mid-term report of the Ministry of Science and Technology Research Project, Project Number: NSC102-2221-E-324-005-MY2 (2014)
3. Tsai TH. Smooth mobile phones, tablets, bad eyesight!, Foundation for National Health Foundation. http://www.twhealth.org.tw/index.php?option=com_zoo&task=item&item_id=845&itemid=21. Accessed 29 May 2018

Musculo Skeletal Disorders (MSDs) Among Algerian Nurses

Houda Kherbache[1(✉)], Lahcene Bouabdellah[1], Mohamed Mokdad[2], Ali Hamaïdia[1], and Abdenacer Tezkratt[1]

[1] Human Resources Development Research Unit (URDRH), Sétif University 2, Setif, Algeria
houdakhe@yahoo.fr, doylettres@yahoo.fr, djnacer90@yahoo.fr, hamaidia@urdrh.com
[2] University of Bahrain, Sakhir, Bahrain
mokdad@hotmail.com
http://www.urdrh.com

Abstract. Musculo-Skeletal Disorders (MSDs) are the leading cause of occupational disease. It is a set of periarticular affections that can affect the upper and lower limbs and spine, more specifically the joints, tendons, muscles and nerves.

The risk factors that trigger MSDs are numerous such as physical constraints, organizational constraints and environmental constraints.

In the proposed paper we will present the results of a short study conducted by our research unit (Research Unit 'Human Resources Development' URDRH - Sétif University 2- Algeria) whose purpose is to determine the various risk factors that may be causing MSDs (MSDs) among nurses care workers in Algeria; in order to deduce prevention strategies.

Our work is based on a sample of 500 members representing different health services and different disciplines. For this purpose, a questionnaire was prepared for this study in hospital-university centers in several Wilayas (department) of Algeria (Batna, Bouira, M'sila, Ouargla, Setif and Constantine). The questionnaire consists of four parts:

- The first dealing with personal data (age, sex, height, weight, seniority etc…)
- The second on the different risk factors
- The third part on the different TMS.
- And finally the fourth on the various details related to the prevention of symptoms and MSDs (modalities, targets, levels: primary, secondary and tertiary prevention).

Keywords: MSDs · Risk factors · Algerian nurses · Prevention

S. Bagnara et al. (Eds.): IEA 2018, AISC 818, pp. 289–297, 2019.
https://doi.org/10.1007/978-3-319-96098-2_38

1 Introduction

Musculo-Skeletal Disorders (MSDs) are the leading cause of occupational disease. They are a set of peri-articular affections that can affect the upper and lower limbs and spine, more specifically the joints, tendons, muscles and nerves.

Work in the health sector is becoming increasingly difficult, especially for paramedics and caregivers in particular. Since they are always at the bedside of their patients, they have to perform delicate tasks that require enormous physical effort, yet they maintain restrictive positions for the spine, the joints and the muscles. Moreover, they often provide night guards; all of this increases the risk of having MSDs.

Statistics tell us that 73–76% of nurses would be affected each year by the problem of back pain in France for example [1]. In Algeria and according to Nafai and Ouaaz [2]: 35.71% of Algerian workers are affected by this disorder, 83% of whom are males, half are under 40; and 10.50% suffer from two or more localizations.

MSDs are defined as all the sensational problems such as pain, cramps, stiffness, heat, tingling, etc., encountered during the practice or the accomplishment of professional tasks. Which affects musculoskeletal structures: muscles, tendons, ligaments, nerves and joints (cartilage, serous bursa…). They can be located at the level of the upper limbs (shoulders, elbows, wrists) than lower limbs (knees), even the neck or the back [3, 4].

These symptoms are due to the "overload" of work of the above-mentioned regions. Intensive use can lead to lesions that, depending on the site of attack, have different and more or less well-known names, such as: tendonitis, when they touch a tendon, carpal tunnel syndrome, when they are located at the level of the canal formed by the bones and ligaments of the wrist or "lumbago" when this lesion is at the bottom of the back [4, 5].

These painful health consequences are caused by many variable factors that need to be analyzed:

- Physical constraints: What is the force exerted? What is the duration of the task? What is the position and how often is the task repeated? Are vibrations produced by the machine used?
- Organizational constraints: Is the activity complex or monotonous? What is the urgency and time to complete the task? How are relations with colleagues, managers, customers treated?
- Environmental constraints: Is the work area hot or cold? Are drafts present? Is there a lot of noise?

A characteristic of MSDs is the slow evolution of symptoms that makes it impossible to determine with certainty the origin of the problem. The tenacious and recurrent nature of the problem is common to these different affections [4, 5].

The main risks of MSDs are numerous. However, the major one are:

Risks Related to Work Positions. Keeping your arms raised above the shoulder height, bending or twisting your wrist, or keeping your back or neck bent forward are uncomfortable positions. Combining these restrictive postures with a lot of effort or the handling of a load further increases the difficulty. The traction or compression on the joints, muscles, tendons and ligaments is high, with the risk of deterioration of these elements. If repeated and prolonged, the risk of developing a musculoskeletal disorder is greatly increased.

Other features will further complicate the situation such as:

- The quality of taking objects (handles, handles…).
- The quality of the tool or furniture.
- The production of vibrations by the tool or the vehicle [3].

Risks Related to the Organization. The difficulty or even the impossibility of planning one's working time or the uneven distribution of breaks concentrates the painful periods. Monotonous work always overloads the same joints. In addition, a messy service, poorly maintained lanes increase the risk of falls and slips, and often force to adopt more restrictive postures. The same is true if there is no work equipment maintenance program. The lack of maintenance of the equipment or the insufficient replacement of the damaged one increases the risks related to the positions, the efforts, the repetitions of the gestures, the vibrations [6].

Risks Related to the Environment. Environmental conditions such as the presence of drafts and cold can increase the risk of suffering from MSDs. For example, cold increases the muscular strength required by forearm muscles and increases tendon strain, which leads to poor tool perception and poorer muscle coordination. Defective lighting makes traveling more risky because of poorer visibility of obstacles and unevenness. Sustained noise disrupts communication and increases feelings of fatigue [6].

From this, we thought in the research unit 'Human Resources Development' URDRH to conduct a study that aims to determine the various risk factors that can cause the onset of MSDs among nurses practicing in hospitals in Algeria, in order to deduce prevention strategies.

Therefore, this study aims at answering the following questions:

- What are the different risk factors of MSDs that affect Algerian nurses?
- What are the most body parts affected by MSDs among Algerian nurses?
- What are the prevention strategies of MSDs used by Algerian nurses?

2 Methodology

Research Method: The descriptive method has been used in view of the nature of the subject treated.

Sample: It consisted of 420 female nurses distributed in 6 Wilayas (Batna: 17.1%, Bouira: 17.1%, M'sila: 22.1%, Ouargla: 10%, Setif: 24.3% and Constantine: 9,3%).

Data Collection Tools: A questionnaire has been developed comprising 20 multiple-choice questions, divided into four axes as follows:

- The first deals with personal data (age, sex, and seniority)
- The second deals with the different risk factors

- The third deals with the most body parts affected by MSDs.
- The fourth deals with the various details related to the prevention of symptoms and MSDs (modalities, targets, levels: primary, secondary and tertiary prevention).

The administration of the questionnaire took place once the reliability and validity criteria were measured and confirmed, (Cronbach alpha reliability: 0.74, Concurrent validity: 0.48).

3 Results and Discussion

3.1 What Are the Different Risk Factors of MSDs that Affect Algerian Nurses?

Risk factors are all the conditions of work situations that can promote the emergence of workplace abnormalities. Like any other functions, careers' professions are not immune to all of these conditions.

The question related to the various risk factors was asked as shown in Table 1.

Table 1. The different risk factors of MSDs that affect Algerian nurses.

Different risk factors of MSDs	Frequency	Percentage
Stress	87	**20.7**
Repetitive gestures	72	17.1
Patients' eagerness	51	12.1
Heavy workload	156	**37.1**
Too much administrative work	9	2.1
Dissatisfaction with own work	45	10.7
Total	**420**	**100.0**

It has been found that both the heavy workload and the stress are the most frequent, with the respective proportions of 37.1% and 20.7%. This tells us that, except for health workers who are aware of the dangerousness of workloads, they also realize that stress is one of the most important risk factors for MSDs. A large proportion of respondents (17.1%) also mentioned repetitive actions that are also a risk factor.

3.2 What Are the Most Body Parts Affected by MSDs Among Algerian Nurses?

In order to know the most common MSDs, our study focuses on topics that represent symptoms of MSDs during the last five years, and to clarify the affected part of the body we asked the question as indicated in the Table 2.

From the Table 2, we can see that the most common MSD in our sample is of course that of the back, and of a less degree and jointly: wrist and shoulders (12.1 and 11.4% respectively). But the unspecified ones represent 30.7% and this indicates

Table 2. The most common body parts affected by MSDs.

The body parts affected by MSDs	Frequency	Percentage
Unspecified	129	**30.7**
Neck	27	6.4
Shoulder	48	11.4
Elbows	18	4.3
Wrist	51	12.1
Back	117	**27.9**
Neck	30	7.1
Total	**420**	**100.0**

something very important: that one third of our sample did not consult otherwise how to explain that they know and that they feel that they have one (or) TMS without being able to determine it? this brings us back to the previous question about risk factors, and the stress response in particular that could be a significant cause of these unspecified MSDs, which could change gait and body part regularly. Stress could also be one of the reasons the worker did not consult, since he is often tired and has little time or desire to consult although the doctor is at hand [3, 4].

Musculoskeletal Pain. Pain is a nociceptive stimulus transmitted by the nervous system, so it is an unpleasant sensory and emotional experience. Regarding the intensity, location, and characteristics of musculoskeletal pain, we asked the following questions: How do you evaluate the pain (s)?

The answers are shown in Table 3:

Table 3. The nature of the musculoskeletal pain.

Answer	Frequency	Percentage
Low pain	126	**30.0**
Average pain	225	**53.6**
Strong pain	69	16.4
Total	**420**	**100.0**

We note that the vast majority have medium or high pain (60%: 53 + 16%) and this poses a problem, both in terms of individual health and for the institution, which accounts for two thirds of its staff suffering of pain that will prevent him from doing his job optimally [6].

At What Level Is This Pain Localized? The answers are shown in Table 4:

For this question, it is concluded that a majority of health personnel have acute pain in the joints (46.4%), followed by acute upper limb (29.3%) and muscle pain. (24.3%) and this agrees well with the results of the previous questions mentioned above. And in order to further determine the nature of these pains in each of the people tested, and to have more details on this subject, we asked the following question that goes further on the characteristics of these pains [2].

Table 4. The location of pain.

Localization of pain	Frequency	Percentage
Diffuse pain in the upper limb	123	**29.3**
Acute pain localized to joints	195	**46.4**
Acute pain localized to the muscles	102	24.3
Total	**420**	**100.0**

What Are the Characteristics of This Discomfort? The answers on this question are shown in Table 5:

Table 5. The characteristics of this discomfort.

Answer	Frequency	Percentage
Unspecified	12	2.9
Pain on palpation	123	**29.3**
Stretching pain	54	12.9
Pain in contraction	93	**22.1**
Pain at rest	63	15.0
Mechanical rhythm pain	45	10.7
Inflammatory rhythm pain	30	7.1
Total	**420**	**100.0**

As is clear on the chart, palpation pain is the most common (29.3%), then comes the contraction pains with the proportion of 22.1%, and this further complicates the problem since it is accompanied with the technical gestures when they do their work, and that means that: either they do not do their work, or worse and worse they do it but in the worst ways [4, 5].

Causes of MSDs. To properly determine the causes of MSDs in our sample, we asked a question about this: What do you think are the causes of this pain (s)?

The answers on this question are shown in Table 6:

Table 6. The causes of pain.

Answer	Frequency	Percentage
Unspecified	63	15.0
Tired	81	**19.3**
overweight	54	12.9
Repetitive gestures	51	12.1
Lack of sleep	21	5.0
Fast walk	3	0.7
Ergonomic problem	48	11.4
Overload	99	**23.6**
Total	**420**	**100.0**

It is evident that overload and fatigue together are the two most frequent causes (with a proportion of 23.6 and 19.3% respectively) according to the workers' comments, and both lead to the same phenomenon: overload work that certainly leads to unbearable fatigue. Then come overweight and repetitive gestures with almost the same percentage (about 12%).

It should be noted that most of the chosen causes are work-related, and are personal among them only the lack of sleep and the overweight which represent a percentage that can be negligible [4, 5].

3.3 What Are the Prevention Strategies of MSDs Used by Algerian Nurses?

To determine the prevention strategies used by our sample to manage the risks of MSDs, The answers on this question are shown in Table 7:

Table 7. Prevention strategies of MSDs used by Algerian nurses.

Question	Answer	Frequency	Percentage
Do you think you are well informed about MSDs?	Yes	180	**42.9**
	No	240	57.1
Do you think you are well informed about how to prevent MSDs?	Yes	168	40.0
	No	252	**60.0**
Do you think you are well informed about modifying daily exercise to avoid MSDs?	Yes	270	**64.3**
	No	150	35.7

It is obvious in Table 7 that more than 50% of individuals are not well informed about MSDs? In addition, 60% of individuals are not well informed about how to prevent MSDs? On the other hand, more than 60% of subjects are well informed about modifying daily exercise to avoid MSDs.

4 Discussion

This study focused on the topic of MSDs. The focus was on three main issues: causes of the problem, its location, and its Prevention.

As to the causes of MSDs, it was found that they are multiple, with the heavy workload being at the top. These results are expected, as nurses in Algeria and in developing countries are doing a lot of work in abnormal conditions. So they are exposed to a lot of MSDs. In large cities such as Setif, Batna, Ouargla, Bouira, M'sila and Constantine, work is much greater and effort is greater and thus more MSDs. These findings are in line with previous researchers' findings either in developing countries [7, 8] or in developed countries [9–14].

With regard to location of MSDs, the back was the area that suffers greatly from MSDs. The fact that the back is the area that is heavily affected by MSDs can be attributed to the daily large number of bending and lifting of the heavy and helpless patients. In addition, the speed required to complete the work contributes to the impact of MSDs on the back. These results are consistent with the results of other researchers [10, 15].

Concerning prevention of MSDs, it has been shown that more than half of the nurses do not have enough information on how to prevent MSDs. Nurses do not start working until after training. It is useful to note that training is very focused on how patients are treated. Although this is a very important topic, it should be pointed out that training on how to overcome work stress and pressure is also important. Since the nurses are not trained on this, the results are that many of them do not know how to cope with MSDs. These results are consistent with the results of previous researchers [4, 16].

5 Conclusion

At the end of this research, we can draw several conclusions:

- The most common MSDs in our sample are those of the back, and of a less degree and jointly: those of the wrist and the shoulders.
- The most common risk factors for MSDs among health care workers are: excessive workload and stress.
- The vast majority of health care workers have moderate or severe pain.
- A majority of 46.4% of health personnel suffer from acute joint pain.
- Pain on palpation is the most widespread among Algerian health staff, then comes in order of frequency pain to contraction.
- The vast majority have moderate or severe pain.
- A majority of health personnel suffer from acute joint pain.
- Regarding the treatment, a proportion of about half (44.3% of respondents) does not take any treatment, and a third uses anti inflammatory.

As for prevention strategies, adopted by health workers in the area of MSDs, the most used are a sort of trick, such as the repeated taking of holidays, or the requests for upgraded positions.

References

1. Maul I, Läubli T, Klipstein A, Krueger H (2003) Course of low back pain among nurses: a longitudinal study across eight years. Occup Environ Med 60:497–503
2. Nafai D, Ouaaz M (2016) MSDs: the case of Algeria. J Occup Med SAMT 20:19–23
3. Bouhourd A (2007) The management of occupational risks related to MSDs: what strategy for the director of care? Memory of the National School of Public Health, Rennes
4. Ghomari O, Beghdadili B, Belabed A, Kandouci AB (2013) Epidemiological surveillance of MSDs in higher sclerosis in enterprises. J Occup Med 20:32–37
5. Bourgeois F, Polin A, Lemarchand C, Faucheux JM, Hubault F, Douillet P, Brun C, Albert E (2006) Troubles musculosquelettiques et travail: quand la santé interroge l'organisation. L'Agence nationale pour l'amélioration des conditions de travail (ANACT), Lyon, France
6. Phajan T, Nilvarangkul K, Settheetham D, Laohasiriwong W (2014) Work-related MSDs among sugarcane farmers in north-eastern Thailand. Asia Pac J Public Health 26(3):320–327

7. Smith DR, Wei N, Zhao L, Wang RS (2004) Musculoskeletal complaints and psychosocial risk factors among Chinese hospital nurses. Occup Med 54(8):579–582
8. Tezel A (2005) Musculoskeletal complaints among a group of Turkish nurses. Int J Neurosci 115(6):871–880
9. Bos J, Kuijer PP, Frings-Dresen MH (2002) Definition and assessment of specific occupational demands concerning lifting, pushing, and pulling based on a systematic literature search. Occup Environ Med 59(12):800–806
10. Trinkoff AM, Lipscomb J, Geiger-Brown J, Brady B (2002) Musculoskeletal problems of the neck, shoulder, and back and functional consequences in nurses. Am J Ind Med 41 (3):170–178
11. Schluter PJ, Turner C, Huntington AD, Bain CJ, McClure RJ (2011) Work/life balance and health: the nurses and midwives e-cohort study. Int Nurs Rev 58(1):28–36
12. Smith DR, Mihashi M, Adachi Y, Koga H, Ishitake T (2006) A detailed analysis of musculoskeletal disorder risk factors among Japanese nurses. J Saf Res 37(2):195–200
13. Louw QA, Morris LD, Grimmer-Somers K (2007) The prevalence of low back pain in Africa: a systematic review. BMC Musculoskelet Disord 8:105
14. Simon M, Tackenberg P, Nienhaus A, Estryn-Behar M, Conway PM, Hasselhorn HM (2008) Back or neck-pain-related disability of nursing staff in hospitals, nursing homes and home care in seven countries–results from the European NEXT-Study. Int J Nurs Stud 45 (1):24–34
15. Feng CK, Chen ML, Mao IF (2007) Prevalence of and risk factors for different measures of low back pain among female nursing aides in Taiwanese nursing homes. BMC Musculoskelet Disord 8:52
16. Israni M, Vyas NJ, Sheth MS (2013) Prevalence of musculoskeletal disorders among nurses. Indian J Phys Ther 1(2):52–55

The Effects of Brushing Practice Using a TBP-Module with Good Real-Life Adaptability

Hiromi Imai[1,2(✉)], Tamiyo Asaga[1], Tetsuo Misawa[3], Ayumi Kimura[2], Sachiko Tsubaki[2], Tomoko Aso[2], and Fusako Kawabe[2]

[1] Graduate School of Chiba Institute of Technology, Narashino-shi, Chiba 2750016, Japan
hiromi.imai@cpuhs.ac.jp
[2] Chiba Prefectural University of Health Sciences, Mihama-ku, Chiba-shi, Chiba 2610014, Japan
[3] Chiba Institute of Technology, Narashino-shi, Chiba 2750016, Japan

Abstract. Objective: The objective of this study was to clarify the learning effect of a tooth brushing practice (TBP) module resembling the oral environment developed to easily practice brushing of patients' teeth.

Methods: The subjects were healthy male students of non-medical fields. The TBP module was coated with artificial plaque. The subjects brushed the module using a tooth brush until they felt that brushing was completed, and the time required for brushing and the degree of residual plaque were evaluated. They repeated this procedure 4 times with about 2-week intervals. The subjects practiced brushing between the 2nd and 3rd experiments (first self-learning) and between the 3rd and 4th experiments (2nd self-learning).

Results: After self-learning, the brushing time shortened and residual plaque decreased.

Discussion: It was suggested that self-learning using the TBP module resembling the oral cavity environment promotes learning of the brushing technique.

Keywords: Brushing technique · Practice module · Self-learning

1 Introduction

The Ministry of Health, Labour and Welfare (MHLW) in Japan reported that pneumonia is ranked third in the cause of death [1] and the most frequent cause of death in persons requiring long-term care [2]. Most of these cases are aspiration pneumonia, which is caused by aspiration of bacteria in the oval cavity with saliva. Yoneyama et al. [3] found that appropriate oral care by specialists leads to prevention of aspiration pneumonia.

Iijima [4] reported that the 8020 (leave 20 teeth at age of 80) Campaign by the Ministry of Health, Labour and Welfare achieved reliable outcomes and that the following task aiming at extension of the healthy life expectancy is prevention of 'oral frailty (weakening of the oral function)'. Oral care improves the physical function by

S. Bagnara et al. (Eds.): IEA 2018, AISC 818, pp. 298–304, 2019.
https://doi.org/10.1007/978-3-319-96098-2_39

preventing health problems and decline of the oral function with aging, playing an important role in prolongation of healthy life expectancy. Nursing professions supporting the 24-hour life process of target persons play the major role.

On the other hand, in the educational content concerning oral care in the current nursing technique education is only cleaning of the oval cavity, and no points of oral care, such as how to load an appropriate pressure and appropriate ways of holding, placing, and moving a brush, are not closely described in text or reference books. Moreover, according to an inquiring survey of researchers, about 90 min at most is used for oral care in the basic nursing education, requiring an educational strategy enabling acquisition of an effective technique within a limited learning time.

With the above background, this study was performed to clarify the learning effect acquired by the use of a TBP-module, with a portable compact size enabling practice for self-learning any time at any place and reproduces a fragile oral cavity environment as an educational strategy from an ergonomic approach.

2 Objective

The objective of this study was to clarify the learning effect of a TBP-module developed to easily practice brushing of patients' teeth (self-learning).

3 Outline of TBP-Module

3.1 Shape

It was prepared to have a portable compact size (see Fig. 1) and comprised of a part of the left lower permanent teeth (Zsigmondy) (lower left No. 3: canine-No. 7: second molar). In addition, a spaced arch, mobile tooth (No. 5: second premolar), and crowding were reproduced. Furthermore, an environment with gingival recession, which is not a normal feature, was reproduced.

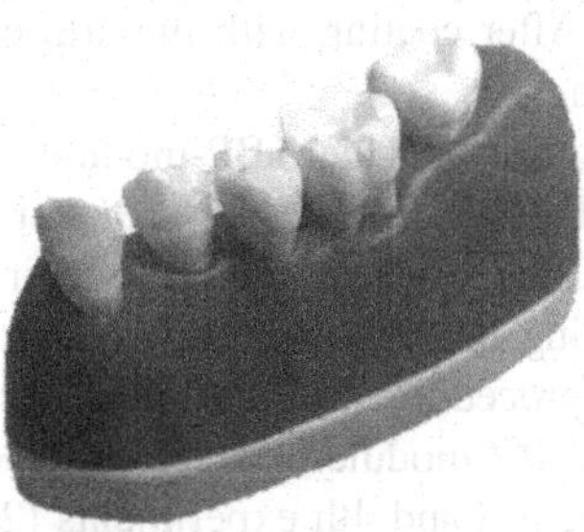

Fig. 1. TBP-module

3.2 Materials

For the material of the teeth, urethane was used. Urethane has a rougher surface than melamine resin adopted in many similar models, and high adhesiveness of artificial plaque can be expected. High adhesiveness of artificial plaque could be expected for urethane because its surface roughness is higher than that of melamine resin adopted in many similar models.

Gingiva was prepared with silicone and molded separately from the teeth to reproduce conditions frequently brushed insufficiently, such as periodontal pockets.

4 Subjects

Subjects were 20 healthy male university students in non-medical field in their twenties who agreed to participate in all experiments.

5 Data Collection Period

The data collection period was October 2017–March 2018.

6 Data Collection Method

The time required for the subjects to brush the artificial plaque-applied TBP-module and the degree of the residual artificial plaque left on the TBP-module were evaluated.

6.1 Experimental Protocol

The experiment was performed according to the following procedure.

(1) Measurement of Brushing Time.
A researcher coated the TBP-module with an artificial plaque (Dental Education Company NISSNI) 3 times. After coating with the artificial plaque, the TBP-module was allowed to dry for 5 min.

After drying, the subjects brushed the TBP-module using a tooth brush (G.V.K MORNIN COMPACT MEDIUM SOFT, K.O. Dental corp.) until they felt that brushing was completed, and the subjects measured their time to complete brushing. A researcher observed brushing throughout the period from initiation to completion. The subjects repeated this procedure 4 times with about 2-week intervals. They practiced brushing using the TBP-module between the 2nd and 3rd experiments (first self-learning) and between the 3rd and 4th experiments (2nd self-learning). The TBP-module and artificial plaque were provided the subjects during this period.

Before self-learning, a researcher explained the following items to the subjects individually taking about 5 min:

(1) How to coat and dry the artificial plaque
(2) How to hold a tooth brush in a pen-grip manner

(3) How to place the tooth brush between the teeth and gingiva and between the teeth
(4) How to move the tooth brush so as to brush each tooth
(5) Appropriate brushing pressure so as not to bend the bristles

A researcher asked the subjects to practice 10 times or more between the 2nd and 3rd experiments and between the 3rd and 4th experiments. The time or frequency of self-learning per day was not limited.

(2) Plaque Control Record (PCR).
After the subjects brushed off artificial plaque coated on the TBP-module, a researcher evaluated the residual plaque on the simulator.

It was evaluated on all 5 teeth of the TBP-module (No. 3–No. 7).

Each tooth was divided into 6 regions (mesial buccal and labial sides, mesial lingual and palatal sides, distal buccal and labial sides, distal lingual and palatal sides, buccal and labial sides, and lingual and palatal sides), and the 30 tooth surfaces (5 teeth × 6 regions) were examined.

The subjects used the same TBP-modules in all experiments. The researcher evaluated the residual plaque and then removed it until no adherent plaque was observed macroscopically.

6.2 Analytical Method

(1) Time Required for Brushing.
The time required for brushing in the 1st–4th experiments was subjected to multiple comparison and significance of differences among the experiments was analyzed.

(2) PCR.
The number of tooth surfaces with residual dental plaque was measured after each of the 1st–4th brushing, and the PCR score for tooth surfaces with residual plaque was calculated using the formula below:

$$\text{PCR score}\,(\%) = \text{number of tooth surfaces with residual dental plaque} \times 100/\text{number of test tooth surfaces}\,(30)$$

The PCR scores calculated in the experiments were subjected to multiple comparison and significance of differences among the experiments was analyzed.

6.3 Ethical Consideration

This study was performed after approval by the Research Ethics Committee of Chiba Prefectural University of Health Sciences (2016-047).

This study was support by JSPS Grant-in-Aid (Scientific Research (C), Grant Number: 17K12082)

7 Result

7.1 Outline of the Subjects

The subjects were 20 male university students aged 20–23 years (mean: 21.6 years old, SD: 1.10). All subjects used a tooth brush for manual brushing in their daily self-care. The frequency of self-practice was 10–13 (mean: 10.65, SD: 0.88) between the 2nd and 3rd experiments and 11–25 (mean: 12.1, SD: 3.14) between the 3rd and 4th experiments.

7.2 Brushing Time

The brushing time that the subjects felt 'brushing was completed' was 184–807 s (mean: 532 s, SD: 161) in the first experiment, 129–809 s (mean: 466 s, SD: 187) in the 2nd experiment, 215–1,017 s (mean: 479 s SD: 227) in the 3rd experiment, and 19–661 s (mean: 408 s, SD: 137) in the 4th experiment (see Fig. 2). On comparison between the first brushing and 4th brushing, the brushing time was significantly shorter.

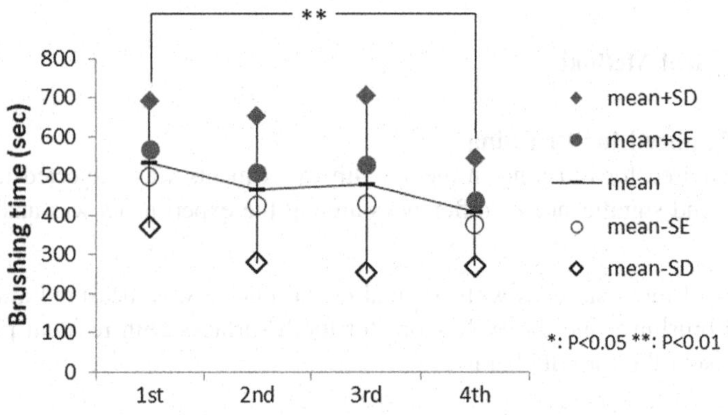

Fig. 2. The brushing time that the subjects felt 'brushing was completed'

7.3 PCR

The PCR score was determined from the amount of the residual dental plaque on the TBP-module after brushing.

The score was 33.3–83.3% (mean: 60.2%) after the first brushing, 26.7–93.3% (mean: 53.2%) after the 2nd brushing, 26.7–63.3% (mean: 43.5%) after the 3rd brushing, and 13.3–50% (mean: 31.0%) after the 4th brushing (see Fig. 3). The PCR score significantly decreased after the 2nd brushing.

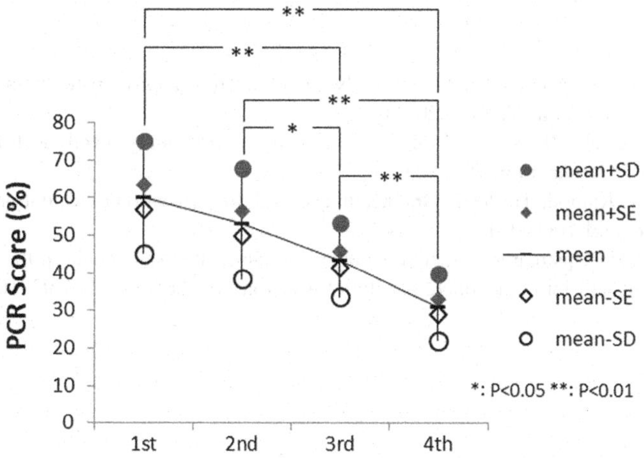

Fig. 3. The PCR score. The residual dental plaque on the TBP-module after brushing.

8 Discussion

After self-learning using the TBP-module reproducing the environment of the oval cavity, the brushing time and the residual plaque significantly decreased.

The subjects had never learned oral care and brushing before. By only short-time explanation by the researcher, they understood the method to brush the teeth of others based on their own brushing technique acquired through their previous experience and acquired the technique by self-learning alone. In addition, the frequency of practice was higher in the 2nd than 1st self-learning period, suggesting that motivation for self-learning persisted. Furthermore, the self-learning kit may have been effective because visible artificial plaque, was removed and practice could be performed within a short time.

By repeating self-learning using the TBP-module, variation of the PCR score decreased and the value also significantly decreased, but the mean score was still about 30% in the 4th brushing.

It was reported that the standard PCR score is 10% or less than 20%, so the technical level required for clinical practice could not be reached within this frequency of self-learning.

However, the PCR score decreased by about 10 points on average after each self-learning period from the 2nd to 4th brushing, suggesting that the standard value can be reached by repeated practice.

It was suggested that self-learning using the compact TBP-module reproducing the oral cavity environment developed for simple practice of brushing patients' teeth improves the brushing technique.

References

1. 2016 Overview of Annual Total of Monthly Vital Statistics (approximate numbers), Ministry of Health, Labour and Welfare (2016)
2. Hasegawa J et al (2013) Place and cause of death in community-dwelling disabled elderly people. Jpn J Geriatr 50(6):797–803
3. Yoneyama T, Kamoda H (2001) Oral health care and the prevention of aspiration pneumonia. Jpn J Gerodontol 16(1):3–13
4. Iijima K (2015) Upstream preventive strategy for age-related sarcopenia in the elderly: why do the elderly fall into inadequate nutrition? Ann Jpn Prosthodont Soc 7:92–101

The Development of Sensor-Based Gait Training System for Locomotive Syndrome: The Effect of Real-Time Gait Feature Feedback on Gait Pattern During Treadmill Walking

Hiroyuki Honda[1(✉)], Yoshiyuki Kobayashi[2], Akihiko Murai[2], and Hiroshi Fujimoto[3]

[1] Graduate School of Human Sciences, Waseda University, 2-579-15, Tokorozawa, Saitama, Japan
hiroyuki.honda@fuji.waseda.jp
[2] Digital Human Research Group, Human Informatics Research Institute, National Institute of Advanced Industrial Science and Technology, 2-8-5 Aomi, Koto-ku, Tokyo 135-0064, Japan
[3] Faculty of Human Sciences, Waseda University, 2-579-15, Tokorozawa, Saitama, Japan

Abstract. The concept of locomotive syndrome was proposed by the Japanese Orthopedic Association; it typifies the condition of reduced mobility resulting from a locomotive organ disorder related to aging. Although several sensor-based gait training systems, which can feedback the gait features in real-time, have been developed for various musculoskeletal disorders, there are no such systems for locomotive syndrome. In this study, we reported how real-time locomotive syndrome related gait feature feedback effects on gait patterns during treadmill walking. 18 healthy participants were assigned into either intervention- or control-group. During 4 sessions (training-session, pre-intervention-session, intervention-session, and post-intervention-session), gait patterns were measured by a motion-capture system. During the intervention-session of the intervention-group, participants received LS-risk-scores made in this study. Meanwhile, they were asked to minimize the LS-risk-scores by modifying their knee joint motion. A two-way-repeated measure ANOVA was conducted on the LS-risk-scores to examine effects of the intervention. When interaction was found, paired t-tests were conducted on the LS-risk-scores and knee angles between the sessions respectively. As a result, the LS-risk-scores were significantly smaller ($p < 0.05$) during the post-intervention-session than the pre-intervention-session in the intervention-group. There were no significant differences on the LS-risk-scores between the sessions in the control-group. Further, in the intervention-group, significant differences ($p < 0.05$) were found between the sessions on the knee angles partially. There were no significant differences between the sessions on the knee angles in the control-group. These results indicate that people can alter their gait pattern if the LS-risk-scores are feedback in real-time.

Keywords: Locomotive syndrome · Real-time visual feedback
Gait training

S. Bagnara et al. (Eds.): IEA 2018, AISC 818, pp. 305–311, 2019.
https://doi.org/10.1007/978-3-319-96098-2_40

1 Introduction

Musculoskeletal disorders, including osteoarthritis and osteoporosis, are major public health problems around the world [1]. In Japan, an estimated 47 million individuals aged over 40 years have musculoskeletal disorders [2]. To increase public awareness of the importance of each individual musculoskeletal disorder, Japanese Orthopedic Association has recently proposed the concept of locomotive syndrome (LS) which is a condition of reduced mobility resulting from a locomotive organ disorder related to aging [3]. It has been reported that physical interventions, such as a squat for strengthening muscle of a lower limb and one leg stand for improving balance, is effective for prevention of LS [4].

Gait training has been reported to be effective for prevention of various musculoskeletal disorders [5, 6]. Nowadays, several sensor-based gait training systems, which feedback gait features in real-time for various musculoskeletal disorders, have been developed [7, 8]. However, there are no such systems for LS. Therefore, we aimed this study to develop sensor-based gait training systems, which feedback gait features in real-time. In this paper, we reported how real-time gait feature feedback effects on gait pattern during treadmill walking.

2 Methods

2.1 Participants

We assigned 18 healthy participants (12 males, 6 females; age 22.7 ± 1.4 years) into either intervention group or control-group.

2.2 Procedure

In a stratified randomized controlled trial, participants were asked to walk on the treadmill (PARAGON 6, HORIZON FITNESS) normally during following 4 sessions: training-session (2-min), pre-intervention-session (2-min), intervention-session (5-min), and post-intervention-session (2-min). They were asked with take a rest (2-min) between the sessions. During the training-session, participants' preferred walking speed (3.7 ± 0.4 km/h) was determined. During all following sessions, participants walked at this speed, and their gait patterns were measured by using a motion-capture system (OptiTrack). During the intervention-session of the intervention-group, participants received LS-risk-scores which were originally made on a basis of a two-step test for assessing locomotive syndrome [9] in this study. The scores were projected on the screen in front of the participants (Fig. 1).

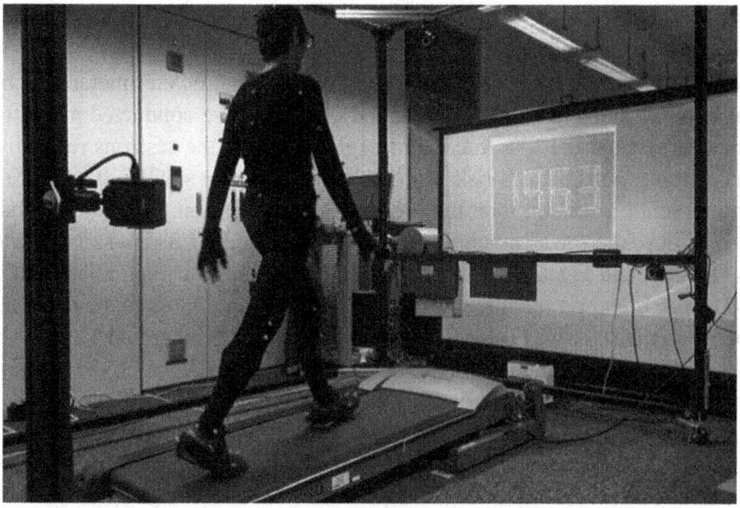

Fig. 1. Real-time visual feedback system

2.3 Data Analysis

Three-dimensional marker positions were collected using the motion-capture system (Motive:Body, OptiTrack; sample frequency 120 Hz) with 8 cameras (Prime 13 W, OptiTrack). The three-dimensional marker positions were filtered with a 4th order low-pass Butterworth filter with a cut off frequency of 6 Hz. Knee joint angles were calculated using Biomech Makersets. The knee joint angles were time-normalized by a gait cycle duration determined based on position of a heel marker and divided into 101 variables ranging from 0 to 100%. The LS-risk-scores were made from right knee joint angles in sagittal plane at four different phases (41%, 72%, 96%, 98%) of locomotion. An algorithm to calculate the scores was originally made from the data of 26 healthy elderly participants (6 belong to LS) in AIST Gait Database 2015 [10] by using discriminant analysis using SPSS (ver.22). All systems were integrated using DhaibaWorks [11] and Motion-dress-up Mirror [12]. Meanwhile, they were asked to minimize the score by modifying their knee joint motion.

2.4 Statistic

A two-way (intervention/control-groups by pre/post-intervention-sessions) repeated measure ANOVA was conducted on the LS-risk-scores, which were calculated from motion-captured data, to examine effects of the intervention. Further, when interaction was found, paired t-tests were conducted on the LS-risk-scores and four knee joint angles which were used to calculate the LS-risk-scores between the sessions respectively. SPSS (ver.22) was used for all statistical analyses.

3 Results

The two-way repeated measure ANOVA revealed significant interaction between the groups and sessions ($p < 0.05$) (Table 1). Therefore, we conducted paired t-test on the LS-risk-scores and the four knee joint angles between the sessions respectively. In the intervention-group, the LS-risk-score was significantly smaller ($p < 0.05$) during the post-intervention-session than the pre-intervention-session. On the other hands, in the control-group, there was no significant difference in the LS-risk-score between the sessions (Table 1 and Fig. 2). Further, in the intervention-group, significant differences ($p < 0.01$) were found on the three timings (72%, 96% and 98%) of the four knee joint angles between the sessions (Table 1 and Fig. 3(a)). On the other hands, in the control-group, there were no significant differences between the sessions on the all timings of the four knee joint angles (Table 1 and Fig. 3(b)).

Table 1. Means and SDs of the calculated LS-risk-scores and four knee joint angles. F-values and P-values for interaction. INT and CON represent the intervention-group and control-group.

Parameters		Group	Pre-intervention-session	Post-intervention-session	Interaction	
			Mean ± SD	Mean ± SD	*F-value*	*P-value*
LS-risk-score		INT	0.1 ± 1.5	2.3 ± 3.0*	5.670	0.030
		CON	−0.1 ± 2.1	−0.2 ± 2.3		
Knee joint angle (degree)	41%	INT	−6.7 ± 3.8	−7.7 ± 3.5	1.692	0.212
		CON	−7.3 ± 5.9	−6.7 ± 4.4		
	72%	INT	−55.5 ± 5.1	−79.3 ± 21.5†	10.960	0.004
		CON	−59.3 ± 8.4	−60.2 ± 7.7		
	96%	INT	−0.8 ± 2.7	−8.8 ± 6.5†	10.768	0.005
		CON	−5.7 ± 8.2	−6.6 ± 8.7		
	98%	INT	−0.7 ± 3.0	−6.3. ± 5.4†	10.367	0.005
		CON	−6.0 ± 7.5	−6.2 ± 7.1		

*Significant differences between the pre- and post-intervention-sessions at $p < 0.05$.
†: Significant differences between the pre- and post-intervention-sessions at $p < 0.01$.

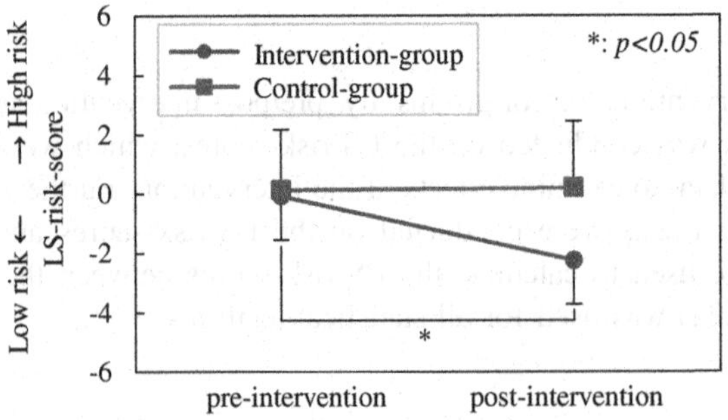

Fig. 2. The LS-risk-scores of the intervention- and control-groups on the pre- and post-intervention-sessions. Positive values indicate high risk for LS.

4 Discussion

As shown in Table 1 and Fig. 2, in the intervention-group, the LS-risk-score was significantly smaller ($p < 0.05$) during the post-intervention-session than the pre-intervention-session. On the other hands, in the control-group, there was no significant difference in the LS-risk-score between the sessions (Table 1 and Fig. 2). These results indicate that people can alter their gait pattern, if the LS-risk-scores are used as real-time gait feature feedback.

Further, as shown in Table 1 and Fig. 3(a), in the intervention-group, significant differences ($p < 0.01$) were found on the three timings (72%, 96%, 98%) of the four knee joint angles between the sessions. On the other hands, as shown in Table 1 and Fig. 3(a), in the control-group, there were no significant differences between the sessions on the all timings of the four knee joint angles. Particularly, in the intervention-group, the post-intervention-session showed an increased approximately 24° of the

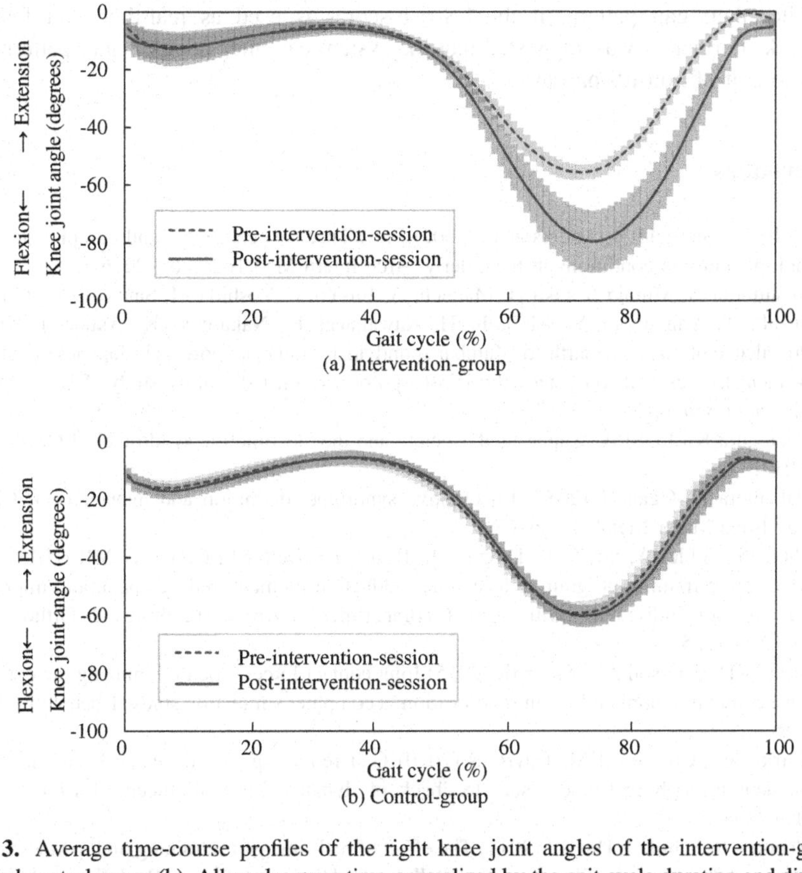

Fig. 3. Average time-course profiles of the right knee joint angles of the intervention-group (a) and control-group (b). All angles were time-normalized by the gait cycle duration and divided into 101 variables ranging from 0 to 100%. Broken lines and solid lines represent the pre-intervention-sessions and post-intervention-sessions respectively. Shaded areas indicate standard deviations. Positive values indicate flexion angle.

knee flexion angle on the 72% during the swing phase than the pre-intervention-session. Further, it has been reported that the gait feature of reduced knee flexion during the swing phase is caused by weakened muscles for hip flexion [13]. The current result and the past finding suggest that the intervention is effect on improving the knee joint angle during swing phase. Therefore, the gait training system, which is made in this study, may be able to alter gait patterns of LS to gait patterns of non-LS partially. However, such alteration is limited only on the certain features of the gait. Therefore, further research is necessary to design intervention which can alter the gait pattern totally.

5 Conclusion

The aim of this study was to develop the sensor-based gait training systems which feedback gait features in real-time. In this study, we reported how real-time gait feature feedback effects on gait pattern during treadmill walking. It was suggested that people can alter their gait pattern, if the LS-risk-scores is used as real-time gait feature feedback. Further, it was suggested that the system can alter people's gait patterns to adapt to gait of non-LS partially.

References

1. Wolff JL, Starfield B, Anderson G (2002) Prevalence, expenditures, and complications of multiple chronic conditions in the elderly. Arch Intern Med 162:2269–2276
2. Yoshimura N, Muraki S, Oka H, Mabuchi A, En-Yo Y, Yoshida M, Saika A, Yoshida H, Suzuki T, Yamamoto Y, Ishibashi H, Kawaguchi H, Nakamura K, Akune T (2009) Prevalence of knee osteoarthritis, lumbar spondylosis, and osteoporosis in Japanese men and women: the research on osteoarthritis/osteoporosis against disability study. J Bone Miner Metab 27:620–628
3. Nakamura K (2008) A "Super-aged" society and the "locomotive syndrome". J Orthop Sci 13:1–2
4. Nakamura K, Ogata T (2016) Locomotive syndrome: definition and management. Clinic Rev Bone Miner Metab 14:56–67
5. Shull PB, Silder A, Shultz R, Dragoo JL, Besier TF, Delp SL, Cutkosky MR (2013) Six-week gait retraining program reduces knee adduction moment, reduces pain, and improves function for individuals with medial compartment knee osteoarthritis. J Orthop Res 31:1020–1025
6. Miller RH, Esterson AY, Shim JK (2015) Joint contact forces when minimizing the external knee adduction moment by gait modification: a computer simulation study. Knee 22(6):481–489
7. Barrios JA, Crossley KM, Davis IS (2010) Gait retraining to reduce the knee adduction moment through real-time visual feedback of dynamic knee alignment. J Biomech 43 (11):2208–2213
8. Noehren B, Scholz J, Davis I (2011) The effect of real-time gait retraining on hip kinematics, pain and function in subjects with patellofemoral pain syndrome. Br J Sports Med 45 (9):691–696

9. Ogata T, Muranaga S, Ishibashi H, Ohe T, Izumida R, Yoshimura N, Iwaya T, Nakamura K (2015) Development of a screening program to assess motor function in the adult population: a cross-sectional observational study. J Orthop Sci 20(5):888–895
10. Kobayashi Y, Hobara H, Mochimaru M (2015) AIST Gait Database 2015. http://www.dh.aist.go.jp/database/gait2015/. Accessed 15 Apr 2018
11. Endo Y, Tada M, Mochimaru M (2014) Dhaiba: development of virtual ergonomic assessment system with human models. In: Proceedings of the 3rd international digital human modeling symposium, Paper vol 58
12. Murai A, Fan K, Tada M Motion-dress-up mirror: real-time motion measurement, analysis, and intervention to human motion. In: Proceedings of the 34th annual conference of the robotics society of Japan (in Japanese)
13. Piazza SJ, Delp SL (1996) The influence of muscles on knee flexion during the swing phase of gait. J Biomech 29:723–733

Affective Appraisal of Hospital Reception Scenes

A. M. M. Maciel[1(✉)], L. L. Costa Filho[2(✉)], and V. Villarouco[2(✉)]

[1] PPGDesign, Universidade Federal de Pernambuco, Recife, Brazil
anamariamaciel@yahoo.com.br
[2] PPGDesign/PPErgo, Universidade Federal de Pernambuco, Recife, Brazil
lourivalcosta@yahoo.com, villarouco@hotmail.com

Abstract. This article describes an affective evaluation of hospital reception scenes and aims to identify dimensions that derive from environmental affection and the physical attributes which most influence this type of judgment. Two hospital reception scenes, typical of private hospitals in the region focused on in this research, one which was judged to have a relaxing quality and the other, exciting, were used as stimuli to collect data from 75 subjects, through a questionnaire, and whose responses were analyzed using graphics and frequency distribution tables. Findings confirm that the scene with relaxing quality raises the perceived affective quality, while the other, with an exciting quality, reduces it. It has also been found that color; furniture and size are the physical attributes which most influence these judgments.

Keywords: Hospital reception · Perceived affective quality
Ergonomics of the built environment

1 Introduction

The empirical evidence shows that buildings can influence health and human well-being, and that the choices made in their design and construction can benefit or harm their users (Andrade et al. 2013; Ulrich et al. 2008).

As such, environmental psychology has been used to study the way in which the physical and spatial context of hospitals influence its users (Lawrence 2002). It should be noted that, according to Beukeboom et al. (2012), during a hospital visit, receptions are the most likely places for stress-generating experiences.

For Maciel, Costa Filho and Villarouco (2018), people tend to be distressed when they are sick and need to look for and wait for medical and hospital care. In these often stressful situations, hospital reception environments whose physical properties exhibit exciting qualities may be inadequate and make the experience even more unpleasant. Conversely, calming-quality environments seem to be more effective in helping people cope with this kind of situation.

In fact, according to Figueiredo (2005), environments judged to be relaxing seem to be more adequate for coping with stressful situations, since they favor a welcoming experience for their users in health care settings.

S. Bagnara et al. (Eds.): IEA 2018, AISC 818, pp. 312–321, 2019.
https://doi.org/10.1007/978-3-319-96098-2_41

Therefore, it is argued here that the experience of people awaiting medical attention in hospital receptions may be helped by the dimensions derived from a calming affective environment since, according to Becjker and Douglas (2008), the environment can communicate values of the organization and create a suitable experience for the patients, as the perceived waiting time is more important than the real time, and has an impact on the level of overall satisfaction.

Villarouco (2011), however, points out that the variables involved in identifying the suitability of an environment are many, which makes the identification of these variables very complex, especially when faced with the approaches used in ergonomics, since this discipline goes beyond just the consideration of physical variables, and focuses also on others from different fields of knowledge, such as environmental psychology.

Studies in the area of ergonomic of the built environment are justified in order to provide empirical information to improve the suitability of hospital reception environment projects to the psychological needs of users, since, according to Nasar (2000), research indicates that the characteristics of the elements of an environment have important impacts on the human experience, being able to evoke strong emotions such as pleasure or unpleasure, to have a stressful or calming effect and to allow inferences about places and people. They may also, according to the author, influence human behavior, so that people are more likely to go and stay in places they perceive favorably and to avoid others which are perceived unfavorablably.

Although evaluative responses, alone, may not predict actual behavior, the combined appraisal of evaluative responses and expected behavior gives a good indication of actual behavior (Nasar 1988). For this reason, in this research, which uses a theoretical reference based on Kaplan's concepts of environmental preference (1988), and Ward and Russell's affective evaluation of the environment (1981), respondents were invited to indicate affective dimensions derived from two scenes of hospital receptions and also identify the physical attributes that influence this type of evaluation.

Affective quality is repeatedly considered an important and salient way in which environments are interpreted and compared to others and occur when a person judges something as having an affective quality. In this way, finding a place relaxing or exciting is to give this place an affective quality. In other words, to say that an environment is relaxing is to say that it can produce calmness. Therefore, the affective quality is a key factor in determining the human response to an environment and cannot be omitted in the evaluation of this environment (Russel 1988).

2 Preference and Environmental Affect

According to Nasar (2008), environmental stimuli, many of them unconscious, shape our feelings, thoughts and behavior. Within this perspective, perceived visual quality has important impacts on the human experience, affecting the worker's productivity, consumer behavior and expected outcomes.

Research on perception sees the perceptual process as inextricably connected with human purposes, and perhaps also with human preferences. If people's reaction to things and spaces depends on their purpose, then understanding environmental preference requires that those purposes be understood first.

As such, environmental preference has been described by Kaplan (1988) as being the product of two underlying processes related to human survival: the need to be involved; and the need for the scene to make sense. The environment must be "involving" to attract human attention, as well as "making sense" so that one can act in it. Complexity and coherence, for this author, play important roles in satisfying these two human needs.

Complexity promotes involvement, and such a relationship has been consistently supported by empirical findings for measures of involvement, such as demand and interest time (Wohlwill 1976). It involves the diversity of elements present in the scene, the number of different visible elements and the distinction between them. Too little complexity is monotonous and boring, too much is chaotic and stressful. A middle level of complexity is the most pleasant (Berlyne 1972; Wohlwill 1976). Hence, the hedonic tone (pleasantness or beauty) has been posited as having an inverted U-shaped relationship to complexity, since the increase in complexity elevates the hedonic tone to some extent and then declines.

For a scene to make sense, it needs unity, patterning, and organization. By aiding understanding, coherence should reduce uncertainty and increase the hedonic tone (Kaplan 1988; Wohlwill 1976). These relationships have been consistently confirmed in empirical research (Nasar 1988). The same author also found that contrast reduces the coherence of a scene.

Regarding evaluations involving psychological concepts, such as those related to perception, cognition, emotion and affection, according to Nasar (1988), they are subjective and have primary reference to either the environment (perceptual/cognitive judgments) or to people's feelings about the environment (emotional/affective judgments). While this type of assessment depends in part on perceptual/cognitive factors (as would ratings on variety of colorfulness of a scene), it is, by definition, an emotional/affective judgment (such as assessing the calmness of a scene), which involves evaluation and feelings.

To be relevant, such judgments must center on the dimensions of evaluation that people actually use in evaluating their surroundings. Ward and Russel (1981) examined this question. Using a variety of research strategies, they consistently found four dimensions to synthesize all the term descriptors of this type of evaluation - which in the English language contains up to 2,000 - namely: pleasant, arousing, exciting, relaxing (their opposites are also considered in the spatial metaphor proposed by the two authors).

According to Ward and Russel (1981), 'pleasing' is a purely evaluative dimension and the stimulus is independent of the evaluative dimension. 'Exciting' and the 'calming' involve mixtures of evaluation and stimulation. Russell explains (1988) that this evidence suggests that affective assessment involves a two-step process. An environment is initially and automatically perceived in terms of pleasant versus unpleasant and arousing versus unarousing. Thus, psychological studies converge towards the likelihood that people experience an exciting environment as more pleasant and arousing than a gloomy one; and relaxing environments as more pleasant and less arousing than distressing ones.

Based on these findings of Ward and Russel (1981), Nasar (2008) argues that people generally prefer and feel calmer in environments that have their elements varying from moderate to high complexity with high coherence (low contrast). They also prefer to visit these types of environments more often than others. Environments with high complexity and low coherence (high contrast) tend to be judged as most exciting. A pleasant environment, however, suggests that the project must have elements of high coherence (low contrast) and moderate complexity. An exciting place would have low coherence (high contrast) and high complexity.

In this context, it is possible to affirm that complexity and coherence (obtained by reducing the contrast levels of the elements in a scene) can influence the affective quality in hospital reception environments in a predictable way. Therefore, a relaxing quality could be increased through moderate complexity and high coherence (low contrast). As empirical evidence, based on Russell's (1988) findings, affective quality will be assessed in this research through various types of affective judgments of two hospital reception scenes - one perceived as having a relaxing quality and another with an exciting quality - typical of receptions in private hospitals in the region where the research was carried out, and evaluated in research developed by Maciel et al. (2018).

The emotional climate of the places, however, must vary to fit the objectives and activities performed there - as the ergonomics of the built environment recommends - always aiming to adapt the places to their users and never the reverse process, which is much more difficult and expensive (Iida 2005).

Ideas about the attributes of affective quality are usually based on the intuition of the designer or on the systemic study of preferences and affective evaluations in relation to environmental attributes. The ideal would be to use a combination of scientific and intuitive knowledge to shape the affective quality of an environment.

Therefore, it should be noted that the evaluation of the affective quality of places is relative to the specific circumstances in which it occurs, including peripheral and previously encountered sites (Russel 1988). The scenes that people perceive as having affective quality therefore depend on the type of scenes they typically encounter in their daily lives. Therefore, an affective assessment of a place will always be relative to the particular circumstances in which it is undertaken.

3 Research Methods and Procedures

The empirical investigation, an exploratory type of field research (Marconi and Lakatos 2003), aimed at the affective evaluation of scenes of hospital receptions, having as its main objectives: (i) to identify the dimensions that derive from environmental affection and (ii) the attributes most influenced by this kind of judgment.

The design of the empirical research has sought to avoid research methods that propose instruments with an emphasis on language, which can bring with it a series of problems, such as the tendency to merge affective and preference evaluations of environments with discourse and the constant need for linguistic analysis.

An alternative method of avoiding such a mishap was the use of a questionnaire, which incorporated two hospital reception scenes typical of private hospitals in the research-focused region - judged to have relaxing (Fig. 1) and exciting (Fig. 2) qualities in an earlier survey undertaken by the present authors - as well as a table with 36 bipolar descriptors (positive/negative) developed by Rohles and Milliken apud Felippe (2014) in support of the proposed evaluation (Table 1).

Fig. 1. Scene 1 hospital reception with a perceived calming quality. **Source.** Maciel et al. (2018).

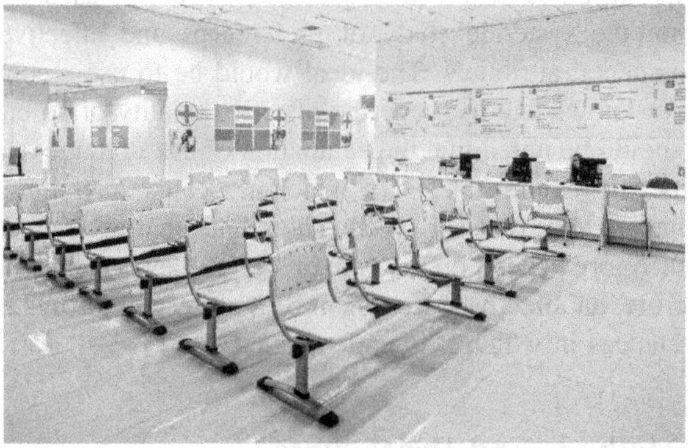

Fig. 2. Scene 2 hospital reception with a perceived exciting quality. **Source.** Maciel et al. (2018)

According to Stamps (1992), when evaluating the visual quality of environments, surveys consistently confirm that reliable results can be obtained by using color photographs, videos, slides, and visual simulations as stimulus elements.

Table 1. Bipolar descriptor terms for environmental assessment.

No	Negative	No	Positive	No	Negative	No	Positive
01	Crowded	10	Uncrowded	19	Uncomfortable	28	Comfortable
02	Confined	11	Spacious	20	Tedious	29	Interesting
03	Monotonous	12	Colorful	21	Exciting	30	Calming
04	Simple	13	Ornamental	22	Stuffy	31	Ventilated
05	Noisy	14	Silent	23	Confusing	32	Organized
06	Ugly	15	Attractive	24	Inelegant	33	Elegant
07	Gloomy	16	Happy	25	Scary	34	Safe
08	Pleasant	17	Pleasant	26	Abandoned	35	Conserved
09	Threatening	18	Welcoming	27	Unfinished	36	Structured

Source. Rohles and Milliken (1981) apud Felippe (2014).

Regarding the research procedures, respondents were initially informed that: (i) the study focused on the visual evaluation of hospital reception environments; (ii) there was no right or wrong answer; (iii) anonymity was guaranteed; (iv) the data collected would only be disclosed in scientific communications. Next, the participant - after identifying himself by sex, age and level of education - was asked to identify seven (7) terms that he judged to be strongly correlated with the two hospital reception scenes presented as stimulus elements. Finally, the same participant was again asked to identify the physical attributes that most influenced the choices of the seven descriptive terms selected.

The treatment of the collected data started with the tabulation and later analysis of the data through graphs and a table of distribution of the frequencies of answers (Marconi and Lakatos 2003), which sought to: (i) identify the main term descriptors of environments most strongly correlated with the two hospital reception scenes used as stimulus elements; (ii) link the physical attributes which most influence this type of judgment, in order to discuss the empirical results.

Research participants were randomly selected on the basis of having recently attended private hospital reception environments in the Metropolitan Region of Recife, capital of the State of Pernambuco, Northeast Brazil. At the end of the period stipulated to close the data collection, there were 75 subjects, most of them women (72.6%), aged between 39 and 48 years (57.5%), and with a level of schooling in higher education (53.4%).

4 Empirical Evidence

Figure 1, a scene judged in the research of Maciel, Costa Filho, and Villarouco (2018) as having a perceived calming quality, is a moderately complex hospital reception environment with high coherence (low contrast) in its physical attributes. Such features of the scene, by elevating the calming quality, would presumably be appropriate to the function that the type of environment plays.

It should be noted that the low contrast characteristic of its physical attributes, as referred to herein, facilitates comprehension, promotes coherence and further reduces uncertainty and increases the hedonic tone (beauty or pleasantness). This finding confirms the theoretical postulates tested.

The fact that it is a moderately complex environment, which is considered the ideal level of stimulation for the type of function of this type of environment, presumably increases the calm. From a theoretical perspective, the calming quality is increased through average complexity, since minimal complexity is postulated as monotonous and boring, while over-complexity is chaotic and stressful.

Figure 2 shows a hospital reception scene with low coherence (high contrast) and high complexity in its environmental attributes, as perceived by the participants in the research of Maciel et al. (2018). This scene was perceived to be the least soothing in terms of quality, and due to these characteristics, inadequate for the function and requirements of this type of environment, being highly distressing in the opinion of the respondents.

In relation to the empirical results obtained, Fig. 3 shows a higher frequency of environmental descriptor terms with a positive character. Scene 1, with a perceived soothing quality, was only perceived to have one negative characteristic, which was monotonous, and which was cited 24 times. The seven most related descriptors, in ascending order, were: spacious (64), comfortable (62), elegant (60), pleasant (58), conserved (57), organized (55), and attractive (53), corroborating both the findings of Maciel et al. (2018) as well as the theoretical postulate that a moderately complex hospital scene with high coherence (low contrast) is appropriate to the function of this type of environment.

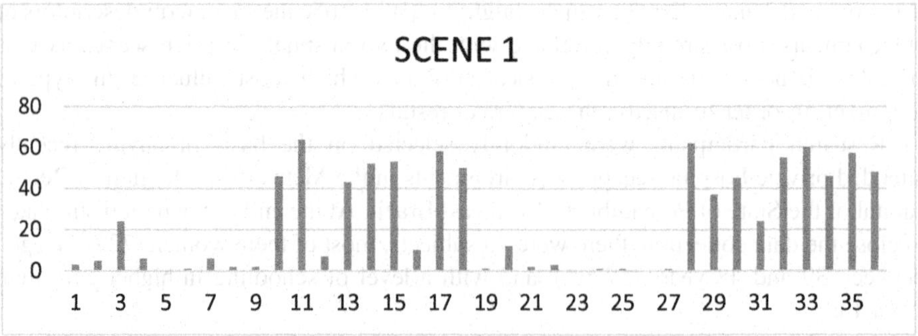

Fig. 3. Evaluative scale for the term descriptors of environments listed in Table 1.

On the other hand, Fig. 4 portends to a frequency of opinions that are clearly more related to terms describing environments which are more negative in character. Among the seven most associated terms, in ascending order of choice, we found: crowded (52), uncomfortable (52), agitated (48), simple (45), noisy (45) and tedious (41). Only with one exception, which does not invalidate the results, does the term color (43) depart from the negative characteristic of the other six.

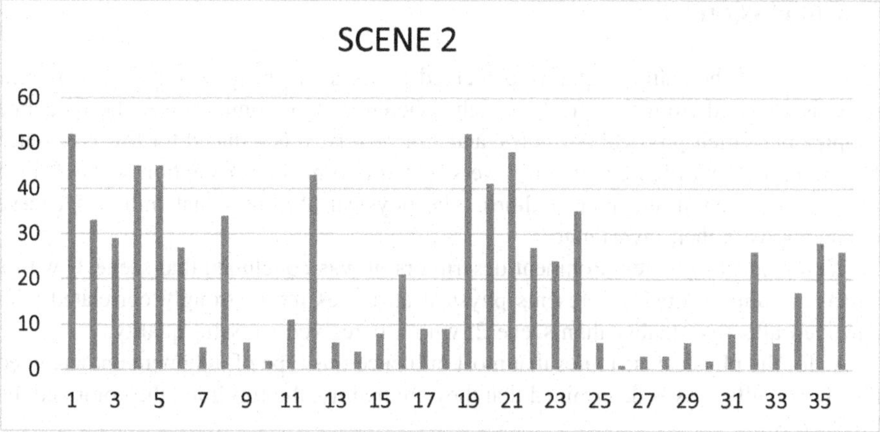

Fig. 4. Evaluative scale for the term descriptors of environments listed in Table 1.

The environmental elements that most influence these types of emotional/affective judgments are summarized in Table 2.

Table 2. Evaluative scores of physical properties (perceptual/cognitive judgments)

1	2	1	2	1	2	1	2	1	2	1	2
Color		Furniture		Dimension		Lighting		Decoration		Other	
22	64	20	127	7	82	8	38	14	65	3	23
85	36	124	19	106	9	92	10	96	17	24	4
6	62	0	96	1	48	3	43	2	40	2	36
74	8	139	28	71	17	93	12	100	10	33	11
187	**170**	**283**	**270**	**185**	**156**	**196**	**103**	**212**	**132**	**62**	**74**

The columns in Table 2 highlight, first, the scenes used; then the physical attribute considered for both scenes; then the scores attributed by the research participants to each of the four groups of nine negative and positive descriptor terms are shown in Table 1; and finally, the total score for the set of 36 environmental descriptors for each physical attribute highlighted by the participants are listed in Table 1.

It was observed it is the furniture, with expressive scores among the elements that obtained greater representativeness, which most influences the perceived affective quality; the color and the spatial dimensioning were also very capable of influencing this type of quality, as well as lighting and decoration.

5 Conclusion

Assuming that the calming quality perceived in scenes of hospital reception environments is elevated from the effect of high coherence (low contrast) and the moderate complexity of their physical attributes, and, conversely, it is reduced for low coherence (high contrast) and high complexity, we selected two hospital reception scenes which reflected these environmental qualities. The physical attributes that influenced these judgments were then investigated.

With respect to the environment descriptors, it was concluded that scene 1, with a perceived calming quality, presents physical attributes more strongly correlated with perceived affective quality than scene 2, with a perceived stressful quality.

As for the physical attributes that most influence this type of judgment on perceived affective quality, it was determined that they are mainly, the furniture, the color and the dimensions.

The empirical results found in this research, however, should not be taken simplistically, since they provide an understanding of the type of stimulus elements presented to the participants, the respondents, the place and time at which the research was carried out.

Acknowledgements. We are grateful for the collaboration received from Gisele Silva, Elizabeth Arruda, Miquelina Gondola and Italo Cabral.

References

Andrade C, Lima L, Pereira C, Fornara F, Bonaiuto M (2013) Inpatients' and outpatients' satisfaction: the mediating role of perceived quality or physical and social environment. Health Place 21:122–132. https://doi.org/10.1016/j.healthplace.2013.01.013

Becjker F, Douglas S (2008) The ecology of the patient visit. J Ambul Care Manag 31:128–141. https://doi.org/10.1097/01.JAC.0000314703.34795.44

Berlyne DE (1972) Ends and meaning of experimental aesthetics. Can J Psychol 26:303–325. https://doi.org/10.1037/h0082439

Beukeboom C, Langeveld D, Tanja-Dijkstra K (2012) Stress-reducing effects of real and artificial nature in a hospital waiting room. J Altern Complement Med 18(4):329–333. https://doi.org/10.1089/acm.2011-0488

Felippe ML (2014) Physical environment and environmental language in the affective stress restoration process in paediatric impatient rooms. PhD Thesis, University of Ferrara

Figueiredo E (2005) Ambientes de saúde – O hospital numa perspectiva ambiental terapêutica. In: Soczka L (ed) Contextos humanos e psicologia ambiental. Fundação Calouste Gulbenkian, Lisboa, pp 303–335

Iida I (2005) Ergonomia: Projeto e Produção. Edgar Blucher, São Paulo

Kaplan S (1988) Perception and landscape: conceptions and misconceptions. In: Nasar JL (ed) Environmental aesthetics: theory, research, and application. Cambridge University Press, New York, pp 45–55. https://doi.org/10.1037/1089-2680.2.2.153

Lawrence R (2002) Healthy residential environments. In: Bechtel R, Churchman A (eds) Handbook of environmental psychology. Wiley, New York, pp 394–412

Maciel AMM, Costa Filho L, Villarouco V (2018) A Qualidade Calmante Percebida em Cenas de Recepções Hospitalares, pp 798–807. In: Blucher, São Paulo. ISSN 2318-6968, https://doi.org/10.5151/eneac2018-058

Marconi MA, Lakatos EM (2003) Fundamentos de metodologia, 5th edn. Atlas, São Paulo

Nasar JL (1988) The effect of sign complexity and coherence on the perceived quality of retail scenes. In: Nasar JL (ed) Environmental aesthetics: theory, research, and application. Cambridge University Press, New York

Nasar JL (2000) The evaluative image of places. In: Walsh WB, Craik KH, Prince RH (eds) Person-environment psychology: new directions and perspectives, 2nd edn. Lawrence Erlbaum Associates, Mahwah, pp 117–168

Nasar JL (2008) Visual quality by design. Haworth Inc, Holland

Russel J (1988) Affective appraisals of environments. In: Nasar JL (ed) Environmental aesthetics: theory, research, and application. Cambridge University Press, New York, pp 120–129

Stamps AE (1992) Perceptual and preferential effects of photomontage simulations of environments: In: Perceptual and motor skills, nº 74. https://doi.org/10.2466/pms.1992.74.3.675

Ulrich R, Zimring C, Zhu X (2008) A review of the research literature on evidence-based healthcare design. Health Environ Des J 1:61–125. https://doi.org/10.1177/193758670800100306

Villarouco V (2011) Tratando de ambientes ergonomicamente adequados: seriam ergoambientes? In: Mont'Alvão C, Villarouco V (eds) Um novo olhar sobre o projeto: A Ergonomia do Ambiente Construído. Teresópolis: 2AB, pp 25–46

Ward L, Russel J (1981) Cognitive set and the perception of place. Environ Behav 13(5): 219–235

Wohlwill J (1976) Environmental aesthetics: the environment as a source of affect. In: Altmann I, Wohlwill J (eds) Human behavior and environment, vol 1, pp 37–86

A Modified Weighted SERVQUAL Method and Its Application on Elder Care House in Taiwan

Chaohua Ko and Chinmei Chou(✉)

Yuan Ze University, No. 135, Yuan-Tung Road, Chung-Li 32003
Taiwan, R.O.C.
Kinmei@saturn.yzu.edu.tw

Abstract. SERVQUAL (SERVice QUALity) instrument is a self-completion questionnaire, containing five dimensions and twenty-two items presented in the form of seven-point Likert scale, is considered the appliance that the most complete to measure service quality. This research proposed a modified weighted SERVQUAL analysis, based on statistical principle, which is able to explore customers' satisfaction combined with importance evaluation on attribute level. The effectiveness of this modified weighted analysis is validated with a comparison of the results of SERVQUAL and weighted SERVQUAL analysis applied to a nursing home study in Taiwan. The results derive from SERVQUAL and weighted SERVQUAL are exactly identical. "Suitable temperature at elderly rooms", "Feel safe and feel at home" and "Medical treatment and doctor visiting are well scheduled" are the first three gap score among all attributes. By applying our modified weighted SERVQUAL analysis, the importance of each attribute could be taken into consideration. "Nurses understand elderly's needs" and "The employees solve the elderly's problem sincerely" are promoted to first two and three gap score, due to their high importance ranking (top 84 and 90% respectively). Our modified weighted analysis demonstrated the usability and discriminatory power of service quality evaluation than original weighted SERVQUAL method. Finally, a number of recommendations and insights to address these shortfalls have been proposed to managers. Service quality cannot be overemphasized because of its strong link to higher customer satisfaction, repurchase, customer loyalty and leads to higher profitability in modern competitive business environment.

Keywords: SERVQUAL · Nursing home · Service quality

1 Introduction

According to the statistics of World Health Organization (WHO), the world ageing population is rapidly growing and its percentage has been predicted will double by 2050. Asia is experiencing faster ageing population growing than any other continent and Taiwan is one of Asia countries with high percentage of ageing population growth [1]. In Taiwan, the percentage of elder was roughly 10% at the end of 2006 and is projected to reach 20% by 2027, and 30% in 2050 [2]. Furthermore, the percentage of

S. Bagnara et al. (Eds.): IEA 2018, AISC 818, pp. 322–330, 2019.
https://doi.org/10.1007/978-3-319-96098-2_42

elder with at least one chronic illness reached a historic high of 65%, and one out of ten needed long-term health care [3]. For elderly, with the increasing needs of long term care, nursing home has become the common solutions to overcome these problems in modern societies [4]. Three main contributions are addressed in this research. Firstly, a validated and reliable SERVQUAL questionnaire was developed specific to nursing home to expose elders' satisfaction on the delivery of services. Secondly, a modified weighted SERVQUAL analysis, based on statistical principle, was proposed to reflect customers' satisfaction on every single attribute without adding any extra burden. Furthermore, the effectiveness and discrimination of this modified analysis are demonstrated with comparing the results of SERVQUAL and weighted SERVQUAL analysis applied to nursing home in Taiwan.

2 Literature Review

2.1 SERVQUAL: The Measurement of Service Quality

SERVQUAL instrument is a self-completion questionnaire, containing five dimensions and twenty-two items presented in the form of seven-point Likert scale, is considered the appliance that the most complete to measure service quality. It is frequently to be used as a diagnostic technique to identify the strengths and weaknesses of service quality. The general validity of the instrument is high enough and has been applied in numerous service industries such as financial services, communication services, higher education, healthcare fields, retail store and IT industries during the past few decades [5–7]. Currently, their conceptualization and measurement of service quality have been received widespread acceptance.

2.2 The Weighted SERVQUAL Measurement and Its Defects

A SERVQUAL measurement weighted by relative importance, named weighted SERVQUAL, can demonstrate how customers will prefer service to be and in order for service providers to make improvements. It enables researchers and managers to discover which dimension is more important to customers as compared to others. The weighted SERVQUAL model avoids the limitation of the original SERVQUAL model in neglecting the relative importance of dimensions, but defects were observed at the same time: (1) The allocation of important points among five SERVQUAL dimensions has formed a burden and trouble to customers. It can be seen from the various cancellation marks on the forms that customers were struggling to allocate the points. Furthermore, once the total points do not add up to 100, how to deal with the questionnaires is another sticky problem. It is not suitable for specific group, such as elders and young children; (2) Customers are hard to access and have identical concepts of reliability, assurance, tangibles, empathy, responsiveness without detail descriptions and specific examples; (3) The importance degree may be not identical among attributes within dimensions, evaluation the relative importance on dimension level may lose critical information.

3 Methodology

3.1 Development of SERVQUAL Questionnaires

In this research, SERVQUAL questionnaire was developed and modified by pilot studies and refer to academic literatures on nursing home. The pilot study has been done with 26 service attributes and test how comprehensible the questionnaire is. Finally, total of 23 service attributes were determined which are consistent with SERVQUAL five dimensions structures. A questionnaire was finally constituted of 5 dimensions and 23 attributes (shown as Table 1).

Table 1. The SERVQUAL attributes applied in this study

Dimension	NO	Code	Service attributes
Tangible	1	T1	Medical instrument and physical facilities are visually appealing
	2	T2	Employees uniform are clean, nice, and neat
	3	T3	Clean, adequate supplies, and well maintained for every rooms
	4	T4	Well lighted for every rooms
	5	T5	Suitable temperature at elderly rooms
	6	T6	Meals served are clean and hygiene
	7	T7	Meals served are delicious
	8	T8	The atmosphere for every rooms are cozy
	9	T9	The scent for every rooms are refreshing
Reliability	10	RL1	Appropriate employees response
	11	RL2	Medical treatment and doctor visiting are well scheduled
	12	RL3	Available and adequate elderly family visiting time
	13	RL4	All elderly activities are well scheduled
	14	RL5	The employees solve the elderly's problem sincerely
	15	RL6	All equipment (AC, TV, radio, light, etc.) work properly
Responsiveness	16	R1	Employees give clear information and understandable
	17	R2	Appropriate and prompt services
	18	R3	Quick medical treatment response when elderly need it
Assurance	19	A1	Feel safe and feel at home
	20	A2	Employees behavior instills confidence in elderly
Empathy	21	E1	Employees are helpful, careful, and friendly
	22	E2	Nurses understands elderly's needs
	23	E2	No discrimination to the elderly

Reliability and validity test was conducted using SPSS 16. Cronbach's Alpha was used in this study to determine the internal consistency or average correlation of items in a survey instrument to gauge its reliability. A number 0.7 was suggested as a standard of Cronbach's Alpha coefficient. In terms of reliability, Cronbach's alphas were 0.877 and 0.911 for both expectations and perceptions respectively. Validity refers to the degree to which evidence and theory support the interpretation of test scores entailed by proposed uses of tests. In this study, Pearson correlation coefficient was used to examine the correlation between total score and subject real performance. Generally, the coefficient value greater than 0.7 indicates strong validity and the coefficient value less than 0.3 indicates weak validity. The higher the coefficient means the better of the data fitness to measurement objective. For expectation and perception questionnaire, all of the service attributes are valid. The content validity of the questionnaire was deemed adequate.

3.2 Participants

Recruitment occurred over a 9-month time period, between January 2016 and October 2016, in nursing home in northern part of Taiwan, including Taoyuan and New Taipei City. All participants stayed in nursing home more than 6 months and already understood the services provided by the nursing home very well. The process of interviewing took around 50 min for each elder.

3.3 Procedure

Service quality and relative importance of attributes should be determined by the customers. Firstly, each participant was given a questionnaire with brief description of five dimensions of SERVQUAL, namely tangibility, reliability, responsiveness, assurance and empathy. Secondly, the participants were asked to divide a total of 100 points among the five dimensions based on the perceived importance of these dimensions to them. The highest points were assigned to the dimension which they perceived as the most important to them and vice versa. Finally, participants were asked to evaluate their expectations and perceptions for the service of nursing home. The question to identify participants' expectation is "how important these services attribute" with 5-point Likert scale ranging from "unimportant at all (1)" to "very important (5)". The question to identify participants' perception is "how do you feel about these services attributes which are already provided", also with 5-point Likert scale method ranging from "very bad (1)" to "very good (5)" (shown as Appendix A).

3.4 Data Collection and Analysis

The gap sores of each service attributes would be figured out by calculating the difference score between participants' perception (P) score and expectation score (E). This step would be conducted per attributes. This calculation does not take the relative importance of quality dimensions into account, named unweighted or original SERVQUAL score. For each attribute of SERVQUAL questionnaires, the SERVQUAL score (SS) is defined as perception score (P) minus expectation score (E), please refer to Eq. 1 as following.

$$SS = P - E \tag{1}$$

The weighting factors (WF) of dimensions would be got by calculating the average score of each dimension evaluated by participants. Then the weighted SERVQUAL scores (WSS) could be got by multiplying the original SERVQUAL score on each attribute with its dimension weighting factors, please refer to Eq. 2 as following [8].

$$WSS = WF * SS \tag{2}$$

WSS: weighted SERVQUAL score on each attribute or dimension.
WF: weighting scores, divide 100 points among five dimensions by participants
SS: SERVQUAL score on each attribute or dimension.

In our modified weighted method, customers' expectation score is not only for comparing with perception score to get gap scope, but also the resource to extract the importance judgments from customers directly. For each SERVQUAL attribute, the relative importance was evaluated more appropriate individually than shared the same points from dimensions for meaningful interpretations. The previous SERVQUAL instruments are constituted of Likert scaling. As it is based on the methodology of ranking, neither calculating means nor standard deviations and not expresses any meaningful results statistically. With sufficient sample sizes, participants' expectation score on attributes will form a normal distribution. The relative importance between attributes could be inferred from the accumulation percentages of this standardized distribution shown as Eq. 3. Then the modified weighted SERVQUAL scores could be got by multiplying the original SERVQUAL score with each modified weighting factor.

$$\mathrm{Y} = f(X|\mu, \sigma^2) = \frac{1}{\sigma\sqrt{2\pi}} e^{\frac{-1}{2}\left(\frac{x-\mu}{\sigma}\right)^2} \tag{3}$$

Y: cumulative probability of specific x
x: the relative importance evaluation on specific attribution
μ: mean of sampling distribution
σ: standard deviation of sampling distribution.

3.5 Data Collection and Analysis

The top three item of perception questionnaire are T9: "The scent for every rooms are refreshing" (mean = 4.00), T3: "Clean, adequate supplies, and well maintained for every rooms" (mean = 3.93) and E1: "Employees are helpful, careful, and friendly" (mean = 3.89), respectively. In term of original SERVQUAL scores, the negative value therefore indicates that the performance of the nursing home was not meeting the expectations of their customers. The top three gap between expectation and perception exist in T5: "Suitable temperature at elderly rooms" (gap = −1.20), A1: "Feel safe and feel at home" (gap = −1.11) and RL2: "Medical treatment and doctor visiting are well scheduled" (gap = −1.02). However, it is imprudent to draw the conclusions from raw SERVQUAL score directly since it did not take the relative importance of the service dimensions into consideration. Thus, the weighted SERVQUAL scores are calculated and shown as following.

The results showed the top three scores of weighted attributes are T5: "Suitable temperature at elderly rooms" (weighted gap = −1.20), A1: "Feel safe and feel at home" (weighted gap = −1.11) and RL2: "Medical treatment and doctor visiting are well scheduled" (weighted gap = −1.02). The negative weighted SERVQUAL score indicates a mismatch between the priorities expressed by the participants and the level of quality delivered by the elder house care.

The modified weighted SERVQUAL scores could be got by multiplying the original SERVQUAL score with each modified weighting factor, attribute by attribute (refer to Eq. 3). The results showed the top three scores of modified weighted attributes are T5: "Suitable temperature at elderly rooms" (modified weighted gap = −1.04), E2: "The employees solve the elderly's problem sincerely" (modified weighted gap = −0.82) and RL5: "The employees solve the elderly's problem sincerely" (modified weighted gap = −0.77).

4 Result and Discussion

By applying SERVQUAL method, "Suitable temperature at elderly rooms (T5)" (Gap = −1.20), "Feel safe and feel at home (A1)" (Gap = −1.11) and "Medical treatment and doctor visiting are well scheduled (RL2)" (Gap = −1.02) are the first three gap score among all attributes. The results derive from weighted SERVQUAL method (WG) are exactly identical with original SERVQUAL scores. By applying modified weighted-SERVQUAL method, "Suitable temperature at elderly rooms (T5)" (MWG = −1.04), "Nurses understands elderly's needs (E2)" (MWG = −0.82) and "The employees solve the elderly's problem sincerely (RL5)" (MWG = −0.77) are the first three gap score among all attributes.

In practical concerns, the evaluation of the weight on dimension levels may be too rough to highlight the importance of attributes. For example, "Nurses understands elderly's needs (E2)" and "The employees solve the elderly's problem sincerely (RL5)" are not place the first three gap score by applying SERVQUAL and weighted SERVQUAL instrument. Once take their attribute importance into consideration, the high importance ranking (top 84% and 90% respectively) make their weighted score promote to first two and three respectively. Assigning identical weighting factor to attributes within dimension may not present the real importance of attributes. Our modified weighting method can present the importance of attributes individually, and improve the sophistication and discriminatory power of service quality diagnosis.

5 Conclusion

There are three main contributions of this research. First of all, a validated and reliable SERVQUAL questionnaire was developed specific to nursing home for managers and researchers to increase customers' satisfaction with the delivery of services. Secondly, a modified weighted SERVQUAL analysis was proposed, based on statistical principle, which is able to explore respondents' satisfaction on attribute level rather than dimension level without any extra burden for respondents. Furthermore, a modified weighted SERVQUAL analysis was proposed to reveal the relative importance between attributes by inferring their accumulation percentages of standardized distribution. It enables managers to grasp customers' satisfaction and importance evaluation of each attribute directly. Thus managers can develop the corresponding coping strategies easily. The effectiveness and discrimination of our modified analysis is demonstrated with a case of comparing the results of SERVQUAL and weighted SERVQUAL analysis applied to nursing home in Taiwan.

Appendix A: SERVQUAL Questionnaire

Expectation Questionnaire: How important these services attributes for you?

Item	Service attributes	Unimportant at all < > Very important				
		1	2	3	4	5
1	Medical instrument and physical facilities are visually appealing					
2	Employees uniform are clean, nice, and neat					
3	Clean, adequate supplies, and well maintained for every rooms					
4	Well lighted for every rooms					
5	Suitable temperature at elderly rooms					
6	Meals served are clean and hygiene					
7	Meals served are delicious					
8	The atmosphere for every rooms are cozy					
9	The scent for every rooms are refreshing					
10	Appropriate employees response					
11	Medical treatment and doctor visiting are well scheduled					
12	Available and adequate elderly family visiting time					
13	All elderly activities are well scheduled					
14	The employees solve the elderly's problem sincerely					
15	All equipment (AC, TV, radio, light, etc.) work properly					
16	Employees give clear information and understandable					
17	Appropriate and prompt services					
18	Quick medical treatment response when elderly need it					
19	Feel safe and feel at home					
20	Employees behavior instills confidence in elderly					
21	Employees are helpful, careful, and friendly					
22	Nurses understands elderly's needs					
23	No discrimination to the elderly					

Perception Questionnaire: How do you feel about these service attributes that already been provided?

Item	Service attributes	Very bad < > Very good				
		1	2	3	4	5
1	Medical instrument and physical facilities are visually appealing					
2	Employees uniform are clean, nice, and neat					
3	Clean, adequate supplies, and well maintained for every rooms					
4	Well lighted for every rooms					
5	Suitable temperature at elderly rooms					
6	Meals served are clean and hygiene					
7	Meals served are delicious					
8	The atmosphere for every rooms are cozy					
9	The scent for every rooms are refreshing					
10	Appropriate employees response					
11	Medical treatment and doctor visiting are well scheduled					
12	Available and adequate elderly family visiting time					
13	All elderly activities are well scheduled					
14	The employees solve the elderly's problem sincerely					
15	All equipment (AC, TV, radio, light, etc.) work properly					
16	Employees give clear information and understandable					
17	Appropriate and prompt services					
18	Quick medical treatment response when elderly need it					
19	Feel safe and feel at home					
20	Employees behavior instills confidence in elderly					
21	Employees are helpful, careful, and friendly					
22	Nurses understands elderly's needs					
23	No discrimination to the elderly					

References

1. Chen C-Y (2010) Meeting the challenges of eldercare in Taiwan's aging society. J Clin Gerontol Geriatr 1(1):2–4
2. Wang WL, Chang HJ, Liu AC, Chen YW (2007) Research into care quality criteria for long-term care institutions. J Nurs Res 15(4):255–283
3. Wang WF, Chen YM (2004) Family experiences of placing the elderly in long-term care facilities. J Long Term Care 8(3):327–344
4. Zinn JS, Aaronson WE, Rosko WE (1993) Variations in the outcomes of care provided in Pennsylvania nursing homes: facility and environmental correlates. Med Care 31:475–487

5. Sohail MS (2007) Determinants of service quality in the hospitality industry: the case of Malaysian hotels. J Account Bus Manag 14(2):64–74
6. Butt MM, de Run EC (2008) Measuring Pakistani mobile cellular customer satisfaction. The ICFAI J Serv Mark 6(1):40–50
7. Niranjan TT, Metri BA (2008) Client-vendor-end user triad: a service quality model for IS/ITES outsourcing. J Serv Res 8(1):123–138
8. Cronin JJ, Taylor SA (1992) Measuring service quality; a re-examination and extension. J Mark 56(3):55–68

Understanding Resilience and Adaptation in the Blood Transfusion Process Using Employee Accounts of Problem Resolution

Alison Watt[1,2](✉), Gyuchan Thomas Jun[1], Patrick Waterson[1], and John Grant-Casey[3]

[1] Loughborough Design School, Loughborough University, Loughborough, UK
A.Watt@lboro.ac.uk
[2] Serious Hazards of Transfusion (SHOT), The UK Haemovigilance Scheme, Manchester, UK
[3] National Comparative Audit of Blood Transfusion, NHS Blood and Transplant, Bristol, UK

Abstract. Blood transfusion is usually considered to be safe and high-quality. Improvement measures have concentrated on assessing adherence to evidence-based guidelines, but this may fail to understand adaptations staff make in a complex, dynamic environment. Three hospitals in England were visited to trial a method for assessing adaptations in the complete vein to vein transfusion process. An open narrative question encouraged staff to describe adaptations voluntarily and a follow up question scored the level of support received from management. Adaptations can be categorised into: (1) preferred adaptations - developments expected to improve the process; (2) forced adaptations - work-arounds and coping strategies when ideal solutions are outside of their control. Preferred adaptations indicate the surrounding system is adequate, but performance/well-being can be improved by adapting. Such adaptations may be good practices and lessons are to be shared. Forced adaptations indicate the surrounding system is inadequate, so people need to adapt to get work done. These may not be desirable, so are unlikely to be suitable for shared learning. Instead the surrounding system probably needs to be assessed or changed. Adaptations are made within staff members' sphere of influence and although managers may have opportunities to make more resilient changes, frontline staff may be forced to make adaptations to get things done. To facilitate the design of more efficient and safer blood transfusion processes, there is a need for better understanding about how to carry out audit and incident investigation or reporting recognising the adaptations and resilience of healthcare staff.

Keywords: Transfusion · Healthcare · Resilience · Adaptation

1 Background

Blood transfusion is usually considered a reliable and safe procedure. Pre-transfusion testing is comprehensive and the risk of viral complication is considered to be very low. Therefore, everything is expected to be satisfactory if staff comply fully with standard

S. Bagnara et al. (Eds.): IEA 2018, AISC 818, pp. 331–338, 2019.
https://doi.org/10.1007/978-3-319-96098-2_43

operating procedures (SOP). When errors are made, which mean patients can be put at risk, the incidents are reportable to the UK haemovigilance scheme, Serious Hazards of Transfusion (SHOT). The focus for improvement has been on incident investigation and assessment of staff adherence to evidence-based guidelines.

However, the process is a complex multidisciplinary procedure, often involving a variety of healthcare professions. SHOT has identified nine major steps in the process from requesting a transfusion and taking a sample from the patient for crossmatching, through all the procedures necessary before transfusing a component to the patient (Fig. 1) [1]. This is known as the vein to vein process, i.e. a sample taken from a patient's vein leading to a transfusion back into the same patient's vein. Different individuals may perform each of these steps and errors at any step can result in patient death or major morbidity.

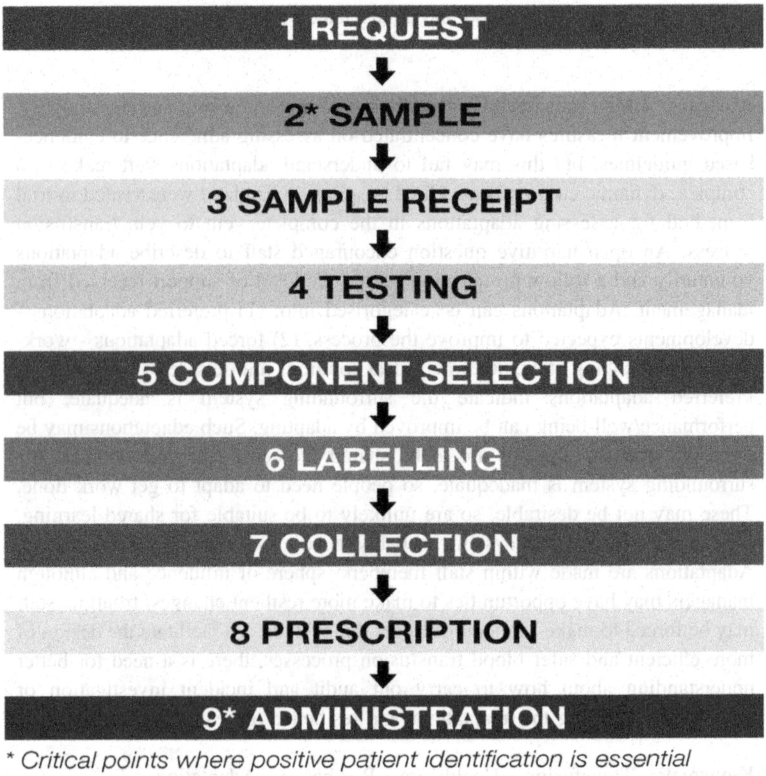

Fig. 1. Diagram of the nine steps in the vein to vein transfusion process, from 2013 Annual SHOT Report (published 2014) [1]

Approaching safety via incident investigation, based on adherence to SOPs often fails to understand the many adaptations staff need to make to deal with this complex and dynamic healthcare environment, so does not identify credible and meaningful information about how to improve the process. There is an urgent need for a better understanding of the adaptation and resilience of healthcare staff in the transfusion process to facilitate the design of more efficient and safer blood transfusion processes.

2 Methods

The aim of this study was to explore a potential new way of discovering the extent to which the processes involved in transfusion are being adapted so this knowledge can be used for process redesign.

Three hospitals in England were visited to assess a method of identifying adaptation and resilience. The complete vein to vein transfusion process was investigated. An open question was adopted, based on previous studies in a hospital dispensary [2, 3] which allows staff to develop their account naturally from identifying a problem to volunteering information on any adaptation(s) used to resolve the issue. Employees working in each of the domains listed in Fig. 1 were interviewed. They represented a range of professions, including doctors, nurses, midwifes and scientists as well as ancillary workers such as phlebotomists, healthcare assistants administrative staff and porters. Each was asked the following question:

- "Please give a short outline of the biggest or most recent difficulty that you have faced when carrying out this procedure and what did you do about the issue?"

Their responses were captured and analysed using these considerations:

- Why and how they adapted? (Fig. 3)
- The efficacy of their adaptations? (Fig. 4)

A follow up question was asked:

"How supportive was your manager/department for how you solved the issue?"

The answers were captured using a Likert scale ranging from 5, very supportive to 1 very unsupportive or N/A for not applicable [4].

3 Results

All employees questioned (n = 37) gave at least one example of a problem/adaptation and several gave more than one (total n = 66). They usually adapted to overcome actual or potential difficulties with processes and to cope with deficiencies in staffing, resources or training. Adaptations were seen at every stage of the nine-step transfusion process (Fig. 2).

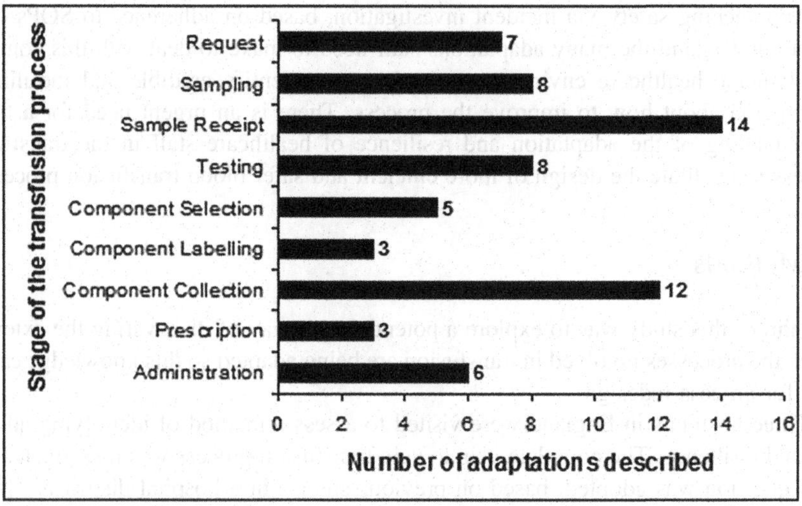

Fig. 2. Number of adaptations described at each stage in the transfusion process

Adaptations are generally made within the sphere of influence of staff members (Fig. 3). The reasons for adaptations can be categorised into:

- preferred adaptations – developments expected to improve the process
- forced adaptations - workarounds and coping strategies when ideal solutions are outside of their control

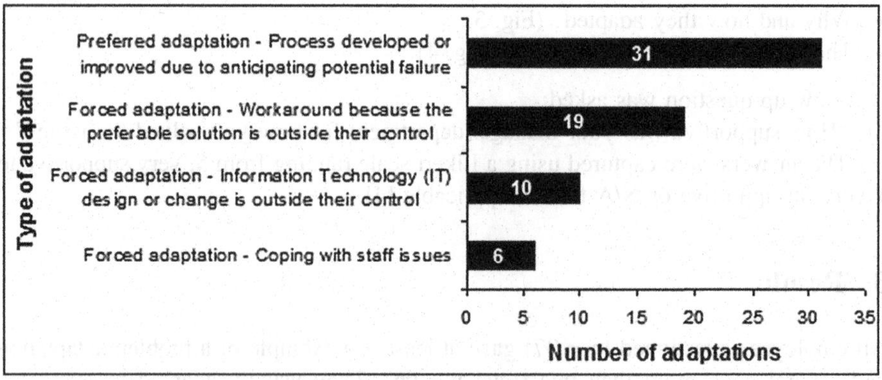

Fig. 3. Type of adaptations made

The types of adaptations usually occurred within aspects that the staff members were empowered to change (Fig. 4) and can be summarised as:

- Agreed process change - where the adaptation was a permanent change made to the standard process
- Added step(s) to the process - One or more extra, often unnecessary, steps were added to the process to compensate for the problem(s) being experienced
- Local process change - where the adaptation was not validated or accepted, but had become custom and practice for a department or an individual

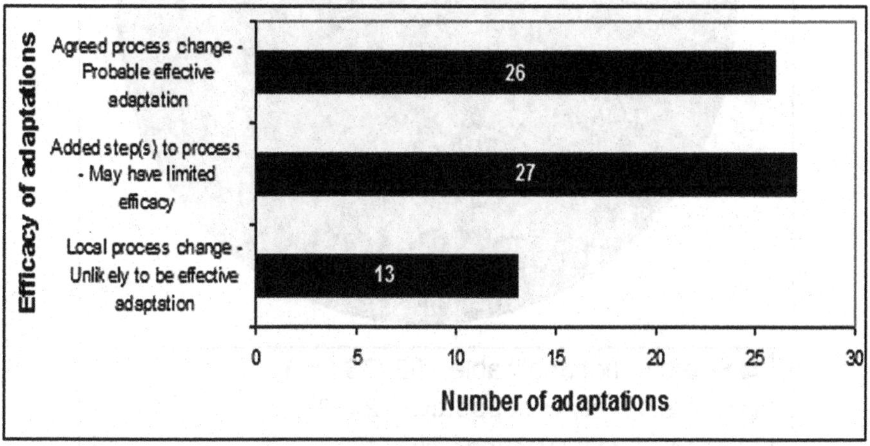

Fig. 4. Efficacy of adaptations made

The scores for question 2, “How supportive was your manager/department for how you solved the issue?” were graded from 5, very supportive to 1, very unsupportive. This was designed to assess whether the adaptations met with managerial or departmental approval, but the most common response was not applicable (N/A), n = 52, showing that adaptations tended to be made without contribution or approval from management or departmental colleagues (Fig. 5).

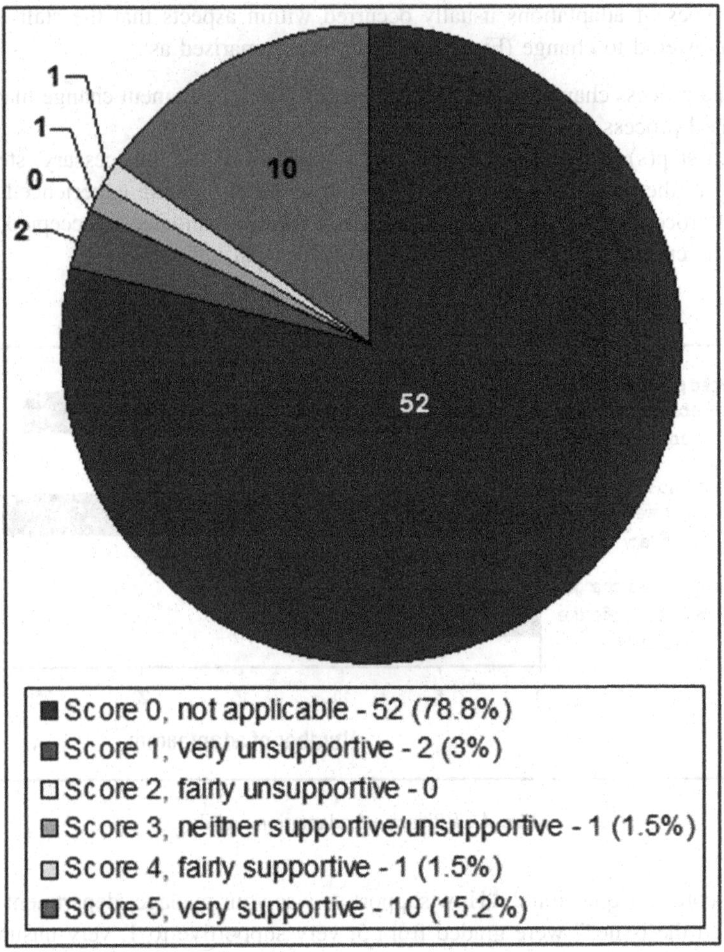

Fig. 5. Assessment of supportiveness of manager/department for the adaptations

4 Discussion and Conclusion

This research is demonstrating a useable method to understand work as imagined compared to work as done. It also enables assessment of the resilience within the transfusion process in any organisation.

The first open question allowed employees to describe their difficulties openly in an unconstrained manner and to present details of any adaptation(s) used to resolve the identified problem. These adaptations were often forced upon the staff members by unexpected problems such as lack of resources and staffing or by work systems, including IT systems, that were not entirely fit for purpose. This may be more likely to happen in healthcare where staff are affected by changeable work system elements and therefore react and adapt to the work system [5]. Forced adaptations indicate the surrounding system is inadequate, so people may need to adapt to get work done. Adaptations that are forced may not be desirable so are unlikely to be suitable for shared learning. Some adaptations were preferred changes, which show the surrounding system is adequate, but a development of the process may improve performance and well-being. Such adaptations may be good practices and lessons are to be shared.

Further analysis of the changes that employees described showed three common themes in how they adapted, i.e. (1) agreed permanent changes; (2) adding step(s) to the process either temporarily or permanently or (3) local process changes which were usually temporary and often not validated or generally accepted. The permanent, agreed changes tended to be most effective and therefore likely to increase resilience. However, even those agreed changes were sometimes not as ideal as staff would have liked. Adding extra steps often resulted in unnecessary extra work, so these adaptations were not particularly effective. Local process changes tended to be the least effective or resilient adaptations and were often driven by variability that had not been foreseen when standard processes were developed. These showed the biggest disparity between work as imagined and work as done [6] and indicate that the surrounding system may need to be assessed or changed.

The second question estimated managerial or departmental support for adaptations. The high number of N/A scores (n = 52) showed little encouragement of resilience, often because managers were not aware of changes implemented. Good support, such as those scoring 5 (n = 10) may indicate if it was an appropriate adaptation. Poor support, e.g. score 1 (n = 2) might indicate concern at the safety implications of the amendment made to the standard process. Participants felt managers may have opportunities to make more resilient changes, while frontline staff may be reluctantly forced to make adaptations to get things done. Under-resourcing of the UK National Health Service (NHS) means adaptations can result from an acceptance that requesting more staff, training or equipment is not realistic. Understanding these drivers, mechanisms and experience of adaptations at different levels suggests possibilities for system redesign and improvement.

Key lessons learned from this research are:

- Adaptations are common in the transfusion process as a means of solving problems and are usually restricted to the sphere of influence of the individual staff member.
- Managers and departmental colleagues are often not aware of or not supportive of these adaptations, which may reduce the opportunities for resilience.

Future work will include collection of more data from hospitals throughout the UK during a large 'Vein to Vein Audit' (V2V), as part of the National Comparative Audit programme. Analysis of answers given to these two questions during the planned V2V audit will highlight areas of the process where adaptations might be made to reduce error in transfusion.

References

1. Bolton-Maggs PHB (ed) Poles D, Watt A, Thomas D (2014) On behalf of the Serious Hazards of Transfusion (SHOT) Steering Group. The 2013 Annual SHOT Report. https://www.shotuk.org/wp-content/uploads/2013.pdf. Accessed 16 May 2018
2. Sujan MA, Pozzi S, Valbonesi C, Ingram C (2011a) Resilience as individual adaptation: preliminary analysis of a hospital dispensary. In: Proceedings of HCP 2011—fourth workshop on human centered processes
3. Sujan MA, Ingram C, McConkey T, Cross S, Cooke MW (2011b) Hassle in the dispensary: pilot study of a proactive risk monitoring tool for organisational learning based on narratives and staff perceptions. Quality and safety in health care, pp.bmjqs-2010
4. Likert R (1932) A technique for the measurement of attitudes. Arch Psychol
5. Carayon P, Wetterneck TB, Rivera-Rodriguez AJ, Hundt AS, Hoonakker P, Holden R, Gurses AP (2014) Human factors systems approach to healthcare quality and patient safety. Appl Ergon 45(1):14–25
6. Braithwaite J, Wears RL, Hollnagel E (eds) (2016) Resilient health care, Volume 3: reconciling work-as-imagined and work-as-done. CRC Press, Boca Raton

Challenges of Disposition Decision Making for Pediatric Trauma Patients in the Emergency Department

Bat-Zion Hose[1,2(✉)], Pascale Carayon[1,2], Peter Hoonakker[2], Abigail Wooldridge[1,2], Tom Brazelton[3], Shannon Dean[3], Ben Eithun[4], Michelle Kelly[2,3], Jonathan Kohler[3], Joshua Ross[3], Deborah Rusy[3], and Ayse Gurses[5]

[1] Department of Industrial and Systems Engineering, University of Wisconsin-Madison, Madison, WI, USA
bhose@wisc.edu
[2] Center for Quality and Productivity Improvement, University of Wisconsin-Madison, Madison, WI, USA
[3] UW School of Medicine and Public Health, Madison, WI, USA
[4] American Family Children's Hospital, University of Wisconsin Hospitals and Clinics, Madison, WI, USA
[5] The Armstrong Institute for Patient Safety and Quality, Johns Hopkins University, Baltimore, MD, USA

Abstract. About 9.2 million children visit the emergency department (ED) in the US annually because of trauma and 20% experience a missed injury. Upon arriving to the hospital, physicians evaluate the child and make the ED disposition decision of whether to admit, operate or discharge. The objective of this study is to report the challenges mentioned by healthcare professionals about ED disposition decision making. We conducted 11 interviews with 12 healthcare professionals and identified 2 challenges of ED disposition decision making. The first challenge was timing of the decision; e.g., an ED nurse explained that a quick decision by physicians is important for providing timely patient care to critically ill children. The second challenge was leadership and team organization; e.g., the OR nurse and surgery resident both mentioned the need to know who to listen to so that they can understand what to do. Analyzing these challenges to ED disposition decision making can help to identify sociotechnical solutions for enhancing team situation awareness.

Keywords: Pediatric trauma · Decision making · Coordination ED disposition · Patient safety · Team situational awareness

1 Introduction

About 9.2 million children visit the emergency department (ED) in the US annually due to traumatic injury [1]. Pediatric trauma care is prone to safety concerns, e.g. missed injuries or delays in care, with 20% of pediatric trauma patients experiencing a missed injury [2]. Effective coordination for pediatric trauma patients when they arrive to the

S. Bagnara et al. (Eds.): IEA 2018, AISC 818, pp. 339–345, 2019.
https://doi.org/10.1007/978-3-319-96098-2_44

ED is critical because many decisions have to be made quickly, often with incomplete information [3].

Pediatric trauma care begins at the scene (i.e. the location of the accident) and continues with the arrival of the child in the ED; it may continue in the hospital. After the child is evaluated in the ED, physicians must decide whether to admit or discharge the patient, which is known as ED disposition [4]. ED disposition must be communicated to other physicians (who may not be present in the ED), nurses, social workers, administrators and others [5]. Delay in or disagreement about ED disposition decision can have negative consequences in the care process and affect patient safety [6].

Pediatric trauma can be characterized by the level of response from the trauma team, which depends on the child's acuity and injury. A level 1 trauma requires a full activation of the trauma team. The American College of Surgeons requires the following to respond to a level 1 trauma: a general surgeon within 15 min of patient arrival, emergency medicine (EM) physician, surgical and EM residents, ED nurses, a laboratory technician, radiology technician, critical care nurse and security officer. A level 2 trauma requires response from a fourth or fifth year surgery resident within 15 min of patient arrival [7]. In a previous study, we identified 53 roles actively involved in care of pediatric trauma patients; some are in the ED: 33 roles for level 1 and 18 roles for level 2 [4]. Such large teams can benefit from team situational awareness (SA) as there is an overlap of informational needs that can change due to the dynamic environment [8].

In this study, using semi-structured interviews, we assessed the challenges of ED disposition making for pediatric trauma and discuss implications for team SA.

2 Methods

2.1 Setting and Participants

This study is part of a larger project aimed at understanding care transitions in pediatric trauma to develop health information technology (IT) design requirements (https://cqpi.wisc.edu/teamwork-and-care-transitions-in-pediatric-trauma/). We conducted 11 interviews with 12 healthcare professionals (1 group interview was conducted with 2 operating room (OR) nurses): 1 EM resident, 2 surgery residents, 2 pediatric intensive care unit (PICU) fellows, 2 ED nurses, 2 OR nurses, 1 PICU nurse, 1 child life specialist and 1 ED coordinator. This study took place at an academic children's hospital in the Midwest of the US. Data collection occurred between May and November 2017. This study was granted IRB exemption from the University of Wisconsin-Madison Institutional Review Board.

2.2 Data Collection Methods

Teams of two human factors researchers conducted semi-structured interviews. Interviews were audio recorded and transcribed by a professional transcription service. The interview guide contained questions about pediatric trauma care transitions, for example, a description of the ED to OR and ED to PICU transitions, roles involved and examples of when the transition went well and went poorly. The average interview duration was 50 min (range: 29–66 min), for a total of 9 h and 6 min. The full version of the interview guide can be found at: http://cqpi.wisc.edu/wp-uploads/2016/08/Pediatric-Trauma-Transitions-Interview-Guide.pdf.

2.3 Data Analysis Methods

The 11 transcripts were uploaded to Dedoose©, a qualitative data analysis software. Two researchers coded the data for work system barriers and facilitators and two other researchers reviewed the coding through a collaborative, consensus-based process. Researchers performed a thematic analysis and identified challenges related to ED disposition decision making [9]. We presented preliminary findings to subject matter experts: attending physicians from pediatric EM, surgery, anesthesia, critical care and hospitalist services, a nursing scholar for the children's hospital and the hospital's pediatric trauma nursing program manager. This is a form of member checking to ensure rigor in qualitative data analysis [10].

3 Results

The thematic analysis produced two themes related to ED disposition decision making: (1) timing of the decision and (2) leadership and team organization (see Table 1).

Table 1. Challenges in ED disposition decision making for pediatric trauma

Challenges	Description
Timing of decision	A quick, timely decision and agreement between physicians about disposition, i.e. what's next for the patient, is needed for care to be provided in a timely manner
Leadership and team organization	A clear, strong leader is needed to provide direction and facilitate agreement about disposition, which improves care and family experience

3.1 Timing of Decision

Eight healthcare professionals mention that, when the ED team makes a quick decision and agrees about next steps for the pediatric trauma patient, timely care is more likely and team members can proceed with next steps of the care plan.

Sometimes a quick decision is needed for sicker patients. One ED nurse mentioned that quick agreement between physicians was especially important for a level 1 patient,

> "*...he was a very critical case anyway...then the decision was made quickly. Sometimes that can be a sticking point, is if the doctors are going, well, are we going to the OR, or are we going to the PICU, or are we going to go somewhere else?*"

Timing of the decision is important for notifying PICU staff, so that they can prepare a bed and the patient can be transitioned quickly, as mentioned by a PICU nurse,

> *"And with most Level 1 s we usually want to get them up pretty quickly, if they don't have to go to the OR, go to a scan. So I'm heading back up to the unit and getting a room ready for them."*

The timing of the decision is influenced by the child's acuity and injury, which can either rush a decision or allow for more time to prepare. A PICU fellow mentioned the difficulty of judging whether the child will get better or worse,

> *"So it's nice when a Level 2 is you have time...when people are rushed...either they over or underestimate their trajectory, that gets a little dangerous. And again, that's hard to judge... because the patient could be changing in time."*

There is more time for level 2 patients because there are fewer decisions about what is immediate to address, as mentioned by an OR nurse,

> *"...there's fewer things you have to deal with...we can fix one and then maybe come back the next day."*

More time for level 2 patients can be helpful to understand the child's condition and prepare for procedures, as mentioned by a surgery resident,

> *"A lot of times we do have the privilege of time...That helps us in a lot of ways because we can prepare for certain procedures."*

The timing of the decision can be affected by the physical presence of physicians in the ED. For level 1 patients, the PICU fellow mentioned that it is helpful to be in the ED to see how sick the child is,

> *"Just an inaccurate assessment...saying that a kid looks okay when they're actually very sick... it's hard to, you know, especially if it's a truly sick trauma patient who has had a lot going on, it's hard to convey all the details. So I think being there is really critical for us."*

For level 2 patients, the attending surgeon is not required to go to the ED. Therefore, the EM resident must communicate with the attending surgeon over the phone. The EM resident described the challenges of decision making when the surgeon has not seen the child,

> *"Just physically seeing the patient can sometimes change surgeons' mind about what they would do...I think there's a little bit more room for communication errors when you can't see what the kid looks like in front of you and when you can't talk to the person face to face or when you can't just go ask a question...and be like I didn't understand that."*

Communication can be less clear and more prone to information loss when physicians are not present in the ED to communicate for level 2 patients.

3.2 Leadership and Team Organization

Ten healthcare professionals mention that the ED disposition decision making process is facilitated by the presence of a clear, strong leader that provides direction and facilitates agreement among the various disciplines involved in patient care.

Strong leadership is needed so that patient care is delivered in a timely manner. A surgery resident mentioned that too many people trying to "call the shots" can lead to uncertainty about who is in charge,

"The things that don't go well are the noise, the chaos, having too many voices involved, and too many people calling the shots, and all that delays patient care...no one knows who is running the code at the point."

An OR nurse mentioned that disagreement among physicians about the ED disposition decision can cause uncertainty about who to listen to,

"And the other half is going, this isn't, you cannot sustain this injury and survive...you quickly get the feeling of that in the room...and who am I supposed to be listening to?...because you don't really know what you need to prepare for...I need a captain here."

Team members may experience uncertainty when the ED disposition decision is not communicated to everyone. An EM resident mentioned the consequences of unclear communication between physicians and nurses,

"...what happened was the lack of very clear communication with the nursing staff in that they weren't clear on exactly what needed to happen before they went up to the OR and things didn't get done that need to get done and it didn't result in any compromise of patient care, but it could have."

Care may be delayed because of poor communication between team members about the ED disposition. A PICU fellow mentioned that the absence of a clear leader to make decisions may result in the child going to the PICU instead of the OR,

"And if it's unclear who is responsible for that patient at the time...and it's unclear whether or not...they go to CT, and they come straight up here, or they should go to CT and back to the patient room... to make sure there's not something that they, instead needs to go to the OR for before coming here. So that has happened a couple of times."

Communication between the trauma surgeon and the rest of the team is important for sharing information so that everyone knows what to do next. For the ED coordinator, knowing the decision and which OR the patient is going to allows her to subsequently admit the patient electronically,

"...communication, making sure that we know which one they're going to...it's the trauma surgeon's job to make sure everyone's on the same page."

When team members experience uncertainty, they are unsure about what they need to do or can tell the family. An ED nurse mentioned how the absence of a clear leader can be distressing for a family,

"It just seems like a lot of the primary nurse's focus gets lost into the flow of the documentation or calling to give reports to people...not having a clear leader sometimes with these peds... Social Work is supposed to be in charge of families, but sometimes I feel like they would like to hear more from the nurse or the doctor."

4 Discussion

When interviewed about the disposition decision making in the ED for pediatric trauma patients, twelve healthcare professionals described challenges related to the timing of the decision as well as leadership and team organization (see Table 1).

The ED disposition decision can affect patient safety and must be made accurately in a timely manner [6]. When the decision is made quickly, there is more time for nurses and ancillary staff to prepare, so that care is not compromised. The child's acuity and injury may impact the timing of the decision as more sick patients require more physicians to be physically present in the ED. When physicians are not together in the ED for level 2 patients, information may be lost in phone communication. Effective and timely decision about ED disposition can facilitate team SA not only among team members involved in the ED, but also team members in the OR and the PICU.

Better leadership and team organization facilitate team SA so that there is communication between not only physicians, but also nurses and ancillary staff about the ED disposition decision. When team members communicate the ED disposition decision, there is less uncertainty about how to prepare the patient and coordinate with the receiving inpatient unit.

Analyzing work system barriers and facilitators to ED disposition decision making can help to identify sociotechnical system solutions for enhancing team SA. Sociotechnical system solutions that enhance team SA in the pediatric trauma care process may also reduce missed injuries and improve timely care, patient family/caregiver satisfaction and clinician satisfaction.

5 Conclusion

Effective, timely and organized decision making about ED disposition can facilitate team SA not only among team members involved in the ED, but also team members in other hospital units, such as the OR and the pediatric ICU. Further research will expand the identification of work system barriers and facilitators in pediatric trauma care transitions, which can help to identify sociotechnical system solutions for improving the care transition process.

Acknowledgements. Funding for this research was provided by the Agency for Healthcare Research and Quality (AHRQ) [Grant No. R01 HS023837]. The project described was also supported by the Clinical and Translational Science Award (CTEAM SA) program, through the National Institutes of Health (NIH) National Center for Advancing Translational Sciences (NCATS), [Grant UL1TR002373]. The content is solely the responsibility of the authors and does not necessarily represent the official views of the funding agencies. We thank the study participants, as our research would not be possible without them.

References

1. Borse NN et al (2009) Unintentional childhood injuries in the United States: key findings from the CDC childhood injury report. J Saf Res 40(1):71–74
2. Peery CL et al (1999) Missed injuries in pediatric trauma. Am Surg 65(11):1067–1069
3. Risser DT et al (1999) The potential for improved teamwork to reduce medical errors in the emergency department. Ann Emerg Med 34(3):373–383
4. Wooldridge AR et al (2018) Complexity of the pediatric trauma care process: implications for multi-level awareness. Cogn Technol Work (submitted)

5. Wooldridge AR et al (2017) Designing and managing healthcare transitions. In: 48th annual conference of the association of Canadian ergonomists, Banff, AB, Canada
6. Calder LA et al (2012) Mapping out the emergency department disposition decision for high-acuity patients. Ann Emerg Med 60(5):567–576
7. Department of Health Services, Division of Public Health, Office of Health Informatics (2018) Wisconsin State Trauma Registry Data Dictionary, version 5.1, p 178
8. Salas E, Prince E, Baker DP (1995) Situation awareness in team performance: implications for measurement and training. Hum Factors 37(1):123–136
9. Robson C, McCartan K (2016) Real world research. Wiley, London
10. Devers KJ (1999) How will we know "good" qualitative research when we see it? Beginning the dialogue in health services research. Health Serv Res 34(5 Pt 2):1153–1188

Estimation Accuracy of Average Walking Speed by Acceleration Signals: Comparison Among Three Different Sensor Locations

Yoshiyuki Kobayashi[1(✉)], Motoki Sudo[2], Hiroyasu Miwa[1], Hiroaki Hobara[1], Satoru Hashizume[1], Kanako Nakajima[1], Naoto Takayanagi[2], Tomoya Ueda[2], Yoshifumi Niki[2], and Masaaki Mochimaru[1]

[1] Digital Human Research Group, Human Informatics Research Institute, National Institute of Advanced Industrial Science and Technology, Waterfront 3F, 2-3-26, Aomi, Koto-ku, Tokyo, Japan
Kobayashi-yoshiyuki@aist.go.jp
[2] Tokyo Research Laboratories, Kao Corporation, 2-1-3 Bunka, Sumida-ku, Tokyo, Japan

Abstract. The ubiquity of wearable sensors now enables us to measure a user's walking speed outside of a laboratory or clinical setting, during activities of daily living. However, this technology is recent, and researchers have yet to determine the locations on the body that produce the most accurate data from these sensors. This study aims to compare the accuracy of average walking speed estimation measured using acceleration data from three body landmarks: wrist, pelvis, and ankle. Estimation models are derived from the gait data of 247 healthy adults using stepwise linear multiple regression analyses. The absolute value of the within-participant mean of errors between actual average walking speed and estimated average walking speed is computed and compered across landmarks. The ankle is the most accurate locations from which to estimate average walking speeds from acceleration signals, whereas the wrist was the least accurate locations. Walking speed is an important measure of health and function, especially in older people, and accurately estimating walking speeds in daily life may be helpful in predicting health outcomes in the elderly.

Keywords: Walking speed · Acceleration · Estimation · Accuracy Landmarks

1 Introduction

Average walking speed is one of the most fundamental gait parameters for predicting an individual's health and function, especially for the elderly [1–3]. Researchers have reported that average walking speeds faster than 1.0 m/s suggest healthier aging, whereas average walking speeds slower than 0.6 m/s increase the likelihood of poor health and function [1, 2].

S. Bagnara et al. (Eds.): IEA 2018, AISC 818, pp. 346–351, 2019.
https://doi.org/10.1007/978-3-319-96098-2_45

Traditionally, average walking speed has been measured and assessed mainly in clinical or laboratory setting by using a stopwatch and a tape measure [4]. Now, however, average walking speeds can be obtained during normal activities of daily living using wearable sensors; a variety of studies have reported the methodology and accuracy of measuring average walking speed using wearable sensors attached various locations of the body [5–9]. However, to the best of our knowledge, these studies have investigated the accuracy of average walking speed estimation using acceleration signals obtained from wearables attached at limited locations on the body (in many cases, from only one location). Furthermore, the methodologies and algorithms used to create estimation models in these studies differ. Therefore, it is difficult to understand whether and how much sensor locations affect the accuracy of average walking speed estimation for a uniform model.

The study compares the accuracy of average waking speed estimation from various sensor locations; we employ acceleration signals obtained from the following three body landmarks [10]: styloid process of the radius (wrist), pelvis (anterior superior iliac spine), and most lateral point of lateral malleolus (ankle). These landmarks are chosen from wearable products currently on the market and from the literature on activity monitoring technology [5–9]. Average walking speed in this study is defined as the average speed of the body's center of mass (COM) during one gait cycle, as in previous studies [11]. We utilize stepwise linear multiple regression analysis to build our estimation models.

2 Methods

2.1 Participants

In this study, we analyzed gait data from 247 healthy adults (114 males and 133 females) aged 20 to 77 included in the National Institute of Advanced Industrial Science and Technology (AIST) gait database [12]. Participant demographics are presented in Table 1. All participants were able to walk independently without assistive devices (e.g., canes, crutches, or orthotic devices), had normal or corrected-to-normal vision, had no history of neuromuscular disease, and lived independently in the local community. Those who had trauma or any orthopedic diseases were excluded. The experimental protocol was approved by the local institutional review board, and all participants gave their written informed consent before participating.

Table 1. Mean (SD) demographics of the participants.

	AGE: yrs	Height: cm	Weight: kg	Walking speed: m/s
All	50.3 (18.8)	162.3 (8.5)	59.2 (10.4)	1.34 (0.16)
Females	49.7 (18.3)	157.3 (6.1)	53.7 (7.8)	1.35 (0.16)
Males	50.9 (19.4)	168.1 (7.0)	65.6 (9.3)	1.32 (0.15)

2.2 Data Collection

Measurements were performed in a room with a straight 10-m path on which the participants could walk. Three-dimensional (3D) positional data were obtained using reflective markers and a 3D motion capture system (VICON MX, VICON, Oxford, UK) using a sampling frequency of 200 Hz. A total of 57 infrared reflective markers were attached by one of three experts (each of them with more than 10 years of experience) in accordance with the guidelines of the Visual 3D software (C-Motion Inc., Rockville, MD, US). We also recorded ground reaction forces (GRFs) using six force plates (BP400600-2000 x 4, and BP400600-1000 x 2, AMTI) sampled at 1 kHz intervals.

The participants were asked to walk barefoot at a comfortable, self-selected speed. Before the walking trials, the positions of the markers were recorded while the participants stood stationary. The participants were then allowed sufficient practice walks to ensure a natural gait. After practicing, ten successful trials in which each participant properly stepped on the force plates were recorded and analyzed.

2.3 Data Analysis

The raw data were digitally filtered using a fourth-order Butterworth filter with zero lag and cut-off frequencies of 10 Hz for the positional data and 56 Hz for the GRFs. The acceleration signals from the eight landmarks, represented in a global coordinate system [6], were calculated using the second derivative method. Gravitational acceleration was added to the vertical component of the calculated acceleration. Acceleration signals were then time-normalized by the gait cycle duration determined from the force plate data and divided into 101 variables ranging from 0 to 100%. Thus, we obtained a dataset of 2424 variables from each trial (i.e., 101 time points, 8 landmarks, and 3 axes). Average walking speed during one gait cycle was defined as the average speed of the body's COM during one gait cycle. COM was computed using the default settings of Visual 3D. Low-pass filtering and variable calculations were also performed using Visual 3D.

2.4 Statistics

In this study, stepwise multiple linear regression analyses were used to build the models for estimating the average walking speed from the acceleration of each landmark. A total of 303 independent variables (101 time-normalized acceleration variables by 3 planes of each landmark) were used to build the models. The dependent variable was the average walking speed in each trial. The inclusion and exclusion criteria for stepwise analysis were set to $p < 0.05$ and $p > 0.10$, respectively. Further, leave-one-subject-out cross-validation was used to build the model and obtain the estimated average walking speed. Therefore, the estimated average walking speed of each participant was obtained from his/her own acceleration data and the model built from the dataset excluding his/her own data.

In the present study, the absolute value of the within-participant mean of errors (AME) between the actual and estimated average walking speeds was computed and used to assess each landmark's estimation accuracy. Further, a one-way analysis of variance (ANOVA) for repeated measures was conducted on the AME to assess the main effect of sensor location (landmark) on average walking speed estimation. Bonferroni's multiple comparison and the point-biserial correlation coefficient r were used for post-hoc analysis if a significant main effect was observed. Because of the large number of participants (n = 247), the differences in the means were considered statistically significant if the $p < 0.05$, $\eta^2 > 0.01$, and $r > 0.10$, indicating a small effect size [13]. All statistical analyses were executed using the SPSS statistical software package (IBM SPSS Statistics Version 24, SPSS Inc., Chicago, IL, USA).

3 Results

Between-participant means (standard deviations) of AME_participant on each landmark were as follows: AME_participant wrist 0.080 (0.061) m/s, pelvis 0.051 (0.040) m/s, and ankle 0.042 (0.035) m/s. Statistical analyses revealed significant differences between wrist and ankle, and wrist and pelvis for AME_participant.

Table 2. Between-participant mean (SD) of parameters among the landmarks

Landmarks	AME_participant: m/s	Adjusted coefficient of determinations	Number of independent variables selected in to the models: mean (SD)
Wrist	0.080 (0.061)	0.534 (0.003)	21.50 (1.48)
Pelvis	0.051 (0.040)	0.795 (0.002)	26.04 (2.89)
Ankle	0.042 (0.035)	0.853 (0.001)	28.04 (1.23)

4 Discussion

This study compared the accuracy of average walking speed estimation from different body landmarks (i.e., sensor locations), with the hypothesis that acceleration signals obtained from the pelvis would provide the most accurate estimates, because of their proximity to the body's COM. Contrary to our initial hypothesis, ankle was the most accurate location from which to estimate average walking speed from the acceleration signals, whereas the wrist was the least accurate locations.

The ankle was the only the landmark for which the mean adjusted coefficients of determination of the estimation models were greater than 0.800, and were the most accurate sensor location for estimating average walking speeds from acceleration signals. Further, the ankle exhibited smallest standard deviations of the adjusted coefficients of determinations of the three estimation models (see Table 2; 0.001 for the ankle). In this study, we used leave one-subject-out cross-validation to build our models. We can interpret these results to mean that we can build models with consistent accuracy regardless of the input dataset if we use acceleration signals from this landmark.

In this study, we used a stepwise technique to select the independent variables for building our walking speed estimation models. On average, only 21 independent variables were selected for the wrist models, whereas more than 28 independent variables were selected for ankle model (Table 2). Generally, the number of independent variables selected for a model is related to the value of its adjusted coefficient of determination. Therefore, these differences may be one reason why the estimation accuracy of the wrist model was relatively low. Zihajezadeh et al. [9], who reported a model to estimate walking speed from wrist acceleration, used principal component analysis to find the direction of the greatest acceleration variation in the horizontal plane to improve estimation accuracy. Applying such an analysis may prove effective to our model as well.

The participants of the present study were 247 healthy adults living within the community and ranging in age from 20 to 77 years old. Their mean (and SD) walking speed was 1.34 (0.16) m/s, as shown in Table 1. The cut-off average walking speed for determining whether an elderly person is fit has been reported as 1.3 m/s [2]; to classify people based on this threshold, models should be as accurate as possible. Therefore, designers, manufacturers, and researchers who utilize sensor systems that can estimate average walking speeds using wearable accelerometers should be aware of the effect of sensor location.

Readers should be aware of several limitations of this study when considering its results. First, of the various methodologies available, this study only used stepwise linear regression analyses to build models. Further research may be needed to understand how different methodologies affect average walking speed estimation using wearables. Second, many studies have demonstrated that 1.0 m/s is the threshold below which the risk of various adverse outcomes is increased [1, 2, 14]. However, because the participants in this study were healthy, community-dwelling adults, their walking speeds were much faster than 1.0 m/s. Thus, further research may be needed to ensure the accuracy of average walking speed measurements for slower subjects.

5 Conclusions

This study compared the accuracy of average waking speed estimated using data from different body landmarks. We found that the ankle were the most accurate sensor locations from which to estimate average walking speeds from acceleration signals, whereas the wrist were the least accurate locations.

Acknowledgements. The authors would like to thank all participants as well as Ms. Rika Ichimura, Ms. Yuko Kawai and Ms. Miho Ono for their support of data acquisition and analyses.

Conflict of Interest. None of the authors have any conflicts of interest associated with this study.

References

1. Cesari M, Kritchevsky SB, Penninx BW et al (2005) Prognostic value of usual gait speed in well-functioning older people—results from the Health, Aging and Body Composition Study. J Am Geriatr Soc 53:1675–1680
2. Studenski S, Perera S, Wallace D et al (2003) Physical performance measures in the clinical setting. J Am Geriatr Soc 51:314–322
3. Fritz S, Lusardi M (2009) Walking speed: the sixth vital sign. J Geriatr Phys Ther 32(2): 46–49
4. Youdas JW, Hollman JH, Albers MJ et al (2006) Agreement between the GAITRite walkway system and a stopwatch-footfall count method for measurement of temporal and spatial gait parameters. Arch Phys Med Rehabil 87(12):1648–1652
5. Yang S, Li Q (2012) Inertial sensor-based methods in walking speed estimation: a systematic review. Sensors 12(5):6102–6116
6. Herren R, Sparti A, Aminian K et al (1999) The prediction of speed and incline in outdoor running in humans using accelerometry. Med Sci Sports Exerc 31(7):1053–1059
7. Schimpl M, Lederer C, Daumer M (2011) Development and validation of a new method to measure walking speed in free-living environments using the Actibelt® platform. PLoS ONE 6(8):e23080
8. Zihajehzadeh S, Park EJ (2017) A Gaussian process regression model for walking speed estimation using a head-worn IMU. In: Conference proceedings of the IEEE engineering in medicine and biology society, pp 2345–2348
9. Zihajehzadeh S, Park EJ (2016) Regression model-based walking speed estimation using wrist-worn inertial sensor. PLoS ONE 11(10):e0165211
10. ISO 7250-1:2017 (2017) Basic human body measurements for technological design—Part 1: Body measurement definitions and landmarks
11. Kobayashi Y, Hobara H, Matsushita S et al (2014) Key joint kinematic characteristics of the gait of fallers identified by principal component analysis. J Biomech 47(10):2424–2429
12. Kobayashi Y, Hobara H, Heldoorn TA et al (2016) Age-independent and age-dependent sex differences in gait pattern determined by principal component analysis. Gait Posture 46:11–17
13. Cohen J (1988) Statistical power analysis for the behavioral sciences, 2nd edn. Lawrence Erlbaum, Hillsdale
14. Abellan van Kan G, Rolland Y, Andrieu S et al (2009) Gait speed at usual pace as a predictor of adverse outcomes in community-dwelling older people an International Academy on Nutrition and Aging (IANA) Task Force. J Nutr Health Aging 13(10):881–889

An Ergonomic Grip Design Process for Vaginal Ultra Sound Probe Based on Analyses of Benchmarking, Hand Data, and Grip Posture

Hayoung Jung[1], Nahyun Lee[1], Soojin Moon[1], Xiaopeng Yang[1], Seungju Lee[2], Junpil Moon[2], Kilsu Ha[2], Jinho Lim[3], and Heecheon You[1](✉)

[1] Department of Industrial and Management Engineering, Pohang University of Science and Technology, Pohang, South Korea
hcyou@postech.ac.kr
[2] Design Group, Samsung Medison Co., Seoul, South Korea
[3] Design Group, Health & Medical Equipment Business Division, Samsung Electronics Co., Seoul, South Korea

Abstract. The present study presents a systematic design process for the ergonomic design of a vaginal probe based on benchmarking, hand data, and grip posture analyses. Five existing vaginal grip designs were compared with each other using subjective measures to identify preferred design features for a new probe grip design. An in-depth analysis of the relationships between grip design variables and hand dimensions was conducted along with the consideration of preferred grip postures of vaginal probe and hand measurements. Two novel vaginal probe grip designs were proposed based on the analysis results of benchmarking, hand data, and grip posture. A validation experiment showed a significant improvement of the hand-data based vaginal grip design compared with the existing designs in terms of subjective satisfaction and wrist flexion.

Keywords: Ergonomic grip design · Grip design process · Ultra sound probe

1 Introduction

The ergonomic design of an ultrasonic probe is needed to improve its usability and prevent work-related musculoskeletal disorders (WMSDs) among health care professionals engaged in sonography. Ultrasonic probe is a device that sends and receives ultrasound signals to the human body during ultrasound examination. Previous studies have reported that 65% to 91% of sonographers experience pains and WMSDs at the neck, shoulder, chest, lumbar, elbow, and wrist [4].

The risk factors of WMSDs experienced by ultrasound technicians include repetitive motion, inadequate posture, static muscle contraction, use of excessive force, prolonged work of ultrasound scanning, and improper design of ultrasound equipment and working environment [1, 3]. Ergonomically designed ultrasound probes can improve ease of use, ease of manipulation, use of proper force, and satisfaction, which can contribute to the prevention of musculoskeletal disorders among sonographers [2, 6].

S. Bagnara et al. (Eds.): IEA 2018, AISC 818, pp. 352–363, 2019.
https://doi.org/10.1007/978-3-319-96098-2_46

An ergonomic evaluation study is conducted to evaluate the physical workload of a sonographer quantitatively during ultrasonic operations and assess the risk of WMSDs. Subjective satisfaction questionnaires, posture and motion analyses, and muscular load (EMG) measurements are collected to evaluate the motion and muscle load during ultrasonic scanning task. Body postures and motions at the shoulder, neck, elbow, and wrist are measured and analyzed by video analysis systems, electrogoniometers, and optoelectronic motion capture systems. The muscular load of an ultra sound probe use is analyzed by measuring EMGs of middle trapezius, supraspinatus, infraspinatus, flexor carpi ulnaris, left extensor digitorum, and left deltoideus anterior muscle [3, 12–14].

Although studies have been conducted to evaluate the workload of ultrasound probe scanning, more research on the preferred design characteristics of the probe handle and the optimal shape based on human body dimensions and grip postures is needed. Burnett and Campbell-Kyureghyn [3] and Village and Trask [14] measured motion, EMG, grip and push forces during ultrasonic operations and analyzed the differences between tasks in physical workload, but the preferred designs of probe were not identified in terms of usability. In contrast, Mazzola et al. [11] and Paschoarelli et al. [12] proposed an ergonomic probe shape and compared the proposed design with those of the existing probes; however, they did not present the design rationale of the proposed design and the corresponding detailed design features in probe size, shape, and weight. The application of anthropometric data of a target population under consideration in the product design process enables to develop a design with better usability [7, 9, 10].

The present study developed a systematic handle grip design process based on benchmarking, hand data analysis and grip posture analysis to improve the usability of a vaginal probe. An improved grip design in terms of usability were proposed and then verified by an ergonomic evaluation in the study. The characteristics of the probe design were analyzed by the benchmarking of existing vaginal probe designs and the preferred design features of the existing designs were analyzed by satisfaction evaluation and an optimal handle grip circumference was derived by applying the hand data of Koreans. The vaginal probe design developed by the ergonomic grip design process in the present study was evaluated in terms of motion, muscular load, grip posture, and subjective satisfaction by ultrasonic probe users in a laboratory environment.

2 Development of Grip Design Process

2.1 Design Analysis of Vaginal Probe

To determine the effect of vaginal probe design on usability, probe designs in different sizes and shapes were analyzed as illustrated in Table 1. Probe designs were scanned using the Artec Eva 3D scanner (Artec Group Inc., Luxembourg) and six probe design dimensions of length, circumference, and angle were measured using the CAD software Alias Automotive 2012 (Autodesk, Inc., USA). The length of vaginal probe and the thickness of the middle grip largely varied with ranges of 307.6–350.8 mm and 32.1–42.6 mm, respectively. The shape of vaginal probe was largely classified into two types, straight (180°) and bent (163.7°–164.0°), by the angle between the head and the grip angle in the side view.

Table 1. The design dimension analysis of vaginal probe (illustrated).

Image			Vaginal probe designs					
Design dimension		Unit	Range	A type	B type	C type	D type	E type
Overall length		mm	307.6–350.8	Small	Large	Medium	Medium	Large
Head-grip angle		° (deg.)	163.7–180.0	Straight	Curved	Straight	Curved	Straight
Grip length		mm	127.3–142.8	Small	Large	Medium	Medium	Large
Thickness of frontal grip		mm	30.3–38.1	Medium	Large	Small	Small	Small
Thickness of middle grip		mm	32.1–42.6	Medium	Large	Medium	Medium	Small
Circumference of middle grip		mm	103.7–118.2	Medium	Large	Medium	Medium	Small

2.2 Design Evaluation for Vaginal Probe

The subjective satisfaction evaluation of five vaginal probe designs was performed to identify the preferred design characteristics of vaginal probe. Seven obstetrician-gynecologists (age = 48.0 ± 11.8 years; career = 19.3 ± 11.3 years; stature = 163.0 ± 5.3 cm) with no history of musculoskeletal disorders and having experience of ultrasonographic examination with vaginal probe for more than 5 years were recruited. The hand length (167.3 ± 6.5 mm) of the participants was statistically similar with the hand length (170.4 ± 7.3 mm) of 20–40 years old female group of Koreans (n = 2,067) from the 6th anthropometric research report (Size Korea, 2010) in terms of mean and variance (t[2, 072] = 1.12, p = 0.26 for mean difference; F[6, 2066] = 1.26, p = 0.58 for variance difference). The vaginal probe designs were evaluated according to length and angle appropriateness, form suitability, pressure dispersion appropriateness, grip comfort, and overall satisfaction using a 7-point scale (1: very unsatisfied, 4: neutral, 7: very satisfied).

Based on the results of the subjective evaluation of the five vaginal probes, the preferred design characteristics of vaginal probe grip were identified. A probe design with the overall length for medium size and the grip length for medium or a large size design was preferred. The straight type of vaginal probe was preferred to the bent type. A probe design with a small size of the frontal grip and a small or medium size of the middle grip was preferred.

2.3 Grip Design Based on Grip Posture and Hand Dimensions

The natural use posture with a vaginal probe grip was analyzed and the human body size data of the fingers were selected to design a preferred handle grip size. Vaginal probe grip postures were identified as two types: normal grip, where the finger and handle are orthogonal based on the power grip; the twisted grip for the inclined finger grip direction. D3 link length and palm-touch link length were selected for the hand dimensions as those related to the circumference of the middle grip.

The preferred grip circumference can be determined as shown in Fig. 1 by the sum of the deformed D3 link, palm-touch link, and clearance and the length deformation due to the twisted angle when the grip posture is changed. In the case of the vaginal probe grip, the D3 link length decreased to 60% of the straight posture as shown in Table 2, and palm-touch link, and clearance were 25% and 10% of the D3 link length, respectively. The preferred grip circumference is determined by the sum of the deformed-D3 length, the palm-touch link length, and the clearance for the normal grip is determined by summing the twisted length with the normal grip circumference at the time of the twisted grip.

The circumferences of vaginal probe grip at the contact points of fingers were derived by applying the finger length ratio and the finger length difference based on the circumference of the middle finger contact point from the index finger to the little finger. US anthropometric measurements data [5] were used for finger length ratio analysis, and the D2, D4, and D5 link length ratios were 99%, 97%, and 78% of the D3 link length, respectively. The optimal grip size for each finger was derived by applying the D2–5 link length ratio to the D3 link length based on the preferred circumference of the vaginal probe grip at the middle finger contact point (103.7 mm).

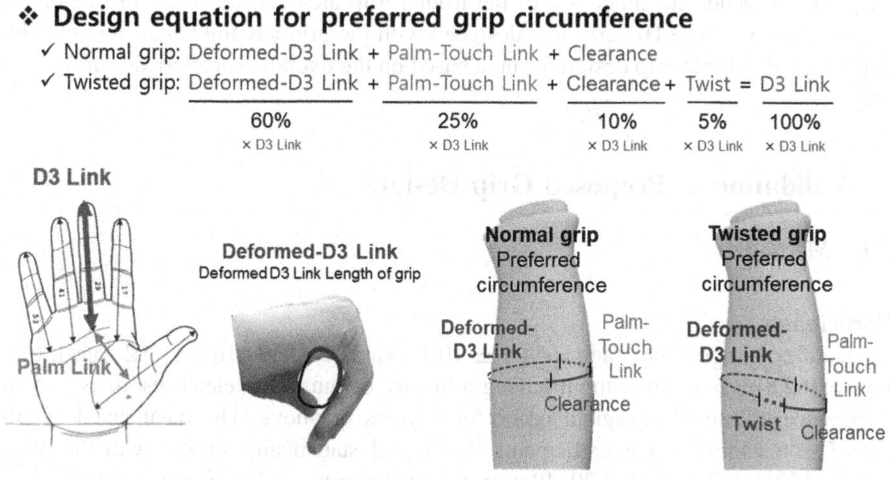

Fig. 1. Design equation of preferred grip circumference.

Table 2. Hand link length deformation analysis.

Participants	S1		S2		S3	
Item	Measurement (mm)	Ratio (%) vs. D3 link	Measurement (mm)	Ratio (%) vs. D3 link	Measurement (mm)	Ratio (%) vs. D3 link
D3 link	98.0	–	96.0	–	106.0	–
Deformed-D3 link	60.0	61.2	58.0	60.4	65.0	61.3
Palm-touch link	24.0	24.5	24.0	25	27.0	25.5
Clearance	9.0	9.2	8.0	8.3	11.0	10.4
Twist	4.0	4.1	4.0	4.2	6.0	5.7
Total	97.0	99.0	94.0	97.9	109.0	102.9
Error rate	1.0%		2.1%		2.9%	

2.4 Design Development for Vaginal Probe

The improved design of the vaginal probe grip was proposed with a benchmarking-based improvement (BM) grip and a hand data-based improvement (HD) grip as shown in Fig. 2. Based on the results of the subjective satisfaction assessment, the BM grip was designed by applying the medium size of overall length and the grip length, the shape of the bend, the small size of the frontal grip area and the medium size of the middle grip area. The HD grip was designed with the grip circumference corrected by applying the optimal grip design formula based on the overall characteristics of the BM grip.

3 Validation of Proposed Grip Design

3.1 Method

Participants

13 female sonographers (age = 44.6 ± 10.1 years; career = 16.8 ± 9.2 years; stature = 162.8 ± 5.0 cm) without having a history of musculoskeletal disorders and an experience of use of a vaginal probe for 5 years or above. The mean hand length (168.5 ± 6.8 mm) of the participants was found statistically similar with the hand length (170.4 ± 7.3 mm) of 20–40 year-old female group of Koreans (n = 2,067) from the 6th anthropometric research report (Size Korea, 2010) in terms mean and variance ($t[2078] = 0.94$, $p = 0.35$ for mean difference; $F[12, 2066] = 1.15$, $p = 0.58$ for variance difference).

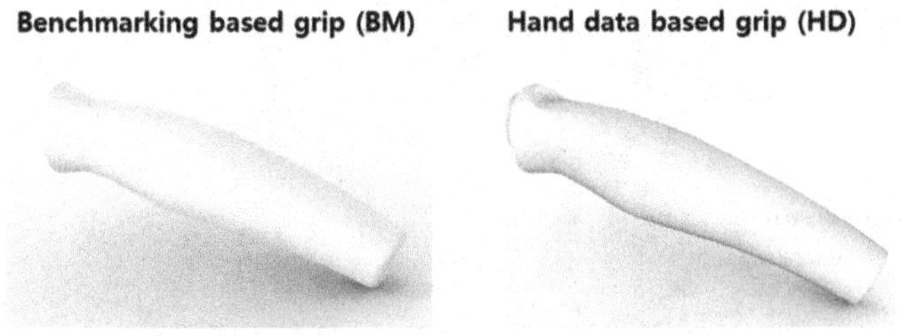

Fig. 2. Proposed vaginal probe grip designs.

Apparatus

The validation experiment of vaginal probe was performed with a medical phantom using two newly proposed probe grips (BM and HD grips) and an existing product. The most preferred probe grip design in terms of subjective satisfaction was selected as superior design product (SP grip, superior probe grip) and the new probe grip designs based on benchmarking and hand data analysis were evaluated along with the SP grip design. For the evaluation of vaginal probe, a phantom for evaluation, which was inspected by the medical staff, was used to implement an environment similar to the actual inspection task. The evaluation phantom was designed to induce a similar vaginal probe task by application of the depth (26 mm) and position guides (engraving depth: 2.0–6.5 mm).

An optical motion analysis system, an EMG measurement device, and a subjective evaluation questionnaire were used to measure changes in wrist motion, muscle activities of the arm and wrist, and subjective satisfaction during the vaginal probe test. The motion analysis system Osprey (Motion Analysis Corp., Santa Rosa: CA, USA; frame rates: 50 Hz) with 10 infrared cameras was installed at various heights around the participants. For the motion analysis 14 reflective markers (front head, rear head, right acromion, left acromion, offset, lateral epicondyle, medial epicondyle, right anterior superior iliac spine (ASIS), left ASIS, sacrum, radius, ulna, 5th metacarpal, and 2nd metacarpal) were attached to the body as shown in Fig. 3. The local coordinate system was defined at the neck, shoulder, waist, elbow, and wrist centers, and the angles of the each body parts (shoulder: flexion/extension, abduction/adduction, internal/external rotation angle; neck: flexion/extension, right/left bending, right/left rotation angle; back: flexion/extension, right/left bending, right/left rotation angle, elbow: flexion/extension angle; wrist: flexion/extension, ulnar/radial deviation, internal/external rotation angle) were measured while using a vaginal probe. The trajectory of the measured reflective marker was filtered using a fourth-order Butterworth filter (cut-off frequency = 5 Hz) to remove noise effects. EMG measurements were performed using the wireless EMG system Telemyo DTS (Noraxon, USA), and nine muscle loads (trapezius, deltoid, extensor digitorum, pronator teres, flexor digitorum, biceps, supinator, FCU, FCR) were measured when using the vaginal probe. Noise of EMG signal was removed using a bandpass filter (lower cut-off frequency: 10 Hz; upper cut-off frequency: 400 Hz) and after rectification, a signal smoothing was performed using root mean square (time window: 400 ms) for amplitude analysis (Fig. 4).

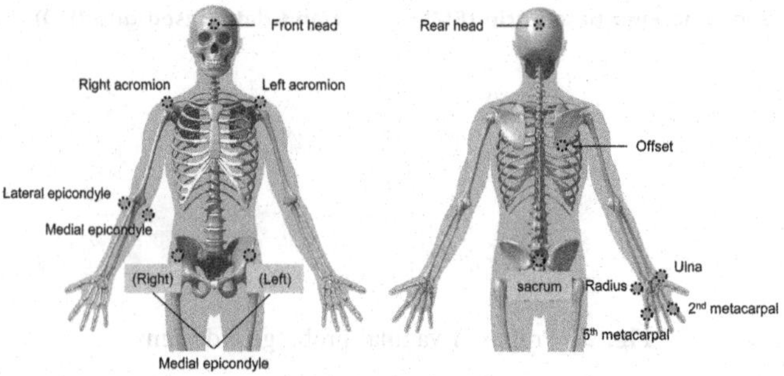

Fig. 3. Reflective markers attached to the whole body.

Subjective satisfaction of each proposed grip was assessed compared to the reference probe grip using a 7-point scale to measure the size (length, thickness, width), angle, and curvature adequacy, form fitness, posture and strength adequacy, grip comfort, and overall satisfaction of vaginal probe (−3: very dissatisfied, 0: neutral, 3: very satisfied).

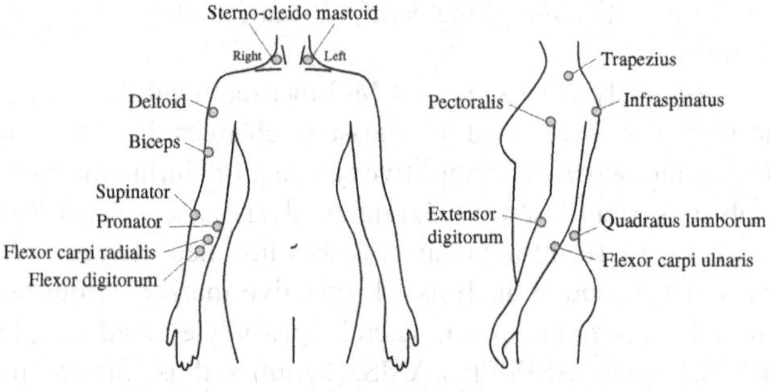

Fig. 4. Attachment of EMG electrodes.

Experiment Procedure

The verification experiment of the effectiveness of vaginal probe design was performed by four steps: (1) preparation of test and body size measurement, (2) measurement of EMG and body motion, (3) evaluation of subjective satisfaction, and (4) post-survey. In the experiment preparation and the human body measurement step, the experimental purpose and method were explained to the participants and informed consent was obtained. The height, weight, hand length, and hand width of the participant were measured after the participants changed with experimental clothing provided, reflective markers, EMG electrodes, and a radio signal transmitter module were attached to the body of the participant. The MVC (maximum voluntary contract) of the participant was measured prior to the experiment. In the main experiment, during the repeated three times of hold, rotate, vertical tilt, and lateral tilt movements, the participant's EMG and body movements were measured simultaneously while the participant was sitting.

A hold motion was performed by holding the probe for 15 s after power grip and insertion into the phantom as shown in Fig. 5a, and rotation motion was performed 90° clockwise from the center of the probe insertion direction in the power grip posture as shown in Fig. 5b. Next, vertical tilt and lateral tilt motions were performed by tilting 30° up/down and left/right in power grip posture as shown in Fig. 5c. The duration of task was controlled from 10 to 15 s per motion and the evaluation and task among the probes were randomized to minimize learning and fatigue effects. In the subjective satisfaction evaluation stage, the satisfaction level of each of the five grip designs (existing three preferred design and the two new designs) was relatively evaluated to the reference probe grip using the evaluation questionnaire based on task performance experience. Finally, a post-survey was conducted on the preference design characteristics of the probe.

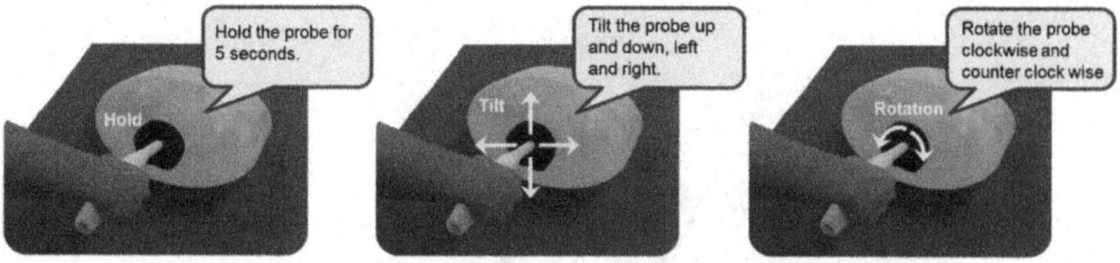

(a) Hold (b) Tilt (c) Rotate

Fig. 5. Use tasks of vaginal probe in ergonomic assessment.

Statistical Analysis

Statistical differences in EMG, body movement, and subjective satisfaction according to vaginal probe design characteristics were analyzed by paired t test at $\alpha = 0.05$. Statistical analysis was performed using MINITAB 14 (Minitab Inc., State College, PA, USA).

3.2 Results

Motion

The improved grip based on hand data (HD grip) was found to improve the wrist motion efficiency in terms of flexion/extension and radial/ulnar deviation compared with the existing superior probe (SP grip). As shown in Fig. 6, the ratio within the average comfortable range of motion (CROM) of the HD grip and the BM grip were 13.5% and 10.3% higher than that of the SP grip in the flexion/extension of the wrist motion, respectively. In the radial/ulnar deviation of the wrist motion, the ratio within the average CROM of the HD grip was 2.5% higher than the SP grip, but there was no statistically significant difference. The ratio within the average CROM of the BM grip was 5.1% lower than the SP grip.

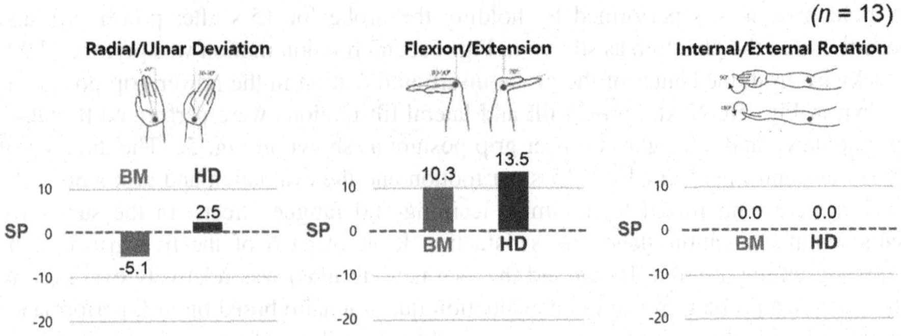

Fig. 6. Comparison of vaginal probe designs in ratio (%) of motion in CROM (comfortable range of motion).

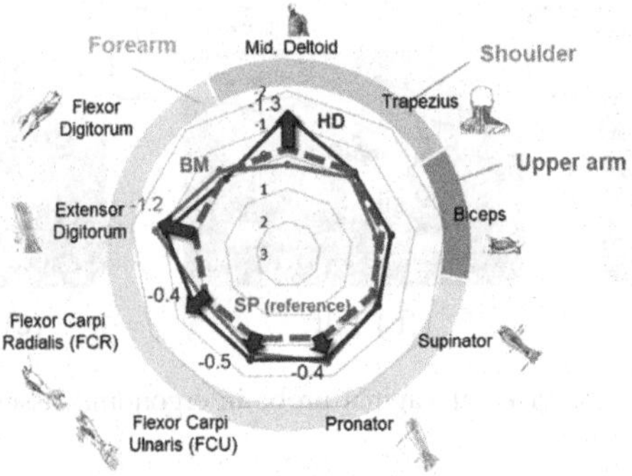

Fig. 7. Comparison of vaginal probe designs by percentage of MVC (%MVC).

Electromyography

The muscle loads of the HD grip and the BM grip were .5% and .2% lower than those of the SP grip, but no statistically significant difference were found. As shown in Fig. 7, % MVC of the HD grip was 1.3%, 0.9%, 0.4%, 0.5% and 0.4% lower in deltoid, extensor digitorum, pronator, FCR and FCU than the SP grip and decreased by 0.5% on average, respectively. When using the BM grip, the % MVC was 1.2% and 0.6% lower in extensor digitorum and pronator than the SP grip, respectively, and decreased by 0.2% on average.

Satisfaction

The HD grip was found to have improved subjective satisfaction with length, curvature and overall usability in comparison with the SP grip. The subjective satisfaction rate of the HD grip and the BM grip were 13.3% and 10.0% higher than that of the SP grip, respectively. As shown in Fig. 8, the subjective satisfaction of the HD grip was improved by 18.3% (Mean = 0.6 point) in length aspect, 15.4% (Mean = 0.5 point) in curvature aspect, and 9.1% (Mean = 0.3 point) in overall usability compared to the SP grip. The subjective satisfaction of the BM grip was increased by 16.5% (Mean = 0.5 point) in length aspect, 15.4% (Mean = 0.5 point) in curvature aspect, and 7.7% (Mean = 0.2 point) in shape aspect compared to the SP grip.

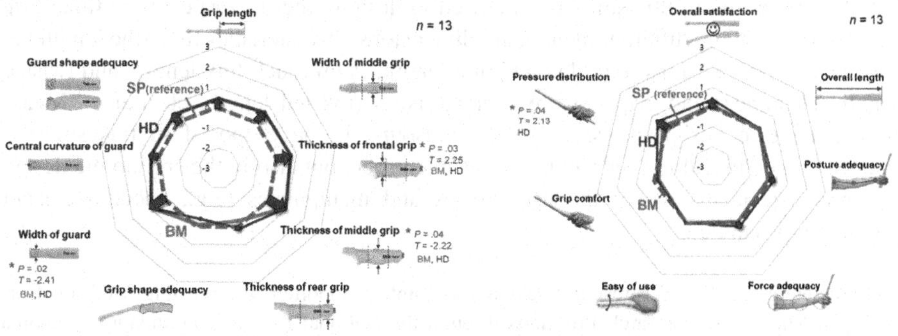

Fig. 8. Comparison of satisfaction evaluation for vaginal probe designs.

4 Discussion

The present study proposed an ergonomic grip design process which includes benchmarking, hand data, and grip posture analysis and developed an improved grip design. A preferred design of vaginal probe was derived using the evaluations of length, circumference, and angle. The proposed vaginal probe was favored in terms of subjective satisfaction with the medium size of the overall length, the shape of the bent, the small size of the frontal part of the grip, and the small or medium size of middle grip part. A design formula using the link length of the middle finger, the link length of palm-touch, clearance, and twisted length based on the preferred circumference of the middle finger contact point and design alternatives were developed based on the benchmarking results and hand data application.

The new design (HD grip) developed through the ergonomic grip design process was found better the most preferred existing design of vaginal probe (SP grip) in terms of motion efficiency and subjective satisfaction. The proposed probe design was evaluated by an ergonomic evaluation protocol, and the comparison of the existing and proposed designs were evaluated in terms of motion, muscular load, and subjective satisfaction for vaginal probe tasks. The HD grip increased 13.5%, 0.5%, and 13.3% in wrist flexion and extension, overall force, and subjective satisfaction, respectively, compared to the SP grip. The BM grip also showed increases in wrist flexion and extension efficiency, overall force, and subjective satisfaction by 10.5%, 0.2%, and 10.0%, respectively. These positive results of the proposed designs support the effectiveness of the proposed grip design process. Both the BM and HD grips were designed to be inversely tapered so that the movement of the wrist flexion required when holding and tilting the wrist could be less than that of the SP grip designed as a tapered shape and the ratios within the average CROM were relatively high.

When the hand data of a particular population under consideration are applied to the proposed design process, the optimal grip design which reflects human body dimension characteristics of the target population can be derived and the design process can be applied to the design of another handle grip. In this study, a vaginal grip design equation was developed by considering the preferential grip circumference of Koreans and hand dimensions characteristics of the Korean and American. The grip design equation proposed in this study can be used to design other race-specific optimal grip designs by applying different racial hand dimensions characteristics (e.g., the length per finger), hand dimensions (e.g., D2–D5 link length, Palm-touch link length, and twisted length, clearance) and grip posture characteristics (twisted length, clearance). Meanwhile, the grip design process of the ultrasound probe proposed in this study is expected to be useful not only in the ultrasonic probe, but also in the ergonomic design of various products which the user grasps and manipulates (e.g., vacuum cleaner handle, tool handle).

Acknowledgements. The present research was jointly supported by Samsung Medison Corporation, Mid-Career Research Programs through the National Research Foundation of Korea (NRF) funded by the Ministry of Education, Science and Technology (NRF-2018R1A2A2A05023299), and the Ministry of Trade, Industry, and Energy (No. 10063384; R0004840, 2017).

References

1. Bernard BP, Putz-Anderson V, Burt SE, Cole LL, Fairfield-Estill C, Fine LJ, Grant KA, Gjessing C, Jenkins L, Hurrell JJ, Nelson N, Pfirman D, Roberts R, Stetson D, Haring Sweeney M, Tanaka S (1997) Musculoskeletal disorders and workplace factor: a critical review of epidemiologic evidence for work-related musculoskeletal disorders of the neck, upper extremity, and low back. National Institute for Occupational Safety and Health (NIOSH), U.S. Department of Health and Human Service, Cincinnati
2. Bohlemann J, Kluth K, Kotzbauer K, Strasser H (1994) Ergonomic assessment of handle design by means of electromyography and subjective rating. Appl Ergon 25(6):346–354
3. Burnett DR, Campbell-Kyureghyan NH (2010) Quantification of scan-specific ergonomic risk-factors in medical sonography. Int J Ind Ergon 40:306–314
4. Evans K, Roll S, Baker J (2009) Work-related musculoskeletal disorders (WRMSD) among registered diagnostic medical sonographers and vascular technologists. J Diagn Med Sonogr 25(6):287–299
5. Gordon C, Churchill T, Clauser CE, Bradtmiller B, McConville JT, Tebetts I, Walker RA (1988) Anthropometric Survey of U.S. Army Personnel: method and summary statistics. U.S. Army NATICK Research, Development and Engineering Center, Natick
6. Harih G, Dolsak B (2014) Comparison of subjective comfort ratings between anatomically shaped and cylindrical handles. Appl Ergon 45:943–954
7. Jeon E, Lee B, Kim H, Park S, You H (2011) An ergonomic design of flight suit pattern according to wearing characteristics. In: Proceedings of the human factors and ergonomics society 55th annual meeting, Las Vegas, NV, USA
8. Korean Agency for Technology and Standard (KATS) (2004) Report on the 6th Size-Korea (Korean Body Measurement and Investigation). Ministry of Knowledge Economy, Republic of Korea
9. Lee W, Jung K, You H (2008) Development and application of a grip design method using hand anthropometric data. In: Proceedings of the 2008 spring joint conference of the Korean Institute of Industrial Engineers & the Korean Operations Research and Management Science Society, Gumi, South Korea

10. Lee W, Jeong J, Park J, Jeon E, Kim H, Jung D, Park S, You H (2015) Analysis of the facial measurements of Korean Air Force pilots for oxygen mask design. Ergonomics 56(9):1451–1464
11. Mazzola M, Forzoni L, D'Onofrio S, Andreoni G (2016) Use of digital human model for ultrasound system design: a case study to minimize the risk of musculoskeletal disorders. Int J Ind Ergon 60:35–46
12. Paschoarelli LC, Oliveira AB, Coury HJCG (2008) Assessment of the ergonomic design of diagnostic ultrasound transducers through wrist movements and subjective evaluation. Int J Ind Ergon 38:999–1006
13. Vannetti F, Atzori T, Matteoli S, Hartmann K, Altobelli G, Molino-Lova R, Forzoni L (2015) Usability characteristics assessment protocol applied to eTouch ultrasound user-defined workflow optimization tool. Proc Manuf 3:104–111
14. Village J, Trask C (2007) Ergonomic analysis of postural and muscular loads to diagnostic sonographers. Int J Ind Ergon 37:781–789

An Ergonomic Evaluation of Convex Probe Designs Using Objective and Subjective Measures

Soojin Moon[1], Hayoung Jung[1], Seunghoon Lee[1], Eunjin Jeon[1], Junpil Moon[2], Seungju Lee[2], Kilsu Ha[2], and Heecheon You[1](✉)

[1] Department of Industrial and Management Engineering, Pohang University of Science and Technology, Pohang, South Korea
hcyou@postech.ac.kr

[2] Design Group, Samsung Medison Co., Seoul, South Korea

Abstract. Use of a convex probe suitable to the hand and operating motion of the probe can contribute to prevention of sonographers from musculoskeletal disorders at work. The present study presents an ergonomic evaluation process customized to convex array ultrasound probe design. Various convex probe designs were evaluated by a mix of nine sonographers and medical doctors in terms of EMG activities of the upper extremity muscles, motion ranges of the upper extremity joints, and subjective satisfaction measures. A randomized controlled testing was administered for the probe designs in a simulation workstation at a designated speed of tilting, pushing, and rotating of convex probe. The subjective satisfaction results were found effective to identify preferred design features in detail, while the EMG and motion analysis results to identify a preferred probe design overall in terms of muscular load at the hand and postural comfort of the forearm.

Keywords: Convex probe design · Ergonomic evaluation
Preferred design feature

1 Introduction

A high prevalence of musculoskeletal symptoms among sonographers has been reported due to elevated exposure to adverse work conditions. Previous studies [2, 4–6] reported significant prevalence rates of musculoskeletal pains among sonographers at the neck (43%–86%), shoulder (29%–84%), upper back (15%–77%), lower back (33%–71%), elbow (5%–57%), and hand/wrist (33%–64%). Risk factors of the musculo-skeletal pains with sonographers include repetitive motion, awkward posture, static muscle contraction, excessive force exertion, prolonged duration of scanning, improper design of device and workstation, insufficient rest, and manual handling of patients [1–3, 9].

S. Bagnara et al. (Eds.): IEA 2018, AISC 818, pp. 364–371, 2019.
https://doi.org/10.1007/978-3-319-96098-2_47

An ergonomic evaluation is needed to identify the desirable design features of an ultrasound probe which can effectively reduce the physical workload of ultrasound task. An ultrasound probe produces sound waves that bounce off body tissues and makes echoes and also receives the echoes and transmits them to a computer for a sonogram. The most common types of ultrasound transducer include linear, convex, phased-array, and endocavity transducers. Few studies have been conducted for the ergonomic design and evaluation of an ultrasound probe to reduce the postural and muscular loads of ultrasound task. Paschoarelli et al. [8] ergonomically designed a linear array ultrasound probe by adding an ergonomic grip and a rotation mechanism at the base of the probe for adjusting their contact areas to the breast to reduce wrist motions during a breast ultrasound scanning task. The proposed probe was compared with two existing probes in terms of wrist motion and subjective satisfaction and concluded that the new probe resulted in less average movement of the wrist, more time within the safe motion range of the wrist, higher acceptance, and lower discomfort than the existing probes. However, detailed features of their proposed probe such as size, shape, and weight were not analyzed in their study.

The present study was intended to identify preferred design features of a convex array ultrasound probe, commonly used for examinations of the abdomen, OB-GYN, and peripheral vasculature, by an ergonomic evaluation. Designs of convex probe in different sizes and shapes were evaluated in a lab environment while simulating ultrasound tasks by health professionals in terms of muscular load, motion, and subjective satisfaction. Preferred design features were suggested for convex probe based on the ergonomic evaluation results.

2 Materials and Methods

Participants

Health professionals (age = 44.6 ± 10.1 years; work experience in sonography = 16.9 ± 9.2 years; hand length = 167.0 ± 6.8 cm) including sonographers and physicians having no history of musculoskeletal disorders participated in the convex probe evaluation. The mean and variance of the participants are not statistically different from those of the corresponding Korean population [7] at $\alpha = .05$. The participants provided informed consent and their participation was compensated.

Apparatus

An EMG measurement system, a motion analysis system, and a subjective satisfaction questionnaire were used in a simulated workstation of sonography (Fig. 1) for the ergonomic evaluation of convex probe design. The participant sitting on a stool (height adjustment range = 440–580 mm).

Simulated motions of pushing and rotation with a convex probe while applying a force of 20 ± 4 N on a silicon abdominal phantom (200 mm from the table) placed on a conventional examination table (length × width × height = 1800 × 730 × 650 mm). The motions of the upper body (wrist: flexion/extension, ulnar/radial deviation, and pronation/supination; elbow: flexion/extension; shoulder: flexion/extension, abduction/adduction, and internal/external rotation; neck: flexion/extension, right/left

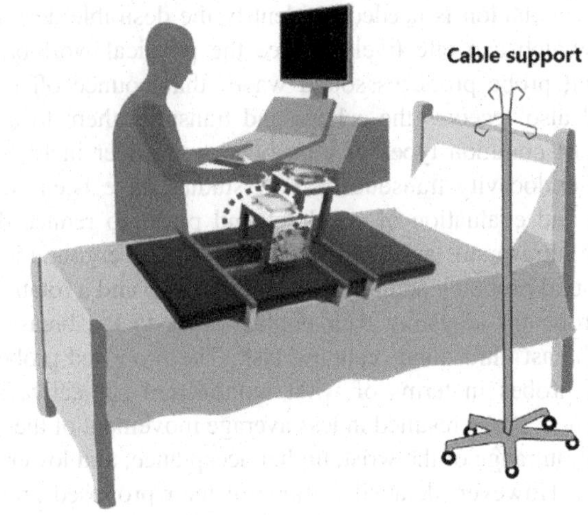

Fig. 1. Sonography simulation workstation

bending, and right/left rotation; back: flexion/extension, right/left bending, and right/left rotation) were measured by 10 infrared cameras of the Osprey motion analysis system (Motion Analysis Corp., Santa Rosa: CA, USA; frame rates: 50 Hz). The outliers of motion measurements collected with a repetition in a probe task cycle were filtered and synchronized, and then the proportion of motion measurements within a comfortable range of motion for each joint motion was calculated as shown in Fig. 2. Next, the muscle activities of the right upper limb and shoulder (thenar muscle, flexor carpi ulnaris, flexor carpi radialis, flexor digitorum, and extensor digitorum) were measured by the wireless EMG system Telemyo DTS (Noraxon, Scottsdale: AZ, USA; frame rates: 1,000 Hz). Noises of EMG were removed using a bandpass filter (10 Hz of lower cut-off frequency and 400 Hz of upper cut-off frequency) and then the filtered EMG signals were rectified, smoothed (time window = 400 ms), and normalized (%MVC) by EMG signals measured at his/her maximum force exertion. Lastly, a subjective questionnaire was used to assess the level of satisfaction with the size, curvature, and shape of the head and grip of a convex probe design in terms of fit to the hand, postural comfort, natural probe manipulation, effective force application, and even pressure distribution using a 7-point scale (−3: very dissatisfied, 0: neutral, 3: very satisfied) compared to a reference probe design designated.

Design of Experiment

Athree-way (probe design × task × grip) within-subject (nested within hand size) design was used for the convex probe design evaluation. Probe design (three probe designs), task (push and rotate), grip (narrow and wide grips), and hand size (small, medium, and large hand size groups) were fixed-effects factors and subject was a random-effects factor in the present study. The three convex probe designs (existing design: ED; benchmarking based design: BM; and hand-data based design: HD; Fig. 3) were prepared using a rapid prototyping machine. Note that the BM and HD designs were developed based on the result of a benchmarking on five convex probe designs and that of the relationship analysis of probe design variables and hand dimensions in a preferred grip as shown in Fig. 4, respectively. The order of probe design was randomized and balanced across the participants.

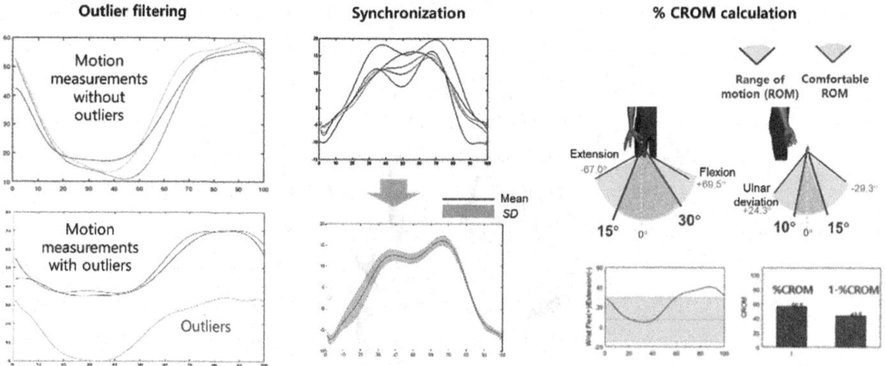

Fig. 2. Analysis process of %CROM (comfortable ROM)

Procedures

The convex probe design evaluation was conducted for two hours per participant in four phases: (1) preparation, (2) practice, (3) main experiment, and (4) debriefing. In the preparation phase, the purpose and procedure of the evaluation were explained, informed consent was obtained, clothing for experiment was worn, electrodes and reflective markers were attached to designated locations on the body, EMG signals of the upper limb and shoulder muscles at the maximum voluntary contraction were collected, and the height, weight, hand length, and hand width of the participant were measured. In the practice phase, the participant was familiarized with the evaluation procedure. In the main experiment phase, while pushing and rotating tasks with each of

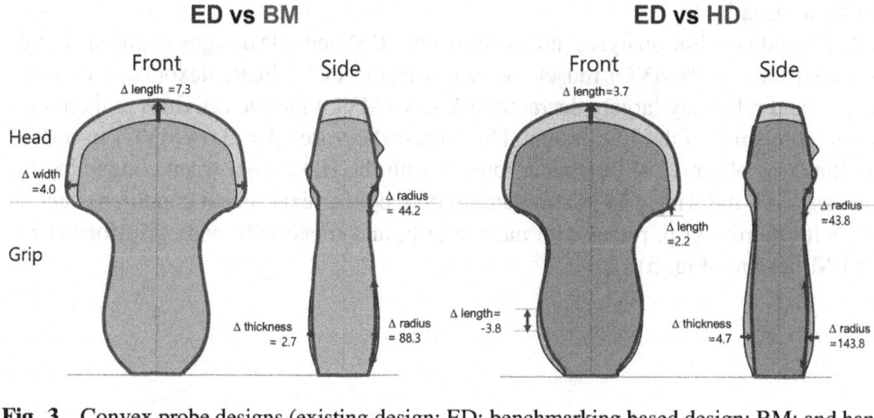

Fig. 3. Convex probe designs (existing design: ED; benchmarking based design: BM; and hand-data based design: HD)

the convex probe designs were simulated using narrow and wide grips at a designated speed by a metronome, the muscle activities and motions of the upper body were measured. After completing the convex probe simulation tasks, the BM and HD designs were compared with the ED design using the satisfaction questionnaire. Finally, in the debriefing stage, the preferred design characteristics of the probe designs were surveyed.

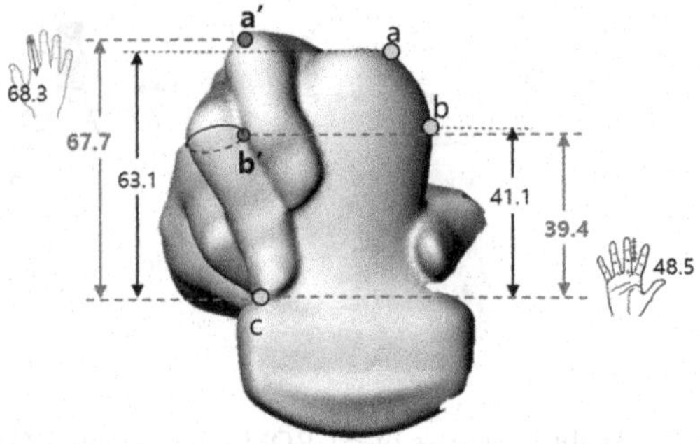

Fig. 4. The relationship analysis between probe design variables and hand dimensions in a preferred grip with a convex probe

Statistical Analysis

Significant factors on EMG, joint motion, and subjective satisfaction were analyzed by ANOVA and then post hoc pairwise comparisons were conducted for significant factors at $\alpha = 0.05$. Statistical analysis was performed using MINITAB 14 (Minitab Inc., State College, PA, USA).

3 Results

Muscular Load

ANOVA and post hoc analyses indicated that the BM and HD designs required slightly decreased ($\Delta < 1.4\%$MVC) muscle activities at the FCU, FCR, flexor, and extensor muscles and relatively largely decreased ($\Delta < 3\%$MVC) muscle activities at the thenar muscle compared to the ED design. The largest decrease ($\Delta = 3.6\%$MVC) in muscle activities was observed at the thenar muscle with the HD design when conducting the push task with the wide grip as compared with the other experiment conditions such as push with narrow grip, rotate with narrow grip, and rotate with wide grip for the ED and BM designs (Fig. 5).

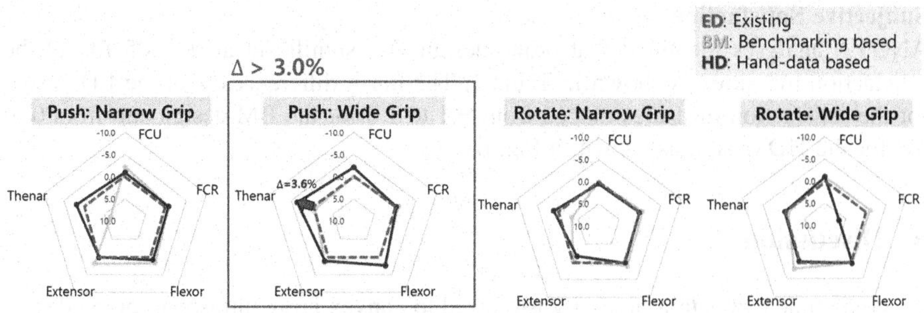

Fig. 5. Comparison of convex probe designs in terms of decrease in %MVC

Table 1. ANOVA on adequacy of grip shape

Sources	SS	df	MS	F_O	*P*-value
Probe design	58.475	3	19.4917	51.22	**<0.01***
HS	0.225	2	0.1125	0.16	0.857
Subject (HS)	5	7	0.7143		
Error	10.2575	27	0.3806		
Total	73.975	39			

Comfortable Motion

ANOVA analyses could not find any statistical differences in %CROM for the upper body by probe design. No post hoc analysis conducted due to the insignificance of probe design.

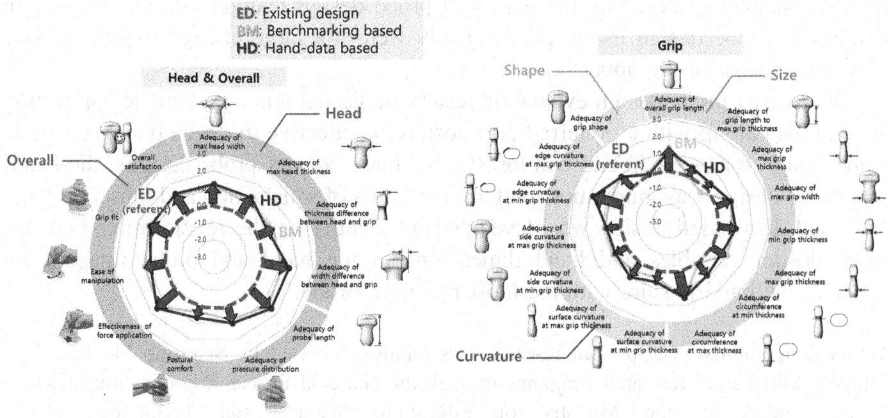

Fig. 6. Comparison of convex probe designs for the satisfaction measures

Subjective Satisfaction

ANOVA analyses identified that probe design was significant at $\alpha = .05$ for all the satisfaction measures as shown in Table 1. The mean differences with the ED design for the satisfaction measures ranged from −.1 to 1.6 for the BM design and from 0 to 1.8 for the HD design as shown in Fig. 6.

4 Discussion

An ergonomic evaluation process customized to convex array ultrasound probe design was established in the present study. A simulated sonography workstation consisting of an examination table, a stool with a height adjustment function, a cable support, and a silicon phantom with a load cell underneath the phantom was prepared. Tasks (push, rotate, tilt, and slide), grip types (narrow and wide grips), a range of force application, and a speed of probe manipulation were specified and controlled in a systematic manner. Lastly, the measurement and analysis protocols of muscle activities, body motions, and satisfaction were established.

The satisfaction assessment method was found effective, cost-efficient, and sensitive to identify the effects of a preferred convex probe design in the present study, while the EMG and motion analysis methods were found specific to identify the effects of a preferred convex probe design in terms of muscular load at the hand and postural comfort of the forearm. The satisfaction assessment method could detect the effects of various probe design features with higher sensitivity than the EMG and motion analysis methods when comparing the three convex probe designs with each other. Next, the EMG analysis method could detect the effect of the probe design based on hand data with higher specificity by identifying a significant decrease in the thenar muscle for pushing with a wide grip out of the five hand-forearm muscles for the experimental conditions of task and grip. Lastly, the motion analysis method was found the least sensitive method in detecting the effects of probe design features because the design changes in probe design in the present study were not large enough to significantly affect the motion of the upper body.

The convex probe design evaluation results suggested that a systematic application of hand data along with a preferred grip posture be effective to develop an ergonomic probe design for better fit and comfort. Of the three convex probe designs, the hand-data based design was found most preferred in terms of satisfaction and muscular load. The hand-data based design was developed by identifying the relationships between probe design variables and hand dimensions in the hand and probe image from scanning the hand posture with the most preferred probe.

Acknowledgements. The present research was jointly supported by Samsung Medison Corporation, Mid-Career Research Programs through the National Research Foundation of Korea (NRF) funded by the Ministry of Education, Science and Technology (NRF-2018R1A2A2A05023299), and the Ministry of Trade, Industry, and Energy (No. 10063384; R0004840, 2017).

References

1. Bernard BP (1997) Musculoskeletal disorders and workplace factors. U.S. Department of Health and Human Services, Washington
2. Burnett DR, Campbell-Kyureghyan NH (2010) Quantification of scan-specific ergonomic risk-factors in medical sonography. Int J Ind Ergon 40:306–314
3. Evans K, Roll S, Baker J (2009) Work-related musculoskeletal disorders (WRMSD) among registered diagnostic medical sonographers and vascular technologists. J Diagn Med Sonogr 25(6):287–299
4. Friesen MN, Friesen R, Quanbury A, Arpin S (2006) Musculoskeletal injuries among ultrasound sonographers in rural Manitoba: a study of workplace ergonomics. Am Assoc Occup Health Nurses J 54(1):32–37
5. Habes DJ, Baron S (2000) Ergonomic evaluation of antenatal ultrasound testing procedures. Appl Occup Environ Hyg 15(7):521–528
6. Horkey J, King P (2004) Ergonomic recommendations and their role in cardiac sonography. Work 22(3):207–218
7. Korean Agency for Technology and Standard (KATS) (2004) Report on the 6th Size-Korea (Korean Body Measurement and Investigation). Ministry of Knowledge Economy, Republic of Korea
8. Paschoarelli LC, Oliveira AB, Coury HJCG (2008) Assessment of the ergonomic design of diagnostic ultrasound transducers through wrist movements and subjective evaluation. Int J Ind Ergon 38:999–1006
9. Punnett L, Wegman DH (2004) Work-related musculoskeletal disorders: the epidemiologic evidence and the debate. J Electromyogr Kinesiol 14(1):13–23

Impact of Emotion Regulation on Mental Health of Japanese University Athletes

Yujiro Kawata[1,2(✉)], Akari Kamimura[1,3], Shinji Yamaguchi[2,4], Miyuki Nakamura[2], Shino Izutsu[5], Masataka Hirosawa[1,2], and Nobuto Shibata[1,2,4]

[1] Faculty of Health and Sports Science, Juntendo University, Inzai, Chiba 270-1695, Japan
yuukawa@juntendo.ac.jp
[2] Graduate School of Health and Sports Science, Juntendo University, Inzai, Chiba 270-1695, Japan
[3] School of Humanities, Wayo Women's University, Ichikawa, Chiba 272-8533, Japan
[4] Institute of Health and Sports Science and Medicine, Juntendo University, Inzai, Chiba 270-1695, Japan
[5] Faculty of Sports and Health Sciences, Japan Women's College of Physical Education, Tokyo 157-0061, Japan

Abstract. Athletes' mental health is a great concern for coaches, supporters, and researchers in the field of sports science. To prevent mental health problems in athletes, emotion regulation strategies are considered effective for coping with stressors. Emotion regulation is defined as the regulation of thoughts or behaviors that influence the emotions. Emotions may influence not only the mental health but also performance in competitive sports. Therefore, we examined the impact of emotion regulation on the mental health of Japanese university athletes. We collected data from 927 Japanese university athletes (535 male and 392 female). We collected information on athletes' demographics, emotion regulation (Emotion Regulation Questionnaire: ERQ), and mental health (Self-report Depression Scale: SDS, and General Health Questionnaire-30; GHQ-30). A regression analysis showed that reappraisal had a significant negative effect on SDS ($\beta = .{-44}$, $p < .001$, $R^2 = .19$) and GHQ-30 scores ($\beta = .{-26}$, $p < .001$, $R^2 = .08$) but suppression did not affect either of them. This indicates that athletes using reappraisal have a good mental health, thereby suggesting that emotion regulation may contribute to the maintenance of athletes' mental health. Thus, we concluded that emotion regulation has an impact on the mental health of Japanese university athletes. Sports coaches and supporters of athletes should pay substantial attention to athletes' emotion regulation strategies to maintain their mental health and enhance their performance.

Keywords: Emotion regulation · Mental health · University athletes

S. Bagnara et al. (Eds.): IEA 2018, AISC 818, pp. 372–382, 2019.
https://doi.org/10.1007/978-3-319-96098-2_48

1 Introduction

Athletes' mental health is a great concern for coaches, supporters, and researchers in the field of sports science. Various types of issues related to poor mental health, such as depression [1], burnout syndrome [2], and eating disorders [3] have been reported. Performance decline leads to not only a loss of opportunity to participate in games/matches as a regular member but also, in some situations, a retirement for the player against their will.

To prevent such problems among athletes, it is very important to develop sufficient support systems to enhance their ability to cope with stress and maintain their mental health. Especially, in Japan, where the Olympics Games will be held in 2020, this is one of the urgent tasks. In fact, in recent years, the number of university athletes covers more than half of Japanese athletes participating in the Olympics Games [4].

In this context, we have been focusing on athletes' mental health and performance [5, 6]. In line with our studies, emotion regulation (a component of resilience) is associated with stress cognition, stress coping, stress responses. That is, athletes using emotion regulation appropriately reported low stress cognition, problem-focused stress coping, and low stress responses, compared to athletes with less emotion regulation. However, the type of emotion regulation strategies frequently used by athletes has not yet been clarified.

Emotion regulation is defined as the regulation of thoughts or behaviors that influence the emotions. To understand the process of emotion regulation, the "process model of emotion regulation" was proposed by Gross [7, 8]. In this model, emotion regulation was conceptualized as two specific regulatory processes: antecedent-focused emotion regulation and response-focused emotion regulation. Antecedent-focused emotion regulation is the strategy used before an emotion appears; while response-focused emotion regulation is the strategy used after an emotion appears.

Cognitive reappraisal strategy has been proposed as a type of antecedent-focused emotion regulation, while expressive suppression strategy has been proposed as a type of response-focused emotion regulation. Cognitive reappraisal strategy is a form of cognitive change that involves construing a potentially emotion-eliciting situation in a way that changes its emotional impact. Expressive suppression strategy is a form of response modulation that involves inhibiting an ongoing emotion-expressive behavior [7, 8].

Antecedent-focused strategies (reappraisal) influence whether or not particular emotional response tendencies are triggered, and are, therefore, expected to have generally positive implications for affective and social functioning. In contrast, response-focused strategies (suppression) influence how emotional response tendencies are modulated once they have been triggered, and are, therefore, expected to have generally negative implications for affective and social functioning [8].

Although the relationship between emotion regulation and mental health of the general population has been studied, this relationship in athletes has not yet been investigated. If sufficient evidence is obtained to suggest that emotion regulations have an impact on the mental health of athletes, it will be useful to develop interventions and establish support systems for the maintenance of good mental health. Emotions influence not only the mental health but also athletes' performance in competitive sports.

Against this background, this study was conducted to examine the effect of emotion regulation on the mental health of Japanese university athletes.

2 Introduction

2.1 Participants

Data were collected from 927 Japanese university athletes (535 males and 392 females; *M* age = 19.8 years, *SD* = 1.20 years). The main sports categories include individual sports and team sports. Their average years of experience in sports events was 9.08 ± 4.42 years.

2.2 Questionnaire Items.

We collected information about athletes' demographics, emotion regulation, and mental health.

Demographic Characteristics. We collected information about athletes' sex, age, academic grade, residential type, individual role in the team, competition level, years of experience in sports events, type of sports event, and sports event category (individual sports and team sports).

Emotion Regulation. We used the Emotion Regulation Questionnaire (ERQ; [9, 10]) to assess emotion regulation. This scale assesses two habitual emotion regulation strategies: reappraisal (6 items) and suppression (4 items). Examples of items are "I control my emotions by changing the way I think about the situation I'm in" (reappraisal) and "I control my emotions by not expressing them" (suppression). Both reappraisal and suppression scales include at least one item asking about regulating negative emotions (illustrated for participants by giving "sadness" and "anger" as examples) and one item asking about regulating positive emotions (illustrated for participants by giving "joy" and "amusement" as examples). Participants responded to each item on a seven-point Likert scale (1 = "disagree" to 5 = "agree"). The mean item score was calculated. A higher score indicates higher tendency to use emotion regulation.

Mental Health. We used the Self-report Depression Scale (SDS; [11]) and the General Health Questionnaire-30 (GHQ; [12]) to assess the mental health. The SDS is a self-report scale to measure depressive symptoms. It has 20 items and assesses a single depression factor. Participants indicated their agreement with each item on a four-point Likert-scale (1 = "disagree" to 4 = "agree"). The overall score was calculated by summing the item scores. A higher score indicates a higher level of depression. The GHQ has 30 items consisting of the following six factors: "general illness," "somatic symptoms," "sleep disturbance," "social dysfunction," "anxiety and dysphoria," and "suicidal depression." Participants indicated their agreement with each item on a five-point Likert scale (1 = "disagree" to 5 = "agree"). The overall score was calculated by summing the item scores. Higher scores indicate poor mental health.

2.3 Procedure

Participants were explained the purpose of the study and given instructions to answer the questionnaire. They took as much time as they needed to respond to the questionnaire. The research staff members distributed the questionnaires to the participants and collected the completed questionnaires. Participation in this study was voluntary. The survey was carried out during participants' free time.

2.4 Ethical Consideration

This study was approved by the Research Ethics Committee at the Faculty of Health and Sports Science, Juntendo University. Prior to the study, we obtained written informed consent from all participants. Each participant was made aware of his or her right to decline cooperation at any time without repercussions, even after consenting to participate. The questionnaire was administered in a quiet university classroom with enough space to ensure that participants' privacy was maintained.

2.5 Analysis

First, we confirmed participants' demographic background to check whether the sample represented Japanese athletes. Then, prior to the main analysis, we identified factors that could confound emotion regulation and SDS and GHQ scores. Finally, to examine the effect of emotion regulation on mental health variables (SDS and GHQ scores), we performed a regression analysis by setting emotion regulation (reappraisal and suppression scores) as the independent variable, mental health (SDS and GHQ scores) as the dependent variables, and demographic characteristics (sex, age, academic grade, residential type, individual role in the team, competition level, experience years of sports event, type of sports event, and sports event category) as covariates. The effect size was calculated. The criteria of the effect size of η^2 in a one-way analysis of variance (ANOVA) were regarded as follows: small effect (η^2 = 0.01), medium effect (η^2 = 0.06), and large effect (η^2 = 0.14). The criteria of the effect size of R2 in regression analysis were regarded as follows: small effect (R^2 = 0.02), medium effect (R^2 = 0.13), and large effect (R^2 = 0.26) [13]. All analyses were performed using SPSS 25.0 (IBM, JAPAN). Statistical significance was set at .05.

3 Results and Discussions

3.1 Demographic Characteristics

Table 1 shows participant's demographic characteristics. We reported the number of participants and the percentages by sex, age, academic grade, residential type, individual role in the team, competition level, years of experience in sports event, type of sports event, and sports event category. Participants' characteristics seem to represent those of Japanese athletes to some extent in terms of the demographic backgrounds.

Table 1. Demographic characteristics of the participants

Attributes	Number (N)	Percentage (%)
Sex		
Male	535	57.7
Female	392	42.3
Age (years)		
18	96	10.4
19	367	39.6
20	192	20.7
21	166	17.9
22	98	10.6
23	8	0.9
Academic grade		
Freshmen (1st year)	411	44.3
Sophomore (2nd year)	210	22.7
Junior (3rd year)	163	17.6
Senior (4th year)	143	15.4
Residential type		
Freshmen dormitory	339	36.6
Club dormitory (single room)	65	7.0
Club dormitory (twin room)	69	7.4
Club dormitory (>triple room)	292	31.5
Living alone	139	15.0
Living with family	21	2.3
Living with friends	2	0.2
Others	339	36.6
Individual role in the team		
Regular member	427	46.1
Non-regular member	500	53.9
Competitive level		
Regional	124	13.4
Prefectural	154	16.6
District	222	23.9
National	389	42.0
International	38	4.1
Years of experience in sports events		
≥5 years	213	23.0
6–10 years	336	36.2
11–15 years	327	35.3
≤16 years	51	5.5

(continued)

Table 1. (*continued*)

Attributes	Number (N)	Percentage (%)
Type of sports event		
Track and field	187	20.2
Soccer	126	13.6
Basketball	67	7.2
Handball	57	6.1
Baseball	55	5.9
Volleyball	53	5.7
Kendo (traditional Japanese martial arts)	42	4.5
Softball	38	4.1
Dancing	34	3.7
Gymnastics	32	3.5
Life-saving	27	2.9
Judo	26	2.8
Futsal	23	2.5
Squash	22	2.4
Swimming	22	2.4
Softball-tennis	16	1.7
Rhythmic gymnastics	15	1.6
Bicycle	14	1.5
Hardball-tennis	14	1.5
Rugby	14	1.5
Triathlon	13	1.4
Badminton	8	0.9
Softball baseball	6	0.6
Ski	4	0.4
Naginata (traditional Japanese martial arts)	4	0.4
Lacrosse	3	0.3
Golf	1	0.1
Table tennis	1	0.1
Fencing	1	0.1
Shorinji Kenpo (traditional Japanese martial arts)	1	0.1
Japanese art of archery	1	0.1
Sports event category		
Individual sports	464	50.1
Team sports	463	49.9

3.2 Effect of Demographic Characteristics on Emotion Regulation and Mental Health

Effect of Demographic Characteristics on Emotion Regulation. Prior to the main analysis, to check the effect of demographic characteristics on emotion regulation, we performed a one-way ANOVA with demographic characteristics as the independent variables and ERQ reappraisal and suppression scores as the dependent variables.

A significant sex difference was found in athletes' suppression scores on the ERQ ($F(1, 925) = 6.58, p < .05, \eta^2 = .013$), wherein male athletes had a higher suppression score than female athletes did, but there was no difference with respect to reappraisal scores. This implies that male athletes have a greater tendency to suppress their emotion than female athletes do. Other culture studies (United States, Germany Italy, and Japan) have also reported the same result [8]. Interestingly, the Japanese population (both male and female) tends to have a higher score on suppression. This might be due to the influence of the Japanese culture. There was a significant age difference in reappraisal scores on the ERQ ($F(5, 921) = 4.08, p < .01, \eta^2 = .030$), wherein older athletes had a higher score on reappraisal than younger athletes did, but there was no difference with respect to suppression scores. This shows that older athletes have a greater tendency to reappraise their emotions than younger athletes do. Similar to age, there was a significant difference in the reappraisal scores on the ERQ ($F(3, 923) = 7.52, p < .001, \eta^2 = .024$) with respect to academic grades, wherein athletes in the upper grades obtained a higher score on reappraisal than athletes in the lower grades did, but there was no difference with respect to suppression scores. This indicates that older athletes have a greater tendency to reappraise their emotions than younger athletes do. There was a significant in for suppression score ($F(6, 920) = 2.60, p < .05, \eta^2 = .020$) in terms of residential type, wherein athletes living with friends had a lower score on suppression than athletes living alone did, but there was no difference with respect to the reappraisal scores. This implies that athletes living with friends have a greater tendency not to suppress their emotions than athletes living alone do. No significant differences in terms of individual role in the team were found. This suggests that having an individual role in the team (regular or non-regular member) does not affect reappraisal and suppression. No significant difference in term of competition level was found. This indicates that competition level does not affect reappraisal and suppression. There was a significant difference in suppression score ($F(3, 923) = 3.80, p < .01, \eta^2 = .019$) in terms of the years of experience in sports events, suggesting that athletes with more experience obtained a higher suppression score compared to athletes with fewer experience, but there was no difference with regard to suppression scores. This suggests that athletes with more experience have a greater tendency to suppress their emotions than athletes with lesser experience do.

Effect of Demographic Characteristics on Mental Health. To confirm the effect of demographic characteristics on athletes' mental health, we performed a one-way ANOVA with demographic characteristics as the independent variables and SDS and GHQ scores as the dependent variables.

There was a significant sex difference in SDS ($F(1, 925) = 33.0$, $p < .001$, $\eta^2 = .038$) and GHQ scores ($F(1, 925) = 36.9$, $p < .05$, $\eta^2 = .034$), wherein female athletes obtained a higher score on the SDS and GHQ than male athletes did. This implies that female athletes have a greater tendency to have depression and poor mental health than male athletes do. A significant age difference was found in SDS ($F(5, 921) = 3.91$, $p < .01$, $\eta^2 = .021$) and GHQ scores ($F(5, 921) = 5.78$, $p < .001$, $\eta^2 = .030$), suggesting that younger athletes had a higher score on the SDS and GHQ than older athletes did. This implies that younger athletes have a higher tendency to have depression and poor mental health than older athletes do. There was a significant difference in the SDS ($F(3, 923) = 9.32$, $p < .001$, $\eta^2 = .029$) and GHQ scores ($F(3, 923) = 5.91$, $p < .001$, $\eta^2 = .019$) with respect to academic grades, suggesting that athletes in the lower grades had a higher score on the SDS and GHQ than athletes in the upper grades did. This implies that athletes in lower grades have a greater tendency to have depression and poor mental health than athletes in upper grades do. No significant difference in term of the residential type was found, suggesting that the residential type does not affect athletes' mental health. There was no significant difference in the SDS and GHQ scores with regard to the individual role in the team. This implies that the individual role in the team does not affect athletes' mental health. There was a significant difference in the GHQ ($F(3, 923) = 9.32$, $p < .001$, $\eta^2 = .029$) but not SDS scores with respect to competition level, suggesting that athletes in the lower level had a higher score on the GHQ than athletes in the upper level. This indicates that athletes in a lower level of competition have a greater tendency to have poor mental health than athletes in higher competition level. There was a significant difference in the SDS ($F(3, 923) = 6.21$, $p < .001$, $\eta^2 = .020$) and GHQ scores ($F(3, 923) = 3.11$, $p < .05$, $\eta^2 = .010$) with respect to years of experience in sports events, suggesting that athletes with lesser experience obtained a higher score on the SDS and GHQ than athletes with more experience did. This implies that athletes with lesser experience have a greater tendency to have depression and poor mental health than athletes with more experience do. There was a significant difference in SDS ($F(1, 925) = 6.48$, $p < .05$, $\eta^2 = .001$) and GHQ scores ($F(1, 925) = 4.47$, $p < .05$, $\eta^2 = .001$) in terms of the sports event category, indicating that athletes in individual sports club obtained a higher score on the SDS and GHQ than athletes in team sports club did. This suggests that athletes in individual sports clubs have a greater tendency to have depression and poor mental health than athletes in team sports clubs do.

Consequently, demographic characteristics, such as sex, age academic grades, residential type, competition level, years of experience in sports events, and the sports event category, might be confounding factors affecting emotion regulation and SDS and GHQ scores. Thus, we considered the demographic characteristics as covariates in the following analysis.

3.3 Impact of Emotion Regulation on Mental Health

To examine the effect of emotion regulation on SDS, we performed a regression analysis considering the above-mentioned confounding factors. The results showed that reappraisal had a significant negative effect on SDS ($\beta = .-44$, $p < .001$, $R^2 = .19$) but suppression had no significant effect on SDS (Figs. 1 and 2). The effect size was medium. This indicates that reappraisal may reduce depressive symptoms in Japanese university athletes. A previous study reported that reappraisal reduces stress responses [8]. The finding of this study is concordant with that of the previous study. This suggests that emotion regulation (reappraisal) could be one of the stress coping strategies of athletes. However, suppression had no effect on the depressive symptoms of Japanese university athletes. Although male athletes engaged in more suppression as an emotion regulation strategy than did female athletes, they need to understand that suppression has no effect on reducing the depressive symptoms.

Next, to examine the effect of emotion regulation score on the GHQ, we performed a regression analysis considering the afore-mentioned confounding factors. The results showed that reappraisal had a significant negative effect on the GHQ score (β = .−26, $p < .001$, $R^2 = .08$) but suppression did not affect the GHQ score (Figs. 3 and 4). The effect size was small. This also indicates that reappraisal may contribute to the maintenance of mental health. This finding is concordant with that of a previous study [8]. This implies that emotion regulation (reappraisal) could be one of the stress coping strategies of athletes. However, similar to the results on depression, suppression had no effect on improving the mental health of Japanese university athletes.

Comparing the effect size of reappraisal scores on the SDS and GHQ, the effect size on the SDS was greater than that on the GHQ. This implies that reappraisal is more effective for reducing depressive symptoms than for maintaining the overall mental health. This discrepancy may come from the fact that the GHQ covers several aspects of mental health (and not just a specific mental health issue such as depression).

Emotion regulation consists of reappraisal and suppression; however, only reappraisal had a positive impact on depression and mental health. In the general Japanese population, reappraisal is correlated with self-esteem, and subjective happiness, while suppression is not [10]. This study suggests that reappraisal may contribute to good mental health. Thus, for athletes, compared to suppression, reappraisal is a more adaptive emotion regulation strategy to reduce depression and improve mental health.

A cross-cultural study reported that reappraisal score was higher than suppression score in different cultural countries [14]. A similar finding was reported by a study in Japan [10]. The present study also showed similar results. Therefore, reappraisal is used more often compared to suppression regardless of the culture. However, it should be noted that for both men and women, the suppression scores of the Japanese population tend to be higher than those of people living in other countries. Suppression is often used by people living in the Japanese culture. The emotion regulation strategies of those living in Japan may have a characteristic pattern with respect to suppression.

This study showed a relationship between emotion regulation and the mental health. However, the causal relationship between emotion regulation and mental health should be clarified. This finding is useful for establishing intervention strategies for better mental health. Further research taking this issue into account will enable the possible applications of emotion regulation research into practice.

Based on the findings of this study, athletes who use reappraisal for stress coping have a good mental health. Emotion regulation may contribute to the maintenance of athletes' mental health.

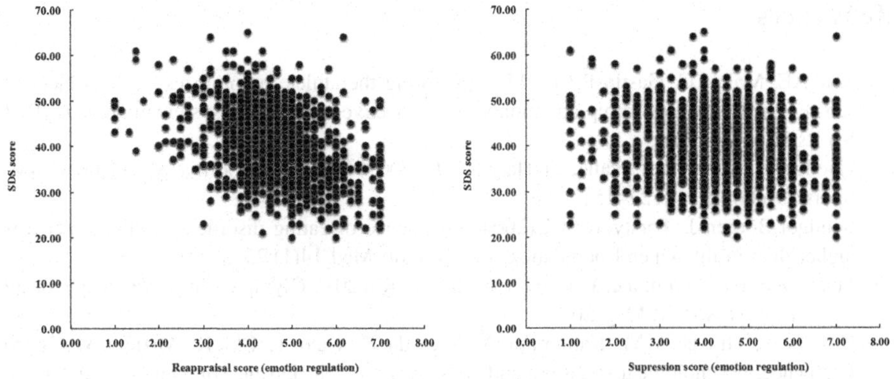

Fig. 1. Scatter plots of reappraisal (left) and suppression (right) and SDS scores

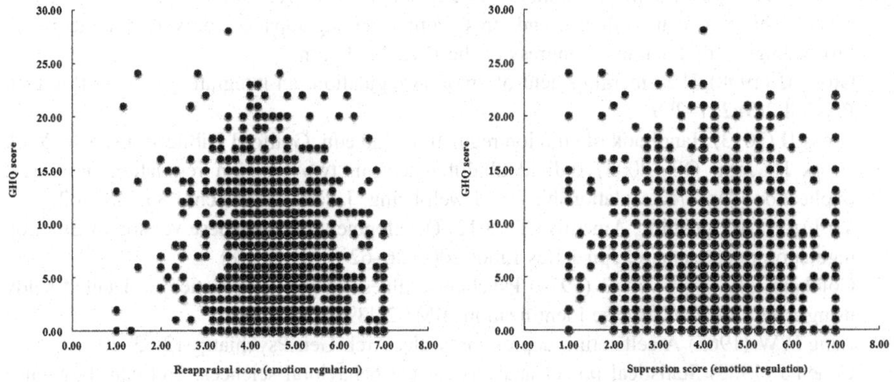

Fig. 2. Scatter plots of reappraisal (left) and suppression (right) and GHQ scores

4 Conclusions

We conclude that emotion regulation has an impact on the mental health of Japanese university athletes. Sports coaches and supporters of athletes should pay substantial attention to athletes' emotion regulation strategies to maintain their mental health and enhance their performance.

Acknowledgements. This study was supported by JSPS KAKENHI Grant Number 18K17802 (PI: Yujiro Kawata). We thank all participants for sharing their time and data used in this study.

References

1. Saw AE, Main LC, Gastin PB (2015) Monitoring the athlete training response: subjective self-reported measures trump commonly used objective measures: a systematic review. Br J Sports Med. 094758
2. Goodger K, Gorely T, Lavallee D, Harwood C (2007) Burnout in sport: a systematic review. Sport Psychol 21(2):127–151
3. Sundgot-Borgen J, Torstveit MK (2004) Prevalence of eating disorders in elite athletes is higher than in the general population. Clin J Sport Med 14(1):25–32
4. Japan Olympic Committee. Japanese medalists in Rio 2016 Olympics. https://www.joc.or.jp/english/. Accessed 28 May 2018
5. Kawata Y, Hirosawa M, Kamimura A, Yamada K, Kato T, Oki K, Mizuno M (2015) Resilience, psychological stressors, and stress responses in Japanese university athletes. In: Yamamoto S, Shibuya M, Izumi H, Shih Y-C, Joe LC, Lim H (eds) New ergonomics perspective: selected papers of the 10th Pan-Pacific conference on ergonomics, Tokyo, Japan. CRC Press, London, pp 231–238
6. Kawata Y, Hirosawa M, Kamimura A, Yamada K, Kato T, Oki K, Mizuno M (2015) Relationship between resilience and stress coping among Japanese university athletes. In: Proceedings 19th Triennial Congress of the IEA, Melbourne
7. Gross JJ (1998) The emerging field of emotion regulation: an integrative review. Rev Gen Psychol 2(3):271–299
8. Gross JJ (2013) Handbook of emotion regulation, 2nd edn. Guilford Publications, New York
9. Gross JJ, John OP (2003) Individual differences in two emotion regulation processes: implications for affect, relationships, and well-being. J Pers Soc Psychol 85:348–362
10. Yoshizu J, Sekiguchi R, Amemiya T (2012) Development of a Japanese version of Emotion regulation questionnaire. Jpn J Res Emot 20(2):56–62 (in Japanese)
11. Goldberg DP, Blackwell B (1970) Psychiatric illness in general practice: a detailed study using a new method of case identification. BMJ 2:439
12. Zung WW (1965) A self-rating depression scale. Arch Gen Psychiatry 12:63–70
13. Cohen J (1988) Statistical power analysis for the behavioral sciences, 2nd edn. Lawrence Erlbaum, Hillsdale
14. Matsumoto D, Yoo SH, Nakagawa S (2008) Culture, emotion regulation, and adjustment. J Pers Soc Psychol 94(6):925

User Centred Development of a Smartphone Application for Wayfinding in a Complex Hospital Environment

Frank Smolenaers[1,2], Tim Chestney[1], Jenny Walsh[1], Simon Mathieson[1], Daniel Thompson[1], Mehmet Gurkan[1], and Stuart Marshall[1,2,3(✉)]

[1] Alfred Health, Commercial Road, Melbourne, VIC 3004, Australia
[2] University of Melbourne, 173 Wilson Ave, Parkville, VIC 3052, Australia
[3] Monash University Department of Anaesthesia and Perioperative Medicine, Level 5, 99 Commercial Road, Melbourne, VIC 3004, Australia
stuart.marshall@monash.edu

Abstract. Finding where to go within a hospital environment can be difficult and frustrating for patients and visitors leading to complaints, missed appointments and inefficiencies. We iteratively developed and tested a navigation application for smartphones, to augment existing wayfinding strategies. The "PowerNav" smart phone app uses indoor location and movement tracking, enabled by low energy Bluetooth beacons (BLE) mounted on the ceiling of hospital corridors. No IT infrastructure is required. A prototyping stage of the interface with user input was followed by a rapid phase user-centred design process involving naïve and experienced users. This culminated in the naturalistic observation of users with the app and ongoing data collection of user experience. The partnering of a public health service and a private developer proved to be a successful venture. Future work will involve accessibility features for visually impaired users and support of additional languages. We believe that future integration of the app with public transport and other travel information will also improve the utility and create a patient centred approach to appointment management from the receipt of the letter to attending the appointment.

Keywords: Wayfinding · Navigation · Hospital · Design

1 Hospital Design and Wayfinding

1.1 The Extent of the Hospital Wayfinding Problem

Hospitals are commonly difficult places to navigate. Metropolitan hospitals tend to be large, tertiary centres housing multiple complex specialty services. Consequently, the buildings and departments can be spread over a large area with patients often needing to access many departments even within the same visit. Newer hospitals may be able to prospectively design departments closer that might be accessed together more frequently, but this is not always possible [1]. Older hospitals provide additional

S. Bagnara et al. (Eds.): IEA 2018, AISC 818, pp. 383–393, 2019.
https://doi.org/10.1007/978-3-319-96098-2_49

challenges and commonly comprise of multiple buildings linked by corridors and outdoor walkways that may be poorly signposted.

Financial and Patient Costs. Poor signage and architectural planning of hospitals invariably leads to inefficiencies and additional cost. These are difficult to quantify in monetary terms because not only are the problems generally hidden, but other factors affect cancelled appointments and late attendances. Nevertheless, in the UK it has been estimated that 8 million outpatient appointments, each costing £120 ($160 USD) were missed in the 2016/2017 financial year alone [2, 3]. If only a conservative 10% of these appointments were missed due to problems finding the clinic this is a cost of over £96 million ($128 million USD) per year, every year. Put another way, the money saved could fund 100,000 cataract operations.

Apart from the financial cost, there is of course also a health cost of cancelled appointments, the risk that patients are 'lost to follow-up' when they need ongoing care, and also the psychological stress of being unable to find the clinics and wards. The deterioration of patients' health leads to disability, untreated and chronic health issues and further social and financial cost to the health system. Furthermore, frustration of patients and visitors adds unnecessary stress to an already difficult time for patients and their families. Stress that could be avoided with improved navigation.

A further consideration with patient and visitor requirements, is the nature of the population attending the hospital. Use of health facilities is predominantly by older adults, with a quarter of over 75 year-olds being admitted to hospital per year, compared with 13% of the whole population [4]. With increased age comes limitation of mobility and reduced acuity of the senses of sight and hearing. This reduction in navigation ability with age is even without taking into account the disease states for which they may be attending the hospital that may lead to further disability. Physical disability may be easily recognized by other individuals and offers of help can be made, but patients with traumatic brain injury, cognitive impairment or psychiatric illnesses might also have specific and not obviously identified impairments of navigation. Patient diversity may also be problematic with language difficulties and the need to cater for a multicultural population.

Unintended Staff Costs. The effects of poor wayfinding on staff are similarly difficult to quantify. In a study in a large tertiary USA hospital in the late 1980s, Zimring suggested up to $220,000 USD ($448 per bed space) was wasted in wayfinding inefficiencies [5]. The majority of this cost was due to an estimated 4,500 h of diverted clinical staff time in dealing with patients and visitors who were disorientated in the hospital environment. A less visible but perhaps more troubling issue is the inability of hospital staff to quickly find their way to emergencies in the clinical environment. There are few studies examining this aspect of wayfinding in hospitals. An interview study examining the non-technical skills of junior doctors working in hospitals after hours found that route planning and wayfinding was an important, and often overlooked competency [6]. As one participant reported, trying to get to a cardiac arrest (crash) call:

> *"Crash calls are often slowed when you don't know where it is. Ward names can give no indication of where they are. Even registrars get lost some times. (Participant 6)"*

The experiences of junior medical staff are particularly problematic, as they are often expected to provide services to a large number of locations across the hospital, occasionally during emergencies and may only be working at the site for a short duration due to the rotational training systems they work within.

1.2 Potential Solutions to the Wayfinding Problem

Design. As noted earlier, the design of a new hospital provides an opportunity to incorporate best practice design techniques and the co-location of departments close to each other that are used together. For instance, an outpatient's department may be located close to the pathology service for blood collection or the radiology department for radiographs (X-rays) required for the fracture clinic. Where this is not possible, strategies such as coloured lines, signs, textures and pictures on floors and walls might help with guidance to related areas and cognitive recognition that you have arrived at the required destination. External cues to the function of buildings and internal cues to the function of areas are important and help guide individuals through the spaces.

Signage. Whilst this is the most obvious method of physical wayfinding in public spaces, it is fraught with complexity and difficulty. Written signs in the local language are clearly important, but the use of icons and symbols can be more effective if they are unambiguous [7].

Maps. These are another obvious solution to indoor navigation, but again, careful design is essential. A UK study showed that many hospital maps had excessive acronyms, lacked detail, or were out of date [8]. Often there are differences between the signage and maps in terms of design, consistent nomenclature and content that are confusing to users.

Symbolism. Additional cues built into the environment can minimize the need for confusing signage. Visual landmarks and motifs help orientation within a space. These features can be utilized to help individuals find specific buildings or regions.

Assistance. One strategy to wayfinding is the use of a concierge system, with an information desk that houses volunteers who can provide ad-hoc wayfinding assistance. This may be with directions from the reception booth, or a concierge accompanying patients and visitors to their destination.

Electronic Devices. Although there are common smart phone outdoor wayfinding applications such as Google Maps, indoor navigation has been limited with electronic wayfinding due to the ability of accurate location and tracking of movement and limitation of the available suitable maps.

Wayfinding Strategy. Wayfinding is a complex function of efficiently moving individuals through spaces, by helping them identify where they are and how to get to where they're going, by aiding decisions and modifying behavior. An effective wayfinding strategy for a building consists of an integrated approach using all of the methods above in a 'legible physical setting' [9].

2 Setting for the Current Study

The setting of this intervention was a large metropolitan hospital in Melbourne, Australia. The Alfred hospital serves a catchment population of over 650,000 and is the referral centre for 14 statewide services. The physical site has sections that are nearly 150 years old and a separate contemporary facility on campus dedicated to day case surgery. This project coincided with the completion of a physical wayfinding strategy (signage and wall maps) and augments the volunteer concierge assistance at major entrances.

The project involved working with a software developer to supplement the existing wayfinding strategy with an additional modality – "Digital Wayfinding". As noted, electronic methods for wayfinding have been fraught with difficulties in pinpointing location and tracking movement. This solution aimed to use low energy Bluetooth beacons in addition to GPS and both integrate with and compliment the newly deployed physical signage solution.

The developer (PowerHealth Solutions) entered a partnership with Alfred Health to further mature the interface and to test it within the Australian Centre for Health Innovation (a division of Alfred Health) and then the broader hospital environment.

3 Methods

An initial meeting was held to discuss the existing prototype of the interface and to establish the acceptability of the solution concept with patients, patient advocates and volunteer concierge staff. Following this, iterative testing was undertaken. A further three phases included (1) testing with naïve users and the initial focus group members on 'typical use scenarios' within the hospital. (2) A naturalistic observation of users asked to download and use the application at the hospital entrance and (3) Ongoing evaluation by the monitoring of use data from the application.

3.1 Phase 1 – Initial Acceptance and Prototype Idea

In this phase, eight potential users consisting of four male and four female patient advocates and volunteers were shown pictures of the prototype interface and asked about the need and initial features of the initial design. The solution, including the low energy Bluetooth beacons, was installed in the Australian Centre for Health Innovation and a demonstration of the application were performed. The duration of the focus group session was a little over an hour. After introductions, a structured set of questions was presented to the group, which examined common problems with wayfinding at the site in general, and then became more specific regarding digital solutions. A short presentation consisting of half a dozen PowerPoint slides of screen captures from the app were displayed to illustrate how such a solution could help.

Need for Electronic Solution. There was a general agreement that a phone or tablet application could improve wayfinding at the Alfred.

It was envisioned that the app could be used independently by the visitors and patients and by the concierge staff at the desk to help explain where to go. Physical maps are currently being used to assist visitors and these could be enhanced, particularly if there was an ability to print the maps out and highlight the route on the app.

Barriers to App Use. Some concerns were raised about the age of the patients and visitors and the ability to use electronic devices. However, a recent survey at the Alfred showed that 40–50% of patients and visitors bring a smartphone with them when they attend the hospital (Fig. 1).

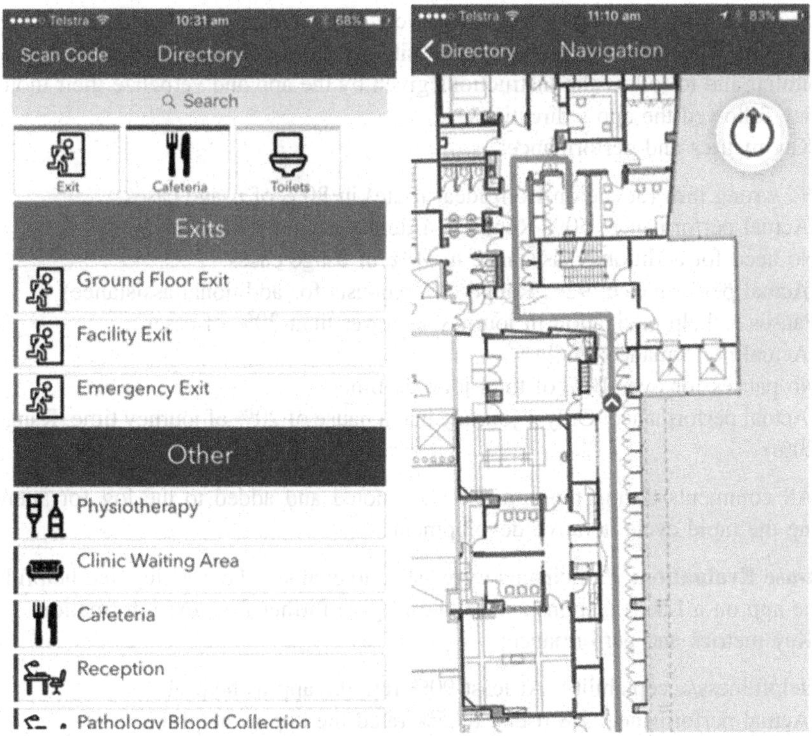

Fig. 1. Screen shots from the initial prototype. On the left is the location selection screen, on the right is the map and navigation track.

3.2 Phase 2 – Testing with Typical Use Scenarios

Prior to this second phase, the low energy Bluetooth beacons were deployed across the hospital site. A further eight participants were recruited (three male and five female) representing patient advocates and volunteer concierges that would be using the app to assist visitors to the site.

This phase was evaluated in terms of downloading the app, using the app within the hospital and post-use evaluation. Key metrics were determined for each of these stages.

Downloading and Activation of the App. The participants downloaded the app from Android "Google Play" or the Apple "App Store". After being escorted to the starting point of their journey, the participants opened the app for the first time and the location information was activated. They had a variety of smart phones in make, model and age.

Key metrics and performance:

- Downloading of app should take less than 1 min in 90% of cases
  (Actual performance: 100%, Range 8–29.3 s)
- Location activation should take less than 30 s in 90% of cases
  (Actual performance: 92%, Range 4–100 s)

Navigation Using the App. Participants were given a location to navigate to using the app. Those who were familiar with the site were asked to imagine that they were unfamiliar and to follow the instructions given by the app and verbalize their thoughts as they followed the app's directions.

Key metrics and performance:

- No wrong turn (deviation from ideal route) in 80% of usage cases
  (Actual performance: 50% Range 0–4 deviations on route)
- No need for additional assistance in 90% of usage cases
  (Actual performance: 92% Range 0–1 requests for additional assistance)
- Pauses to help navigation in journey in fewer than 20% of cases
  (Actual performance: 25%)
- No pauses for over 10% of total journey time
  (Actual performance: Only 1 journey had a pause of 20% of journey time. Range 3–20%)

All comments during the app use were noted and added to the log for attention during the rapid cycle iterative development.

Post-use Evaluation. Participants were asked to evaluate the usability and helpfulness of the app on a Likert response scale and to give further free text information.

Key metrics and performance:

- Helpfulness/acceptability: At least 90% rate the app as helpful
  (Actual performance: 7/8 users, 87.5% rated the app as helpful)
- Ease of use: At least 90% report easy or very easy to use. (Actual performance: 62.5% report easy or very easy to use with the rest being neutral)

Specific problems were noted in the car parking area, outdoor areas between buildings and in tracking the movement of the individual. Smoothing of the path, rotation of the map and the symbols on the interface were found to be confusing for some users. The use of shading and colour for outdoor areas was suggested by the users as a solution and the level of detail on the map was generally agreed to be confusing.

3.3 Phase 3 – Naturalistic Testing with Naïve Users

Once the feedback from the Phases above was implemented, along with further app refinement generally, we progressed to a Phase 3 study. Five ad-hoc participants were recruited at one of two main entrances to the hospital. Demographic questions were asked and the participant given a smart phone with the app pre-installed to help them find their way to their intended destination. Performance was measured against defined metrics and transcription of their comments was recorded concurrently.

Navigation using the app was observed by the researcher

Key metrics and performance:

- No wrong turn (deviation from ideal route) in 80% of usage cases
  (Actual performance: 60% Range 0–2 deviations on route)
- No need for additional assistance in 90% of usage cases
  (Actual performance: 80% Range 0–1 requests for additional assistance)

Post-use Evaluation Participants were asked to evaluate the usability and helpfulness of the app on a Likert response scale and to give further (free text information see below)

Key metrics and performance:

- Helpfulness/acceptability: At least 90% rate the app as helpful
  (Actual performance: 5/5 users, 100% rated the app as helpful)
- Ease of use: At least 90% report easy or very easy to use. (Actual performance: 100% report easy or very easy to use)

Problems were again noted with the level of detail and lack of colour in the maps.

One test patient, recovering from an acute brain injury and accompanied by his mother, was readily recruited. The ease of navigation and sheer excitement of the patient to interact with the app led his mother to be overcome with emotion. She was thrilled that her son could manage to navigate to his own appointment – a real sense of clinical improvement and independence.

4 Final Design Solution and Discussion

4.1 Visual Design

Following the feedback from earlier versions a second developer was engaged to redesign the electronic maps for the app to be clearer and include colour and shading, to distinguish inside from outside and paths from rooms. Additional features were added such as icons to identify, the position of the nearest toilet facilities, café, shops and information points (Fig. 2). Lag during movement was minimized by modification of tracking algorithms and location names were standardized and searchable. Furthermore, the forward path was highlighted in Blue with a Blue arrow indicating direction, whilst the path behind the user's journey changes to white. This colour scheme was chosen after consulting a colour blind male volunteer as it provided him the best discrimination.

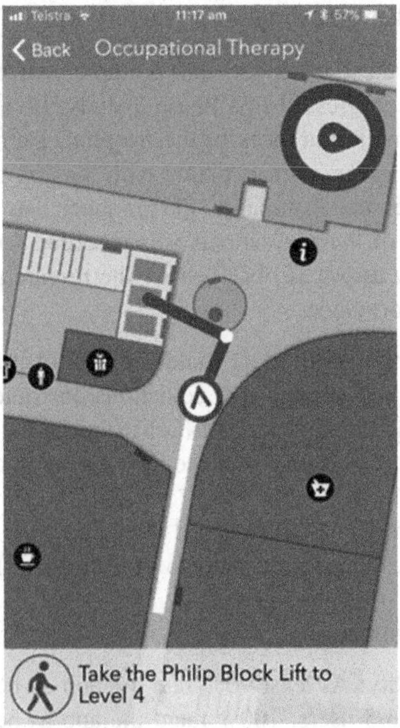

Fig. 2. Design of the interface at the time of public release

4.2 Feedback from Users

At the end of every journey, the app has a simple survey to solicit user feedback on their experience with the app and any comments (Fig. 3).

The solution also includes the ability to message a family member or friend with your current or planned destination (e.g. being admitted to a ward from emergency, or meet at the cafeteria) such that the receiver can simply click on an embedded link in the message and launch the app and navigate to the sender easily.

Fig. 3. Simple app rating and feedback survey

4.3 Integration with Appointments and Existing Wayfinding

Further improvements stimulated by this project included the standardized naming of clinical areas and the process for sending clinic appointment letters. The appointment letters will include a visual QR type code (Aztec Code) that can be scanned from within the app to pre program the location details of the appointment for the app and get the user on their way (Fig. 4).

A "Favourites" or "Recent Destinations" list will help the user to find their way back to areas such as where their car is parked or to return to the clinic waiting room after diagnostic procedures, for example.

There is an aspiration to provide navigation 'from home to appointments' using public transport, parking and traffic information with advice on when to depart to minimize the time stress for patients needing to attend appointments.

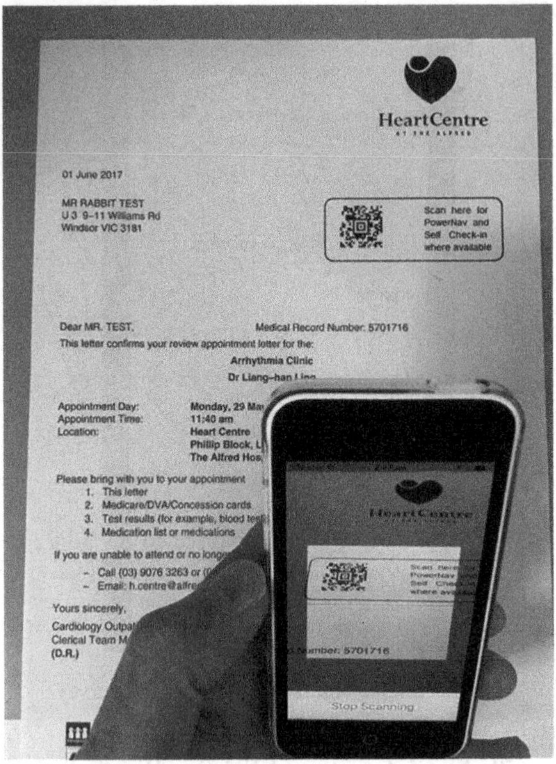

Fig. 4. Aztec scanning code on Patient appointment letter to launch app and set destination

5 Conclusion

During this project, an Australian public, government funded health service, partnered with smart phone app (PowerHealth Solutions) and wayfinding map developers (ID/Lab) to improve patient and visitor wayfinding in a complex hospital. This electronic system of wayfinding is an adjunct modality to the other more traditional methods of wayfinding in the hospital, in keeping with best practice that a strategy for wayfinding rather than a single intervention is more effective. The early outcome data is very encouraging with good uptake and feedback since the 30th November 2017 launch. Patient Experience Survey data will be analysed periodically to determine if the expected improvement in satisfaction is seen and attributable to the adjunct digital wayfinding modality. Further analysis using efficiency data and user feedback from within the app will continue to guide the refinement of the technology.

References

1. Cooper R (2000) Wayfinding for health care: best practices for today's facilities. AHA Press, Washington
2. Bryant B (2014) NHS England using technology to beat cost of missed appointments. https://www.england.nhs.uk/2014/03/missed-appts/. Accessed 27 May 2018
3. Matthews-King A (2018) NHS appointment no-shows cost health service £1bn last year, in independent. London
4. ABS (2011) Australian social trends: health services use and patient experience. Australian Bureau of Statistics, Canberra
5. Zimring C (1990) The costs of confusion: non-monetary and monetary costs of the Emory University Hospital Wayfinding System. Georgia Institute of Technology, Atlanta
6. Brown M et al (2015) A survey-based cross-sectional study of doctors' expectations and experiences of non-technical skills for out of hours work. BMJ Open 5:e006102
7. Calori C, Vanden-Eynden D (eds) (2015) Signage and wayfinding design: a complete guide to creating environmental graphic design systems, 2nd edn. Wiley, Hoboken
8. Pinchin J (2015) Getting lost in hospitals costs the NHS and patients, in Guardian. London
9. Carpman J, Grant M (eds) (1993) Design that cares: planning health facilities for patients and visitors, 2nd edn. American Hospital Publishing, Chicago

Assessing Symptoms of Excessive SNS Usage Based on User Behavior: Identifying Effective Factors Associated with Addiction Components

Ploypailin Intapong[1(✉)], Sarompom Charoenpit[2], Tiranee Achalakul[3], and Michiko Ohkura[1]

[1] Shibaura Institute of Technology, Tokyo 135-0064, Japan
nb15508@shibaura-it.ac.jp,
ohkura@sic.shibaura-it.ac.jp

[2] Thai-Nichi Institute of Technology, Bangkok 10250, Thailand
sarompom@tni.ac.th

[3] King Mongkut's University of Technology Thonburi, Bangkok 10140, Thailand
tiranee.ach@mail.kmutt.ac.th

Abstract. Social Networking Sites (SNSs) have exploded as a type of popular communication, suggesting exponential appeal. Unfortunately, one reason for their rise is the potential of excessive usage, which leads to negative consequences that are associated with addiction. In this research, we assessed the symptoms of excessive SNS usage by studying user behavior in SNSs. We employed the modified Internet Addiction Test (IAT) and the modified Bergen Facebook Addiction Scale (BFAS) to reflect addictive behaviors. We previously developed a data collection application and experimentally collected data from undergraduates in Thailand. In this article, we clarify the factors associated with addiction components (e.g., salience, mood modification, tolerance, withdrawal, conflict, and relapse), which are reflected by the questions of IAT and BFAS. We analyzed questionnaire and Facebook data by various methods. Our analytic results identified the effective factors associated with addiction components. Then we employed the Support Vector Regression (SVR) for evaluation. The outcome of our research can be applied for developing prevention strategies to increase the awareness of excessive SNS usage.

Keywords: Social networking site · SNS · SNS addiction
Addiction components

1 Introduction

Social Networking Sites (SNSs) have become an incredibly popular type of communication through which groups of people virtually meet, interact, and share similar interests [1]. In January 2018, the use of SNSs dramatically rose 13% from 2.8 billion to 3.2 billion active users, an astonishing total that equals 42% of the world's population [2]. This phenomenon suggests their exponential appeal. One reason, however, for such a rise is the potential of excessive SNS usage. Previous research argued that

S. Bagnara et al. (Eds.): IEA 2018, AISC 818, pp. 394–406, 2019.
https://doi.org/10.1007/978-3-319-96098-2_50

excessive SNS usage leads to such negative consequences as relationship problems [1] (e.g., disconnection from reality and damaged family relations), performance problems [1, 3] (e.g., reduced work productivity, lower academic performance, and poor time management), health-related problems [4] (e.g., eyestrain, headaches, body pain, and lack of sleep), and emotional problems [5] (e.g., negative effects on emotional health and brain development). Excessive and compulsive use of SNSs leads to symptoms associated with addiction [4] and have been described as addiction components with regards to behavioral addictions [4, 6].

In this article, we clarify the factors associated with addiction components. We employed the modified version of the Internet Addiction Test (IAT) and the Bergen Facebook Addiction Scale (BFAS) for measuring SNS addiction. The IAT and BFAS questions reflect addiction components, which consist of common addictive symptoms. We analyzed questionnaire and Facebook data to clarify the risk factors associated with these addiction components by various methods, including basic statistics, regression analysis, and decision tree analysis. The analytic results of these methods highlighted the factors associated with addiction components. We combined these results and selected the factors that might be effective for dealing with addiction components and employed the Support Vector Regression (SVR) classifier with the selected factors for evaluation.

The outcome of this research will be applied to develop prevention strategies to increase the awareness of excessive SNS usage.

2 Background

2.1 Our Previous Studies

We conducted this research to assess the symptoms of excessive SNS usage by studying the behaviors of SNS users. We divided our research into the following three main stages.

Stage 1 Collect SNS User Behavior Data. We previously designed and developed a data collection application as a tool for collecting SNS data from questionnaires and SNSs themselves [7, 8]. We organized our questionnaire that gathered SNS user experiences in three parts: (1) personal information, (2) SNS usage, and (3) SNS addiction. In the third part, we employed the Thai versions of IAT [9] and BFAS [5] and modified them for SNSs to reflect addictive behaviors. We implemented Facebook and Twitter quizzes that asked such questions as "How often do you tweet?" When users completed the quizzes, their Facebook and Twitter data were directly retrieved through SNS APIs [10].

We also experimentally collected data from undergraduates at the Thai-Nichi Institute of Technology (TNI), Thailand [11].

Stage 2 Clarify Characteristics of SNS Usage and Relationships with SNS Addiction. We statistically analyzed the obtained data from Stage 1 by various methods to clarify the characteristics of SNS usage and their relationships with SNS addiction [11, 12]. We separately analyzed the data from the questionnaires, Facebook, and Twitter. The analytic results identified candidates of effective factors that differentiate excessive users from normal ones [11, 12].

Stage 3 Assess Symptoms of Excessive SNS Usage. The analytic results in Stage 2 can be used for classifying users who are at risk for addiction. In this stage (explained below), we clarify the risk factors that identify the symptoms of excessive SNS usage and that can be applied for developing appropriate prevention strategies for individual users.

2.2 Addiction Components

SNS addiction shares similarities with other behavioral addictions [1, 4], which also have common addictive symptoms called addiction components [6]. Several screening questionnaires about SNS addiction exist in the literature [1]. In this study, we measured SNS addiction with two tests: the Internet Addiction Test (IAT) and the Bergen Facebook Addiction Scale (BFAS). The IAT and BFAS questions reflect these addiction components.

The IAT questions identify the following addictive symptoms [13, 14]:

1. **Salience:** Addicts feel preoccupied with the Internet, hide such behavior from others, display a loss of interest in other activities and/or relationships, and feel bored or depressed without the Internet.
2. **Excessive use:** Addicts engage in excessive behavior and compulsive usage and have difficulty controlling their time online. High ratings also suggest that addicts become depressed, stressed, or angry when such use is restricted.
3. **Neglecting work:** Performance and productivity decreased due to the amount of time spent online. Addicts may also hide or lie about such time.
4. **Anticipation:** Addicts think about being online and feel compelled to use the Internet when they are offline.
5. **Lack of control:** Addicts have trouble managing their time online. Family, friends, and co-workers complain about the amount of time a potential addict is spending online.
6. **Neglecting social life:** Addicts form new relationships with online users to cope with problems and/or reduce mental tension and stress.

The BFAS items reflect the following addictive symptoms [6, 15]:

1. **Salience:** SNS use becomes the most important activity in a person's life, leading to preoccupations and obsessions. SNS use tends to dominate the behaviors, the thoughts, and the feelings of addicts.
2. **Mood modification:** Addicts use SNSs to make themselves feel better, to alter their moods, and to create feelings of pleasure. SNS activities are mood-altering.
3. **Tolerance:** Addicts increase the amount of time they spend on SNSs to achieve the same feelings and mental states that occurred in their initial usage phases.

4. **Withdrawal:** This refers to the unpleasantness that occurs when SNS use is discontinued, slashed, or restricted.
5. **Conflict:** SNS use causes relationship problems: (1) personal relationships (family and friends), (2) working and education lives, and (3) other social activities.
6. **Relapse:** This refers to the failure to avoid using. Addicts quickly return to excessive behaviors after periods of control.

3 Effective Factors Associated with Addiction Components

3.1 Dataset

We experimentally collected data from 467 undergraduate volunteers from various universities in Thailand. After data cleaning, we had questionnaire data from 374 participants (80.09%), Facebook data from 221 (47.32%), and Twitter data from 74 (15.85%). Due to the small amount of Twitter data, we only used the questionnaire and Facebook data for clarifying the effective factors associated with each addiction component that is reflected by the questions of IAT and BFAS. There are 49 variables (Table 2): 27 from questionnaires, which are categorical data, and 22 from Facebook, which are continuous data.

3.2 Method

Since the data type of the questionnaires and Facebook variables are different, we separately analyzed their data.

1. The relationship among the questionnaire variables was analyzed by Cramver's V and the relationship among the Facebook variables was analyzed by a Spearman's correlation analysis.
2. To clarify the effective factors associated with addiction components, we employed various methods. Figure 1 illustrates our method for clarifying the effective factors associated with addiction components.

A T-test and ANOVA were used to examine the differences between the questionnaire variables and the scores of each addiction component. A Spearman's correlation analysis clarified the relationships among the Facebook variables and the scores of each addiction component.

Curve estimation is the process of constructing a curve, or mathematical function that has the best fit to a series of data points. We used a curve estimation to examine the relationship between variables and the scores of each addiction component.

Regression analysis examined the relationships between the sets of variables and the scores of each addiction component. A forward stepwise method was used with four different criteria for entry and removal: Akaike Information Criterion (AICC), F statistics, Adjusted R-squared, and Average Squared Error (ASE).

We also used a decision tree analysis to examine the relationships between the sets of variables and the scores of each addiction component. The CHAID and Exhaustive CHAID algorithms were used.

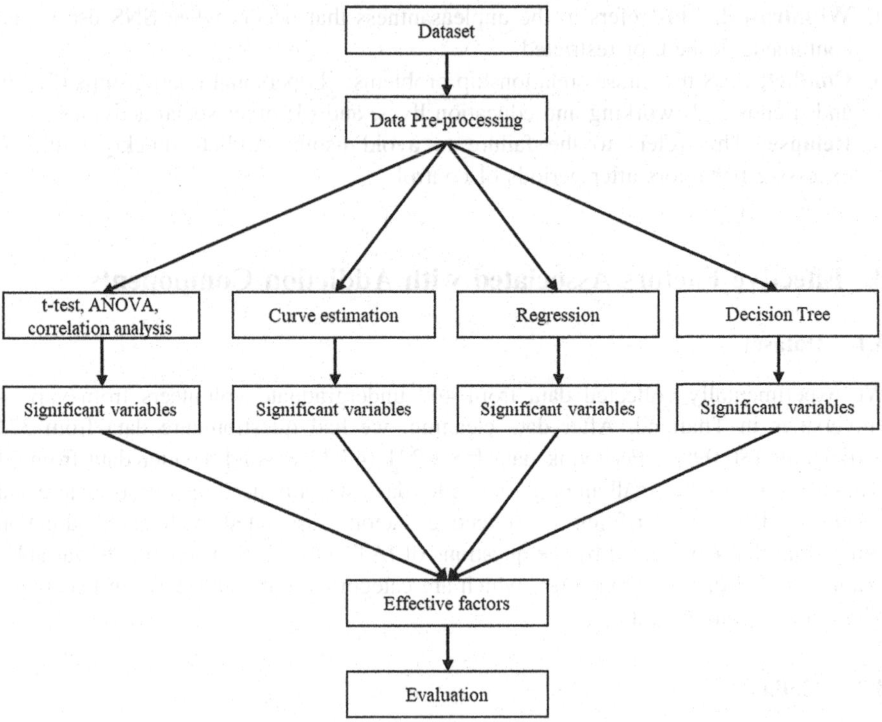

Fig. 1. Method for identifying effective factors associated with addiction components

Then we combined the analytic results of each method and selected the effective factors. Finally, we evaluated the selected factors using Support Vector Regression (SVR) to confirm which effective factors were associated with addiction components.

3.3 Results

The Cramer's V results indicated that questionnaire variables were independent of each other. However, the results of Spearman's correlation analysis indicated that some Facebook variables are dependent.

To clarify the effective factors associated with addiction components, we employed the following methods:

1. Basic statistics: T-test, ANOVA, and correlation analysis
2. Curve estimation
3. Forward stepwise method with AICC criterion
4. Forward stepwise method with F statistics criterion
5. Forward stepwise method with adjusted R-squared criterion
6. Forward stepwise method with ASE criterion
7. CHAID algorithm
8. Exhaustive CHAID algorithm

Table 1. Significant variables associated with IAT addiction components from results of methods 1–8

Variables	IAT addiction components					
	Salience	Excessive use	Neglecting work	Anticipation	Lack of control	Neglecting social life
Purpose						
Finding information					1	1
Playing games						1, 3, 4, 5, 7, 8
Making new friends						5, 7, 8
Keeping in touch				7, 8		
Expressing identity						
Sharing experiences	7, 8	7, 8		7, 8		7, 8
Killing time				7, 8		
SNS usage						
Time spent	1	5, 6	3, 5	1	5	5, 6
Frequency of use		1, 2, 3, 4, 5, 6, 7, 8	5	2, 3, 6, 7, 8	1, 2, 3, 5, 6, 7, 8	1
Length of use	1, 2, 3, 4, 5, 6	1, 3, 4, 5	3, 4, 5, 6	1, 3, 4, 5	1, 3, 4, 5, 6	1, 3, 4, 5, 6
Usage period						
06:00–09:00				7, 8		
09:00–12:00	1, 2, 5, 6	1, 2	1, 2, 3, 4, 5, 6	1		
12:00–13:00	1, 3, 5	1, 2, 3, 5, 6	1, 2, 3, 4, 5	1	1, 2, 3, 4, 5, 6	1, 2, 3, 4, 5
13:00–18:00			7, 8	1, 2, 3, 4, 5, 6, 7, 8		
18:00–24:00	1, 2, 3, 4, 5, 6	1, 2, 3, 4, 5, 6, 7, 8	1, 2, 7, 8		1, 2, 3, 4, 5, 6, 7, 8	1
After midnight						
Location						
Home	2, 3, 4, 5, 6, 7, 8	3, 5, 6	3, 4, 5, 6, 7, 8	7, 8	2, 3, 4, 5	
University	7, 8		7, 8	7, 8		3, 5, 6
Walking			7, 8			5
In vehicles				3, 4, 5, 6		

(*continued*)

Table 1. (*continued*)

Variables	IAT addiction components					
	Salience	Excessive use	Neglecting work	Anticipation	Lack of control	Neglecting social life
Activity						
Viewing feed						
Viewing friend's page	1, 2, 3, 4, 5, 6, 7, 8	1, 2, 5, 7, 8	1, 2, 3, 4, 5, 6, 7, 8	1, 2, 3, 5, 7, 8	1, 2, 3, 4, 5, 7, 8	1, 2, 3, 4, 5, 7, 8
Posting				1, 7, 8		
Commenting				1, 2, 7, 8		1
Updating profile		1, 2, 3, 4, 5, 6	1, 2, 3, 5, 6	1	1	1, 2, 3, 5
Messaging		7, 8				
Playing games		7, 8				
Facebook usage						
Friends	1	1	6		1, 2	1, 2, 5
Time spent	2, 3, 5, 6	2, 3, 4, 5, 6	2, 3, 4, 5, 6, 7	1, 2, 3, 4, 5, 6	5, 6	2
Length		2	2	1, 2		1, 2, 6
Frequency				7, 8		2
Sessions						1, 2, 3, 4
Posts		5, 6	5, 6	6	3, 5, 6	1, 7, 8
Comments	3, 5	5			5	1, 2
Replies	1	1				1, 2
Ratio of usage period						
06:00–09:00	2	7, 8		5	2, 7, 8	
09:00–12:00	5, 6		6			
12:00–13:00	1	1, 2, 5, 6			6	
13:00–18:00	5	5, 6		8	5	6
18:00–24:00	5	6			2, 6	5, 6
After midnight	2, 3, 4, 5				2, 3, 5	
Types of posts						
Status	1	1, 5, 6	1, 5, 6		3, 4, 5, 6	1, 2
Photos	6	6	6		3, 5, 6	1
Videos		6				5, 6
Links	6			6		1, 2, 3, 4

(*continued*)

Table 1. (*continued*)

Variables	IAT addiction components					
	Salience	Excessive use	Neglecting work	Anticipation	Lack of control	Neglecting social life
Ratio of posts						
Status	1, 2, 3, 4, 5	1, 2, 3, 4, 5, 6, 7, 8	1, 2, 5, 6	6	1, 2, 3, 4, 5, 6, 7, 8	1, 2, 3, 4, 5, 6, 7, 8
Photos	3, 5, 6	3			2, 3, 4, 5	
Videos	1, 2, 6, 7	5	2, 3, 4, 5, 6, 7	2, 6	6	1, 2, 5, 6
Links	2, 6, 7	5, 6, 8		6	6	2, 5, 6

Table 2. Significant variables associated with BFAS addiction components from results of methods 1–8

Variables	BFAS addiction components					
	Salience	Mood modification	Tolerance	Withdrawal	Conflict	Relapse
Purpose						
Finding information						
Playing games			7, 8	1	1, 2, 3, 4, 5, 7, 8	2, 3, 4, 5, 6, 7, 8
Making new friends		7			7, 8	
Keeping in touch						
Expressing identity			1, 2, 5, 7, 8			
Sharing experiences		7	7, 8			7, 8
Killing time	1, 2, 3, 4, 5, 6, 7, 8					
SNS usage						
Time spent	1	1, 2, 5	2	3, 5, 6, 7	1, 2, 3, 4, 5, 6, 7, 8	7, 8
Frequency of use	1	1	1, 2, 3, 4, 5		6, 7, 8	
Length of use	2, 3, 4, 5	1, 2, 3, 4, 5, 6, 7	3, 4, 5, 6		3, 4, 5, 6, 7	1, 3, 4, 5

(*continued*)

Table 2. (*continued*)

Variables	BFAS addiction components					
	Salience	Mood modification	Tolerance	Withdrawal	Conflict	Relapse
Usage period						
06:00–09:00		7		7, 8		
09:00–12:00		1		2, 5, 7, 8	2, 3, 4, 5, 7, 8	1, 2, 3, 4, 5, 6, 7, 8
12:00–13:00	1	1, 3, 5, 6, 7	1, 5, 6, 7, 8	3, 4, 5, 6	1, 2, 3, 5, 6	
13:00–18:00		1			1, 7, 8	1, 3, 4, 5, 7, 8
18:00–24:00	1, 3, 5	1, 7	1, 7, 8	1, 7, 8	1, 6, 7, 8	
After midnight						
Location						
Home	1			3, 4, 5	3, 5, 6	7, 8
University	3, 5, 6	3, 5, 6	3, 4, 5	6	3, 4, 5	3, 5
Walking			1, 2		1	5
In vehicles			3, 5, 6			5
Activity						
Viewing feed				7, 8		
Viewing friend's page		1	1, 6	1, 3, 5, 7, 8	1	5
Posting		1	1			
Commenting		1	1, 2, 3, 4, 5, 7, 8		1, 7, 8	
Updating profile		1	1			1, 2, 3, 4, 5, 6, 7, 8
Messaging						7, 8
Playing games		1		7, 8		
Facebook usage						
Friends	1, 2, 7	2, 5, 6	1, 2, 3, 5	1, 7, 8	1	5, 6
Time spent	1, 2, 3, 4, 5	1, 2, 3, 4, 6	1, 2	2, 6	2, 3, 4, 5, 6	3, 5, 6
Length	1, 2	1, 2	2, 7	2, 5	2	6
Frequency	1, 2, 3, 5, 6	5	2, 5	5	5, 6	
Sessions	1, 2	1, 2	1, 2, 5	1, 2, 3, 4, 5	1, 2, 5	5
Posts	1	1	1, 5, 7, 8	1	1, 7	5, 6
Comments	1, 2, 3, 4, 5, 6	2	2	1, 2	5	
Replies	1, 2, 3, 4, 5, 6, 7, 8	1, 2	1, 2, 3, 4, 5	1, 2	1, 6	5

(*continued*)

Table 2. (*continued*)

Variables	BFAS addiction components					
	Salience	Mood modification	Tolerance	Withdrawal	Conflict	Relapse
Ratio of usage period						
06:00–09:00	2, 3, 4, 5, 6, 7	1, 2, 3, 4, 5, 6, 7, 8	1, 2, 3, 4, 5	6, 7, 8	2	6
09:00–12:00	7, 8	6	6		6	
12:00–13:00	2, 5, 8			6	6	6
13:00–18:00	2, 7		3, 5	2, 6, 8	5	
18:00–24:00	1, 2, 3, 4, 5, 6	2	3, 5, 6	2, 3, 4, 5, 6		2
After midnight	2	2, 5	2	2, 5, 7, 8	5	
Types of posts						
Status	2	1, 5	1, 2	1, 2	1	6
Photos	2	1, 2	1, 5	1	1	6
Videos	3, 5	3, 5, 8			5, 6	
Links	7, 8	8	5			
Ratio of posts						
Status	1, 2, 5, 6	1, 2, 3, 5, 6	1, 2, 7	1, 7, 8	1, 5	6
Photos	2, 5, 6, 7, 8	2, 3, 5		2, 6	2, 5, 6	2, 6
Videos	6	2	8	6, 8		
Links	2, 3	2, 6	6		3, 6	3, 5

The analytic results of these methods identified the significant variables associated with IAT addiction components (Table 1) and BFAS addiction components (Table 2). Variables that have at least two methods with significant results were candidates for effective factors associated with addiction components.

For the IAT addiction components, the common variables associated with all the addiction components were length of use and viewing friend's page. The common variables associated with any five addiction components were the usage period from 12:00–13:00, accessing SNSs from home, time spent on Facebook, and the ratio of posting status on Facebook.

For the BFAS addiction components, the common variable associated with all the addiction components was time spent on Facebook. The common variables associated with any five addiction components were length of use, the usage period during 18:00–24:00, accessing SNSs from university, number of Facebook's friends, number of sessions, number of replies, the ratio of posting status, and the ratio of posting photos.

3.4 Evaluation

We employed the SVR classifier for our evaluation. We trained the model with the selected factors from the previous section to confirm the relationships between those factors and the addiction components. The evaluation results are shown in Tables 3 and 4. The correlation measures the strength of the relationship between the selected factors and each addiction component. Mean Absolute Error (MAE) averages the absolute differences between the predictions and actual values. The number of factors is the selected elements used for training the models.

The results of the trained model show a high correlation between the selected factors and each addiction component. However, the SVR models did much better on the training set than on the test set, which suggests an overfitting problem. Further evaluations with an alternative method will confirm the effective factors associated with addiction components.

Table 3. Results of training SVR model with selected factors for IAT addiction components

IAT addiction component	Correlation	MAE	Factors
1. Salience	0.891	0.315	16
2. Excessive use	0.903	0.272	20
3. Neglecting work	0.867	0.347	16
4. Anticipation	0.981	0.144	17
5. Lack of control	0.720	0.542	16
6. Neglecting social life	0.788	0.440	21

Table 4. Results of training SVR model with selected factors for BFAS addiction components

BFAS addiction component	Correlation	MAE	Factors
1. Salience	0.711	0.492	21
2. Mood modification	0.622	0.458	18
3. Tolerance	0.968	0.170	25
4. Withdrawal	0.809	0.431	23
5. Conflict	0.950	0.190	21
6. Relapse	0.901	0.293	15

4 Discussion and Conclusion

This research studied the behaviors of SNS users to clarify their characteristics of usage and relationships with SNS addiction for assessing the symptoms of excessive SNS usage. SNS addiction shares similarities with other behavioral addictions [1, 4], which also have common addictive symptoms called addiction components [6]. In this article, we employed various methods and combined their results to clarify the factors associated with addiction components, which are reflected by the questions of IAT and BFAS.

IAT was originally developed for measuring Internet addiction, and BFAS was developed for assessing Facebook. We modified them for measuring SNS addiction. Based on our previous studies [11], our finding identified similar classification results and a positive correlation between them. However, the IAT and BFAS questions reflect different addiction components [6, 13–15].

To clarify the effective factors associated with addiction components, we recruited undergraduates in Thailand and statistically analyzed their questionnaire and Facebook data by various methods. The analytic results found significant variables associated with IAT and BFAS addiction components as well as common significant variables. We combined these results and selected variables that have at least two methods with significant results as candidates of effective factors associated with addiction components. Even though the candidates of effective factors were different for each addiction component, some were shared, and common variables were associated with both IAT and BFAS addiction components. Then we evaluated the selected factors with an SVR classifier. The results of the trained model show good correlation between the selected factors and each addiction component.

Regarding our results, the factors associated with each addiction component are different. Therefore, prevention strategies and treatment should be developed based on risk factors. Based on our finding, the risk factors (e.g., time spent, frequency of use, length of use, period of use) can be prevented by reducing time online, scheduled specific use times, and finding alternative activities. In addition, the goal of prevention should be managing usage rather than prohibiting it [16]. Prevention strategies will fail unless the SNS users themselves admit their problems and seek help.

Empirical research suggests that generational and cultural differences exist in many aspects of SNS usage and addiction [1]. Our current study, which was limited to college-aged Thai SNS users, explored the factors that correlate with SNS addiction. It explained our analysis methods and obtained factors related to SNS addiction, which can be applied to future studies.

References

1. Kuss DJ, Griffiths MD (2011) Online social networking and addiction—a review of the psychological literature. Int J Environ Res Public Health 8(9):3528–3552
2. We Are Social. http://wearesocial.net. Accessed 20 Feb 2018
3. Al-Menayes JJ (2015) Dimensions of social media addiction among university students in Kuwait. Psychol Behav Sci 4(1):23–28
4. Andreassen CS, Torsheim T, Brunborg GS, Pallesen S (2012) Development of a Facebook addiction scale. Psychol Rep 110(2):501–517
5. MPh MM (2015) Validation of the Thai version of Bergen Facebook addiction scale (Thai-BFAS). J Med Assoc Thai 98(2):108–117
6. Griffiths MD (2005) A components model of addiction within a biopsychosocial framework. J Subst Use 10(4):191–197
7. Intapong P, Achalakul T, Ohkura M (2016) Collecting data of SNS user behavior to detect symptoms of excessive usage: design of data collection application. In: International symposium of affective science and engineering (ISASE 2016), Tokyo, Japan, 21–22 March 2016

8. Intapong P, Achalakul T, Ohkura M (2016) Collecting data of SNS user behavior to detect symptoms of excessive usage: development of data collection application. Adv Ergon Model Usability Spec Popul 468:88–99
9. Weerachatyanukul S (2015) Effect of internet addiction on students' academic performance of the second year students. HCU J 18(36):47–63
10. Intapong P, Achalakul T, Ohkura M (2016) Collecting data of SNS user behavior to detect symptoms of excessive usage: technique for retrieving SNS data. TNI J Eng Technol 4 (2):14–19
11. Intapong P, Charoenpit S, Achalakul T, Ohkura M (2017) Assessing symptoms of excessive SNS usage based on user behavior and emotion: analysis of data obtained by questionnaire. In: International symposium of affective science and engineering (ISASE 2017), Tokyo, Japan, 20–21 Mar 2017
12. Intapong P, Charoenpit S, Achalakul T, Ohkura M (2017) Assessing symptoms of excessive SNS usage based on user behavior and emotion: analysis of data obtained by SNS APIs. In: Meiselwitz G (eds) Social computing and social media. Human behavior, SCSM 2017, pp 71–83
13. Young KS (1999) The research and controversy surrounding internet addiction. Cyber Psychol Behav 2(5):381–383
14. Young KS (2017) IAT manual. https://www.flexiblemindtherapy.com/internet-gaming-and-other-problematic-screen-use-in-individuals-with-asd.html. Accessed 20 Dec 2017
15. Kuss D, Griffiths M (2017) Social networking sites and addiction: ten lessons learned. Int J Environ Res Public Health 14(3):311
16. Echeburua E, de Corral P (2010) Addiction to new technologies and to online social networking in young people: a new challenge. Adicciones 22(2):91–95

Fighting Panic with Haptics!

Sara Yxhage(✉) and Hanna Gustafsson

Department of Product and Production Development,
Chalmers University of Technology, Gothenburg, Sweden
sara.yxhage@outlook.com

Abstract. This paper presents some of the findings from the master's thesis project *Fighting Panic with Haptics!* (Gustafsson and Yxhage 2016) which aimed to answer the question of if and how one could make a product to help people suffering from panic attacks and panic disorder. Through the use of design and human factors tools such as personas, user journeys and user testing we came to the conclusion that it is possible to create a physical product that can help against panic attacks. Furthermore, our findings suggest that such a product should be designed to help the persons suffering from panic attacks break their internal focus by including the world around them since internal focus on thoughts and bodily reactions was found to be one of the core triggers of the attacks. Moreover, we could conclude that design theories and methodologies are effective tools also when trying to solve complex problems outside of the conventional product design sphere.

Keywords: Human factors · Design methodology · Design for mental health
Panic attack

1 Background

Panic attacks is an issue that affects many people, around one third of all people have experienced an attack during the past year alone, according to an estimation by Carlbring and Hanell (2011). A panic attack can be described as a period of intense fear, and the symptoms range from accelerated heart rate, shortness of breath, nausea, fear of losing control or going crazy and a fear of dying (Semple and Smyth 2013). Furthermore, Carlbring and Hanell (2011) explains panic attacks as a fight or flight reaction to an imaginary threat. The fight or flight reaction can be catalysed by for example stress, hyperventilation, fear of normal bodily reactions and misinterpretations of bodily symptoms. Some persons have recurrent panic attacks, and they often have a persistent worry about having another attack or worry about the consequences of having panic attacks. These can be diagnosed with panic disorder.

The available treatment options are different therapies and medications, and most commonly cognitive behavioural therapy and antidepressants or anxiolytics are used, either separately or in combination (Semple and Smyth 2013). In addition to the available treatment options, there might be room for other ways of helping people who struggle with panic attacks. We have investigated whether the designer's approach could be successful.

S. Bagnara et al. (Eds.): IEA 2018, AISC 818, pp. 407–416, 2019.
https://doi.org/10.1007/978-3-319-96098-2_51

As designers we have the ability of making complex issues tangible, both in our working process and in the resulting products. In the master's thesis *Fighting Panic with Haptics!* (Gustafsson and Yxhage 2016) we took on the task of investigating if a product could help against panic attacks, and if so, how it could be designed. This paper will explain how we used our concretization skills in our process and final product concepts in order to show the unique ways in which the designer's approach can help in such issues.

2 Questions and Objectives

This paper presents parts of the master's thesis project *Fighting Panic with Haptics!* (Gustafsson and Yxhage 2016) that intended to investigate the possibilities of creating a product that in some way helps against panic attacks. This included (1) investigating what problem, if any, we as designers could, and should tackle when creating a product against panic attacks, (2) investigating in what way this problem could be solved or minimized, and (3) develop one or more product concepts with the potential to help the user during a panic attack. This paper will focus on the objective (1) and (2), please see the research results in its entirety for the third objective in *Fighting Panic with Haptics!* (Gustafsson and Yxhage 2016).

3 Method and Theory

The project followed the structure of the ACD^3 framework (Bligård 2015) which divides the design process in to five main phases that one goes through in iterations. These phases are *Effect, Usage, Architecture, Interaction and Element* and they help to break down the problem by narrowing the scope from wide and open in the effect stage, to narrow and more precise in the element stage. Furthermore, the ACD^3 framework supports one in defining requirements and designs at a correct abstraction level at the right time. The focus in this project was to define the first three steps of the product development process, i.e., the *Effect* that the product should have on the sociotechnical system, the desired *Usage* of the product, and the product's *Architecture*.

As a part of the ACD^3 framework (Bligård 2015), we identified a problem in the panic attack that could be possible and successful for us as designers to solve and defined what desired effect a future product should have. These are necessary steps to ensure we create a product that is meaningful for the user. In order to do this, we needed to learn more about the origin of panic attacks, how they are experienced and what existing methods and strategies help the sufferers today.

Data was collected by studying medical literature such as *Ingen panik, frifrån panik ochångestattackeri 10 steg med kognitivbeteendevetenskap* (Carlbring and Hanell 2011) and *Oxford Handbook of Psychiatry* (Semple and Smyth 2013).

Having a designer's approach, the personal experiences of panic attacks are of highest concern. Thus, in addition to the literature study, a questionnaire survey with 116 respondents as well as 13 in-depth interviews with people suffering from panic attacks were performed. Moreover, we interviewed four experts within the area - a behaviourist, a psychotherapist, a psychologist and a psychiatrist.

The data collected from the interviews, the questionnaire survey and literature study were then synthesized and concretized by creating personas. Personas are used as a tool within user-centred design and they are descriptions of imaginary people who showcase archetypical characteristics of a user group. The user groups are based on a categorisation of common goals, motivations and behaviours (Cooper et al. 2014).

In addition, emotion journey maps were co-created with the interviewees in order to get an understanding of the phases and experience of a panic attack. The emotion journey map was inspired by a customer journey map which is a tool used to understand and address customer needs and pain points by visualising the process that a person goes through to accomplish a goal (Kaplan 2016).

After the data was synthesized, we identified the main problem for us to solve along with the desired effect a future product should have. The desired effect was used as a stepping stone for the later stages of the project, in which we defined the usage and architecture. By exploratory studies with sensory input and ideation of how the product should be used, we created concepts which were evaluated, in a non-panic situation, with people with experience of panic attacks.

4 Findings

This section describes some of the core findings made in the project *Fighting Panic with Haptics!* (Gustafsson and Yxhage 2016).

4.1 Personas - Making Large Data Personal and Tangible

In contrast to the medical descriptions of persons suffering from panic attacks, our personas were not mainly created by grouping symptoms or underlying conditions leading to panic attacks into different characters. Instead, we created the personas based on the interviewees' experiences and needs in the attacks according to design methodology.

This lead to five different personas with different attitudes, behaviours, triggers of panic attacks and experiences of panic attacks. Each persona was described by a persona story, intending to convey the difficult emotions of panic attacks in a gripping way to make the person taking part of it empathise with the persona. In this way we concretized our findings of the user study into relatable, yet comprehensive stories that respected the interview and questionnaire participants' integrity.

Using personas enabled us to transform the data into a tool for us to use in the subsequent design process. We analysed the needs for the different personas, finding needs not related to specific symptoms or medical conditions but rather to personality traits and personal experiences (see Table 1), thus opening up the product creation process to more than e.g. symptom relief. The personas followed us throughout the process and we used them to validate our ideas and final concepts.

Table 1. The needs of the personas

Persona	Attribute of persona	Need	Product's function
Heart attack Henry 	Thought he was having a heart attack the first time, still fears it and over interprets fast heartbeats	To face his fears and stop with his avoidant behaviour. Concretise therapy	Guide through and remove focus from the heartbeat symptoms
Ashamed Ashley 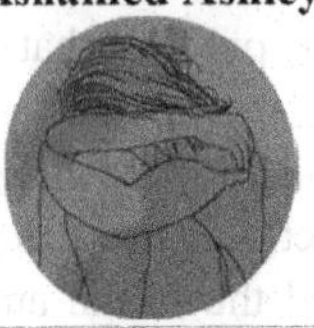	Afraid of making a fool of herself. Panic attacks manifested by a strong urgency of needing to go to the bathroom	Support, compassion, helping hand to get through a panic attack. Dare to go to school	Companion and guide through
Exhausted Emilia 	History of exhaustion disorder and started getting panic attacks. Accepts herself as she is now	Support in doing exercises to maintain mental health such as ACT and mindfulness	Help with staying in the present during an attack
Losing control Lisa 	Feels insecure and does not believe she can handle situations. Needs safety persons. Thinks she will lose her mind	Break the feeling of derealisation and depersonalisation. Help in getting free from safety persons	Snap out, give consistent feedback to action
Detached Daniela 	Feels detached and disoriented during an attack, hates the feeling	Break the feeling of derealisation and depersonalization. Recover control over her body	Snap out

4.2 Emotion Journey Maps - Helping People Make Their Emotions and Experiences Tangible

In order to understand the stages of a panic attack better, the interviewees were asked to draw their emotional journey in a panic attack. The participants described how they felt before, during and after a panic attack, both in terms of what sensory impressions, experiences and thoughts they had (heart palpitations, a feeling of losing control, etc.) as well as how much panic they felt. The emotional state was mapped on the vertical axis, ranging from feeling good, to complete panic, and on the horizontal axis the time was mapped, from before the panic attack starts to after it has happened.

Generally, the interviewees felt normal or slightly more anxious before the first signs of a panic attack started to appear in *phase 1: pre-panic attack*. Then there was usually an initial symptom such as trouble breathing, heart palpitations or worrisome thoughts (e.g. thoughts of being trapped in a confined space, being publicly shamed or fear of losing control) that scared them which we refer to as *phase 2: symptom scare*. At this stage, a common strategy was to try to calm oneself down in order to avoid having a panic attack. Sometimes it worked, and the panic attack could be fended off altogether, but other times the short relief of calmness was interrupted by a complete fall into panic, and they end up in *phase 3: panic attack*. In this phase the interviewees described the symptoms being very intense, and that this stage of complete panic and anxiety could last from a couple of minutes up to an hour. For those who experienced long panic attacks, the most intense panic would come and go in waves. After a panic attack, in *phase 4: recovery* and *phase 5: post panic attack*, the participants described a feeling of being very tired, drained of energy as well as ashamed.

The outcome of the emotion journey mapping gave us the insight that the needs are shifting during a panic attack, and thus that one single future product probably will not be effective throughout a panic attack (see Fig. 1). Similar to the personas, the emotion journey map followed us throughout the project as a means to validate different product ideas.

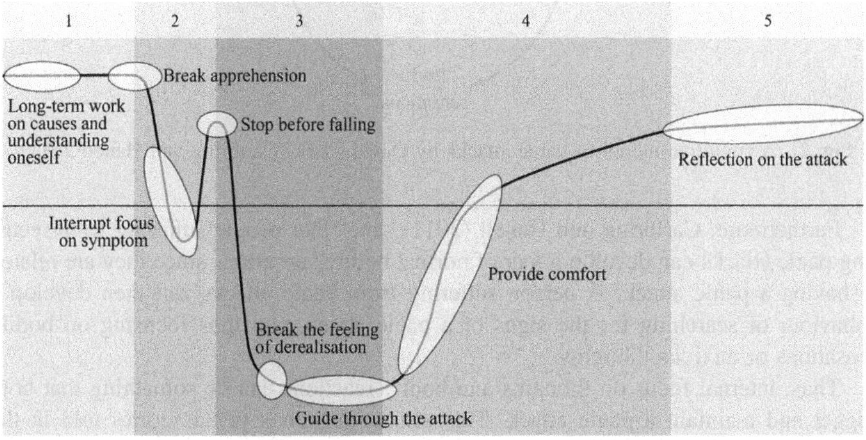

Fig. 1. Possible ways for products to help in the different phases of panic attacks: 1 pre-panic attack, 2 symptom scare, 3 panic attack, 4 recovery and 5 post panic attack.

4.3 The Desired Effect - Refocus

From the personas and the emotion journey maps we had learned that the needs in a panic attack differs both based on the personality of the sufferers as well as in different stages of the panic attack. We had also found a wide range of symptoms and underlying causes of panic attacks. However, we were still in search of a common denominator to serve as the main problem for us to solve.

A pivotal piece of the puzzle was found in David Clark's cognitive model of panic attacks (Carlbring and Hanell 2011). Clark's model propose that panic attacks originates from catastrophic misinterpretation of normal bodily sensations such as heart palpitations, loss of breath or simply getting a strange feeling. Clark's model is illustrated in Fig. 2 and implies that an internal or external trigger (e.g. heart palpitations, a scary or uncomfortable thought, or being in a big crowd if you are suffering from agoraphobia) is being perceived as a threat which causes the body and mind to react with anxiety. The anxiety then causes more bodily sensations, which are interpreted as catastrophic, and the perceived threat grows bigger, causing even more anxiety and bodily reactions. This vicious circle of anxiety then culminates in a panic attack.

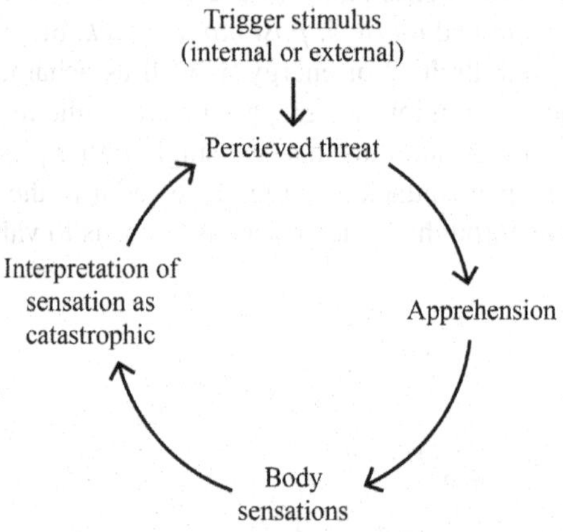

Fig. 2. A cognitive model of panic attacks by David Clark (Carlbring and Hanell 2011).

Furthermore, Carlbring and Hanell (2011) states that people suffering from recurring panic attacks can develop a fear of normal bodily sensations since they are related to having a panic attack. A person suffering from panic attacks can then develop a behaviour of searching for the signs of a panic attack, and thus focusing on bodily sensations or anxious thoughts.

Thus, internal focus on thoughts and bodily reactions can be something that both trigger and maintain a panic attack. This also corresponds to the stories told in the emotion journey maps, manifested by the symptom scare phase that is the starting point of the panic attack. The internal focus is therefore a problem that seems to be universal for people suffering from panic attacks, independently of symptoms manifested during the panic attack, and the underlying reasons for having the attacks. Some existing therapy methods for panic attacks, such as mindfulness, help you to consciously refocus and being present in the moment (1177.se 2012), thus strengthening the thesis of this approach being effective.

Since the focus on thoughts and bodily reactions was universal, we had found our common denominator. As designers, we could see potential here to create a product that could be helpful. We could therefore answer research question number one and defined our main problem that we wanted to solve to be the internal focus on thoughts and bodily reactions.

Another key factor emerged in the literature and interview studies, a term called safety behaviour. A safety behaviour is in short to rely on a certain behaviour, person or object to handle a difficult situation. This is seemingly helping the person in the situation, but relying on a safety behaviour, person or object is rather helping to maintain the focus on the threat, fuelling the fear and contributing to the vicious panic attack cycle. We saw a risk of creating a product that could be a safety behaviour and therefore concluded that the product should not completely block out the user from the panic attack, but rather help the user to include the world around them to reduce the internal focus. The philosophy of not pushing away, but to rather accept the emotions that are hard to handle is found in the Acceptance and Commitment Therapy (ACT) (psychologytoday.com 2011). The strategies used in ACT became a source of inspiration for the creation of a product for us.

Making a person refocus from the inside to the external world can be done in different ways. As our senses is the interface to the world around us, we saw that a sensory impression could be useful for refocusing, and as designers, we have the competence of creating impressions for the senses that affect our emotions and thoughts. Thus, the desired effect of a future product, as well as the answer to research question number two, was *to break the internal focus on thoughts and bodily reactions by including the external world around the user in the present. This should be done by giving the user a sensory input.*

4.4 Fighting Panic with Haptics!

Through ideation, prototyping, testing and evaluation with people with experience of panic attacks in a non-panic situation, three concepts emerged.

The concepts were all based on haptic impressions due to the long lasting yet not too intense distracting effect. Figure 3 shows how different impressions and activities affects the experience of pain, which was used as an analogue to a panic attack during exploratory testing.

However, the three concepts differed in terms of what persona they suited and at what phase of the panic attack they were deemed by the test participants to be the most effective. The concepts were based on three different principles: *distraction* to break the internal focus on symptoms and thoughts, *comfort* to recover faster and relieve feelings of shame, emptiness and sadness and finally, *externalisation* to show empathy for oneself and create a kinder inner dialogue. These concepts were designed into three products: *the distracting stone, the comforting hand,* and *the externalization birdie*, please see *Fighting Panic with Haptics!* (Gustafsson and Yxhage 2016) for more information about the concepts (Fig. 4).

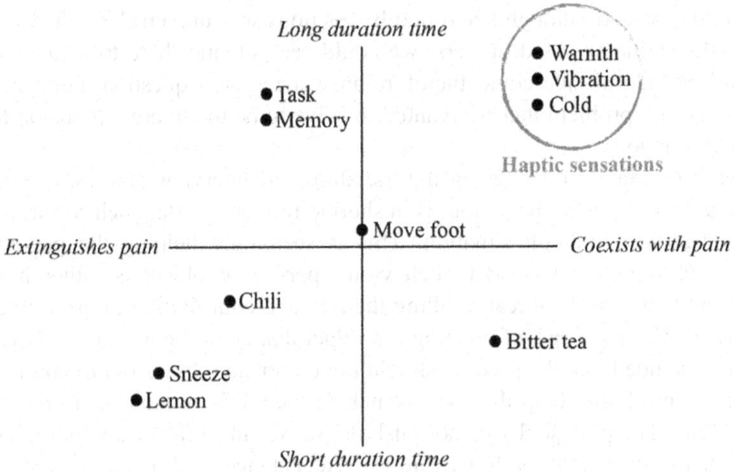

Fig. 3. The effects of tested stimuli concerning their duration time and relation to pain.

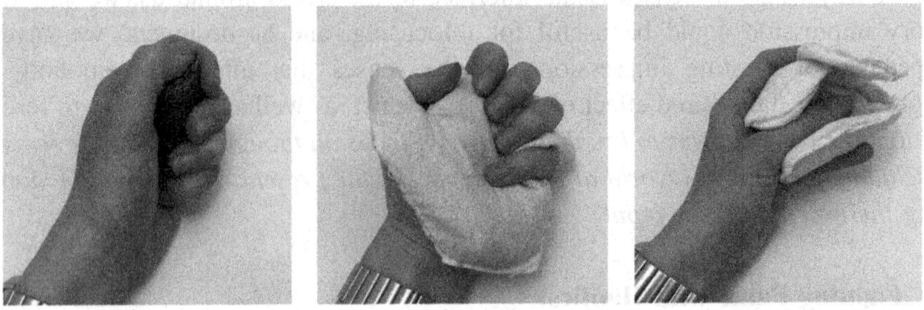

Fig. 4. The three concepts, *the distracting stone, the comforting hand* and *the externalization birdie.*

The stone focusing on providing distraction was deemed as being most effective in the initial stage of a panic attack, while the comforting hand was deemed to be most helpful in the later stages of an attack. The externalization birdie on the other hand did not have a clear focus area but was rather deemed to be able to help throughout the attack. The place for the concepts in the panic attack cycle is shown in Fig. 5. Aside from being more or less effective in different stages of the panic attack cycle, the different concepts also suit different kinds of personas and needs.

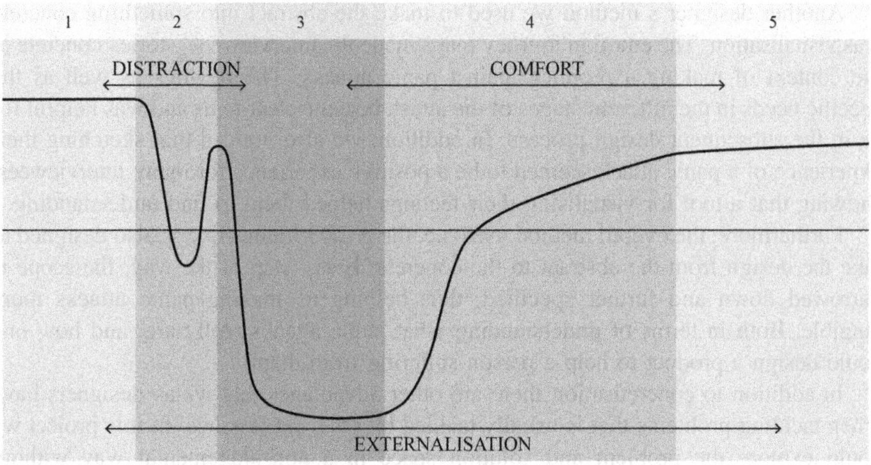

Fig. 5. The potential of products in the different phases of a panic attack, 1 pre-panic attack, 2 symptom scare, 3 panic attack, 4 recovery and 5 post panic attack.

5 Discussion

Panic attacks can be excruciating for the person suffering from them and it has so far been up to professions such as medicine and psychology to find solutions that help the sufferers. The master's thesis *Fighting Panic with Haptics!* (Gustafsson and Yxhage 2016) instead gives the designer's approach to finding solutions to these issues. What turned out to be one of the most useful designer skills was the ability to make the abstract into something concrete and we used this in many different forms.

The personas made the data from the user study tangible into five persona stories that resonated with the participants. User-centred design is sometimes called empathic design and it is evident that describing the experiences of panic attacks through the persona stories gives a greater impact to the reader than e.g. reading a list of symptoms. For us as designers, categorising the experiences in personas made the results from the user study easier to grasp and we found who we were designing for which enabled us to focus on their needs rather than e.g. medical conditions. This gave us a more open approach to solving the experience of panic attacks, rather than a specific medical condition, opening up the product development process and end results. By not being limited to medical categorisations, it is left to the user to use the product as they want, whether it is for a panic attack, a PTSD attack or to find calmness and manage stress with the products. This non-judgemental and empathic approach was met with appreciation by the participants and led to empathic product concepts.

Another designer's method we used to make the abstract into something concrete was visualisation. The emotion journey maps made the interviewees' stories concrete in the context of making a product against panic attacks. The feelings as well as the specific needs in the different stages of the attack became clear to us and was helpful for us in the subsequent design process. In addition, we also noticed that sketching their experience of a panic attack seemed to be a positive experience for many interviewees, showing that a tool for visualising their feelings helped them to find understanding.

Furthermore, the overall method used, i.e. the ACD3 framework, is also designed to take the design from the abstract to the concrete. Every step of the way, the scope is narrowed down and further specified, thus helping us making panic attacks more tangible. Both in terms of understanding what panic attacks really are, and how one could design a product to help a person suffering from them.

In addition to concretisation there are other advantages that we as designers have when tackling problems that is usually tackled by other professions. In this project we could explore the problem and solution space in a non-judgemental way without preconceived notions, but instead based our findings on the experiences of the users which we categorised into personas and emotion journey maps. For subjects such as mental health issues, having a non-judgemental approach is especially important as there are many taboos with both society and the sufferers blaming themselves for their problems. We believe that the non-judgemental way of investigating the problem space and creating solutions makes the designer's approach a successful approach for the issue of panic attacks and that it has the potential of being successful when applied to other mental health issues as well.

References

Bligård L-O (2015) ACD3 utvecklingsprocessen ur ett människa-maskinperspektiv, andra upplagan. Teknisk rapport, nr 96, Chalmers Tekniska Högskolan, inst. för Produkt- och produktionsutveckling, Gothenburg

Carlbring C, Hanell Å (2011) Ingen panik, fri från panik och ångestattacker i 10 steg med kognitiv beteendevetenskap, 2nd edn. Natur & Kultur, Stockholm, pp 21, 47–53, 70–73

Cooper A et al (2014) About face: the essentials of interaction design, 4th edn. Wiley, Indianapolis, pp 62–69

Gustafsson H, Yxhage S (2016) Fighting panic with haptics! Chalmers University of Technology, Gothenburg

Kaplan K (2016) When and how to create customer journey maps. Nielsen Norman Group. https://www.nngroup.com/articles/customer-journey-mapping

Psychologytoday.com (2011) Acceptance and commitment therapy—a mindful way to treat disorders. https://www.psychologytoday.com/blog/two-takes-depression/201102/acceptance-and-commitment-therapy. Accessed 24 May 2016

Semple D, Smyth R (2013) Oxford handbook of psychiatry, 3rd edn. Oxford University Press, Oxford, pp 358–362

1177.se (2012) Mindfulness. http://www.1177.se/Vastra-Gotaland/Tema/Halsa/Stress/Mindfulness/. Accessed 24 May 2016

Prevalence of Death Anxiety Among Emergency Medical Services (EMS) Staffs in Bam and Kerman Cities, IRAN. 2016

Akram Sadat Jafari Roodbandi[1], Vafa Feyzi[2], Farzaneh Akbari[3], Zeinab Rasouli Kahaki[4(✉)], Azar Asadabadi[5], and Nilofar Azadbakht[6]

[1] Research Center for Health Sciences, Institute of Health, Shiraz University of Medical Sciences, Shiraz, Iran
ergonomic.jafari@gmail.com

[2] Department of Occupational Health, Health Center of Saghez, Kurdistan University of Medical Sciences, Sanandaj, Iran
vafa.faizi@yahoo.com

[3] Student Research Committee, School of Public Health, Bam University of Medical Sciences, Bam, Iran
Farzaneh120374@gmail.com

[4] Ergonomics Department, Institute of Health, Shiraz University of Medical Sciences, Shiraz, Iran
zeinab.rasouli96@gmail.com

[5] Bam University of Medical Sciences, Bam, Iran
azar.asadabadi2009@gmail.com

[6] Member of the Student Research Committee of Kerman University of Medical Sciences, Kerman, Iran
niloufar.azadbakht@gmail.com

Abstract. Introduction: Death Anxiety includes the prediction of death or the fear of the death and death process in person or important individuals in a person's life, which can have a detrimental effect on the quality of life of individuals. Death anxiety has a high prevalence in some occupations such as EMS staff, due to occupational exposure to injured patients and dying patients. The purpose of this study was to determine the prevalence of death anxiety in EMS staff in Bam and Kerman covered by university of medical sciences hospitals.

Methods: This cross-sectional study was conducted in 2016. Sampling method was census in Bam and Kerman cities. Data were collected using Templar death anxiety scale (DAS, 1970). DAS is a tool that determines the attitude of the subjects about death. This scale is a self-report questionnaire consisting of 15 questions with yes/no choice answer. A higher score indicates higher death anxiety. SPSS software was used for statistical analysis with a significance level of 0.05.

A. S. J. Roodbandi—MSc in Occupational Health, PhD by Research Student.V. Feyzo—MSc in Occupational Health. F. Akbari—BSc of Occupational Health. Z. R. Kahaki—MSc in Ergonomics, PhD Student of Ergonomics. A. Asadabadi—MSc in Biostatistics. N. Azadbakht—MSc of Health Services Management.

S. Bagnara et al. (Eds.): IEA 2018, AISC 818, pp. 417–423, 2019.
https://doi.org/10.1007/978-3-319-96098-2_52

Result: 242 people participated in this study that 83.1% (n = 201) were male and 70.2% (n = 170) were married. Mean of age and job experience were 31.5 ± 6.32 and 7.2 ± 4.91 respectively. T-test showed that increasing age leads to a decrease in death anxiety (P-value = 0.003) death anxiety had significant relationship with experience of presence in earthquake zone and death of first-degree family member and death of friends and neighbors in Bam earthquake in 2003.

Conclusion: Encounter in major natural disasters such as earthquakes probably causes changes in concepts related to death and death anxiety in EMS staff. Therefore, further studies are recommended.

Keywords: Death anxiety · Emergency medical services · Earthquakes

1 Introduction

Anxiety disorders are one of the most common psychiatric disorders worldwide. Anxiety is a reaction to an unknown, intrinsic, vague and unconscious and uncontrollable source, and is associated with many factors. "Death anxiety" is a multidimensional concept and its definition is difficult and often defined as fear of dying yourself and others. In other words, "death anxiety" includes the prediction of death and the fear of the death and death process of important individuals in the person's life which can have a damaging effect on the quality of life of individuals [1].

Social psychologists believe that Social psychologists believe that anxiety is obtained through learning and can decrease the satisfaction of life of individuals [2]. On the other hand, people's ability to deal with anxiety is different. Anxiety can reduce physical and mental strength and cause many physical and mental problems [3]. Having a little anxiety about death is normal and it makes life worthwhile, but if this anxiety becomes too severe, it will undermine the person's adaptability and efficiency [4].

Yousefzadeh et al. (2014) studied the relationship between death anxiety and burnout in emergency medical staff. The results indicated that the relationship between death anxiety and burnout in the emergency medical staff was significant ($P < 0.01$) and findings have confirmed the direct effect of death anxiety on increased burnout in the group [5].

Death anxiety is one of the human tensions and has a high prevalence in some occupations, including emergency medical personnel, due to the association with the injured and the dying patients. According to the medical emergency is a community-based health management system that is coordinated with the entire health care system and staff are the most important asset of it [6].

Considering that medical emergencies are a community-based health management system that is coordinated with the entire health care system, and employees are the most important capital of the system [6], as well as taking into account the critical role of the medical emergencies The aim of this study was to determine the prevalence of death anxiety in emergency medical emergencies (115) under cover of Bam and Kerman University of Medical Sciences.

2 Method

Methods: This cross-sectional study was conducted in 2016. Sampling method was census in Bam and Kerman cities. Data were collected using Templar death anxiety scale (DAS, 1970). DAS is a tool that determines the attitude of the subjects about death. This scale is a self-report questionnaire consisting of 15 questions with yes/no choice answer. A higher score indicates higher death anxiety. SPSS software was used for statistical analysis with a significance level of 0.05.

The study population included all emergency medical services (115) from Bam and Kerman University of Medical Sciences Who were ready to participate in health research.

Data were collected using Templar Death Anxiety Questionnaire (DAS, 1970), which is a widely used and valid questionnaire on death anxiety. In this questionnaire, the attitude of the subjects is determined by the death anxiety [7–9].This scale is a self-report questionnaire consisting of 15 questions (yes-no). The answer to this question reflects the presence of death anxiety in individuals and the scores range from 0 to 15, with higher scores (higher than average, scores 8) indicating higher death anxiety.

This questionnaire was evaluated by Rajabi and Bahrani (2001) in Iran, which reported a reliability of 60% in Split Half method and reported an internal consistency coefficient of 73%. Templer (1970) obtained the re-test coefficient of the scale of 0.83.

This questionnaire was evaluated by Rajabi and Bahrani (2001) in Iran, which reported reliability of Split Half method and an internal consistency coefficient, 60% and 73%, respectively [10]. Studying the reliability of DAS, Templer (1970) reported 0.83 coefficients for test-retest reliability and 0.76 for internal consistency coefficient.

The data was collected by referring the researcher to EMS staff workplace. They completed the questionnaire. The researcher referred to EMS work place and after the explanation, they completed the questionnaire with full satisfaction. It should be noted that the completion questionnaire was carried out at times when there is no interference with their work and mission.

The collected data were analyzed by SPSS software version 18 and analyzed at 0.05. Simple single-variable monitors such as chi-square and t-test were used.

3 Result

In this study, 242 people participated that 83.1% were male and 70.2% were married. 175 of them (72.3%) were from Kerman and the rest were from Bam.

The mean age and work experience were 31.5 ± 6 and, 7.2 ± 4.9 respectively. Other information is given in Table 1.

It was found that 48.3% (117) of the participants in the study suffered from death anxiety. Using Chi-square test, there was a significant difference in the incidence of death anxiety in two cities of Kerman, Bam.

The chance of anxiety for deaths of people who live in Bam is 0.424 times higher than those who live in Kerman (Table 2).In other words, life in Bam is less likely to reduce death anxiety by 58%.

Table 1. Frequency and percentage of qualitative demographic variables.

Variables		Frequency (percent)
City	Kerman	175 (72.3)
	Bam	67 (27.7)
Gender	Man	201 (83.1)
	Female	41 (16.9)
Marital status	Single	72 (29.8)
	Married	170 (70.2)
Level of education	Diploma and associated degree	106 (43.8)
	Bachelor's degree and higher	136 (56.2)
Job	Nurse	141 (58.3)
	Nurse assistance	12 (4)
	First responder	44 (18.2)
	Technician	28 (11.6)
	Dispatch	17 (7)
Service location	Urban emergency	87 (36)
	Road emergency	98 (40.5)
	Control center	42 (17.4)
	Urban and road emergency	15 (6.2)
Shiftwork schedule	Yes	206 (85.2)
	No	36 (14.8)
Anti-anxiety/sedative drug	Yes	11 (4.5)
	No	231 (95.5)
Enjoying of job	Yes	216 (89.3)
	No	26 (10.7)

Table 2. Correlation coefficient of age and work experience with death anxiety in emergency and EMS staff.

Variable		Correlation coefficient		p-value	
Death anxiety	Yes	Age	Work experience	Age	Work experience
	No	−0.154	−0.048	0.017	0.46

The Chi-Square test revealed that emergency workers who had been involved in the Bam earthquake in 1382 were significantly less anxious than those who were not present at the time in Bam (P = 0.003).

The results of the Chi-Square test showed that the participants in the study who have seen the death of their first-degree family (father-mother-sister-brother-wife-child) in the 2003 Bam earthquake, or have seen the deaths their friends and neighbors, had death anxiety lower than other emergency and EMS staff. This relationship was significant and had p-value = 0.013 and p-value = 0.005 respectively.

Using T-test, it was found that increasing age decreases the death anxiety and this relationship was statistically significant (P-value = 0.003) but no significant relationship was found between work experience and death anxiety (P-value = 0.78) (Table 3).

Table 3. Relationship between death anxiety and age and work experience in emergency and EMS staff (independent T-test).

Variable			Mean	SD	t	p-value
Age	Death anxiety	Yes	30.54	5.7	2.4	0.017
		No	32.48	6.74		
Work experience	Death anxiety	Yes	7.16	5.1	−0.152	0.87
		No	7.24	4.6		

The results showed that there is a significant relationship between death anxiety and qualitative variables such as the city of service, the experience of Bam earthquake in 2003, the history of the death of the first degree family and the history of the death of friends and neighbors in the Bam earthquake in 2003. And from the quantitative variables, only the age variable has a significant effect on the depended variable.

Table 4. Qualitative variable with death anxiety.

Variable	χ^2	p-value
Marital status	0.38	0.53
Gender	0.937	0.33
City	4.51	0.03
Service location	5.27	0.15
Anti-anxiety/sedative drug	0.66	0.41
Enjoying of job	0.42	0.51
Education	0.2	0.9
Second job	1.31	0.85
Shiftwork schedule	2.62	0.75

4 Discussion

The present study aimed to investigate and identify the factors affecting the death of Emergency and EMS staff. The results of the study showed that 48.3% (n = 117) had death anxiety. Which showed a significant difference between the two cities of Kerman and Bam. Bagherian et al. [11] also concluded that Bam nursing students had less death anxiety than Kerman nursing students. The reason for this could be the 2003 earthquake experience, and the death history of family members and friends and neighbors. Because studies have shown that people who have seen death of their loved ones are less anxious than death [12, 13]. In addition, we can mention the differences in culture and customs in the two cities of Kerman and Bam. And the third factor can be different working conditions in terms of incident rate and different workloads in these two cities. According to the results of this study, Rohi et al. [14] conclude that nurses' work places have a significant relationship with their death anxiety. Also, the results of death anxiety in other cities and countries indicate the difference in death rate between cultures and different working conditions [15].

In this study, age is known as a factor affecting the level of death anxiety, so Christina et al. [16] and Deffner et al. [17] also refer to the same result in their study which, the degree of death anxiety decrease when age grows. But Masoudzadeh et al. [18] stated that age and death anxiety have no significant relationship with each other. Peter et al. [19] also found that younger nurses had more experience of death anxiety than older nurses.

In our study, there was no significant relationship between the work experience of the Emergency and EMS staff and death anxiety (Table 2). The reason may be that there are always faced with dying patients and death anxiety is always high, and death anxiety don't decrease by increasing work experience [20]. Hojjati et al. [21] stated that people with more work experience had less anxiety, which could be due to increased experience and normalization of the subject.

Relationship between gender variables, education, second job, marital status, etc. was not observed with death anxiety (Table 4). This is in line with other studies. In the study of Agha Jani et al. (2010), there was no significant relationship between death anxiety with variables such as age, sex, education level, workshift scugual, employment status, having children, shifting nurses and observing the death of patients [22].

5 Conclusion

Periodic screening and educational interventions are needed to empower people against anxiety.

Acknowledgement. This study was funded by the Bam Medical University. Thanks and appreciation from Bam University of Medical Sciences and all the people who have collaborated in this research.

References

1. Aghajani M, Raisi M, Heidari F (2013) The relationship between Quran and religious believes with death anxiety in heart patients. 3th provincial congress on Quran and health (in Persian)
2. Sharafaldin H (2009) Relationship of social anxiety and social support with subjective well-being in female students. J Women Cult 1(2):48–58
3. Rabie Siahkali S, Avazeh A, Eskandari F, Khalegh doost Mohamadi T, Mazloom S, Paryad E (2011) A survey on psychological and environmental factors on family anxiety of the hospitalized patients in intensive care units. Iran J Crit Care Nurs 3(4):175–180 (in Persian)
4. Hobfoll SE, Vaux A (1993) Social support, social resources and social contextinal. In: Codberger, S.Brezitz(end), nand book of stress. MACM, New York

5. Yusefzade I, Eshaghi M, Hojjati A, Zamanshoar E (2014) An investigation on intermediary role of self- determined needs in relationship of death anxiety with job burnout in emergency staff. Int Res J Appl Basic Sci 8(3):388–392
6. Saberi Nia A, Nekouei Moghadam M, Mahmoudi Meymand F (2013) Identify stressful factors causing dissatisfaction in pre-hospital emergency personnel in Kerman. Paramed Fac Tehran Univ Med Sci (Payavard Health) 6(6):489–497 (in Persian)
7. Ghasempour A, Sooreh J, Tohid Seid Tazeh Kand M (2012) Predicting death anxiety on the basis of emotion cognitive regulation strategies. Knowl Res Appl Psychol 13(2):63–70
8. Feliu T, Balle M, Sese A (2010) Relationships between negative affectivity, emotion regulation, anxiety, and depressive symptoms in adolescents as examined through structural equation modeling. J Anxiety Disord 24:686–693
9. Werner KH, Goldin PR, Ball TM, Heimberg RG, Gross JJ (2011) Assessing emotion regulation in social anxiety disorder: the emotion regulation interview. J Psychopathol Behav Assess 33:346–358
10. Rajabi GH, Bohrani M (2002) Item factor analysis of the death anxiety scale. J Psychol 5 (20):331–344 (in Persian)
11. Bagherian S, Iranmanesh S, Dargahi H, Abbaszadeh A (2009) Nurses' attitudes toward caring of dying patients in Cancer Center and Vali-e- Asr hospital in Tehran. J Nurs Midwifery (Razi) Kerman Univ Med Sci 9(1&2):8–14 (in Persian)
12. Wessel EM, Rutledge DN (2005) Home care and hospice nurses' attitudes towards death and caring for dying patients. J Hosp Palliat Nurs 7:212–218
13. Gama G, Vieira M, Barbosa F (2012) Factors influencing nurses' attitudes toward death. Int J Palliat Nurs 18(6):267–273
14. Rohi M, Dadgri F, Farsi Z (2015) Death anxiety in nurses working in critical care units of AJA hospitals. Mil Caring Sci 2(3):150–157
15. Aghajani M, Valiee S, Tol A (2010) Death anxiety amongst nurses in critical care and general wards. Iran J Nurs: IJN 23(67):59–68 (in Persian)
16. Christina A, Christian B (2010) The relationship of death anxiety with age and psychosocial maturity. J Psychol 130(2):141–144
17. Deffner JM, Bell SK (2005) Nurses' death anxiety comfort level during communication with patients and families regarding death, and exposure to communication education. J Nurs Staff Dev 21(1):19–21
18. Masoudzadeh A, Setareh J, Mohammadpour R, Modanloo kordi M (2008) A survey of death anxiety among personnel of a hospital in Sari. J Mazandaran Univ Med Sci 18(67):84–90 (in Persian)
19. Peters L, Cant R, Payne S, O'Connor M, McDermott F, Hood K, Morphet J, Shimoinaba K (2013) How death anxiety impacts nurses' caring for patients at the end of life: a review of literature. Open Nurs J 7:14
20. Sadeghi H, Hoseinzade M, Mehrabi F, Bahrami M, Frouzan R (2017) Death anxiety in students of medical emergency and emergency technicians of Sabzevar in 2013. J Sabzevar Univ Med Sci 24(6):71–79 (in Persian)
21. Hojjati H, Hekmati Pour N, Nasrabadi T, Hoseini S (2015) Attitudes of nurses towards death. J Health Care 17(2):146–153
22. Aghajani M, Valiai S (2010) P01-148-"Death anxiety" in special and general ward's nurses. Eur Psychiatry 25:356

The Role of Ergonomics in the Design of an Intravenous Therapy (IV) Set for Neonatal Intensive Care

Armagan Albayrak[1(✉)], Myra Vreede[1], Luuk Evers[2], and Richard Goossens[1]

[1] Delft University of Technology, Delft, The Netherlands
a.albayrak@tudelft.nl
[2] University Medical Centre Utrecht, Utrecht, The Netherlands

Abstract. Intravenous therapy (IV therapy) refers to a treatment in which medication and nutrition is provided patients via a system that infuses these fluids directly into the patient's bloodstream. Patients that are admitted to intensive care facilities, often require infusion of multiple fluids as physical systems fail to operate as intended. IV therapy administers drugs and nutrition into the patient's bloodstream by connecting tubing, through which fluids can be transported, to the vein. The smallest fluctuations in dosages or drug administration delays have severe consequences for the new-born infants in the Neonatal Intensive Care unit. This study focuses on the development of an innovative IV set based on a new principle for merging various drugs and nutrition, called "Tulive".

A user-centred approach was applied during the whole design process, including analysis of the complex healthcare context, literature study, observations & interviews with the different stakeholders and mainly its key user the 'nursing staff', contextual design models & methods, generative sessions and multiple user feedback sessions, etc.

Main themes that were uncovered include: product-user communication in terms of simplistic & unambiguous use cues, a sense of feedback regarding product related actions on multiple levels, user confidence to positively influence product interaction and inconveniences with fluid flow distinction.

Multiple design iterations have been performed where the needs, expectations and limitations of the user in context played a central role and motivated the design decisions made during the process. All these insights were translated and integrated in the final design proposal, "Tulive". A name that relates to the aesthetic design, the tulip shape that aims to conceal the complexity and suggest operation in a user-friendly way.

The value of user-centered design lies in the increased product acceptance that can be achieved after new product implementation. In designing for healthcare, barriers that impede successful product acceptance can most often be attributed to the fact that majority of the medical products are not designed according to the needs, expectations and limitations of the medical professionals. Centralizing the intended user of the product during the design phases and allowing user input and contextual factors to shape the design in a systematic way will help evade these barriers and support product acceptance of any innovation.

Keywords: User-centered design · Ergonomics · Healthcare innovation

S. Bagnara et al. (Eds.): IEA 2018, AISC 818, pp. 424–436, 2019.
https://doi.org/10.1007/978-3-319-96098-2_53

1 Introduction

1.1 A Subsection Sample

The Neonatal Intensive Care Unit (NICU) is the hospital department that cares for pre-term infants or new-born that have difficulty surviving by their own mechanisms [1]. Inside incubators, that simulate the environment inside the womb, these infants are connected to several systems that measure and control bodily parameters. These include heart monitors, respiratory devices and an Intravenous Therapy (IV therapy) system [1].

Specific patient characteristics do not always influence product design of medical products. However, due to the underdeveloped state of the skin and the highly limited internal volume, NICU patients, require an intra venous system that is tailored to their needs.

IV therapy refers to a treatment in which medication and nutrition is provided to the patient via a system that infuses the fluids directly into the patient's bloodstream. Patients that are admitted to intensive care facilities, often require infusion of multiple fluids because physical systems fail to operate as intended. Administering multiple fluids into the patient using IV therapy with a single point of entry is referred to as "multi-infusion". Multi infusion is realized by merging various infusion flows, containing different fluids to compose a stream of fluids that will eventually enter the patient's vein. Merging multiple infusion lines is accommodated by an IV set with multiple inlets and a single outlet. IV sets connect the fluids to the catheter that enters the patient's vein. In NICU patients, the vessels of the forearm, which is used as an entry point in adults and children, will tear from catheter insertion. Therefore, the umbilical orifice serves as the entry point in most cases. Automated systems are used to administer the different fluids encased in bags or syringes. For an overview of the IV system (see Fig. 1).

Drug infusion in neonates, is characterized by extremely low infusion volumes. A pre-term infant that weighs 1 kg, can take in 60 ml of fluid on its first day. This results in a flow rate of 2,5 ml/h with which the totality of fluids can be infused into the patient. In neonatal intensive care, common flow rates lie between 0,5 ml/h and 10 ml/h [1]. To put that into perspective, the adult ICU employs flow rates of 10 ml–2 l/h, depending on the fluid to be infused.

An IV set (See Fig. 2) typically makes a distinction between inlets to connect critical medication and non-critical medication or nutrition. Critical medication is defined as drugs that have a direct effect on cardiac parameters such as heart rate and blood pressure. These critical fluids need to be administered at extremely low flow rates and are therefore connected to the IV set as close to the entry point as possible. The distant inlets are used for non-critical medication and nutrition that can be administered at higher flow rates. They allow the comfort of connection outside of the incubator. The third type of inlet is specifically reserved for fluids that consist of large molecules, such as lipids, as they cannot pass through the filter.

Multi-infusion has been a subject of research for many years. Dosage errors of pump administered drugs and the chemical reactions of interacting fluids are two of the most commonly depicted aspects. Within the field of multi-infusion, the actual merging

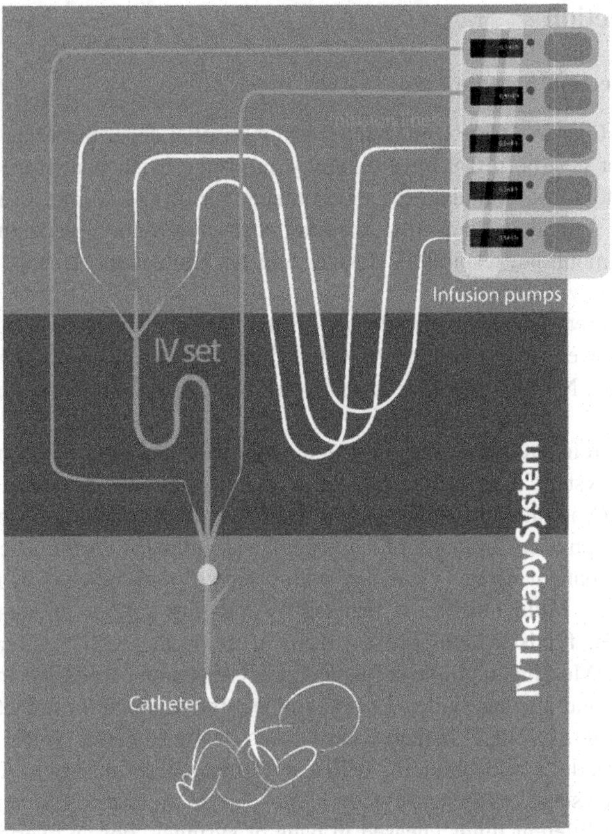

Fig. 1. A schematic overview of the intravenous therapy system in neonatal intensive care.

of fluids by the IV set has deserved little attention until the last five years. In 2009, the department of clinical physics of University Medical Centre Utrecht (UMCU) decided to investigate the characteristics of the flow within the IV sets. The outcome of In-Vitro studies showed undesirable *flow rate fluctuations* and *prolonged standardization time* of drug administrations created by the confluence of multiple fluids.

As a result of these two factors the intended administration set by healthcare professionals and the actual administration to the patient are unequal. This discrepancy creates both hazards in patient safety and implicates a limited sense of control for the medical professionals that operate and rely on the IV set. The flow rate fluctuations that cause inaccurately administered dosages and the delays that impede immediate administration can be attributed to two phenomena. Merging various fluids through the currently used branching system causes ***back-flow*** into infusion lines that carry no fluids or fluids with smaller flow rates (see Fig. 3). These differences in flow rates also create a ***pressure threshold***, the need for pressure build-up inside the infusion lines with smaller flow rates before confluence of the two streams can be realized (see Fig. 3).

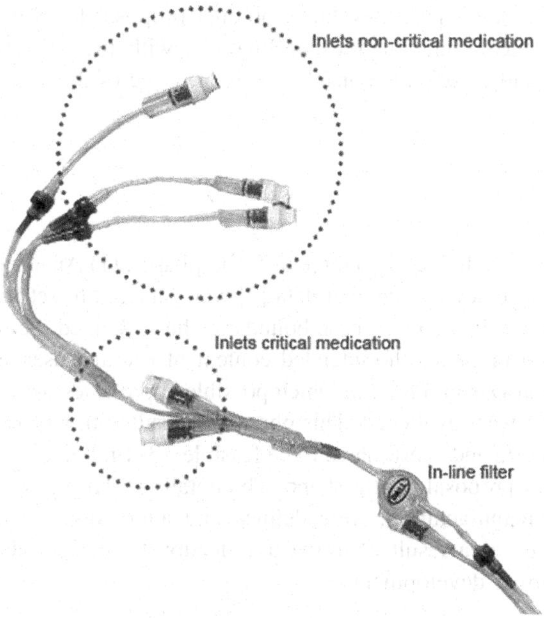

Fig. 2. Standard issue IV set.

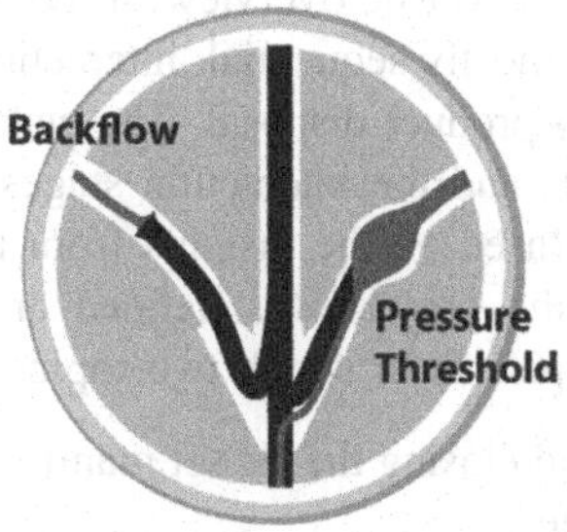

Fig. 3. The undesired effects of converging infusion fluids in conventional IV sets.

Both phenomena cause delayed and inaccurate administration and call for an improved manner of merging various streams to eliminate these consequences.

Based on the in-vitro studies UMCU had developed a concept the so-called 'Innofuse' which is based on the principle of pre-filling the inlets of low flow rate fluids. Innofuse adds a manifold, a rigid component with integrated fluid channels, to the IV set that allows the user to retain throughput of fluids to the patient for the purpose of pre-filling the channels up to the confluence point. Throughput can be restored after pre-filling and newly added low flow-rate fluids will enter the main flow almost immediately. Proof of principle was gained by laboratory testing. Proof of concept required a product proposal applying the principle as well as incorporation the requirements of the context, user needs and expectations.

The aim of this study is to develop a product proposal for a new IV set which integrates the technical functionalities of Innofuse with the requirements from the context, the users and ergonomic considerations of usage of the IV set in the specific context of NICU.

2 Methods

This design project was divided into three different phases. (1) Analysis Phase in which parallel to literature review, contextual data [2] was gathered to set the boundaries for the project (list of requirements). These boundaries have defined the scope and design space for a solution that suits the intended context of use and user needs & expectations. (2) Contextual Design Phase in which possible approaches were investigated and preliminary designs were evaluated. This phase has resulted in a product proposal that meets the main criteria and is estimated to be feasible. (3) Embodiment Phase, referring to the final product proposal taking shape. The details regarding development of the design, including manufacturing, were defined and alternations to the final product proposal were made as a result of target user feedback sessions along with recommendations for further development.

2.1 Analysis Phase

The different stages of human-product interaction were mapped in a product journey. The product journey provided a holistic overview of the stakeholders that interact with the IV set at some level. The mostly sequential, interactions define product interaction stages from product origin to product disposal. The stakeholders were defined by an extensive stakeholder analysis and the interaction stages through NICU observations and stakeholder interviews. Based on the insights from literature and contextual data collection [2] the scope of the project was defined and design opportunities were identified resulting in four topics from which two topics are included in this study;

(1) Solutions for opening and closing the IV set manifold and the distinction between the open and closed state.
(2) Solutions for providing feedback on the completion of the pre-filling action.

2.2 Contextual Design Phase

The opportunities identified in the previous phase were investigated further in this phase using the so-called ‘morphological chart’ method. Morphological chart is a method to generate ideas in an analytical and systematic manner. Usually, functions of the product are taken as a starting point [3]. Using this method allows for generating creative solutions for specific problems without the interference of feasibility boundaries caused by looking at the complete design challenge.

Based on the morphological chart multiple concepts were developed from which one proposal was selected for further elaboration in the next phase. The selected design proposal met the main criteria set in the analysis phase, matched the requirements of the use context and was aligned with users needs & expectations.

2.3 Embodiment Phase

The selected design proposal was elaborated further in this phase by focusing on three levels of use, namely:

- How to guide correct use?
- How to impede incorrect use?
- How to indicate throughput of critical medication?

The design should accommodate (by use cues) showing the user whether throughput to the patient is enabled. Although the key user is expected to be cautious in using the pre-filling function, it is necessary to show the user the distinction between the pre-filling and administrating states. To determine what aspects of the design can help trigger the correct positions of throughput for the correct actions, a creative session was set up. Based on the input of this creative session three different prototypes were developed and evaluated by the key users.

A simulated NICU context and a representative tasking script were used to conduct structured user testing. The goal of the user test was to evaluate the different prototypes by answering the following questions:

- What is the user's attitude towards the different prototypes and what drives their reasoning selecting one over the other?
- Is product use clear from the design?
  - Open the leaves to connect to the IV set
  - Close the leaves to start pre-filling
- Do the instructions suffice?

Three true scale prototypes were used so the tasked interactions could be thoroughly simulated. For the three different prototypes; color, shape and sign as variations on the selected design proposal (see Fig. 4).

The different prototypes and the context of the user test is shown underneath (see Fig. 5).

Whether the three prototypes evoked intuitive use and understanding of operation was tested during the sessions. This cognitive aspect was measured by tasking the participants with representative actions to be performed with the IV set. A brief instruction manual for the product proposal was made available to the participants and as they were asked to perform the tasks, their interaction was recorded. The product use was evaluated through video analysis of prototype handling.

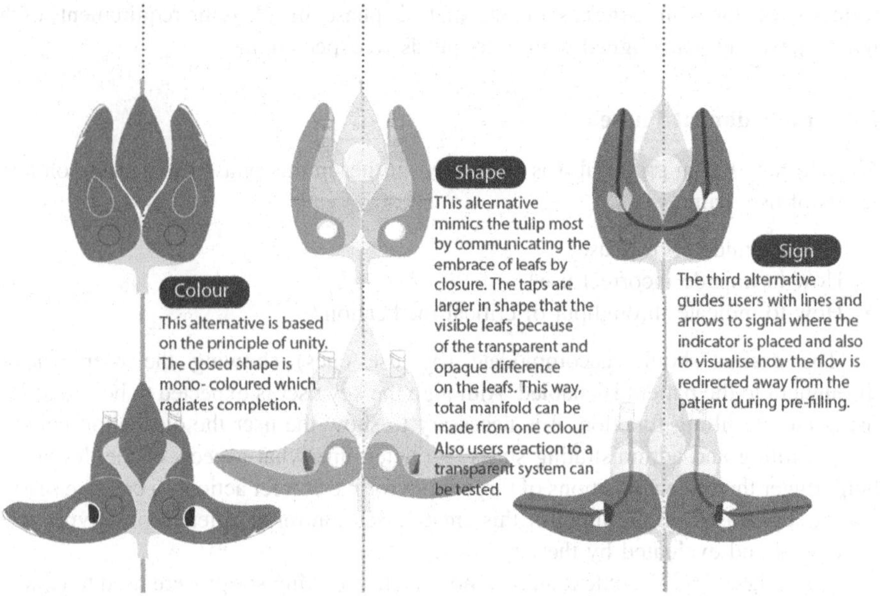

Fig. 4. Design proposals for prototypes.

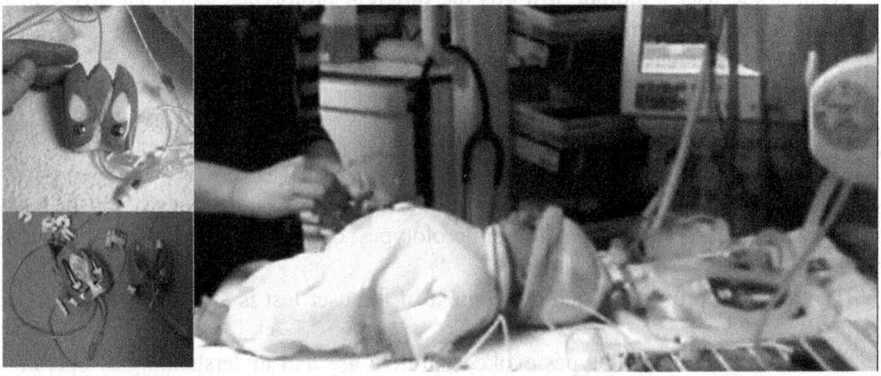

Fig. 5. Left: prototypes used in feedback session. Right: NICU simulated context (skills-lab).

3 Results

3.1 Analysis Phase

Stakeholder Analysis. The scope of the project was divided into three levels. The application field, the context and the product use. A graphical overview of the scope was made (see Fig. 6).

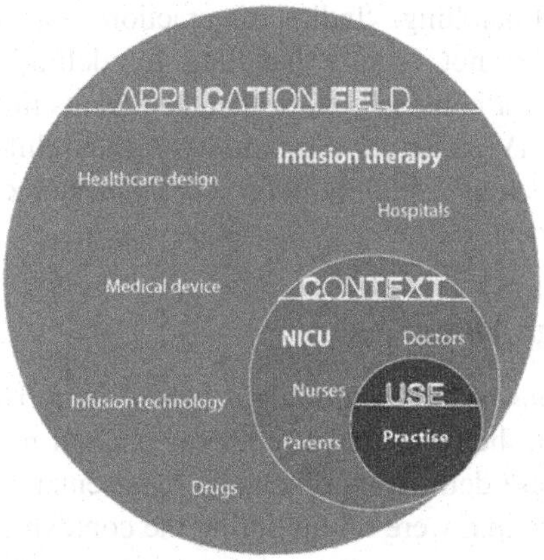

Fig. 6. Scope of the study.

Within the context three main stakeholders are represented. Personas were created based on stakeholder interviews. The NICU nurse was identified as the key user.

Product Journey. A product journey was defined describing the different interactions with the product performed by different stakeholders (see Fig. 7). It distinguishes different use phases starting with the selection of the product for purchase, when it exists on the market to when the final user disposes of the product. The different stages mentioned in this overview were mainly derived from analysis of the existing procedure protocols (protocols of UMCU used in 2014) for medical staff regarding infusion sets. The stage "pre-filling" is highlighted because this is the main task of the key user and the focus of this project.

According to the product journey shown in Fig. 7 the key user, has the most interactions with the IV set and therefore positioned at the top of the list. The interactions were divided into three categories. "Direct interaction" referring to the handling

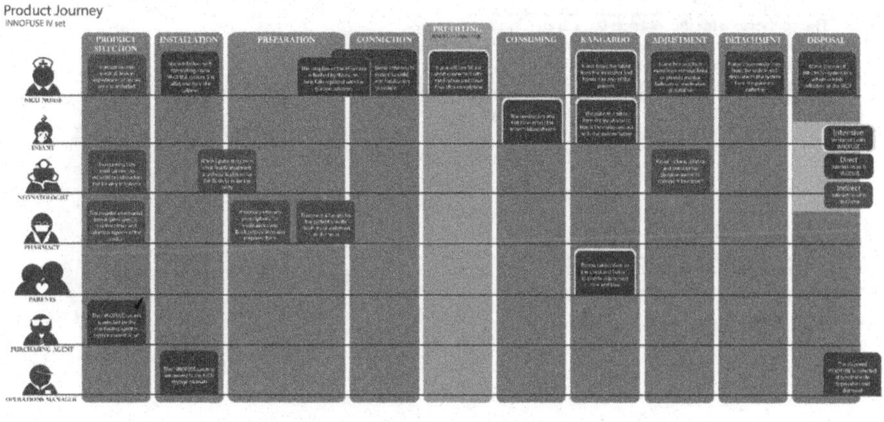

Fig. 7. Product Journey IV set.

of the IV set or physical handling. "Indirect interaction" was defined as an interaction through senses that does not require handling by definition, for example visual assessment of the IV set for consultation, or that the parents in context were confronted with the presence of the IV set in close proximity to their child. A third category, was called "Intensive interaction", highlighting the key interactions that significantly influence product design.

3.2 Contextual Design Phase

Combinations of solutions of the morphological chart comprised several concepts that were evaluated based on the program of requirements and matched to the user needs and expectations that were defined. A schematic representation of the iteration phases and the design guidelines that were set up during the contextual design phase is shown in Fig. 8.

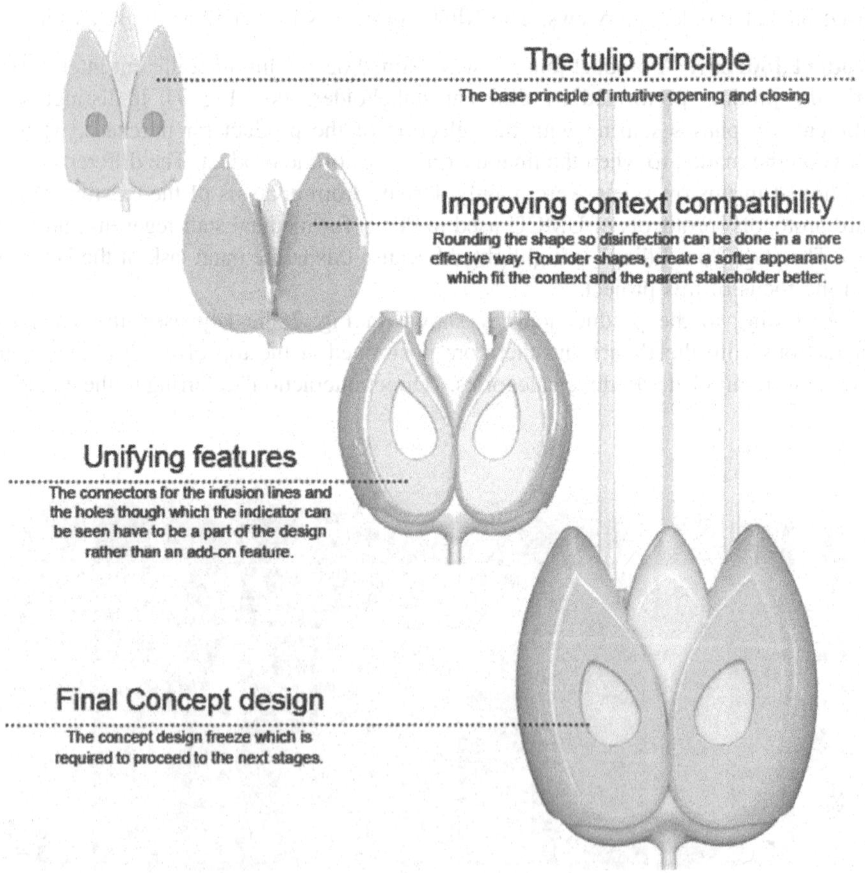
The tulip principle
The base principle of intuitive opening and closing
Improving context compatibility
Rounding the shape so disinfection can be done in a more effective way. Rounder shapes, create a softer appearance which fit the context and the parent stakeholder better.
Unifying features
The connectors for the infusion lines and the holes though which the indicator can be seen have to be a part of the design rather than an add-on feature.
Final Concept design
The concept design freeze which is required to proceed to the next stages.

Fig. 8. Schematic representation of the design iteration phases.

3.3 Embodiment Phase

An impression of the user testing can be found in Fig. 9.

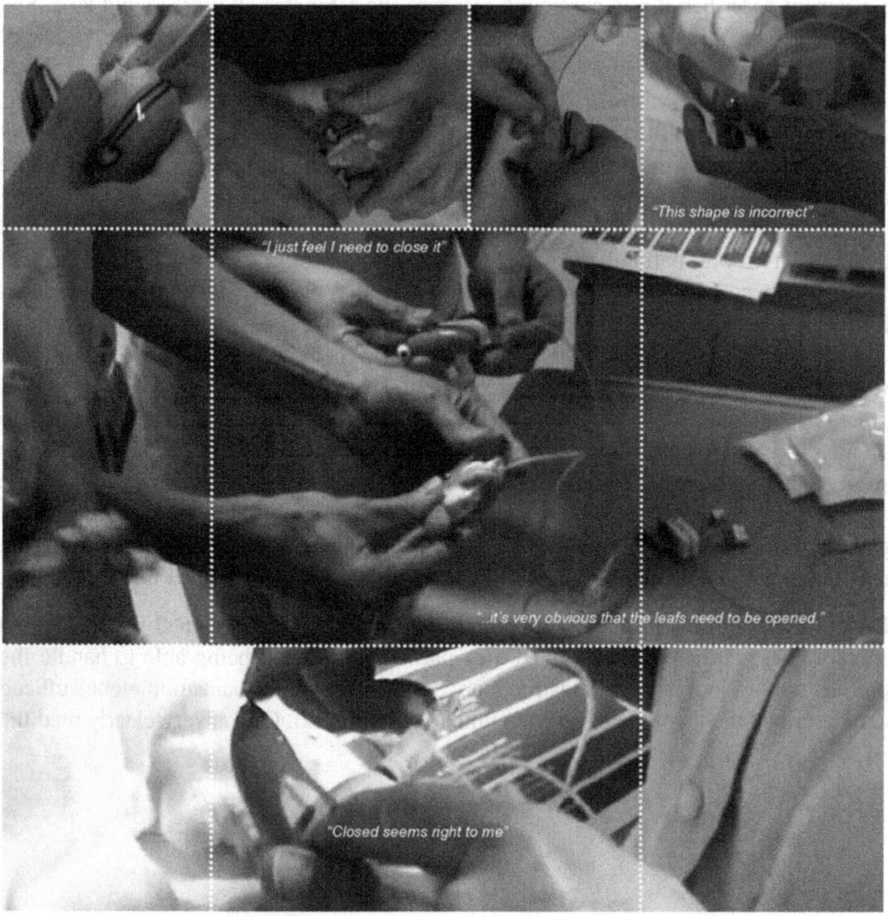

Fig. 9. Impression of the user testing session.

The user test included 23 participants. The results of the user test were clustered according to the goals defined for the user testing.

What is the User's Attitude Towards the Different Prototypes and What Drives Their Reasoning Selecting One over the Other? The participants were asked to pick up the models first to assess which one they would choose to use. Eight out of thirteen participants expressed their preference selecting the COLOUR model (see Fig. 4). None of these participants described their reasoning behind this selection to be influenced by the color scheme particularly but all used phrases that could be reduced to communicating a sense of simplicity.

Is Product Use Clear from the Design?
Opening the leaves to connect to the IV set. Without the specific task to connect tubing to the critical inlets, 80% of all users opened at least one of the leaves. Most users very clearly opened the leaf by pushing it towards the point that it could move no more. One participant described it as a correcting measure she estimates all users would take. She stated: "I think that if you started pre-filling with the leaf not really opened, you will open it after a while because you will look for the indicator."

Closing the leaves to start pre-filling. Some of the participants described a cautious attitude towards pre-filling. One nurse described she would feel anxious to perform that action because she would feel she is pushing the medication towards the patient. The three other participants in this session felt seemingly less anxiety towards pre-filling and stated they understood that this was what the leaves are for. One participant stated that he felt it was particularly clear that pre-filling would take place with the leaf opened. Others described they would make sure they would not close it before performing this action.

The purpose of the indicator as a feedback mechanism for pre-filling completion was eventually understood by all participants. A preventative measure taken in current practice would have to be stopped with the introduction of this product proposal. There seemed to be a positive attitude towards the de-coloring of the indicator, and inquiry showed that participants estimated that this signal would suffice.

Users did describe the desire to manually prefill the manifold as the tactile feedback gained from pushing the plunger of the syringe would be an extra conformation for the user that pre-filling is completed, and medication has not been pushed into the patient.

Do the Instructions Suffice? The way the instructions were designed was perceived as very clear. Participants explained they particularly valued being able to handle the product next to looking at the instructions. Some stated the visualization alone sufficed and described they would not take the time to read the text. Others very clearly read the

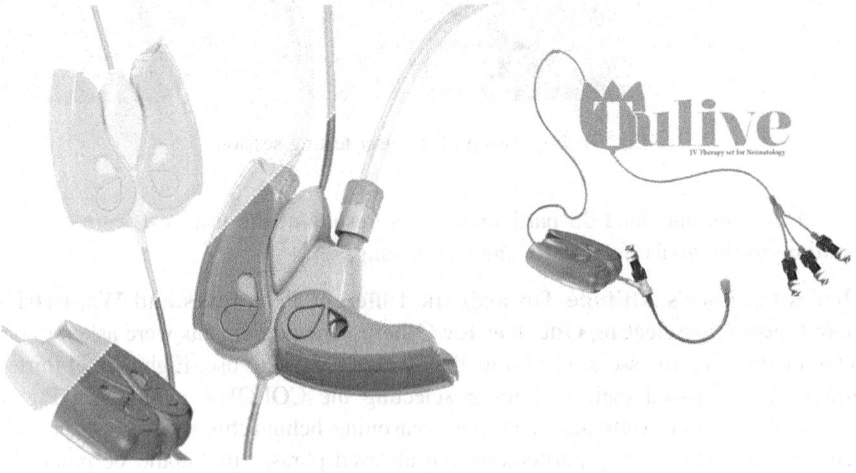

Fig. 10. Final design proposal "Tulive".

text. Because of this division, the assumption was made that both visual and literal instruction would be necessary.

All the feedback and insights from the user test and program of requirements has resulted in a final design iteration and the so-called "Tulive" product proposal was created (see Fig. 10).

4 Discussion

The aim of this study was to develop a product proposal for a new IV set which integrates the technical functionalities of Innofuse with the requirements from the context, the users and ergonomic considerations of usage of the IV set in the specific context of NICU.

This aim could only be fulfilled by understanding the complexity of the technical functionalities of the IV set of Innofuse, the high demands of safety protocols of NICU, the interactions between a qualified nurse and a very vulnerable patient and integrating these insights in a product proposal by applying a user-centered design approach.

Since the key users, in this case NICU nurses, are highly specialized in their profession guided by safety protocols it is necessary to empathize with their context, needs and expectations. These insights will shape the boundaries of the design space. In order to increase the product acceptance at the end the solution should match seamlessly to the values, norms and workflow of the key user. This was ensured by using a user-centered design approach where the key user is the source of all input regarding human-product interaction throughout the whole design process.

A systematic approach as applied in the product journey allows the designer to identify the stakeholders and their interaction with the product in the different phases defined in the timeline. It is holistic and creates an overview but also enable the designer to make choices.

Looking more specifically to the project it can be concluded that principles of ergonomic considerations were applied throughout the whole project incorporated in the design process. Questions like how the design can guide the user for correct use, impede incorrect use and indicate throughput of critical medication were design challenges related to cognitive ergonomics. Considerations of physical ergonomics regarding the limited available space in the incubator, entangling of the different tubes, physical contact between the pre-term infants and their parents, etc. were considered in this the design project. These could not be discussed in this paper due to format limitations.

All the aspects discussed above show the necessity of a holistic approach when designing in the medical field. Defining the different stakeholders, key users and the scope of the project, collecting contextual data next to the literature study, including the feedback of the key users/stakeholders throughout the whole design process makes it possible to understand the complexity of the healthcare context. Learning the language of the medical professionals to communicate including protocols & cultural aspects was part if this approach. To synthesize all these insights and incorporate them into a solution demand an analytical and integrated approach.

The value of user-centered design lies in the increased product acceptance that can be achieved after new product implementation. In designing for healthcare, barriers that impede successful product acceptance can most often be attributed to the fact that majority of the medical products are not designed according to the needs, expectations and limitations of the medical professionals. Centralizing the intended user of the product during the de-sign phases and allowing user input and contextual factors to shape the design in a systematic way will help evade these barriers and support product acceptance of any innovation.

Acknowledgment. The authors acknowledge the contribution of the Innofuse team comprised of Brechtje Riphagen, Luuk Evers, Pim Mijers, and master student Marijn Lardinois for the studies which were performed and as a proof of principle was delivered for the start of this project.

References

1. Auckland District Health Board Newborn Services Clinical Guideline. http://www.adhb.govt.nz/newborn/Guidelines/Nutrition/Nutrition.htm. Accessed 05 Mar 2015
2. Beyer H, Holtzblatt K (1999) Contextual design: defining customer-centered systems. Elsevier Science & Technology, New York. Author, F., Author, S., Author, T.: Book title. 2nd edn. Publisher, Location (1999)
3. van Boeijen AGC, Daalhuizen JJ, Zijlstra JJ, van der Schoor RSA (2013) Delft design guide. BIS Publishers, Amsterdam

System Diagrams for Healthcare Incident Investigation: Ease of Understanding and Usefulness Perceived by Healthcare Workers

Gyuchan Thomas Jun(✉) and Patrick Waterson

Loughborough Design School, Loughborough University, Loughborough, UK
G.Jun@lboro.ac.uk

Abstract. There is a growing awareness of the problems of root cause analysis in analyzing patient safety incidents. The need has been highlighted for the use of alternative systemic accident analysis methods, but it is unknown how applicable they are to healthcare. This study aims to evaluate how healthcare workers perceive the usability and utility of three alternative system diagrams for patient safety incident analysis: AcciMap; Hierarchical Control Structure Diagram (HCSD); Causal Loop Diagram (CLD). A two-hour workshop was carried out and how twenty-one healthcare workers perceive the applicability of these methods were captured by asking their level of agreement (five Likert scale) with statements on ease of understanding, usefulness for patient safety incident analysis. In terms of ease of understanding, both AcciMap and CLD were considered equally positive by 80% of participants, while HCSD was considered much less easy to understand (35% positive). In terms of utility, AcciMap was considered more positive (75–90%) than the other two diagrams (44–65%). Consequently, the participants' intention to use system diagram was in order of the following preference: AcciMap (60%); CLD (25%); HCSD (15%). However, the participants shared their concern that if the use of any system diagram requires significant amount of additional resources (expertise, time, etc.), it will be very challenging to apply them in healthcare.

Keywords: Patient safety incident · AcciMAP · STAMP
Causal loop diagram

1 Background

There is a growing awareness of the problems of root cause analysis (RCA)-based patient safety incident investigation in healthcare. For example, the five whys framework and fishbone diagrams within the RCA promote a flawed reductionist view, which can easily result in a blame culture and action plans focusing only on staff retraining [1, 2]. The need has been highlighted for the use of alternative systemic accident analysis methods [3], so this study aims to evaluate how healthcare workers perceive the usability and utility of three alternative system diagrams for the systemic

S. Bagnara et al. (Eds.): IEA 2018, AISC 818, pp. 437–440, 2019.
https://doi.org/10.1007/978-3-319-96098-2_54

analysis of patient safety incidents: AcciMap; Hierarchical Control Structure Diagram (HCSD); Causal Loop Diagram (CLD).

2 Methods

A two-hour workshop was carried out to introduce three alternative system diagrams (origin, concept, structure and examples) and evaluate how healthcare workers perceive the usability and utility of them. responses using a questionnaire and group discussion. Twenty healthcare workers (twelve clinicians, six healthcare managers and three improvement researchers) attended the workshop with average 19 year working experience in healthcare and 90% of them previously involved in healthcare incident investigations. The four aspects were evaluated using the five Likert scale by asking their level of agreement with the statements on the following four aspects: (*i*) *ease of understanding;* (*ii*) *usefulness for analysing contributing factors;* (*iii*) *usefulness for identifying recommendations;* (*iv*) *intention to use*. Ethical approval was given by Loughborough University Ethics Human Participants Sub-Committee.

3 Results

Figure 1 shows ease of understanding, usefulness for analysing contributing factors, usefulness for identifying recommendations and intention to use of each system diagram perceived by 21 healthcare workers.

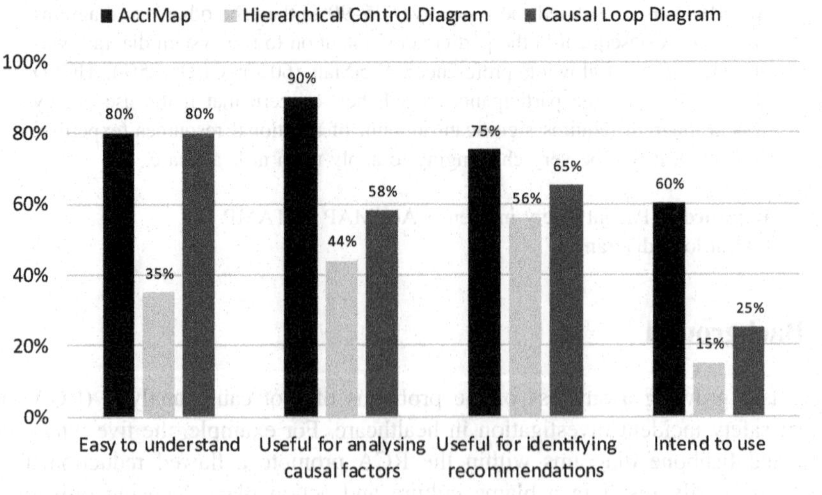

Fig. 1. Ease of use, usefulness and intend to use perceived by 21 healthcare workers

In terms of ease of understanding, both AcciMap and CLD were considered equally positive by 80% of participants, while HCSD was considered much less easy to understand (35% positive). In terms of utility, AcciMap was considered more positive (75–90%) than the other two diagrams (44–65%). Consequently, the participants' intention to use system diagram was in order of the following preference: AcciMap (60%); CLD (25%); HCSD (15%).

The participants' qualitative comments for each diagram include: (i) AcciMap is considered useful particularly for carrying out high-level aggregated analysis of a cluster of incidents and highlighting common causation, but need some more clarity on how to define interactions (strength, certainty, hypothesized, etc.); (ii) HCSD is considered useful since the same control structure can be reliably reused once created, but its control-focused structure is considered unfit for healthcare where complexity is managed by professionalism, not by control; (iii) CLD is was seen useful particularly for understanding in-depth dynamic interactions of certain contributing factors, but not necessarily all the contributing factors.

The participants also made the following general comments. First, more practical evidence of effectiveness of these methods will be required before considering using it more widely in practice. Once convinced the effectiveness, they would need more training or software tools for the efficient application of them first. If the use of any system diagram requires significant amount of additional resources (time, expertise, etc.), practical challenges might outweigh the potential benefits of using them. Besides, they shared a concern that the use of any system diagram might help identify system-level issues, but some people in the organisation might be more reluctant to admit them, which is beyond the methodological issue.

4 Discussion and Conclusion

The results of this study show that AcciMap is the most promising. The participants perceived it to be the easiest to understand, most useful for both analysing causal factors and identifying recommendations. Consequently 60% of the participants showed their intention to use it. At the same time, concerns and comments shared by the participants demonstrate that some system diagrams can improve the analysis quality of patient safety incident investigations, but it is more than methodological issues as presented by Lundberg et al. [4]. A limitation of this study is that the findings are based on the participants' perception since it was not possible to provide them an opportunity to apply the system diagrams to real incident analysis. Given the participants' positive response to AcciMap, the application of AcciMap to real patient safety incidents would provide more in-depth insights into the applicability of AcciMap.

References

1. Peerally MF, Carr S, Waring J, Dixon-Woods M (2016) The problem with root cause analysis. BMJ Qual Saf 26(5):417–422
2. Kellogg KM, Hettinger Z, Shah M, Wears RL, Sellers CR, Squires M, Fairbanks RJ (2016) Our current approach to root cause analysis: is it contributing to our failure to improve patient safety? BMJ Qual Saf 26:381–387
3. Leveson N, Samost A, Dekker S, Finkelstein S, Raman J (2016) A systems approach to analyzing and preventing hospital adverse events. J Patient Saf
4. Lundberg J, Rollenhagen C, Hollnagel E (2010) What you find is not always what you fix-how other aspects than causes of accidents decide recommendations for remedial actions. Accid Anal Prev 42(6):2132–2139

Evaluation of the Remote Control Affordance of Medicalized Bed for People with Mental Disabilities Getting Older (PDO)

Chibaudel Quentin[1(✉)], Lespinet-Najib Véronique[1], Durand Karima[2], Piant Laurence[3], and Piant Frédéric[3]

[1] Équipe Cognitique et Ingénierie Humaine (CIH), Laboratoire de l'Intégration du Matériau au Système (IMS) – UMR CNRS 5218; École Nationale Supérieure de Cognitique (ENSC), Institut Polytechnique de Bordeaux, Talence, France
quentin.chibaudel@ensc.fr

[2] Association pour le Développement et la Gestion des Équipements Sociaux, Médico-sociaux et Sanitaires (ADGESSA), Paris, France

[3] Groupe Conseil, Gestion et Prestation de Dispositifs Médicaux (CGPDM), Martignas-sur-Jalle, France

Abstract. Thanks the progress of medicine, the life expectancy for people with mental disorder has increased. A new population is appearing: people with mental disorder getting older (PDO). They need more care than general population. But they face with more difficulties, especially with medical devices. The aim of this article is to study the affordance of a medicalized bed remote control for people with mental disabilities getting older. 10 PDO participated to the evaluation. Data show they have lots of difficulties in using the remote of the bed: it is a very complex tool for them and they have access to too many functionalities in comparison with their needs. We propose adaptations to improve the affordance and the ergonomic of this remote control. We propose recommendation on the remote itself but also the remote operating manual.

Keywords: Mental disorder · Aging · Remote control · Ergonomics Affordance

1 Introduction

1.1 People with Mental Disorder Getting Older (PDO)

A major societal challenge today is the access to care for people with mental disabilities. Even more so for those who are aged: the PDO. According to the Blanc's report [3], the appearance of this new population is "*a true challenge*" for society. A PDO is someone who lived with a disability (whatever the specific nature of the handicap or the cause) before facing the effects of ageing [2, 8]. That is, the disability preceded old age. As pointed out, "*there is in fact an accumulation of the degenerative problems associated with age on the pre-existing conditions. The handicaps caused by chronic diseases which can occur as a part of normal ageing become, "adding handicap to handicap"*" [2]. Similar to the population at large, the PHA population

S. Bagnara et al. (Eds.): IEA 2018, AISC 818, pp. 441–448, 2019.
https://doi.org/10.1007/978-3-319-96098-2_55

experienced a remarkable increase in life expectancy during the 20th and 21st centuries. In 1983, Carter and Jancar showed that between 1930 and 1980 the life expectancy for mentally disabled males and females increased by 38.4 and 37.8 years, respectively [5]. This increase in life expectancy means that the proportion of PHA among the overall population has also increased. A study showed that in 2002 France had 635,000 PHA, of whom 267,000 were over 65, that is 42% of the PHA population [9]. In comparison, in the rest of the world only 16% of the population was over 65 in 2002 [9]. This newly emerging population faces great difficulties in accessing quality health care.

1.2 Access to Care for PDO

Despite a greater need for health care than the rest of the population, PHA have a harder time obtaining health care [10, 16, 18]. People with disabilities are in overall poorer health than others, as evidenced by higher morbidity and mortality rates [21]. They have the same health problems as the rest of the population, but are more susceptible to chronic diseases such as diabetes and hypertension. This might be explained by high risk behavior: little or no physical activity, higher use of tobacco and/or alcohol than the rest of the population [12, 23] and difficulties in accessing prevention messages (messages are too complex, inaccessible, etc.). Despite their extra needs, PHA are marginalized in their access to care [13] and receive less preventive care [15, 17]. As has been repeatedly demonstrated [14, 16, 20], this population faces many challenges which makes it very difficult to access quality health care. Many factors contribute, and primary among these is the organization of the French health care system.

1.3 Situation in France

The French system is divided into two separate and distinct areas. On one side is mental disabilities which engenders associated social problems, and on the other side is ageing and its associated health problems [24]. This distinction is based on age alone: a person under 60 will live in housing for the disabled while a person over 60 lives in housing for the aged [19]. That creates an iniquity to access to care for this population [22]. Because of this dichotomy, care professionals are trained separately and use different tools. Professionals in nursing homes learn to work with the aged as they face e.g., loss of autonomy or dementia and use medical devices adapted for them. Professionals working with people with disabilities in e.g., supported housing, occupational or job services are trained to aid based on the handicap, in e.g., performing daily activities and use medical devices adapted to them. Medical devices are adapted either for aged people either for people with mental disabilities. This system raises one question: how can medical devices be adapted for PDO?

1.4 Medical Devices

The French public health code (art. L. 52211-1 and R. 5211-1) defines a medical devices as "*any apparatus, appliance, software, material, or other article—whether used alone or in combination, including the software intended by its manufacturer to be used specifically for diagnostic and/or therapeutic purposes and necessary for its proper application—intended by the manufacturer to be used for human beings*". One purpose is to diagnose, monitor, treat or compensate for an injury or a disability, whatever the situation of the person. Medical devices do not take in account the specificity of a person. They are very general and adapted to situation. As a result, monitoring of this population is not adapted [6].

1.5 Purpose of This Study

Medical devices are not adapted. And especially not for people with mental disabilities getting older. The aim of this work is to study a specific device (a medicalized bed) and understand what ergonomics recommendations can be made to adapt them to PDO.

2 Methodology

Based on field analysis, 5 personas of PDO were proposed [7]. Resident were selected according their affiliation with the personas. To put in practice tests, establishment were visited and PDO were met. There were two aims. The first was to explain them the objective of the study. The second was to know the architecture of the establishment to ask them to realize action the usually do. Protocol tests were established from these personas and these observations. They were conceived in order to measure the usability of the bed and its user experience. Usability is defined by the norm ISO 9241-11 as the extent to which can be used by specified user to achieve specified goals with effectiveness, efficiency and satisfaction in specified context of use. Quantitative and qualitative data were collected in order to measure the usability of the medicalized bed.

3 Results

3.1 Description of the Sample

Tests were realized in retirements homes Aquitaine, in France in 2017. 10 people with mental disabilities getting older realized tests in 2 retirement homes. The average age was 65 years old.

3.2 Difficulty to Use the Medical Device

During test, quantitative data were collected. Resident were asked to explain how many actions they could realize (5 actions could be realized with this medicalized bed). They were also asked to evaluate the difficulty to realize the action asked. Possible answers were from 0 (really easy to achieve) to 5 (really hard to achieve). And they were asked how they preferred to use the bed from 0 (use the bed alone) to 5 (use the bed with someone's help).

The Fig. 1 illustrate results for these three questions according each resident.

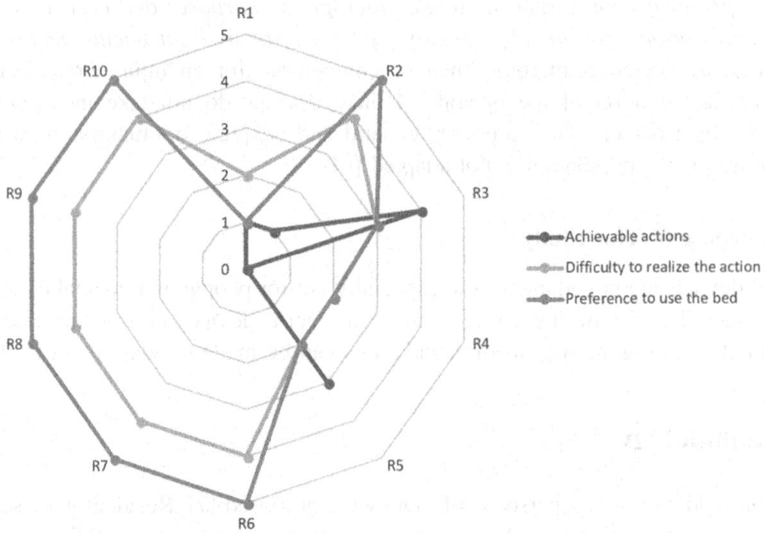

Fig. 1. Preference to use the bed according action that can be realized and the difficulty to achieve them

6 resident estimates it is hard to realize actions with the medical device. All of them prefer to use it with someone's help. Only one know he can do one action with the bed. This results shows that the medicalized bed is really hard to use for PDO. That's why they need someone's help. The bed is too hard for them to be used. And particularly one element: the remote control.

3.3 Lack of Affordance for the Remote Control

The remote control is very hard to use for PDO. They do not understand how it works. 4 of the participants decided to stop the test because they did not understand the remote control, even when we tried to explain them. They felt frustrated and sad because of that.

For those who could use the remote, they made a lot of mistakes. For example, one task was to lower the part for the legs. When the resident was asked "*are you lowing down your legs?*", the answer was "*yes*" whereas he was upping it. This residents knows how to use only one button: the one to lower the bed. Indeed, this action is realized every day. The remote control proposes too many actions in comparison with the use in the everyday resident's life.

Another resident could not explain which button to use to realize an action. He knew what he was asked to do but he could not achieve the action. According to him, "it is hard to do that".

Globally, the remote control affordance is insufficient: its ergonomics need to be improved.

4 Perspectives

4.1 Modifications of the Remote Control

The Fig. 2 is a picture of the remote control tested.

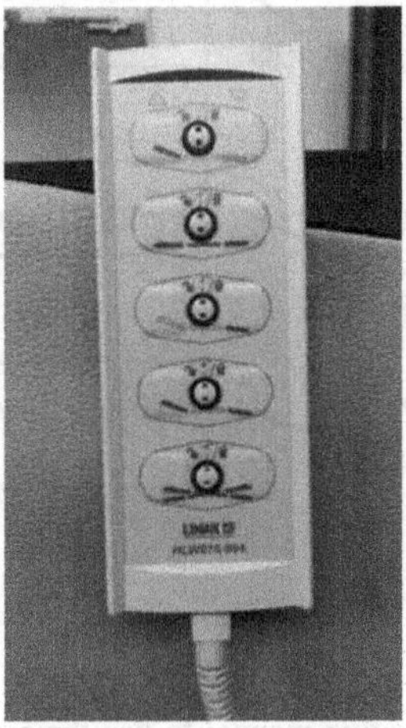

Fig. 2. Picture of the remote control tested

Pictograms are not clear. The bed is divided in three parts. Each button has a consequence on one of the parts. It is very hard to identify the part of the bed concerned. Plus, arrows to indicate the movement are very hard to distinguish. They must be colored and specific to each button. They could be added on each button with a high contrast to be easily visible.

However, the remote control is part of the medicalized bed. To be sold, it must be certified by the European Union with the certification "CE". If one part of the device is modified, everything must be tested again. It is a very long and expensive process. It is hard to put in practice these pieces of advice. Another element is easier to modify: the operating manual.

4.2 Modifications of the Operating Manual

The Fig. 3 shows the current operating manual for the remote control tested[1].

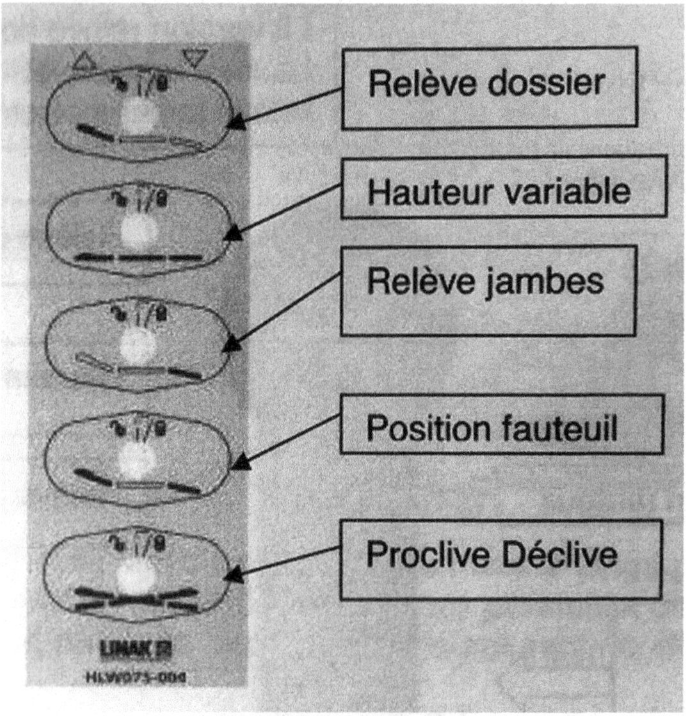

Fig. 3. Operating manual to use the remote control

Some requirements can be made to make this illustration easier to understand.

First of all, semantic must be more accessible. For example, the words "*proclive*" and "*declive*" respectively ("*trendelenburd*" and "*reverse-trendelenburg*") are very uncommon in the usual language and are very hard to be understood. And especially for PDO.

Then, indications must be clearer. The second propositions indicates that the buttons can be used for "*hauteur variable*" ("*variable height*"). What eight is variable? And does that influence the position of the bed? This is not clear.

Finally, differences between the button are not clear. For example, the first button (from the bottom) is indicated as "*relève dossier*" ("*increase backrest*", the third button is indicated as "*relève jambes*" ("*increase leg rest*") and the fourth one is indicated as "*position fauteuil*" ("*wheelchair position*"). On the draws, we can see that the "wheelchair position" concerns the same element as "increase backrest" and "increase leg rest".

[1] The operating manual is available only in French. Translations are proposed.

Differences between these action are hard to understand. And, once again, especially for a PDO who has difficulties in understanding the different functions available.

5 Conclusion

5.1 Results of the Study

This work is about the accessibility to care for people with mental disorder getting older. In this study, we selected a specific medical device (a medicalized bed) and we tested with PDO. 10 people participated to the evaluation. They all live in retirement homes in Aquitaine, France. We have shown that this population has a lots of difficulties in using the remote control of a medicalized bed. As a result, they often ask for help and monopolize caregivers. It is hard to modify the remote control itself. We propose some requirement to make the operating manual more accessible. We hope to give them more autonomy and give more time to caregivers to monitor residents.

5.2 Limit of This Study

The sample is limited: 10 people is insufficient to propose general adaptations. Results may have bias. But, according the specificity of the population, it is an acceptable sample. These results represent a first step in our work. We now have to put them in practice with more participants to have a representative sample.

5.3 Future Perspectives

Now, we need to put in practice our requirements and test them again. Plus, universal design principles, also called "*design for all*" [4] or "*barrier free environment*" [11], are applied in this work. This methodology considers that principles applied to very frailty users will include less frailty users. In other words, if our simplifications are relevant for people with mental disabilities getting older, they will be relevant for other people: people getting older, people with dementia like Alzheimer's disease… We will have to test them also on this kind of people.

References

1. Audry A, Ghislain JC (2009) Le dispositif médical: «Que sais-je?» n° 3858. Presses universitaires de France
2. Azéma B, Martinez N (2005) Les personnes handicapées vieillissantes: espérances de vie et de santé; qualité de vie. Revue française des affaires sociales 2:295–333
3. Blanc P (2006) Rapport d'information Fait au nom de la commission des Affaires sociales (1) sur l'application de la loi n° 2005-102 du 11 février 2005 pour l'égalité des droits et des chances, la participation et la citoyenneté des personnes handicapées. Sénat session extraordinaire, 2007
4. Burzagli L, Emiliani PL, Gabbanini F (2009) Design for all in action: an example of analysis and implementation. Expert Syst Appl 36(2):985–994

5. Carter G, Jancar J (1983) Mortality in the mentally handicapped: a 50-year study at the Stoke Park group of hospitals (1930–1980). J Mental Defic Res 27:143–156
6. Chibaudel Q, Lespinet-Najib V, Durand K, Piant L, Piant F (2016) Access to care for people with mental disabilities getting older through the study of the accessibility to medical devices: the situation in France. In: Proceedings of healthcare systems ergonomics and patient safety (HEPS), Toulouse, 2016
7. Chibaudel Q, Lespinet-Najib V, Durand K, Piant L, Piant F (2017) Personne en situation de handicap mental avançant en âge (PHA): établissements de profils psychologiques à l'aide de la méthode des personas. Sfp2017: 58 ème congrès annuel de la société française de psychologie; 30 août – 1 September 2017, Nice, France
8. CNSA (2010) Aide à l'adaptation et à la planification de l'offre médico-sociale en faveur des personnes handicapées vieillissantes
9. DREES (2002) Les personnes handicapées vieillissantes: une approche à partir de l'enquête HID. 204
10. DREES (2011) Comptes nationaux de la santé 2011, s.l., Ministère de l'économie, ministère de la santé, ministère du travail (coll. «Etude et statistiques»)
11. Goodman-Deane J, Ward J, Hosking I, Clarkson PJ (2014) A comparison of methods currently used in inclusive design. Appl Ergonom 45(4):886–894
12. Havercamp SM, Scandlin D, Roth M (2004) Health disparities among adults with developmental disabilities, adults with other disabilities, and adults not reporting disability in North Carolina. Public Health Rep 119(4):418–426
13. Jacob P (2013) L'accès aux soins et à la santé des personnes handicapées: un droit citoyen. Gestion hospitalière 531:636–637
14. Jensen PM, Saunders RL, Thierer T, Friedman B (2008) Factors associated with oral health-related quality of life in community-dwelling elderly persons with disabilities. J Am Geriatr Soc 56(4):711–717
15. Kancherla V, Braun KVN, Yeargin-Allsopp M (2013) Dental care among young adults with intellectual disability. Res Dev Disabil 34(5):1630–1641
16. Lengagne P, Penneau A, Pichetti S, Sermet C (2015) L'accès aux soins courants et préventifs des personnes en situation de handicap en France. Tome 1 Résultats de l'enquête Handicap-Santé Ménages
17. Lengagne P, Penneau A, Pichetti S, Sermet C (2014) L'accès aux soins dentaires, ophtalmologiques et gynécologiques des personnes en situation de handicap en France. Une exploitation de l'enquête Handicap-Santé Ménages
18. OMS (2011) Rapport Mondial sur le Handicap
19. Penneau A, Pichetti S, Sermet C (2015) L'accès aux soins courants et préventifs des personnes en situation de handicap en France. Tome 2 Résultats de l'enquête Handicap-Santé volet Institutions
20. Pezzementi ML, Fisher MA (2005) Oral health status of people with intellectual disabilities in the southeastern United States. J Am Dent Assoc 136(7):903–912
21. Prince M, Patel V, Saxena S, Maj M, Maselko J, Phillips MR, Rahman A (2007) No health without mental health. The Lancet 370(9590):859–877
22. Ramos-Gorand M, Rapegno N (2016) L'accueil institutionnel du handicap et de la dépendance: différenciations, conséquences territoriales et parcours résidentiels. Revue française des affaires sociales 4:225–247
23. Rimmer JH, Rowland JL (2008) Health promotion for people with disabilities: Implications for empowering the person and promoting disability-friendly environments. Am J Lifestyle Med 2(5):409–420
24. Roy D (2016) Les personnes âgées et handicapées en France et les politiques publiques d'accompagnement. Revue française des affaires sociales 4:21–33

Safety I and Safety II for Suicide Prevention – Lessons from How Things Go Wrong and How Things Go Right in Community-Based Mental Health Services

Gyuchan Thomas Jun[1(✉)], Aneurin Canham[1], Fabida Noushad[2], and Satheesh Kumar Gangadharan[2]

[1] Loughborough Design School, Loughborough University, Loughborough, UK
G.Jun@lboro.ac.uk
[2] Leicestershire Partnership NHS Trust, Leicester, UK

Abstract. Prevention of patient suicide is a major challenge for mental health services. This study applied both safety I and safety II approaches to gain an understanding of the detection and response process for suicide prevention in community mental health care in order to compare/contrast outputs from each approach. For safety I, 41 suicide incident reports were analysed using a systemic analysis approach. For safety II, interviews with 20 community-based mental health practitioners and managers were conducted asking their know-hows to successful suicide risk detection and response. The five key issues found from the Safety I approach were: (i) an inherent weakness in the interactions between patient and clinician with the presence of uncertainty in the risk detection; (ii) Poor patients' engagement with services; (iii) Reliance on patients self-presenting in crisis and declining the offered support options; (iv) Delay in treating new patients; (v) Coordination, communication and process issues. On the other hand, the safety II approach revealed a complex decision-making process with the presence of uncertainty and trade-offs between patient clinical need, patient desire, legal and procedural obligations, and resource considerations. It also revealed a strong theme on the importance of peer-support. The results of this study indicate that safety II approach provides valuable insights into how to strengthen the system performance without challenging systemic issues, while system I approach identifies systemic issues and raise questions how to address them. These findings suggest the potential benefit of applying both approaches to quality and safety improvement in healthcare.

Keywords: Safety II · Mental health · Suicide prevention

1 Background

More than 5,000 people committed suicide in 2016 in the UK and suicide is the leading cause of death among young people aged 20–34 years [1]. Prevention of patient suicide is a major challenge for mental health services. A current focus of suicide prevention is in risk assessment methods which are used to identify risk factors and initiate appropriate treatment. However, risk assessment does not remove the uncertainty around the

S. Bagnara et al. (Eds.): IEA 2018, AISC 818, pp. 449–452, 2019.
https://doi.org/10.1007/978-3-319-96098-2_56

potential for suicide [2]. This study applied both safety I and safety II [3] approaches to gain an understanding of the detection and response process for suicide prevention in community mental health care in order to compare and contrast outputs from each approach.

2 Methods

The case study was carried out on the community-based mental health care services of a mental health trust in England. The teams involved include community mental health teams (ongoing support for complex and serious mental health problems), crisis teams (urgent support for a mental health crisis) and psychological service teams (support through psychological therapy). For safety I, forty-one root cause analysis-based suicide incident reports produced between 2015 and 2016 were re-analysed using an alternative systemic analysis approach - Systemic Theoretic Accident Modelling and Process (STAMP) [4].

For safety II, interviews with 20 community-based mental health professionals (3 managers, 11 crisis team staff, 6 community team staff) were conducted asking their know-hows to successful suicide risk detection and response.

Ethical approval was given by Loughborough University Ethics Human Participants Sub-Committee and the research was granted approval by the Health Research Authority.

3 Results

Figure 1 represents a hierarchical safety control structure diagram of suicide prevention and five key issues identified in the analysis of the incident reports (safety I).

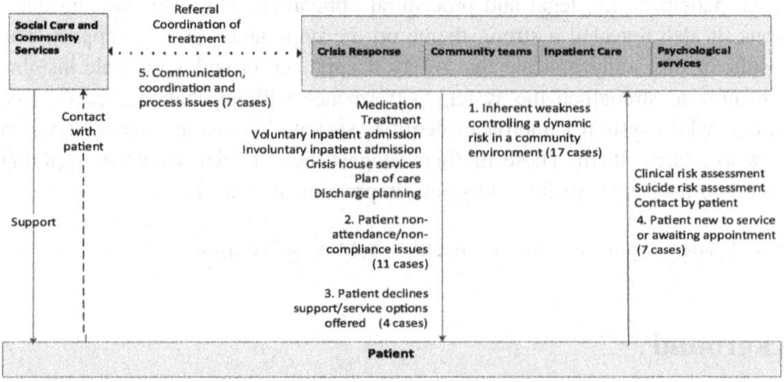

Fig. 1. Simplified hierarchical safety control diagram of suicide prevention

The five main weaknesses in the control structure were identified as:

i. An inherent weakness in the interactions between patient and clinician with the presence of uncertainty in the risk detection (17 cases)
ii. Poor patients' engagement with services including non-attendance and non-compliance (11 cases)
iii. Reliance on patients self-presenting in crisis and declining the offered support options (4 cases)
iv. Delay in treating new patients, with suicides occurring while on waiting lists or having only had initial assessments (7 cases)
v. Coordination, communication and process issues within services interrupting patient care (7 cases)

On the other hand, the interviews with staff (safety II) revealed complex decision-making elements displaying three main considerations and trade-offs that healthcare professionals need to deal with. All this is considered within a level of uncertainty and ambiguity with the difficulties in detecting risk, prediction human behaviour and the dynamic nature of the risk.

i. Being patient-centred in terms of clinical need and patient desires;
ii. A resource consideration in terms of their individual emotional, time, experience and skill resource and the resources available within services;
iii. A legal and procedural responsibility with constraints from laws and regulations.

The interview findings also revealed a strong theme on the importance of peer-support for successful suicide risk detection and responses.

4 Discussion and Conclusion

Safety I approach identified patient engagement issues and highlighted a problem to a care model reliant on patients adhering to care plans and presenting at times of crisis. Two questions were also raised as to whether the system has the resources to accommodate different patient needs and how services can fit to patient desire. On the other hand, safety II approach found the importance of peer-to-peer learning and support for successful detection and response to suicide risk.

The Safety II finding in this study is in line with what Vincent and Amalberti's finding [5]. In an overview of safety approaches in safety critical industries, they have classified three main approaches to safety for healthcare: ultra-adaptive, high reliability and ultra-safe. In many respects, the safety approach in community mental health care is comparable to the ultra-adaptive model, in which risk is embraced and risk management is heavily reliant on the judgement, adaptability and resilience of individuals. This model prioritises adaptation and recovery strategies, power is given to experts, with safety improvements coming through peer-to-peer learning, shadowing, acquiring professional experience and personnel having awareness regarding their own limitations [5].

The results of this study indicate that safety II approach provides valuable in-sights into how to strengthen the system performance without challenging systemic issues, while system I approach identifies systemic issues and raise questions how to address them. These findings suggest the potential benefit of applying both approaches to quality and safety improvement in healthcare.

References

1. Office for National Statistics: Suicides in GB (2016) Registrations—ONS. https://www.ons.gov.uk/peoplepopulationandcommunity/birthsdeathsandmarriages/deaths/bulletins/suicidesintheunitedkingdom/2016registration. Accessed 28 May 2018
2. Mulder R (2011) Problems with suicide risk assessment. Aust N Z J Psychiatry 45(8):605–607. https://doi.org/10.3109/00048674.2011.594786
3. Hollnagel E (2014) Safety-I and safety-II: the past and future of safety management. Ashgate Publishing, Farnham
4. Leveson N (2004) A new accident model for engineering safer systems. Saf Sci 42(4):237–270. https://doi.org/10.1016/S0925-7535(03)00047-X
5. Vincent C, Amalberti R (2016) Safer healthcare. Springer, Cham. https://doi.org/10.1007/978-3-319-25559-0

Users' Awareness of Alarm Settings

Analyzing the User Interface of a Physiological Monitor

Kathrin Lange[1(✉)], Armin Janß[2], Siyamak Farjoudi Manjili[2], Miriam Nowak[1], Wolfgang Lauer[1], and Klaus Radermacher[2]

[1] Federal Institute for Drugs and Medical Devices [BfArM], Bonn, Germany
kathrin.lange@bfarm.de

[2] Chair of Medical Engineering in the Helmholtz-Institute for Biomedical Engineering, RWTH Aachen University, Aachen, Germany

Abstract. Whether or not a physiological monitor will issue an alarm in a certain condition depends largely on the particular settings of the device. To be able to safely use the monitor, the user has to be aware of these settings. We addressed a potential contribution of interface design to users' awareness of devices settings by analyzing a monitor's user interface. Based on a previous analysis of incident reports, we selected the following functions for further analysis: Deactivating individual alarms (SpO_2), changing alarm limits (arhythmia), muting alarm volume and completely disabling the measurement of a particular parameter (blood pressure). We applied two different methods of assessing interface usability: an analytical approach supported by the formal-analytical method mAIXuse and an empirical approach, i.e. observing whether expert users in a simulated care setting were aware of the various causes underlying alarm failures. In the simulation study, seven experienced intensive care nurses took part. The mAIXuse analysis showed that detecting altered alarm limits, the muting of alarms and the disabled measurement should be somewhat less easy than detecting the deactivation of an alarm. With regard to comprehension, alarm limits and disabled measurement were judged to be inferior to muting and blocking of alarms. During the usability-test, not a single user identified the muted alarm and the altered alarm limits as potential causes for the absence of an audible alarm without being prompted. By contrast, three of seven nurses directly recognized that the SpO_2 alarm was blocked. The results of the formal-analytical analysis and the simulation study will be compared and conclusions regarding the contribution of the monitor display to users' awareness of alarm settings will be discussed.

Keywords: Physiologic monitoring alarms · User computer interface
Awareness

1 Introduction

1.1 To Safely Use Alarms, Users Have to Be Aware of Device Settings

Intensive care nurses have to be aware of the vital status of the patients assigned to them. The monitoring of vital functions is assisted by specific devices (physiological

S. Bagnara et al. (Eds.): IEA 2018, AISC 818, pp. 453–464, 2019.
https://doi.org/10.1007/978-3-319-96098-2_57

monitors) and largely automated. The device measures and interprets vital parameters and provides an alarm to alert the user, if it infers the presence of a critical physiological condition, e.g. insufficient oxygen saturation. In addition, devices monitor for technical conditions that compromise their intended functioning, for instance the detachment of an electrode. Theoretically, the automatic monitoring of vital parameters and use of alarms allow care givers to immediately become aware of critical conditions without watching the patient uninterruptedly. This is necessary, because there are situations, in which nurses have to rely on alarms – at least for a certain while. Examples include the need to care for a second patient or performing other tasks off the patient's bed. To be able to rely on alarms in these situations, nurses have to be aware of the factors affecting whether or not a certain physiological condition will trigger an alarm.

1.2 Users of Monitoring Devices May Not Always Be Sufficiently Aware of Alarm-Related Settings

There is evidence that awareness for alarms and alarm behavior of physiological monitors is not as straightforward as it should be, as a recent analysis of incident reports regarding "alarm failure" in physiological monitors that were issued to the Bundesinstitut für Arzneimittel und Medizinprodukte [Federal Institute for Drugs and Medical Devices] (BfArM) shows [9]. Between 2009 and 2015, 184 alarm-related incidents with physiological monitors were reported. The alarm-related malfunctions were missing alarms (144/184), ceasing alarms (21/184), wrong alarms (16/184) and alarms not saved (3/184). Crucially, only a small proportion of these malfunctions had a technical cause (16/184). Rather, more than half of the reported problems (101/184) reflected incorrect assumptions or expectations regarding device functioning. Often, users were not aware of the complex conditions that determined the specific alarm behavior or whether the monitor would issue an alarm, at all.

Table 1 gives an overview of the monitor settings mainly involved in the reported incidents. In another 58 cases the root-cause could not be identified, 9 cases were due to failures during implementation or maintenance.

Table 1. Selected monitor settings that have contributed to missing alarms that were interpreted as device failure and reported to the BfArM, see [9] for details

1. Deactivation of alarm
2. Suspension of alarm
3. Alarm volume
4. Alarm routing
5. Alarm-related interactions at other components in network
6. Adaptation of monitor settings to characteristics of specific patient groups (e.g. pacemaker)

1.3 Users' Awareness of Alarm-Related Device Settings Depends on Many Factors

Particularly in a complex work environment such as critical care, users' awareness of device behavior or device settings may be influenced by a variety of factors (see also Fig. 1). Two of these seem pivotal: Firstly, users need a precise mental model of the device. Secondly, they have to be able to retrieve and actively use this knowledge if necessary. The users' mental model of the device (i.e. device knowledge) includes, for instance, its functionality, its individual components and their interplay, and the external factors that may influence device functioning [7]. This requires an adequate representation of the respective information in long term memory. For active use, the information has to be retrieved and maintained, which is related to attentional control and working memory functions.

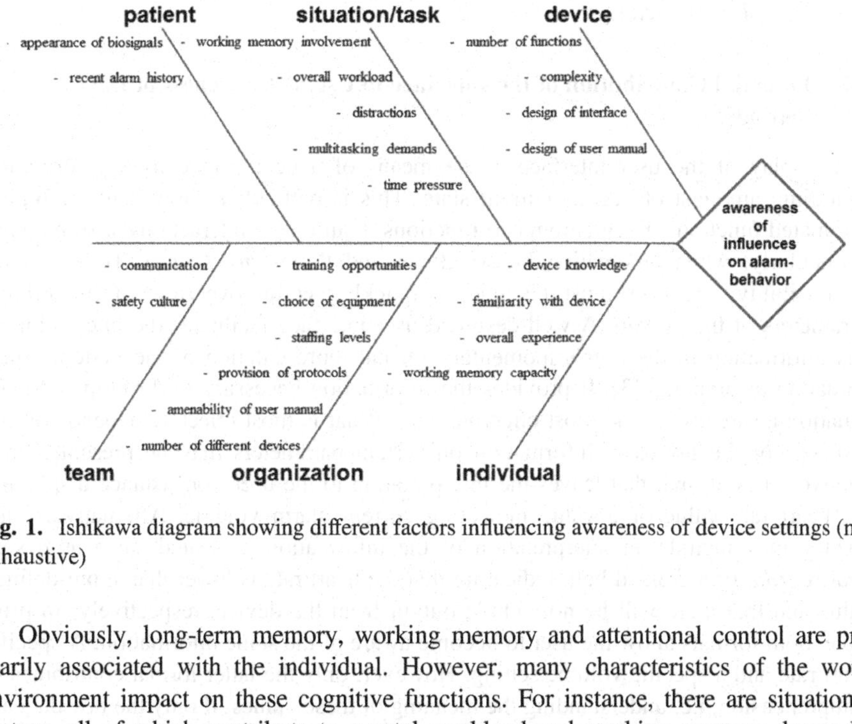

Fig. 1. Ishikawa diagram showing different factors influencing awareness of device settings (not exhaustive)

Obviously, long-term memory, working memory and attentional control are primarily associated with the individual. However, many characteristics of the work environment impact on these cognitive functions. For instance, there are situational factors, all of which contribute to mental workload and working memory demands, such as the complexity or difficulty of the current task(s), the frequency of distractions, the multitasking demands, or the presence of time pressure. Organizational influences include training opportunities, staffing levels, standards and protocols regarding device settings or the choice of equipment (regarding both individual devices and the composition of the device pool). Examples for team factors are communication (e.g. of changes to device settings) or the safety culture on the ward (how strictly does the personnel adhere to protocols). Even characteristics of the patient may have an effect, e.g. the specific appearance of the measured bio-signals or the recent alarm-history.

Finally, the device itself has to be considered, for instance the number and complexity of functions, the precise algorithms used, the design of the user manual or – crucial for the present paper - the design of the user interface.

Evidently, there are many interdependencies between the individual factors depicted in Fig. 1 (not shown). For instance, it is safe to assume that an individual's device knowledge (a feature of the individual) is influenced (amongst other things) by the number and complexity of functions of the device (a device property), by the frequency and quality of the training (an organizational factor), and by whether only a single device model is used for the purpose in question (also an organizational decision). For related discussions of these topics see also [8, 9]. Likewise, users' awareness of specific alarm settings may depend on the interaction between device knowledge, attentional orienting and the design of the user-interface (a device property) and may be expected to be further mediated by the current demands on working memory. In the present paper, we focus on the user-interface and its suitability for enhancing users' awareness of alarm-settings.

1.4 Potential Contribution of the Interface to Users' Awareness of Device Settings

The display of the user interface is the means of a device to convey information regarding, amongst others, its current state. This is particularly important for highly automated functions. Even automated functions require user interactions at some point (particularly, when automation breaks down and things go wrong). To take over responsibility, the user must be able to quickly get an overview of the critical parameters of the system. A well-designed user interface facilitates the integration of this information in the user's momentary mental representation of the system– their "situation awareness" [3]. It provides the information necessary to develop adequate situation awareness in the most effective way. What is most effective depends on the demands of the use-case. Information on certain parameters may be presented in a relatively raw format that leaves the interpretation to the user, for instance displaying the heart rate value or a value indicating current alarm volume. Alternatively, the display may include an interpretation of the information presented, such as a sign *Bradycardia* or a crossed bell, indicating that the heart rate is lower than a pre-defined value and that there will be no audible output from the device, respectively. In principle, both formats allow the user to become aware of the same information: A specific heart rate and a specific volume setting. However, only the latter format contributes to comprehension, i.e. understanding the meaning of these values. If only the raw data are given, they have to be combined with knowledge stored in long-term memory to attain the same level of awareness or understanding. While this may be effortlessly achieved by expert users in standard situations, greater difficulty may be experienced by novices or if mental demands are particularly high, thus loading on working memory. Another important aspect, which may be particularly important for complex systems, is the challenge associated with the need to deal with large numbers of variables and the interdependencies between these variables. Of course, users have to be provided complex information. However, overly complex displays compromise the perceptual salience of the information that is actually presented [2, 4].

The user interface of a physiological monitor is only one facet of the entire "device-patient" system a critical care nurse has to keep in mind. The display of a physiological monitor includes graphical and numerical representations of the bio-signals measured, together with their respective alarm settings (e.g. status of alarm or alarm limits). Complexity results from the fact that a typical device monitors a large number of different vital functions. Moreover, for each of these functions, there are many parameters influencing alarm behavior, only some of which are explicitly displayed. Our goal was to investigate how well displays of currently used physiological monitors support users' awareness of alarm-related device settings. We started by investigating a particular monitor model of a manufacturer frequently used in German hospitals, for which several alarm-related incident reports were issued. Based on an analysis of these reports [9], we chose four settings for evaluation: muting alarm volume, disabling the measurement of a bio-signal, blocking the alarm for a particular bio-signal, and changing alarm limits of a bio-signal. These settings differ with respect to whether they affect all or only selected bio-signals and alarm-conditions and how the current setting is displayed by the device (Fig. 2 and Table 2). We evaluated how well the monitor's user interface supports awareness for these functions, both using a formal analytical approach (mAIXuse) and by empirically assessing how easily users become aware of the individual settings if they search for the reason why an expected alarm did not sound.

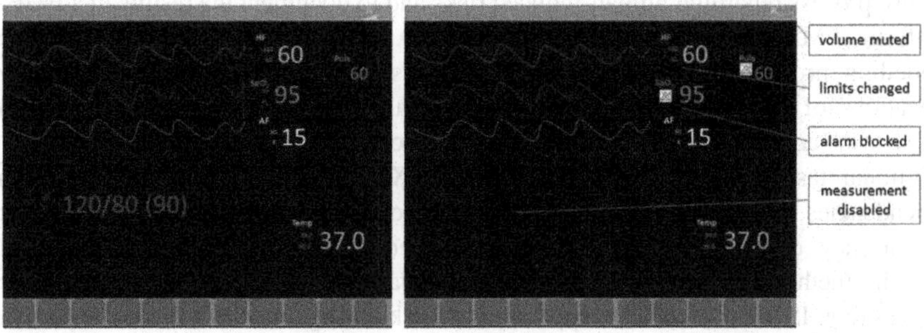

Fig. 2. Overview of the alarm-related monitor settings addressed in the present study before and after changes were performed

2 Methods

2.1 Formal-Analytical Approach: mAIXuse

In the framework of the usability evaluation of the graphical user interface (GUI) of the patient monitor, which has later been used for the interaction-centered usability tests, the mAIXuse method has been conducted initially, in order to get further impressions regarding the usage deficits. In a two day workshop the GUI of the patient monitor has been evaluated with the mAIXuse methodology at the Chair of Medical Engineering.

Table 2. Overview of alarm-related settings selected for evaluation

	Bio-signals affected	Alarm conditions affected	Other indicator of alarm condition?	What indicator on display?	Absolute indicator?
Alarm volume	All	All	Visual	Symbol	Yes
Measurement of bio-signal	Selected	All	No	Absence of readings/values	Not applicable
Blocking of alarms	Selected	All	No	Symbol next to parameter value	No (position on screen)
Alarm limit	Selected	Selected	Not applicable	Limit values next to parameter value	No (position on screen, semantic relation)

The formal-analytical mAIXuse methodology [5] for usability evaluation and human risk analysis is based on a two-folded approach. On the one hand, HiFEM provides a task-specific modelling structure (user- and system-tasks) with additional integrated temporal relations and on the other hand the methodology enables to analyze the use process regarding human-induced risks and to document the results in a FMEA-conform (FMEA - Failure Modes and Effect Analysis) data sheet. Within the task modelling process, the investigator gains a systematic overview of the high-level operations of the user, the system and especially the Human-Machine-Interaction. These tasks are subdivided into "Perception", "Cognition" and "Action". In contrast to traditional task analyses [1], within the mAIXuse modelling not only hierarchic dependencies are presented, but with the help of temporal relations (e.g. sequence, concurrency, disabling, sharing, choice, etc.) coequal tasks are linked with each other [6]. The methodology is based on several standards for the development of medical devices (e.g. ISO 14971:2013 - Application of risk management to medical devices and IEC 62366:2015 - Application of usability engineering to medical devices). The approach can be used from early developmental stages up to the validation process.

In the present study, we assessed perceptual, cognitive, and motor qualities of display elements underlying selected alarm-related monitor settings within the framework of the mAIXuse analysis. In order to compare the results of the formal-analytical approach and the usability-evaluation, the same device settings were chosen for both approaches: turning off alarm volume, deactivating measurement of blood pressure, blocking the SpO_2 alarm and changing alarm limits.

The mAIXuse analysis provides failure causes (in the perceptional and the cognitive level of the human information processing) and potential failures (in the motoric action) within use process steps and classifies these potential failures and causes according to various human error taxonomies. In the subsequent human risk analysis the parameters *causes*, *failures* and *consequences* (which characterize the risk priority number) are documented in a FMEA standard sheet. Finally, countermeasures for the different parameters of the risk priority number are systematically identified and documented as well.

2.2 Empirical Approach: User Performance in a Simulated Use Scenario

To empirically investigate, how awareness of alarm settings is supported by the monitor's user interface, we simulated a use scenario, in which participants were asked to identify reasons for the absence of audible alarms, given a deterioration of the patient and certain device settings. Seven nurses from two different intensive care units of University Hospital Aachen took part after giving their informed consent. All had experience with the monitor that was used (between 1 month and 10 years) and no one had used models by other manufacturers before.

The setting consisted of two rooms at the Chair of Medical Engineering, simulating patient rooms. Each was equipped with a physiological monitor, syringes, and dressing material. Room 1 was equipped with a manikin that could be used for performing the task "changing dressings". Note, however, that a fully functional manikin, allowing for the simulation of alarm conditions, was not available (this will be discussed below).

During the test scenario, participants were asked to perform a selection of tasks typically performed on an ICU. They started with the initial control of the patient and the monitor, which is mandated at the hospital at the beginning of a shift. Performance of the initial control served as a manipulation to evaluate, whether the participants engaged in the simulation. As a next step, participants took blood, performed a blood gas analysis and started to change the dressings on patient A (see also Fig. 3). Triggered by a syringe pump alarm (played from a recorded sound), they were required to switch to room 2 to perform a different task, there (calculate medication dose, prepare syringe, administer drug).

While the participant was absent from room 1, the experimenters changed several alarm-related monitor settings (see also Fig. 3):

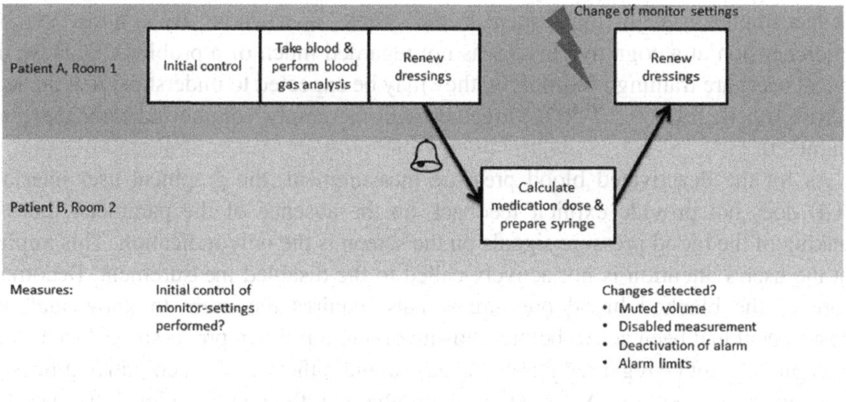

Fig. 3. Procedure of use simulation

a. Alarm volume was changed from 7 to 0
b. Measurement (and display) of blood pressure was deactivated
c. The alarm indicating oxygen de-saturation was blocked (i.e. neither visual nor acoustic alarms would be presented)
d. Alarm limits for heart rate were altered (lower limit from 50 to 30, upper limit from 120 to 190)

Having addressed the alarm in room 2, the participants returned to room 1 and resumed changing dressings. They were told that they had been away for 15 min to indicate the possibility that a third person might have altered device settings in the meantime. Shortly before finishing their task and while facing the patient rather than the device, the participants were told that the patient's status had deteriorated, but that the device had not alarmed. At this point, they were asked to search for potential causes of this absence of alarms.

3 Results

3.1 mAIXuse

The formal analytical assessment separately evaluates the contribution of perceptual, cognitive, and motor functions to using a particular display element for a selected task. For the present purpose (awareness of alarm settings), we focused on the perceptual (detection) and the cognitive (comprehension) evaluation.

The mAIXuse analysis came to the conclusion that the symbols indicating alarm volume settings are difficult to detect. Likely causes for the low salience are the small symbol size as well as the insufficient color coding (light grey) of the symbol associated with volume settings, per se. As for the qualitative transition implied by a volume setting of "zero", this is not associated with a prominent symbol change (e.g. in color or contrast) to capture the user's attention. This may be regarded a hazard, because when alarm volume is turned off, only visual alarms are available. Not becoming aware of this fact implies a major impairment to the system, as a whole. By contrast, symbol comprehension at a cognitive level was not regarded much of a problem. Because the clinical users are training accordingly, they may be expected to understand that the icon (a crossed bell, IEC 60417-5576) together with the empty volume indicator represent volume "0".

As for the deactivated blood pressure measurement, the graphical user interface (GUI) does not provide explicit feedback on the absence of the parameter, i.e. the blanking of the blood pressure signals on the screen is the only indication. This implies that the user's attention is not actively called to the disabled measurement. Becoming aware of the blanked blood pressure signals requires the users to know that this measurement had been active before, thus involving top-down processing. Hence, even detection may not be regarded purely perceptual but rather involves cognitive processes like memory or attention. As an additional problem at the cognitive level, the user has to understand all the implications of the disabled measurement signals ("no measurement", "no alarms"). Again, because of the complete absence of any explicit symbol, the GUI does not provide any cue to facilitate awareness. At a basic perceptual level, the system's feedback regarding the blocking of the SpO_2 alarm can be easily detected. However, the mAIXuse analysis yielded some difficulty understanding the meaning of the symbol at a cognitive level. Two causes may be responsible for this. Firstly, the symbol representing deactivation of alarms was judged to be ambiguous. Secondly, the mode of alarm deactivation for the respective parameter/signal presentation is not distinct. Confusingly, the vital signs presentation is continuously displayed in an unmodified way. Countermeasures e.g. would be to visually mark the signal itself (e.g. with a different signal color or different background color), the blocked parameter itself or the complete display area (with a border strip), where the signal is presented. However, similar to the icons indicating the muted volume, trained users may be regarded familiar with the icons indicating "blocked alarms" (crossed alarm symbol, IEC 60417-5319) and may, therefore, effortlessly decode its meaning.

Perception of alarm limits was evaluated to be difficult for the user due to the small size and the light green color. The combination of the selected color with the black background is not recommended according to usability design guidelines. Additionally, the system allows setting alarm limits to unreasonable values and does not provide visual feedback if this is the case, i.e. the user interface does not offer any cue to facilitate comprehension. As a consequence, in order to understand the implication of the specific alarm limits, users have to actively relate the displayed value with their stored knowledge.

3.2 Device-Related Performance in the Use Simulation

Participants engaged in the simulation situation somewhat differently: The majority (4/7) thoroughly checked the initial settings of the monitor and even changed certain values. 2 of the 7 participants extensively viewed the settings, but did not alter anything. One participant only performed a cursory control.

With respect to the alterations in alarm-related monitor settings, detection differed between functions: Almost all participants detected the blocking of the SpO_2 alarm either directly or upon being prompted and still a majority noticed directly or after a prompt that blood pressure was no longer measured (Table 3). By contrast, not a single participant detected the altered limits of the arrhythmia alarm or the muted alarm volume – and 4 of 7 participants never found these potential causes for missing alarms.

Table 3. Number of participants, who detected changes at the different stages

Detection of ...	Directly	Prompted	Never
Muted alarm volume	0	3	4
Disabled blood pressure measurement	3	1	3
Blocked SpO_2 alarm	3	3	1
Changed limits of arhythmia alarm	0	3	4

4 Discussion

We aimed at assessing how well the interfaces of currently used physiological monitors support users in obtaining awareness of the device's alarm-related settings. These settings are important, because they determine whether or not a particular physiological or technical condition will result in an audible alarm. Full awareness of these settings determines the users' situational expectations for alarms and, therefore, their capability to safely use the alarm function. Triggered by device issues identified in incident reports, we performed an initial assessment of how selected alarm-settings are displayed by one particular model of a physiological monitor. The procedure presented here may serve as a blueprint for following-up on universal issues regarding device use identified by incident reports.

4.1 Comparing the Results of the Formal Analytical Assessment and the Simulation Study

Perceptual Aspects of Selected Display Parts: Detection of Relevant Settings
For three of the four settings, the formal analytical assessment yielded problems with salience: the muted alarms, the alarm limits and the disabled measurement. For muted alarms and alarm limits, reasons are the small size, the low luminance and the low color contrast of the respective elements of the display. For the disabled measurement, information is not actively presented on the display, i.e. attention cannot be exogenously drawn to the absence of measurement. Only the display element indicating the blocking of alarms was considered to be salient, particularly because of its high contrast and luminance. In line with this assessment, the blocked alarms were almost always detected by the participants of the simulation study, whereas more than half of the participants never recognized that alarms may not have sounded because the alarms were muted or because alarm limits had been changed.

Of course, display salience is not the only factor to determine visual information sampling. After debriefing, several participants indicated that volume settings are never changed at their hospital. Therefore, they did not expect this parameter to vary. It may be hypothesized that this affected their top-down attentional control and information sampling strategy, i.e. they simply did not check for the possibility of a muted alarm. This is a particularly adverse constellation, because attending to a potentially critical element of the display is supported by neither bottom-up nor top-down factors. Whereas the volume settings may not be of critical value, per se, at least the complete muting of sound may be indicated with greater salience (e.g. by changing its contrast or color), to overcome the hazard associated with missing this critical information. In a similar vein, the high detection rate for "blocked alarms" may not have been caused by a particularly salient display but rather by the participants' expectations, based on their previous experience: In fact, the only participant, who did not identify that alarms were blocked, had the least experience as an intensive care nurse.

Cognitive Aspects of Selected Display Parts: Comprehension of Relevant Settings
The degree to which a display element captures the user's attention is just one aspect of display usability. To achieve full situation awareness, users not only have to perceive

the relevant elements of the environment – they have to be able to effortlessly draw the correct conclusions, i.e. comprehend the meaning of their percepts. As discussed above, this should be particularly easy, when the display element in question includes the complete information necessary for interpretation. Comprehension may require more effort, if the displayed element has to be related to other information on the display or to knowledge from long term memory. We regard the general muting of alarms and the blocking of visual and acoustic alarms of a particular parameter to be examples of the former: Both are indicated by a single symbol - a crossed bell (IEC 60417-5576) and a crossed alarm-symbol (IEC 60417-5319), respectively. Once the meaning of these symbols is learned, it may be almost automatically derived. This is an intuitive process for the crossed bell, whereas some learning may be necessary for the crossed alarm symbol. However, because both signs are defined in IEC 60417 and are used frequently on medical devices, professional clinical users should not have difficulty with comprehension.

Comprehending the actual meaning of a particular pair of alarm limits (or changes to them) may not be achieved as easily. These values have to be interpreted relative to, for instance, default values or patient-specific criteria. To conclude that the current settings of the device are the reason why an expected alarm does not sound, users may not rely on a single element of the display but rather have to relate the current parameter value to the alarm limits. In a similar vein, to conclude that no information at all is available for a parameter, which is not displayed on the screen, is not directly evident from this absence: Users need to possess a well-developed mental model of this function and they must be able to recall the respective information from long-term memory. The latter is a question of their current mental workload. Notably, for both the disabled measurement and the changes to alarm limits, there were several participants (3 and 4, respectively), who never recognized these settings as potential causes for the absence of an alarm, although they had been asked several times, if the settings were safe for further monitoring. Whether this was due to the limited salience of the respective display parts (see above) or suggests difficulties in comprehending the associated displays cannot be concluded from the present data.

4.2 Limitations of the Current Study and Future Directions

Our current study was only a first step towards investigating the different factors that contribute to users' alarm-specific situation awareness. While our study revealed some interesting points, several aspects will have to be addressed by future research. For instance, it would be interesting to compare different device models of the same or different manufacturers to identify, which specific features of display design may particularly contribute to awareness of settings. Another aspect relates to the ecological validity of the simulation. There is some evidence that participants did engage in the situation, since all but one thoroughly conducted the initial control. The simulated tasks were typical for those performed on an ICU, as was the interruption and call to a different patient, such that one patient could no longer be monitored visually (see also [10]. Similarly, four of the seven participants indicated that situations, where colleagues changed device settings without notification, are conceivable, one had even personal experience. Nevertheless, future studies should increase the ecological validity of the simulation. This may include using several networked monitors, increasing overall workload and time pressure or presenting typical ICU-noise to approximate perceptual and cognitive demands of the simulation to those of a typical work environment [10]. Noticing the changed settings was measured using a graded system of verbalized conscious detection (direct identification, prompted identification, no identification). Future studies should make use of additional measures to less obtrusively assess awareness of device settings, e.g. by analyzing eye movements or pupil dilation to assess the distribution of users' attention and their information sampling behavior, respectively, see also [11, 12].

References

1. Diaper D, Stanton NA (2004) The handbook of task analysis for human–computer interaction. Lawrence Erlbaum Associates, London
2. Endsley MR (1996) Automation and situation awareness. In: Parasuraman R, Mouloua M (eds) Automation and human performance: theory and applications. Lawrence Erlbaum, Mahwah, pp 163–181
3. Endsley MR (1988) Design and evaluation for situation awareness enhancement. Proc Hum Factors Soc Annu Meet 32:97–101
4. Endsley MR (2011) Designing for situation awareness: an approach to user-centered design. CRC Press, Boca Raton
5. Janß A, Lauer W, Radermacher K (2009) Bewertung sicherheitskritischer Systeme im OperationssaalEvaluation of Risk-Sensitive Systems in the OR. i-com Zeitschrift für interaktive und kooperative Medien 8:32
6. Janß A, Radermacher K (2014) Usability first. Bundesgesundheitsblatt - Gesundheitsforschung - Gesundheitsschutz 57:1384–1392
7. Kluwe RH (2006) Informationsaufnahme und Informationsverarbeitung. In: Zimolong B, Konradt U (eds) Ingenieurpsychologie. Hogrefe, Göttingen, pp 35–70
8. Lange K, Brinker A, Nowak M et al Patientengefährdungen durch Gerätediversität? Diskussion eines Risikofaktors anhand der Ergebnisse zweier Befragungen an deutschen Kliniken. Der Anästhesist (in press)
9. Lange K, Nowak M, Neudörfl C et al (2017) Umgang mit Patientenmonitoren und ihren Alarmen: Vorkommnismeldungen liefern Hinweise auf Probleme mit Gerätewissen. Zeitschrift für Evidenz Fortbildung und Qualität im Gesundheitswesen 125:14–22
10. Lange K, Nowak M, Zoller R et al (2016) Boundary conditions for safe detection of clinical alarms: an observational study to identify the cognitive and perceptual demands on an Intensive Care Unit. In: de Waard D, Brookhuis KA, Toffetti A, Stuiver A, Weikert C, Coelho D, Manzey D, Ünal AB, Röttger S, Merat N (eds) Human factors & user experience in everyday life, medicine, and work. Proceedings of the human factors and ergonomics society Europe chapter 2015 annual conference. ISSN 2333-4959 (online), pp 195–208
11. Steelman KS, Mccarley JS, Wickens CD (2011) Modeling the control of attention in visual workspaces. Hum Factors 53:142–153
12. Wickens CD (2015) Noticing events in the visual workplace: the SEEV and NSEEV models. In: Szalma JL, Scerbo MW, Hancock PA, Parasuraman R, Hoffman RR (eds) The Cambridge handbook of applied perception research. Cambridge University Press, Cambridge, pp 749–768

Specific Risks Related to Robotic Surgery: Are They Real?

Luca Moraldi(✉), Giuseppe Barbato, and Andrea Coratti

Azienda Ospedaliero-Universitaria Careggi, Florence, Italy
luca.moraldi@gmail.com

Abstract. Robotic surgery, in recent years, has had an exponential increase, both as a number of procedures and as new indications to a robotic surgical approach. With the massive use of the robotic approach in surgery, the safety of this device and monitoring of any malfunctions play an increasingly important role. The goal of this literature review is to analyze different causes of unexpected complications or potential errors in the Da Vinci system, to increase patient and operator safety in the future. The complication rate directly related to robotic malfunction is very low, approximately from 0,02% to 4,97%. Splitting the data across studies MAUDE based and single or multi center experience will notice immediately that the rate is lowest among the MAUDE based (from 0,02% to 0,61% vs from 2,39% to 4,97%). Malfunctions are uncommon and the need to abort or convert to another modality is rare. Most importantly, while mechanical and electronic errors can happen, they do not appear to impact surgical outcomes or patient safety.

Keywords: Robotic surgery · Da Vinci system · Malfunctions

1 Introduction

The use of robotic-assisted surgery has greatly expanded worldwide since it was first approved in 2000 [1]. During the last years, both the number of procedures performed using the Da Vinci system and the number of Da Vinci systems installed in hospitals have exponentially expanded [2]. Surgical robots allow surgeons to conduct minimally invasive, but complex procedures, with better visualization, enhanced precision and better dexterity compared to laparoscopy. The Da Vinci system provide 3-dimensional magnified view of the surgical operations and translate the surgeon's hand, wrist, and finger movements into precisely engineered movements of miniaturized surgical instruments inside the patient's body. Robotic surgery permit surgeons to perform complex surgical tasks through tiny incisions with high accuracy. Compared to open surgery, robotic surgery results in less pain and a very small scars. Like other minimally invasive techniques, robotic surgery provides a cosmetic benefit compared to open approaches. The rapid development in technology and improvement in surgeon skills have allowed robot-assisted procedures, such as single site surgery and natural orifice transluminal endoscopic surgery (NOTES), to result in minimal impact on cosmetic outcome and reduce postoperative pain. As briefly described, applications of robotic surgery are constantly expanding with increasing complexity procedures that

S. Bagnara et al. (Eds.): IEA 2018, AISC 818, pp. 465–470, 2019.
https://doi.org/10.1007/978-3-319-96098-2_58

require accurate tools while maintaining safety of the patient and of the operators as the cornerstone of every procedure. The aim of this literature review is to analyze different causes of unexpected complications or potential errors in the Da Vinci system, in order to avoid adverse events in the future.

2 Materials and Methods

The purpose of this study is a literature review that focus on determining the safety of robot-assisted surgery including eventual malfunctions during robotic procedures.

In the US, when an unexpected event or a malfunction occur, the hospital is required to report these incidents to the manufacturer, whoever in turn is required to report those incidents to the Food and Drug Administration (FDA). The FDA then creates a deidentified report of the incidents in the Manufacturer and User Facility Device Experience (MAUDE) database [3, 4].

The MAUDE database is a public collection of "suspected medical device-related" adverse event reports, submitted by mandatory (user facilities, manufacturers, and distributors) and voluntary (health care professional, patients, and customers) reporters to the FDA. This research included studies relative to single or multi-centric experience in robotic procedures across major surgical centers. This study is based on an online search performed using Medline, PubMed, EMBASE and Cochrane. The keywords used were: "robotic surgery OR robot-assisted surgery OR Da Vinci OR intuitive" AND "malfunction OR adverse event OR device failure OR complication OR safety". We further expanded our search by reviewing the bibliographies of related studies for potential relevant papers. All studies published on robotic surgery, with a focus on specific risks of robot assisted surgery, were fully and deeply reviewed.

Inclusion criteria were: analysis of the MAUDE database or institutional experience about malfunction occurring during robotic procedures.

Exclusion criteria were: comparison or inclusion of other minimally invasive laparoscopic surgical techniques.

All the adverse events are divided based on relationship of event to device malfunction and adverse event severity.

3 Results

The complication rate directly related to robotic malfunction is very low, approximately from 0,02% to 4,97%. Splitting the data across studies MAUDE based and single or multi center experience will notice immediately that the rate is lowest among the MAUDE based (from 0,02% to 0,61% vs 2,39% to 4,97%).

Rajih et al. [5] made a single center review of 1228 robotic multispecialty procedures performed from 2012 to 2015 with 61 failures (4.97%) but no adverse event, no conversion to open technique but only surgical delay. The most common error was related to pressure sensors in the robotic arms indicating out of limit output (2.04% - 25/1228). In this report, the incidence of robotic malfunction rate during surgery was

3.7%. Surgical delay was reported only in one patient, robotic malfunction occurrence does not affect patient safety or surgical outcome (Table 1).

Table 1. Analysis of the adverse event on robotic surgical systems reports in MAUDE database and authors experience.

Author	Total robotic procedures	Adverse events	Period	Type of study	Percentage
Agcaoglu [8]	223	10	2008–2011	Mono	4,48
Rajih [5]	1228	61	2012–2015	Mono	4,97
Kim [10]	1797	43	2005–2008	Multi	2,39
Chen [11]	400	14	2005–2011	Mono	3,50
Gupta [6]	1057000	2837	2009–2012	MAUDE	0,27
Cooper [13]	1180000	245	2000–2012	MAUDE	0,02
Lucas [12]	435000	1914	2003–2009	MAUDE	0,44
Alemzadeh [7]	1750000	10624	2000–2013	MAUDE	0,61

Gupta et al. [6] identified 2837 events for MAUDE analysis from 2009 to 2012. The authors classified the events by severity in 75% as mild, 18% as moderate, 4% as severe. The 81.4% of the moderate (Deviation from planned procedure without change in surgical outcome) and severe (Significant intra-operative deviation from planned procedure requiring aggressive intervention) adverse events are not related to malfunction but to patient's baseline health, or surgeon's error.

Alemzadeh [7] analyzed all the adverse events reported to MAUDE database from 2000 to 2013. 8061 device malfunctions were reported. The number of injury per procedure have stayed constant over the years. Despite widespread adoption of robotic assisted surgery, a non negligible number of technical difficulties and complications are still being experienced during operations. As a consequence of the increasing use of the complex robot-assisted procedures in surgical specialties there are a higher number of adverse events.

Furthermore, Alemzadeh identified five major categories of device and instrument malfunctions:

- system errors and video/imaging problems (7.4%)
- falling of the broken/burnt pieces into the patient's body (14.7%)
- electrical arcing, sparking, or charring of instruments (10.5%)
- unintended operation of instruments (10.1%)

Alemzadeh found that despite a relatively high number of reports, the vast majority of procedures were successful and did not involve any adverse event and the number of injuries/death per procedure has stayed relatively constant since 2007.

Agcaoglu [8] made a single center review of 223 cases with 10 failures (4,5%) from 2008 to 2011 but none of them led to adverse patient consequences or conversion to open technique. Most of the malfunction in their series could have been discovered and managed earlier with a comprehensive evaluation of the system before the incision. The author compares the data of the Talamini [9] study and those of his own study highlighting how the surgical team learning curves and technical improvement of the equipment shows a significantly lower conversion rate (16% vs 2.3%).

Kim [10] analyze a multi center urology review of 1797 cases with 43 failures (2.4%). Chen [11] perform a single center review from 2005 to 2011 analyzing 400 cases with 14 failures but no adverse event (3,5%). Both studies report overlapping results with the data of similar studies in the literature.

4 Discussion

Surgical robots allow surgeons to conduct minimally invasive, but complex procedures, with better visualization, enhanced precision and better dexterity compared to laparoscopy. The surgeon works at an operating microscope, personally controls the vision, uses a "third" hand to manage exposure and traction, has an assistant who keeps the operating field exposed and clean. This "condition" is profoundly different from the laparoscopic one and allows to work with extreme precision even in very complex technical contexts like enucleation of a small tumor in the uncinate process or in the head of the pancreas. And it is obvious, therefore, that even a complex gesture, such as the control of a bleeding, is more feasible and effective than in laparoscopy. From this derives, in general, the reduction of blood loss and the conversion rate in open reported for robotics in the field of complex interventions.

Robotic assisted surgery carries specific risks of an advanced computerized system, such as the possibility of mechanical failure. The da Vinci surgical system is a computer-assisted device with three main integrated subsystems:

(1) the surgeon console, which is the control center of the system
(2) the patient-side cart, including the robot with articulated mechanical arms
(3) vision cart, which contains supporting hardware and software components, including the electrical surgical unit

The possibility of a robotic malfunction can potentially occur in any of these subsystems.

Lucas et al. [12] made a retrospective review of MAUDE database from 2003 to 2009 categorizing malfunctions as problems in the console, patient side cart, camera, instruments and cannula accessories. Adverse events caused by malfunction are categorized as aborted procedure, injury or death. Patient injury did not change with year of surgery (0,5–5,4% of malfunction) and open conversion declined.

A severe limitation of these MAUDE based studies is within the data itself [12, 13] utilizing the MAUDE database. The number of events relies on voluntary reporting by the surgical team. The degree of underreporting of device malfunctions cannot be determined [14]. Single or multi center analysis of incidence or malfunctions in robotic surgery reveals higher reports then the MAUDE based review, due to the possible underreporting of the adverse events in the MAUDE database and the overreporting of the annual number of robotic procedures in the manufacturing company investor presentations, the estimated rates of adverse events per procedures should be conservative and represent the lower bounds on the prevalence of events. An additional limitation of the database is the mode of recording data into the database because they are entered in a standardized case report which adds translational bias. This study is not without limitations because of its retrospective design, it is more likely to suffer from selection, misclassification, or information bias. There is no method to ascribe causality between an adverse event during a robot-assisted procedure with a surgical complication, doing it with coded parameters is an even more difficult task. Various authors who have addressed the issue before us have approached the problem with different methods. The Clavien–Dindo scale is the most widely accepted classification of surgical complications, but it is not adequate to fully describe complications that are associated with robot-assisted procedures. As complex medical devices become more ubiquitous in the operating room, it is important to characterize complications associated with these devices as well as to use a standardized method to ascribe causality. A standardized classifier would allow for an easier comparison of events across disciplines and institutions, and it would allow us to standardize reporting by which complications associated with surgical systems can be attributed to device failure. While the robotic surgical systems have been successfully adopted in many different specialties the overall numbers of injury and death events per procedure have stayed relatively constant over the years. As robotic surgery grows, it is important that the true incidence of complications that occur with the system be known to ensure continued safe innovation. The decline in robotic malfunction provoked by the surgeon observed in this report suggests the need of a dedicated experienced team to minimize robotic malfunctions [15–17].

5 Conclusions

Malfunctions are infrequent, and the need to abort or convert to another modality is rare. Most importantly, while mechanical and electronic errors can happen, they do not appear to impact surgical outcomes or patient safety. Overall, the unrecoverable fault rate with the technology is low. However, we need to continue to critically evaluate this outcome. Although the structural limits of the MAUDE database, even in Europe and Asia, the exponentially spreading of robotic making desirable to establish an international database that includes all the malfunctions over the world. In conclusion robotic surgery offers many potential benefits with low risks related robotic malfunctions, but it is important the correct training of surgical team.

References

1. Hussain A, Malik A, Halim MU, Ali AM (2014) The use of robotics in surgery: a review. Int J Clin Pract 68(11):1376–1382
2. Annual report 2016. http://www.annualreports.com/HostedData/AnnualReports/PDF/NASDAQ_ISRG_2016.pdf
3. MAUDE: Manufacturer and User Facility Device Experience. U.S. Food and Drug Administration. http://www.accessdata.fda.gov/scripts/cdrh/cfdocs/cfMAUDE/search.CFM
4. Gurtcheff SE (2008) Introduction to the MAUDE database. Clin Obstet Gynecol 51(1):120–123. https://doi.org/10.1097/GRF.0b013e318161e657
5. Rajih E (2017) Error reporting from the da Vinci surgical system in robotic surgery: a Canadian multispecialty experience at a single academic centre. Can Urol Assoc J 11(5): E197
6. Gupta P (2017) Development of a classification scheme for examining adverse events associated with medical devices, specifically the DaVinci surgical system as reported in the FDA MAUDE database. J Endourol 31(1):27–31
7. Alemzadeh H, Raman J, Leveson N, Kalbarczy Z (2016) Adverse events in robotic surgery: a retrospective study of 14 years of FDA. PLoS ONE 11(4):e0151470
8. Agcaoglu O (2012) Malfunction and failure of robotic systems during general surgical procedures. Surg Endosc 26(12):3580–3583
9. Talamini M (2002) Robotic gastrointestinal surgery: early experience and system description. J Laparoendosc Adv Surg Tech Videosc 12(4):225–232
10. Kim WT (2009) Failure and malfunction of da Vinci surgical systems during various robotic surgeries: experience from six departments at a single institute. Laparosc Robot
11. Chen CC (2012) Malfunction of the da Vinci robotic system in urology. Int J Urol. https://doi.org/10.1111/j.1442-2042.2012.03010.x
12. Lucas SM, Pattison E, Sundaram CP (2012) Global robotic experience and the type of surgical system impact the types of robotic malfunctions and their clinical consequences. An FDA MAUDE review. BJU Int 109(8):1222–1227
13. Cooper M, Ibrahim A, Lyu H, Makary MA (2015) Underreporting of robotic surgery complications. J Health Qual 37(2):133–138
14. Andonian S (2008) Device failures associated with patient injuries during robot-assisted laparoscopic surgeries: a comprehensive review of FDA MAUDE database. Can J Urol 15(1):3912–3916
15. Dubeck D (2014) Robotic-assisted surgery: focus on training and credentialing. PA Patient Saf Advis 11(3):93–101
16. Catchpole KR (2018) Diagnosing barriers to safety and efficiency in robotic surgery. Ergonomics 61(1):26–39
17. Ashwin N (2017) Sridhar training in robotic surgery—an overview. Urosurg Curr Urol Rep 18:58. https://doi.org/10.1007/s11934-017-0710-y

From Causality to Narration: The Search for Meaning in Accident Analyzes

Ciccone Elodie(✉), Cuvelier Lucie, and Decortis Françoise

Université Paris 8, 2 rue de la Liberté, 93526 Saint-Denis, France
ciccone.elodie@gmail.com

Abstract. This paper aims to reflect on safety trainings. Based on psychologic and ergonomic research, we want to see how the narrative approach and the use of stories can help to improve current safety trainings. After a reflexion about the current place of narratives in safety and the possible ways of development, we will present a study on a training based on simulation in anesthesia. This study aims to show the importance of the narrative approach in relation to causal approach.

Keywords: Safety training · Narrative approach · Simulation training

1 Introduction

Our research aims to develop innovative safety trainings in the medical field. It is conducted in partnership with a pediatric resuscitation anesthesia service. In this service, simulator trainings have been conducted for 10 years. The demand we answer is to improve these trainings. Our research hypothesis is that one possible way to improve safety trainings is to move from a causal (classical) approach to a narrative approach. This hypothesis is based on the evolution of our way of thinking about safety. Over the last few years, a number of limitations have been pointed out regarding risks and accidents causal analyses (Lundberg et al. 2009). Models for thinking about safety have shifted from a linear (cause-and-effect) view to a more systemic one, where we try to understand the impact of all the components of the system and where we take more into account resilience capacity of systems (Hollnagel et al. 2006).

Admittedly, the models evolved, but the tools used to analyze accidents and the principles underlying the design of training remain based on causal analytical approaches. The search for the causes of accidents is always at the heart of our safety procedures (for example, the causal-tree method). In this context, how can we improve our safety tools and training? Which theories, different from the causal approach, can help us to understand accidents and develop new trainings? Researches in psychology and ergonomics about narrative activity and stories give pathways for moving forward. After presenting them (part 1), we will explore the possible links between these notions and the current accident analysis tools (part 2). Then we will illustrate our point with a research we conduct currently that aims to explore empirically these links (part 3).

S. Bagnara et al. (Eds.): IEA 2018, AISC 818, pp. 471–478, 2019.
https://doi.org/10.1007/978-3-319-96098-2_59

2 The Fundamental Role of Narrative Activity and Reflexivity in Development

As Bruner points out (2006), our relationship to the world is first of all narrative. Individuals learn, grow, and develop themselves through stories that play a fundamental role in the construction and transmission of a culture. Anyway, for Ricoeur (1983), our first relation to language is not that we speak but that we listen. This "narrative gift" as Bruner calls it is innate and a source of development from the youngest age. Researchers interested in adult development are also thinking about how narrative can support development. Pastré (2011) explains that in all learning there is a part that comes from us and a part that comes from others, transmitted through stories, conveying lived experiences.

On the other hand, research into professional didactics has focused on the role of reflection and reflexivity in the development of skills in the field of education (Donnay and Charlier 2008, Faingold 2015; Perrenoud 1998). Reflection on a lived situation consists in reconstructing what happened. Reflexivity, on the other hand, consists of taking a distance from this situation. This distancing is a source of development, as Perrenoud (1998) explains. "Insofar as singular action is accomplished, thinking about it only makes sense in the aftermath to understand, learn, integrate that happened. Reflection is not limited to an evocation, but goes through a critique, an analysis, a connection with rules, theories or other actions possible or attested in a similar situation" (p. 3). The narrative approach allows a better understanding of the entry into the reflection on work situations involving humans and their environment. Donnay and Charlier 2008, for example, use this link between language and reflexivity, they explain that the level of language used by individuals helps to identify the moments of reflexivity, for example, if we talk about a situation that we just lived, we are less in the reflexivity than when we put in link this situation with others lived (to give an example of similar situation already lived). This reflection on the situations is a base of sharing (proposes ways of doing, lived experiences…).

The narrative approach has only recently been mobilized by ergonomics. At first, work on the narrative approach took off in the field of child-oriented ergonomics (Decortis 2013) and in the field of cultural mediation (Bationo-Tillon 2013). Other aspects of ergonomics focus on the use of narratives, particularly those of training, didactics and education (Beaujouan 2011; Perrenoud 1998; Donnay and Charlier 2008, Faingold 2015). In the field of safety, studies recognizing the importance of stories are also beginning to emerge (Beaujouan and Daniellou 2012; Marchand and Falzon 2015).

3 Are Stories Currently Excluding Form Safety Training?

3.1 A Willingness for Objectivity

Analyses of past events, especially of adverse events (e.g., accidents, incidents, failures, errors, malfunctions) are seen as essential prerequisites to manage safety in organizations. These accident analyzes are at the heart of safety training: they are given as examples, feed the exercises, are references for the design of scenarios on simulators, etc. But the relationship to these past events is always in an analytical, argumentative and under the guise of objectivity.

A range of tools (such as causal trees) drawn from fields including human factors, ergonomics and safety science, are used to establish scientifically how and why an incident occurred in an attempt to identify how it, and similar problems, might be prevented from happening again. These accident analysis tools and the principles that underlie safety training are mostly based on research into the causes of accidents. For example, if an accident occurs, analysis processes based on the search for causes is used to understand it and to prevent a similar accident from happening again. Or, during a training exercise (for example a simulator-based scenario), a debriefing is used to understand the possible failure in the management and to identify the causes.

These causal models are supposed to be objective because they are based on facts. But, in practice, it is shown that the purely objective identification of facts and the establishment of certain causal links between these facts is strictly impossible. As Lundberg, Rollenhagen and Hollnagel point out in 2009, when we analyze accidents, we find what we are looking for. Interpretations and inferences are always present. Thus, the same event, analyzed by different experts and/or in different contexts give rise to multiple explanations. This explanation is therefore always somehow a narrative of an event by one or more narrators (Bruner 1996).

3.2 New Training Practices: A Way to Narrative Activity

3.2.1 New Training Practices: A Window on Narrative Activity

In addition to this impossibility of achieving objectivity, the work done in ergonomics shows that the safety of the systems is also based on «adaptive safety», that is to say on a set of know-how and rules of the trade (Daniellou et al. 2011; Nascimento et al. 2014). This concrete, operational knowledge arises from diverse disciplines and is based upon the experience of individuals and work groups. Hence, it includes local variations that specialists are not aware of. These local knowledges, brought by users at the sharp-end (workers, users, organizers, managers, supervisors, etc.) is essentially transmitted via the narratives.

In companies, stories are used for a long time to manage safety. For example, the storytelling (Salmon 2007) which promotes the companies' mark is also a way to promote safety. Old employees tell about their experience to younger, but this practice is often informal and not necessary controlled by management. These stories may relate real experiences or myths. Recently, new training devices aim to make these managed safety practices more visible and to leave more room for these singular knowledge transmissions. The new trends in management sciences even advocate the institutionalization of discussion time (Detchessahar 2011, 2013; Falzon 2008; Mulholland et al. 2005). In safety sciences, these devices take different names: «work debates spaces» «constructive cause analysis» or «activity sharing space» for example (Cuvelier et al. 2017; Rocha et al. 2015; Thellier et al. 2015; Van Belleghem 2016). Their aim is to facilitate the dialogue of views and knowledge in order to transform the conditions of people's activity and the people themselves.

The challenge is now to develop a true "engineering of discussion", namely a system whereby discussion is no longer considered as one means among others to serve the stakeholders, but rather as the very subject of the intervention (Detchessahar 2013; Rocha et al. 2015; Van Belleghem and Forcioli Conti 2015). Current research perspectives are focused upon the need to understand and develop this type of mechanism and provide tools for it. In safety sciences, current work is being undertaken to investigate the possibility of developing other risk analysis mechanisms in this new constructive perspective (Rocha et al. 2015; Van Belleghem 2016). The creative and developmental power of "narratives activity" and "story" is a promising framework to get down this path (Bruner 1996; Clot and Santiago-Delefosse 2004; Decortis 2013).

4 The Example of Simulator Training in Anesthesia

The study we are going to present now is conducted in partnership with anesthetists in an intensive care unit and it concerns a simulator communication training. In this study we are interested in post-simulation debriefing moments to study the place of stories in this formation. The questions we are trying to answer are: Are there narratives in these existing trainings? If yes, which stories are transmitted? Why? How? How are they exploited? Do the trainers' analyzes and interventions aim to encourage and exploit the stories?

4.1 Presentation of the Training Device

This training in communication was developed by pediatric anesthetists. It aims to improve the ability of caregivers (physicians and nurses) to communicate with patients and their families to improve care and reduce the possibility of medical error related to miscommunication. The training concerns on the one hand the interns in pediatric anesthesia and intensive care and on the other hand the nurses of the same service. Interns are trained by physicians (supervisor) and nurses by other nurses (colleagues). Psychologists also participate in these trainings as trainers. After a theoretical contribution on the psychological defense mechanisms that parents can show when their child is hospitalized, the trainees are led to participate in 4 simulated cases. Interns must announce difficult news to parents, nurses meet families during their care round. Simulated parents are played by actors.

4.2 Data Collection and First Results

We attended two interns training sessions and two nurses training sessions. In these four sessions, we filmed the post-simulation debriefings, i.e. 16 films equivalent to about 10 h of videos. The analysis of these films are in progress.

The first analyzes show that in these formations, the narrative activity is at two levels:

- that of the reconstitution of the simulated situation (extract 1)
- that of the story of an event actually lived in the past (extract 2).

Extract 1

Intern 1: we do not know if it's uncomfortable, we know that she saw that in the script, or is it the fact that she waited without hearing from her child, we know it was a trauma for her to see what happened in the emergency room. We arrive in front of this mother, we know that in emergencies it happened, that it's been an hour that she is waiting and we took care of his child but we do not know what made her in this state there we know a little but we do not know what specific episode is that it's the fact that she stayed alone so as she starts to talk it allows to engage the conversation
Psychologist: But before she started talking is what you spotted, there was something between them early in the interaction and with you when the father arrived, what did he tell you?
Intern 2: You talked about the child he went straight to his wife
Psychologist: And what did he say?
Intern 2: Are you okay?
Psychologist: He told them "wait" by raising your hand and he took care of his wife while you may have been at that moment in a desire to talk about saying because you've been waiting also since a long moment of silence in front of this woman on the ground, you were spectators of her distress and the dad he arrived you started to "we will explain" and in fact he stopped you, you went account?

In this sequence we see the hypotheses emitted by the intern 1 to try to identify the situation in front of which it is located. We also see the role of the psychologist trainer on the reconstruction of the facts, if the interns had cut little part of the story to make it clearer, the psychologist allows them to highlight events to which he had not paid attention and reconstruct the whole story to explain the reaction of the father to his arrival.

Extract 2

Psychologist: Maybe these parents were not very very finally that their reaction was moderate there are probably parents more difficult to manage and at the same time the first difficulty is the mother who removes the electrode, **is there something in your experience, is there reactions that promote communication or that, on the contrary, there are** …
Trainer: How do you react?
Nurse 1: In general there are parents who say the noise it bothers but the noise is made for that the alarms are strong because we need but there are parents who react badly who want to lower the sound or stop the thing but no we can not
Psychologist: And in this case suddenly you explain what the point is?
Nurse 1: We explain it's surveillance yes
[…]
Nurse 2: Yes, there are a few parents who set on the scope
Nurse 1: There it was that the baby didn't exist, for that mom there was only the machine but the baby she never took him in the arms it was weird like relationship
Nurse 3: I had Dad put sat on his finger because it sounded too much
Trainer: How do we react to this?

Nurse 3: So a little surprise because suddenly it had already rang a first time I was introduced me a second time and then I made him understand finally especially there he had saturation problems including so there is a little complicated «yes but it sounds and then it annoys me" "oh yes but you must not do it because it is very dangerous for your child there you know he has small problems in his saturation"
Nurse 1: He's the one we're watching, you're not

In this extract, the psychologist trainer attempts to relate the simulated situation to situations already lived. In this way, the trainees explore situations in which they meet similar problems. This allows them to enrich the plot and make other assumptions about how the situation might have happened and share ways of doing things. This narrative of lived situations related to simulated situations allows participants to create or consolidate rules collectively.

Finally, the important point that can be noted is that this training modality with post-simulation debriefings and trainers that relaunch both to flesh out the plot and to link simulated situations with lived situations allows to get out of a purely causal analysis of the «what happened» situations to move to a more narrative mode that explores more possibilities. The story, unlike the analysis of the causes, includes a plot (Pastré 2005b). This means that it has two dimensions: the causal sequence dimension and the contingency dimension. It is characteristic of the story, its richness: the plot contains a mixture of causal relationships, of finality and chance relation (Pastré 2005a). This narrative aspect is also favored by the presence of comedians. For example, in extract 3 which follows, we see that the actor who plays the father describes the scene as it happened and explains what he would have liked as a father as intervention of physicians.

Extract 3

Father: I could hear a summary that I had been very short, what I would have liked you to do to me is "your child stopped to respire and then suddenly he is fine" "Ah you see everything is fine" here I would have understood better while there "I will take since the departure, the bronchiolitis …" I make fun because it was actually the father was in a problematic very … after that it was still very you were open and we managed to make it talk and make it sort

The fact that the "father" can express his feelings on the situation is also a way to explore the possibilities, in this time not only from the point of view of the caregiver but by integrating the point of view of their interlocutors.

5 Conclusion: Two Ways of Mind to Articulate

According to Bruner (2000), there are two types of cognitive functioning, "two modes of mind". Each one represents a particular way of ordering experience and learning. The first form is argumentation. It aims to establish formal and empirical evidence. It makes it possible to define universal conditions of truth and to state laws in the form "if … then …". It is this form of mind that underlies all current safety practices. The second form is the narrative practice described above. For Bruner, these two modes of mind are complementary and irreducible to each other. To grasp the rich diversity of mind, it is necessary to retain these two modes of mind: "any attempt to reduce one of these two modes to another, or to ignore one for the benefit of the other inevitably result in not being able to grasp the rich diversity of mind » . The rest of our project seeks to define practical modalities that promote concrete articulation in the safety training of these two modes of mind.

References

Bationo-Tillon A (2013) Ergonomie et domaine muséal. Activites. https://doi.org/10.4000/activites.752

Beaujouan J (2011) Contributions des récits professionnels à l'apprentissage d'un métier. Le cas d'une formation d'ergonomes. Thèse de doctorat de l'Université de Bordeaux

Beaujouan J, Daniellou F (2012) Les récits professionnels dans une formation d'ergonomes. Le travail humain 75(4):353–376. https://doi.org/10.3917/th.754.0353

Bruner J (1996) The culture of education. Harvard University Press, Cambridge

Bruner J (2000) Culture et modes de pensée: l'esprit humain dans ses oeuvres. Retz, trad. Bonin, Y, Paris

Bruner J (2006) La culture, l'esprit, les récits. Enfance 58(2):118–125

Clot Y, Santiago-Delefosse M (2004) La perspective historico-culturelle en psychologie. Présentation. Bulletin de psychologie 469(57–1):3–4

Cuvelier L, Benchekroun H, Morel G (2017) New vistas on causal-tree methods: from root cause analysis (RCA) to constructive cause analysis (CCA). Cogn Technol Work. https://doi.org/10.1007/s10111-017-0404-8

Daniellou F, Simard M, Boissières I (2011) Human and organizational factors of safety: state of the art. Retrieved from Toulouse. http://www.foncsi.org/fr/publications/collections/cahiers-securite-industrielle/human-and-organizational-factors-of-safety-state-of-the-art-1/human-and-organizational-factors-of-safety-state-of-the-art

Decortis F (2013) L'activité narrative dans ses dimensions multi instrumentée et créative en situation pédagogique. Activités 10(1):3–30. http://www.activites.org/v10n31/v10n31.pdf

Detchessahar M (2011) Santé au travail: quand le management n'est pas le problème mais la solution…. Revue française de gestion 214:89–105

Detchessahar M (2013) Faire face aux risques psycho-sociaux: quelques éléments d'un management par la discussion. Négociations 1(19):57–80. https://doi.org/10.3917/neg.019.0057

Donnay J, Charlier E (2008) Apprendre par l'analyse de pratiques: initiation au compagnonnage réflexif (2ème éd. revue et augmentée). Presses universitaires de Namur, Namur

Faingold N (2015) Un dispositif universitaire centré sur un travail réflexif d'exploration du vécu subjectif: formation et transformation de soi. Rech Form 80:47–62. https://doi.org/10.4000/rechercheformation.2499

Falzon P (2008) Enabling safety: issues in design and continuous design. Cogn Technol Work 10:7

Hollnagel E, Woods D, Leveson N (2006) Resilience engineering: concepts and precepts. Ashgate, Aldershot

Marchand AL, Falzon P (2015) Usages pédagogiques de la pratique anecdotale dans la formation à la gestion des risques. Psychologie du Travail et des organisations 21(3):248–259

Mulholland P, Zdrahal Z, Domingue J (2005) Supporting continuous learning in a large organization: the role of group and organizational perspectives. Appl Ergon 36(2):127–134. https://doi.org/10.1016/j.apergo.2004.09.009

Nascimento A, Cuvelier L, Mollo V, Dicioccio A, Falzon P (2014) Constructing safety: from the normative to the adaptive view. In: Falzon P (ed) Constructive ergonomics. CRC Press, Boca Raton, p 294

Pastré P (2005a) Apprendre par la résolution de problèmes: le rôle de la simulation Apprendre par la simulation. Octarès, Toulouse, pp 17–40

Pastré P (2005b) Apprendre par la simulation: de l'analyse du travail aux apprentissages professionnels. Octarès, Toulouse

Pastré P (2011) La didactique professionnelle. Approche anthropologique du développement chez les adultes

Perrenoud P (1998) De la réflexion dans le feu de l'action à une pratique réflexive. Université de Genève, Faculté de psychologie et des sciences de l'éducation (repris dans Perrenoud, Ph. Développer la pratique réflexive dans le métier d'enseignant. Professionnalisation et raison pédagogique. Paris: ESF, 2001)

Ricoeur P (1983) Temps et Récit - Tome 1: L'intrigue et le récit historique. Seuil, Paris

Rocha R, Mollo V, Daniellou F (2015) Work debate spaces: a tool for developing a participatory safety management. Appl Ergon 46(Part A(0)):107–114. https://doi.org/10.1016/j.apergo.2014.07.012

Thellier S, Falzon P, Cuvelier L (2015) Construction of an "activity sharing space" to improve healthcare safety. In: 32nd annual conference of the European Association of Cognitive Ergonomics (ECCE 2015), Warsaw, Poland, 1–3 July

Van Belleghem L (2016) Eliciting activity: a method of analysis at the service of discussion. Le Travail Humain 79(3):285–306

Van Belleghem L, Forcioli Conti E (2015) Une ingénierie de la discussion? Chiche! 50ème congrès de la SELF, Paris les 23, 24, 25 septembre 2015

Things Falling Through the Cracks: Information Loss During Pediatric Trauma Care Transitions

Peter Hoonakker[1(✉)], Abigail Wooldridge[1], Bat-Zion Hose[1], Pascale Carayon[1], Ben Eithun[1], Thomas Brazelton III[1], Shannon Dean[1], Michelle Kelly[1], Jonathan Kohler[1], Joshua Ross[1], Deborah Rusy[1], and Ayse Gurses[2]

[1] University of Wisconsin-Madison, Madison, WI, USA
peter.hoonakker@wisc.edu
[2] Johns Hopkins University, Baltimore, MD, USA

Abstract. Pediatric trauma is one of the leading causes of morbidity and mortality in children in the USA. Several clinical teams converge to support trauma care in the Emergency Department (ED); the most severe trauma cases often need surgery in the operating room (OR) and/or are admitted to the pediatric intensive care unit (PICU). These care transitions can result in loss of information or transfer of incorrect information, We interviewed 18 clinicians about communication and coordination during care transitions between the ED, OR and PICU. Clinicians completed a short questionnaire about patient safety during transitions. Results show that, although many services and units involved in pediatric trauma work well together, important patient care information may be lost in the transitions. To safely manage transitions of this fragile, unstable, complex population, we need to better manage the information flow during these transitions.

Keywords: Pediatric trauma · Teamwork · Care transitions · Patient safety

1 Introduction

Pediatric trauma is one of the leading causes of morbidity and mortality in children in the USA. Every year, nearly 10 million children are evaluated in emergency departments (EDs) for traumatic injuries, resulting in 250,000 hospital admissions and 10,000 deaths. Traffic accidents are by far the leading cause of death for children in the USA [1]. The nature of traumatic injuries creates work characterized by incomplete or nonexistent information (e.g., non-identified patient) and, thus, ambiguity about the patient's conditions and treatment [2]. Prior to arriving to the ED, patients are triaged into trauma levels based on perceived injury severity and mechanism: level 1 trauma cases are most urgent with potential immediately life-threatening injuries while level 2 trauma cases are still critical but less likely to have life-threatening injuries. Depending on the trauma level, several clinical teams (including emergency medicine, anesthesiology, surgery, and pediatric critical care) converge to help support trauma care in the ED. This collocation in

S. Bagnara et al. (Eds.): IEA 2018, AISC 818, pp. 479–488, 2019.
https://doi.org/10.1007/978-3-319-96098-2_60

the ED can help to support communication and coordination of team members. Pediatric trauma cases often need surgery in the operating room (OR) and/or admission to the pediatric intensive care unit (PICU). These transitions can result in loss of information or transfer of incorrect information, which can negatively affect the child's care. The multiple roles involved in pediatric trauma care increase care complexity and can pose challenges to communication and coordination. The number of personnel involved in (pediatric) trauma cases and the potential number of transitions is impressive; up to 53 different roles can be involved and children may experience up to 24 transitions between different services and units before discharge [3].

2 Background: Care Transitions

There are many types of care transitions, e.g., between shifts of healthcare professionals, hospital units, or care settings. In this study, we focus on inter-departmental care transitions, in particular transitions between the ED and OR, the OR and PICU and the ED and the PICU. Inter-departmental transitions carry a high risk because of transfer in at least three domains: provider, department, and physical location, which do not have to occur simultaneously [4]. Care transitions include handoffs, "a transfer of care involving a transfer of information, responsibility, and authority between clinicians" [5]; care transitions between hospital units can increase risk to the patient if information is lost [6, 7]. Effective handoffs are necessary for communication and coordination of successful patient care activities that improve care quality and safety [2, 5]. The literature has identified issues with information transfer during handoffs: omission, lack of data, delayed information, inaccuracies, poorly organized data and information overload [2, 8]. Horwitz et al. [4] examined the transition of adults from ED to inpatient care; they found that 29% of survey respondents reported that a patient of theirs experienced a near miss or an adverse advent after a transition. Bigham et al. [9] showed that 26% of handoffs in 23 children's hospitals were linked to a handoff-related care failure before the implementation of a standardized hand-off, which decreased these failures to 8%. The study also showed that ED to inpatient (IP) handoffs had some of the highest failures (37% before implementation). The standardized handoff tool reduced ED to IP unit handoff failures to 13% [9]. However, little is known about transitions in pediatric trauma care and factors associated with information flow problems.

The goal of this study is to understand the different transitions and the barriers to the flow of information. If we better understand the transitions and their barriers, we can design health information technology that supports the transitions. Eventually, better supported care transitions should lead to fewer missed injuries, less delay in care, more patient and family involvement, less clinician frustration and more satisfaction.

3 Methods

3.1 Setting and Sample

This study took place in an academic hospital in the Midwestern USA. Healthcare professionals, i.e. nurses, physicians and other support staff, in the ED, OR and PICU were involved in the study; physicians, i.e. attendings, fellows and residents, are on the emergency medicine, surgery, anesthesia or pediatric critical care services. At the hospital, pediatric trauma care is initiated in the ED where a specialized pediatric trauma team cares for the patient. After care provided in the ED, the patient may be transferred to another unit or discharged. We conducted interviews on transitions of care between the ED and OR, OR and PICU, and ED and PICU with 18 people (see Table 1).

Table 1. Interviewee and survey respondent characteristics

#	Service/unit	Job Title	Transitions	Duration
1	ED	ED RN	ED-OR, ED-PICU	0:56
2	ED	ED RN	ED-OR, ED-PICU	1:06
3	ED	Coordinator	ED-OR, ED-PICU	0:37
4	ED	Child life specialist	ED-OR, ED-PICU	0:55
5	ED	Resident	ED-OR, ED-PICU	0:54
6	Anesthesia	Attending	ED-OR, OR-PICU	0:51
7	Anesthesia	Resident	ED-OR, OR-PICU	0:50
8	Anesthesia	Certified Nurse Anesthetist	ED-OR, OR-PICU	0:57
9	OR	OR nurse manager	ED-OR, OR-PICU	0:44
10	OR	Care team leader	ED-OR, OR-PICU	0:44
11	OR	Circulating RN	ED-OR, OR-PICU	0:54
12	OR	Director of surgical services	ED-OR, OR-PICU	1:02
13	Surgery	Resident	ED-OR, OR-PICU	0:53
14	Surgery	Resident	ED-OR, OR-PICU	0:58
15	PICU	PICU RN/care team leader	ED-PICU, OR-PICU	1:04
16	PICU	Care team leader	ED-PICU, OR-PICU	0:45
17	PICU	Fellow	ED-PICU, OR-PICU	0:49
18	PICU	Fellow	ED-PICU, OR-PICU	0:29
	Total			14:44

3.2 Procedure

We used a semi-structured interview guide to ask about care transition processes between the ED, OR and PICU, including communication and coordination. After the interview, clinicians filled out a short questionnaire about patient safety during transitions (response rate: 100%). The study protocol was reviewed and approved by the University of Wisconsin-Madison Internal Review Board (IRB).

3.3 Data Collection and Analysis

Interviews. An interview guide was developed to collect data on the different transitions, and barriers and facilitators to these transitions. Participants were asked to describe the transition and provide an example of a transition from ED to OR (or OR to PICU, or ED to PICU) that went well and an example of a transition that went poorly. All questions were asked for both level 1 and level 2 patients. All interviews were audio-recorded, transcribed, uploaded to a qualitative analysis software (Dedoose®). The interviews lasted between 30 min and little more than an hour (average duration: 51 min). In total, the interviews lasted nearly 15 h. One interview was conducted with two interviewees (interviewee #9 and #10 in Table 1). The interview data analysis in Dedoose® focused on text that indicated problems with information flow in the transitions. The interviews contained 22 excerpts with these problems. The analysis process identified three themes for barriers to information transfer: (1) information is not available, (2) information is lost and (3) incorrect information is transferred.

Survey. Clinicians filled out a short version of the AHRQ Hospital Survey on Patient Safety Culture (HSPC) [10] about patient safety during transitions. HSPC contains 11 questions about care transitions and communication, coordination and cooperation. Three questions about hospital management and patient safety were not used in this project. Two (of the 8) questions are aimed at shift changes; the wording in these questions was changed to transitions between units. Survey data were entered in SPSS©. Frequencies and percentages were calculated for the individual items.

4 Results

4.1 Survey

Results of the survey show that, despite many services and units involved in pediatric trauma cooperating well together during trauma cases (89% of respondent agree with that statement, see Fig. 1), "Important patient care information is often lost when transitioning patients between units" (44% of respondents agree) and "Things 'fall between the cracks' when transferring patients from one unit to another" (50% of respondents agree).

4.2 Interview

Results of the qualitative analyses of the interview data identified three themes: (1) information not available or accessible, (2) information is not transferred, (3) transferred information is incomplete, and (4) transferred information is not accurate or incorrect; see Table 2.

Information Is Not Available or Not Accessible. Information that is not available or not accessible can be a problem for pediatric trauma. For example, because the patient may be unconscious or the parents have not arrived to the ED yet, information about the patient is unavailable. ["*A lot of times you don't know a history on the patient, or*

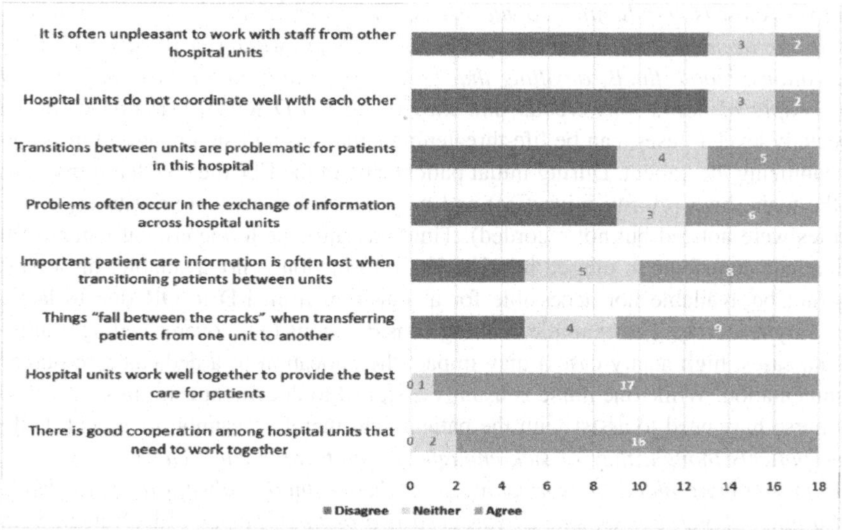

Fig. 1. Perceptions of care transitions (N = 18)

Table 2. Problems with information transfer during pediatric trauma transitions

Problem with information flow	Description	ED-OR	OR-PICU	ED-PICU
1. Information not transferred because the information is not available or not accessible	Sending party does not have the information and can therefore not transfer information to receiving party, or sending party does have information, but information is not (yet) recorded or not accessible and can therefore not be transferred to receiving party	6	2	3
2. Information is not transferred because of omission	Information is available, but information does not get transferred from the sending to receiving party	1	1	1
3. Information transfer is incomplete or loss of information during transfer	Information is available but during transfer from sender to receiver, information gets lost	1	3	3
4. Inaccurate information transferred	Correct information is available, but during transition from sender to receiver, information becomes inaccurate	0	0	1
Total		8	6	8

Note: The numbers in the table refer to the number of excerpts *what you know is very limited, so vital signs, the history of the event, so, you know, is this a car accident or a burn or whatever, the lines, what you've given as far as medications, blood, fluids, anything that basically is going to have an impact on the next couple of hours, any trends that they've had*" ED nurse]. Trauma cases, and especially level 1 cases, can be life-threatening situations and significant effort is spent on stabilizing the patient. During initial patient care in the ED, the care team may focus on the main, high priority injury(ies) and may miss other low(er) priority injuries (or injuries were noticed but not recorded). Time pressure due to the critical nature of the patient can contribute to missed injuries, i.e., information is not available. Information may not be available nor accessible for a transition from ED to OR due to lack of necessary resources. For example, staffing considerations (e.g., overcrowding, multiple trauma cases, high acuity cases) may impact the allocation of a dedicated resource to documentation. While one nurse is usually assigned to document care, in severe cases, this nurse may need to assist with the patient, and therefore cannot document. [*"With some patients though, they're sick enough that that nurse can't chart. They have to jump in and hang blood or, you know, so in theory, there's always a nurse charting and a nurse doing, but sometimes you need two or three nurses"* ED Nurse]. Sometimes there is not enough time to enter all available information in the EHR (information overflow), which can be aggravated by EHR usability issues. For example, information may be available, but not recorded due to distraction, and is 'lost'. ["*And a lot of times it's like the critical information, you get a distraction, and then it goes from your memory. But like I said, my experience has been the very severe cases"* ED Nurse].

Information Is Available But Is Not Transferred. Sometimes, the information is available, but is not transferred. In some situations, the receiving unit is not notified about a patient's transfer before their arrival on the unit. This is sometimes caused by an adult ED team taking care of the pediatric patient and not being familiar with procedures, contacts, etc. for pediatric patients [*"... But the trauma team that takes care of the patient in the ED is the adult trauma team. And then they, oftentimes the kid doesn't need surgery, so they'll just admit them directly to the PICU or to the floor, and then they become a pediatric general surgery patient, but we're not always told that the patient even exists"*, Anesthesia intern]. The receiving unit then does not receive information about the patient. ["*...It's like I didn't even realize this patient was on the unit. The surgeon is nowhere to be found, and the anesthesiologist has started a case, went home... And so in that scenario, it's really just a lot of loss of things, a lot of loss of information that was, you know, they never passed on or maybe never documented or and then wasn't verbally passed onto our team. So that's kind of the worst"*, PICU Fellow].

Incomplete Information or Loss of Information. The sending unit may have information but fails to transfer it adequately to the receiving unit, for example because the people involved in the information exchange have changed, e.g., over shift change. The clinical team on the receiving unit may have specific questions; if they try to contact the physician or nurse from the sending unit after shift change, the individual who cared for the patient may not be available. The same issue may happen with shift change in the receiving unit. ["*And that's a pretty detailed report we would give to the* *floor nurse or the ICU nurse. The only hard thing about it is that sometimes, depending on how long the patient is going to be in the OR, it may be a change of shift before that patient gets up to the unit. So then it's like kind of playing telephone with having to make sure they all get the right information, if they have to pass the information to a next nurse"*, ED RN]. Several interviewees mentioned information loss when the information is exchanged only through written documentation. Verbal communication often contains more details than written communication, and the opportunity for questions and feedback is missing from non-verbal communication. ["*…I think just the words that you use to describe them are different, verbally versus just on paper. You could, and then it's a conversation, right, and you could ask questions. So when you get, have a chance to have a conversation with someone, it's just, there's just that much more information that gets passed on. There's less likely things get missed… So there's just a lot of things that aren't communicated just by the written word. To a lesser extent, the same can happen when the information is exchanged over the phone, and not face-to-face"* PICU Fellow]. Information may also be lost when information is only transferred from one role in the sending unit (e.g. nurse) to the same role (nurse) in the receiving unit rather than communication between all roles. Information that is important to the other roles of the care team (e.g. physician, physical therapist, etc.) may not be transmitted.

Inaccurate Information. Few interviewees made comments about incorrect or inaccurate information during a transfer. One example is when the patient has received a preliminary diagnosis in the ED, but that diagnosis needs to be confirmed by diagnostic tests, e.g., imaging and labs. If the patient is transferred with a preliminary diagnosis before test results have been received, the results may arrive to the ED and not be communicated to the receiving unit, resulting in a failure to confirm/reject the preliminary diagnosis or additional, unnecessary testing. Another example is information that is incorrectly entered in the electronic system. For example, a bed may have been assigned to a patient, but upon arrival in the receiving unit, the bed is already taken by another patient. ["*It might just be whoever, coordinators or something might have put the wrong thing in or, so it physically won't let us go any further if a patient is already in that bed in the PICU. So again, I just call and say, hey, I've got a, you know, bed 48 down but it says it's occupied, and they'll go, oh, yeah, they should go in here"*, ED Coordinator].

5 Conclusion and Discussion

5.1 Conclusion

Pediatric trauma care transitions are complex with vulnerable patients and many different hospital services, units and clinicians involved. To safely manage the transition of this fragile and complex patient population, we need to better manage the information flow during these transitions. Results confirm the potentially considerable information loss during pediatric trauma transitions. Although the many services and units involved in pediatric trauma work well together during trauma cases, important patient care information can get lost when transitioning patients between units and

"Things 'fall between the cracks' when transferring patients from one unit to another". Results of our interview analysis show that during the transitions, information may not be available or accessible, incomplete, or inaccurate. It is not always possible to strictly separate these three categories from each other. They are not independent and what happens in an early stage can have an effect on later stages of the care process. For example, a level 1 pediatric trauma patient gets to the ED and the ED physician suspects a certain medical condition and orders laboratory tests to confirm the preliminary diagnosis. However, before the test results are available, the patient (with the preliminary diagnosis) has been transferred to the OR and then to the PICU. The tests are ordered in the ED and therefore the results are sent to the ED. Sometimes this information is forwarded to the next service(s), but sometimes it does not. The preliminary diagnosis, even if the test results would have changed this diagnosis, may remain the same during the patient trajectory over the different units and services. Information problems in the transitions between the ED and OR or PICU are often about unavailable/inaccessible information. In trauma cases, the information may not be available and in pediatric trauma cases, the child often cannot provide information. The potential limited time and personnel resources in the ED during a trauma case, and the fact that the diagnostic process has not been completed yet compound to the problem. Finally, the often hectic and stressful nature of a trauma case means that a standardized handoff between ED and OR may not be beneficial. Results show further that information loss occurs in the transitions between OR and PICU and ED and PICU. These results confirm the study by Zakrison et al. [11] that showed that in nearly 25% of patients, injuries were missed in the transition from ED to PICU, and in 48% of patients, information discrepancies occurred. Interestingly, the transition from OR to PICU is a team handoff that should reduce information loss. Some studies have demonstrated the benefits of team handoffs [12], but there are also some negative consequences. First, team handoffs require much effort and coordination. Second, when team handoffs are not well organized, and for example, the physician and nurse handoff take place at the same time, participants can have problems focusing. Research is needed to explore advantages and disadvantages of team handoffs.

5.2 Discussion

Several "mechanisms" are proposed to improve care transitions, including standardizing the process, organizational change and health information technology (IT) support.

Standardized Handoff to Support the (Individual) Handoff. Several tools have been developed to support care transitions such standardized tools (e.g. SBAR), mnemonics (memory aids), checklist, handoff tools (for example I-pass) and other health IT tools. Some of these tools have been successful in information transfer during transitions. However, most of the literature focuses on inpatient transitions and not on transitions from the ED to inpatient units. More research is needed to study handoffs between ED and hospital services/units, and how these handoffs can be improved.

Organizational Change to Support the Transitions. Organizational changes have been implemented to support care transitions. Certified level 1 trauma centers require personnel from different services and units to be present during level 1 and 2 trauma cases in the ED. When attending physicians and sometimes nurses from surgery, anesthesiology, OR and the PICU are present during a trauma case in the ED, the need for a handoff is greatly reduced: the receiving unit/service already has the information needed to take care of the patient. While this organizational intervention may have benefits, there are also some drawbacks. First, it requires resources in time and personnel. Second, during a trauma case, the large number of people impede identifying individuals role, let alone what information they need and/or possess. The trauma team response does not necessarily impact the OR to PICU transition, and so different ways of organizing that handoff have been developed, e.g., the team handoff. Research should evaluate the impact of this organization.

Information Technology to Support the Transitions. Health information technology can help in addressing information loss. Several studies have shown that health IT can support handoffs [13]. For example, I-pass is an evidence-based package of interventions that has been shown to reduce communication problems during handoffs [14, 15].

Many solutions to address information flow problems in care transitions have been relatively successful. However, it may be difficult to apply these solutions to *trauma care*. The high-speed environment of trauma care does not always allow for successful application of these solutions. Our study aims to better understand the communication and information transfer issues, and formulate design requirements for a health IT tool that can support the clinicians in these fast-paced, sometimes life-threatening situations.

5.3 Study Limitations

The study took place in one academic hospital, which makes it difficult to generalize results to other hospitals. Further, we only examined transitions from ED to OR, OR to PICU and ED to PICU, but not other transitions such as the transitions from ED, OR or PICU to floor. During the interviews, we asked clinicians about barriers and facilitators to transitioning pediatric trauma patients from one service/unit to another, and not specifically about problems with information flow. This resulted in a limited number of excerpts for analysis. Future studies should specifically focus on the transfer of information during the transitions, to better understand what information is missing, incomplete or inaccurate. Finally, survey results are based on only 18 respondents, which also makes it difficult to generalize the results.

Acknowledgements. Funding for this research was provided by the Agency for Healthcare Research and Quality (AHRQ) [Grant No. R01 HS023837]. The project described was supported by the Clinical and Translational Science Award (CTSA) program, through the National Institutes of Health (NIH) National Center for Advancing Translational Sciences (NCATS), [Grant UL1TR002373]. The content is solely the responsibility of the authors and does not necessarily represent the official views of the funding agencies. We thank the study participants, as our research would not be possible without them.

References

1. CDC. https://www.cdc.gov/injury/images/lc-charts/leading_causes_of_death_age_group_2015_1050w740h.gif
2. Apker J, Mallak LA, Gibson SC (2007) Communicating in the "gray zone": perceptions about emergency physician hospitalist handoffs and patient safety. Acad Emerg Med 14:884–894
3. Wooldridge AR, Carayon P, Hoonakker P, Hose B-Z, Ross J, Kohler J, Brazelton T, Eithun B, Kelly M, Dean S, Rusy D, Gurses A (2017) Understanding team complexity in pediatric trauma care. Human Factors and Ergonomics in Healthcare, New Orleans
4. Horwitz LI, Meredith T, Schuur JD, Shah NR, Kulkarni RG, Jenq GY (2009) Dropping the baton: a qualitative analysis of failures during the transition from Emergency Department to Inpatient Care. Ann Emerg Med 53:701.e704–710.e704
5. Abraham J, Kannampallil T, Patel VL (2014) A systematic review of the literature on the evaluation of handoff tools: implications for research and practice. J Am Med Inform Assoc 21:154–162
6. American Academy of Pediatrics Committee on Emergency Medicine (2016) Handoffs: transitions of care for children in the Emergency Department. Pediatrics 138:1–12
7. Joint Commission (2012) Joint Commission Center for Transforming Healthcare releases targeted solutions tool for hand-off communications. Joint Commission perspectives. Joint Commission on Accreditation of Healthcare Organizations 32, 1, 3
8. Arora VM, Johnson JK, Lovinger D, Humphrey HJ, Meltzer DO (2005) Communication failures in patient sign-out and suggestions for improvement: a critical incident analysis. Qual Saf Health Care 14:401–407
9. Bigham MT, Logsdon TR, Manicone PE, Landrigan CP, Hayes LW, Randall KH, Grover P, Collins SB, Ramirez DE, O'Guin CD, Williams CI, Warnick RJ, Sharek PJ (2014) Decreasing handoff-related care failures in children's hospitals. Pediatrics 134:e572
10. Sorra J, Gray L, Streagle S, Famolaro T, Yount N, Behm J (2016) AHRQ Hospital survey on patient safety culture: user's guide. Agency for Healthcare Research and Quality (AHRQ), Rockville
11. Zakrison TL, Rosenbloom B, McFarlan A, Jovicic A, Soklaridis S, Allen C, Schulman C, Namias N, Rizoli S (2016) Lost information during the handover of critically injured trauma patients: a mixed-methods study. BMJ Qual Saf 25:929–936
12. Joy BF, Elliott E, Hardy C, Sullivan C, Backer CL, Kane JM (2011) Standardized multidisciplinary protocol improves handover of cardiac surgery patients to the intensive care unit. Pediatr Crit Care Med 12:304–308
13. Bernstein J, MacCourt DC, Jacob DM, Mehta S (2010) Utilizing information technology to mitigate the handoff risks caused by resident work hour restrictions. Clin Orthop Relat Res 468:2627–2632
14. Starmer AJ, Landrigan CP (2015) Changes in medical errors with a handoff program. N Engl J Med 372:490–491
15. Joint Commission: Inadequate Hand-Off Communication (2017). Sentinel Alert Event, pp 1–6

Ergonomic Factors Triggering Risk in the Pharmacotherapy Process Carried Out by Nurses

Izabela Witczak[1(✉)], Janusz Pokorski[2], Anna Kołcz-Trzęsicka[3], Joanna Rosińczuk[4], and Łukasz Rypicz[1]

[1] Division of Economics and Quality in Healthcare, Department of Public Health, Faculty of Health Sciences, Wroclaw Medical University, 5 Bartla Street, 51-618 Wroclaw, Poland
izaeuro@wp.pl

[2] Department of Ergonomics and Physiological Effort Physics, Jagiellonian University, Collegium Medicum in Krakow, 20 Grzegórzecka Street, 31-531 Kraków, Poland

[3] Division of Rehabilitation in Movement Dysfunctions, Department of Physiotherapy, Faculty of Health Sciences, Wroclaw Medical University, 2 Grunwaldzka Street, 50-355 Wroclaw, Poland

[4] Department of Public Health, Faculty of Health Sciences, Wroclaw Medical University, 5 Bartla Street, 51-618 Wroclaw, Poland

Abstract. The pharmacotherapy process consists, among other things, of storing medicines, ordering a medicine by a doctor, dissolving and administering a drug to a patient, and is carried out by nurses. This process is very important for the safety of patient treatment. Nurses should also feel safe when carrying out such an important professional activity. The analysis of potential triggers of risk will make it possible to develop preventive measures in order to limit adverse events related to this. The study covered 305 nurses employed in hospitals and primary healthcare centres in Poland. The study used a diagnostic poll method, which was developed by the authors of the poll. The poll identified nine psychosocial and organisational ergonomic factors that could trigger risks in the pharmacotherapy process carried out by nurses. The nurses recognised the following factors with a very high risk in the pharmacotherapy process: unreadable medical orders (82%), poor communication between the physician and nurse (55%), time pressure during nurse duties (70%), drug preparation for patients combined with other activities performed by nurses (74%), lack of orders by the physician to use individual/specific solvents for drugs (63%) or shift work causing psychophysiological fatigue (52%). The professional activity of nurses is associated with ergonomic factors that trigger numerous significant risks in the pharmacotherapy process. Nurses do not have a full sense of safety in the implementation of pharmacotherapy, and thus patients may also feel threatened.

Keywords: Pharmacotherapy process · Nurses · Risk · Ergonomic factors

S. Bagnara et al. (Eds.): IEA 2018, AISC 818, pp. 489–497, 2019.
https://doi.org/10.1007/978-3-319-96098-2_61

1 Introduction

By analysing processes performed in health facilities (both in hospitals as a whole and in primary healthcare centres), it should be stated that nurses play various roles in individual processes due to their interdisciplinary profession - they play a dominant role when, for example, they care for patients, while their role is participatory e.g. during the diagnostic or pharmacotherapy processes. It is very often that their role is crucial in order to ensure the quality of provided services as well as patients' safety. The professional activity of nurses features a great many issues of an ergonomic nature that may have a potential impact on the occurrence of adverse events during the pharmacotherapy process. They include:

- work in shifts, at night, or overtime;
- considerable exposure to biological, chemical, or physical factors;
- time pressure, stress, interpersonal conflicts;
- drugs with similar names and very similar packaging [1].

It is obvious that people prefer to work in places where it is safe, the use of which is simple, and which contribute to health and efficiency, and this is the result of following ergonomics principles. Ergonomics focuses on people and constitutes one of the largest and most valuable resources of every organisation [2].

The risk of adverse events associated with the pharmacotherapy process is multi-factorial and depends not only on the nature of the patient's disease, but also on the nature of diagnostic, therapeutic, or nursing care measures. This risk also results from the so-called errors made by medical staff (the cause of which is various, e.g. personal conflicts, poor interpersonal communication etc.), failures of medical equipment, mistakes in procedures as well as from poor technical and organisational solutions in terms of ergonomics [3].

Medication errors have been studied and discussed for many years and numerous times. They are defined as a failure in treatment that leads or may lead to patient harm [4]. Extensive papers on medication errors were published as early as in the 1970s, e.g. the results of the study by A. Chapanis, who emphasises the importance of system errors that have to be identified and eliminated [5]. When analysing risks when it comes to the pharmacotherapy process, it should be noted that a medication adverse event is made up of adverse drug reactions and medication errors [6].

Medication errors are among the most common medical errors in the healthcare system and constitute between 10 and 18% of all medical errors. These errors may lead to serious consequences resulting in patient harm, or even in his/her death [7]. About 44,000–98,000 patients in the USA are estimated to die as a result of medication errors, i.e. more than those who die of breast cancer or in road accidents [7–9].

The pharmacotherapy process has a high risk of adverse events and it is nurses who play a crucial role in it. It may be divided into four stages to which leaders responsible for the correct implementation of each stage may be assigned (Table 1). Nonetheless, the highest risk is associated with the last two stages for which nurses are responsible. These stages include drug preparation for patients and administering a drug to them [10].

Table 1. Pharmacotherapy process stages

Pharmacotherapy process stage	Leader (person responsible for the implementation of the stage)
Drug ordering	Physician
Drug dispensing	Pharmacist
Drug preparation	Nurse
Drug administration	Nurse

2 Materials and Methods

The study involved a group of 305 nurses working in hospitals or primary healthcare centres in Poland. The age bracket of the nurses participating in the study was 22–59. People in the 40–59 age bracket were the most populous (73%), while the least populous age bracket was 50–59.

The majority of the studied nurses - 69% - obtained secondary education, whereas 31% of the respondents obtained higher education (BSc - 20%, MSc - 11%).

The largest percentage of the nurses (43%) have a 24–35-year working experience as a nurse. A considerable proportion of them have a long working experience as a nurse, i.e. 36 years or more (33%). The remaining nurses (24%) have a 1–24-year working experience as a nurse.

A diagnostic poll method with the use of authors' own poll was used to carry out the study. The poll consisted of 5 closed-ended questions and a matrix to evaluate selected ergonomic factors that trigger risks in the pharmacotherapy process. The aforementioned matrix contained a detailed description of the selected ergonomic factors that have an impact on the safety of the pharmacological process. These factors are as follows:

(a) unclear or unreadable medical orders;
(b) poor communication between the physician and nurse in terms of changes in medical prescriptions;
(c) time pressure when nurses are on duty;
(d) method for drug preparation for patients combined with performing other nursing care duties based on the adopted work organisation (telephones, current diagnostic orders etc.);
(e) ignorance of the list of drug substitutes by nurses;
(f) no orders to use specific solvents for individual drugs by a physician;
(g) work in shifts resulting in physiological fatigue;
(h) poor access to training on the effects, side effects and adverse effects of drugs administered to patients;
(i) preparation of cards with the first name and surname of a patient and the name and dose of a drug (different writing by different persons).

The results were analysed in statistical terms. A five-point scale of risk was use to evaluate the above-mentioned ergonomic factors:

1. marginal risk;
2. low risk;
3. medium risk;
4. significant risk;
5. very significant risk.

The poll was validated by measuring reliability with the use of Cronbach's alpha. The alpha for the entire sheet was 0.86. The results were analysed in statistical terms using

Excel 2010 spreadsheet tools. The quantitative study was carried out using the Statistica 10.0 PL (STATSOFT) software.

3 Results

The study reveals that as many as 61% of the nurses regard the safety of the pharmacological process in their healthcare centre as medium. Only 12% of the respondents said that the above-mentioned safety is very high, 23% of them regard it as low, and 4% of the nurses have no opinion on it.

The nurses were asked who, in their opinion, faces the highest risk of committing an error during the pharmacological process. The results are as follows: 88% of the respondents answered that it is the nurse who faces the highest risk of committing such an error; 11% of them believe it is the physician, and 1% of the nurses said it is the pharmacist.

The analysis of further data shows that 53% of the nurses believe that the introduction of electronic medical records will not eliminate medication errors. 23% of the respondents are of a different opinion. They believe that this introduction will eliminate the risk of committing medication errors to a certain extent. On the other hand, 15% of the nurses said they have no opinion on this matter.

The nurses were also asked whether the patient or a person authorised by the patient should be informed about committing an error in drug or drug dose administration. According to 87% of the nurses, the patient or his/her family should be informed about it, 8% of the respondents replied that they should not, and 5% of them claim it does not matter.

Over 50% of the nurses said that it is the physician who should inform the patient or his/her family about committing a medication error. 35% of the respondents, i.e. more than one third, replied that it is the responsibility of the chief physician, 10% of them claimed that the representative of the Commissioner for Patients' Rights should do it, and only 2% of the nurses replied that it is the nurse who should inform the patient or his/her family about a medication error.

The nurses were also asked to evaluate the risk of committing a medication error using individual factors in the risk matrix. Detailed results are presented in Table 2. Over 80% of the respondents considered *unclear or unreadable medical orders* to be a factor that triggers risk in the pharmacological process (it was ranked the highest on a five-point scale of risk). Only 2% of the nurses rated the aforementioned factor as marginal or low.

Table 2. Matrix for risk assessment in the pharmacotherapy process

Factors that trigger risks in the pharmacological process	Direction of risk growth →				
	Marginal (1)	Low (2)	Medium (3)	Significant (4)	Very significant (5)
Unclear or unreadable medical orders	2%	2%	6%	8%	82%
Poor communication between the physician and nurse in terms of changes in medical prescriptions	1%	1%	14%	28%	56%
Time pressure when nurses are on duty	1%	1%	11%	17%	70%
Method for drug preparation for patients combined with performing other nursing care duties based on the adopted work organisation (telephones, current diagnostic orders etc.)	1%	2%	6%	17%	74%
Ignorance of the list of drug substitutes by nurses	3%	4%	15%	25%	53%
No orders to use specific solvents for individual drugs by a physician	2%	2%	11%	22%	63%
Work in shifts resulting in physiological fatigue	3%	3%	14%	28%	52%
Poor access to training on the effects, side effects and adverse effects of drugs administered to patients	2%	3%	18%	27%	50%
Preparation of cards with the first name and surname of a patient and the name and dose of a drug (on a medicine tray)	4%	4%	15%	25%	52%

Another factor that may trigger risks in the pharmacotherapy process was *poor communication between the physician and nurse in terms of changes in medical prescriptions* - 56% of the nurses regard it as very significant, 28% of them as significant, and 14% of them as medium. The impact of *time pressure when nurses are on duty* was also evaluated and the results are as follows: as many as 70% of the nurses considered this factor to be very significant when it comes to triggering risk in the pharmacological process. When asked about the risk of committing a medication error considering the *method for drug preparation for patients combined with performing other nursing care duties based on the adopted work organisation,* the nurses claimed that it is very significant - it was evaluated in this way by 74% of the respondents in fact. As many as 53% of the nurses considered the risk associated with the *ignorance of the list of drug substitutes by nurses* to be very significant, while one fourth of them considered it to be significant. According to 63% of the respondents, *no orders to use specific solvents for individual drugs by a physician* are associated with a very high risk of committing a medication error. More than half of the nurses believed that *work in shifts resulting in physiological fatigue* is a factor that triggers a very high risk in the pharmacological process, while over 25% of them considered it to be significant. According to 50% of the nurses, *poor access to training on the effects, side effects and adverse effects of drugs administered to patients* is associated with a very high risk of committing a medication error. Preparation of cards with the first name and surname of a patient and the name and dose of a drug on medicine trays was the last evaluated factor - over 50% of the nurses considered it a factor that triggers a very high risk in the pharmacological process.

4 Discussion

There is a considerable number of reports on medication errors which are part of adverse events. Adverse events associated with drug administration are divided into avoidable (events caused by an error) and unavoidable (e.g. unpredictable allergic reactions in patients). Nurses face the highest risk in the pharmacotherapy process as they carry out two out of four stages of the pharmacological process, i.e. drug preparation and administration [10].

Creation of optimal conditions to carry out the pharmacological process based on a high pro-quality culture of an organisation should be a priority to healthcare centre/hospital management in order to reduce medication errors to a minimum, and thus potential adverse events. This translates into the safety of patients as well as medical staff providing medical services. The literature uses the concept of a "victim" in the context of adverse events. The patient is the first victim, the nurse is the second, and the third victim is the organisation in which the adverse event occurred. Therefore, it is vital to provide support for nurses who are the so-called second victim. The support should not only be psychological, but also include the implementation of preventive measures in the organisation to avoid medication errors in future [6, 9, 11].

The study reveals that 88% of the nurses believe that they themselves face the highest risk of committing a medication error. This also confirms reports available in foreign literature. They reveal that the implementation of two last pharmacological process stages, i.e. drug preparation and administration, is crucial in ensuring patient safety. Possible medication errors may still be noticed at this stage and corrective measures may be implemented to avoid potential adverse events associated with pharmacotherapy.

Reports on the studies carried out in the Western countries indicate a great deal of ergonomic factors that trigger potential risks in the pharmacotherapy process. The so-called risk triggers include for example nursing care for many patients simultaneously, hurry, time pressure, work in shifts affecting the biological clock, family issues, or poor work organisation. Design ergonomics in the everyday work of nurses should also be emphasised. It turns out that there are numerous drugs with a similar name but a different use. Drug packaging is another practical ergonomic issue. There is a large group of drugs with almost identical packaging; there are also drugs that differ in dose but have the same packaging. Practical ergonomics plays a crucial role in the safety of the pharmacotherapy process. Research conducted in Australia reveals that as many as 56% of medication errors result from similar drug names and packaging [12].

The impact of work in shifts by nurses on patient safety in the pharmacological process is emphasised all over the world [6]. Fatigue associated with circadian rhythm disturbance caused by work in shifts, i.e. at night and in early morning, may contribute to a high number of adverse events related to pharmacotherapy [3].

The study shows that over 50% of the nurses indicated work in shifts resulting in physiological fatigue is a very significant factor that triggers risks in the pharmacological process, while over 25% of the respondents considered it to be a significant factor.

In her review paper, Hewitt [1] emphasises that the occurrence of medication errors is affected by, in particular, incorrect calculation of drug doses, lack of knowledge by medical staff, fatigue, number of nurses on duty, lack of experience, inadequate working environment, and inappropriate medical equipment. In addition, the author raises the issue of the legibility of handwritten medical order sheets completed by physicians.

Meanwhile, the study conducted by the authors of this paper reveals that 82% of the nurses considered unclear or unreadable medical orders are associated with a very high risk in the pharmacological process. What is more, 53% of the nurses claimed that the introduction of electronic medical records will not eliminate factors triggering risks of medication errors completely.

Risk management in the pharmacotherapy process in healthcare centres/hospitals should have an effect on building a safe organisational culture among the staff providing medical services. This should make it possible to achieve a high level of safety and should include: responsibility for your own actions and your colleagues' that have an impact on patient safety or increased collective prudence (reporting errors, adopting measures that eliminate the cause, analysing and drawing conclusions, implementing corrective and preventive measures).

There is a considerable number of reports on medication errors and safe pharmacotherapy which might serve as training material for medical staff. In addition, development of risk management systems and systems for reporting errors and medical events might constitute an excellent knowledge base to be used for preventive purposes.

5 Conclusions

The profession of a nurse is associated with a great many ergonomic factors that trigger very significant risks in the pharmacotherapy process. Nurses do not feel entirely safe during the implementation of the pharmacological process, and thus patients may feel at risk as well. The presented results of this study indicate that these factors may as a result lead to adverse events ("not this patient, not this drug, not this dose" errors etc.). Hurry (e.g. in the event of emergency), multitasking, work interruption while performing tasks by nurses, and fatigue have an impact on the increase in the risk of committing medication errors.

This study reveals that the following actions should be implemented in order to limit medication errors:

(a) improvements to interpersonal communication (between the physician, nurse, and patient);
(b) improvements to the drug ordering rules, i.e. handwriting has to be legible, determination of the drug administration route as well as drug dose and drug administration frequency;
(c) changes in the organisation of nursing care work in order to ensure that nurses responsible for drug preparation and administration are not involved in other professional activities while on duty;
(d) nursing staff training based on professional literature and scientific reports as well as on practical rules, e.g. 7 rights of drug administration, i.e. right medication, right route, right time, right dose, right patient, right documentation, and right education.

Risk management in healthcare centres/hospitals should contribute to building a safe organisational culture that involves numerous qualities and behaviour patterns of staff providing medical services. This should lead to the achievement of a high level of safety and should include: responsibility for your own actions and your colleagues' that have an impact on patient safety or increased collective prudence (reporting errors, adopting measures that eliminate the cause, analysing and drawing conclusions, implementing corrective and preventive measures).

Predicting potential sources of risk may pose severe difficulties in risk management. When it comes to medical services, these sources are specific and their enormous diversity makes the risk management capacity to be limited.

References

1. Hewitt P (2010) Nurses' perceptions of the causes of medication errors: an integrative literature review. Med Surg Nurs J 19(3):159–167
2. Springer T (2007) Ergonomics for healthcare environments. Knoll, Inc., East Greenville, pp 11–16
3. Jurek TM, Świątek B, Golema W (2010) Adverse events and medical error. In: Pokorski J, Pokorska J, Złowodzki M (eds) Medical error, ergonomic conditions, Kraków, pp. 142–147

4. Aronson JK (2009) Medication errors: what they are, how they happen, and how to avoid then. QJM Int J Med 102(8):513–521
5. Carayon P, Xie A, Kinafar S (2014) Human factors and ergonomics as a patient safety practice. BMJ Qual Saf 23:196–205
6. Volpe CRG, Pinho DLM, Stival MM, de Oliviera G, Karnikowski M (2014) Medication errors in a public hospital in Brazil. Br J Nurs 23(11):552–559
7. Fathi A, Hajizadeh M, Moradi K, Zandian H, DezhKameh M, Kazemzadeh S, Rezaei S (2017) Medication errors among nurses in teaching hospitals in the west of Iran: what we need to know about prevalence, types, and barriers to reporting. Epidemiol Health 39:1–7
8. Pop M, Finocchi M (2016) Medication errors: a case-based review. AACN Adv Crit Care 27 (1):5–11
9. Mahmoud EE, Abu SJ (2018) Nurse as second victims after adverse event. J Altern Perspect Soc Sci 9(1):58–83
10. Kavanagh C (2017) Medication governance: preventing errors and promoting patient safety. Br J Nurs 26(3):159–165
11. Mayo AM, Duncan D (2004) Nurse perceptions of medication errors: what we need to know for patient safety. J Nurs Care Qual 19(3):209–217
12. Shrivastav AS, Sachdeva PD (2018) Medication error: its type, causes, and strategies to avoid them. Pharm Sci Monit 9(1):404–415

Evaluation of the Symmetry of Lower Limbs Symmetry Loading and Body Composition as Elements of Monitoring of Health-Related Behaviours Among Professionally Active Nurses

Anna Kołcz-Trzęsicka[1(✉)], Izabela Witczak[2], Piotr Karniej[3], Anna Pecuch[1], and Łukasz Rypicz[2]

[1] Division of Rehabilitation in Movement Dysfunctions, Department of Physiotherapy, Faculty of Health Sciences, Wroclaw Medical University, 2 Grunwaldzka Street, 50-355 Wroclaw, Poland
anna.kolcz-trzesicka@umed.wroc.pl
[2] Division of Economics and Quality in Healthcare, Department of Public Health, Faculty of Health Sciences, Wroclaw Medical University, 5 Bartla Street, 51-618 Wroclaw, Poland
[3] Division of Organization and Management, Department of Public Health, Faculty of Health Sciences, Wroclaw Medical University, 5 Bartla Street, 51-618 Wroclaw, Poland

Abstract. The study purpose was to assess the body composition and distribution of foot forces on the ground in nurses. A group of 72 participants was qualified in the study including professionally active nurses (study group, n = 29) and nursing students (control group, n = 43). The lower limb load assessment was carried out using the baropodometric platform, while the body composition was measured using the electrical bioimpedance device. Moreover, a brief questionnaire designed by the authors was carried out three months after research completion. The mean visceral fat index in the study group was significantly higher than in the control group, respectively 5.48 kg and 1.79 kg ($p < 0.0001$). The mean total body water was significantly lower in the study group, than in the control, respectively 49.06% and 54.56% ($p < 0.0001$). The forefoot overload was more frequent in both groups and the centre of gravity was shifted in the same direction. The ground peak pressure point was higher in the control group than in the study, respectively 67.4% and 55.2% without statistically significant difference ($p > 0.05$). It was observed that 97.1% of respondents considered the body composition measurement to be useful and 68.6% considered the information given in the study as important for changing everyday habits related to work ergonomics and lifestyle. Regular education in the field of workplace ergonomics and health-promoting behaviours should increase the awareness of employees in the healthcare sector and can significantly improve the quality of comfort and functioning during professional activities.

S. Bagnara et al. (Eds.): IEA 2018, AISC 818, pp. 498–510, 2019.
https://doi.org/10.1007/978-3-319-96098-2_62

Keywords: Nurses · Body composition · Bioimpedance analysis
Lower limbs load symmetry · Baropodometric assessment
Workplace ergonomics · Health-promoting behaviours

1 Introduction

The professional duties of people working in the healthcare system as female or male nurses are varied and require concentration and commitment as well as precision and reliability. There are many workplace factors that potentially limit the proper implementation of professional activities [1]. These include chemical and biological factors but also, in many cases, threats resulting from disorders of the musculoskeletal system. Functional disorders manifest through pain, motor discomfort and fatigue, which often, as the preserved state, lead to the occurrence of employee injuries and absenteeism. These abovementioned factors limit the effectiveness of nurses' work and contribute to committing errors and increasing the risk of accidents at work [2, 3]. Principles of work ergonomics, including room ergonomics where varied activities are carried out, and education in the field of movement ergonomics of the body during work performance are currently of significant importance in improving working conditions and preventing employee absenteeism [4].

The nursing profession is qualified in the literature as the second largest in terms of physical load [5]. In the search for new solutions to this problem, promoting ergonomic behaviours and minimising the loads within the musculoskeletal system and the entire body have become one of the basic issues of scientific research conducted in this professional group [6, 7]. The time of a single nursing activity performed by hospital nurses is about 30 s; however, the body position adopted at that time is the anteversion of the entire body silhouette. The consequence of this type of work-related behaviour—that is, the adoption of a strained position, both under static and dynamic conditions—is a pain, which, in consequence, provokes degenerative changes [8, 9].

The most important threats to which nurses are exposed include working conditions (working in strained and unnatural positions), workplace diversity and specificity (working on the move and in a hurry as well as workplace chemical and biological threats) and shift work (nonstandard working time and night work) [3]. The risk factors and causes of chronic pain include ignorance of the work ergonomics principles or lack of their implementation, lack of regular physical activity and poor nutrition (consequently leading to metabolic disorders) as well as the use of painkillers as the only way of analgesic treatment. Also, stress (emotional burden and exposure to aggressive behaviour) and work monotony can contribute to the deepening of pain and general exhaustion of the body [10].

According to the European Commission, as presented in the 2011 publication *Occupational Health and Safety Risks in the Healthcare Sector* [11], about 10% of people working in the European Union (EU) are employed in the healthcare and social care sectors, many of whom carry out their professional duties in hospitals. These people are particularly exposed to various risks. Currently, EU legislation on health and safety at work takes into account most types of such risks. However, due to simultaneous combinations of various risks and the fact that it is a high-risk sector, the discussion was undertaken at the EU level on the need to adopt the necessary solutions in terms of an approach to the work environment in order to improve the health and safety of hospital staff [11].

Currently, the number of defined civilization and occupational diseases is increasing, and these are determined on the basis of objective research results and expert opinions. Health risks that are directly related to performing professional activities are defined and specified in the ergonomic laboratories [12]. However, there are also professional groups in the healthcare sector, such as paramedics, for which, due to the fact that they are quite 'young professions', the health risks resulting directly from professional activities are still not specified in detail [13]. The most frequently described general health threats for employees in the healthcare sector are those related to an overload of the movement system, effects of long-term persistent stress and improper nutrition [14, 15].

The load on the musculoskeletal system is mainly due to the body position at work, the type and size of the external load, the frequency of repetitions of a particular activity and the duration of the load. It is the most widespread work-related health problem in Europe, affecting millions of employees [16]. It is the source of anxiety for an individual employee, but also for the management staff of the medical entity because of the economic consequences. In some countries, dysfunction of the musculoskeletal system accounts for 40% of the cost of compensation for employees and has reached 1.6% of gross domestic product (GDP). It should be emphasised that overloading of the musculoskeletal system, resulting from improper employee behaviour, reduces the profitability of healthcare entities and increases social costs [17–19].

A 2014 report by the European Agency for Safety and Health at Work (EU-OSHA) [20] focused on one of the main features and at the same time the main task of people employed in the healthcare sector: solicitude for the patient. However, in the same document, the authors emphasised that the improper performance of professional activities carries a high risk for the safety and health of employees. The main goal of preventive actions was thus determined: focusing on the need to convince people that, in order to achieve and maintain high-quality patient care, the safety and health of employees in the healthcare sector in the workplace must be treated with absolute priority. Therefore, it is necessary to implement training and education based on theory but also practical ones that will eliminate the inappropriate unergonomic behaviours of employees in the healthcare sector. Healthcare is a rapidly growing sector, which, in the near future, should offer increasing employment opportunities due to the observed tendency of an ageing EU society. Expanding the scope of services aims to meet the demand for better-quality care and the growing demand for personal care services, but properly trained and ergonomically working staff is the basis of the whole worldwide healthcare system [20].

Human involvement to perform one or more tasks is defined as the degree of workload. In professionally active nurses, the degree of workload is high because of the number and variety of activities they perform every day. Pain complaints resulting from the musculoskeletal system load directly affect the quality of life (QOL). The way individual employees feel about and cope with the workload is also due to their personal predispositions manifested in their family situation, health status and lifestyle [2].

Holistic approaches to employee health and monitoring health behaviours in order to develop appropriate mechanisms focused on maintaining health is becoming of great value. Previous research has indicated that there is a relationship between the body mass index (BMI) and the intensity of perceived overload pain—as the BMI value increases, the intensity of pain sensation increases [21, 22]. One of the non-invasive, simple, reliable and relatively quick techniques to measure body composition as part of the prevention of civilization and occupational diseases is the BIA. This method consists of measuring the body composition based on the detection of differences in electrical resistance of tissues. BIA can be used in both healthy and sick people of all ages. It is widely used for predicting the risk of cardiovascular and metabolic diseases, during research in eating disorders and in sports medicine [23–25].

For people working in the healthcare sector in dynamic conditions and requiring a long-term standing position, a useful examination may also be the assessment of the static and dynamic postures, including parameters such as lower limb symmetry, pelvic tilt and gait pattern. The asymmetry of the lower limbs load under static conditions and during movement action may result in perceived functional limitations, both during professional activities and daily activities [26]. Also, holding the wrong body posture during professional activities (abnormal tension of the postural muscles, shifting the centre of gravity towards the front or back and excessive pressure on one one-side foot or part of the foot) may result in overloads of the skeletal and muscular system and induce pain [27, 28].

The aim of the study was to determine the composition of the body and to assess the distribution of foot forces on the ground in both groups professionally active nurses and nursing students as an educational exercise for expanding knowledge about health-related behaviours.

2 Materials and Methods

The entire study procedures were performed in the research room at the Division of Rehabilitation in Movement Dysfunctions, Department of Physiotherapy, Faculty of Health Sciences, Wroclaw Medical University, Poland. Signed informed consent forms were obtained from all patients who were included in the study. The study protocol was approved by the independent Bioethics Committee of the Wroclaw Medical University, Poland (approval no. KB–232/2016). This study was conducted under a research project funded by a statutory grant of the Wroclaw Medical University for maintaining research potential (no. ST.E.060.17.020).

A group of 72 people was qualified to participate in the study. A total of 29 professionally active nurses were assigned to the study, and the qualification criterion was job seniority with professional experience not shorter than 10 years. The control group consisted of 43 nurses who were second-degree nursing students (master studies).

The research procedure included the body composition measurement of the BIA using the Tanita MC-780MA device (Tanita Corporation, Tokyo, Japan). This examination was carried out in accordance with the manufacturer's instructions, with all safety recommendations and methodological guidelines [29]. Four hours before the test, no food or drink was consumed, and no physical effort was made. Alcohol consumption was limited to 48 h before the test. Based on the collected results recorded in the generated report, the following parameters were recorded: metabolic age [years], visceral fat index (VFI) [kg] and total body water (TBW) [%]. During the test, the examined person was standing with bare feet on the platform, ensuring their proper contact with the metal electrodes and simultaneously holding two electrodes with their hands. During testing, the lower limbs could not touch each other, and the upper limbs were positioned in abduction (about 45° from the torso). During this measurement, a very slight electrical current sent from metal electrodes passed quickly through the tissues. Electrolytes and water contained in the human tissues work as an electric conductor, where fat tissue with a small percentage of water limits its flow (higher resistance), and muscle tissue containing a lot of water allows the electrical signal to increase (lower resistance). As the water content in the body decreases, the bioelectric impedance ratio increases. An important role in the BIA is played by the TBW because any changes in the water content will have a significant impact on the bioelectric impedance ratio and body fat content. In the body composition of body mass, a fat-free part which consists of bones, viscera and muscles can be marked as well as adipose tissue, which covers 10–20% of the body weight of an adult human.

The assessment of the lower limbs load and the plantar pressure was carried out using a baropodometric platform FreeMED Professional FM6050 (Sensor Medica, Rome, Italy). This examination was also carried out in accordance with the manufacturer's instructions, with all safety recommendations and methodological guidelines [30]. Each participant was in the vertical position, standing comfortably, but immobile on both feet, looking at the horizon, arms along the body, without any weight in their pockets or any type of movement during the assessment, which lasted 20 s. The device measures were based on the baropodometric technique, which consists of comparing the pressure of individual parts of the tested person's foot while standing on the platform. Such a test can be performed while standing, walking or running, and the obtained recordings are analysed by dedicated analytical computer software. In addition, the software enables the generating and reporting of 3D visualisation, which allows a more accurate assessment of the morphological structure of the feet and the symmetry of the lower limbs load. The research report data concerned the forefoot and hindfoot load [%], the distribution of the ground pressure between the left and the right side [%], the centre of gravity (COG) position [cm] and the ground peak pressure point [%].

Additionally, during the study, all nurses were informed about the importance and manner of the proper lower limbs loading as well as the adoption of a correct postural pattern while performing professional activities. Furthermore, three months after the study was finalised, a brief questionnaire designed by the authors consisting of six questions was carried out by the nurses. Nurses were asked about the usefulness of the present study and potential improvable changes in their everyday habits.

Statistical analysis was performed using STATISTICA software version 12 (StatSoft, Inc., USA) and Microsoft Excel 2007. For measurable variables, arithmetic means, standard deviations and range of variations (extreme values) were calculated. Comparisons of results for these variables between groups were calculated using the Mann–Whitney U test. Comparisons of qualitative variables were made using the chi-square test.

The level $\alpha = 0.05$ was used for all comparisons, and the obtained p-values were rounded up to three decimal places for the purpose of this study.

3 Results

The mean nurses' age from the study group was 45.2 years (professionally active). However, the participants in the control group were 22.7 years (master's degree students).

In all cases, the p-value indicated a very significant heterogeneity of the study groups regarding achieved results ($p < 0.05$). Detailed data for the body composition in both groups are presented collectively in Table 1.

Table 1. Comparison of the results of the body composition for both groups.

Data	Parameter	Study group	Control group	P-value
BA [years]	M	45.21	22.65	0.000
	Min–Max	32–55	22–27	
	SD	5.67	1.11	
MA [years]	M	41	19.21	0.000
	Min–Max	26–66	12–39	
	SD	11.35	9.19	
VFI [kg]	M	5.48	1.79	0.000
	Min–Max	2–11	1–8	
	SD	2.26	1.63	
TBW [%]	M	49.06	54.56	0.000
	Min–Max	40.4–60.2	41.7–62.4	
	SD	4.15	5	

Abbreviations: BA – biological age; MA – metabolic age; VFI – visceral fat index; TBW – total body water: M – mean value; Min – minimum value; Max – maximum value; SD – standard deviation; p-value – statistical significance

The mean values of the MA in both groups were at a lower level than the mean biological age (BA) of the nurses. However, when the results were analysed individually, the MA was not always lower than the BA, which indicates metabolic abnormalities of some subjects. In some nurses, the MA was determined at a much higher level than the BA. Similarly, the VFI was at a very different level individually, which indicates systemic metabolic disorders. However, it should be pointed out that the mean value of the VFI in both groups was within the normal range. A lower value of the TBW in the study group indicates poorer hydration of nurses with longer professional experience (see Fig. 1).

Data obtained during the measurement using the baropodometric platform did not show a significant heterogeneity of the subjects between the two compared groups. However, it is worth noting the tendency of occurrences in the study group, with higher

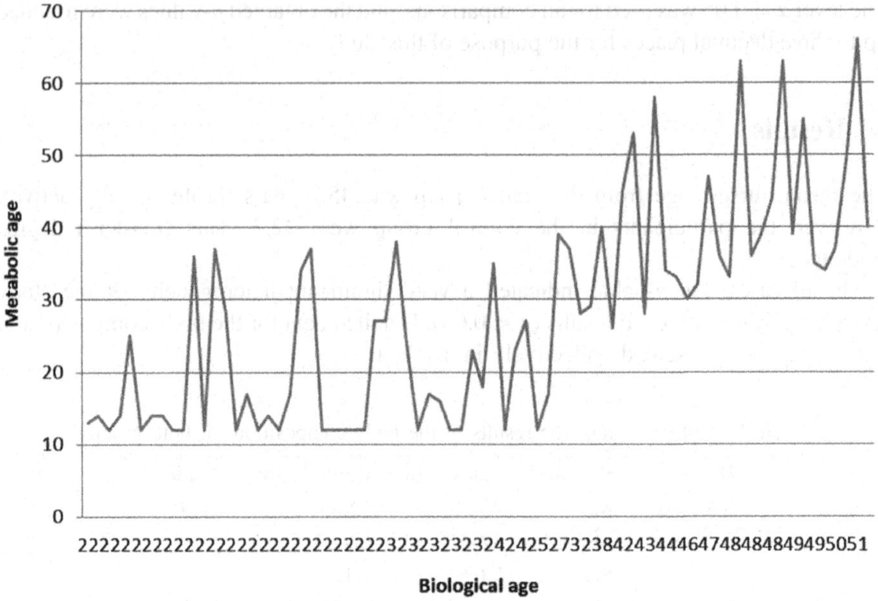

Fig. 1. Individual comparison of results for biological and metabolic age in nurses.

extreme values and more frequent abnormalities in the lower limbs load. The range of percentages for the forefoot and hindfoot load was comparable; nevertheless, a distinct tendency for overloading the forefoot was observed in both groups. Nurses from the study group achieved higher results of the forefoot load (the maximum forefoot load was 85% of the overall load), compared to the nurses from the control group (the maximum forefoot load was 73% of the overall load). Detailed data are presented in Table 2.

The percentage of people in both groups who load the limbs evenly was comparable and amounted to 6.9%. Also, the range of overloads on both right and left sides indicated a similar level in both groups and was ranged between 41%–57% and 43%–60% for the right lower limb overall load in the study and control groups, respectively and was 43%–59% and 40%–57% for the left lower limb overall load in the study and control groups, respectively. Detailed data are presented in Table 2.

Assessment of the COG degree oscillation, which is resultant of the feet pressure forces on the ground, in the majority of the respondents indicated the forward COG shift, with a result of 93% in the study group and 81.4% in the control group. The ground peak pressure point was mainly located in the hindfoot in both groups, i.e. 67.4% in the control group and 55.2% in the study group. This parameter was within the normal range and there were no statistically significant difference between groups ($p = 0.292$).

Three months after research completion, a questionnaire was carried out by the nurses. Its aim was to assess the suitability of the conducted study and the impact of the study on raising awareness of ergonomics and work safety behaviours. It was observed

Table 2. Comparison of the results of the lower limbs load for both groups.

Data	Parameter	Study group	Control group	P-value
FF [%]	M	53.34	52.35	0.931
	Min–Max	41–85	29–73	
	SD	10.03	9.59	
HF [%]	M	46.66	47.65	0.931
	Min–Max	15–59	29–71	
	SD	10.03	9.59	
COG [cm]	M	2.11	1.93	0.383
	Min–Max	0.27–6.57	0.04–7.24	
	SD	1.37	1.46	
L side [%]	M	50.38	50.21	0.535
	Min–Max	40–57	43–59	
	SD	3.77	4.21	
R side [%]	M	49.62	49.79	0.535
	Min–Max	43–60	41–57	
	SD	3.77	4.21	
L/R angle [°]	M	4.24	2.56	0.036
	Min–Max	0–12	0–9	
	SD	3.42	2.49	

Abbreviations: FF – forefoot; HF – hindfoot; COG – centre of gravity; L – left; R – right; M – mean value; Min – minimum value; Max – maximum value; SD – standard deviation; p-value – statistical significance

that 97.1% of respondents considered the body composition test to be useful, while the remaining 2.9% found it difficult to assess. In the study group, 94.3% of the subjects assessed a measurement with Tanita for the symmetry of lower limb load as useful, while the remaining 5.7% admitted that it was difficult to say.

The next question concerned the impact of the obtained results using BIA on the change of eating habits. A group of 51.4% of respondents try to nourish in a thoughtful way and take into account greater water consumption. While 22.9% initially changed their habits, over time they returned to their previous eating habits. Unfortunately, in the case of 25.7%, the study did not have any impact on this issue (see Fig. 2).

Moreover, a group of 68.6% of respondents considered the information given in the study results regarding body composition and symmetry of the lower limb load as important for changing everyday habits related to work ergonomics and lifestyle. The remaining 31.4% admitted that it was difficult to say. A group of 71.4% of nurses confirmed that instructional information presented during the course related to the principles of workplace ergonomics and everyday activities and those regarding the proper activation of postural muscles are needed and justified. In turn, 28.6% of respondents admitted that it is difficult to say

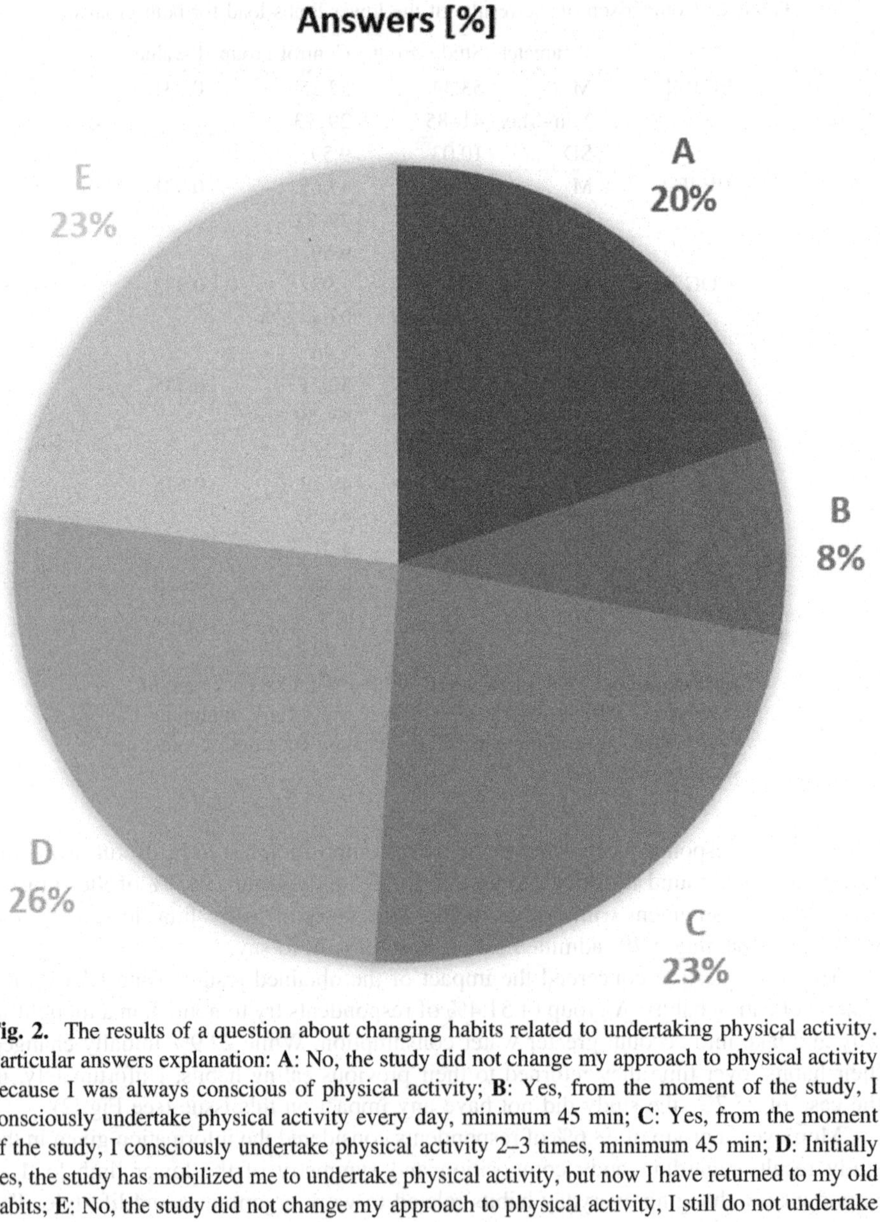

Fig. 2. The results of a question about changing habits related to undertaking physical activity. Particular answers explanation: **A**: No, the study did not change my approach to physical activity because I was always conscious of physical activity; **B**: Yes, from the moment of the study, I consciously undertake physical activity every day, minimum 45 min; **C**: Yes, from the moment of the study, I consciously undertake physical activity 2–3 times, minimum 45 min; **D**: Initially yes, the study has mobilized me to undertake physical activity, but now I have returned to my old habits; **E**: No, the study did not change my approach to physical activity, I still do not undertake any physical activity

4 Discussion

A study by Bilski and Sykutera [31] showed that 73% of nurses suffer from pain syndromes of the spine or joints. In this study, 81% of nurses showed poor knowledge of the legal foundation determining the permissible mass of lifted weight during professional activities. Moreover, only 31% of nurses used exercises of the spine muscles as a means for pain reduction. According to Kułagowska et al. [5], with the increasing interest of researchers in the ergonomics of nurses during their professional activities, the amount of evidence for the occurrence of spinal overload symptoms has increased. The subject of musculoskeletal overloads in the group of professionally active nurses seems to be important and worthy of interest. An important aspect of the prevention of the musculoskeletal overloads is to absorb the knowledge about basic principles of workplace ergonomics and the specificity. Our own study and a summary survey show that it is an important element of preventative physiotherapy to provide information to healthcare providers on the subject of lower limb load and ergonomic principles regarding proper body position during professional activities (also by intentionally activating appropriate postural muscles and proper foot loading).

A study conducted by Maciuk et al. [8] showed that 81% of surveyed nurses feel pain in their lower spine (low back pain). Due to the numerous overloads and accumulated microdamages in the musculoskeletal system in nurses, the number of preventing methods of overload caused by professional activities should increase, and undoubtedly, this subject is worth promoting in medical professions.

Also, Sakowski [32] conducted research among 200 nurses who were asked to indicate areas of education that need to be expanded or supplemented. Frequently mentioned elements were the promotion of health, the implementation of health promotion programmes and the creation of pro-health programmes. The use of appropriate diagnostic, research and informative tools and methods in health-promoting programmes is an important element that can improve the functioning of medical personnel in their workplaces.

The legitimacy of the implementation of preventive examinations for the purpose of health monitoring is justified by Malczyk and Krzonkalla-Bartnik [33], who provided data on the nutritional status of 315 inhabitants of rural areas of southwestern Poland. Analysis of body composition showed irregularities in over 50% of participants. Too high-fat tissue values (VFI) were observed, and this result was characteristic of every fourth examined person. Low body water values (TBW) were also reported, and this abnormal result was observed in every sixth studied person. The best nutritional status was characterised by people from the youngest age group, in the range of 19–30 years. According to the authors, it is advisable to monitor the nutritional status of the society in order to verify the effectiveness of prevention of civilization diseases and promote a healthy lifestyle, referring to regular physical activity and healthy diet. Monitoring of individual factors allows for effective selection of resources in health-promoting activities.

5 Conclusions

The tendency of forefoot overloading and forward COG deflection increases with the length of the professional experience. It may be a consequence of the lack of proper postural muscle tension and be associated with the wrong movement pattern during professional activities in the workplace, the reduction of the proper muscular tension or the use of inappropriate footwear.

It is, therefore, worth implementing regular education in the field of workplace ergonomics, which should increase the awareness of employees in the healthcare sector. Such actions give the chance to improve the functional capabilities of healthcare employees as well as to increase a deliberate control for deep muscle activity, which can lead to a normal load on both feet.

Preventive actions of a research and educational nature, the implementation of work ergonomics and preventive physiotherapy principles should constitute the purpose for eliminating inappropriate movement patterns in professional groups who are strongly exposed to musculoskeletal system overloads. Also, body composition analysis is a useful measurement tool that allows controlling changes in the physical state of the body. Therefore, BIA devices for body composition analysis, such as Tanita, can be successfully used as a tool for monitoring the implementation progress of the prevention principles of civilization diseases, including those related to professional activities.

References

1. National Research Council (US) Committee (2011) Environmental health and safety management system. National Academies Press (US), Washington
2. Kuriata E, Felińczak A, Grzebieluch J, Szachniewicz M (2011) Occupational hazards and the workload of nurses employed at the hospital. Part II. Nurs Public Health 1(3):269–273
3. Carayon P, Xie A, Kianfar S (2014) Human factors and ergonomics as a patient safety practice. BMJ Qual Saf 23(3):196–205
4. Kisakye AN, Tweheyo R, Ssengooba F, Pariyo GW, Rutebemberwa E, Kiwanuka SN (2016) Regulatory mechanisms for absenteeism in the health sector: a systematic review of strategies and their implementation. J Healthc Leadersh 8:81–94
5. Kułagowska E (2009) Musculoskeletal system load in operating room nurses and its determinants. Pract Med 60(3):187–195
6. Ribeiro NF, de Fernandes RCP, Solla DJF, Santos Junior AC, de Sena Junior AS (2012) Prevalence of musculoskeletal disorders in nursing professionals. Braz J Epidemiol 15 (2):429–438
7. Moreira RFC, Sato TO, Foltran FA, Silva LCCB, Coury HJCG (2014) Prevalence of musculoskeletal symptoms in hospital nurse technicians and licensed practical nurses: associations with demographic factors. Braz J Phys Ther 18(4):323–333
8. Maciuk M, Krajewska-Kulak E, Klimaszewska K (2012) Self-assessment of low back pain incidence in professionally active nurses. Probl Hyg Epidemiol 93(4):728–738
9. Rezaee M, Ghasemi M (2014) Prevalence of low back pain among nurses: predisposing factors and role of work place violence. Trauma Mon 19(4):e17926

10. Davis KG, Kotowski SE (2015) Prevalence of musculoskeletal disorders for nurses in hospitals, long-term care facilities, and home health care: a comprehensive review. Hum Factors 57(5):754–792
11. European Commission (2013) Directorate-General for Employment, Social Affairs and Inclusion: Occupational health and safety risks in the healthcare sector-guide to prevention and good practice. Publications Office of the European Union
12. Masoudi Alavi N (2014) Occupational hazards in nursing. Nurs Midwifery Stud 3(3):e22357
13. Donnelly E, Siebert D (2009) Occupational risk factors in the emergency medical services. Prehospital Disaster Med 24(5):422–429
14. DiMaria-Ghalili RA, Mirtallo JM, Tobin BW, Hark L, Van Horn L, Palmer CA (2014) Challenges and opportunities for nutrition education and training in the health care professions: intraprofessional and interprofessional call to action. Am J Clin Nutr 99 (5):1184–1193
15. Kris-Etherton PM, Akabas SR, Douglas P, Kohlmeier M, Laur C, Lenders CM, Levy MD, Nowson C, Ray S, Pratt CA, Seidner DL, Saltzman E (2015) Nutrition competencies in health professionals' education and training: a new paradigm. Adv Nutr 6(1):83–87
16. Lunde L-K, Koch M, Knardahl S, Wærsted M, Mathiassen SE, Forsman M, Holtermann A, Veiersted KB (2014) Musculoskeletal health and work ability in physically demanding occupations: study protocol for a prospective field study on construction and health care workers. BMC Public Health 14:1075
17. Dall TM, Gallo P, Koenig L, Gu Q, Ruiz D (2013) Modeling the indirect economic implications of musculoskeletal disorders and treatment. Cost Eff Resour Alloc 11(5):5
18. Moar JMR, Alvarez-Campana JM, Míguez JL, González LML, Ramos DG (2015) Comparative study of the relevance of musculoskeletal disorders between the Spanish and the European working population. Work (Reading, Mass) 51(4):645–656
19. Bevan S (2015) Economic impact of musculoskeletal disorders (MSDs) on work in Europe. Best Pract Res Clin Rheumatol 29(3):356–373
20. Report ERO (2014) Current and emerging occupational safety and health (OSH) issues in the healthcare sector, including home and community care. Publications Office of the European Union, Luxembourg
21. Zdziarski LA, Wasser JG, Vincent HK (2015) Chronic pain management in the obese patient: a focused review of key challenges and potential exercise solutions. J Pain Res 8:63–77
22. Boughattas W, Maalel OE, Maoua M, Bougmiza I, Kalboussi H, Brahem A, Chatti S, Mahjoub F, Mrizak N (2017) Low back pain among nurses: prevalence, and occupational risk factors. Occup Environ Med 05:26
23. Barbosa-Silva MCG, Barros AJD (2005) Bioelectrical impedance analysis in clinical practice: a new perspective on its use beyond body composition equations. Curr Opin Clin Nutr Metab Care 8(3):311–317
24. Bera TK (2014) Bioelectrical impedance methods for non-invasive health monitoring: a review. J Med Eng 2014:381251
25. Khalil SF, Mohktar MS, Ibrahim F (2014) The theory and fundamentals of bioimpedance analysis in clinical status monitoring and diagnosis of diseases. Sensors 14(6):10895–10928
26. Wang J, Cui Y, He L, Xu X, Yuan Z, Jin X, Li Z (2017) Work-related musculoskeletal disorders and risk factors among Chinese medical staff of obstetrics and gynecology. Int J Environ Res Public Health 14(6):562
27. Hodder JN, Holmes MWR, Keir PJ (2010) Continuous assessment of work activities and posture in long-term care nurses. Ergonomics 53(9):1097–1107

28. Freitag S, Seddouki R, Dulon M, Kersten JF, Larsson TJ, Nienhaus A (2014) The effect of working position on trunk posture and exertion for routine nursing tasks: an experimental study. Ann Occup Hyg 58(3):317–325
29. Sergi G, De Rui M, Stubbs B, Veronese N, Manzato E (2017) Measurement of lean body mass using bioelectrical impedance analysis: a consideration of the pros and cons. Aging Clin Exp Res 29(4):591–597
30. Rosário JLP (2014) A review of the utilization of baropodometry in postural assessment. J Bodyw Mov Ther 18(2):215–219
31. Bilski B, Sykutera L (2004) Determinants of musculoskeletal system load and their health effects among nurses from four Poznan hospitals. Occup Med 55(5):411–416
32. Sakowski P (2010) Self-assessment of tasks and roles of occupational medicine service (OMS) nurses in the polish system of workers' health protection. Occup Med 61(5):561–572
33. Malczyk E (2017) K: evaluation of the nutritional status and composition of the inhabitants of Lower Silesia and Opole villages. Gen Med Health Sci 23(4):250–256

Design and Development of a Medical Device (Artificial Ganglion) for Aids in the Treatment of Lymphedema

Pilar Hernandez-Grajeda(✉), Alberto Rossa-Sierra, and Gabriela Durán-Aguilar

Facultad de Ingeniería, Universidad Panamericana, Prolongación Calzada Circunvalación Poniente 49, 45010 Zapopan, Jalisco, Mexico
phernand@up.edu.mx
http://www.up.edu.mx

Abstract. In 2016, according to the National Institute of Statistics and Geography the incidence of breast malignancy among the population aged 20 years and over is 14.80 new cases per 100,000 people. In women, it peaks in those of the 60–64 age group (68.05 for every 100,000 women in that age group). Worldwide, it is estimated that each year 1.38 million new cases are detected and there are 458,000 deaths due to this cause, being this type of cancer with higher incidence among women [1]. Depending on the type of cancer detected, the treatment is different. Doctors indicate, however, that when there is surgery or radiation in the procedure, lymph nodes are removed or damaged resulting in the majority of cases the disease is known as *lymphedema*. Through the studies of the American Cancer Society [2] it can be observed that "lymphedema is produced by surgery, radiation or cancer". In this work a literature review will be presented in order to focus on the need of considering Human Factors and Usability evaluations during the first stages of the design process for a new artificial ganglion that may help breast cancer survivors with lymphedema to improve their well-being.

Keywords: Medical devices · Artificial ganglion · Lymphedema
User-centered design · Usability · Human factors · Systems approach

1 Introduction

Each year, 1.38 million new cases of breast malignancy are detected worldwide [3]. The treatment a patient receives after being diagnosed with breast cancer could include surgery, radiotherapy, chemotherapy, hormone therapy and directed therapy [4]. These treatments cause side effects on the patient, also called late effects, and some of them appear months or years after the end of the treatment [5]. Lymphedema is one of these late effects, and it occurs when the lymph is not able to circulate through the body as it should [6].

Medical devices are being used in an increasingly wide range of settings, and there have been calls for an acceleration of the integration of Human Factors in patient safety [7]. For example, in the US the FDA requires developers to apply Human Factors

S. Bagnara et al. (Eds.): IEA 2018, AISC 818, pp. 511–519, 2019.
https://doi.org/10.1007/978-3-319-96098-2_63

principles throughout the development process to identify, understand and address use-related hazards [8]; in the European Union, a medical device manufacturer is required to be reducing the risk of use error due to the ergonomic features of the device and the environment in which the device is intended to be used [9].

Although it is not easy to identify the direct relationship between a medical device and its outcome [7] because the consequent behavior of interest, such as long term health behavior, could be temporary distant from the point of interaction with the device, there's been an increased appreciation of the importance of user issues in medical device design, with research focusing mainly on the links between device design, poor usability, human error and patient safety [8]. The importance of Human Factors in medical devices development is recognized in many publications and enforced by international guidelines, but pervasion of industrial development processes by Human Factors principles and trained staff are still in a state of infancy [10].

In this work a literature review will be presented in order to focus on the need of considering Human Factors and Usability evaluations during the first stages of the design process for a new artificial ganglion that may help breast cancer survivors with lymphedema to improve their well-being. We will study the current design process used at medical environments, and we will set the ideal workflow and recommendations for defining the user's goals with the new artificial ganglion and their ideal state of comfort in both short and long term.

2 Background

2.1 Breast Cancer and Lymphedema

There are 1.38 million new cases of breast malignancy detected each year worldwide, and there are 458,000 deaths due to this cause, being this type of cancer the one with higher incidence among women [3]. According to the World Health Organization (WHO), breast cancer is the most frequent and increasing type of cancer in women in both developed and developing countries.

The breast is composed of glands called lobules and thin tubes, that are able to produce milk, and thin tubes called ducts that carry the milk from the lobules to the nipple. The breast tissue also contains fat and connective tissue, lymph nodes and blood vessels [11]. The most common type of breast cancer is ductal carcinoma, which starts in the duct cells. Breast cancer can also start in the cells of the lobules and in other tissues of the breast. Ductal carcinoma in situ is a condition in which abnormal cells are found in the lining of the ducts, but that did not spread outside the duct. Breast cancer that has spread from where it started in the ducts or lobules to the surrounding tissues is called invasive breast cancer [12]. Cancer cells could spread in the body through the tissue, through the lymphatic system or through the blood; the more lymph nodes found with breast cancer cells during a surgery, the greater is the probability of finding cancer in other organs of the body. Usually, during the surgery doctors remove one or more lymph nodes to check for the spreading cancer.

The treatment a patient receives after being diagnosed with breast cancer could include surgery, radiotherapy, chemotherapy, hormone therapy and directed therapy [4].

These treatments cause side effects on the patient, also called late effects, and some of them appear months or years after the end of the treatment [5]:

- Inflammation of the lung after radiotherapy directed to the breast; in particular, when chemotherapy is administered at the same time.
- Lymphedema in the arm; in particular, when radiotherapy is administered after lymph node dissection.
- Higher cancer risk in the other breast for women younger than 45 who receive radiation therapy to the chest wall after a mastectomy.

One possible long-term side effect of a lymph node surgery is swelling in the arm or chest, called lymphedema. It occurs when the lymph is not able to circulate through the body as it should [6]. Because any excess fluid in the arms normally returns to the bloodstream through the lymphatic system, the removal of the lymph nodes sometimes blocks the drainage of the arm, which causes the accumulation of this fluid. Up to 30% of women who undergo an axillary lymph node dissection have lymphedema.

2.2 Medical Devices

Medical devices are fundamental to healthcare and patient safety [10]. According to ISO 13845, a medical device can be defined as any instrument, apparatus, implement, machine, appliance, implant, in vitro reagent or calibrator, software, material or other similar or related article, intended by the manufacturer to be used, alone or in combination, for human beings for one or more of the specific purpose of:

- Diagnosis, prevention, monitoring, treatment or alleviation of disease.
- Diagnosis, monitoring, treatment, alleviation or compensation for an injury.
- Investigation, replacement, modification, or support of the anatomy of a physiological process.
- Supporting of sustaining life.
- Control of conception.
- Disinfection of medical devices.

2.3 Design Process in Medical Environments

Well-designed medical devices will assure a high quality care for patients, being firstly clinically effective and safe, but also meeting the needs of the people that will use them and that will be treated by them [8]. But this is not an easy task, since medical environments involve a complex interaction between regulations, a highly diverse user base, a multitude of established procedures and a vast body of underlying science, all of which must be factored into any medical device design process [13]. There is a need to consult a broad range of specialists during the design and development process.

Given the complexity of medical environments and the need to design within this context, there have been some approaches of translating existing design or engineering methods and tools to include medical knowledge in the development process. Recent studies examining the relationship between healthcare professionals and equipment providers have found an avoidance of consultation with patients or less senior staff, and a tendency to make design modifications on the basis of intuition rather than more formal approach of user testing [14]. The first stages of a design process are crucial on the development of a new medical device, so the early elicitation of the final user views ensures that the findings from the clinical applications and both the clinical and patient users are correctly identified and they will be incorporated into prototype design more easily and with less cost [8]. Design teams are typically not composed of medical domain experts and, therefore, they often lack detailed knowledge of potential user or use environments; the involvement of clinicians and potential device users in development and evaluation is costly in terms of resources but it is ultimately critical to the functional and economic success of a medical product [15].

Effective design thinking can facilitate the delivery of products, services processes and environments that are intuitive, simple to understood, simple to use, convenient, comfortable and consequently less likely to lead to accidental misuse, error or accidents [16]. Design can also reduce the like hood and consequences of error. There have been different methods used on the first stages of a new medical device design process: interviews, observations, focus groups, workshops, Human Factors analyses and prototype design; all of them must consider the full range of applications and users for the devices, even the ones that appeared unlikely [8].

Hagendorn et al. [13] presented a framework called *Concept Ideation Framework for Medical Device Design*, developed in order to accomplish three interrelated tasks aimed at improving the design and innovation process and to managing medical knowledge for incorporating it into the early phases of engineering design:

1. Unify a high level understanding of medical concepts, practices and resources with detailed engineering descriptions of their functional characteristics, as well as a repository of similarly annotated design solutions.
2. Facilitate automated reasoning both within each domain, as well as across domains, enabling high level inferences not immediately available in any individual field.
3. Create a basis for identifying analogous solutions to an engineering problem in a domain agnostic way, so that a designer can incorporate methods and innovations made in other medical field into a medical device design.

3 Human Factors on Medical Devices Development

3.1 Defining the Final User of a Medical Device

The ultimate aim of any medical device is to improve the well-being of the person receiving diagnosis, treatment of medication [7], but it's also important to remember that the final user of this device may not necessarily be the patient. A device may be achieving its goal if it allows a procedure to be completed more quickly, and thus minimizing discomfort for both the patient and the non-patient user. When we refer to improving the user's well-being, or in other words, to the succeed of a medical device, it's understood that there are other factors included among the improvement of health: the device could be getting the user to be more independent during his treatment, or it could be minimizing or slowing down the negative progress or impact of a disease or condition. At the hospitals, there's often a purchasing committee for new medical devices, that typically represents a range of interested parties: end users, power users, trainers, pharmacy staff, and those responsible for the management and maintenance of the equipment [17].

3.2 The Impact of a Poor Design Process of a Medical Device on the Final User

Both the design and use of medical devices affect patient safety and the quality of care. For medical devices, communication deficits during the design process have been shown to result in the wrong device being developed or purchased [9]. Normally, the process of purchasing a medical device is driven by engineering standards and the emphasis on functional requirements, rather than those relating to social or organizational needs [17]. There have been great improvements worldwide in order to get better software and hardware development, and to create new information technologies and biotechnology, but this effort will never be enough if the medical device manufacturers don't take the final user as the greater reason behind every other effort. In other words, a medical device exists because it aims to be a solution for a specific need of a particular user with well detected characteristics.

3.3 Human-Centered Design in a Medical Context

Human-centered design can be defined as a focus on the critical human issues throughout the design and development process so that the inevitable trade-offs between human, commercial and technical issues can be made in a balanced way [7]. The benefits of this approach are wide: identification of new ideas, paradigms and design directions, providing better experiences for users and reducing complains. There are many challenges faced by those who work at the design and development process of a medical device when implementing human-centered design, i.e. understanding the users and their situation or providing adequate justification for the adoption of a user-centered approach [18]. Feedback about user requirements has the longer-term potential to inform manufacturers about user needs and to raise the importance of usability within the development process [17]. Normally, with a human-centered approach, before starting a new design process the design team wants to ensure that their research with the user focused on the right topics, asked the right questions and produced data that could be easily understood and implemented into development.

A good understanding of the relationship between design and the user behavior can support the design itself and provide evidence for the importance of considering human-centered design when aiming to influence or change user behavior in a medical context [7]. Failing to adequately study the potential user of a device at the beginning of development may result in assumptions being made about them, and these assumptions may soon become accepted and unquestioned, and if they are false or incomplete then the device will be developed and evaluated on incorrect info [8]. This has serious implications not just for the safety of the new device, but also for the commercial success of the device.

4 The Design Process for Developing an Artificial Ganglion for the Treatment of Lymphedema

4.1 A Systems Approach to Understand the Complexity of the Problem

It is crucial for the success of the new medical device to not consider device design in isolation from the organizational or social context, and to consider the range of stakeholders or final users of the device. By taking a systems approach and understanding the complexity of implementation and use of medical devices, effective, efficient and satisfying device design should be an achievable goal. In order to provide specific and useful advice to device manufacturers it is necessary to understand in more detail the relationship between the consequence of a particular artefact and the way in which it is designed. For example, we should consider the interaction between the user and the device in a systems context, and develop techniques that allow us to provide specific device insight findings and guidance, and understand what the impact of this specific device design could be in the context of a medical system [7].

When developers had not correctly understood the context of use of their device and had not anticipated likely error situations, there will probably incidents where the design of a medical device had potentially contributed to the incident. This can also happen at instances where the devices had not been designed with the user's expectations in mind, with errors occurring when the device did not function as the user had expected. So the authors of previous studies, as Martin et al. [8] suggest that the key to the success is to include participants for all clinical departments during the design process, identified as potentially benefiting from the device and not restricting the work to those that had already been identified as the target users, so that the problems, needs and difficulties are as easy as possible to identify. In seeking to identify the impact of device design on overall consequences of use, it is crucial that the context of use of a device is acknowledged [7], as it says on the ISO 13407 *Context of use*, encompassing the users and other stakeholder groups, the characteristics of the users, the tasks of the users and the environment of the system.

In the process of understanding the context and problems of the users or patients with lymphedema, in order to define the best design direction for the development of a new device that decreases the side effects of the lymphedema, we must consider the context of use of the designed device, the range of users and its consequence of use, in an attempt of fully understand the impact of the design on the resulting user behavior and well-being. This user-centered approach must be taken into account during the early stages of development of the new device for two purposes: to validate and refine the concept for the new device, as well as to collect user information, experiences and preferences, and to determine the value of the information collected and the impact of this on the product development pathway [8]. To specifically meeting the needs of the user, we must be aware of the capabilities and working patterns of clinical users, the needs and lifestyles of the patients, the environment in which the device will be used and the system of which it will be part.

4.2 Human Factors and Usability as a Guide During the Design Process

Human Factors is a term applied to the application of theory, principles, data and methods to design in order to optimize human well-being and overall system performance [9]. In a medical environment, it seeks to ensure that medical devices that are placed on the market are usable and safe [19]. The discipline of Human Factor has demonstrated that if a device is well designed then this will have positive implications for usability [7]. Approaching the design with Human Factors and Usability Engineering has proven to be an effective means to enhance performance-related outcomes such as fewer errors, less time and lower mental effort [10]. It has also been used to collect input from a wide array of stakeholders, support multidisciplinary communication, support reconciliation of viewpoints [17], and being used as a validation test for performance requirements, such as efficiency and safety [10].

Since Usability is the level to which a product can be used by specified users to achieve specified goals with effectiveness, efficiency and satisfaction, in a specific context of use, and an improved usability may contribute to the quality of patient and staff experience as well as improving safety, cost time and reliability [17], the artificial ganglion designed on the subsequent stages of this work must consider the level of compliance desired for the following usability components:

- *Effectiveness*, as the accuracy and completeness with which users achieve specified goals.
- *Efficiency,* as the resources expended in relation to the accuracy and completeness with which users achieve goals.
- *Satisfaction,* as the freedom from discomfort, and positive attitudes to the use of the product.

In order to define the user's goals with the new artificial ganglion and their ideal state of comfort, we must form a varied team with those involved in the medical environment: nurses, doctors, pharmacists, biomedical technicians, quality improvement staff, unit managers, patients, trainers and patient's family. This emphasizes that the consideration of usability needs to accommodate multiple perspectives and adopt a holistic approach [17].

5 Conclusion

Design and development of new medical devices is shared across multiple disciplines, and there is a lack of common ground between disciplines, with the risk of leading to a misunderstanding of the reference points at the firsts stages of the design process. Medical devices are being used in an increasingly wide range of settings, and there have been calls for an acceleration of the integration of Human Factors in patient safety [9].

During the design process of the new artificial ganglion for users with lymphedema, we must be clear about how Human Factors are a constant process of the wider product development. If the usability recommendations are taken seriously during the early stages of the design process, it will me more likely that time and money will be saved, but most important, a user-centered perspective will lead to an artificial ganglion that is more effective, efficient and that gives satisfaction to the user.

In further work, we will define who the final user of the artificial ganglion will be, how is the environment around the user, who are the stakeholders that will also interact at some point with the device and which tasks will the artificial ganglion be desired to perform, so that the functional features of the new device will be based on the user needs and characteristics as a result of a systems approach.

References

1. Instituto Nacional de Estadística y Geografía (2016) Estadísticas a propósito del día mundial de la lucha contra el cáncer de mama. http://www.inegi.org.mx/saladeprensa/aproposito/2016/mama2016_0.pdf. Accessed 05 June 2017
2. American Cancer Society (2016) Qué es lo que causa linfedema vinculado con el cáncer. https://www.cancer.org/es/tratamiento/tratamientos-y-efectossecundarios/efectos-secundarios-fisicos/linfedema/que-es-linfedema.html. Accessed 17 July 2017
3. Organización Mundial de la Salud (2016) Organización Mundial de la Salud. http://www.who.int/disabilities/care/es/
4. National Cancer Institute (2017) Aspectos generales de las opciones de tratamiento. https://www.cancer.gov/espanol/tipos/seno/paciente/tratamiento-seno-pdq#section/_52. Accessed 10 June 2017
5. National Cancer Institute (2017) Efectos Secundarios. https://www.cancer.gov/espanol/tipos/seno/paciente/tratamiento-seno-pdq#section/_52. Accessed 15 July 2017
6. American Cancer Institute (2016) Cirugía de ganglios linfáticos para el cáncer de seno. https://www.cancer.org/es/cancer/cancer-de-seno/tratamiento/cirugia-del-cancer-de-seno/cirugia-de-ganglios-linfaticos-para-el-cancer-de-seno.html. Accessed 16 June 2017
7. Sharples S, Martin J, Lang A, Craven M, O'Neill S, Barnett J (2012) Medical device design in context: a model of user-device interaction and consequences. Displays 33:221–232
8. Martin JL, Clark DJ, Morgan SP, Crowe JA, Murphy E (2012) A user-centred approach to requirements elicitation in medical device development: a case study from a industry perspective. Appl Ergon 43:184–190
9. Vincent CJ, Li Y, Blandford A (2014) Integration of human factors and ergonomics during medical device design and development: it's all about communication. Appl Ergon 45:413–419
10. Schmettow M, Schnittker R, Schraagen JM (2017) An extended protocol for usability validation of medical devices: research design and reference model. J Biomed Inf 69:99–114
11. National Cancer Institute (2017) Anatomía de la mama femenina. Se muestran el pezón, la aréola, los ganglios linfáticos, los lóbulos, los lobulillos, los conductos y otras partes de la mama. https://www.cancer.gov/espanol/tipos/seno. Accessed 20 June 2017
12. National Cancer Institute (2017) El cáncer se disemina en el cuerpo de tres maneras. https://www.cancer.gov/espanol/tipos/seno/paciente/tratamiento-seno-pdq#section/_24. Accessed 17 June 2017
13. Hagedorn TJ, Grosse IR, Krishnamurty S (2015) A concept ideation framework for medical device design. J Biomed Inf 55:218–230
14. Money A, Barnett J, Kuljis J, Craven M, Martin J, Young T (2011) The role of the user within the medical device design and development process: medical device manufacturers' perspectives. BMC Med Inf Decis Mak 11:1

15. Shah SGS, Robinson I (2007) Benefits of and barriers to involving users in medical device technology development and evaluation. Health Care 23:131–137
16. Clarkson P, Buckle P, Coleman R, Stubbs D, Ward J, Jarret J (2004) Design for patient safety: a review of the effectiveness design in the UK health service. J Eng Des 15:123–140
17. Vincent CJ, Blandford A (2017) How do health service professionals consider human factors when purchasing interactive medical devices? A qualitative interview study. Appl Ergon 59:114–122
18. Vincent C, Blanford A (2011) Designing for safety and usability: user-centered techniques in medical device design practice. In: Proceedings of the HdES 55th annual meeting. Sage, Las Vegas, pp 793–797
19. Furniss D, Masci P, Curzon P, Mayer A, Blandford A (2014) 7 Themes for guiding situated ergonomic assessments of medical devices: a case study of an inpatient glucometer. Appl Ergon 45:1667–1668
20. Organización Mundial de la Salud, Noviembre 2013. http://www.who.int/mediacentre/factsheets/fs384/es/

The Design and Use of Ergonomic Checkpoints for Health Care Work

Kazutaka Kogi[1(✉)], Yumi Sano[1], Toru Yoshikawa[2], and Setsuko Yoshikawa[3]

[1] Ohara Memorial Institute for Science of Labour, Tokyo 151-0051, Japan
k.kogi@isl.or.jp
[2] National Institute of Occupational Safety and Health, Kawasaki 214-8585, Japan
[3] Japanese Red Cross College of Nursing, Tokyo 180-8618, Japan

Abstract. Recent participatory approaches for improving health care workplaces clearly indicated the usefulness of applying broad-ranging ergonomic measures feasible in local situations of health care facilities. In order to facilitate the application of these measures, ergonomic checkpoints in health care work were compiled based on these recent approaches. Comparing the multifaceted improvements achieved by health care workers with those from participatory approaches in other sectors, ergonomic aspects applicable to health care work were identified. Ergonomic checkpoints for use by health care work were drafted by using the established format presented by the International Labour Office (ILO) in collaboration with the International Ergonomics Association (IEA). Ten technical areas were confirmed as important ergonomic action areas for health care work. They included materials handling, machine safety, person transfer, workstations, physical environment, hazardous agents, welfare facilities and work organization. Typical low-cost ergonomic measures in these areas were compiled into 60 ergonomic checkpoints by utilizing the ILO/IEA format. The validity of these checkpoints was tested by pilot application of action tools based on them. Main contributing actors leading to successful applications of the proposed checkpoints were simple procedures for multifaceted risk reduction, emphasis placed on locally feasible improvements and the use of action-oriented tools such as locally. adapted action checklists. The new checkpoints compiled by examining recent participatory approaches were found useful for improving health care workplaces.

Keywords: Health care work · Ergonomic checkpoints
Participatory approach

1 Introduction

The effective use of ergonomic checkpoints covering major technical areas important for improving the working environment in each specific occupation is widely known. This awareness has been raised through the collaboration of the International Labour Office (ILO) and the International Ergonomic Association (IEA) by developing and widely applying ILO/IEA Ergonomic Checkpoints for small industries and ILO/IEA

S. Bagnara et al. (Eds.): IEA 2018, AISC 818, pp. 520–527, 2019.
https://doi.org/10.1007/978-3-319-96098-2_64

Ergonomic Checkpoints in Agriculture [3, 4, 6, 7], followed by ILO Stress at Work Prevention Checkpoints [5]. These checkpoints are providing good impetus for participatory workplace improvements in many countries. A recent trend is to apply participatory steps for stress prevention at work [2, 10, 12, 16, 18, 20].

The need to apply ergonomic measures for improving safety and health of health care and nursing personnel is increasingly recognized. Recent participatory approaches for improving health care workplaces indicate the usefulness of applying broad-ranging ergonomic measures feasible in each local situation [1, 13, 19]. In order to promote the application of these measures, it is useful to compile ergonomic checkpoints in health care work based on these recent approaches.

2 Methods

Participatory steps taken for improving health care work were reviewed [13, 19]. Comparing the multifaceted improvements achieved by health care workers with those from participatory approaches in other sectors [3, 4, 6, 7, 9, 11, 14], ergonomic aspects applicable to health care work were examined. Typical low-cost improvements applicable to health care work were extracted based on recent experiences in health care facilities.

Ergonomic checkpoints useful for improving health care work were then compiled by using the established format presented by the ILO in collaboration with the IEA [4, 17]. The illustrations corresponding to these checkpoints were created by taking into account the practical types of improvements undertaken by health care workers [1, 13, 19].

The validity of these checkpoints was tested by pilot application of action tools based on these checkpoints [8, 15, 20].

3 Results

3.1 Ergonomic Areas Relevant to Improving Health Care Facilities

In editing the new checkpoints applicable to health care settings, the working group has collected good ergonomic practices in these settings and identified commonly relevant technical areas. The newly proposed ergonomic checkpoints in health care work thus cover a broad range of ergonomic measures. These broad-ranging technical areas are covered also by the sample action checklist presented in the annexes of the new checkpoints. It is useful to compare these measures with the composition of the IEA/ILO Ergonomic Checkpoints (ILO, 2010). Table 1 compares the distribution of the items in different technical areas among the action checklists used in WISE (small enterprises), WIND (agriculture) and POSITIVE (various industries) programs as well as the Mental Health Action Checklist and the new human are work checklist.

These checklists for participatory workplace improvement activities commonly cover the main technical areas within the ILO/IEA Ergonomic Checkpoints. The Mental Health Action Checklist additionally covers the areas corresponding to communication, emergency preparedness and other areas including social support. In the case of the

Table 1. Relations between the IEA/ILO Ergonomic Checkpoints and items in action checklists.

Technical area in the IEA/ILO Ergonomic Checkpoints		Number of corresponding items in the checklist				
		Small enterprises	Agriculture	Trade union initiative	Mental health action checklist	Health care work
Materials handling		9	6	7	1	3
Workstations and tools		5	6	8	2	3
Machine safety		7	5	6	1	3
Physical environment		11	6	11	3	6
Welfare facilities		8	5	6	1	2
Work organization		3	3	4	2	3
Not in the IEA/ILO Checkpoints	Communication	2	11	12	7	1
	People transfer					3
	Infection control					3
	Preparedness				4	3
	Other areas				9	
Total number in the checklist		45	42	54	30	30

sample health care work checklist being newly tested, all these technical areas are covered since its emphasis is placed on multifaceted ergonomic and stress-related aspects. It also includes check items specific to health care settings such as people transfer and infection control. These wide areas covered by the health care work checklist reflect the broad-ranging good practices in health care work collected by the HES working group. It is therefore reasonable that the new health care work checkpoints cover the broad technical areas indicated in Table 2. The ten technical areas of the proposed checkpoints include materials storage and handling, machine and hand-tool safety, people transfer, workstations, physical environment, hazardous agents, infection

Table 2. Ten technical areas identified for ergonomic checkpoints in health care work.

Technical area	Examples of measures
1. Materials storage and handling	Racks, carts, labels
2. Machine and hand-tool safety	Guards, safety devices, wiring
3. People transfer	Mobile equipment, lifters
4. Workstations	Reach, height, coding
5. Physical environment	Lights, indoor climate, partitions
6. Hazardous substances and agents	Isolation, shielding, labelling
7. Infection control	Hand hygiene. Safe procedures, protective gear
8. Welfare facilities	Toilets, resting, meetings
9. Emergency preparedness	Plans, anti-harassment measures, extinguishers
10. Work organization	Schedules, stress prevention

control, welfare facilities, emergency preparedness and work organization. As indicated by the examples of measures, these ten technical areas are concerned with physical, mental and social aspects relevant to safety and health risks in health care services. It is noteworthy that the technical areas covered in the case of improving health care facilities are broader than in the case of traditional participatory ergonomics programs.

Examples of the workplace improvements achieved by works in selected health care facilities also indicate that these technical areas are relevant to their experiences. While the majority of improvements were usually done in work methods and physical environment, many improvements were also conducted to improve people transfer, infectious control, emergency preparedness and work schedules. It is clear that the checkpoints should address organizational and psychosocial aspects of health care work.

3.2 Types of Low-Cost Actions for Improving Workplace Environment of Health Care Work

The experiences in applying action checklists for small enterprises, agriculture and trade union activities and the Mental Health Action Checklist clearly show the importance of focusing on low-cost improvement actions that have real impact on reducing work-related risks in each work setting. These experiences are well taken into account in assembling the 60 checkpoints for the ten technical areas.

A special attention has been paid to selecting low-cost improvement actions that reflect basic ergonomic principles in respective technical areas. Table 3 shows these practical and basic ergonomic principles incorporated in the typical low-cost actions of the compiled checkpoints. These basic ergonomic principles may help achieve the reduction of multifaceted work-related risks.

Table 3. Ten technical areas identified for ergonomic checkpoints in health care work.

Technical area	Basic ergonomic principles	Examples of the impact
A. Materials handling	Organized storage, easy transport	Less physical demands
B. Machine-tool safety	Guarding, safe equipment/wiring	Reduced accident risks
C. People transfer	Proper transfer procedures	Reassured safe procedures
D. Workstations	Natural posture, efficient operations	Efficient work with less load
E. Physical environment	Good climate, isolating hazard sources	Comfortable, barrier-free space
F. Hazardous agents	Proper isolation/shielding, labels	Reduced contacts with hazards
G. Infection control	Safe procedures, specific protection	Reduced risks of infection
H. Welfare facilities	Sanitary/resting facilities, recreation	Refreshing effects and relations
I. Preparedness	Emergency/anti-harassment planning	Shared emergency/violence steps
J. Work organization	Restful schedules, better communication	Shared plans for better teamwork

The application of these basic ergonomic principles can lead to improvements in physical, mental and social aspects of health care work. These improvements reflecting basic ergonomic principles may promote the stepwise progress in each work situation.

3.3 Participatory Action-Oriented Steps for Improving Health Care Work

The planning of low-cost improvements is shown to be facilitated through the use of an action checklist listing typical improvement actions extracted from the compiled ergonomic checkpoints. Examples of the check items included in the action checklist are shown in Fig. 1.

Technical area	Typical check item and its illustrated example		Do you propose action?
A. Materials storage and handling	Use multi-level racks and small containers to minimize manual transport of materials.		[ ]No [ ]Yes- └[ ]Priority
B. Machine and hand-tool safety	Establish safe handling procedures of sharps and use safety devices and safe disposal containers.		[ ]No [ ]Yes- └[ ]Priority
C. People transfer	Utilize safe lifting or transferring devices when lifting of the person is involved.		[ ]No [ ]Yes- └[ ]Priority
D. Workstations	Adjust the working height for each worker at elbow level or slightly below it.		[ ]No [ ]Yes- └[ ]Priority
E. Physical environment	Use partitions, curtains and other arrangements for protecting privacy of persons cared.		[ ]No [ ]Yes- └[ ]Priority
F. Hazardous substances and agents	Label and store properly containers of hazardous chemicals to communicate warnings and to ensure safe handling.		[ ]No [ ]Yes- └[ ]Priority
G. Infection control	Ensure regular and proper use of personal protective equipment adequate for protecting potential infections.		[ ]No [ ]Yes- └[ ]Priority
H. Welfare facilities	Provide refreshing resting facilities and, for night shift workers, restful napping facilities.		[ ]No [ ]Yes- └[ ]Priority
I. Emergency preparedness	Establish emergency plans to ensure correct emergency operation, easy access to facilities and rapid evacuation.		[ ]No [ ]Yes- └[ ]Priority
J. Work organization	Arrange working schedules avoiding excessive work hours and securing enough resting periods and short breaks.		[ ]No [ ]Yes- └[ ]Priority

Fig. 1. Typical check items incorporated in an action checklist corresponding to the proposed ergonomic checkpoints in health care work.

The action checklist asks each participant to tick either 'No' or 'Yes' to the question 'Do you propose action?' for each check item. The answer 'No' means that the action is not proposed because that action is already implemented or not applicable. The answer 'Yes' means that the action is proposed for improving the existing conditions. The participants may select some of the 'Yes' items as 'Priority' items. The checklist can be used to identify good practices and propose necessary actions.

It is important that these basic ergonomic principles in multiple technical areas correspond to low-cost types of ergonomic improvements also in health care work. This fact is confirmed by the pilot activities in healthcare and nursing workplaces. The action checklists used in these pilot activities contain low-cost ergonomic improvement actions in all these technical areas.

The participatory steps for applying the ergonomic checkpoints by using an action checklist may consist of the four stages in Fig. 2. The initial steps are for learning local good practices and organizing group work to propose locally feasible options for improving existing workplace conditions. As in other participatory programs, the participants then agree on immediate improvements that are implemented and reported.

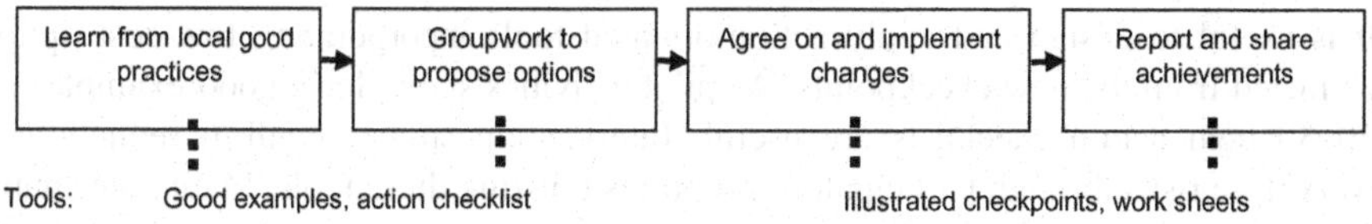

Fig. 2. Participatory steps proven valid for applying the proposed ergonomic checkpoints in health care work.

The use of the action checklist can facilitate these participatory steps, since the participants can easily understand the types of actions to be taken and select priority actions. In this way, the use of the action checklist can facilitate the participatory steps leading to concrete results. The participants can plan and implement locally feasible improvements selected from multiple technical areas. The action checklist can help workers understand feasible options and plan actions indicated in the action checklist.

3.4 Supporting Improvement Steps at Varied Health Care Services

These results suggest the importance of utilizing the proposed ergonomic checkpoints as practical means of facilitating participatory steps for proposing and implementing improvements in varied health care services. The main contributing factors of the reported results may include simple procedures for multifaceted risk reduction, a clear focus on feasible actions and the locally adjusted use of action tools. The use of the ergonomic checkpoints makes it easy to design a locally adapted action checklist. This merit is confirmed through pilot activities for participatory steps in health care and nursing services.

The effectiveness of these programs has been confirmed by intervention studies on participatory programs applying ergonomic checkpoints. These intervention studies indicate that the application of action checklists through participatory programs can reduce workload and health disorders [11], decrease musculoskeletal disorders [13] and reduce stress scores and improve work performance [8, 15]. Multifaceted outcomes in stress reduction can result from participatory steps in varied jobs [20]. It is evident that the participatory application of ergonomic checkpoints can lead to reduced workload and improved mental health among workers. It is useful to organize participatory programs utilizing these action checklists.

The sustainability of the participatory activities aimed at feasible multifaceted improvements is also proven by the follow-up results reported in these studies. It is suggested that the wide application of the ergonomic checkpoints in health care work, combined with the use of locally adjusted action checklists, may contribute to the promotion of workplace improvement activities in many countries.

4 Conclusions

It is useful to design and apply action-oriented tools incorporating low-cost options extracted from the new checkpoints. As pilot activities show, local good examples and 20–30 item action checklists are useful. Intervention studies confirm reductions in workplace risks through participatory programs utilizing these tools. Main contributing factors leading to positive achievements are (a) simple procedures, (b) a clear focus on locally feasible improvements and (c) the use of action-oriented toolkits. It is recommended to design and use the ergonomic checkpoints in health care work by compiling multifaceted ergonomic measures.

References

1. Bourbonnais R, Brisson C, Vinet A, Vezina M, Abdous B, Gaudet M (2006) Effectiveness of a participative intervention on psychosocial work factors to prevent mental health problems in a hospital setting. Occup Environ Med 63:335–342
2. Dul J, Bruder R, Buckle P, Carayon P, Falzon P, Marras WS, Wilson JR, van der Doelen B (2012) A strategy for human factors/ergonomics: developing the discipline and profession. Ergonomics 55:377–395
3. International Labour Office (2004) WISE: work improvement in small enterprises: package for trainers. International Labour Office Subregional Office for East Asia, Bangkok
4. International Labour Office (2010) Ergonomic checkpoints, 2nd edn. International Labour Office, Geneva
5. International Labour Office (2012) Stress prevention at work checkpoints. International Labour Office, Geneva
6. International Labour Office (2014) Global manual for WIND: work improvement in neighbourhood development. International Labour Office, Geneva
7. Kawakami T, Kogi K, Toyama N (2004) Participatory approaches to improving safety and health under trade union initiative – experiences of POSITIVE training program in Asia. Ind Health 42:196–206

8. Kobayashi Y, Kaneyoshi A, Yokota A, Kawakami N (2008) Effects of a worker participatory program for improving work environments on job stressors and mental health among workers: a controlled trial. J Occup Health 50:455–470
9. Kogi K (2008) Facilitating participatory steps for planning and implementing low-cost improvements in small workplaces. Appl Ergon 39:475–481
10. Kogi K (2012) Practical ways to facilitate ergonomics improvements in occupational health practice. Hum Factors 54:890–900
11. Kogi K, Kawakami T, Itani T, Batino JM (2003) Low-cost work improvements that can reduce the risk of musculoskeletal disorders. Intern J Ind Ergon 1:179–184
12. Kogi K, Yoshikawa T, Kawakami T, Lee MS, Yoshikawa E (2016) Low-cost improvements for reducing multifaceted work-related risks and preventing stress at work. J Ergon 6:147. https://doi.org/10.4172/2165-7556.1000147
13. Lee JE, Kim SI, Jung HS, Koo JW, Woo KH, Kim MT (2009) Participatory action-oriented training for hospital nurses (PAOTHN) program to prevent musculoskeletal disorders. J Occup Health 51:370–376
14. Scott P, Kogi K, McPhee B (2010) Ergonomics guidelines for occupational health practice in industrially developing countries. University of Darmstadt, Institute for Ergonomics, Darmstadt
15. Tsutsumi A, Nagami M, Yoshikawa T, Kogi K, Kawakami N (2009) Participatory intervention for workplace improvements on mental health and job performance among blue-collar workers: a cluster randomized controlled trial. J Occup Environ Med 51:554–563
16. Yoshikawa E (2013) Concept analysis of a participatory approach to occupational safety and health. Sangyo Eiseigaku Zasshi 55:45–52
17. Yoshikawa T, Kawakami N, Kogoi K, Tsutsumi A, Shimazu M, Nagami M, Shimazu A (2007) Development of a mental health action checklist for improving workplace environment as means of job stress prevention. Sangyo Eiseigaku Zasshi 49:127–142
18. Yoshikawa T, Kogi K (2010) Roles in stress prevention of good practices for workplace improvements and the use of action support tools. Job Stress Res 17:267–274
19. Yoshikawa T, Kogi K, Kawakami T, Osiri P, Arphorn S, Ismail NH, Van Chin P, Khai TT, Koo JW, Park JS, Toyama N, Mitsuhashi T, Tsutsumi A, Nagasu M, Matsuda F, Mizuno Y, Sakai K (2006) The role of participatory action-oriented training in building an Asian network for occupational safety and health of health care workers. J Sci Labour 83:182–187
20. Yoshikawa T, Yoshikawa E, Tsuchiya M, Kobayashi Y, Shimazu A, Tsutsumi A, Odagiri Y, Kogi K, Kawakami N (2013) Development of evidence-based medicine guidelines for improving the workplace environment by means of primary job stress prevention. Job Stress Res 20:135–145

Simulating the Impact of Patient Acuity and Nurse-Patient Ratio on Nurse Workload and Care Quality

Sadeem M. Qureshi[1(✉)], Nancy Purdy[2], and W. Patrick Neumann[1]

[1] Human Factors Engineering Lab, Department of Mechanical and Industrial Engineering, Ryerson University, Toronto, ON M5B 2K3, Canada
{slqureshi,pneumann}@ryerson.ca

[2] Daphne Cockwell School of Nursing, Ryerson University, Toronto, ON M5B 2K3, Canada
npurdy@ryerson.ca

Abstract. Increased patient acuity and nurse-patient ratios are associated with increased nurse workload and deteriorated care quality – quantifying this change is a challenge. A novel approach to nurse focused simulation approach was conceived, to quantify the effects of changing nurse-patient ratios and patient acuity in terms of care quality and nurse workload. The demonstrator model was run on different levels of patient acuity (present-case, −10%, +10%, +20%, +30%), and nurse-patient ratio (one nurse assigned to 2, 3, 4, 5, 6 patients). Inputs to the model were: real patient-care task data, workflow process sequence, and physical layout. Outputs included: nurse workload in terms of walking distance, care delivery time, and task in queue, and care quality indicators including missed care delivery time and missed care. The model was able to quantify: as nurse-patient ratios decreased and patient acuity increased, nurse workload increased and care quality deteriorated. In comparison to the base case, walking distance increased up to 18%; care delivery time up to 40%; task in queue up to 354%, missed care delivery time increased up to 354%; and missed care increased up to 253%.

Keywords: Process quality improvement · Healthcare ergonomics
Discrete Event Simulation

1 Introduction

This paper explores the novel application of Discrete Event Simulation (DES) to examine two trends in healthcare: nurse-patient ratios and patient acuity, and their effects on nurse workload and care quality.

1.1 Workload Drivers: Nurse-Patient Ratio and Patient Acuity

One of the challenges hospital managers face is to stay on a budget and have adequate levels of staff that deliver high quality of care (Seago et al. 2001). In addition to this,

S. Bagnara et al. (Eds.): IEA 2018, AISC 818, pp. 528–535, 2019.
https://doi.org/10.1007/978-3-319-96098-2_65

hospital managers have to cope with increased demands from an aging population, rising healthcare technologies cost and complex treatment processes (Letvak and Buck 2008). Since the cost of nurses is the highest budget item, hospital managers may become tempted to reduce the workforce directly, despite the increase in nurse workload.

Another driver of workload is patient acuity. To address the increased healthcare demands, newer polices have been conceived to improve patient throughput by discharging patients earlier, leading to increased patient acuity levels throughout the unit that leads to increased workload for nurses (Daly and Brennan 2009). Negative effects arising from increased workload include: absenteeism, overtime, burnouts, work related musculoskeletal disorders (MSD) and injury (Hughes 2008). There is a lack of tools to help hospital managers understand these issues in qualitative terms, to support better decision making. Hence, a tool is needed that can proactively predict the effects of changing patient acuity levels and nurse-patient ratios – Discrete Event Simulation (DES) is a potential tool.

1.2 Discrete Event Simulation (DES)

Current approaches such as real-life trial and error, force workers to be exposed to untested work environments that can be hazardous and expensive (Howard et al. 2003). DES is a tool that can virtually assess the effects of technical design and policy changes on the performance and outcomes of the system in question.

DES is the process of representing complex systems as an ordered event sequence in which variable(s) change at a discrete sets (Banks et al. 2005). It is an operational research technique used to assess and predict the efficiency of a proposed or an existing system (Jun et al. 1999). DES has been effectively used for the analysis of system design alternatives and business modeling. It is a widely used tool in manufacturing and service industries to examine emergent behaviors (Dode et al. 2016; Greasley and Owen 2018; Perez et al. 2014).

In the domain of healthcare, DES has been used in healthcare to model patient flow in peri anesthesia units, the emergency department, spatial layouts to examine wait times, and modelling operations in pharmacies (Choudhary et al. 2010; Gunal and Pidd 2010; Mohammadi and Shamohammadi 2012). However, these studies have been limited to modelling patient flow, and do not focus on workload for nurses delivering care. Attention to workload and wellbeing issues in DES modelling has largely been limited to manufacturing (Dode et al. 2016; Neumann and Medbo 2016; Perez et al. 2014). Previous work has introduced an approach to modelling nurses' workloads and the process of care delivery instead of the traditional 'patient as a product flow through creation of Simulated Care Delivery Unit (SCDU) model (Qureshi et al. 2016, 2017). In this study, we are extending that approach to include attention to both patient acuity and nurse-patient ratios.

The aim of the study is to extend the SCDU model, simulating the delivery of care in acute care hospital units (Qureshi et al. 2016, 2017), by developing a modelling capability to test both patient acuity and nurse-patient ratio. Furthermore, the model is then used to quantify the impact of changing these factors on nurse workload and care quality.

2 Methods

The Simulated Care Delivery Unit (SCDU) model is created on DES environment software – Rockwell (ARENA). The SCDU model simulates the working environment of nurses and the patient care delivery activities and records the effects of changing technical design factors on nurse workload and care quality. Model inputs include: work demands, workflow process, physical layout. Outputs are recorded separately for nurse workload, in terms of cumulative walking distance, number of tasks in queue, care delivery time, and care quality indicators including number of missed care tasks, missed care delivery time. These are elaborated below:

2.1 Model Design and Inputs

Specifying Work demands entails the essential details of the daily patient care tasks that a nurse performs. This was obtained from an in-patient unit from a teaching hospital in Toronto, Canada, as part of nurse workload report generated using a management information system called GRASP (Grace-Reynolds Application and Study of PETO). The Ontario Ministry of Health and Long Term Care (MOHTC) has mandated the use of MIS to measure workload (Farrington et al. 2000). This dataset includes: Task name, such as medication, vital signs etc.; Task frequency, how frequent a care delivery task is performed; and Time duration, a frequency weighted average of the standardized time as reported by GRASP, was used to calculate time duration for each task.

Workflow process is the operating logic programmed in the SCDU model. The operating logic consists of task priorities, task schedules (either random or fixed times e.g. patient discharge at 11am), nurse priorities of which task is required next, task location (nurse station or patient bedside). This was defined after consultation with the subject matter expert a Registered Nurse (RN) in Canada, on the research team with 25+ years of practical and research experience.

Physical layout describes the overall floorplan of the in-patient unit. It was designed using Microsoft Visio. The physical layout consists of the following information: total beds, types of room (single, double, or quad beds), distance between patient beds and nurse station, and bed assignment. The layout will affect travel distances as nurses move between beds.

These inputs were programmed in the SCDU model, where they were simulated at using a combination of different design parameters – staffing ratio and patient acuity.

2.2 Model Outputs

Nurse workload is assessed using: *care delivery time*, total time spent by nurse in delivering care; *walking distance*, the total distance walked by a nurse in one shift; *task in queue* is a type of a mental workload indicator (Potter et al. 2009) and, it is a stack of pending tasks for a nurse.

Care Quality is assessed using: *missed care*, the number of care delivery tasks not completed at the end of shift; *missed care delivery time*, the time that would be required to complete these tasks.

2.3 Experimental Design and Analysis

Staffing ratio entails the number of patients assigned to one nurse. In this study, the following staffing ratios were explored: one nurse assigned to 2, 3, 4, 5 and 6 patients respectively. The most common actual nurse-patient ratios are 1:4, 1:5 and 1:6. In this study, we aimed our experimental conditions to span the range of all possible acuity scenarios and nurse-patient ratios.

Patient acuity is the level of sickness of a patient. In this study, patient acuity is operationalized by the intensity of care (frequency of care delivery tasks and/or time duration). The following levels of patient acuity were tested: baseline case, −10%, +10%, +20% and +30% of the baseline case. This simulation range was chosen to span a realistic operating range. In reality, −10% of the baseline case for patient acuity may not exist as newer policies have been conceived to discharge patients earlier, to improve system throughput. Hence, +10% and +20% increase in patient acuity levels are more realistic, given the new policies of discharging patient earlier than previous. However, a 30% increase in patient acuity may conceivably be expected, given the increase in acuity levels of patients in recent years (Needleman 2013).

Modelling Experiment – The demonstrator model was run using a combination of different levels of patient acuity (baseline case, −10%, +10%, +20% and +30% of the baseline case) and, nurse-patient ratio (1 nurse assigned to 2, 3, 4, 5 and 6 patients respectively). The SCDU model was tested on 25 different conditions – a combination of all different staffing ratio and patient acuity levels. Each trial was run for 365 days with data collected for each day separately. A warm up period of 21 days was applied using the method of Hoad et al. (2008) which resulted in 344 simulated shift-data for each condition. A full factorial ANOVA and Post Hoc analysis (Tukey's test) were conducted for all 25 conditions.

3 Results

A nurse focused DES modelling approach has been developed that demonstrates the ability to assess the impact of changing nurse-patient ratio and patient acuity on nurse workload and care quality. These are presented below:

3.1 Nurse Workload Results

In this study, the impact of changing patient acuity and nurse-patient ratio on nurse workload is assed using: 'task in queue', 'walking distance' and 'care delivery time'. *Task in queue* – An increasing pattern can be observed as a range of 1 to 33 tasks were always in queue for nurses for all 25 trials. *Care delivery time* – A ceiling effect can be observed as for most cases, the nurse spent a constant 11.7 h out of 12 h in care delivery, excluding walking time. The *walking distance* increased during high patient acuity levels and nurse-patient ratios, in comparison to lower nurse-patient ratio and patient acuity levels.

3.2 Care Quality Results

In this study, care quality is represented using: ‘missed care delivery time’ and ‘missed care’. *Missed care delivery time* – An increasing pattern can be observed as a range of 1 to 27 h were spent in performing care delivery tasks that were not completed before the end of shift. *Missed care* – A linear increase can be observed where a range of 2 to 79 tasks were not completed before the end of shift.

A full factorial analysis (ANOVA) showed a statistical significant difference ($p < 0.05$) was observed for *missed care, missed care delivery time, task in queue* except for *walking distance*, for baseline & 10% increase in baseline; 20% & 30% increase in baseline case; 10% & 30% increase in baseline case, and *care delivery time* 5 & 6 beds; 4 & 5 beds. Tukey’s test was selected as a Post Hoc test that concluded statistical significant difference for all cases of *missed care, missed care delivery time, task in queue* except for *walking distance*, for 10% & 30% increase in baseline; 20% & 30% increase in baseline case, and *care delivery time* 5 & 6 beds; 4 & 6 beds; 4 & 5 beds.

4 Discussion

In this paper a novel approach to nurse-focused DES modelling was presented, that includes patient acuity and nurse-patient ratio quantitatively. This nurse focused DES modelling approach simulates the delivery of care by nurse caregivers. There has been a lack of attention to nurse workload in DES models, as traditional approaches have been limited to modelling ‘patients as a product flow’.

4.1 Nurse Workload

The demonstrator model shows increased mental workload. For most trials, 30 to 45 tasks were always observed in the nurse task queue. These ‘stacked’ tasks create increased mental workload (Potter et al. 2009). Furthermore, nurses spent more than three quarters of the shift were spent by nurses in delivering care for patients, similar to Hendrich et al. (2008). Nurse walking distance decreased as patient acuity levels and nurse-patient ratio increased because the physical layout of the unit used in this study had an optimistic bed assignment – all beds assigned to patient were next to each other. In real world scenarios, bed assignment is contingent upon nurse skillset, bed availability, treatment priority, acuity level, and resource utilization (Schmidt et al. 2013). Therefore, bed assignment isn’t always optimistic and further research is needed to address this complicating issue.

4.2 Care Quality

Ausserhofer et al. (2014) compared uncompleted nursing care delivery task for hospitals across 12 European countries, and were found consistent with the output from the demonstrator model i.e. a range of 1 to 80 tasks left undone. The time to perform these tasks were 8 to 26 h, which may not be the case in real world scenarios as nurses tend to work much faster than the standardized times reported in GRASP systems. In reality, nurses under time-pressure may be forced to skip some lowest priority documentation tasks and/or entering some documentation task data on GRASP systems.

4.3 Methodological Modeling Issues

Further validation of the demonstrator model is needed. The model was built in consultation with only one subject matter expert, using optimistic bed assignments and obtaining patient data from a hospital that used GRASP system to record data; GRASP uses standardized times that does not address the differences between a novice and an expert worker, and attention to the patient acuity effects on time. Future work includes a field study, using realistic bed assignment and introducing biomechanical load and risk factors in the model. Further sensitivity testing is required to understand critical variables affecting model, such as different bed assignment. The current work represents an early stage in the development of this technology. Future work should also include physical workload aspects related to nurse injury risk, as well as fatigue aspects that may be related to human error rates. Further validation is required.

5 Conclusion

This paper presents a novel approach to nurse-focused DES modelling, that includes patient acuity and nurse-patient ratios quantitively. The demonstrator model simulated the delivery of care of nurses, to quantify the impact of changing nurse-patient ratio and patient acuity on nurse workload and care quality indicators. Results of the demonstrator model showed that as nurse-patient ratios decreased, and patient acuity increased, nurse workload increases and care quality deteriorated. In comparison to base case, walking distance increased up to 18%; care delivery time up to 40%; task in queue up to 354%, and missed care delivery time up to 354%; missed care up to 253%. Although the model shows good potential, further model development, testing, and validation studies are still needed.

References

Ausserhofer D, Zander B, Busse R, Schubert M, De Geest S, Rafferty AM, Ball J, Scott A, Kinnunen J, Heinen M, Sjetne IS, Moreno-Casbas T, Kózka M, Lindqvist R, Diomidous M, Bruyneel L, Sermeus W, Aiken LH, Schwendimann R (2014) Prevalence, patterns and predictors of nursing care left undone in European hospitals: results from the multicountry cross-sectional RN4CAST study. BMJ Qual Saf 23(2):126. LP-135. http://qualitysafety.bmj.com/content/23/2/126.abstract

Banks J, Carson JSI, Nelson NL, Nicol DM (2005) Discrete-event system simulation, 4th edn. Prentice Hall International Series in Industrial and Systems Engineering, Upper Saddle River

Choudhary R, Bafna S, Heo Y, Hendrich A, Chow M (2010) A predictive model for computing the influence of space layouts on nurses' movement in hospital units. J Build Perform Simul 3(3):171–184. https://doi.org/10.1080/19401490903174280

Daly BJ, Brennan CW (2009) Patient acuity: a concept analysis. J Adv Nurs 65(5):1114–1126. https://doi.org/10.1111/j.1365-2648.2008.04920.x

Dode P (Pete), Greig M, Zolfaghari S, Neumann WP (2016) Integrating human factors into discrete event simulation: a proactive approach to simultaneously design for system performance and employees' well being. Int J Prod Res 54(10):3105. https://doi.org/10.1080/00207543.2016.1166287

Farrington M, Trundle C, Redpath C, Anderson L (2000) Effects on nursing workload of different methicillin-resistant Staphylococcus aureus (MRSA) control strategies. J Hosp Infect 46(2):118–122. https://doi.org/10.1053/jhin.2000.0808

Greasley A, Owen C (2018) Modelling people's behaviour using discrete-event simulation: a review. Int J Oper Prod Manag. https://doi.org/10.1108/IJOPM-10-2016-0604

Gunal MM, Pidd M (2010) Discrete event simulation for performance modelling in health care: a review of the literature. J Simul 4(1):42–51. https://doi.org/10.1057/jos.2009.25

Hendrich A, Chow MP, Skierczynski BA, Lu Z (2008) A 36-hospital time and motion study: how do medical-surgical nurses spend their time? Permanente J 12(3):25–34. http://www.pubmedcentral.nih.gov/articlerender.fcgi?artid=3037121&tool=pmcentrez&rendertype=abstract

Hoad K, Robinson S, Davies R (2008) Automating warm-up length estimation. In: Proceedings - winter simulation conference, pp 532–540. https://doi.org/10.1109/WSC.2008.4736110

Howard SK, Gaba DM, Smith BE, Weinger MB, Herndon C, Keshavacharya S, Rosekind MR (2003) Simulation study of rested versus sleep-deprived anesthesiologists. Anesthesiology 98(6):1345–1355. Discussion 5A. https://doi.org/10.1097/01.sa.0000101124.76490.aa

Hughes RG (2008) Patient safety and quality: an evidence-based handbook for nurses. Agency for Healthcare Research and Quality, US Department of Health and Human Services. https://doi.org/AHRQ Publication No. 08-0043

Jun JB, Jacobson SH, Swisher JR (1999) Application of discrete-event simulation in health care clinics: a survey. J Oper Res Soc 50(2):109–123. https://doi.org/10.1057/palgrave.jors.2600669

Letvak S, Buck R (2008) Factors influencing work productivity and intent to stay in nursing. Nurs Econ 26(3):159–165

Mohammadi M, Shamohammadi M (2012) Queuing analytic theory using WITNESS simulation in hospital pharmacy. Ijens.Org (06):20–27. http://www.ijens.org/Vol_12_I_06/123806-9494-IJET-IJENS.pdf

Needleman J (2013) Increasing acuity, increasing technology, and the changing demands on nurses. Nurs Econ 31(4):200

Neumann WP, Medbo P (2016) Simulating operator learning during production ramp-up in parallel vs. serial flow production. Int J Prod Res 7543(August):1–13. https://doi.org/10.1080/00207543.2016.1217362

Perez J, de Looze MP, Bosch T, Neumann WP (2014) Discrete event simulation as an ergonomic tool to predict workload exposures during systems design. Int J Ind Ergon 44(2):298–306. https://doi.org/10.1016/j.ergon.2013.04.007

Potter P, Wolf L, Boxerman S, Grayson D, Sledge J, Dunagan C, Evanoff B (2009) An analysis of nurses' cognitive work : a new perspective for understanding medical errors. Int J Healthcare Inf Syst Inform 4(3):39–52. http://www.igi-global.com/viewtitlesample.aspx?id=3978

Qureshi SM, Purdy N, Neumann WP (2016) Predicting nursing workload using discrete event simulation. In: Proceedings of the Association of Canadian Ergonomists (ACE) conference 2016: harnessing the power of ergonomics, Niagara Falls, ON, Canada, 18–20 October 2016

Qureshi SM, Purdy N, Neumann WP (2017) Simulating the impact of patient acuity on nurse workload and care quality. In: Joint proceedings 48th annual conference of the Association of Canadian Ergonomists (ACE) & 12th international symposium on human factors in organizational design and management (ODAM), Banff, Canada, July 31–August 3 2017, pp 232–238

Schmidt R, Geisler S, Spreckelsen C (2013) Decision support for hospital bed management using adaptable individual length of stay estimations and shared resources. BMC Med Inform Decis Mak 13(1):1–19. https://doi.org/10.1186/1472-6947-13-3

Seago JA, Ash M, Spetz J, Coffman J, Kevin G (2001) Hospital registered nurse shortages: environmental, patient, and institutional search. Health Serv Res 36:5(October):831–852. http://onlinelibrary.wiley.com/journal/10.1111/%28ISSN%291475-6773/issues

Physical and Physiological Synchrony Between Care Worker and Care Recipient During Care Operation

Tsuneo Kawano[1(✉)], Yukie Majima[2], Yasuko Maekawa[3], Mako Katagiri[4], and Atsushi Ishigame[2]

[1] Setsunan University, Neyagawa, Osaka 572-8508, Japan
kawano@mec.setsunan.ac.jp
[2] Osaka Prefecture University, Sakai, Osaka 599-8531, Japan
[3] Kagawa University, Kita-gun, Kagawa 761-0793, Japan
[4] Osaka Research Institute of Industrial Science and Technology, Osaka 594-1157, Japan

Abstract. Musculoskeletal disorders are a widespread affliction among nurses at nursing-care facilities in general. Nursing proficient skills are supposed to make the disorders reduce and also to lead to tender care for the care recipients. Tacit knowledge such as "proficient skills" and "knacks" in nursing care skills seems not to be applied by care worker alone but by the interaction between care worker and care recipient. The purpose of this study is to examine the physical and physiological synchrony between care worker and care recipient during care operation. It is also to reveal the relationship between the synchrony and physical loads, and the relationship of cerebral activities between them. Experiments of the care operations were carried out in the care worker-recipient pairs. In the experiments the care recipient gave two kinds of roles. One was to act a physically unimpaired person, and the other was to act a bedridden person that was unable to move on his/her own. The body motions, muscle loads, and cerebral activities of the care worker and care recipient were measured using two sets of motion capture, EMG device, and near-infrared spectroscopy (NIRS), respectively. Experimental results indicated that the muscle loads of care worker decreased when the temporal changes of physical and physiological data of the care worker and care recipient got into synchronization. Furthermore, in that case the cerebral activities of the care recipient were synchronized with those of the care worker.

Keywords: Care operation · Physical synchrony · Physiological synchrony
Physical load · Cerebral activity

1 Introduction

Musculoskeletal disorders are a widespread affliction among nurses at nursing-care facilities in general. Back or neck-pain-related disability of nursing staff is mainly attributed to physical and psychosocial risk factors [1, 2]. Nursing proficient skills are supposed to make the disorders reduce and also to lead to tender care for the care

S. Bagnara et al. (Eds.): IEA 2018, AISC 818, pp. 536–543, 2019.
https://doi.org/10.1007/978-3-319-96098-2_66

recipients. Tacit knowledge such as "proficient skills" and "knacks" in nursing care skills seems not to be applied by care worker alone but by the interaction between care worker and care recipient.

The purpose of this study is to examine the physical and physiological synchrony between care worker and care recipient during care operation. And it is also to reveal the relationship between the synchrony and physical loads, and the relationship of cerebral activities between them. Our previous study has successfully attempted to examined the inter-brain synchronization between nurse and patient during drawing blood [3].

In this study two kinds of care operations were adopted as nursing care skills. One was the operation in which the care worker moved the care recipient from supine position to sitting position on the bed. The other was the operation in which the care worker transferred the care recipient from the bed to the wheelchair. Experiments of the care operations were carried out in the care worker-recipient pairs. In the experiments the care recipient gave two kinds of roles. One was to act a physically unimpaired person, and the other was to act a bedridden person that was unable to move on his/her own. The body motions, muscle loads, and cerebral activities of the care worker and care recipient were measured using two sets of motion capture, EMG device, and near-infrared spectroscopy (NIRS), respectively.

2 Experimental Methods

2.1 Participants

In the experiment participants were seven persons, i.e. four males and three females. Two males of them were students for engineering, and played the role of the care recipients (22.5 ± 0.7 yrs, 171.5 ± 2.1 cm height, 67.5 ± 3.5 kg weight). The other males of them played the role of the care workers (28.5 ± 7.8 yrs, 174.5 ± 0.7 cm height, 74.5 ± 7.8 kg weight). One of them was a student for nursing. The other was a student for mechanical engineering, i.e. inexperienced person for nursing. Three females were working nurses (46.3 ± 1.5 yrs, 160.7 ± 3.5 cm height, 55.0 ± 8.7 kg weight). They played the role of the care workers. Their years of experience for nursing were 23.0 ± 3.0 yrs.

The authors explained ethical considerations to the participants and obtained informed consent before the study.

2.2 Care Operations

In this study two kinds of care operations were adopted as nursing care skills. One was the operation in which the care worker moved the care recipient from supine position to sitting position on the bed. The other was the operation in which the care worker transferred the care recipient from the bed to the wheelchair. Experiments of the care operations were carried out in the care worker-recipient pairs. In the experiments the care recipient gave two kinds of roles. One was to act a physically unimpaired person, and the other was to act a bedridden person that was unable to move on his/her own.

2.3 Kinesiological Measurements

The care motions, muscle loads, and cerebral activities of the care worker and care recipient were measured using two sets of motion capture, EMG device, and near- infrared spectroscopy (NIRS), respectively.

Figure 1 shows a photo of one scene in a care operation. The care motions of care worker and care recipient were simultaneously captured using wearable motion capture (MVN, Xsens). The participants attached 17 inertial sensors to the body parts with straps and belts. 3D coordinates of 23 points of the body were captured at up to 240 Hz.

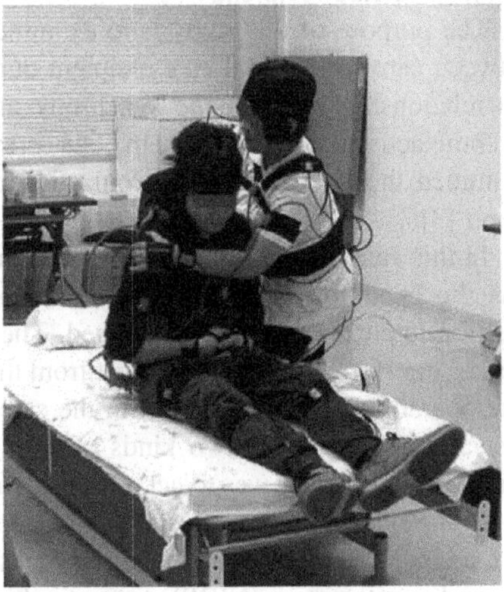

Fig. 1. Simultaneously capturing care motions and measuring EMG for care worker and care recipient using wearable motion capture (MVN, Xsens) and EMG device (SYNA ACT MT-11, NEC).

In order to investigate the relation between care motions and muscle loads of care worker and recipient, surface electromyography (EMG) activities were recorded using a conventional telemeter system including a bio amplifier (SYNA ACT MT-11, NEC). Sampling rate was 1 kHz. Surface electrodes were placed on the rectus femoris, rectus abdominis muscle, erector spinae muscles, and biceps as shown in Fig. 2. The surface EMG raw data were high pass filtered with cut-offs of 30 Hz by a second-order Butterworth filter to remove motion artifact and electrocardiographic interference.

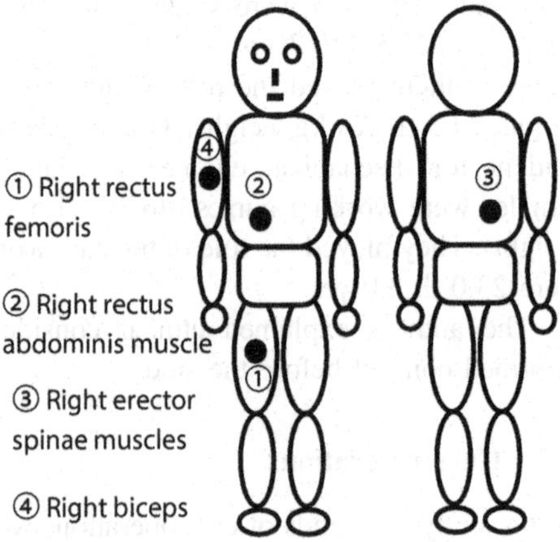

Fig. 2. The muscles for EMG measurement

The cerebral activities of the care worker and care recipient during the care motions were investigated for their brain synchrony. The relative changes in oxygenated and deoxygenated hemoglobin concentration at prefrontal cortices of two participants were simultaneously measured using NIRS (Wearable Optical Topography, WOT-100, HITACHI). Figure 3 shows a measuring scene of cerebral activities of the care worker and care recipient.

3 Results and Discussions

3.1 Body Motions

Figure 4 shows an example of relationships between the body motions of two participants who is a physically unimpaired person and a bedridden person during care operations on the bed. Every temporal changes of distances were calculated from the origin to every points measured by the motion captures. Cross correlations of the temporal changes of distances between two participants were calculated in every time intervals of approximately 2.2 s. Averages of absolute of the cross correlations are shown in Fig. 4. As a result, it is found that almost body points in the case of a physically unimpaired person have higher correlations than in the case of a bedridden person.

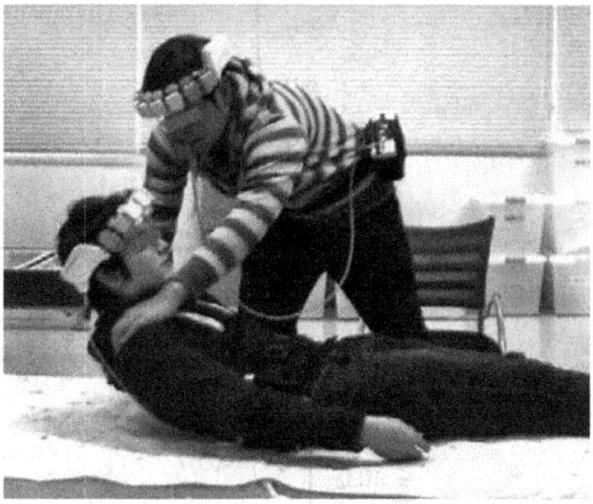

Fig. 3. Simultaneously measuring cerebral activities of the care worker and care recipient using wearable optical topography based on NIRS (WOT-100, HITACHI)

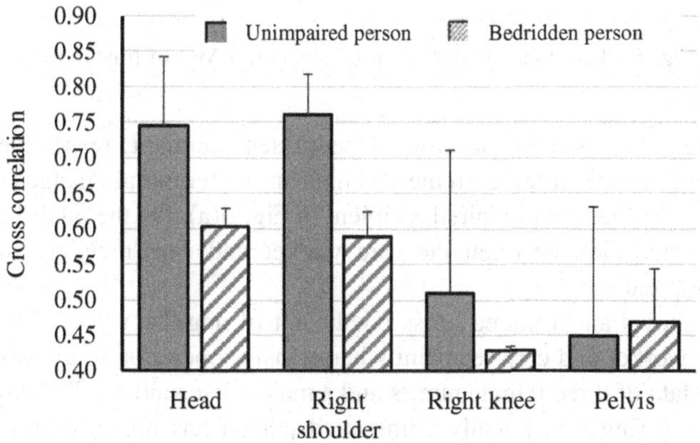

Fig. 4. Cross correlations in changes of each joint movement between cooperative and uncooperative operation. Care worker is a male student.

3.2 EMG

Figure 5 shows an example of the temporal changes of EMG during the care operations on the bed. The graphs show envelope changes in total absolute EMG of four muscles as shown in Fig. 2. Figures 5(a) and (b) show the change in the case of physically

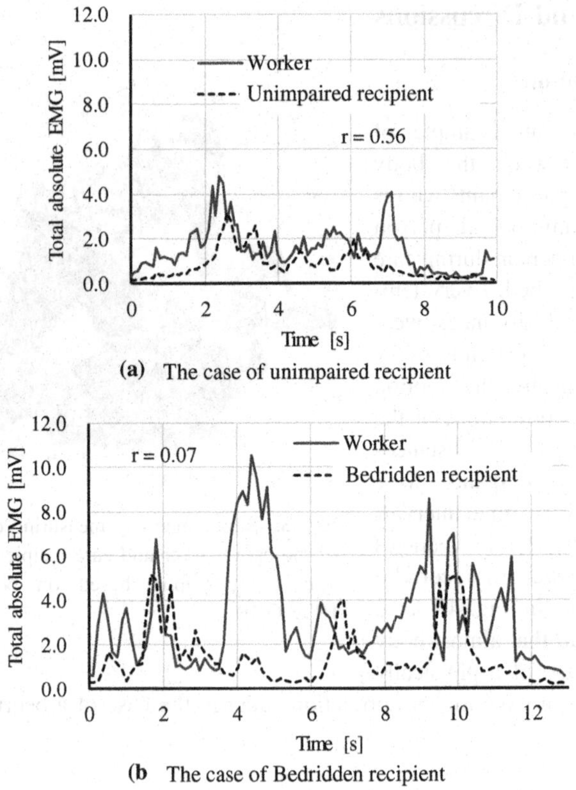

Fig. 5. Envelope changes in total absolute EMG of four muscles.

unimpaired recipient and in the case of bedridden recipient, respectively. The care worker and care recipient are extremely similar in the temporal changes of total EMG in the case of physically unimpaired recipient in Fig. 5(a). On the other hand, Fig. 5(b) shows little correlation between the care worker and care recipient in the case of bedridden recipient.

Figure 6 shows an averaged cross coefficient of angular velocity of lumber joint between care worker and care recipient when transfer operation to the wheelchair. It is based on the data of three female nurses and a male nurse student. The cross coefficient in the case of caring a physically unimpaired person has higher correlations significantly than in the case of caring a bedridden person.

Figure 7 shows an averaged muscle loads of care worker's low back (right erector spinae muscles) when transfer operation to the wheelchair. It is also based on the data of three female nurses and a male nurse student. The muscle loads of care workers in the case of caring a physically unimpaired person is significantly smaller than in the case of caring a bedridden person. From the results of Figs. 6 and 7 it is found that the muscle loads of low back become small as the cross coefficient of angular velocity of lumber joint between care worker and care recipient is small.

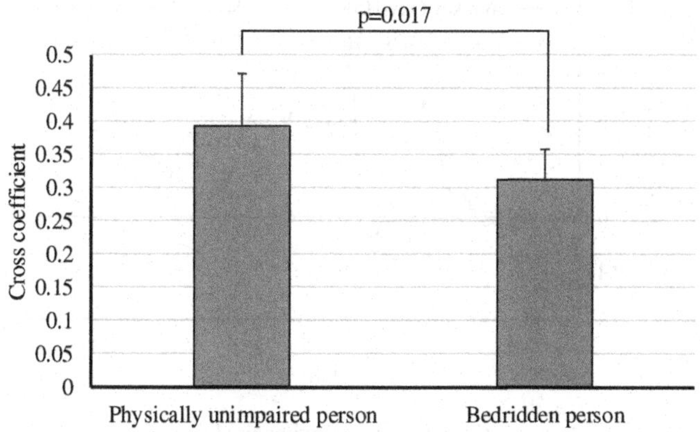

Fig. 6. Cross coefficient of angular velocity of lumber joint between care worker and care recipient when transfer operation to the wheelchair. (Average of three female nurses and a male nurse student.)

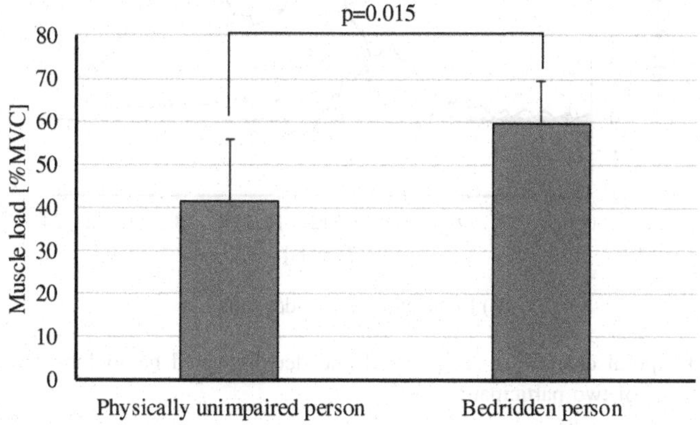

Fig. 7. Muscle loads of care worker's low back (right erector spinae muscles) when transfer operation to the wheelchair. (Average of three female nurses and a male nurse student.)

3.3 NIRS

The relation of brain activities between two participants were investigated [4]. Figure 8 shows the temporal changes in oxygenated and deoxygenated hemoglobin concentration at prefrontal cortices of two participants. In the figure the oxygenated and deoxygenated hemoglobin concentration are referred to as [Oxy-Hb] and [Deoxy-Hb], respectively. Their values are related to brain activation. As the motion of care worker begins, [Oxy-Hb] increases and blood oxygenation-level dependent (BOLD) response is existed [5].

In the case of caring a physically unimpaired person, the worker's [Oxy-Hb] increases primarily and the recipient's [Oxy-Hb] follows by approximately 4 s behind.

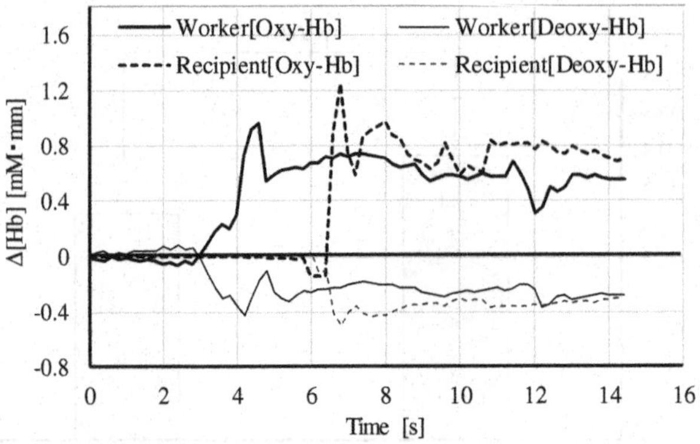

(a) In the case of unimpaired person

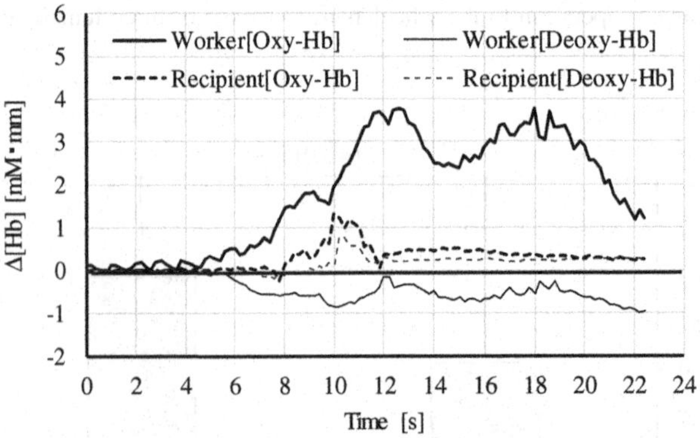

(b) In the case of bedridden person

Fig. 8. The temporal changes in oxygenated and deoxygenated hemoglobin concentration at prefrontal cortices of two participants

The temporal changes of [Oxy-Hb] and [Deoxy-Hb] of two participants show a strong correlation. On the other hand, in the case of caring a bedridden person their brain activities show unsynchronized changes. As a result, it is suggested that muscle loads become large when the brain activities are not synchronized.

4 Conclusion

This study was aimed to investigate the behavioral synchrony and physical loads of care worker and care recipient. Experiments were carried out to evaluate their interaction and their physical loads, and the relationship between cerebral activities and their loads. In this study two kinds of care operations were adopted as nursing care skills, e.g. shifting in posture on bed and transferring from bed to wheelchair.

Furthermore, in the experiments the care recipient gave two kinds of roles. One was to act a physically unimpaired person, and the other was to act a bedridden person that was unable to move on his/her own.

Main remarks are as follows:

(1) In the case of caring a physically unimpaired person, the motions between care worker and care recipient were synchronized. The changes of EMG and brain activities are also synchronized.
(2) It is suggested that muscle loads become large when the temporal changes of EMG of two participants are not synchronized.
(3) Muscle loads seem to become large when the body motions of two participants are not synchronized.
(4) In the case of caring a physically unimpaired person, the cerebral activities of the care recipient were synchronized with those of the care worker.

Acknowledgements. This work was supported by a JSPS Grant-in-Aid for Scientific Research (B) (Grant-#16H05571).

References

1. Luime JJ, Kuiper JI, Koes BW, Verhaar JA, Miedema HS, Burdorf A (2004) Work-related risk factors for the incidence and recurrence of shoulder and neck complaints among nursing-home and elderly-care workers. Scand J Work Environ Health 30(4):279–286
2. Simon M, Tackenberg P, Nienhaus A, Estryn-Behar M, Conway PM, Hasselhorn HM (2008) Back or neck-pain-related disability of nursing staff in hospitals, nursing homes and home care in seven countries—results from the European NEXT-Study. Int J Nurs Stud 45(1):24–34
3. Kawano T, Majima Y, Maekawa Y, Katagiri M, Ishigame A (2016) Inter-brain synchronization between nurse and patient during drawing blood. In: Proceedings of the 9th international joint conference on biomedical engineering systems and technologies (BIOSTEC 2016), vol 5. HEALTHINF, pp 507–511
4. Funane T, Kiguchi M, Atsumori H, Sato H, Kubota K, Koizumi H (2011) Synchronous activity of two people's prefrontal cortices during a cooperative task measured by simultaneous near-infrared spectroscopy. J Biomed Opt 16(7):1–10. 077011
5. Mullinger KJ, Mayhew SD, Bagshaw AP, Bowtell R, Francis ST (2014) Evidence that the negative BOLD response is neuronal in origin: a simultaneous EEG–BOLD–CBF study in humans. Neuroimage 94:263–274. Author, F (2016) Article title. Journal 2(5):99–110

Information for Users as a Key Element to Delivering a Better Healthcare Service

Zuli T. Galindo-Estupiñan[1(✉)], Carlos Aceves-Gonzalez[1,2], John A. Rey-Galindo[1,2], and Elvia Luz Gonzalez-Muñoz[1,2]

[1] University of Guadalajara, 44250 Guadalajara, Jalisco, Mexico
zuli.galindo@alumnos.udg.mx

[2] Ergonomics Research Centre, University of Guadalajara, 44250 Guadalajara, Jalisco, Mexico

Abstract. Design of healthcare services should consider the diversity of capabilities and needs of users with the purpose of delivering an efficient, effective and satisfactory service. The information delivery by service providers is a key element of the system that allows that users have a clear idea about how to use the service but also it has an effect on the perceived quality. The aim of this study was to identify the information needs of users in two stages in an external consultation service of Neurology. The study compiled 11 interviews with staff members and 32 questionnaires with users (patients and caregivers). Results show that overall patients were satisfied with the information received during the analysed stages. Nevertheless, the users manifested the need of receiving more and clearer information about waiting times, who is attending them and medical and administrative procedures that they have to follow before and after consultation. The results also emphasize the need to provide information not only face to face but using another communication channels to support the information delivery. It is important to concentrate efforts on designing service information considering users' needs and abilities, especially the limited ability that the most vulnerable users might have.

Keywords: Service design · Information design · User capabilities

1 Introduction

Health is one of the aspects that affect physical well-being, which is directly related to the quality of life of people [1]. Nowadays, diverse health problems affect the population worldwide [2], therefore medical attention is a high priority service. Health services are complex and dynamic systems [3] due to the diversity of users and the multiple personal interactions that occur during the provision of the service [4].

Information is a fundamental element that facilitates users, staff members and the organization to communicate effectively and efficiently during the provision of the

Z. T. Galindo-Estupiñan, C. Aceves-Gonzalez, J. A. Rey-Galindo and E. L. Gonzalez-Muñoz—Master in Ergonomics.

S. Bagnara et al. (Eds.): IEA 2018, AISC 818, pp. 544–552, 2019.
https://doi.org/10.1007/978-3-319-96098-2_67

service. The channels of transmission of information, as well as the people who are involved in the flow of information during the interaction with patients, have already been the subject of research [5]. Likewise, the number of transitions in care, the gap in the flow of information, communication, the coordination of problems and effectiveness during the provision of information to the user have been raised as a research topic [6]. It has also been studied, the loss of information regarding pending examinations at the time of discharge of the patient. In the advice given to patients and family members, a communication deficit is recognized, as well as in the transfer of information between patients and staff. Similarly, in the literature, it is recognized that the information provided by health professionals is an aspect that affects the satisfaction of the user [7]. However, the information has not been investigated as a transversal element in the communication process that occurs during the provision of the service.

On the other hand, the importance of exploring the Latin American context is due to the potential increase in the demand for health services due to the increase in the adult population over 60 years that is expected by the phenomenon of demographic change [8]. The trend shows an accelerated growth of this population in developing countries [2] and consequently, older people would be frequent users of the healthcare services because they are more vulnerable due to age conditions [9].

Among the main needs of older adults is that they require more information to use health services in order to take care of themselves [10]. Additionally, it is important to point out that information is a fundamental element to properly use a service, so it is important to consider users with reduced motor, sensory and cognitive capacity due to age [11].

This case of the study was carried out in the external consultation service that provides the Neurological Department in a healthcare institution in Mexico. The patients of this service suffer from chronic diseases like Epilepsy, Parkinson, Multiple Sclerosis, Dementia and other neurological pathologies. The consultation service provides care for a large percentage of older adults but also some young adults. Ergonomics as a discipline and the design of inclusive services [12] was used to study the diversity of users and the interactions with the different elements of the system. The aim of this study was to identify the information needs of users in two stages in external consultation service.

2 Method

2.1 Semi-structured Interviews

Eleven semi-structured interviews were conducted with staff members: neurologists, medicine residents, medical assistants, nurse, and information module staff. The selection of participants was made for convenience among those staff members who were linked to the external consultation service. Semi-structured interviews were conducted due to their flexibility for data collection [13]. In the interview four themes were established to guide the questions [14], however, for the interest of this study, the results related to the needs of the external consultation service are presented, particularly those related to the delivery of information. A semantic inductive thematic analysis was used to analyse the data collected in the staff interviews, in order to identify, analyse and report the patterns that are presented in the reports of the participants' reality [15].

2.2 Questionnaires

The interview was conducted to 32 users of the outpatient service of the Department of Neurology. The respondents were patients and in some cases, the companions were the ones who answered the questions asked when the patients could not do it due to their diminished cognitive capacity. The participants were randomly selected during different days of external consultation.

The questionnaire of ten questions was used to identify the importance of the information required before and after the medical consultation, as well as the reasons behind their answers. The participants agreed to participate prior to the start of the questionnaire. On the occasions that the participant could not understand the statement of the question, the researcher reformulated the question to obtain the required answers. During the interviews, some participants provided additional information for what was noted in the observations section. People who participated in this study were informed about the procedure to be performed and their doubts (if any) were clarified before signing the informed consent and the participation itself.

The data collected were transcribed to a data processor. Each question was coded and a descriptive analysis was carried out. In relation to the qualitative responses, they were grouped into categories and the frequency with which they were repeated was recorded.

3 Results

Two stages of the external consultation service were explored [16]. These stages correspond to the moments in which there is a greater exchange of information (stages 4 and 6 according with [16]). It is the moment in which the users interact directly with the medical assistant in charge of providing care to all the medical offices. Table 1 shows the results of the analysis of the interviews with the staff members.

On the other hand, during the pilot test of the questionnaires, it was identified that elderly users or those with a low educational level had difficulty answering the questionnaire using the numerical scales. However, responding affirmatively or negatively to the questions was easier for everyone. The average age of the participants was 44.3 (SD 13.64), of which 23 were women and 9 were men. At least 75% of patients attended accompanied, because they had mobility problems (5), loss of vision (2), medical prescription (15), to feel accompanied (2). Table 2 shows the number of participants who answered yes or no to each question and the corresponding justifications. The average time in which patients arrived early for the consultation was 54.6 min (SD 42.59).

Table 1. Thematic analysis from staff interviews

Stage	Service staff's information needs	Mean to deliver the information
Stage 4- Arriving at the consultation module	- Identify the patient that is arriving - Information about patients who have an appointment - Knowing the dynamics of patient care according to the doctors' requirements: by order of arrival or by schedule - Nurses require information about the medical speciality they have been cited to determine the procedure they must perform, special care according to the illness (inability to speak or move)	- Verbal information (for the most part) and very little written - Patient's card with the identification number and name - Handwritten paper with the vital signs data
Stage 6- Return to the consultation module	- Know if the doctor has assigned a subsequent appointment and how long should be scheduled - Know what procedures they requested from the patient - Information about places inside the hospital (also done by nurses, information assistants, security guards and other staff members)	- Verbal information mostly and a little written - Communication between medical staff - Paper was written by hand with the time in which the patient should be given the subsequent appointment, sometimes lost by the patient - Data of the subsequent appointment written on the card

4 Discussion and Conclusion

The aim of this study was to identify the information needs of users in two stages in external consultation service. The results show the existence of three information needs in these stages of the service from the perspective of staff and patients: (1) demand for attention during the delivery of verbal information, (2) information about waiting times and (3) information from support for a better use of the service.

Table 2 identifies the users' need to make eye contact to be sure that they are being looked after. This aspect is the basis of communication and social interaction [17], and it is a key element during the search for information [18] since it allows to obtain feedback on the reactions of people [19]. Another element noted was the preference for requesting information about the patient and the medical appointment. This gives patients the security of being in the right place, which reduces uncertainty, a key attribute of the delivery of information [20]. It is important, then, to propose a protocol of care that allows generating greater empathy and certainty in the information given to patients.

On the other hand, users require information on the hours in which they will be served in a real-life context. However, as shown in Table 1, this information about

Table 2. Satisfacción sobre la información recibida en las etapas 4 y 6

Question	Number of answers	Reasons behind the answers
Is it important that the medical assistant look at you when you arrive at the attention module?	Yes = 31	- To feel that I am not ignored/Because I know the medical assistant is paying attention/it makes me feel more secure - I know the medical assistant has my case in mind - Because the medical assistant recognizes the patient health conditions so can be more considered
	No = 1	- It is enough if the medical assistant answers my questions
Is it important that the medical assistant ask for your name when you arrive at the appointment?	Yes = 29	- It is important that the medical assistant can check my name in the list and follow with the process - To avoid confusion with the name of somebody else - The medical assistant can identify who is the patient and the correct arrival time - Because the medical assistant can register each patient correctly - I feel doubtful when I arrive at the consultation module, that question makes me feel in the correct place
	No = 3	- Because my name is written on my patient card was write and scheduled in a list for that day
Is it important that the medical assistant asks you for the name of the physician who you have an appointment with?	Yes = 32	- Because the medical assistant needs to know which physician will attend me - Because the medical assistant can correct if there is any error - Because the list may have inconsistencies - Because the medical assistant can confirm that my physician is there - To let the physician know that the patient has arrived
Do you consider that receiving adequately the indications for the taking of vital signs is important?	Yes = 32	- Because it is part of the patient care protocol - To not be guessing where to go - To know which is the next step of the process

(*continued*)

Table 2. *(continued)*

Question	Number of answers	Reasons behind the answers
		- Because I don't have knowledge of space - Then I can be sure where the place I have to go
Would you like to receive more information when you arrive at the appointment?	Yes = 16	- When I will be attended - How long I will be waiting - If the doctor is there or if he/she has already arrived at the consultation - A guide with information about other specialities when the patient is transferred to another service - Protocols about the procedures for each disease - About the location of consultation modules
	No = 16	- Without comments
Have you ever been missing or lost an appointment for being late?	Yes = 9	- I forgot that had an appointment or I was confused with the date - Because I had work to do - Because of the traffic - I did not attend because I thought my health was fine
	No = 23	- Without comments
Is it important that the medical assistant confirm verbally the date of the next appointment?	Yes = 26	- To keep the idea of the date given by the medical assistant - To verify not having another commitment or appointment - To see if there is a match between what the medical assistant says and what he/she writes - Because if not I forget my appointments
	No = 6	- I have total time availability

patient care times is uncertain for the staff, and therefore it is difficult to determine the approximate waiting time for patients for the consultation. This situation also produces an accumulation of people at the waiting rooms [21], which could be a source of uncertainty, frustration, and stress on patients [22], this affects the satisfaction of users with the service [7, 22, 23] and subsequently, stress could increase on staff members [24]. In the context of the study, it was identified that users arrive with different ranges of time in advance of the consultation and there is a delay in the hours assigned to be attended [16], which affects the variation of the waiting time [21]. The consultation time established in this medical centre is 20 min for first-time consultations and 15 for medical follow-up, which according to doctors is insufficient to treat patients with neurological diseases such as dementia, Parkinson's or multiple sclerosis [16]. Cayirli et al. [25] report that in the literature most of the scheduling systems for medical appointments consider that patients are homogeneous. However, the allocation of consultation hours should reflect the different types of patients who use the service and the variation depending on the needs of care of one of them [24]. Additionally, it should be considered that the consultation time can be a limitation for doctors to carry out an adequate diagnosis and follow-up process. This could lead to errors that contribute to patients having to increase their permanence in the service or the unnecessary use of laboratory resources, among others. For this reason, future research could define patient profiles according to the time requirements, physical and cognitive abilities of patients to propose more inclusive and real-time schedules.

Even with a specific appointment allocation schedule close to the real, it should be considered that waiting times can never be completely eliminated [23]. Therefore, the need to design a strategy to provide information on waiting times to the user is still a relevant aspect to improve service provision, since it allows managing the perception of waiting time and user expectations and improving the satisfaction of them [26].

The responses of staff and users of the service allowed to identify the need for information on the location of the different places in the hospital and support information to correctly use the service. This coincides with the literature, which highlights the implementation of wayfinding as an essential element in a health service [27–30]. As Bopp [20] states, service providers must deliver information before the service is provided to help users understand how to enter the system and how it operates.

Finally, it should be pointed out the need that the staff members have in the transfer of information between the medical assistant and the doctor during the assignment of subsequent appointments and the schedule in which the patients will be attended. The absence of effective mechanisms for the transfer of information can be solved with the implementation of participatory ergonomics strategies to involve the staff members in the generation of ideas to improve or change communication mechanisms or useful information products. This way, the flow of information can be efficient [31] and errors and times in the attention process are reduced.

In conclusion, it can be said that this work shows the effect of the interactions of one stage of the service in subsequent ones. In this particular case, the absence of information in the early stages of the service on procedures and the location of hospital sites is required by users in later stages causing loss of time, disorientation and dissatisfaction. Likewise, inappropriate face-to-face interactions can generate greater uncertainty in patients and make them believe that they are doing something wrong. Another key aspect is the importance of letting users know the waiting time. The importance of this information should be socialized among the staff to understand the advantages of providing this information for a better service delivery.

This study, exploring the aspects related to the external health consultation service, allows evidencing the improvement opportunities in two stages of the service. The complexity and extension of this system can be studied from the perspective of the needs of those most vulnerable users in order to propose interventions in the design of the service to benefit all stakeholders and thus improve the quality of the service.

Acknowledgments. This study is part of the research project "Evaluation of Clinic of Cognitive Deterioration and Dementia of the Department of Neurology using the Inclusive Service Design Approach" with funds from the Mexican Government through PRODEP. Additionally, we are grateful with the Department of Neurology of the CMNO, IMSS for authorizing the execution of this research, as well as the collaboration of the staff members.

References

1. Felce D, Perry J (1995) Quality of life: its definition and measurement. Res Dev Disabil 16 (1):51–74
2. United Nations, Department of Economic and Social Affairs, Population Division (2013) World population ageing 2013
3. Effken J (2002) Different lenses, improved outcomes: a new approach to the analysis and design of healthcare information systems. Int J Med Inform 65(1):59–74
4. Carayon P, Bass EJ, Bellandi T, Gurses AP, Hallbeck MS, Mollo V (2011) Sociotechnical systems analysis in health care: a research agenda. IIE Trans Healthc Syst Eng 1(3):145–160
5. Waterson P (2014) Health information technology and sociotechnical systems: a progress report on recent developments within the UK National Health Service (NHS). Appl Ergon 45 (2, Part A):150–161
6. Hignett S, Carayon P, Buckle P, Catchpole K (2013) State of science: human factors and ergonomics in healthcare. Ergonomics 56:1491–1503
7. Sauceda-Valenzuela AL, Wirtz VJ, Santa-Ana-Téllez Y, de la Luz Kageyama-Escobar M (2010) Ambulatory health service users' experience of waiting time and expenditure and factors associated with the perception of low quality of care in Mexico. BMC Health Serv Res 10(1):178
8. United Nations, Department of Economic and Social Affairs, Population Division (2002) World population ageing: 1950-2050
9. United Nations, Department of Economic and Social Affairs, Population Division (2015) World population ageing 2015
10. Barrett J (2005) Support and information needs of older and disabled older people in the UK. Appl Ergon 36(2):177–183
11. Tenneti R, Johnson D, Goldenberg L, Parker RA, Huppert FA (2012) Towards a capabilities database to inform inclusive design: experimental investigation of effective survey-based predictors of human-product interaction. Appl Ergon 43:713–726
12. Aceves-Gonzalez C (2014) The application and development of inclusive service design in the context of a bus service. Loughborough University, Doctoral thesis
13. Stanton N, Salmon PM, Rafferty LA (2013) Human factors methods: a practical guide for engineering and design. Ashgate Publishing, Ltd., Burlington
14. Robson C, McCartan K (2011) Real world research. Wiley, Hoboken
15. Braun V, Clarke V (2006) Using thematic analysis in psychology. Qual Res Psychol 3:77–101
16. Galindo-Estupiñan ZT, Aceves-Gonzalez C, Ortiz G, Rey-Galindo J, Mireles-Ramirez M (2017) Information characteristics in the operation of a healthcare service from the staff perspective. In: International conference on applied human factors and ergonomics. Springer, Cham, pp 303–313
17. Senju A, Johnson MH (2009) The eye contact effect: mechanisms and development. Trends Cogn Sci 13(3):127–134
18. Kleinke CL (1986) Gaze and eye contact: a research review. Psychol Bull 100(1):78

19. Argyle M, Dean J (1965) Eye-contact, distance and affiliation. Sociometry 28:289–304
20. Bopp KD (1989) Value-added ambulatory encounters: a conceptual framework. J Ambul Care Manag 12(3):36–44
21. Fetter RB, Thompson JD (1966) Patients' waiting time and doctors' idle time in the outpatient setting. Health Serv Res 1(1):66
22. Thompson DA, Yarnold PR (1995) Relating patient satisfaction to waiting time perceptions and expectations: the disconfirmation paradigm. Acad Emerg Med 2(12):1057–1062
23. Davis MM, Heineke J (1998) How disconfirmation, perception and actual waiting times impact customer satisfaction. Int J Serv Ind Manag 9(1):64–73
24. Harper PR, Gamlin HM (2003) Reduced outpatient waiting times with improved appointment scheduling: a simulation modelling approach. OR Spectr 25(2):207–222
25. Cayirli T, Veral E, Rosen H (2006) Designing appointment scheduling systems for ambulatory care services. Health Care Manag Sci 9(1):47–58
26. Thompson DA, Yarnold PR, Williams DR, Adams SL (1996) Effects of actual waiting time, perceived waiting time, information delivery, and expressive quality on patient satisfaction in the emergency department. Ann Emerg Med 28(6):657–665
27. Carpman JR, Grant MA (2016) Design that cares: planning health facilities for patients and visitors, vol 142. Wiley, Hoboken
28. Devlin AS, Arneill AB (2003) Health care environments and patient outcomes: a review of the literature. Environ Behav 35(5):665–694
29. Pati D, Harvey Jr. TE, Willis DA, Pati S (2015) Identifying elements of the health care environment that contribute to wayfinding. HERD Health Environ Res Des J 8(3):44–67
30. Ulrich RS (2001) Effects of healthcare environmental design on medical outcomes. In: Design and health: proceedings of the second international conference on health and design. Svensk Byggtjanst, Stockholm, pp 49–59
31. Hendrick HW, Kleiner BM (2005) Macroergonomics, theory, methods, and applications. CRC Press Taylor & Francis Group, Boca Raton

Ergonomics Intervention Project in Undergraduate Physical Therapy Program. A Curricular Innovation Approach

Cerda Díaz Leonidas[1], Rodríguez Carolina[1], Cerda Díaz Eduardo[1], Olivares Giovanni[1], and Antúnez Marcela[2](✉)

[1] Ergonomics and Biomechanical Laboratory, Department of Physical Therapy, Faculty of Medicine, University of Chile, Santiago, Chile
[2] Department of Education in Health Sciences, Faculty of Medicine, University of Chile, Santiago, Chile
mantunez@med.uchile.cl

Abstract. Competency-based education in ergonomics was incorporated into the physical therapy undergraduate educational program in the University of Chile in 2009 in conjunction with the construction of the graduation profile. The course "Ergonomic Intervention Project" is developed in the last year of the career (5 years of duration). The aim of this non-experimental, cross-sectional analytical design study was to evaluate the academic performance of the course and the students' perceptions regarding the pedagogical and disciplinary domain. Forty-three students took the course. The academic performance was recorded and a survey was applied in order to evaluate pedagogical domain, disciplinary domain and general aspects. Academic performance average was 5.94 (min 5.28 max 6.51) on a scale of 1 to 7 with all the students approved. Thirty-three students answered the survey sent. Pedagogical dimension obtained a 2.87 score and the disciplinary domain a 3.41 score (Likert 1–4). Regarding the evaluation of the general aspects of the course, 73% was satisfied/very satisfied with the performance of the faculty team, 87% declared to know the evaluation criteria of the subject in a timely manner, 91% considered requirement of the course adequate. Student perception in the pedagogical and disciplinary domain were satisfactory. Teaching-learning strategies based on the experiential learning cycle, in context guided by expert teachers considering an educational competencies model, allow the habilitation in ergonomics themes in undergraduate students.

Keywords: Undergraduate education · Ergonomics · Curricular innovation

1 Introduction

1.1 Institutional Academic Undergraduate Degree Context in Physical Therapy

The educational model of the Universidad de Chile and the Faculty of Medicine have been considered for the construction of the formation plan for the career of physical therapy and the determination of competencies that contribute to the profile of the

S. Bagnara et al. (Eds.): IEA 2018, AISC 818, pp. 553–562, 2019.
https://doi.org/10.1007/978-3-319-96098-2_68

graduates within a process of curricular innovation. Competencies of the undergraduate program are organized in five domains, which must act in an integrated manner to allow graduates to perform as pertinent, responsible, highly qualified professionals, committed to the needs of the country, along with an institutional seal that sets them apart. The five domains of competencies that are described in the current formation plan are Health and Study of the Human Movement; Investigation; Public Health and Management; Introduction to Teaching, and Transversal Generic Domain.

In the organization of the Department of Physical Therapy today, there are eight specific lines of development, in which the teachers must organize their undergraduate courses, articulating their domains of competencies in order to fulfill the declares profile of the graduate. The line of ergonomics is one of these subjects.

The process of teaching and learning of the students within this formation plan by competencies is based on active learning, on the foundation of the knowledge of structure and function of human anatomy integrated with the pertaining psychological and social aspects.

1.2 Ergonomic Competencies in Undergraduate Physical Therapy Students

Ergonomics is "the scientific discipline concerned with the understanding of interactions among humans and other elements of a system and the profession that applies theory, principles, data and methods to design in order to optimize human well-being and overall system performance" (IEA 2001). In this sense, an undergraduate education with professors that stimulate critical and reflexive thinking, integrating the physical and mental existence of the human being and his environment, would allow them to solve problems associated to the efficiency and effectiveness of human tasks (Barbosa and Pinheiro 2012a, b).

The competencies domains associated with undergraduate Ergonomics courses were constructed on the design of an infrastructure that allowed the organization of a progressive learning sequence of relevant content and disciplinary aspects. After this design, three specific courses were identified, based on the expected realization of each course, the associated learning goals and the pertaining strategies of teaching-learning, along with its evaluations. The Ergonomics courses based on competencies have been monitored since their launch and their development is evaluated in cycles, considering self-regulation and flexible learning options (Gruppen et al. 2012).

This proposal composes an innovative and unprecedented paradigm in the disciplinary area of professional formation in undergraduate ergonomics when we consider the health needs of society, which represent a challenge for academics involved in the development of teaching in ergonomics.

The branch of ergonomics in the Department of Physical therapy is composed of academic and disciplinary experts. The coursed related to Ergonomics have a representation of eight credits (27 h per credit) within the study plan and it considers the proposal of curricular innovation as a response to current disciplinary ergonomic needs. These courses are: (1) Course of Analysis of the relationship of the person to the environment, two credits; (2) Course of Ergonomic Evaluation, three credits;

(3) Course of Ergonomic Intervention Project, three credits. This takes place at the end of the career which lasts five years in its entirety.

The educational goal of the third course, the Ergonomic Intervention Program, is to promote critical thinking of patient-environment interaction in an ergonomic evaluation in order to promote interventions in a real-life setting. The associated learning outcomes are to analyze an ergonomic intervention, to identify main associated variables and propose a systematization of interventions in specific clinical areas.

The final competency defined for the course is to "apply ergonomic principles and criteria in different stages and levels of intervention of the physical therapist in human activity in areas of day to day activities such as work, recreational and within a clinical field and, as an associated subfield, identifying and evaluating the conditions of the relationship between a person and his or her environment, taking into account the physiological, biomechanical, histological, anatomical, sensorial, anthropometric, physical, environmental, dimensional and physio-cognitive factors.

On the other hand, there is an evaluation on a yearly basis, that is carried out by the students in order to revise the activity planning and programming as well as a proposal, carried out by the professors, for improvements, which transforms this course program into a dynamic structure around the definition of ergonomic competencies that were initially proclaimed (Gruppen et al. 2012).

Even though core competencies have been defined in ergonomics for work performance (Williams, IEA 2001), these can be initiated at an undergraduate level to allow a progressive articulation to postgraduate to answer to the continuing formation and the professional challenges that the ergonomic discipline involves in the particular context of Chile. Core knowledge, as well as basic applications of this core knowledge, the correct use of tools, processes and system integration, are the levels considered for the construction of courses in ergonomic (see Fig. 1). The focalization and advance in each level depend on the selection of content by the teachers for each course, as will the learning objectives that are set and the educational context.

The levels that are considered according to the educational needs of the undergraduate students, and the professional perspectives that the graduate must meet, correspond to a core knowledge and basic application, allowing the possibility to advance in the levels of tools and processes and system integration, in the formation courses that follow the graduation of that study plan. It will serve as a foundation for the specialization that follows.

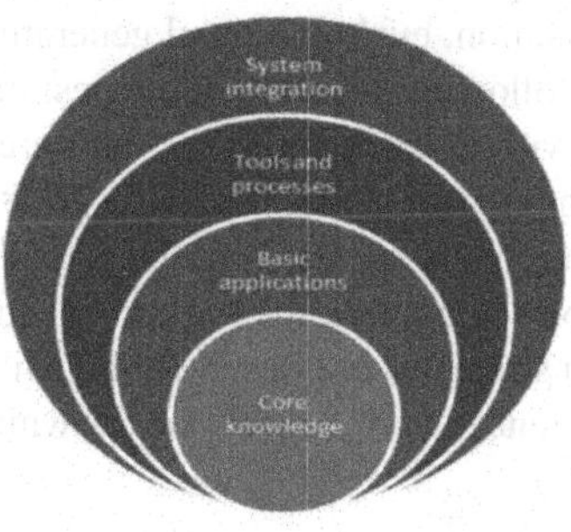

Fig. 1. Levels of knowledge in ergonomics education (Bures 2015).

1.3 Learning, Teaching and Evaluation Strategies in Ergonomic Intervention Project Course

The content prepared by the professors has the basic structure that follows the educational purposes proposed by the IEA (IEA 2001; Bridger 2012), in the two first levels:

Core knowledge: In this course, the knowledge and competencies of the two courses that have come before, in the branch of Ergonomics, are put into use in a way that we can identify work systems, carry on an analysis of a task and the relationship of a person with his or her environment and/or tools (first course). Following this, comes an ergonomic evaluation (as seen in the second course), which will be integrated into the third course in the process of an ergonomic intervention project in a clinical, work or day to day aspect, in order to improve the quality of people's lives in any health-related situation. It also integrates competencies of the clinical courses of the career, in order to generate integral physical therapy evaluations in the context of an ergonomic intervention, which is the foundation for the future professional's performance.

Basic Applications: they are founded on the use of basic knowledge so as to analyze and evaluate, using quantitative and qualitative criteria, the conditions and elements with which a person or a group of people interact in a real context. This will be the focus of the ergonomic intervention project, in order to generate a proposal to improve the conditions of the environment, interface adjustments and over life quality of the people.

The third course, Ergonomic Intervention Project, is in the last year of the career of physical therapy. It has three credits which correspond to 19 in-person hours and 62 non-present hours. It has two modules: the first is called, "Fundamentals for the development of Intervention Projects" and has a duration of three weeks which total 16 in-person hours and 14 non-present hours, and presents tools in audiovisual communication for the Intervention Project, as well as the conceptual foundation for the structure of the project, alternative configuration analysis of intervention systems and strategies, with learning strategies from lectures, non-present independent work, workshops and case-based learning (De Miguel Díaz 2005).

The second module is called "Intervention Project Development", it has a duration of 14 weeks and it is structured in a way that the student can draft and carry out the development of a project of ergonomic intervention in a real case in a clinical context, determining criteria for evaluation, evaluating and generating an intervention proposal. This module counts on the following learning strategies: external practice in a clinical field, self-studies, tutorials with assisted feedback, non-present independent work and procedure logbook (De Miguel Díaz 2005; Williams 2008).

The strategies for evaluation that are used by stages in the project are rubrics for the procedure logbook, a brief presentation of the pre-project and for the oral defense of the final project. The evaluation strategies are based on the Miller framework for assessing competencies in health, science, and education (Downing and Yudkowsky 2009; Gruppen et al. 2012) (Fig. 2).

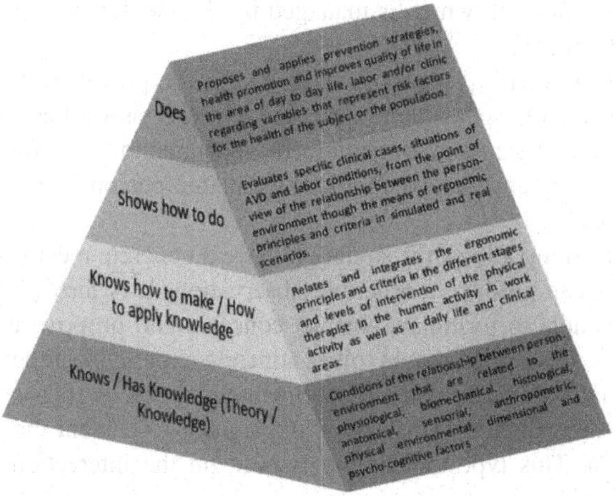

Fig. 2. Levels of competencies. Evaluation in ergonomics (Miller framework)

The comprehension of the learning strategies to solve, in an effective manner, the problems that are presented in ergonomics and in prevention, can contribute to the process of transition from the undergraduate program to the professional practice (Adam et al. 2014). The course involves strategies of teaching and learning individually and in small groups. The strategies of teaching and learning in small groups is based on the principles of experiential learning as well as the principles of constructivism, which proposes that people construct their own learning through experience, solving problems in a context of complexity given by their cognitive development, guided by a professor that oversees and serves as a sort of scaffold of knowledge that is built on a known foundation (Bures 2015; Taylor and Hamdy 2013).

The strategies used in a small collaborative group are (1) Case-based learning: Students must carry out the processing of the information among the members of the group for the resolution of real problems, utilizing the material prepared by the teachers and their independent work, stimulating the intrinsic motivation by delimiting interventions in a real context and within their own culture (Pintrich 2012). (2) Workshops: Based on the analysis of real life and simulated cases, students are guided by a team of teachers to develop approaching strategies to deliver a problem-solving proposal that applies to different scenarios. They are given and shown the diverse use of audiovisual communication tools to design digital solutions (blueprints, environments and the design of 3D tools), which has already been a formative and summative evaluation in the first module.

Another methodology used in the course is the Procedure logbook. This is enclosed at the beginning of the second module, so as the student can establish the different phases in the development "Ergonomic Intervention Project" in the procedure logbook. The second module coincides with the beginning of the student's first clinical internship, which is how the project focuses on a read clinical case, where the student must identify a problem and define it in his or her procedure logbook under the item: "Problem Statement based on a clinical case". Next to this item, they must state their project objective (General Objective and Specific Objectives). In the course of this phase, the students can lay out their doubts by means of a digital classroom (a platform arranged by the university), which is managed by the mentor professor who has been assigned for each student.

The development of this process is stipulated in the procedure logbook recording the daily or weekly progress of the project on a timeline as well as the problems they are faced with and each step and decision made. This method is evaluated with a grade when it is sent digitally and also when it is explained in person to the mentor, who will then give a feedback.

Experiential learning allows the student to have a concrete means of experience, on which the student can produce a reflexive observation. It is through these reflections that the student is able to formulate abstract concepts and propose appropriate generalizations. For this, this cycle –that involves concrete experience, reflexive observation, abstract conceptualization and active experimentation– needs to consolidate itself by being applied to new situations that allow the student to cement the learning (Taylor and Hamdy 2013). This type of learning is built on the interaction in small groups (Seen in Fig. 3).

Students finalize their course with the presentation of their intervention projects in front of a teaching commission and are evaluated with a rubric with criteria for specific tasks, previously validated in content by the faculty team.

Fig. 3. Images in reference to work in small groups.

2 Methods

An analytic, non-experimental, transversal study was performed. 43 students fulfilled the course, 48.9% men and 51.1% women with an average age of 22 years (min 22 years, max 30 years). Academic performance was recorded, and an electronic survey was applied in order to evaluate general aspects, pedagogical and disciplinary domains. The analyses were performed using SPSS V. 21.

3 Results

Academic performance average was 5.94 (min 5.28 max 6.51) on a scale of 1 to 7 and all the students approved. 33 of 43 students answered the survey sent by academic electronic platform. In a Likert Scale from 1 to 4, a 2.87 score was obtained in the pedagogical domain and 3.41 in the disciplinary domain. In regards with general aspects, 73% of the students felt satisfied/very satisfied with the faculty team performance, 87% described themselves as familiar with the criteria used for evaluation, 91% considered requirement of the course as adequate.

Table 1. General aspects evaluation and pedagogical – disciplinary domain

General Aspects

	Categories			
	Very Unsatisfied	*Unsatisfied*	*Satisfied*	*Very Satisfied*
In general terms, I feel satisfied with the performance of the faculty team	3%	24%	67%	6%

Evaluations

	Categories			
	Strongly Disagree	*Disagree*	*Agree*	*Strongly Agree*
I was given knowledge of the criteria of evaluation of the subject in a timely manner	3%	10%	52%	35%
	Too low	*A little low*	*Adequate*	*Too high*
In general terms, I consider that the level of demand of the course is:	0%	0%	91%	9%

Pedagogical Dimension

	Categories			
	Strongly Disagree	*Disagree*	*Agree*	*Strongly Agree*
The teaching methodology used in the course favored my learning and participation.	6%	25%	56%	13%
During the course, there were moments where I was presented with appropriate and challenging problems.	6%	15%	67%	12%
The activities and examples used in the course were closely related to the tasks of my profession.	3%	3%	81%	13%
The evaluations applied were coherent with the purpose and methodology of the course.	6%	3%	66%	25%
The feedback on the evaluations allowed me to reinforce my learning.	10%	23%	52%	16%
The faculty was clear when exposing, answering questions or clarifying doubts that I had.	6%	6%	66%	22%

Disciplinary Domain

	Categories			
	Strongly Disagree	*Disagree*	*Agree*	*Strongly Agree*
In my opinion, the faculty showed domain over the subjects treated.	0%	0%	53%	47%
In my opinion, the bibliography of the course is pertinent and updated.	0%	6%	52%	42%

Regarding pedagogical dimension, 69% considered that methodologies promote participation and learning; 79% considered reflection instances were promoted when appropriate and challenging problems were raised; 94% considered activities and examples used were related with the profession requirements. 93% considered that the evaluations applied were coherent with course purpose and methodology; 68% considered that the received feedback allowed them to reinforce the learning process and 88% that the teachers were clear in the resolution of doubts.

Regarding to the disciplinary domain, 66% recognized teacher's commitment with students learning processes, 100% considered that teachers showed a complete management of knowledge of topics, and 96% considered the bibliography as relevant and updated (Table 1).

4 Discussion

This study presents the planning of competencies in ergonomy in a undergraduate course, the strategies of teaching and learning utilized and the evaluations carried out in a context of curricular innovation. The perception of the students as to the pedagogical and disciplinary domains of the course was satisfactory and all the students achieved the approval of the course, demonstrating an effective process of education by competencies.

There is a correspondence between the proposal of the authors regarding the international recommendations in education in Ergonomics (Bures 2015; IEA 2001), allowing for the preparation of the students in the career of Physical therapy of the University of Chile to fulfill their profile of competent graduates to analyze persons that present a certain degree of alteration in their functionality in an integral way and to make decisions for their reinsertion in their activities on a day to day basis and also their work lives.

Education with innovative methods, with perspectives that involve the progressive development of the content of professionals, focused on the needs of the community, that promote a proactive stand and that uses assessment strategies relative to the educational objectives proposed to allow the students to face their problems in context, to solve them in an effective way considering the context of complexity that it involves (Barbosa and Pinheiro 2012a, b). The physical therapist has a fundamental role in the rehabilitation of persons that suffer different work-related injuries and diseases. Prendushi has highlighted the importance of physical therapy students give to the subject of work disability, as an interesting subject of study (Prendushi 2016), as it is a relevant aspect to approach in the contexts that involve the teaching of evaluation and basic procedures in Ergonomics.

The process of effective teaching and learning in the course of ergonomic intervention involves the exposure to real problems, allowing for the mobilization of knowledge for its resolution, utilizing the necessary core knowledge to carry out basic applications in a relevant manner, coinciding with the importance assigned to the structure of the competency-based programs (Furniss et al. 2017).

The collaborative work that involves the methodology of small groups inside the course, along with the autonomy of the students to chose their cases and resolve them, stimulates the intrinsic motivation, the disposition for learning, self-directed and self-regulated learning, fulfilling the principles of andragogy as a concept of permanent education (Knowles et al. 2012). Teachers within the process act as enablers in a context of social constructivism (Taylor and Hamdy 2013), promoting the consolidation of a significant educational experience.

Inside of the methodologies of small groups, the use of a logbook as an instrument for keeping records and for evaluation allows for the participation of the student in an active manner in his or her process of formation, due to the assigned deadlines that the tutor designates to complete specific tasks and, in the way, analyze the impact of learning and achieve an effective communication with students (Barrios Castañeda et al. 2012). The logbooks have been proposed as educational strategies in ergonomy on an expert level (Williams 2008; Barrios Castañeda et al. 2012). In this study, they have been set forth in accordance with the level of the educational objectives to achieve an effective monitoring of the work carried out.

5 Conclusions

The course of ergonomic intervention project contains active learning strategies in small groups like case-based learning, workshops and procedure logbooks lead by skilled teachers in a supervised context which promote ergonomic competency achievements by undergraduate students in a learning process based on constructivism. Students fulfilled the competencies regarding the application of principles and ergonomic criteria in the different stages and possible intervention levels. The process of teaching, learning and student assessment was consistent with the principles of the experimental learning cycle and effective practice-based learning.

References

University of Chile (2018) Educative model, Santiago, Chile. http://www.libros.uchile.cl/717

Faculty of Medicine, University of Chile (2012) Educative model, Santiago, Chile. http://decsa.med.uchile.cl/wp-content/uploads/Modelo-Educativo.pdf

De Miguel Díaz M (2005) Modalidades de enseñanza centradas en el desarrollo de competencias: orientaciones para promover el cambio metodológico en el espacio europèo de educación superior. Servicio de Publicaciones, Universidad de Oviedo

Gruppen LD, Mangrulkar RS, Kolars JC (2012) The promise of competency-based education in the health professions for improving global health. Hum Resour Health 10(1):43

Adam K, Strong J, Chipchase L (2014) Readiness for work injury management and prevention: important attributes for early graduate occupational therapists and physiotherapists. Work 48 (4):567–578

Barbosa LH, Pinheiro MHC (2012a) Teaching ergonomics to undergraduate physical therapy students: new methodologies and impressions of a Brazilian experience. Work 41(Supplement 1):4790–4794

Barbosa LH, Pinheiro MHC (2012b) The challenges of interdisciplinary education and its application on teaching ergonomics. Work 41(Supplement 1):5456–5458

Barrios Castañeda P, Ruiz LA, González Guerrero K (2012) The logbook as a monitoring and evaluation instrument-formation of residents in the ophthalmology program. Inv Andina 14(24):402–412

Bridger RS (2012) An international perspective on ergonomics education. Ergon Des 20(4): 12–17

Bures M (2015) Efficient education of ergonomics in industrial engineering study program. Procedia-Soc Behav Sci 174:3204–3209

Downing SM, Yudkowsky R (2009) Assessment in health professions education. Routledge, New York

Furniss D, Curzon P, Blandford A (2017) Exploring organizational competencies in Human Factors and UX project work: managing careers, project tactics and organizational strategy. Ergonomics, 1–52

Hignett S, Jones EL, Miller D, Wolf L, Modi C, Shahzad MW, Catchpole K (2015) Human factors and ergonomics and quality improvement science: integrating approaches for safety in healthcare. BMJ Qual Saf 24(4):250–254

Prendushi H (2016) The attitudes of physiotherapy students toward occupational medicine. Eur Sci J ESJ 12(36)

Williams C (2008) In search of ergonomics expertise. Doctoral thesis, © Claire Williams

Wilson JR (2000) Education and recognition of ergonomists. In: Proceedings of the human factors and ergonomics society annual meeting, vol. 44, no. 33. SAGE Publications, Los Angeles, pp 6–100

Knowles M, Holton E, Swanson RA (2012) The adult learner. The definitive classic in adult education and human resource development. Elsevier

Pintrich PR (2003) A motivational science perspective on the role of student motivation in learning and teaching contexts. J Educ Psychol 95(4):667–686

Taylor DC, Hamdy H (2013) Adult learning theories: implications for learning and teaching in medical education: AMEE guide no. 83. Med Teach 35(11):e1561–e1572

International Ergonomics Association: Professional Standards and Education Committee, Version 2 (2001, October) Full version of core competencies in ergonomics: units, elements and performance criteria. https://www.iea.cc/project/2_Full%20Version%20of%20Core%20Competencies%20in%20Ergonomics%20Units%20Elements%20and%20Performance%20Criteria%20October%202001.pdf

Relationship Between Acceptance of Workforce Diversity and Mental Health Condition Among Japanese Nurses

Yasuyuki Yamada[1,2(✉)], Takumi Iwaasa[1], Takeshi Ebara[3], Teruko Shimizu[4], and Motoki Mizuno[1,2]

[1] Juntendo University Graduate School of Health and Sports Science, 1-1, Hiragagakuendai, Inzai-shi, Chiba 270-1695, Japan
yayamada@juntendo.ac.jp

[2] Juntendo University School of Health and Sports Science, 1-1, Hiragagakuendai, Inzai-shi, Chiba 270-1695, Japan

[3] Nagoya City University Graduate School of Medical Sciences, Nagoya, 1, Kawasumi, Mizuho-cho, Mizuho-ku, Nagoya-shi, Aichi 467-8601, Japan

[4] Daiyukai Daiichi Hospital, 1-6-12, Hagoromo, Ichinomiya-shi, Aichi 491-8851, Japan

Abstract. Diversity management has been expected not only to progress productive nursing service, but also to promote well-being among Japanese nurses. However, Japanese conventional studies have not shown enough statistical evidences. Hence, this study examined the relationship between perception of workforce diversity and mental health condition among Japanese nurses. Through the internet research, we collected a total of 1,031 valid data (male = 217, female = 814). This study constructed eighteen original items to assess the acceptance levels of diversity elements; seniority, managerial position, clinical experience, employment history, academic background, generation, gender, nationality, role orientation, employment pattern, license, personality, health condition, family situation, work and life priorities, hometown, nursing ability and work motivation. Each diversity element was evaluated by four acceptance levels; (1) refusing diversity (Resistance), (2) ignoring diversity (Assimilation), (3) valuing diversity (Separation) and (4) utilizing diversity (Integration). Mental health condition was assessed by the 12-item General Health Questionnaire (GHQ-12, high-stress ≧ 6 point). As the results of a logistic regression analysis, higher acceptance of the managerial position, personality and work and life priorities were negatively related with high-stress. Higher acceptance of the employment pattern and work motivation were positively related with high-stress. These results indicated that diversity management was one of the effective approaches to improve mental health condition among Japanese working nurses.

Keywords: Diversity management · Mental health · Nursing

S. Bagnara et al. (Eds.): IEA 2018, AISC 818, pp. 563–567, 2019.
https://doi.org/10.1007/978-3-319-96098-2_69

1 Introduction

Since nursing shortage has been chronical social issue in Japan, the nursing organizations have tried to develop new human resources, e.g. education of male nurses, invitation of foreign nurses and acceptance of various working styles. From a conceptual framework of diversity management (DM) [1], these challenges could be regarded as not only solution of the manpower shortage, but also strategies of productive performance and innovation to enhance the product or service quality, health and safety, and financial indices. Accordingly, the researchers have discussed about the effective DM approaches [2, 3], guidelines [4, 5] and good practices [6–9]. On the other hand, especially in Japan, empirical study was not fully progressed. Under these circumstances, we conducted descriptive study and categorized some diversity elements observed in the nursing workplace as a first step for the empirical study in Japan [10]. However, it was not unclear that what kind of diversity elements strongly contributed to progress the organizational and individual outcome. As a next step, this study evaluated acceptance levels of diversity elements in the Japanese nursing workplace and examined the relationships with working nurses' mental health condition.

2 Methods

2.1 Participants

Study participants were Japanese nurses, those who were monitors of Macromill inc (Tokyo) internet research company. Through the internet research, we collected a total of 1,031 valid data (male = 217, female = 814).

2.2 Measurements

To assess the acceptance levels of diversity, we designed original items through the preliminary study. This scale dealt with diversity elements included in the seniority, managerial position, clinical experience, employment history, academic background, generation, gender, nationality, job orientation, employment pattern, license, personality, health condition, family situation, work and life priorities, hometown, nursing ability and work motivation. On the basis of theoretical study of DM [11], this study evaluated acceptance levels of diversity from four stages; (1) refusing diversity (Resistance), (2) ignoring diversity (Assimilation), (3) valuing diversity (Separation) and (4) utilizing diversity (Integration).

Mental health condition was evaluated by the 12-item General Health Questionnaire (GHQ-12). The GHQ-12 is a self-administered questionnaire [12] with 4-point scale; (1) less than usual, (2) no more than usual, (3) fairly more than usual, (4) or much more than usual. On the basis of the 0-0-1-1 method, this study calculated the total scores (range = 0–12 points) and categorized high-stress group (standard = 0–5 points, high-stress = 6–12 points).

2.3 Data Analysis

This study conducted a logistic regression analysis (LRA) to examine the relationships between acceptance levels of diversity elements and mental health condition with adjustment of the age, sex, license, marriage and managerial post. Adjusted OR means the strength of relationship between acceptance of diversity elements and high-stress group in GHQ-12.

2.4 Ethics and Conflict of Interest (COI)

This study was approved by the ethical committee of Juntendo University Graduate School of Health and Sports Science. Moreover, there was not potential COI to disclose with this study.

3 Results

The results of LRA were shown in Table 1. In this research, progress of acceptance levels of managerial position diversity (Resistance-Assimilation: OR = 0.52, 95% CI = 0.30–0.92, Resistance-Separation: OR = 0.44, 95%CI = 0.24–0.81, Resistance-Integration: OR = 0.34, 95%CI = 0.17–0.69), personality (Resistance-Integration: OR = 0.37, 95%CI = 0.16–0.87) and work and life priorities (Resistance-Separation: OR = 0.37, 95%CI = 0.19–0.71) were negatively related with high-stress condition. On the other hand, higher acceptance levels of the diversities of employment pattern (Resistance-Assimilation: OR = 1.82, 95%CI = 1.03–3.23, Resistance-Integration: OR = 2.58, 95%CI = 1.29–5.15) and work motivation (Resistance-Integration: OR = 3.71, 95%CI = 1.63–8.44) were positively related with high-stress condition. The significant relationships were not observed in other diversity elements.

4 Discussions

In this study, acceptance of managerial position was negatively related to high-stress condition. This result indicated that understanding, respecting and encouraging each other's differences among manager, chief and staff nurses were meaningful to construct less stressful working conditions. Moreover, intervention in the interaction between manager and chief nurses' leadership and staff nurses' followership will be effective to improve the stressful conditions. Utilization of the personality traits also negatively related with high-stress condition. Hence, working nurses have to deal with co-workers' personality traits including an assertiveness, aggression, slow learning, awkwardness and withdrawal to accomplish of team nursing with lower stress. Additionally, acknowledgement of work and life priorities negatively related with high-stress condition. Namely, to keep good work-life balance and good mental health condition, mutual support among work-oriented nurses and family-oriented nurses is necessary. Meanwhile, diversity elements of employment pattern and work motivation were positively related to high-stress condition. Although active utilization of part-time nurses has been regarded as effective DM strategy in Japan, it was not appropriate in the context of mental health management. Likewise, while securing working nurses is undeniably important, employment and acceptance of unmotivated negative nurses can be regarded as a risk of stressful working conditions.

Table 1. Relationships between acceptance levels of diversity elements and stress outcome led by the Logistic Regression Analysis.

Diversity elements	Acceptance levels	Adjusted OR (95%CI)	Diversity elements	Acceptance levels	Adjusted OR (95%CI)
1. Seniority	Resistance		10. Employment pattern	Resistance	
	Assimilation	1.14 (0.64 - 2.05)		Assimilation	1.82 (1.03 - 3.23) *
	Separation	0.85 (0.46 - 1.58)		Separation	1.36 (0.76 - 2.43)
	Integration	1.29 (0.63 - 2.63)		Integration	2.58 (1.29 - 5.15) **
2. Managerial position	Resistance		11. License	Resistance	
	Assimilation	0.52 (0.30 - 0.92) *		Assimilation	1.25 (0.70 - 2.21)
	Separation	0.44 (0.24 - 0.81) **		Separation	1.39 (0.76 - 2.55)
	Integration	0.34 (0.17 - 0.69) **		Integration	1.01 (0.52 - 1.96)
3. Clinical experience	Resistance		12. Personality	Resistance	
	Assimilation	0.73 (0.40 - 1.33)		Assimilation	0.77 (0.45 - 1.30)
	Separation	0.83 (0.43 - 1.60)		Separation	0.95 (0.53 - 1.72)
	Integration	0.73 (0.34 - 1.54)		Integration	0.37 (0.16 - 0.87) *
4. Employment history	Resistance		13. Health condition	Resistance	
	Assimilation	0.97 (0.55 - 1.74)		Assimilation	1.28 (0.63 - 2.61)
	Separation	0.96 (0.52 - 1.78)		Separation	1.16 (0.54 - 2.47)
	Integration	0.78 (0.37 - 1.64)		Integration	1.27 (0.49 - 3.28)
5. Academic background	Resistance		14. Family situation	Resistance	
	Assimilation	0.75 (0.47 - 1.20)		Assimilation	1.15 (0.58 - 2.28)
	Separation	0.70 (0.42 - 1.16)		Separation	1.19 (0.58 - 2.43)
	Integration	0.97 (0.51 - 1.84)		Integration	1.06 (0.46 - 2.47)
6. Generation	Resistance		15. Work and life priorities	Resistance	
	Assimilation	1.08 (0.61 - 1.91)		Assimilation	0.63 (0.34 - 1.17)
	Separation	0.97 (0.54 - 1.73)		Separation	0.37 (0.19 - 0.71) *
	Integration	0.54 (0.26 - 1.15)		Integration	0.42 (0.17 - 1.06)
7. Gender	Resistance		16. Hometown	Resistance	
	Assimilation	0.80 (0.46 - 1.40)		Assimilation	1.07 (0.61 - 1.87)
	Separation	0.93 (0.55 - 1.58)		Separation	1.29 (0.73 - 2.29)
	Integration	1.15 (0.63 - 2.08)		Integration	0.86 (0.40 - 1.89)
8. Nationality	Resistance		17. Nursing ability	Resistance	
	Assimilation	1.14 (0.56 - 2.32)		Assimilation	1.14 (0.65 - 2.02)
	Separation	1.06 (0.51 - 2.18)		Separation	0.88 (0.47 - 1.65)
	Integration	1.69 (0.73 - 3.90)		Integration	1.39 (0.61 - 3.14)
9. Role orientation	Resistance		18. Work motivation	Resistance	
	Assimilation	1.23 (0.66 - 2.26)		Assimilation	1.18 (0.71 - 1.96)
	Separation	1.10 (0.58 - 2.08)		Separation	1.34 (0.73 - 2.47)
	Integration	1.50 (0.74 - 3.05)		Integration	3.71 (1.63 - 8.44) *

Adjusted OR: Odds Ratio adjusted by sex, marriage, managerial post, license, age, 95%CI: 95 percent confidence interval of adjusted OR, Resistance: Refuse the difference of diversity, Assimilation: Ignore the difference of diversity, Separation: Valuing the difference of diversity, Integration: Utilizing the difference of diversity, *p<.05, **p<.01.

Acknowledgement. This work was supported by JSPS KAKENHI Grant Number 15K16177.

References

1. Nishii LH, Özbilgin MF (2007) Global diversity management: towards a conceptual framework. Int J Hum Res Manuf 18(11):1883–1894
2. Slovensky DJ, Paustian PE (2005) Preparing for diversity management strategies: teaching tactics for an undergraduate healthcare management program. J Health Adm Educ 22(2):189–199
3. Wallace PE Jr, Ermer CM, Motshabi DN (1996) Managing diversity: a senior management perspective. Hosp Health Serv Adm 41(1):91–104
4. Fry B (2011) Thriving in the workplace: a nurse's guide to intergenerational diversity. Canadian Federation of Nurses Unions
5. AONE (2007) Guiding principles for diversity in health care organizations nurse leader 5(5):17–24
6. Wisconsin Center for Nursing, Inc. (2013) Enhancing diversity in the nursing workforce: a report by the Wisconsin Center for Nursing diversity taskforce
7. National Advisory Council for Nursing Education and Practice (2013) Achieving Health Equity through Nursing Workforce Diversity. U.S. Department of Health and Human Services, Health Resources and Services Administration: Washington, DC
8. Kreitz PA (2008) Best practices for managing organizational diversity. J Acad Lib 34(2):101–120
9. Bunton W (2000) Best Practices in achieving workforce diversity, benchmarking study of the U.S. Department of Commerce and Vice President Al Gore's National Partnership for Reinventing Government
10. Yamada Y, Mizuno M, Shimizu T, Asano Y, Iwaasa T, Ebara T (2017) Elements of workforce diversity in Japanese nursing workplace. In: Advances in social and occupational ergonomics. Advances in intelligent systems and computing, vol 487. Springer, Heidelberg
11. Taniguchi M (2005) Diversity Management. Hakuto-Shobo, Tokyo
12. Goldberg DP, Hillier VF (1979) A scaled version of the General Health Questionnaire. Psychol Med 9(1):139–145

Design of a Dynamic Hand Orthosis for Stroke Patient to Improve Hand Movement

Yung-Chang Li[1,2], Shu-Zon Lou[1,2], Kwok-Tak Yeung[1,2], Ya-Ling Teng[1,2], and Chiung-Ling Chen[1,2(✉)]

[1] Department of Occupational Therapy, Chung Shan Medical University, Taichung, Taiwan
joelin@csmu.edu.tw
[2] Occupational Therapy Room, Chung Shan Medical University Hospital, Taichung, Taiwan

Abstract. Motor recovery of hand in most of the stroke survivors is poor. Hand rehabilitation therefore continues to be a major focus for occupational therapy. Only some stroke survivors are candidates for constraint-induced movement therapy and the robots for robot-assisted therapy are expensive to purchase. Therefore, it is necessary to develop a low cost hand rehabilitation device for use in home or a therapeutic environment. The purposes of this study were to design of a firm, durable and adjustable hand orthosis and to examine the usability of the dynamic hand orthosis for persons with stroke. Splinting materials, 3D prints and hardware materials were used to design a firm and durable dynamic hand orthosis. Then user satisfaction questionnaire, hand function test and motion analysis system were used to examine the usability of the dynamic hand orthosis in persons with stroke. The results show that the persons with stroke satisfied with the orthosis. When wearing the orthosis the pick-up and release hand function improved immediately. The range of motion of metacarpophalangeal joint from hand open to pulp grip also increased. Future research is needed to evaluate the effects of the dynamic hand orthosis intervention on hand movement and function in both chronic and acute stroke.

Keywords: Hand orthosis · Stroke · Hand rehabilitation · Hand function

1 Introduction

Stroke is a common and highly debilitating disease. Most of stroke survivors experience unilateral motor deficit that leads to chronic upper extremity impairment. Only 38% of stroke survivors have some dexterity in the paretic arm after 6 months [1]. Two-thirds of patients still suffer from profoundly impaired dexterity, which significantly impacts the individual's ability in activities of daily living [2]. Restoration of hand function is a major rehabilitation and health care challenge.

Functional grasp/release is an important and basic function for various movements in daily life. However, most of the stroke patients regain the ability to flex the fingers but they cannot voluntarily extend their fingers. The reasons of limited finger extension include inappropriate activity in flexors interferes with voluntary extension movements [3],

S. Bagnara et al. (Eds.): IEA 2018, AISC 818, pp. 568–575, 2019.
https://doi.org/10.1007/978-3-319-96098-2_70

inability to active extension movement due to weakness in extensors [4], abnormal co-contraction of flexors during voluntary extension tasks [5, 6] and impairment in inter-joint coordination and finger fractionation [7].

Studies have found that intense active repetitive movement practice can enhance the strength and functional use of the affected hand [8]. There are several approaches to facilitate hand motor recovery such as constraint-induced movement therapy [9], functional electrical stimulation [10] and robot-assisted therapy [11]. However, only some stroke survivors are candidates for constraint-induced movement therapy and the robots for robot-assisted therapy are expensive to purchase. Therefore, it is necessary to develop a low cost hand rehabilitation device for use in home or a therapeutic environment.

The purposes of this study were to design of a firm, durable and adjustable hand orthosis and to examine the usability of the dynamic hand orthosis for persons with stroke.

2 Methods

Splinting materials (low temperature orthoplast), 3D prints and hardware materials were used to design a firm and durable dynamic hand orthosis. Then satisfaction questionnaire, hand function test and motion analysis system were used to examine the usability of the dynamic hand orthosis in four persons with stroke.

2.1 Orthosis Design

The four components of the dynamic hand orthosis are forearm base, finger gutter, elastic outrigger mechanism and resistance adjustment system. The design of the orthosis is shown in Fig. 1.

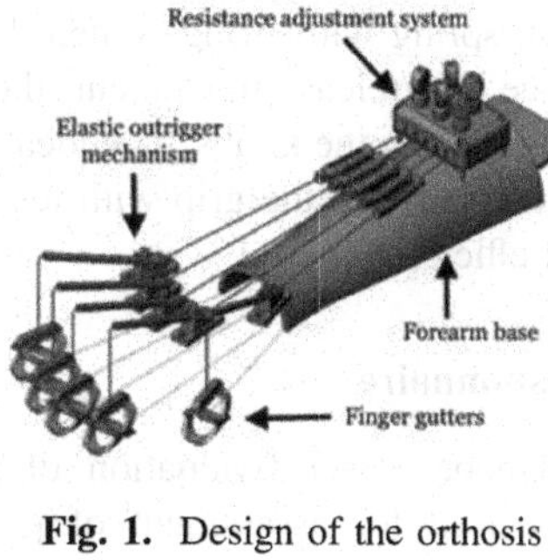

Fig. 1. Design of the orthosis

Forearm Base

The low-temperature orthoplast with 3.2 mm thickness was used to fit the shape of forearm. This orthoplast is made from polycaprolactone material, which is resistant to over-stretching and retains sufficient elasticity during application to achieve optimum moulding and comformability. The forearm base comfortably contacts with skin and provides a stable base to attach the following outrigger and adjustment system. It can withstand the external force and high numbers of grip-release repetitions without detachment.

Finger Gutter
We used the ORFITCAST material to mold the finger in a slight flexion position. The gutters designed to restrict distal interphalangeal (DIP) and proximal interphalangeal (PIP) joints flexion movements led to an increase in metacarpophalangeal (MP) joint flexion angle when performing pulp grip movement. This kind of movements mimics the normal movement pattern of pulp pinch.

Elastic Outrigger Mechanism
The elastic outrigger mechanism was composed of strings, hard springs with 3D printed base and soft springs. The strings passing through the hard springs are like tendon passing through the sheath, which link the finger gutters to soft springs.

The hard and soft springs provide resistance for finger flexion and assistance for finger extension. When gripping movement initiate, the soft springs distend first with lighter resistance, the patient can initiate the movement easier. The hard springs place individual finger into the proper 90° line of pull. As gripping movement continued, the hard springs guide alignment of the fingers allowing them to move in the right direction. When patients try to release grip, both springs will produce counter-force to assist MP joint extension.

Resistance Adjustment System
The outrigger with spring system provides resistance for finger flexion and assists in opening the hand. In addition, we can adjust the resistance of the springs individually according to the muscle strength and flexor muscle tone of the fingers. Because of the flexor muscle tone may be different for each finger and may change from day to day, or even at different times during the same day. The individual finger strength may also be different.

On the forearm base, we design a small box which holds the five knobs steady. Each knob connects to the soft spring with string. When the persons with stroke who wearing the orthosis to practice grip/release movement, they have to learn and try to adjust the tension of springs by turning the knobs. The ideal tension of springs for each finger means that the person can perform the grip with resistance and release the grip with assistance smoothly and efficiently.

2.2 User Satisfaction Questionnaire

The Taiwanese version of Quebec User Evaluation of Satisfaction with Assistive Technology (QUEST-T) was used for assessment of patient satisfaction with the orthosis. The items of the QUEST-T include size, weight, adjustments, safety, durability, simplicity of use, comfort and effectiveness. Each item is scored on a 5-point ordinal satisfaction scale, i.e., 1. not satisfied at all; 2. not very satisfied; 3. more or less satisfied; 4. quite satisfied; and 5. very satisfied. Space for comments is provided next to each item.

2.3 Hand Function Test

The participants were asked to pick up and release a 5 cm^3 block with and without the orthosis repeatedly within 2 min. The number of times of pick-up and release was calculated to compare the speed of pulp grip/release movement with and without the orthosis.

2.4 Motion Analysis

The Motion Analysis Corporation Eagle system was used to measure the joint movements of fingers. Figure 2 shows the reflexive markers placed on the hand with and without wearing the orthosis. The participants were asked to open the hand as much as possible and then perform pulp grip. The range of motion (ROM) of MP, PIP and DIP was measured with and without the orthosis.

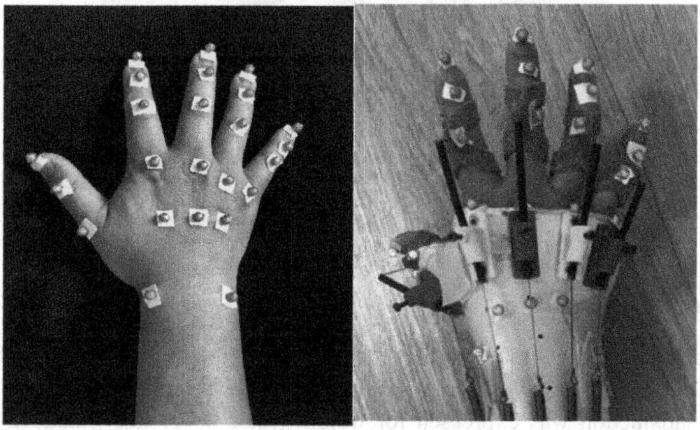

Fig. 2. The reflexive markers placed on the hand without and with the orthosis.

3 Results and Discussion

Previously, the static hand splints were used to reduce hand spasticity. Recently, the dynamic hand orthosis was suggested to applied in hand rehabilitation. There is evidence to demonstrate that intensive massed and repeated practice may be necessary to modify neural organization [12]. We design the dynamic hand orthosis for persons with stroke to practice grip and release movement repeatedly.

Four men with stroke participated in the usability test. The mean age (±SD) at test was 44.8 ± 21.3 years, and onset duration was 51 ± 28.5 months. One participant suffered from right hemiplegia and 3 from left hemiplegia. All of them were right handed. The persons' Brunnstrom stage of motor recovery on hand were one in III, two in III-IV and one in IV stage.

3.1 Dynamic Hand Orthosis

This dynamic hand orthosis is a custom fabricated, non-electrical, mechanical orthosis. We use low temperature orthoplast to make a forearm base, mounted with an elastic outrigger system which connected with a resistance adjustment for individual finger (Fig. 3). The objective of this orthosis was to help patients to reopen their fingers after functional grip.

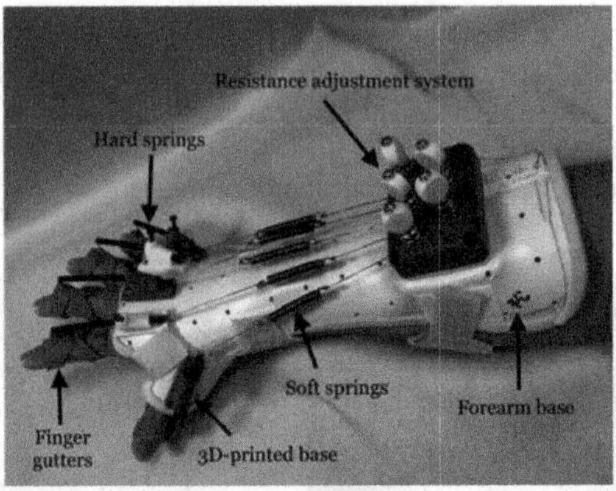

Fig. 3. The dynamic hand orthosis.

3.2 User Satisfaction

The participants indicated that they were satisfied with the dynamic hand orthosis. The mean total score of the eight items (8–40) was 29.8. Mean score of each item was 3.72. The most satisfaction was expressed for weight (rated 4.75) size (rated 4.5) and followed by adjustments (3.75), simplicity of use and effectiveness (rated 3.5). The subject rated 3.25 on durability and safety items, rated 3 on comfort. The subject's comments were about the poor durability of the strings and 3D printed base, and discomfort of the finger gutter material.

The authors will find more durable strings and softer finger gutter materials as well as change 3D printed orientation from vertical to horizontal direction to increase the strength of the base.

3.3 Hand Function Test

Figure 4 shows the participants pick up and release the block without and with the orthosis. One participant could not perform the test and one participant completed the test with assistant by sound hand without the orthosis. They all could completed the test with the orthosis. The mean repetitions of participants completed picking up and releasing the block was 6.5 without the orthosis and 14 with the orthosis within 2 min. Because the spasticity of finger flexors interferes with voluntary extension, the speed of movement was slower without orthosis. When wearing the orthosis, the participants extended the fingers quickly with assistance of the orthosis.

Fig. 4. The subject picks up and releases the block without and with the orthosis

3.4 Motion Analysis Outcome

Table 1 shows the ROM of MP, PIP and DIP joints in all fingers from hand open to pulp grip without orthosis and with orthosis. The results demonstrated that the ROM of MP joint increased when wearing the orthosis. The participants can open their finger with much more range as expected. The orthosis has the biomechanical advantage in allowing pulp grip and release for stroke subject with spasticity.

For PIP and DIP joints, the ROM from hand open to pulp grip decreased when wearing the orthosis. The results consist with the expected outcome of the design. We use the finger gutters to restrict the flexion movements of DIP and DIP joints. It was postulated that the finger flexor tone would be reduced by keeping the joints from too much flexion movement. Research has suggested that spasticity can be prevented and treated by prolonged muscle stretching [13].

Table 1. Range of motion in MP, PIP and DIP joints from hand open to pulp grip

Finger joint	Without orthosis (degree) Mean (SD)	With orthosis (degree) Mean (SD)
Thumb MP	16.9(5.4)–33.2(12.1)	13.3(10.9)–37.4(15.9)
Index MP	28.9(3.2)–49.9(14.7)	6.0(3.5)–44.9(14.2)
Middle finger MP	27.2(7.5)–52.1(22.0)	10.2(9.0)–50.3(11.7)
Ring finger MP	15.4(6.3)–47.7(26.8)	5.3(3.8)–37.1(13.2)
Little finger MP	14.3(5.2)–35.6(27.7)	8.5(8.3)–43.9(29.1)
Thumb IP	17.5(18.3)–51.5(31.1)	12.6(8.4)–30.9(11.5)
Index PIP	28.9(19.5)–73.7(4.5)	23.1(9.3)–42.7(3.9)
Middle finger PIP	26.4(19.4)–63.1(7.9)	30.6(12.0)–45.8(7.5)
Ring finger PIP	35.8(22.9)–72.2(8.6)	33.9(14.4)–53.6(14.6)
Little finger PIP	38.2(22.6)–69.8(7.6)	23.1(14.3)–53.7(15.6)
Index DIP	24.3(15.2)–57.0(14.2)	26.3(9.4)–47.2(15.0)
Middle finger DIP	11.7(11.8)–54.9(18.2)	36.9(9.3)–57.7(15.0)
Ring finger DIP	7.1(4.9)–35.4(8.5)	24.4(13.2)–50.6(21.9)
Little finger DIP	12.2(7.1)–54.2(18.1)	26.5(11.1)–44.2(18.0)

4 Conclusion

In this paper, the design and usability test of the dynamic hand orthosis was presented. The findings of the usability test suggest that the dynamic hand orthosis can applied to persons with stroke to enhance the extension movement of MP joint and hand function immediately. Future study would be warranted in order to demonstrate the effects of the dynamic hand orthosis as an intervention to improve hand movement and function of acute and chronic stroke survivors.

Acknowledgment. This research was funded by the Ministry of Science and Technology in Taiwan (MOST 106-2314-B-040-018).

References

1. Kwakkel G et al (2003) Probability of regaining dexterity in the flaccid upper limb: impact of severity of paresis and time since onset in acute stroke. Stroke 34(9):2181–2186
2. Feigin VL et al (2003) Stroke epidemiology: a review of population-based studies of incidence, prevalence, and case-fatality in the late 20th century. Lancet Neurol 2(1):43–53
3. Kamper DG et al (2003) Relative contributions of neural mechanisms versus muscle mechanics in promoting finger extension deficits following stroke. Muscle Nerve 28(3):309–318
4. Kamper DG et al (2006) Weakness is the primary contributor to finger impairment in chronic stroke. Arch Phys Med Rehabil 87(9):1262–1269
5. Kamper DG, Rymer WZ (2001) Impairment of voluntary control of finger motion following stroke: role of inappropriate muscle coactivation. Muscle Nerve 24(5):673–681

6. Cruz EG, Waldinger HC, Kamper DG (2005) Kinetic and kinematic workspaces of the index finger following stroke. Brain 128(Pt 5):1112–1121
7. Carpinella I, Jonsdottir J, Ferrarin M (2011) Multi-finger coordination in healthy subjects and stroke patients: a mathematical modelling approach. J Neuroeng Rehabil 8:19
8. Liepert J et al (2001) Motor cortex plasticity during forced-use therapy in stroke patients: a preliminary study. J Neurol 248(4):315–321
9. Tarkka IM, Pitkanen K, Sivenius J (2005) Paretic hand rehabilitation with constraint-induced movement therapy after stroke. Am J Phys Med Rehabil 84(7):501–505
10. Quandt F, Hummel FC (2014) The influence of functional electrical stimulation on hand motor recovery in stroke patients: a review. Exp Transl Stroke Med 6:9
11. Sale P et al (2014) Recovery of hand function with robot-assisted therapy in acute stroke patients: a randomized-controlled trial. Int J Rehabil Res 37(3):236–242
12. Nudo RJ et al (1996) Neural substrates for the effects of rehabilitative training on motor recovery after ischemic infarct. Science 272(5269):1791–1794
13. Smania N et al (2010) Rehabilitation procedures in the management of spasticity. Eur J Phys Rehabil Med 46(3):423–438

Muscle-Training Effect Associated with Scaler Operation

Tamiyo Asaga[1,2](✉), Tomoko Aso[2], and Tetsuo Misawa[3]

[1] Graduate School of Chiba Institute of Technology, Narashino, Japan
tamiyo.asaga@cpuhs.ac.jp
[2] Chiba Prefectural University of Health Sciences, Chiba, Japan
[3] Chiba Institute of Technology, Narashino, Japan

Abstract. Objective: In this study, since the gripping of the scaler is based on the gripping of the writing instrument, we examined the effect of training of the correct writing instrument's gripping motion on the operation of the scaler.
Methods: Training of holding a pen with a grip was performed in subjects whose thumb did not properly contact with the pen. In the measurement of EMG, we measured the first dorsal interdigital muscle and the flexor pollicis brevis muscle in the scaler manipulation and measured the daily grasping and correct grasping motion, the forearm rotational motion of the scaler operation and the finger flexion exercise before and after the training went.
Results: RMS of the flexor pollicis brevis muscle significantly increased during usual and proper pen-holding motions and that of the first dorsal interosseous muscle significantly increased in scaler operation. In addition, the flexor pollicis brevis muscle activity level in scaler significantly decreased operation during bending and extending the fingers after training. It was suggested that training of positioning and the way of applying the force of the thumb could be performed leading to a reduction of the muscle activity level by daily use of writing instruments with a grip.
Discussion: A strong correlation was noted between proper pen-holding motion operation and scaler operation-induced forearm rotation and RMS of the flexor pollicis brevis muscle, suggesting that holding a pen with a grip serves as training strengthening the flexor pollicis brevis muscle.

Keywords: Scaler · Training · Electromyography

1 Introduction

Scaling is to remove dental plaque and dental calculus from the tooth surface.

It is a basic treatment of periodontal disease performed by a dental hygienist and it is important for prevention, and accurate and effective operation of the scaler is required.

The basic scaler-holding method is a modified pen grasp to hold writing instruments, in which a scaler is held by 3 fingers: the thumb, index finger, and middle finger (see Fig. 1). In scaling, by slightly rotating the blade of the scaler by pushing and pulling the thumb, it is possible to operate by adapting it to the curved surface of a complicated teeth.

S. Bagnara et al. (Eds.): IEA 2018, AISC 818, pp. 576–581, 2019.
https://doi.org/10.1007/978-3-319-96098-2_71

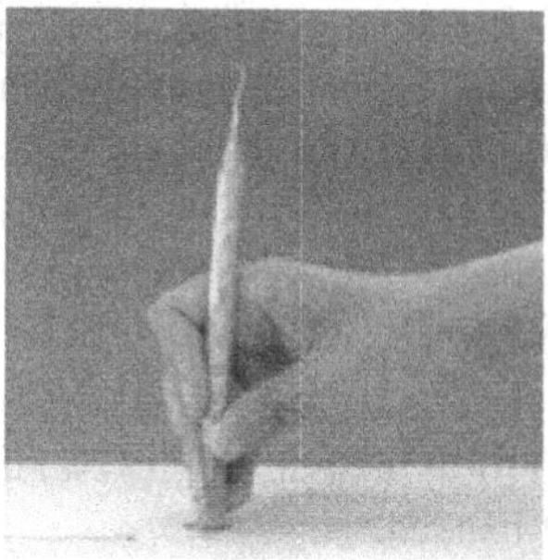

Fig. 1. Scaler grasping method

In addition, attaching the fingertips of thumb and index finger to the scaler, can sensitively sense the condition of the tooth surface. Asaga et al. [1] and Ioku et al. [2] reported from studies of the method of grasping writing instruments of university students that there are many students holding writing instruments in contact with the base of the thumb and holding the thumbs like holding their thumbs overlap their index finger.

Asaga et al. reported that they told students to hold a scaler like holding a pen, and for students who do not hold a pen with 3 fingers (the thumb, index finger, and middle finger), sufficient explanation of the positions of the thumb, index finger, and middle finger is important.

In this study, the effect of training of proper pen-holding motion for scaler operation was investigated.

2 Subjects and Methods

2.1 Subjects

The subjects were 19 female students of the Dental Hygiene Department of a university who agreed to cooperate all experiment.

2.2 Methods

Holding style of pen in all subjects were photographed and classified as follows (see Fig. 2).

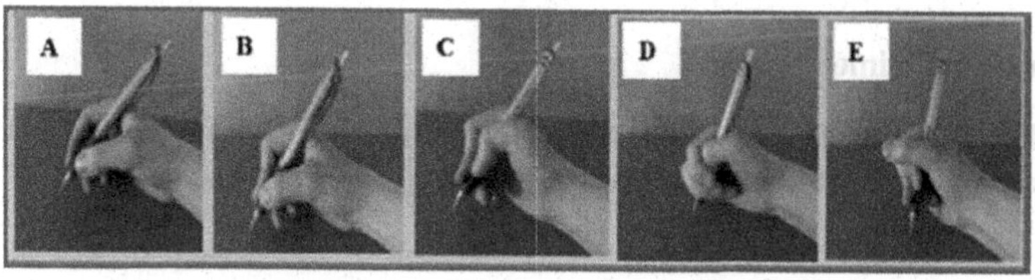

Fig. 2. How to hold a writing instrument

Style A: The joint of the index finger with an arch shape with mild flexion position
Style B: The thumb and index finger are strongly bending.
Style C: Supporting the pen shaft with the thumb side face and index finger
Style D: The index finger is hanging over on the thumb.
Style E: the Thumb overlaps the index finger
Style F: How to hold that does not fall under a to b

The pen-holding method was classified 3 subjects (16%) in Style-A, 1 (5%) in Style-B, 4 (21%) in Style-C, 0 (0%) in Style-D, 7 (37%) in Style-E and 1(5%) in Style-F. And three (16%) were left-handed.

Electromyogram (EMG) measurements were performed on subjects classified as type C, E, and F and whose thumbs important for scaler operation were not in proper contact with the pen.

Electromyograms were inductively recorded from the first dorsal interdigital muscle (FDI) and flexor pollicis brevis muscle (FPB) involved in the scaler maneuver before and after training of the pen holding movement.

As the criterion for calculating the muscle activity level, the thumb and index finger were pushed so that both fingers did not separate from each other in the opposite position, and the pressing force was set to the maximum value (100%).

For training of pen-holding motion, a pen attached with Punyu Grip (KUTSUWA), in which depressions are present at the proper finger positions, was used. The duration of its use was 2 months and the subjects recorded the time of its use daily. Electromyogram was measured during mark sheet-filling motions in the longitudinal direction with a pen held usually and properly.

Electromyogram of scaler operation, wrist forearm motion and finger flexing movement using Skelton 9.0 mm (YDM Corporation) were measured.

For electromyogram measurement, EMG Master KM-104 (Mediarea Support) was used. Electrodes were attached to FDI and FPB in parallel to the muscle fiber and electromyograms were derived by bipolar lead. The acquired electromyograms were analyzed using a data recording/analysis system, ML846PowerLab4/26 (Bioresearch Center). The root mean square value (RMS) of data of measurement sections (2 s) with stable electromyogram waveforms was calculated.

3 Results

3.1 Pen-Holding Method

The thumb important for scaler operation did not properly contact with a pen in 12 subjects, accounting for 63% of all subjects.

3.2 Training Time

The daily time of using a pen with a grip for training was 4 h maximum and 30 min minimum, and the mean use time over 2 months per person was 42 min.

3.3 RMS

In the 12 subjects whose thumb did not properly contact with a pen, the mean RMS of the maximum pressing force at initiation was 0.304 mV ± 1.109 in FDI and 0.156 mV ± 0.164 in FPB, and those at completion were 0.362 mV ± 0.148 and 0.298 mV ± 0.170, respectively.

The mean RMS in usual and proper holding motions in FDI and FPB and those during forearm rotation in scaler operation and bending and extension of the fingers are shown in Table 1.

Table 1. RMS measurement result

Season	Measured muscle	Maximum pressing value	How to hold everyday	How to hold it right	Operation of the scaler	
					Wrist forearm motion	Finger flexing movement
Start time	FDI	0.304 ± 0.109	0.149 ± 0.085	0.106 ± 0.033	0.087 ± 0.033	0.095 ± 0.032
	FPB	0.156 ± 0.164	0.089 ± 0.054	0.073 ± 0.055	0.104 ± 0.085	0.117 ± 0.107
End time	FDI	0.362 ± 0.148	0.129 ± 0.051	0.104 ± 0.030	*0.128 ± 0.057	0.110 ± 0.055
	FPB	0.298 ± 0.170	*0.156 ± 0.080	*0.150 ± 0.089	0.140 ± 0.070	0.130 ± 0.062

FDI: first dorsal interosseous muscle FPB: flexor pollicis brevis muscle RMS ± SD *$p < 0.05$

RMS of FPB significantly increased during usual and proper pen-holding motions using the grip, and that without the grip tended to increase ($r = 0.091$). RMS of FDI significantly increased wrist forearm motion in scaler operation ($p < 0.05$).

The muscle activity level was calculated from the mean RMS of the maximum pressing force. The FPB muscle activity level finger flexing movement in scaler operation was 72.3% before initiation and it significantly decreased to 51.5% at 2 months after initiation ($p < 0.05$) (see Fig. 3).

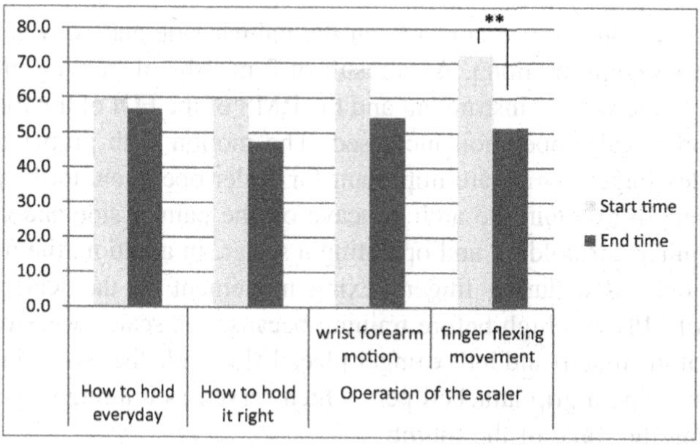

Fig. 3. Flexor pollicis brevis muscle (FPB) Muscle activity amount against pushing pressure

3.4 Relationship Between Pen-Holding Motion and Scaler Operation

In FPB, correlations were noted between the proper holding motion using the grip and wrist forearm motion in scaler operation (r = 0.587). A high correlation was found between the grip operation of an accurate holding method without using a grip and the forearm rotation operation (r = 0.828) of the scaler operation, the grasping motion of the correct holding method and the finger flexion and retraction motion (r = 0.841) ($p < 0.01$) (Table 2).

Table 2. Relationship between pen-holding motion and scaler operation

		Correct holding		Operation of the scaler	
		Using grip	No grip	Wrist forearm motion	Finger flexing movement
Correct holding	Using grip	1	0.667**	0.587**	0.411
	No grip	0.667**	1	0.828**	0.841**
Operation of the scaler	Wrist forearm motion	0.587**	0.828**	1	0.897**
	Finger flexing movement	0.411	0.841**	0.897**	1

In FDI, no correlation was noted between the proper holding motion and wrist forearm motion (r = 0.255) or finger flexing movement (r = 0.210) in scaler operation. There was a high correlation between the wrist forearm motion of the scaler operation and the finger flexing movement (r = 0.845) ($p < 0.01$).

4 Discussion

The hands form an arch shape concave on the palmar side and convex on the dorsal side to adapt to holding motion3. As a result of 2 months of training, the FPB of the grip operation of the writing instrument and the RMS of the FDI of the forearm rotation movement of the scaler operation increased. The motion with proper contact of the thumb and index finger, which are important for scaler operation, to the pen improved the muscle strength to retain the arch concave on the palmar side and convex on the dorsal side required for holding and operating a scaler. In addition, the muscle activity level of FPB decreased during finger flexing movement on the scaler. The muscle activity level of FPB was high before training because the scaler was strongly held by the fingertips of the thumb and index finger placed alongside the scaler, but it decreased after constantly using a grip-attached pen, which served as training of positioning and way of applying the force of the thumb.

A strong correlation was noted between the motion in the proper way of holding and wrist forearm motion by holding a scaler and RMS of FPB during, finger flexing movement suggesting that training of the proper pen-holding motion using the grip strengthened FPB.

5 Conclusion

(1) RMS of FPB significantly increased during the usual and proper pen-holding motions and that of FDI significantly increased during in wrist forearm motion scaler operation.
(2) The muscle activity level of FPB during finger flexing movement significantly decreased.
(3) In FPB, a strong correlation was noted between the proper holding motion and wrist forearm motion in scaler operation and between the proper holding motion and finger flexing movement.

It was suggested that the proper pen-holding motion using the grip serves as training of scaling operation.

References

1. Asaga Tamiyo, Asou Tomoko et al (2016) Relationship of the muscle activity of dental hygiene students when holding writing instruments/scalers. J Jap Soc Dent Hyg Educ 7 (2):97–102
2. Ioku Kazuki, Yamamoto Misako et al (2006) A study on the writing posture of university students. J Hyogo Soc Phys Educ Sports Sci 5:21–26
3. Takei J (ed) (2013) Visual encyclopedia of mechanisms of muscles and joints, Seitosha Co., Ltd., Tokyo, 70–71

Design, Understanding and Usability Evaluation of Connected Devices in the Field of Health: Contribution of Cognitive Psychology and Ergonomics

Noémie Chaniaud(✉), Emilie Loup-Escande, and Olga Megalakaki

EA7273 CRP-CPO - Université de Jules Verne Picardie, 80000 Amiens, France
Noemie.chaniaud@u-picardie.fr

Abstract. EHealth appears to be both an economic solution and a way to overcome the increase in health needs due to the ageing of the population and the increase of diseases. The presented project is aimed at developing individualized monitoring of patients who have undergone surgery and have been supported by a medical information system and connected devices. These connected systems have many benefits for healthcare system, but they also have disadvantages, relating to the complexity of their use. Their evaluation must be done in real work situations with health professionals and patients with different profiles (age, needs, skills, literacy, pathology, and expert/novice in technologies …). The aim of this work is to propose an integrative theoretical and methodological framework for a standardized evaluation of usability (including *efficiency, effectiveness, satisfaction* and *understanding*) in e-health in a user-centered design context. To do this, below is presented a state-of-the-art on usability evaluation. Finally, this work presents insights on how ergonomics and cognitive psychology approaches can contribute to greater understanding and the use of these devices in healthcare (patients, caregivers, family careers).

Keywords: EHealth · Usability · Understanding

1 Introduction

The demographic and economic landscape of health in most developed countries has to meet numerous challenges. The shortage of health care professionals and ageing populations is leading to an increase in pathologies (e.g. chronic diseases) cause a worrying congestion of hospitals. Thus, the healthcare system is turning to alternative economic and technological solutions commonly referred to as eHealth, in order to meet societal needs. EHealth (or "connected health") is defined as "*an emerging field in the intersection of medical informatics, public health and business, referring to health services and information delivered or enhanced through the Internet and related technologies*" [p. 1, 1]. The patient can perform his own care and send information from home to the hospital. Consequently, these medical devices keep patients connected to hospitals and thus have a decongestive impact on health care services. Studies [1, 2] even report that these tools improve the quality of health care. One of the reasons

S. Bagnara et al. (Eds.): IEA 2018, AISC 818, pp. 582–591, 2019.
https://doi.org/10.1007/978-3-319-96098-2_72

why eHealth makes the patient active and efficient by the self-management of their own health.

Connected devices have many benefits for the healthcare system, but there are also current disadvantages. These emerging technologies have confronted hurdles such as; the complexity of their use, the reliability, data security and the resulting socio-economic transformations [3]. Consequentially many projects are emerging to design and evaluate these medical devices. According to Borycki and Kushniruk [4], ergonomic evaluations should be mandatory. Designers have an important responsibility and face a crucial challenge.

This is the case of the Smart Angel project. This collaborative project is co-designing a patient monitoring system by involving all the project's stakeholders (research laboratory, companies and hospitals). This medical device includes interfaces (for patients and nurses) but also connected objects. This project has therefore produced a doctoral thesis to assist the design and ergonomic evaluations of its interfaces. The aim of this research is to propose improvements to medical devices to promote their usability and avoid rejection by patients or healthcare workers. To do this, reliable and holistic model and effective methods such as user-centered design methods must be provided to develop a system that is accessible to the social and cognitive specificities of each patient. It is also necessary to evaluate these medical devices on different ergonomic dimensions, such as the usability, as well as psychological dimensions as understanding dimension. This last dimension will be studied to understand the characteristics that influence it, such as health literacy or the design of the user manual.

The purpose of this paper is to propose thoughts in order to design a forthcoming holistic usability model. Ultimately, this theoretical and methodological framework will make it possible to design devices from a user-centered perspective.

2 Design and Evaluation of Usability in Field of EHealth

Since users have been at the center of the design process, many approaches to design and evaluation have emerged. Ergonomic assessments are progressively included into a number of design phases. The most widespread of evaluations in ergonomics are made through three famous dimensions that are acceptability, usefulness and usability. We focus below on the usability dimension because an important need emerges from literature in the field of health. The success of implementation in these systems highly depends on the degree of usability.

2.1 Usability

Usability is defined by ISO [5] based on three distinct aspects: *effectiveness*, the accuracy and completeness with which users attain certain aims; *efficiency*, the relation between the accuracy and the resources expended in attaining them; and *satisfaction*, the users' comfort and assessment of interaction. The relationships between these three aspects are still poorly understood, and studies have not yet proved any correlation between these three dimensions [6]. That's why, the usability has to be assessed in its whole, i.e. by separately assessing its three dimensions [7] but also by using a combination of them [8].

A large number of evaluations already exist: (1) inspections via experts such as heuristic evaluations [7] or (2) usability tests with future users reproducing tasks representative of the system.

These assessments can be done during and after system design. There are four kinds of usability assessment:

- The subjective evaluation: the aim of this evaluation is to detect usability problem with a subjective point of view [9]. An example of its methods is the interview, the questionnaire and or the focus group.
- The objective evaluation: its purpose is to obtain results independently of user's personals opinions [10]. Observations or eye tracking can be used.
- The summative evaluation: the final goal is to deliver a certification. Da costa et al. [11] proposes a certification for medical technologies (IEC 62 366). This standard includes both evaluations in the usability assessment processes: risk management and quality management.
- The formative evaluation: This is an assessment to detect problems in the design phase [12].

2.2 Usability in eHealth

The methods presented below are also applied to the eHealth fields but studies report significant difficulties around the usability of these devices. The main problem reported by systematic reviews on the usability of health technologies is the lack of a framework and standardization of usability studies [14]. One of the major challenges in the development of medical devices is the design of user friendly interfaces. To enable health tools to be more adapted, the usability dimension must be present throughout the design process. This requires during the early design phases to include usability experts [15] but also usability tests with the target population [8].

It has been demonstrated in other fields, such as website design, the usability of a technology can be important because it can improve practices, the customer fidelity and thus be more economical for companies [15]. In that respect, a developed design is highly important for home medical equipment because users may have cognitive specificities which may include; pain, disability, high stress, limited training, lack of knowledge or support [p. 3, 16] quoted in [17]. Moreover, in the medical field as Kortum and Peres [18] claim it: "a lack of usability can cost lives!". Usability is therefore critical in the design of medical devices. Despite these warnings, these technologies are not to this day suitable enough. For example, Fairbanks et al. [19] have shown that a defibrillator cannot be used by untrained people because it can injure the patient (or worse).

Studies by Kortum and Peres [18] have evaluated the subjective usability of many health devices such as the blood glucose monitor (69,59 SUS score), blood pressure monitor (73,56) and even the pregnancy test kit (63,74) by offering a simple online SUS (system usability scale) test to a large sample of students (N = 271). The authors compared these results with those of Kortum and Bangor [20] on domestic equipment.

They showed that a microwave (86 SUS score) or Gmail (84) scored significantly higher in subjective usability than the most usable medicals devices, the thermometer (80,53). Even if the authors evaluate only one of the usability dimensions, they show the lack of satisfaction about these tools. Thus, we can deduce that conventional usability models are not sufficient for eHealth.

2.3 Usability and Health/eHealth Literacy

Another dimension to take into account especially in the field of Health is the health literacy. The Medical Library Association [21] define health information literacy as "*the set of abilities needed to recognize a health information need, identify likely information sources and use them to retrieve relevant information, assess the quality of the information and its applicability to a specific situation, and analyze, understand, and use the information to make good health decisions.*" Systematic studies [22] have shown that patients with good literacy have better health outcomes and are more successful in recovery. The patient has indeed a better understanding of his treatment and of the package leaflet of the medicinal product. To address the lack of patient literacy, the designer must provide patients with enough information for them to understand these medical devices. This is what Monkman and Kushniruk [23] propose in their model called "The Consumer Health Information System Adoption Model" specifically applied to eHealth. To build it, they consider that understanding a system results in two factors: usability and health literacy.

In others studies, with the same model, the authors evaluate various devices (Diagnostic Imaging [24], blood pressure application [25]) with eleven heuristics specifically designed with the health and eHealth literacy dimension. To assess these devices, they review each page and give them a degree of severity according to each heuristic of their model. they argue that the low literacy level of some users should be taken into account in device design because these individuals may use health data differently or even misinterpret it. Devices should therefore help individuals to better understand the system by providing glossaries, or explanations about blood pressure for example [26].

Although the link between health literacy and usability seems obvious, it has never been proven. The authors [24–26] propose a model based on the heuristics aforementioned but no usability test with end-users. In addition, to perform user tests, it is also necessary to know when it is wise to do them. The first thing to do when designing a device as complex as a medical device is to take users and their needs into account in the design process.

2.4 Usability and User Needs

There is currently consensus on the idea that technology must meet the user's needs [27]. This requires the adoption of user-centered design methods [28]. There are many examples of approaches in the scientific literature [28, 29]. These studies based on user-centered design methods have shown that the created devices have greater usability because they take into account end-user constraints. This also makes it possible to propose universal recommendations for these devices [29]. The purpose is to identify these constraints, requirements, needs, preferences and cognitive or social specificities for future users. Several studies propose user-centered design models [29–31]. They take the form of iterative processes where at the end of each iteration, an evaluation is set up. Patients, caregivers, health professionals and others are involved in these assessments, adding further specificity to future the study of health assessment models. Thus, they must be collaborative.

Health personnel is often missing out on eHealth education and literature. Studies show that the focus is predominantly laid onto the patient in terms of self-management or self-monitoring and patient self-sufficiency. However, Andersen et al. [32] demonstrate through their data collected between therapists and patients the importance of collaboration on patient motivation. This should not be called self-monitoring but co-monitoring [33]. Behind each medical device, there are caregivers taking care of the patient's health. This pair must be part of the design process and involves a collaborative and evaluative approach. The great complexity of these objects requires a close the follow-up of a professional. Unfortunately, this cannot always be the case. The patient should therefore have the equivalent skills as a health professional. Obviously, this is idealistic, and this is where ergonomic methods allow us to approach this idea.

2.5 Synthesis: Usability, User Needs and Health Literacy

We have seen above that previous studies have focused on current usability failures. These failures are due to the lack of a reliable and validated usability model applied to eHealth characteristics. However, there is a consensus that health literacy is a fundamental component of usability. Even if this link has never been proven, this relationship has emerged from user needs studies. Are there other crucial dimensions to be taken into account to improve usability? We postulate that the basic user need is understanding a dimension that includes health literacy but also other dimensions such as the understanding of content.

3 Discussion: How to Improve Usability in eHealth?

One of the main complexities when designing medical technologies is their adaptation to all user profiles. Even within the elderly people, there are many specificities much broader than for a young user. There is indeed a multitude of generations, with a diversity of cognitive and physical abilities, health status, background, and technological experience. There is a special discipline for the applied design of technology targeting the elderly, called "gerontechnology" [34].

Having presented difficulties met with health devices, these are not necessarily related to the ease of use of the system. This may be hypothetically due to a lack of health literacy [26]. The blood pressure monitor, for example, may be more difficult for older people to use, not only because these users are less familiar and comfortable with these devices, but also because it requires prior knowledge about blood pressure to interpret the data [26], and because users must remember the sequence of tasks to use them correctly [17]. Thus, there is an important need for the design of a new usability model in eHealth taking into account all the dimensions presented above. We propose below some tracks for this new model, which will be our future object of study.

3.1 Understanding and eHealth

Understanding is difficult to define because this cognitive process is relative to its context. It can be linked to an object such as eHealth technologies, to a content, and to a situation. Understanding mobilizes knowledge that requires to be restored and manipulated. When we talk about understanding in the field of connected health, several issues arise. The understanding is done at different levels, these include; does the user understand what he has to do to use the object, does he understand the interest of using this object, is the user able to interpret the measurements given by the medical device? To answer these questions, we have to wonder what skills and knowledge are needed to use a health device? It is important to understand users' abilities and resources to engage in technology [35]. To do that, Kayser et al. [36] have developed a method based on the concept of Kushiniruk and Turner [37] of the "user-task-context" matrix. This matrix includes three dimensions: task dimension (for example the complexity and difficulty of the task), user dimension (includes age, gender, motivation, illness, and computer experience), and finally health context dimension (for example the geographical location, and level of emergency). The cause of the misunderstanding must be diagnosed. Is it related to the individual's lack of knowledge or to the object itself? This is why we differentiate between understanding of content and individual related understanding such as literacy.

Understanding of Content: The Evaluation of Procedural Documents
Usability already includes a dimension called understandability, with the frequency of using help and documentation or with the time spent using help and documentation [38]. This usability sub-dimension assesses procedural documents that also play an important role in understanding the material. This "understanding of content" is particularly developed in the field of language processing in cognitive psychology [39]. For application of this field, Mykityshyn, Fisk and Rogers [17] demonstrated that it was advisable to focus on video presentations during training sessions on these devices, especially for the elderly, because it avoids cognitive overload. These cognitive assessments should be systematic to check if the misunderstanding does not come from the user manual [40].

Understanding and Users: Literacy and Engagement
The specificity in the health field is that the individual may well know how to use a connected object or health application, but not be aware of his inability to understand what he is doing and why he is doing it. To agree to use a connected health device, the user has to be engaged, and for this purpose, he must understand it. This idea can be found in the characteristics of commitment [41]. Greene and Hibbard [42] have shown (on a sample of 25,047 adult patients) that activation and engagement would have a potential impact on health outcomes even though these studies require further investigation. Barcello et al. [43] explain that if active patients have better outcomes than inactive patients it is because they are able to make sense of their care experience.

To allow patients to be engaged, we must therefore give them all the available tools, and to makes sense of their disease [41]. For this, they should be instructed to a good level of education and have sufficient knowledge of health, anatomy and well-being (this means providing them with a good level of health literacy). Conversely, we can take into consideration the low levels of health literacy of users during the design stages and include specific aids tailored for each in order to interpret the instructions [23]. The current problem is that no model is really reliable and validated to prove the relationships between literacy and usability.

The standardized evaluation of these devices becomes a priority. Recent studies [32, 33] have attempted to bring the understanding dimension into their usability experience. Global understanding [32] has been integrated as a set of questions (QU) into the usability technique, CUT (cooperative usability testing). In this study, understanding cannot be fully appreciated, because it is addressed through only subjective evaluation (interviews).

We therefore postulate that the *comprehension* dimension needs a framework and to be integrated into usability models at the same level as the three dimensions of usability (*effectiveness, efficiency, satisfaction*) in the field of eHealth (see Fig. 1.). The understanding could be evaluated on specific criteria: criteria related to the device such as the user manual, but also criteria relating to users' skills such as their health literacy level, their commitment, and their ease of use of technologies.

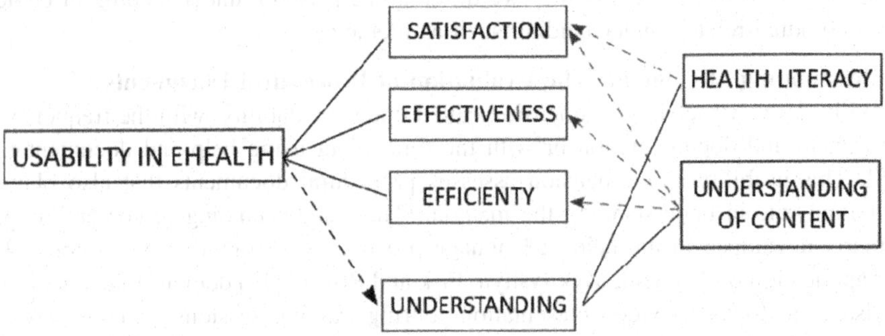

Fig. 1. Prospecting for the new usability model including understanding

4 Conclusion

EHealth is a particular field of study that has specific design and evaluation essential needs. To manage these needs, designers must rely on a user-centered approach, focusing on the user interface of the product. However, systematic usability studies reveal a lack of standardization in current findings because there is no reliable and validated usability model to include all the specificities of the end-user in the field of eHealth. This model has to take into account the use needs such as the level of health literacy, the user profile (patients vs medical professionals) and the context of use (home vs hospital).

During the manipulation of home medical devices, the patient has to be trained in a multitude of skills. We consider that patient understanding is fundamental to the use the device and will include many variables, one of which being health literacy. On the other hand, the patient acknowledgment will largely be dependent on the quality of the presentation, and the formatting of the user manual. All the points raised above narrow down the study towards a clear link between understanding and usability.

Future studies will attempt to better understand the relationship between these two fields, usability through these three dimensions (*efficiency, effectiveness, satisfaction*) and *understanding* including these specific dimensions prospected above. To this end, we propose two research topics. The first direction will aim to demonstrate the links between usability and understanding relating to the individual (i.e. literacy level). Whilst the second topic will focus on usability based on the understanding of the material presented (such as the user manual).

Acknowledgments. This project is financially supported by FEDER, EVOUCARE and by the Programme d'Investissements d'Avenir (PIA) under the programme of «Projets de Recherche et Développement Structurants pour la Compétitivité» PSPC 5.

References

1. Eysenbach G (2001) What is e-health? J Med Internet Res 3(20):1–2
2. Sharry J, Davidson R, McLoughlin O, Doherty G (2013) A service-base evaluation of a therapist-supported on line cognitive behavioral therapy program for depression. J Med Internet Res 15(6). https://doi.org/10.2196/jmir.2248
3. Kjeldskov J, Graham C (2003) A review of mobile HCI research methods 9(8):317–335. https://doi.org/10.1007/978-3-540-45233-1_23
4. Borycki E, Kushniruk A (2005) Identifying and preventing technology-induced error using simulations: application of usability engineering techniques. Healthc Q 8:99–105. https://doi.org/10.12927/hcq.17673
5. ISO 9241-11:1998 (1998) Exigences ergonomiques pour travail de bureau avec terminaux à écrans de visualisation (TEV) – Partie 11: lignes directrices relatives à l'utilisabilité
6. Frøkjær E, Hertzum M, Hornbæk K (2000) Measuring usability: are effectiveness, efficiency, and satisfaction really correlated? In: Proceedings of the SIGCHI conference on human factors in computing sytems, CHI 2000, SIGCHI 2000, pp 345–352. The Netherlands, The Hague. https://doi.org/10.1145/332040.332455
7. Nielsen J (1993) Usability engineering. Academic Press, Cambridge
8. Jaspers M (2009) A comparison of usability methods for testing interactive health technologies: methodological aspects and empirical evidence. Int J Med Inf 78(5):340–353. https://doi.org/10.1016/j.ijmedinf.2008.10.002
9. Nishiuchi N, Takahasi Y (2015) Objective evaluation method of usability using parameters of user's fingertip movement. In: Gavrilova M, Tan C, Saeed K, Chaki N, Shaikh S (eds) Transactions on computational science XXV. Lecture notes in computer science, LNCS, vol 9030, pp 77–89. Springer, Heidelberg. https://doi.org/10.1007/978-3-662-47074-9_5
10. Bernsen N, Dybkjaer L (2009) Multimodal usability. Springer, Berlin

11. Bras Da Costa S, Beuscart-Zéphir M, Bastien J, Pelayo S Usability and safety of software medical devices: need for multidisciplinary expertise to apply the IEC 62355. Stud Health Technol Inf 216:353–357 (2007)
12. Hartson R (2003) Cognitive, physical, sensory and functional affordances in interaction design. Beh Info Tech 22(5):315–338. https://doi.org/10.1080/01449290310001592587
13. Peute L, Driest K, Marcilly R, Bras Da Costa S, Beuscart-zephir M, Jaspers MA (2013) A Framework for reporting on human factor/usability studies of health information technologies. Stud Health Technol Inf 194:54–60. https://doi.org/10.3233/978-1-61499-293-6-54
14. Peute L, Spithoven W, Bakker P, Jaspers M (2008) Usability studies on interactive health information systems; where do we stand? eHealth Beyond the Horizon 136:327–332. https://doi.org/10.3233/978-1-58603-864-9-327
15. Marcus A (2005) User interface design's return on investment: examples and statistics. In: Bias R, Mayhew D (eds) Cost justifying usability: an update for the internet age, pp 17–39. Elsevier, Amsterdam. https://doi.org/10.1016/B978-012095811-5/50002-X
16. Klatzky RL, Kober N, Mavor A (1996) Safe, comfortable, attractive, and easy to use: improving the usability of home medical devices. National Academy, Washington, DC. https://doi.org/10.17226/9058
17. Mykityshyn AL, Fisk AD, Roger WA (2002) Learning to use a home medical device: mediating age-related differences with training. Hum Factors 44(3):354–364. https://doi.org/10.1518/0018720024497727
18. Kortum PT, Peres C (2015) Evaluation of home health care devices: remote usability assessment. JMIR Hum Factors 2(1):1–9. https://doi.org/10.2196/humanfactors.4570
19. Fairbanks RJ, Caplan S (2004) Poor interface design and lack of usability testing facilitate medical error. Joint Comm J Qual Saf 30(10):579–584. https://doi.org/10.1016/S1549-3741(04)30068-7
20. Kortum PT, Bangor A (2013) Usability ratings for everyday products measured with the system usability scale. Int J Hum-Comput Inter 29(2):67–76. https://doi.org/10.1080/10447318.2012.681221
21. Medical Library Association: Health Information Literacy. http://www.mlanet.org/resources/healthlit/define.html. Accessed 21 April 2018
22. Berkman ND, Sheridan SL, Donahue KE, Halpern DJ, Crotty K (2011) Low health literacy and health outcomes: an updated systematic review. Ann Intern Med 155(2):97–107. https://doi.org/10.7326/0003-4819-155-2-201107190-00005
23. Monkman H, Kushniruk AW (2015) eHealth literacy issues, constructs, models, and methods for health information technology design and evaluation. KM&EL 7(4):1–6
24. Griffith J, Monkman H (2017) Usability and eHealth literacy evaluation of a mobile health application prototype to track diagnostic imaging examinations. Stud Health Technol Inf 234:150–155
25. Monkman H, Griffith J, Kushniruk AW (2015) Evidence-based heuristics for evaluating demands on eHealth literacy and usability in a mobile consumer health application. Stud Health Technol Inform 216:358–362. https://doi.org/10.3233/978-1-61499-564-7-358
26. Monkman H, Kushniruk AW (2013) A health literacy and usability heuristic evaluation of a mobile consumer health application. Stud Health Technol Inf 192(1):724–728. https://doi.org/10.3233/978-1-61499-289-9-724
27. Piau A, Campo E, Rumeau P, Vellas B, Nourhashémi F (2014) Aging society and gerontechnology: a solution for an independent living? J Nutr Health Aging 18(1):97–112. https://doi.org/10.1007/s12603-013-0356-5
28. Greenhalgh T, Russel J (2010) Why do evaluations of eHealth programs fail? An alternative set of guiding princiaples. PLoS Med 7(1000360). https://doi.org/10.1371/journal.pmed.1000360

29. Matthew-Maich N, Harris L, Ploeg J, Markle-Reid M, Valaitis R, Ibrahim S, Gafni A, Isaacs S (2016) Designing, implementing, and evaluating mobile health technologies for managing chronic conditions in older adults: a scoping review. JMIR mHealth and uHealth 4 (2). https://doi.org/10.2196/mhealth.5127
30. Schnall R, Rojas M, Bakken S, Brown W, Carballo-Dieguez A, Carry M, Gelaude D, Moley J, Travers J (2016) A user-centered model for designing consumer mobile health applications. J Biomed Inf 60:243–251. https://doi.org/10.1016/j.jbi.2016.02.002
31. LeRouge C, Wickramasinghe N (2013) A review of usercentered design for diabetes-related consumer health informatics technologies. J Diabetes Sci Technol 7(4):1039–1056. https://doi.org/10.1177/193229681300700429
32. Andersen SB, Rasmussen CK, Frøkjær E (2017) Bringing content understanding into usability testing in complex application domains - a case study in eHealth. Des User Experience Usability Theor Methodol Manage 10288:327–341. https://doi.org/10.1007/978-3-319-58634-2_25
33. Huvila I (2016) Taking health information behaviour into account the design of e-health services. Finnish J eHealth eWelfare 8(4):153–165
34. Bjering H, Curry J, Maeder A (2014) Gerontechnology: the importance of user participation in ICT development for older adults. Stud Health Technol Inform 204:7–12
35. Karnøe Knudsen A, Kayser L (2016) Validation of the eHealth literacy assessment tool (eHLA). Inter J Integr Care 16(6):1–8. https://doi.org/10.5334/ijic.2897
36. Kayser L, Kushniruk A, Osborne RH, Norgaard O, Turner P (2015) Enhancing the effectiveness of consumer-focused health information technology systems through eHealth literacy: a framework for understanding users' needs. JMIR Hum Factors 2(1):1–13. https://doi.org/10.2196/humanfactors.3696
37. Kushniruk A, Turner P (2012) A framework for user involvement and context in the design and development of safe e-Health. Stud Health Technol Inform 180:353–357
38. ISO/IEC TR 9126-2 (2003) Software engineering: product quality Part 2: External metrics
39. Jamet E (2008) La compréhension des documents multimédias: de la cognition à la conception. Solal, Marseille
40. Ganier F, Querrec R (2012) TIP-EWE: a software tool for studying the use and understanding of procedural documents. IEEE Trans Prof Commun 55(2):106–121. https://doi.org/10.1109/TPC.2012.2194600
41. Graffigna G, Barello S, Riva G (2013) Technologies for patient engagement. Health Aff 32 (1172). https://doi.org/10.1377/hlthaff.2013.0279
42. Greene J, Hibbard JH (2012) Why does patient activation matter? an examination of the relationships between patient activation and health-related outcomes. J Gen Intern Med 27:520–526. https://doi.org/10.1007/s11606-011-1931-2
43. Barello S, Triberti S, Graffigna G, Libreri C, Serino S, Hibbard J, Riva G (2016) EHealth for patient engagement: a systematic review. Front Psychol 6(2013):1–13. https://doi.org/10.3389/fpsyg.2015.02013

The Effectiveness of Cardiac Rehabilitation Patient Education Through Various Media

Chi-Ping Tseng[1], Yu-Lin Lai[1], and Eric Wang[2](✉)

[1] Department of PMR, National Taiwan University Hospital Hsin-Chu Branch, NO. 25, Lane 442, Sec. 1, Jingguo Rd., Hsinchu 300, Taiwan, R.O.C.
[2] Department of IEEM, National Tsing Hua University, No. 101, Section 2, Kuang-Fu Road, Hsinchu, Taiwan, R.O.C.
mywangeric@gmail.com

Abstract. Background: Ischemic heart disease (IHD) is one of the leading cause of death globally for the past 15 years according to the updated data from World Health Organization in January 2017. Deaths caused by IHD increased by an estimated 41.7% from 1990 to 2013. Therefore, in addition to medical procedure improvement, continuing efforts should be put in the prevention and control of heart disease by actively educating people and promoting self-health management. For patients with coronary artery disease, cardiac rehabilitation may reduce its recurrence and mortality. By practicing the cardiac rehabilitation, the rates of subsequent coronary events would be decreased and the morbidity and the total mortality would be reduced as well. Patient education is also an important part of the cardiac rehabilitation, while its effectiveness through various media may be different. **Purpose:** To explore and compare the effectiveness of cardiac rehabilitation patient education among three types of media, text only, text and pictures, and video. **Methods:** This study was divided into two stages. The first stage was the testing of validity of "Cardiac Rehabilitation Awareness Scale (CRAS)" done by five experts. The second stage was the test of the effectiveness of various media by 90 subjects using the CRAS. The subjects were divided into three groups and each group used one type of patient education medium. The subjects carried out the tests according to five different age levels. **Results:** 1. The validity of "Cardiac Rehabilitation Awareness Scale" was confirmed as a measurement tool of the effectiveness of cardiac rehabilitation patient education. 2. There was no significant difference in the effectiveness of patient education of the three media ($P = 0.6001$) but "video" received highest subjects preference. **Conclusion:** It is suggested that currently for clinical application, the "text and pictures" material may be used to educate the IHD patients for its convenience, low cost and preference. With the development of information and communication technology, the "video" may be the better option in the future.

Keywords: Cardiac rehabilitation · Patient education · Media

S. Bagnara et al. (Eds.): IEA 2018, AISC 818, pp. 592–599, 2019.
https://doi.org/10.1007/978-3-319-96098-2_73

1 Introduction

1.1 Background

Ischemic heart disease (IHD) is at the first place among the Top 10 causes of death worldwide in 2015 according to the data from World Health Organization updated in January 2017. There are 8.76 million deaths caused by IHD (15.5% of 56.4 million total death toll) in 2017. IHD is also one of the leading causes of death globally in the last 15 years [1]. Since the medical expenditure for IHD in the US went up to $108.7 billion in 2011–2012 [2], the American Heart Association (AHA) proposed a new concept of cardiovascular health, characterized by seven metrics including diet quality, physical activity, smoking, body mass index, blood cholesterol, blood pressure and blood glucose. The goal of this initiative is to improve the cardiovascular health of all Americans by 20% and to reduce death from IHD by 20% by 2020 [3].

Hence, we should continue our efforts in the prevention and control of heart disease by actively educating people and promoting self-health management. For IHD patients, it is important to reduce their recurrence and mortality.

Cardiac rehabilitation can improve the quality of IHD patient's life and reduce mortality rate by 36% to 38% [4]. It plays a very important role in the secondary prevention of IHD [5]. The focus of secondary prevention is to change the behavior and lifestyle of patients through education to reduce mortality and re-hospitalization rates [6]. Patient education is an important part of the cardiac rehabilitation. It has been proven to effectively promote healthy behavior change and health-related quality of life [7].

Researches on patient education are broadly divided into four categories: (1) Improving the health literacy of at-risk groups. (2) Improving communication between patients and medical professionals. (3) Changing the health care system. (4) Changing the design of educational media [8]. In this study, we will focus on the scope of category 4.

In addition to the commonly used text only or text and pictures educational media, the patient education video has become more and more common in recent years. Although medical professionals have more choices of various media while conducting patient education, it would be important to know which of the media is most effective for the patient education. If the most appropriate medium is used for the patient education, improving health condition could be expected.

1.2 Purpose

To explore and compare the effectiveness of cardiac rehabilitation patient education among three types of media: text only, text and pictures, and video.

2 Methods

This study was divided into two stages. The first stage was the testing of validity of "Cardiac Rehabilitation Awareness Scale (CRAS)" done by five experts. The second stage was the test of the effectiveness of various media by 90 subjects using the CRAS. The subjects were divided into three groups and each group used one type of patient education medium. The subjects carried out the tests according to five different age levels.

2.1 The First Stage

As shown in Fig. 1, three types of media (text only, text and pictures, and video) for cardiac rehabilitation patient education were used in the first stage of the study. There were three parts in our educational media: "the purpose of cardiac rehabilitation", "the exercise program in cardiac rehabilitation" and "the adjustment and control of coronary artery disease risk factors". The content of the patient education materials of the three media is the same.

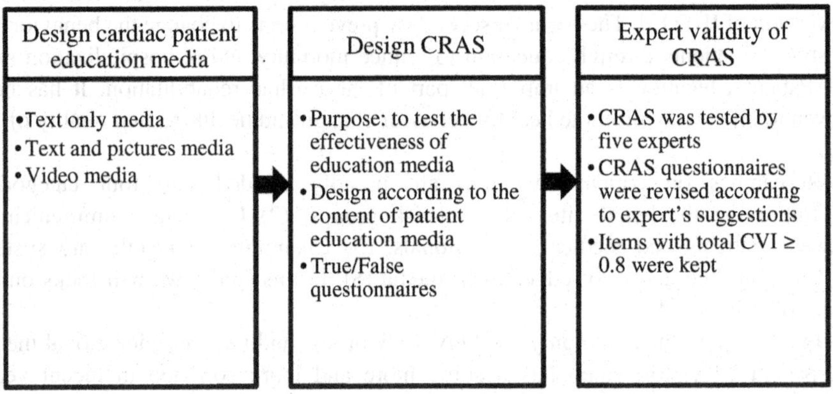

Fig. 1. Flow diagram of the first stage

The "Cardiac Rehabilitation Awareness Scale (CRAS)" was designed according to the content of cardiac rehabilitation patient education. The validity of the CRAS was confirmed by five experts who were two cardiologists, one rehabilitation physician, and two physical therapists. The experts judged the breadth and appropriateness of the questionnaires in CRAS. Meanwhile, we gave three types of educational media to experts for their references.

According to the experts' suggestions, the CRAS questionnaire were revised with the items of a Total CVI (Content Validity Index) of 0.8 or higher were kept. The CRAS was used for measuring the effectiveness of cardiac rehabilitation patient education materials with different media.

2.2 The Second Stage

Ninety (90) subjects were recruited through advertisement at the second stage of this study. All the subjects were informed about the purpose and procedures of our study, and a consent form was signed before participation. The study protocol was approved by the local IRB (Institutional Review Board). Then the subjects filled in the form of personal data including sex, age, education level, actual information sources and preferred information sources.

At first, all subjects were allocated to their individual age group (20–34y/o, 35–44y/o, 45–54y/o, 55–64y/o, ≥ 65y/o) then sequentially assigned to three media groups (text only, text and pictures, video) (Fig. 2). The subjects in each group read one type of patient educational medium. After 10 min of reading/watching, the subjects played specified mental poker card game for 5 min. Then, they answered the CRAS questionnaires. The flow diagram of this stage is shown in Fig. 3.

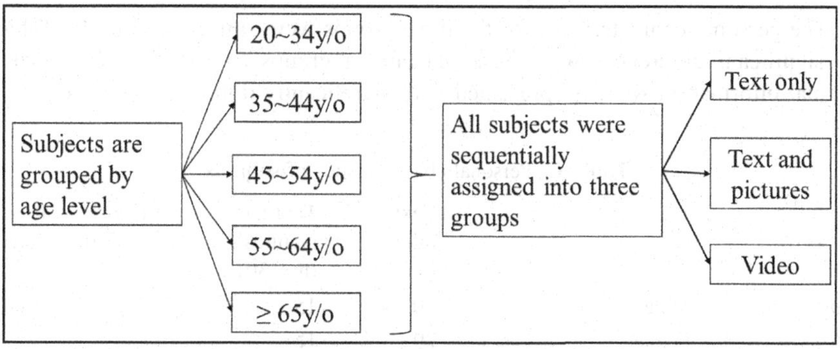

Fig. 2. Grouping process of subjects

3 Data Analyses and Results

3.1 The Validity of "Cardiac Rehabilitation Awareness Scale"

The Total CVI of CRAS was calculated, and its value was 1. All of the questionnaires in CRAS were retained. The CRAS was used to test the effectiveness of cardiac rehabilitation patient education.

3.2 The Effectiveness of Cardiac Rehabilitation Patient Education of the Three Media

There were 90 subjects participated in this study. The majority of our subjects were female (57.78%), and most of our subjects were 20–34 y/o (34.44%). About their education level, most of them were university graduates (47.78%). Last, most of our subjects got information from the Internet (61%), while their preferred information sources were "video" (43.33%) and "text and pictures" (35.56%).

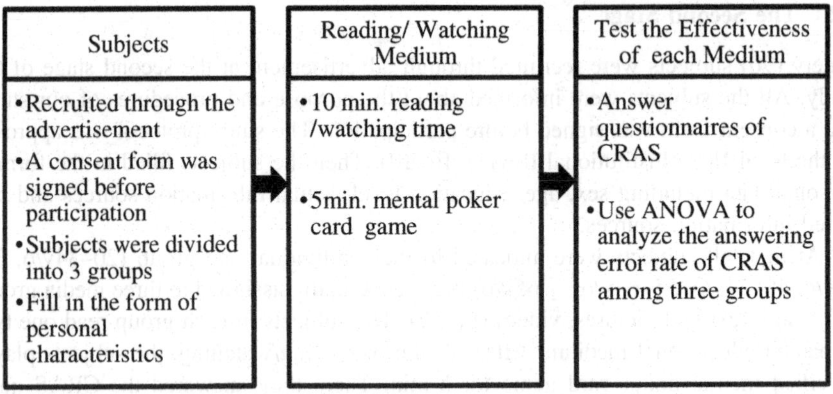
Subjects
•Recruited through the advertisement
•A consent form was signed before participation
•Subjects were divided into 3 groups
•Fill in the form of personal characteristics
Reading/ Watching Medium
•10 min. reading /watching time
•5min. mental poker card game
Test the Effectiveness of each Medium
•Answer questionnaires of CRAS
•Use ANOVA to analyze the answering error rate of CRAS among three groups

Fig. 3. Flow diagram of the second stage

The personal characteristics for the three groups were summarized in the Table 1. No significant differences were observed among groups ($p > 0.05$) except that the "Actual information sources" presented with significantly difference ($p < 0.05$).

Table 1. Personal characteristics of subjects

		Text only (n = 30)	Text and pictures (n = 30)	Video (n = 30)	P value
Sex	Male	10	15	13	0.4209
	Female	20	15	17	
Age	20–34y/o	11	10	10	0.9440
	35–44y/o	4	5	5	
	45–54y/o	8	7	8	
	55–64y/o	2	4	5	
	≥ 65 y/o	5	4	2	
Education level	Uneducated	0	0	0	0.7392
	Elementary	2	0	1	
	Junior High	1	1	0	
	High School	5	3	3	
	University	13	13	17	
	Master/Doctor	9	13	9	
Actual information sources	Newspapers/magazines	1	5	4	0.0212
	Television	10	3	4	
	Internet	14	18	22	
	Others	5	3	0	
Preferred information sources	Text	2	8	8	0.2203
	Text and pictures	14	9	9	
	Video	13	13	13	

There was no significant difference in the effectiveness of cardiac rehabilitation patient education through analysis of variance (ANOVA) among these three subject groups (Table 2). However, the percentage of "wrong + uncertain" in the text and pictures group was lower. The education level was a significant factor (P = 0.0023) that affected the percentages of "wrong answer" through linear regression analysis. The higher the education level was, the less the percentages of "wrong answer" was.

Table 2. The effectiveness of cardiac rehabilitation patient education: result of CRAS

	Text only	Text and pictures	Video	P value
Percentages of "wrong answer"	10.93%	8.52%	8.33%	0.3617
Percentages of "uncertain"	3.89%	3.52%	4.81%	0.8002
Percentages of "wrong + uncertain"	14.81%	12.04%	13.15%	0.6001

4 Discussion

4.1 Personal Characteristics of Subjects

Since our subjects were recruited around the university campus, most of our subjects were 20–34 y/o (34.44%). For education level, more than 80% of these participants were above university graduates. However, the education level above university graduates of Taiwanese was 43.31% during 2014–2016 [9]. Compared with the above data, the education level of our subjects was even higher obviously. The major information resources of our subjects were from the Internet (61%). The possible reason behind that was our subjects were most from higher education group and younger generation. Why the preferable information source was "video" (43.33%) and "text and pictures" (35.56%)? It was because that the available information sources mostly came with "video" or "text and pictures" in addition to "text" content. Pure text content usually can't attract people's attention.

4.2 The Effectiveness of Cardiac Rehabilitation Patient Education

There was no significant difference in the effectiveness of cardiac rehabilitation patient education among these three subject groups. Our result was similar as the study result from Frieske and Park (1999). In their study, they found there were no significant differences on the recall of content among young age (20–40y/o) and old age (60–80y/o) for various media (print, audio or TV) [10]. However, our result was different with the study result from Harskamp [11]. Harskamp's study found the teenager's (16–17y/o) learning effectiveness through "video" was better than through "text" and "text and pictures". But Mayer's (2005) study toward university students found the learning effectiveness through "text" and "text and pictures" was better than through "video" [12].

According to the paper from Wilson and Wolf (2009), the advantages and disadvantages of enhanced print and video-based health material was listed [8]. We could expect that the effectiveness of video medium with manipulation procedures was better.

Since our cardiac rehabilitation patient education did not contain manipulation procedures, there were no significant differences between video medium and other media. Besides, we also expected bad effectiveness if we applied text only medium to low literacy people because of their difficulties of reading and comprehending text material. There were 95% of our subjects have high school or higher degrees, so poor effectiveness of low literacy didn't apply to our study.

Our study indicated around 80% of our subjects preferred "video" and "text and pictures". Although video and text and pictures media didn't have obvious learning effectiveness advantages compared with text medium in our study, learning with preferred media would yield out better performance. Based on the cognitive theory of multimedia learning, visual and verbal presentation would enhance active learning to integrate people's prior knowledge and then become long-term memory [13]. Therefore, people could learn to change their behavior and lead to a healthier life.

4.3 Methods Limitations

The CRAS was used to measure the effectiveness of cardiac rehabilitation patient education in this study. However, since no pre-test for learning was conducted before the subjects were reading education media, hence we couldn't really assure whether it's the learning effect or not. In the future study, we should conduct pre-test to confirm the effectiveness of learning.

4.4 Clinical Application

The results of this study show that there is no significant difference in the effectiveness of cardiac rehabilitation patient education through three types of media. However, the longest time and most human resources are required for creating video contents while less efforts for "text and pictures" medium and the least efforts for text only medium. As regard to the execution convenience for clinical staffs, video medium needs a device to play, while the "text only" and "text and pictures" media need papers only. At present, most patients with coronary artery disease (CAD) are elders usually with less education; the text only medium is a burden for them and reduces the willingness of learning. Text and pictures medium of the patient education can enhance the reader's comprehension and memory, while video medium may cause difficulties in the execution of playing video for elderly patients. To sum up, we suggest to choose the text and pictures medium to educate the CAD patients to learn cardiac rehabilitation knowledge.

5 Conclusion

This study investigated the effectiveness of cardiac rehabilitation patient education through various media. The result showed that there is no significant differences on the effectiveness among three types of media (text only, text and pictures and video). For now, it may be acceptable to use "text and pictures" medium for practical implementation of cardiac rehabilitation patient education. With the highest subjects preference and the development of the information and communication technology, the "video" medium may be the better option in the future.

References

1. WHO (2017) The top 10 causes of death 2017/01. http://www.who.int/mediacentre/factsheets/fs310/en/
2. Benjamin EJ et al (2017) Heart disease and stroke statistics-2017 update a report from the American heart association. Circulation 135(10):E146–E603
3. Mozaffarian D (2016) Heart disease and stroke statistics-2016 update: a report from the american heart association. Circulation 133(15):E599–E599 (vol 133, pg e38)
4. Rauch B et al (2016) The prognostic effect of cardiac rehabilitation in the era of acute revascularisation and statin therapy: a systematic review and meta-analysis of randomized and non-randomized studies - the cardiac rehabilitation outcome study (CROS). Eur J Prev Cardiol 23(18):1914–1939
5. Giannuzzi P et al (2003) Secondary prevention through cardiac rehabilitation - position paper of the working group on cardiac rehabilitation and exercise physiology of the european society of cardiology. Eur Heart J 24(13):1273–1278
6. Boyde M et al (2015) What have our patients learnt after being hospitalised for an acute myocardial infarction? Aust Crit Care 28(3):134–139
7. Aldcroft SA et al (2011) Psychoeducational rehabilitation for health behavior change in coronary artery disease a systematic review of controlled trials. J Cardiopulm Rehabil Prev 31(5):273–281
8. Wilson EAH, Wolf MS (2009) Working memory and the design of health materials: a cognitive factors perspective. Patient Educ Couns 74(3):318–322
9. Education level of the age above 15 years old statistical information network of the Republic of China 2017. http://win.dgbas.gov.tw/dgbas04/bc4/manpower/year/year_t1-t23.asp?table=5&ym=1&yearb=103&yeare=105&out=1
10. Frieske DA, Park DC (1999) Memory for news in young and old adults. Psychol Aging 14(1):90–98
11. Harskamp EG, Mayer RE, Suhre C (2007) Does the modality principle for multimedia learning apply to science classrooms? Learn Instr 17(5):465–477
12. Mayer RE et al (2005) When static media promote active learning: annotated illustrations versus narrated animations in multimedia instruction. J Exp Psychol-Appl 11(4):256–265
13. Mayer RE (2003) The promise of multimedia learning: using the same instructional design methods across different media. Learn Instr 13(2):125–139

Occupational Hazards: Awareness and Level of Precautions Among Physiotherapists in Selected Health Institutions in Lagos, Nigeria

Udoka C. Okafor[1(✉)], Gboyega A. Awe[2], Nicholas S. Oghumu[1], Ayomide C. Adeniyi[1], and Ganiyu O. Sokunbi[3]

[1] Physiotherapy Department, College of Medicine, University of Lagos, Lagos, Nigeria
udochris@yahoo.com, satghumu@yahoo.com, aadeniyi@gmail.com, udochris@cmul.edu.ng

[2] Physiotherapy Department, Lagos University Teaching Hospital, Lagos, Nigeria
gboyegaawe@gmail.com

[3] Physiotherapy Department, Bayero University Kano, Kano, Nigeria
ganiyusokunbi@gmail.com

Abstract. Background: Exposure to Occupational hazards can be very detrimental to the health of workers. Physiotherapists are healthcare professionals with immense burden to deliver on musculoskeletal and other aspects of health of patients and clients. Exposure to Occupational hazards can affect the discharge of their duties and negatively affect the quality of healthcare delivery. Identification, precaution and control of occupational hazards in the workplace are imperatives to the health and well-being of physiotherapists.

Methodology: This analytical cross-sectional survey involved 112 physiotherapists who were evaluated using a 52-item questionnaire which sought information on bio-data, job content, physical, mechanical/ergonomic and psychosocial health hazards, precautionary measures, physiotherapy specific hazards and influence of occupational hazards. Data were analysed using Statistical Package for Social Sciences version 20 and summarised using descriptive and inferential statistics of bar charts, pie charts, mean, standard deviation, and Chi Square.

Results: Forty-nine (43.8%) respondents agreed that they suffer musculoskeletal symptoms during their work while thirty-seven (33%) respondents agreed that they are often exposed to communicable diseases such as Tuberculosis, HIV, and Hepatitis. However, less than half (38.4%–48.2%) of the respondents disagreed that occupational hazards led to their frequent musculoskeletal symptoms, lack of motivation, reduced job output, permanent injury or disability, and frequent days off work. Also, 33–35.7% respondents agreed that they had either a pre-employment training or pre-employment entrance health examination when newly employed. There was no significant association between highest educational attainment and level of precaution. However, a significant association ($p < 0.05$) was found between duration of time spent at work a day and level of precaution

S. Bagnara et al. (Eds.): IEA 2018, AISC 818, pp. 600–608, 2019.
https://doi.org/10.1007/978-3-319-96098-2_74

Conclusion: The opinion of physiotherapists garnered from this study concerning occupational hazard is an indication of their knowledge and perception of occupational hazard and their practice of Occupational Health and Safety. Good knowledge of occupational hazard will influence the level of precaution of physiotherapists and also encourage the healthcare system in general to promote Occupational safety at work.

Keywords: Occupational hazard · Awareness · Precaution · Physiotherapist

1 Introduction

The burden of occupational injuries and work-related diseases and disorders is of major importance at workplaces given the various occupational hazards workers are exposed to and the economic cost associated with it (Takala et al. 2014). Occupational hazard is the risk, harm, or danger that an individual is exposed to at the workplace with concomitant occupational diseases from such exposures (Awodele et al. 2014). Exposure to Occupational hazards can be very detrimental to the health of workers. Yet workers must work because work has its positive health-promoting effects, as the financial dividend provides the worker with the basic necessities of life (Snyder et al. 2011; Awodele et al. 2014). There is the belief that a reciprocal and interactive relationship exists between workers and their work environment (Isah et al. 1997). The knowledge of these interactions between work and health is fundamental in understanding and practicing occupational health and safety which is often overlooked (Aliyu and Shehu 2006; Eyayo 2014). Hence, the World Confederation for Physical Therapy advocates for the right of physiotherapists to a safe and healthy practice environment that assures their own health and safety and that of their patients/clients (World Confederation for Physical Therapy 2011).

Physiotherapists are healthcare professionals with immense burden to deliver on musculoskeletal and other aspects of health of patients and clients. Exposure to Occupational hazards can affect the discharge of their duties and negatively affect the quality of healthcare delivery. Physiotherapists are known to be prone to WMSDs due to the nature of their work which is often repetitive, labour intensive and involving direct contact with patients (Chartered Society of Physiotherapy 2001). Previous studies have reported the life-time prevalence of WMSDs among physiotherapists in different populations; 68% in the United Kingdom (Glover et al. 2005), 55% (West and Gardner 2001); 91% (Cromie et al. 2000) in Australia, and 85% in Turkey (Salik and Ozcan 2004). Low back pain (LBP) is the most common WMSD among physiotherapists (West and Gardner 2001) with career and annual prevalence of LBP among physiotherapists in the United Kingdom reported as 68% and 58% respectively (Glover et al. 2005); 49% in Canada (Mierzejewski and Kumar 1997), and 70% in Kuwait (Shehab et al. 2003).

Several factors such as age, years of work experience and cultural values have been attributed to WMSDs among physiotherapists (Cromie et al. 2002; Salik and Ozcan 2004; Glover et al. 2005). Notably among these factors is the suggestion that the cultural values of physiotherapists may make it difficult for practitioners to avoid the risks of WMSDS during their work (Cromie et al. 2002). This is because these cultural values are generic and unique to physiotherapy. Despite the wealth of information on WMSDs among physiotherapists around the world, little seems to be known about the occupational hazards associated with physiotherapy practice in Nigeria. Identification, precaution and control of occupational hazards in the workplace are imperative to the health and well-being of Nigerian physiotherapists as applicable to their counterpart globally. Therefore, this study sought to know the influence of occupational hazards and level of precautions among physiotherapists in selected health institutions in Lagos, Nigeria.

1.1 Materials and Method

This cross-sectional survey used purposive sampling technique to recruit one hundred and thirty (130) respondents who are physiotherapists from various secondary and tertiary health institutions within Lagos metropolis. Ethical approval was sought and obtained from the Health Research and Ethics Committee of the Lagos University Teaching Hospital (LUTH) and from the appropriate authorities in selected health institutions, prior to the commencement of the study. Informed consent was also sought from all respondents prior to questionnaire administration. Physiotherapists with minimum of one year working experience post internship training were included whereas non-practicing physiotherapists and interns were excluded from the study. The study utilized questionnaire instrument titled; 'Occupational Hazards: Awareness and Level of Precautions among Physiotherapists in Selected Health Institutions in Lagos, Nigeria, adapted from the Occupational Health Hazards Questionnaire and modified by a five-man focus group comprising physiotherapy lecturers and clinicians. The questionnaires were administered to 130 physiotherapists at selected health institutions in Lagos State and consists of 3 sections, A, B and C. Section A captured information relating to socio-demographic data such as age, gender, highest educational attainment, marital status, religion, tribe, year qualified as a physiotherapist, nature of job, area of specialization, when current job began and duration of work per day; Section B contains the job content while Section C contains health hazard identification such as physical health hazards, ergonomic health hazards, psychosocial health hazards, physiotherapy specific hazards, the influence of occupational hazards and the precautionary measures taken.

1.2 Data Analysis

Out of the 130 questionnaires administered, 12 respondents did not return theirs, 6 questionnaires were not valid for analysis while 112 were analyzed. Data were analyzed using Statistical Package for Social Sciences version 20 and summarized using descriptive statistics of bar charts, pie charts, mean, and standard deviation. Chi Square test was used to test the association between variables.

2 Results

A response rate of 86.15% was obtained from 112 validly returned questionnaires. Socio-demographic distribution of respondents were age range (23–58 years), gender (male: 58%, female: 42%), tribe (75.9% Yoruba, 18.5% Igbo, others 5.4%), area of specialization (0.9% Burns and Trauma, 6.3% Cardiopulmonary, 0.9% Ergonomics, 0.9% Medicine, 16.1% Neurology, 0.9% Neurology/Orthopaedics, 34.9% Orthopaedics, 14.3% Orthopaedics and Sport, R6.3% Paediatrics, 0.9% Public health, 0.9% Sports and Rehabilitation, and 0.9% in Women's health), years of experience and nature of job (92.9% were clinical practice physiotherapists while 17.1% were academic physiotherapists).

The result revealed divergent views of respondents on identification of health hazards. While 44.6% disagreed that the noise level in their workplace is relatively high, 46.4% strongly agreed that loss of hearing could result from exposure to loud noise, 47.3% respondents agreed that their job function have to do with working with machineries that have high temperature, 42% agreed that extreme heat could cause muscle cramp, 49.1% strongly disagreed that their work place shakes as a result of vibration from workplace, 50% agreed that vibration could cause disorder and fatigue in the spine, 56.3% agreed that their work place is adequately lighted, 47.3% agreed that inadequate illumination could affect the eyes, 50% agreed that radiations could be emitted as they perform their job function, and 52.7% agreed that radiation can cause cancer and premature skin ageing.

On Ergonomic health hazards, respondents equally expressed divergent views. Whereas 52.7% agreed that they sometimes took awkward postures when working, 31.3% disagreed that they sometimes work at height, 46.4% agreed that they stand for a long time while performing their job function, 29.5% disagreed that they lift heavy objects manually, 53.6% disagreed that they sit most times when they are on duty, 36.6% agreed that their work is repetitive and monotonous, 49.1% strongly agreed that ergonomic hazards could cause deformity of one's body, and 59.8% strongly agreed that ergonomic hazards could cause musculoskeletal problems.

Respondent also showed divergent views on psychological health hazards as 43.8% agreed that their workload is very challenging, 42.9% disagreed that they would like to be transferred to another unit or department, 46.4% disagreed that they work in isolation, 39.3% strongly disagreed that they are constantly talked down by their supervisors, 41.1% strongly disagreed that they are faced by some kind of aggression and harassment in their place of work, and 43.8% agreed that psychological hazard could cause hypertension, anxiety or boredom.

Our findings further showed that one third of respondents either agreed or disagreed on precautionary measures. Whereas 33% agreed that they had a pre-employment training when newly employed, 35.7% strongly agreed that they had a pre-employment entrance health examination when newly employed, 35.7% disagreed that their employer periodically calls for a health examination monitoring or surveillance on their employees, 38.4% disagreed that Personal Protective Equipment (PPE) provided by management is adequate and appropriate, 34.8% disagreed that their employers periodically sent the employees for training to update and upgrade their efficiency and effectiveness, 33.9% disagreed that there is a first aid box in their workplace, 25.9% strongly disagreed that there is a Health Safety and Environment (HSE) Policy that is duly signed by the Managing Director in their work station, 28.6% strongly disagreed that implementation of the HSE policy is taken seriously by management, 32.1% agreed that the management is completely committed to the health and well-being of their workers, 33% disagreed that there is a very functional and active Occupational Health Safety System in their place of work. Respondents equally expressed divergent views on physiotherapy specific hazards in their work places. Whereas 50% agreed that their job involves a lot of lifting during patient care, 49.1% agreed that their job involves a lot of musculoskeletal maneuvers or manual therapy, 43.8% agreed that they suffer musculoskeletal symptoms during their work, 61.6% agreed that they are often tired at the end of the day's job, 47.3% agreed that their job involves gymnasium exercise activities, 42% agreed that they spend many hours engaging in physical exercises with their patients, 48.2% agreed that they engage in private practice physiotherapy outside their routine hospital work, 42% agreed that their work surfaces in private practice physiotherapy settings are usually smooth and adequate, 59.8% agreed that they are experienced in the clinical skills of manual handling. Less than half of the respondents disagreed on the influence of occupational hazards on job content: 42% disagreed that their job output is reduced as a result of occupational hazards, 40.2% disagreed that their motivation is dwindled as a result of occupational hazards, 48.2% disagreed that occupational hazards led to their frequent days off work, 50% disagreed that occupational hazards caused them to be assigned to a particular duty specification, 38.4% disagreed that occupational hazards led to their frequent musculoskeletal symptoms, 43.8% disagreed that occupational hazards led to their permanent injury or disability. However, 33% respondents agreed that they are often exposed to communicable diseases such as Tuberculosis, HIV, Hepatitis.

The study found significant association ($p < 0.05$) between duration of time spent at work per day and level of precaution whereas no significant association ($p > 0.05$) was found between the level of influence of occupational hazard and level of precaution (Table 1); between job content and level of precaution (Table 1) and between the level of precaution and each of year of qualification and highest educational attainment (Table 2). A significant association ($p < 0.05$) was found between job content and each of duration of time a day and highest educational attainment (Table 3). However, no significant association was found between job content and each of year of respondent's qualification and resumption of current job (Table 3).

3 Discussion

This study sought to determine the awareness and level of precautions among physiotherapists in selected health institutions in Lagos state, Nigeria. The finding that more male physiotherapists participated in the survey is a reflection of the population from which the sample was drawn and is contrary to the findings in a previous study by Bork et al. (1996). This result is understandable since unlike in Europe and America, the physiotherapy profession in Nigeria is male dominated.

Table 1. Chi-square analysis for duration of time spent, level of influence of occupational hazard a day, job content and level of precaution

Variables	Level of precaution				
	Low n(%)	High n(%)	Total n(%)	X^2	p
Duration of time a day					
1–8	80(84.2)	15(15.8)	95(100.0)	5.88	0.02
>8	10(58.8)	7(41.2)	17(100.0)		
Level of influence of occupational hazard					
Low influence	54(48.2)	16(14.3)	70(62.5)	1.22	0.27
High influence	36(32.1)	6(5.4)	42(37.5)		
Job content					
Low	7(7.8)	83(92.2)	90(100.0)	2.16	0.14
High	4(18.2)	18(81.8)	22(100.0)		

* Significance set at $p \leq 0.05$. Key: n = frequency; % = percentage; X^2 = Chi-square; p = p value

Table 2. Chi-square analysis for year of qualification, highest educational attainment, and level of precaution

Variables	Level of precaution				
	Low n(%)	High n(%)	Total n(%)	X^2	p
Year of qualification					
1–5	23(79.3)	6(20.7)	29(100.0)	0.15	0.93
6–10	22(78.6)	6(21.4)	28(100.0)		
>10	49(81.8)	10(18.2)	55(100.0)		
Highest educational attainment					
B.PT	54(79.5)	14(20.5)	68(100.0)	10.34	0.07
DPT	0(0.0)	1(100.0)	1(100.0)		
M.Sc	31(81.6)	7(18.4)	38(100)		
PhD	5(100.0)	0(100.0)	5(100.0)		

* Significance set at $p \leq 0.05$. Key: n = frequency; % = percentage; X^2 = Chi-square; p = p value

Despite the high prevalence of occupational hazards, limited knowledge was found among physiotherapists in Lagos, Nigeria. With further probing, about half the study population were able to identify the hazardous working conditions. Overall, physiotherapists did not adequately recognize the serious health and safety implications of occupational exposures as one third of respondents either agreed or disagreed on precautionary measures thus showing inconsistencies on their perception on precautionary measures towards work place hazards. This finding is in agreement with Senthil et al. (2015) who reported non-compliance by health care workers to standard health precautions and general lack of awareness about occupational safety.

Table 3. Chi-square analysis for duration of time a day, year of qualification, highest educational attainment, year respondent started current job, and job content

Variables	Job content				
	Low n(%)	High n(%)	Total n(%)	X^2	p
Duration of time a day					
1–8	6(6.4)	89(93.7)	95(100.0)	8.68	0.00
>8	5(29.4)	12(70.6)	17(100.0)		
Year of qualification					
1–5	3(10.3)	26(89.7)	29(100.0)	0.067	0.97
6–10	3(11.1)	25(89.3)	28(100.0)		
>10	5(9.1)	50(90.9)	5(100.0)		
Highest educational attainment					
B.PT	4(4.8)	63(95.2)	67(100.0)	11.82	0.04
DPT	1(100.0)	0(0.0)	1(100.0)		
M.Sc	5(13.2)	33(86.8)	38(100)		
PhD	1(20.0)	4(80.0)	5(100.0)		
Year respondent					
Started current job					
1–5	4(11.1)	32(88.9)	36(100.0)	0.54	0.76
6–10	4(12.9)	27(87.1)	31(100.0)		
>10	2(7.1)	26(92.9)	28(100.0)		

* Significance set at $p \leq 0.05$. Key: n = frequency; % = percentage; X^2 = Chi-square; p = p value

While one third of physiotherapists agreed to have undergone pre-employment training when newly employed, same one third of physiotherapists disagreed to have been periodically examined by their employers on health examination surveillance. This highlights the need for management of various institutions to examine physiotherapists by carrying out both pre-employment assessment, regular screening and also periodically call for examination of their employees. This is important to improve standard of practice. Similarly, this study revealed that Personal Protective Equipment (PPE) provided by management is inadequate and not appropriate according to one third of the respondents. A previous study in Tanzania by Rongo et al. (2004) agrees with this finding. The gross inadequacy of adherence to occupational safety measures could be as a result of lack of awareness on the importance of essential equipment on prevention of occupational diseases.

Less than half of the respondent disagreed on influence of occupational hazards on their job content. Notably, half of the respondents disagreed that occupational hazards caused them to be assigned to a particular duty specification. This finding agrees with a similar study by Adegoke et al. (2008) where majority of physiotherapist did not change their area of practice and did not leave the profession due to their WMSDs. Again, one third of physiotherapists agreed that they are often exposed to communicable diseases such as hepatitis and tuberculosis. This assertion is in agreement with a previous study by Awodele et al. (2014) in the paint industry which reported that 90%

of the respondents have symptoms relating to occupational hazard. This can be as a result of negligence both on the part of the workers and also attitude of management to occupational health. Prevention of these exposures are important in ensuring a safe working environment for physiotherapists, It also complements healthcare system, patients safety and infection control efforts. Thus, it is important that physiotherapists are safe in patient care and health care delivery and should be aware of the practices that promote patient and workers safety.

There was no significant association between year of qualification and level of precaution as all respondents who qualified between 1983 and 2015 reported low level of precaution. Same findings were obtained between highest educational attainment and level of precaution, level of influence of occupational hazard and level of precaution and also job content and level of precaution. Significant association was found only between duration of time spent a day and level of precaution. The level of precaution is an indication of level of compliance with the Occupational Health and Safety Management System. Thus respondents with low level of precaution have low level of compliance with the Occupational Health and Safety Management System.

3.1 Conclusion

Identification of Occupational Health Hazards, the awareness of physiotherapists on the health hazards, the risk associated with them and the effectiveness of the occupational health practices are crucial in the promotion, protection and rehabilitation of the health and well-being of physiotherapists working in health institutions. From this study, it could be deduced that there is low level of precaution and low level of compliance to occupational health and safety management. There is therefore need for management of various institutions to holistically engage in health promotion of physiotherapists on occupational health awareness and practice.

3.2 Recommendation

Occupational Health and Safety Management System is an integral part of workers' management policy and as such should be paramount in their practice by physiotherapist. Staff and management of any healthcare institution should not be seen lacking in this area.

References

Adegoke O, Akodu K, Oyeyemi L (2008) Work-related musculoskeletal disorders among nigerian physiotherapists. BMC Musculoskelet Disord 9:112

Aliyu AA, Shehu AU (2006) Occupational hazards and safety measures among stone quarry workers in Northern Nigeria. Niger Med Pract 50:42–47

Awodele O, Popoola D, Alade A (2014) Occupational hazards and safety measures amongst the paint factory workers in Lagos State. Safety Health Work 5:100–111

Bork BE, Cook TM, Rosecrance JC, Engelhardt KA, Thomason ME, Wauford IJ, Worley RK (1996) Work-related musculoskeletal disorders among physical therapists. Phys Ther 76:827–835

Chartered Society of Physiotherapy (2001) Health and Safety Briefing Pack. No 11 Work-Related Strain Injuries (musculoskeletal disorders). CSP, London

Cromie JE, Robertson VJ, Best MO (2000) Work-related musculoskeletal disorders in physical therapists: prevalence, severity, risks and responses. Phys Ther 80:336–351

Cromie JE, Robertson VJ, Best MO (2002) Work-related musculoskeletal disorders and the culture of physical therapy. Phys Ther 82(5):459–472

Eyayo F (2014) Evaluation of occupational health hazards among oil industry workers. J Environ Sci Toxicol Food Technol 8(12):22–53

Glover W, McGregor A, Sullivan C, Hague J (2005) Work-related musculoskeletal disorders affecting members of the chartered society of physiotherapy. Physiotherapy 91:138–147. https://doi.org/10.1016/j.physio.2005.06.001

Isah EC, Asuzu MC, Okojie OH (1997) Occupational health hazards in manufacturing industries in nigeria. J Community Med Primary Health Care 9:26–34

Mierzejewski M, Kumar S (1997) Prevalence of low back pain among physical therapists in Edmonton, Canada. Disabil Rehabil 19(8):309–317

Rongo LMB, Barten F, Msamanga GI, Heederik D, Dolmans WMV (2004) Occupational exposure and health problems in small-scale industry workers in Dares Salaam, Tanzania: a situation analysis. Occup Med 54:42–46

Salik Y, Ozcan A (2004) Work-related musculoskeletal disorders: a survey of physical therapists in Izmir Turkey. BMC Musculoskelet Disord 5:27

Senthil A, Anandh B, Jayachandran P, Thangavel G, Josephin D, Yamini R, Kalpana B (2015) Perception and prevalence of work-related health hazards among health care workers in public health facilities in Southern India. Int J Occup Environ Health 21(1):74–81

Shehab D, Al-jarallah K, Moussa MA, Adham N (2003) Prevalence of low back pain among physical therapists in Kuwait. Med Principles Pract 12:224–230

Snyder CR, Lopez SJ, Pedrotti JT (2011) Positive psychology: the scientific and practical explorations of human strengths. Sage Publications Inc., Thousand Oaks

Takala J, Hämäläinen P, Saarela KL, Yun LY, Manickam K, Jin TW, Heng P, Tjong C, Kheng LG, Lim S, Lin GS (2014) Global estimates of the burden of injury and illness at work in 2012. J Occup Environ Hyg 11(5):326–337

West DJ, Gardner D (2001) Occupational injuries of physiotherapists in north and central Queensland. Aust J Physiotherapy 47:179–186

World Confederation for Physical Therapy (2011) Policy statement: occupational health and safety of physical therapist, 2–5

Barriers and Facilitators in Implementing a Moving and Handling People Programme – An Exploratory Study

Hannele Lahti[1](✉), Kirsten Olsen[2], Mark Lidegaard[2], and Stephen Legg[2]

[1] University of Eastern Finland, 70210 Kuopio, Finland
hannelah@student.uef.fi
[2] Massey University, Palmerston North 4474, New Zealand

Abstract. Health care workers, including nurses, have one of the highest musculoskeletal injury incidence rates of any profession, especially work-related back injuries. The majority of musculoskeletal disorders (MSD) in health care are caused by moving and handling of people (MHP). In order to reduce MSDs due to MHP, some national health care sector authorities worldwide have developed intervention programmes or guidelines that can be used by their health care organisations. However, very few of the national interventions have been evaluated for their efficacy or impact. In those that have, the effort to reduce the incidence of MSDs caused by MHP has been largely unsuccessful, often because of barriers. This study aimed to identify what barriers, and facilitators, existed in health care organisations in relation to implementation of the New Zealand MHP guidelines. It was found that implementing a MHP programme especially requires sufficient resources for training and strong support based on evidence, or legislation. Management support for equipment purchase or maintenance is also essential, as is sufficient training for changing the culture of the workplace. It is also important to assess staff MHP knowledge and practices. MHP programmes need to be designed to be suitable for different sectors such as age care, and different users such as a foreign work force.

Keywords: Guidelines · Evaluation · Interventions

1 Introduction

Health care workers, including nurses, have one of the highest musculoskeletal injury incidence rates of any profession, especially work-related back injuries [1]. The majority of musculoskeletal disorders (MSDs) and related injuries in health care are caused by moving and handling of people (MHP) [1, 2]. In order to reduce MSDs due to MHP, some national health care sector organisations worldwide have developed intervention programmes [2] or guidelines. However, very few of the national interventions have been evaluated for their efficacy or impact [3]. In those that have been evaluated, the effort to reduce the incidence of MSDs caused by MHP has been largely unsuccessful [1]. This study identifies what barriers and facilitators exist in health care organisations in relation to implementation of the New Zealand MHP guidelines.

S. Bagnara et al. (Eds.): IEA 2018, AISC 818, pp. 609–618, 2019.
https://doi.org/10.1007/978-3-319-96098-2_75

In an effort by New Zealand's Accident Compensation Corporation to help the country's health care sector reduce its incidence of musculoskeletal discomfort, pain and injury, the New Zealand Moving and Handling People Guidelines (MHPG) were published in 2012. They include seven core skills and competencies and physical resources, and four organisational systems for MHP. The seven core skills and competencies and physical resources were: 'risk assessment', 'MHP techniques', 'MHP training', 'organising training', 'MHP equipment', 'equipment management', and 'facility design and upgrading'. The four organisational systems were: 'policy and programme planning', 'workplace culture', 'monitoring and evaluation', and 'audits'. In addition to these, the MHPG include the sections of 'introduction', 'why moving and handling programmes are needed' and 'bariatric clients', resulting altogether in 14 sections [4].

Multifactor interventions based on a risk assessment programme have been found to be successful in reducing patient handling related risk factors [1]. According to an evidence review by Humrickhouse and Knibbe [5], the most effective workplace intervention for preventing manual patient handling related injuries is now thought to be a structured safe patient handling (SPH) program including training, ergonomic intervention and the use of mechanical devices.

Identified barriers from the literature can be divided in barriers for implementing evidence or guidelines in practice and barriers for using SPH technology. Dogherty et al. [6] identified the barriers for implementing evidence into practice with nurses. The barriers were lack of engagement such as management support or interest, lack of resources such as equipment, funding, time or human resources, conflicting interests for example between staff and the facilitator, constraints in team functioning and lack of evaluation and sustainability. Engkvist [7] divided the barriers in patient, facility and organisation related factors. Patient related factors included for instance emergency situations and language barriers, facility related factors constraints in equipment or space and organisation related factors, lack of time and staff. Krill et al. [8] assessed barriers and attitudes toward SPH and the need for SPH equipment. Factors that were negatively affecting for the attitudes were lack of "No Lift" -policy, adequate equipment or space, lack of follow-up training, patient-related factors and lack of time for SPH. Silverstein et al. [9] identified barriers for the introduction and integration of SPH legislation into organisations. Barriers were related to facilities such as lack of storage place for equipment or lack of space for handle equipment, resources for example lack of staff, funding or competency in using the equipment, and patient-related factors such as limitation for patient independence or refusal to use the equipment. Two years after implementing the SPH legislation, lack of on-going funding and training, difficulty to maintain awareness and age of the equipment were decreasing the sustainability of the process.

For using SPH technology, identified barriers were physical, such as lack of access for the equipment, lack of storage space or space to maneuver equipment, incompatibility of equipment, lack of time or staff [10] lack of equipment, lack of staff or training [11], constraints in maintenance or storage [12] and inappropriate or unavailable equipment [13]. Knowledge and skills related barriers identified were lack of training [10, 11] and confidence in using the equipment or patient assessment [10]. Barriers could also have concerned unit culture such as different patient handling practices [10] or uncertain roles at the workplace [12] or patient related factors, for example patient's refusal to use the equipment [10, 11].

The aims of the present study were to identify barriers and facilitators in implementing the seven core skills and competencies and physical resources, and four organisational systems of the MHPG in healthcare organisations in New Zealand, and to identify possible solutions to overcome identified barriers.

2 Methods

Ten individuals that had used the MHPG were purposively selected to participate in semi-structured interviews based on a questionnaire that had been used in the project: *Assessment of the uptake and impact of the ACC New Zealand Moving and Handling People Guidelines (2012)* [14]. Data collected from the interviews was transcribed and analysed using data based inductive content analysis [15].

2.1 Selection of Interview Participants

The participants for interview were selected based on a questionnaire that was sent to 3,052 persons in the health care sector [3]. Of the 689 respondents, 144 indicated their willingness to be re-contacted with respect to participation in further research. Of these, 61 had work roles identified to be in the key target group for the MHPG: 'Physiotherapist or Occupational Therapist', 'Management', 'Ward or Unit Manager', 'Occupational Health and Safety Manager', 'MHP Coordinator', 'Health and Safety Representative', and 'Policy Maker' and worked either in 'Public Hospital', 'Private Hospital', 'Hospice', or 'Residential and Aged Care'. From these 61 persons, 28 potential interviewees were selected to be contacted for the present study. These individuals were sequenced according to their level of the MHPG uptake and contacted in priority order. In order to identify the 'high uptake' users, a MHPG usage-score was created based on their responses to the 2016 questionnaire. The questionnaire identified the use of the seven core skills and competencies and physical resources, and the four organisational systems from the MHPG. The more they had used the seven core skills and competencies and physical resources and the four organizational systems, the higher the usage-score was. The potential interviewees were sent an invitation e-mail and information sheet and asked for their permission to be interviewed. The number of participants selected for interview was limited to ten, due to time limitations.

Participants worked either in public hospital, age care, residential care, home care, hospice or training and education. Seven of them worked in multiple sectors and three of them were contracted in different units to organise MHP training. All the contracted participants were physiotherapists and worked in companies that employed less than 20 people. All seven other participants worked in companies with 20 to 100 and more employees. Two of them were nurses, five of them were physiotherapists and one of them was an occupational therapist. Two of the participants did not report a certain occupation but had a role as MHP coordinator or advisor, educator and health and safety representative, and occupational health and safety manager, coordinator or advisor. Six of the participants also reported they had a role in MHP coordinator or advisor and five of them as a health and safety representative.

2.2 Semi-structured Interviews and Analysis of the Data

Participation in the semi-structured interview was voluntary and entirely confidential. Each interview lasted 45–60 min and was conducted by telephone or skype (and voice recorded with the permission of each participant) during September–November in 2017. Interviewees were sent the interview schedule before the interview. The interview schedule comprised questions about the use and implementation of the MHPG, questions about the 14 sections of the MHPG and individualised questions specific to each interviewee based on their answers in the 2016 questionnaire [3]. Individual questions were based on the sections the interviewee had used and aimed to find out what barriers and facilitators there were in relation to the implementation of the section.

Recorded interviews were transcribed individually, checked by the members of the research group, revised and sent to the interviewee for final factual verification and approval. The final approved transcripts were then analysed using inductive content analysis [15]. In this way, the main barriers and solutions (i.e. ways to overcome barriers) and the facilitators were identified.

3 Barriers for Implementing a MHP Programme and Ways to Overcome Them

Five major barrier themes were identified. They related to: the implementor (the interviewee), the MHPG, the organisation's resources, the workplace culture, and the organisation's requirements (Fig. 1).

For implementor related barriers, lack of influence in the organisation was mentioned most often. This included lack of influence in staffing, facility design, policy planning, workplace culture, budgeting and equipment storage or maintenance. All three of the contracted worker interviewees mentioned lack of influence in the organisation as a barrier. Other barriers related to the implementor were implementor's or trainer's individual factors such as lack of qualifications as a trainer. There was only one specific mention of a way to overcome an implementor related barrier. One of the interviewees mentioned trying to overcome the barrier of lack of influence in the organisation's practices by referencing evidence from the literature, even though that did not lead to any changes in the organisation.

For MHPG related barriers, the most often mentioned issues were that guidelines were too generic for special needs and language barriers. Identified special needs included age care, residential care and dementia patient, bariatric patients, and patients in risk of pressure injuries. This barrier was also mentioned by all three of the contracted worker interviewees. By language barriers interviewees meant that the MHPG were difficult to understand by patients and their families, and by the foreign workforce. Other barriers related to the MHPG were lack of compatibility with the organisations' practices or parallel programmes, lack of practical instructions for the implementation and lack of legal requirements for following the MHPG. There were

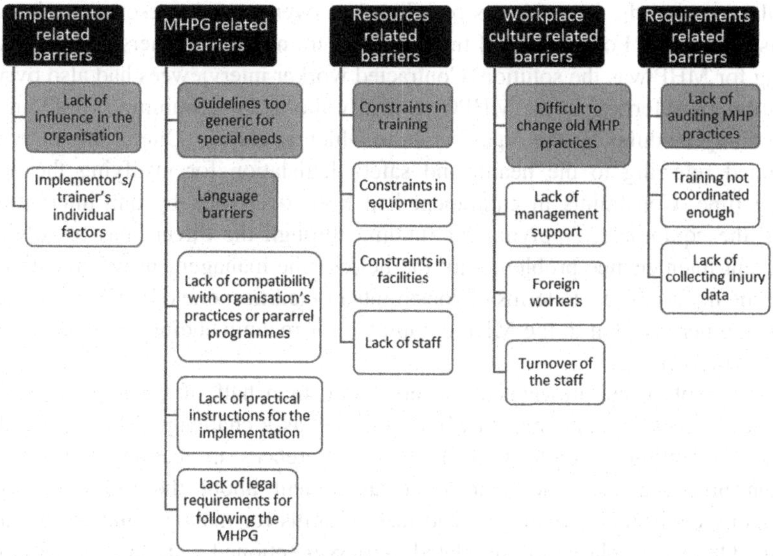

Fig. 1. Themes for barriers for implementing a MHP programme (black background) and their subthemes (white background) with the main barriers (grey background)

several ways that the interviewees mentioned that MHPG barriers could be overcome. For the barrier of the MHPG being too generic for age care, one of the contracted worker interviewees had overcome the barrier by adjusting the guidelines for the focus workplace and by using techniques learned from other organisations. For the lack of practical instructions for implementing the MHPG into an organization, two of the interviewees had used feedback for care staff, adapting themselves in the situation and starting with small and easy steps, and using variety of other resources as instructions, including adding the teaching the body control during MHP. One of the interviewees mentioned that there were too many constraints concerning the full use of MHPG. She mentioned that cooperation with other trainers had helped to overcome the barrier. There were a few concerns about language barriers for foreign workforce but also for the patients and their families. These challenges were overcome by using different formats of the MHPG, such as pictures and DVD, and by making instructions clearer and more basic.

For resources related barriers, most of the interviewees mentioned constraints in training. This included lack of training staff, lack of training time, and lack of funding for training. Constraints in training was acknowledged by all three of the contracted worker interviewees. Senior carers' lack of knowledge of MHP, and lack of possibility to force staff to attend the training were mentioned by one of the contracted worker interviewees. Other identified resources related barriers were constraints in equipment and facilities, and lack of staff. Constraints in equipment included lack of funding, lack of suitable equipment for the patient, and constraints in equipment maintenance. Constraints in facilities included lack of space for the equipment use, lack of cooperation between building designers and ward staff, and lack of funding. There were several ways that the interviewees had found to overcome the resources related constraints in training. For the lack of training time, training the trainers or using external training for MHP was the solution. Contracted worker interviewees had also overcome the barrier by referring to the MHPG to justify the need of training or building their own training modules for the organisation to which to choose. One of the interviewees mentioned referring to the health and safety legislation for justifying the need of training time. Constraints in equipment had been overcome by one interviewee by renting the equipment. Applying for funding through the direct boss or solving the equipment maintenance problem with the help of the management was mentioned by two of the interviewees. In terms of constraints in facilities, one of the interviewees had been using her position in the MHP committee for recommending the facility upgrade based on the MHPG.

For workplace culture related barriers, more than half of the interviewees mentioned difficulties in changing old MHP practices as a challenge. This included difficulties in adapting to uniform techniques, differences in learning the techniques between nurses and physiotherapists, disagreement about the techniques or risk assessment, environmental change, and lack of skills for manual handling of bariatric patients. Other workplace culture related barriers mentioned were lack of management support, foreign workers and turnover of the staff. In terms of foreign workforce, reluctancy for reporting issues and lack of MHP knowledge were mentioned. This barrier has been overcome by introducing the preceptor system for supporting new staff. In order to overcome the barrier for changing old work practices or cultural change at the workplace, three of the interviewees emphasised the importance of sufficient practice in SPH techniques. One of the interviewees also composed a song for reminding the MHP techniques. For the disagreement about the techniques between staff members, one of the interviewees used the MHPG as a reference to justify the technique. One of the interviewees overcame the barrier for implementing the cultural change to the workplace by on-going training, introducing the new equipment and by requiring mandatory patient risk assessment.

Requirements related barriers included lack of auditing the MHP practices, lack of coordination in training and lack of collected injury data from the organisation. Lack of auditing the MHP practices was mentioned most often and mostly by the contracted worker interviewees who were unable to measure staff competency, monitor the training or trainer's qualifications, or measure the effectiveness of the training in the focus organisation. In order to overcome the barrier for lack of hours to measure staff competency, one of the contracted worker interviewees had shared the training responsibility with the Champion nurses.

According to the interviews, two of the biggest barriers for implementing the MHPG into an organisation were resources related barriers; constraints in training and equipment. There were several ways identified to overcome these barriers: training the trainers, using external training, referring to the MHPG or legislation and building own training modules in case of constraints in training, and in terms of equipment constraints, renting the equipment or having the support from the management for equipment purchase or maintenance. Other barriers that more than half of the interviewees mentioned were implementor's lack of influence in the organisation and difficulty to change old MHP practices at the workplace. For overcoming the barrier for lack of influence in the organisation, even though it did not lead to any changes, referencing to the evidence from the literature was used as a solution. For changing the old MHP practices, sufficient practice and a MHP song to remind about the practices were used.

4 Facilitators for Implementing a MHP Programme

Six major facilitator themes were identified. They related to: the implementor, the MHPG, resources, requirements, the organisations' attitude, and training (Fig. 2).

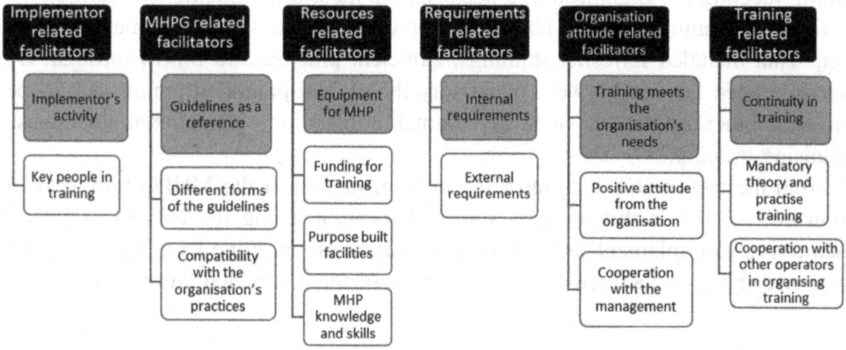

Fig. 2. Themes for facilitators for implementing a MHP programme (black background), their subthemes (white background) and identified main facilitators (grey background)

Implementor required facilitators included implementor's activity related factors and identifying key people for responsible in training. Four of the interviewees emphasised the importance of implementor's or trainer's activity in the implementation. That included trainer's confidence and persistence in programme implementation, trainer's own knowledge and education in MHP and the relationship between the staff and the trainer. Two of the contracted worker interviewees mentioned the trainer's persistence and own knowledge important in the implementation of the MHP programme. Two of the contracted worker interviewees also emphasised the importance of key people for training in the organization. Nine of the ten interviewees mentioned using MHPG as a reference for justifying the need of change in the organisation as a facilitator in the guidelines implementation. The MHPG were used as a reference for example for equipment purchase or creating safe working environment. One of the contracted worker interviewees used the MHPG as a reference for her own training manuals.

From the resources related facilitators, appropriate equipment for MHP was experienced the most important facilitator. That also included funding for and the team purchasing the equipment. Other resources related facilitators were funding for training, purpose-built facilities and MHP knowledge and skills. Two of the contracted worker interviewees mentioned paying the staff to attend the training as facilitators for the implementation of the MHPG.

Internal and external requirements were also experienced as facilitators for the MHPG implementation. Internal requirements in this context mean competency assessments, reporting the incidents and accidents, observation of MHP practices, monitoring the attendance in the trainings and regular updates for the patients' transfer plans. External requirements in this context mean Health and Safety legislation, environmental change and external auditing.

Organisation attitude related facilitators included training that meets the organisation's needs, positive attitude from the organization and cooperation with the management. These facilitators were emphasised by contracted worker interviewees, training meeting the organisation's needs as the most often mentioned facilitator.

From the training related facilitators, continuity in training was mentioned most often. That included refresher trainings, sufficient practice and follow-through. Other training related facilitators were mandatory theory and practice training and cooperation with other operators, such as external trainers or occupational therapists, in organising training.

From the factors that facilitated the implementation of the MHPG into an organisation one facilitator should have a special mention: using the MHPG as reference when trying to implement the change in an organisation. Nine from ten interviewees mentioned this as a facilitator, so did all three of the contracted worker interviewees.

5 Discussion

Findings from this study follow the results from the earlier literature about the facilitators and barriers of implementing MHP programmes into health care organisations, especially in terms of facilitators. Koppelaar et al. [16] mentioned patient related facilitators such as patient's comfort and safety as a facilitator, which was interestingly not mentioned by the interviewees in this study. Whereas using guidelines a reference, internal and external requirements, continuity in training and mandatory theory and practice training, were identified as facilitators in the interviews in this study and were not mentioned in the previous studies. Elnitsky et al. [17] emphasises the importance of the characteristics, skills and activities of the programme implementor, which were partly also mentioned by four of the interviewees in this study. Earlier literature also identified facilitators for MHP equipment use, which was not the aim of this study, but adds an interesting point of view because MHP equipment use is an important part of the MHPG. The literature emphasises concerns of staff personal health [10, 11] and patients' safety and mobility as facilitators for equipment use. External requirements were also mentioned, such as requirements by law [11], as well as promotional efforts [12]. In terms of barriers for implementing the MHP programme into an organisation, the previous literature emphasises the importance of patient related factors such as refusal for using the equipment [7–11]. Patient related factors were not mentioned in the interviews of this study. Other factors that were identified in the earlier literature, but did not come up in this study, were conflicting interests between staff and implementors and constraints in team functioning [6]. As identified in this study, previous literature also acknowledges barriers related to resources, workplace culture and requirements such as lack of auditing. In addition to the earlier findings, this study adds implementor's lack of influence in the organisation and factors related to MHPG as barriers for the MHP programme implementation.

6 Conclusions

Implementing a MHP programme especially requires sufficient resources for training and strong support based on evidence, or legislation. Management support for equipment purchase or maintenance is also essential, as is sufficient training for changing the culture of the workplace. It is also important to assess staff MHP knowledge and practices. MHP programmes need to be designed to be suitable for different sectors such as age care, and different users such as a foreign work force.

References

1. Nelson A, Matz M, Chen F, Siddharthan K, Lloyd J, Fragala G (2006) Development and evaluation of a multifaceted ergonomics program to prevent injuries associated with patient handling tasks. Int J Nurs Stud 43(6):717–733
2. Hignett S (2003) Intervention strategies to reduce musculoskeletal injuries associated with handling patients: a systematic review. Occup Environ Med 60(9):e6
3. Lidegaard M (2016) Uptake and impact of the ACC New Zealand moving and handling people guidelines. Ph D confirmation report. Massey University (2016)
4. Accident Compensation Corporation. https://www.acc.co.nz/assets/provider/acc6075-moving-and-handling-peopleguidelines.pdf. Accessed 27 May 2018
5. Humrickhouse R, Knibbe HJ (2016) The importance of safe patient handling to create a culture of safety: an evidential review. Ergon Open J 9(1):2742
6. Dogherty EJ, Harrison MB, Graham ID, Vandyk AD, Keeping-Burke L (2013) Turning knowledge into action at the point-of-care: the collective experience of nurses facilitating the implementation of evidence-based practice. Worldviews Evid-Based Nurs 10(3):129–139
7. Engkvist I-L (2008) Back injuries among nurses – a comparison of the accident processes after a 10-year follow-up. Saf Sci 46(2):291–301
8. Krill C, Staffileno BA, Raven C (2012) Empowering staff nurses to use research to change practice for safe patient handling. Nurs Outlook 60(3):157–162
9. Silverstein B, Schurke J (2012) Washington state department of labor and industries' SHARP program. Safety and health assessment and research for prevention. Washington State Department of Labor & Industries
10. Kanaskie ML, Snyder C (2018) Nurses and nursing assistants decision-making regarding use of safe patient handling and mobility technology: a qualitative study. Appl Nurs Res 39:141–147
11. Olkowski BF, Stolfi AM (2014) Safe patient handling perceptions and practices: a survey of acute care physical therapists. Phys Ther 94(5):682–695
12. Schoenfisch AL, Pompeii LA, Myers DJ, James T, Yeung YL, Fricklas E, Lipscomb HJ (2011) Objective measures of adoption of patient lift and transfer devices to reduce nursing staff injuries in the hospital setting. Am J Ind Med 54(12):935–945

13. Wardell H (2017) Reduction of injuries associated with patient handling. Aaohn J 55 (10):407–412
14. Olsen K, Lidegaard M, Legg S (2016) Assessment of the uptake and impact of the ACC New Zealand moving and handling people guidelines (2012). Report. Stage 2, Uptake and use. Part A, Descriptive analysis of questionnaire findings. Centre for ergonomics, Occupational safety and health, School of public health, Massey University, New Zealand
15. Vaismoradi M, Turunen H, Bondas T (2013) Content analysis and thematic analysis: implications for conducting a qualitative descriptive study. Nurs Health Sci 15(3):398–405
16. Koppelaar E, Knibbe H, Miedema H, Burdorf A (2009) Determinants of implementation of primary preventive interventions on patient handling in healthcare: a systematic review. Occup Environ Med 66(6):353–360
17. Elnitsky CA, Powell-Cope G, Besterman-Dahan KL, Rugs D, Ullrich PM (2015) Implementation of safe patient handling in the U.S. veterans health system: a qualitative study of internal facilitators' perceptions. Worldviews Evid Based Nurs 12(4):208–216

Does Exposure from the Use of Mobile Phones Affect Lateralised Performance and Mood in a Visual Task?

Joanna E. N. Fowler(✉) and Janet M. Noyes

School of Experimental Psychology, University of Bristol, Bristol BS8 1TU, UK
psjenf@bristol.ac.uk

Abstract. There has been increasing use of mobile phones and concerns about exposure to radiofrequency electromagnetic fields (RF EMF). The aim of this study was to investigate lateralised effects of exposure on cognitive performance. Forty-eight healthy young males and females performed a dot detection task in three phone conditions, active right, active left and inactive; the completion of a well-being test followed this. Results show that reaction times (RT) are faster when the phones are switched off. In addition, RT is faster for the active phone in the right visual field when it is positioned over the right hemisphere than when the phone is positioned over the right hemisphere and in the left visual field. Also, when the active phone is over the left hemisphere, participants report greater sadness compared to the active phone over the right hemisphere. Previous studies on assymetrical brain function suggest deactivation of the left hemisphere increases negative affect. The combination of the results in the current study suggest a potential mood effect is operating.

Keywords: Radiofrequency electromagnetic fields (RF EMF)
Cognitive performance · Well-being

1 Introduction

Mobile phone ownership and use has increased over the last decade. Figures from the Pew Research Centre in the United States show that all young people between 18–29 years owned a mobile phone and 94% of those are smartphones [1]. Mobile phones are ubiquitous devices and are widely used for their multi-faceted functionality, portability and, are considered to be an essential tool in everyday life. However, there has been concern over possible health implications of use. These include acoustic neuromas [2] parotid gland tumours [3], and effects on the brain [4].

It has been suggested that the effects of radiofrequency electromagnetic fields (RF EMF) emitted from the use of mobiles phones could be having an effect on cognitive performance in humans. A review by Zhang et al. [5] assessed the results of recent studies on neurocognitive functions [6, 7] and recommended further study on the effect of RF EMF exposure on brain function and cognitive performance. Many previous studies have investigated performance in visual tasks.

S. Bagnara et al. (Eds.): IEA 2018, AISC 818, pp. 619–627, 2019.
https://doi.org/10.1007/978-3-319-96098-2_76

Freude et al. [8] carried out a complex and demanding visual monitoring task. Although no significant effect on cognitive performance was found, a significant decrease in preparatory slow brain potentials during exposure to RF EMF from the phone was recorded. Jech et al. [9] found that reaction time (RT) was significantly shortened by 20 ms in response to all target stimuli exposure in an adapted visual oddball task significantly enhanced. The positivity of event related potentials (ERPs) in response to target stimuli in the right hemifield of the screen was significantly enhanced. A significant difference was found to RF EMF exposure in a visual masking task compared to a control group Rodina et al. [10]. Mortazavi et el. [11] showed decreased RT for acute short-term exposure in a visual search task [11] whilst Vecchio et al. [12] found increased RT to 'go' stimuli in the post exposure period in a visual, go/no go, task. These studies show that RF EMF effects have been found to affect brain physiology in visual tasks and some studies have found cognitive performance is affected.

Differential effects have found between the hemispheres. Croft et al. [13] investigated the effects of RF EMFs from mobile phones in an auditory task and found that exposure altered the resting EEG, decreasing delta (1–4 Hz activity) in right hemisphere sites relative to the left hemisphere sites. Ferreri et al. [14] found that intracortical excitability curve was significantly modified during exposure, with short intracortical inhibition (SICI) being reduced and intra cortical facilitation (ICF) being enhanced in the acutely exposed brain hemisphere. The results demonstrated that mobile phone exposure modifies brain excitability. These studies provide further support for the effect of mobile phone exposure on brain physiology. Hamblin et al. [15] found that mobile phone exposure may affect neural activity with differential effects in the right and left hemisphere in an auditory oddball task. Eliyahu et al. [16] and Luria et al. [17] identified a lateralized effect of RF EMF conditions in cognitive performance. The results of these studies show that there is possible evidence for differential effects in cognitive performance and/or brain physiology in the right and left hemispheres of the brain.

In the past, it was thought that damage to the left hemisphere was more likely to lead to depressed behaviour compared to similar damage to the right hemisphere [18, 19]. For example, the closer the damage to the frontal pole of the left frontal lobe [20], the more severe depressive symptoms were. It was thought that damage to the left frontal region resulted in a deficit in behaviour and experience with a lowered threshold for the experience of sadness and depression. More recent studies and reviews [21] however have questioned whether the negative emotions of sadness and depression are more specific to the left hemisphere, as findings have shown that lesions anywhere in the right hemisphere can cause difficulties with perception of emotional expressions (for example, [22]).

The aim of this experiment was to find out if the presence of RF EMF exposure from a mobile phone might affect cognitive performance in a visual task and have differential effects in the right and left hemisphere. A further hypothesis was that the lateralized effect would be related to well-being.

2 Method

2.1 Experimental Design

There were three phone conditions. The first phone condition was an active phone on the right side of the head (active right) and an inactive phone on the left. The second phone condition was an active phone on the left side of the head (active left) and an inactive phone on the right. The third condition had an inactive phone on the right side of the head and an inactive phone on the left (both off). This was the control or sham condition. There were 24 combinations of active right, active left and both phones in the first block of trials and a further 24 combinations in the second block. A visual detection task was designed with 72 dots appearing equally in the left and right visual fields and a single blind, counterbalanced crossover design was used. Six trials were run with a well-being test developed by the Bristol Well Being Group [23]. This was administered at the end of each trial to assess the mood of participants.

2.2 Participants

Forty-eight university students, 24 females and 24 males, were tested in all phone conditions. Ages ranged from 18 to 29 years ($M = 21.04 \pm 2.98$). Participants had used mobile phones for 8.5 years on average ($M = 8.56 \pm 2.71$ with a range of use of 2–16 years). Mean number of texts sent were 38 (38.28 ± 36.28) and mean number of calls made were 2 (2.29 ± 2.55). The group ($n = 48$) was made up of 67% describing themselves as heavy mobile phone users and 33% as light users. Participants were asked on which side of the head they usually held their phone: 79% said the right side of the head, 19% the left side of the head and 2% both sides. Participants had to be neurologically healthy to participate. The University Ethics Committee had approved the study.

3 Materials

A recording, not heard by the participants, was played from a laptop to the phone to create an active phone condition. Three GSM 900 MHz Nokia 130 (RM-103) phones were used. The phones had a maximum specific absorption rate (SAR) of 1.28 kW/kg over 10 g when held against the head. Participants could not tell if the phone was active or not. This was verified were at the end of the experiment. A socket that is usually attached to the connection to make hands free operation of the phone possible had been cut. A temperature sensor was attached to the back of the active phone with matt black tape so that the amount of heat created by the phone could be measured. The time of day was also recorded. Demographic and well-being data was collected in questionnaires. At the beginning of the experiment, participants were asked about their health, health behaviour, mobile phone usage behaviour and personality traits. At the end of each of the six trials, participants completed an online well-being scale which focuses on mood, alertness, and engagement.

3.1 Procedure

Participants sat 50 cm away from the computer monitor. The center of the monitor was positioned at eye level and the monitor screen was positioned vertical to the viewer. Phones were positioned on either the right or left posterior side of the head in the area of the occipital lobe. They were kept in place by a headband. The antennae on the phones were on the outside of the phone, positioned 2 cm away from the top of the phone and over the area of the bottom of the posterior lobe, the side of the temporal lobe and the top of the occipital lobe.

Participants were requested to fixate on a central stimulus, a cross, at the middle of the screen. Whenever the experimental stimulus, a dot, appeared in either the left or right of the central stimulus on the screen, two keys had to be pressed at the same time, as quickly as possible. The importance of responding "as quickly as possible" but without making mistakes, was emphasised to participants. There were two practice sessions before the six visual task trials began. One practice session was for the visual search task and the other was for the well-being scale. It was important that all participants maintain a similar level of alertness and fixated on the center of the screen. They were asked to try not to look left or right and were informed that if they looked in a half of the screen when a target occurred in the other side, they might miss the target. The keys were "X" and ".". Both hands were used when a response was made, and participants were informed of the importance of using both hands to respond to each detection on each trial. Three participants were excluded as outliers as trials were rejected if they were 2SD ± M. Data was therefore analysed for 45 participants, 23 males and 22 females.

4 Results

A significant difference in reaction time in the phone condition, $F(2, 88) = 3.24$, $p = .044$ was found in a 3-way ANOVA for field, block (trial). This indicates that the phone condition is making a difference to how participants are performing in the task, and RF EMF exposure could be affecting performance. No effect of visual field, $F(1, 44) = 2.53$, $p = .119$ or block were found, $F(1, 44) = 1.85$, $p = .181$. The performance of participants is not affected by the visual field as a variable on its own, and the lack of statistical significance concerning block indicates there is not a learning curve effect. Further analyses indicated no interaction between field and block, $F(1, 44) = 0.22$, $p = .645$, but a significant difference was found between visual field and phone, $F(1, 44) = 3.56$, $p = .031$. No interaction between block and phone was found, $F(2, 88) = 1.17$, $p = .314$, or between field and block and phone, $F(2, 88) = 0.13$, $p = .874$. A pairwise comparison between the active right condition and the phones off condition, the active left and the phones off condition showed a significant mean difference, 6.11, 95% C.I. = (0.069–12.149), $p = .047$, where the active phone on the left yielded slower response times than the inactive phones. The p value was Bonferroni adjusted. It was shown that RT is faster when the phones are not on (Condition 3), in the right visual field, when the active phone positioned over the right hemisphere (Condition 1) and in the right visual field compared to when the phone is in the active phone condition over the right hemisphere is in the left visual field. When the active phone is positioned over the left hemisphere (Condition 2), RT is slower in the right visual field. This is represented in Fig. 1.

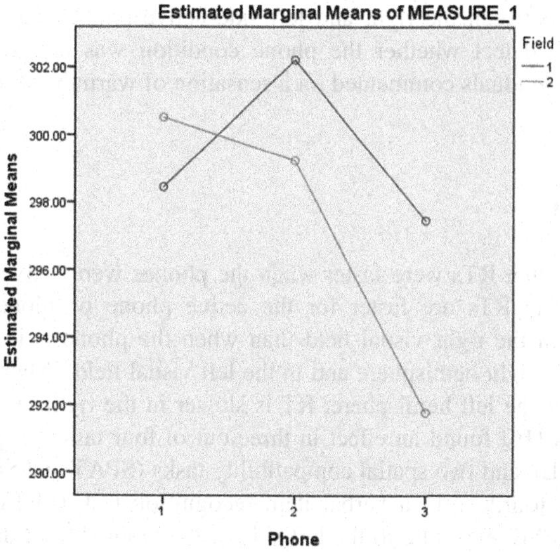

Fig. 1. Reaction times of participants in the right and left visual fields in the different phone conditions

4.1 Well-Being Data

Well-being data was analysed using a 2-way ANOVA with two levels of 'block' and three levels of 'phone'. There were three mood variables of interest: boredom, sadness, and physical fatigue. A significant effect of boredom was found, $F(2, 88) = 3.24$, $p = .044$. A test of within subject contrasts revealed no significance for the active phone over the right hemisphere (Condition 1) when compared to the inactive phone/control condition (Condition 3), $F(1, 44) = 3.07$, $p = .087$. The result is significant when the phone positioned over the left hemisphere (Condition 2) is compared to the inactive phones/control condition (Condition 3), $F(1, 44) = 6.75$, $p = .022$. These results show that participants were less bored when the phones are active, and when the active phone was over the left hemisphere. A significant effect of phone condition on sadness was found, $F(2, 88) = 4.23$, $p = 0.17$. Participants report feeling greater sadness over the left hemisphere (Condition 2) when the active phone over the right hemisphere (Condition 1) is compared with the active phone over the left hemisphere (Condition 2), $F(1, 44) = 7.59$, $p = .008$. The result survives Bonferroni modification. A significant effect was shown for block and phone, $F(2, 88) = 7.08$, $p = .001$. Further comparisons showed significance for the right hemisphere when compared to the inactive/control condition, $F(1, 44) = 7.24$, $p = .010$ and for the left hemisphere when compared to the inactive/control condition, $F(1, 44) = 13.74$, $p = .001$. Participants are less physically fatigued when the phone is active.

4.2 Temperature

Temperature data was analysed using a one-way ANOVA. It was predicted that there would be an increase of temperature in the active phone condition. Analysis showed that there was an increase of 0.5 °C. Further analysis showed that there was a very small difference between the right and left side active phone condition. Participants were not able to detect whether the phone condition was active, inactive, or off, although three individuals commented on a sensation of warmth from where the phone was on the head.

5 Discussion

The results show that RTs were faster when the phones were active in the visual dot detection task. The RTs are faster for the active phone positioned over the right hemisphere and in the right visual field than when the phone is in the active phone condition over the right hemisphere and in the left visual field. When the active phone is positioned over the left hemisphere, RT is slower in the right visual field.

Eliyahu et al. [16] found an effect in three out of four tasks: a spatial item recognition task (FACE), and two spatial compatibility tasks (SPAT and SIMON). No effect was found for a fourth task, a verbal item recognition task (LETTER). The effects showed that RF EMF exposure to the left side of the brain slowed down the left-hand response time, in the second later part of their experiment. This left sided difference was not, however, thought to be due to a hemisphere dependence, as the task functions affected were related to activities of both hemispheres.

In a further experiment by this team, it was found that the average RT of right hand responses with left side RF EMF exposure was significantly longer than those of right side exposure and the control condition [17]. The task performed was a spatial working memory task and the significant results were found only in the first two blocks out of a total of 12 blocks of trials. Again, it was not possible to identify a clear explanation for these results. Differences in the right and left hemisphere functions were considered but it was thought there were other reasons, for example, phone type and model, the positioning of the phone for the experiment, exposure methodology, exposure time, type of cognitive task. The results of Eliyahu et al. [16] and Luria et al. [17], however, are in line with the result in the current experiment. If taken together and with reference to research on cerebral brain asymmetry that supports a right hemisphere advantage for visuospatial analysis [24–26] these results might suggest that different causal mechanisms are operating in the right and left hemisphere. It is suggested they are differentially affected by mobile phone exposure.

Jech et al. [9] carried out a visual oddball experiment with patients with narcolepsy-cataplexy. The team concluded that the RF EMF from a mobile phone may suppress excessive sleepiness and improve performance while completing the task. It was observed that RF EMF effects occurred in situations when the target stimulus appeared exclusively in the right hemifield. This was thought to reflect an interaction of the RF EMF with the right hemisphere function. It might indicate either a direct stimulating effect of the RF EMF to the right hemisphere neurons or the mobile phone exposure might produce an inhibition of processes which normally reduce the processing of information coming from the left hemisphere. The mobile phone exposure significantly heightened the amplitude without affecting the endogenous complex latency. So, instead of a change in the rate of processing for the target stimulus, there was an improvement in performance while the P3a wave was being generated. These studies, along with the current experiment, give further support to a lateralised effect of mobile phone exposure.

A temperature increase of 0.5 °C was found in the active phone condition. Hyland [27] suggested that the body's thermoregulatory homeostasis mechanisms can accommodate a temperature rise as long it does not exceed 1 °C. Three participants commented on a sensation of warmth from the phone but could not determine whether the phone was active, inactive, or off when asked at the end of the experiment. So, the effect of temperature in the active condition was not possible to identify clearly.

Regarding well-being, participants reported greater sadness when the active phone was over the left hemisphere than when the active phone was over the right hemisphere. They were also less bored when the phones were active especially when the active phone was over the left hemisphere. Finally, participants demonstrated less physically fatigue when the phones were active.

It would seem that the effect of the mobile phone exposure creates a mood effect and especially over the left hemisphere. Brain activation patterns associated with emotional states have been investigated for pleasant and unpleasant emotions [28]. George et al. and Canli et al. [28, 29] found transient sadness and happiness affected different brain regions in divergent directions and not merely opposite activity in identical brain regions. Transient sadness significantly activated bilateral limbic and paralimbic structures (cingulate, medial prefrontal and mesia temporal cortex) as well as brain stem, thalamus, and caudate/putamen. In contrast, transient happiness had no areas of significantly increased activity but was associated with increased cerebral blood flow especially in the right prefrontal and bilateral temporal region. In the current experiment, participants seem to be stimulated by the phone being positioned over the left hemisphere by showing less boredom, but at the same time, experience greater sadness. They are also stimulated by the phone in the active condition, being less physically fatigued.

In conclusion, the RT is faster for the active phone over the right hemisphere and in the right visual field than when the phone is in the active phone condition over the right hemisphere and in the left visual field. When the active phone is positioned over the left hemisphere, RT is slower in the right visual field. Also, when the active phone is over the left hemisphere, participants report greater sadness compared to the active phone over the right hemisphere. Previous studies on asymmetrical brain function suggest deactivation of the left hemisphere increases negative affect. The combination of the results here suggests a potential mood effect is operating. Future research could investigate the lateralized effects of mobile phone exposure and mood effects. One suggestion might be to replicate this experiment using neuro-imaging techniques, so that areas of the brain that are affected by RF EMF exposure can be identified.

References

1. Pew Research Centre, Mobile Technology Fact Sheet, February 2018. www.pewinternet.org/factsheet/mobile. Accessed 30 May 2018
2. Hardell L, Carlberg M, Soderqvist F, Hansson Mild K (2008) Meta-analsis of long-term mobile phone use and the association with brain tumours. Int J Oncol 32:1097–1103
3. Sadetaki S, Chetrit A, Jarus-Hakak A et al (2009) Cellular phone use and risk of benign and malignant parotid gland tumors – a nationwide case control study. Am J Epidemiol 164 (7):637–643
4. Utton T (2017) Radiation from mobiles may lead to brain damage. http://www.dailymail.co.uk/health/article-124179/Radiation-mobiles-lead-brain-damage.html. Accessed 16 Nov 2017
5. Zhang J, Sumich A, Wang G (2017) Acute effects of radio frequency electromagnetic fields emitted by mobile phone on brain function. Bioelectromagnetics 38(5):329–338
6. Lv B, Chen Z, Wu T, Shao Q, Yan D, Ma L, Lu K, Xie Y (2014) The alteration of spontaneous low frequency oscillations caused by electromagnetic fields exposure. Clin Neurophysiol 125:277–286
7. Goshn R, Yahia-Cherif L, Hugueville L, Ducorps A, Lemarechal J-D, Thuroczy G, de Seze R, Selmaoui B (2015) Radio-frequency signal affects alpha band in resting electroencephalogram. J Neurophysiol 113:2753–2759
8. Freude G, Ullsperge S, Eggert S, Ruppe I (2000) Microwaves emitted by cellular telephones affect human slow brain potentials. Eur J Appl Physiol 81:18–27
9. Jech R, Sonka K, Ruzicka E, Nebuzelsky Bohm J, Juklickova M, Nevsimalova S (2001) Electromagnetic field of mobile phone affects visual event related potential in patients with narcolepsy. Bioelectromagnetics 22(7):519–528
10. Rodina A, Lass J, Ripulk J, Bachmann T, Hinrikus H (2005) Study of low level microwave field by method of face masking. Bioelectromagnetics 26:571–577
11. Mortazavi SMJ, Rouintan MS, Taeb S, Dehghan N, Ghaffarpanah AA, Sadeghi Z, Ghafouri F (2012) Human short-term exposure to electromagnetic fields emitted by mobile phones decreases computer-assisted visual reaction time. Acta Neurol Belg 112:171–175. https://doi.org/10.1007/s13760-012-0044-y
12. Vecchio F, Buffo P, Sergio S, Iacoviello D, Rossini PM, Bablioni C (2012) Mobile phone emission modulates event related desynchronization of alpha rhythms and cognitive-motor performance in healthy humans. Clin Neurophysiol 12:121–128
13. Croft RJ, Chandler AP, Burgess AP, Barry RJ, Williams JD, Clarke AR (2002) Acute mobile phone operation affects neural function in humans. Clin Neurophysiol 113:1623–1632
14. Ferreri F, Curcio G, Pasqualetti P, Gennaro L, Fini R, Fossini PM (2006) Mobile phone emissions and human excitability. Acta Neurol 60:188–196
15. Hamblin DL, Wood AW, Croft RJ, Stough C (2004) Examining the effects of electromagnetic fields emitted by GSM mobile phones on human event-related potentials and performance during an auditory task. Clin Neurophysiol 115:171–178
16. Eliyahu I, Luria R, Hareven R, Margaliot M, Nachshon M, Gad S (2006) Effects of radiofrequency radiation emitted by cellular telephones on the cognitive functions of humans. Bioelectromagnetics 27:119–216
17. Luria R, Eliyahu I, Margaliot MH, Meiran N (2009) Cognitive effects of radiation emitted by cellular phones: the influence of exposure side and time. Bioelectromagnetics 30:198–204
18. Gainotti G (1972) Emotional behaviour and hemispheric side of lesion. Cortex 8:41–55

19. Sakeim HA, Greenberg MS, Weiman AL, Gur R, Hungerbuhler JP, Geschwind N (1982) Hemispheric asymmetry in the expression of positive and negative emotions. Arch Neurol 39:210–218
20. Robinson RG, Kubos KL, Starr LB, Rao K, Price TR (1984) Mood disorders in stroke patients: importance of location of lesion. Brain 107:81–93
21. Harmon-Jones E, Gable PA, Peterson CK (2010) The role of asymmetric frontal cortical activity in emotion-related phenomena: a review and update. Biol Psychol 84:451
22. Kolb B, Taylor L (1991) Affective behaviour in patients with localised cortical excisions: role of lesion site and side. Science 214:89–90
23. Bennett K, Magee K, Pleydell-Pearce C (2017) Bristol well being group, state well being scale, work in progress
24. Davidoff JB (1982) Studies with non-verbal stimuli. In: Beaumont JG (ed) Divided visual field studies of cerebral organisation. Academic Press, London
25. Young AW, Ratcliffe G (1982) Visuospatial abilities of the right hemisphere. In: Young AW (ed) Functions of the right cerebral hemisphere. Academic Press, London
26. Rhodes G (1985) Lateralised processes in face recognition. Br J Psychol 76:249–271
27. Hyland GJ (2000) Physics and biology of mobile telephony. Lancet 356:1833–1836
28. George MS, Ketter TA, Parekh PI, Howitz B, Herscovitch P, Post RM (1995) Brain activity during transient sadness and happiness in healthy women. Am J Psychiatry 152:341–351
29. Canli T, Desmond JE, Zhao Z, Glover G, Gabrieli JDE (1998) Hemispheric asymmetry for emotional stimuli detected with fMRI. Neuroreport 9(14):3233–3239

Comparison of Muscle Activity in Young People and Seniors During Hip Adduction and Abduction Movements According to Backrest Angle

Juyoung Na[1,2](✉), Jaesoo Hong[1](✉), KwangTae Jung[2](✉), ByeongHee Won[1](✉), and JongHyun Kim[1](✉)

[1] Biomedical System & Technology Group, Korea Institute of Industrial Technology, Cheonan, Korea
{njy,jshong94,bhwon,ddalki}@kitech.re.kr
[2] Department of Industrial Design Engineering, KOREATECH, Cheonan, Korea
ktjung@koreatech.ac.kr

Abstract. The purpose of this study was to analysis patterns of changes in muscle activity and subjective assessment at various backrest angles during lower-limb rehabilitative exercises across different age groups to determine the appropriate backrest angles during exercise. Elderly people experience difficulties in normal walking due to their reduced muscle strength, balance, and flexibility caused by aging. Although equipment has been developed for muscle strengthening and rehabilitative exercises, studies on exercise postures and chair backrests are relatively insufficient. In this study, participants performed hip adduction and abduction rehabilitative exercises. Electromyogram test and a subjective questionnaire survey were conducted for 15 healthy adults in their 20s and 4 healthy elderly persons aged 65 years or older. Differences in muscle activity level according to backrest angle were observed. A significant difference in angle preference was found between young and old participants ($p < 0.05$). Results of this study may be used as ergonomic data to develop rehabilitative exercise equipment.

Keywords: Backrest angle · Rehabilitation · Exercise posture

1 Introduction

Physical functions deteriorate more drastically in old age than in other life stages. A previous study has reported that muscle strength will decrease with age [1–3]. Muscle strength, balance, and flexibility are essential factors for efficient and stable walking. In older adults, spinal changes and poor posture due to educed flexibility and muscle strength can lead to unstable gait patterns. Irregular gait rhythms or asymmetrical gaits are observed to have poor postures. The cause of such poor posture is usually weakened hip abductor muscle or leg length discrepancy [4]. Musculoskeletal changes due to aging not only induce changes in gait function, flexibility, and muscle strength, but also reduce changes in range of joint motion. James and Parker [5] have

S. Bagnara et al. (Eds.): IEA 2018, AISC 818, pp. 628–634, 2019.
https://doi.org/10.1007/978-3-319-96098-2_77

reported that the range of joint motion decreases with age. Kramer et al. [6] have reported that older women have 61% hip abductor muscle strength compared to younger women. Reduced hip abductor muscle strength significantly affects gait function, stability, and balance. The hip adductor muscle is associated with sphincters. Weakened hip adductor muscle can eventually lead to urinary incontinence.

Urinary incontinence also refers to involuntary urination. Urinary incontinence not only affects physical functions, but also affects psychological health. It is associated with reduced quality of life [7]. Therefore, hip joint rehabilitative exercises are crucial for older adults. Rehabilitative exercises can directly and preemptively prevent disabilities [8]. In response to the need of older adults for rehabilitative exercises, it is necessary to develop rehabilitative exercise equipment that can strengthen hip adductor and abductor muscles.

Exercise posture is an important factor that affects exercise outcome. The role of a chair backrest is to attenuate stresses exerted on the vertebral column by relaxing the erector spinae musculature while maintaining lumbar lordosis and increasing comfort [9].

The backrest angle directly determines the posture of the user during exercise. It can significantly affect the user's overall satisfaction with exercise experience as backrest is the part of the exercise equipment that comes in direct contact with the user. Although chair backrests have been studied in various fields, most studies have focused on chairs for office use or car seats. Carcone and Keir [10] have studied backrest pressure, spinal posture, and comfort during the use of chairs in a computer workstation. Shibata and Maeda [11] have studied backrest steepness to prevent back pain in occupational drivers. Although many studies have been conducted on chairs, research on chairs used in rehabilitative exercises with chair backrests is relatively insufficient which hinders the development of rehabilitative exercise equipment for lower limbs.

Therefore, the objective of this study was to determine optimal backrest angles for rehabilitative purposes by conducting electromyogram (EMG) test at different backrest angles during hip adductor and abductor muscle exercises. A subjective questionnaire survey was also conducted.

Please note that the first paragraph of a section or subsection is not indented. The first paragraphs that follows a table, figure, equation etc. does not have an indent, either.

2 Methods

2.1 Subjects

Participants were divided into the young group (YG), which consisted of 15 healthy young adults in their 20s, and the senior group (SG), which consisted of 4 older adults aged 65 years or older (Table 1). Patients with serious chronic disease, musculoskeletal pain, musculoskeletal disease, mental disorder, or skin allergy, and those who were deemed unsuited for research participation by our researchers, were excluded. Participants who provided voluntary consent to research participation were included in this study.

Table 1. Characteristics of subjects.

Group	Sex	Number	Age (years)	Height (cm)	Weight (kg)
Young group	Male	9	23.6 ± 3	176.9 ± 5.1	71.7 ± 7.8
	Female	6	26.3 ± 1	163.3 ± 5.9	53.3 ± 6.2
Senior group	Male	4	72.75 ± 3.5	168.25 ± 5.7	64.5 ± 3.4

2.2 Test Protocol

Changes in the level of muscle activity according to the backrest angle during lower-limb rehabilitative exercises were observed, and a subjective questionnaire survey was conducted. An experimental chair with an adjustable backrest (backrest angle could be adjusted by 10°) was prepared for this study. Five different backrest angles (90, 100, 110, 120, and 130)° were set as the experimental variables (Fig. 1). Muscle activity was measured using the Telemyo DTS wireless EMG device (Noraxon Co. USA) during the experiment, participants performed hip adduction and abduction exercises. Measurements were done in the dominant leg. The adductor longus was measured during hip adductor exercises. The tensor fascia latae and gluteus medius were measured during hip abductor exercises.

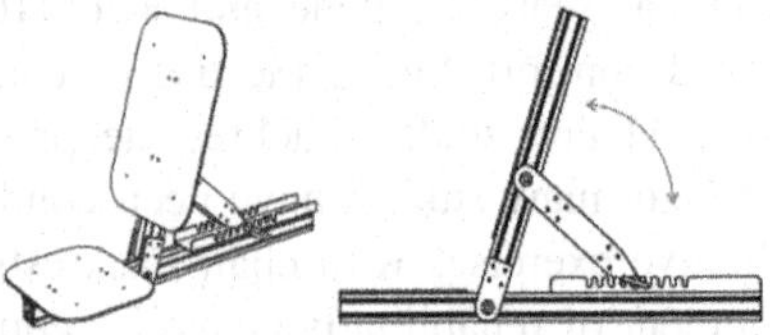

Fig. 1. Backrest-adjusting equipment produced for the experiment.

Before the experiment, the participants first received an explanation of the study, and were asked to fill out an experiment assessment form. After sensors were attached to the participants, the participants performed simple warm-up exercises, and had their maximum voluntary isometric contraction (MVIC) measured. After a 20-min break, the main experiment was begun. Measurement was performed while the participants were properly seated on the rehabilitative exercise equipment, with their hands crossed over their chest. The exercise load ranged from (40 to 60)% of the maximum muscle strength. It has been reported that exercise loads within (40–60)% of the maximum muscle strength are advisable for older adults, and that in this range of loads, older adults can still communicate during exercise [4]. The participants performed adduction exercises, followed by abductor exercises. The backrest angle was randomly set between (90 and 130)°. The participants repeated the exercise (3–5) times per angle, and after each exercise, were given a 3-min break. During the break, they subjectively assessed on a 5-point Likert scale if the backrest angle at which they had just exercised was appropriate or not. The same procedure was used for both experimental groups (SG, YG).

2.3 Data Analysis

EMG data were normalized using a bandpass finite impulse response filter of (10–350) Hz to eliminate noise. The average values of three repeated measurements obtained at each angle were analyzed. They were also converted to %MVIC for analysis. To determine if there were significant differences in the data between the two age groups, a t-test was performed using SPSS 12.0 (SPSS Inc., Chicago, IL, USA). The level of statistical significance was set at $p < 0.05$.

3 Results

3.1 Data Analysis of Hip Adduction

The adductor longus activity was measured during adduction exercises, and a significant difference was observed between the SG and the YG ($p < 0.05$). Relatively low muscle activity in the SG was observed at (110 and 120)° compared with other angles, while low muscle activity in the YG was observed at (100 and 110)° (Fig. 2). In the subjective assessment, the most preferred angles in the SG were (110 and 120)°, while in the YG they were (100 and 110)°. The two groups showed differences in the level of muscle activity and in the preferred backrest angle. Both groups showed the highest muscle activity at (90 and 130)°, which participants in both groups rated as the most uncomfortable angles. It is deemed that as exercise postures become more uncomfortable, excessive muscle use occurs, because the backrest angle is not appropriate.

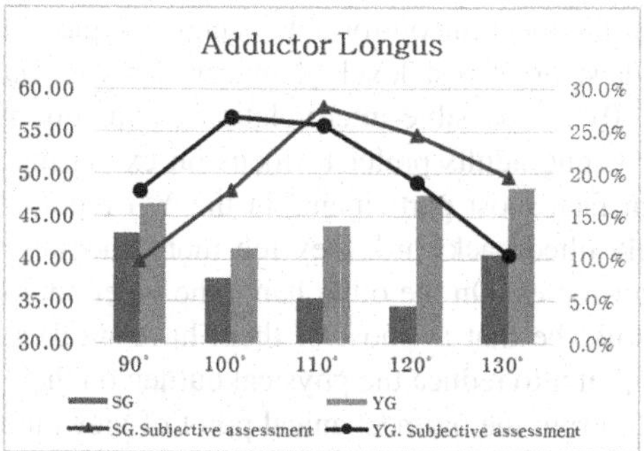

Fig. 2. Muscle activity in hip adduction and subjective assessment analysis.

3.2 Data Analysis of Hip Abduction

The tensor fascia latae and gluteus medius activities were measured during abduction exercises. A significant difference in the tensor fascia latae activity was observed between the SG and the YG ($p < 0.05$). No significant difference was observed for the gluteus medius ($p > 0.05$). In both groups, as the backrest angle increased, the level of tensor fascia latae activity increased. In the SG, as the backrest angle increased, the level of gluteus medius activity slightly increased, and in the YG, the level decreased, or did not change at all. In the subjective assessment, the SG preferred (100 and 120)° the most, while the YG preferred (100 and 110)° the most. In both groups, the highest and lowest levels of muscle activity were observed at (90 and 130)°. In the subjective assessment, these two angles were rated as the most inadequate; therefore, they are not adequate for exercise (Fig. 3).

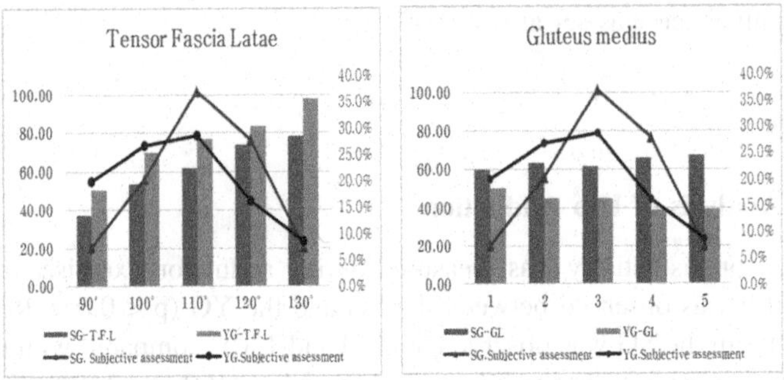

Fig. 3. Muscle activity in hip abduction, and subjective assessment analysis.

4 Discussion

A significant difference in the activity level of the adductor longus and tensor fascia latae muscles was observed during adduction and abduction exercises between the two groups ($p < 0.05$). No significant difference was observed for the gluteus medius. Both groups responded that the most uncomfortable backrest angles were (90 and 130)°. At these angles, the highest or lowest level of muscle activity was observed. The YG preferred (100 and 110)°. A possible interpretation of this observation is that as they have healthy bodies, young adults prefer to focus on exercising the leg muscles with their back straight. In fact, most participants in the YG reported that as the backrest angle was increasingly tilted backward, they felt more uncomfortable, and their muscles had to exert larger forces. On the other hand, the older adults (SG) most preferred (110 and 120)°. It could be that as most of them have weaker lower-limb and back muscles, older adults prefer to reduce the physical burden on their body by leaning their upper body backward. From the biomechanical point of view, it has been reported that changes in the backrest angle and posture can have crucial effects on the body. A study has also reported that leaning the body backward, or tilting the chair seat by >15°, can effectively distribute the body pressure in the hip area [12]. The pressure on the spinal disc is reported to decrease as the backrest angle increases. At 120°, only 50% of the disc pressure at 90° was observed. Therefore, increasing the backrest angle decreases disc pressure [13]. In the present study, the participants responded that they could exercise with more ease as their backs were comfortable at 120° than at 100°, and that they had difficulty moving their legs starting at 130°. Significant differences were observed in the activity level of the adductor longus during the adduction exercises, and in the activity level of the tensor fascia latae during the abduction exercises at all angles. However, no significant changes were observed for the gluteus medius.

A previous study reported that as the backrest tilts backward, the shearing force in the hip area increases [14]. Therefore, it is possible that changes in the gluteus medius strength according to angle changes were cancelled by the increasing shearing force.

Based on the muscle activity levels measured in this study, it is difficult to determine whether or not changing the backrest angle is effective for exercise. By analyzing muscle activities at different angles and subjective satisfaction levels, we could determine the adequate backrest angles. These findings may be used as research data for determining the appropriate postures for older adults during the use of rehabilitative equipment.

5 Conclusions

This study compared the differences in changes in muscle activity at different backrest angles during adduction and abduction exercises and the subjective assessment results between experimental groups, and determined the adequate backrest angles. Backrest angle is an important factor of exercise posture, which in turn is important for those trying to maintain a steady exercise routine. Research on the appropriate exercise postures for older adults during the use of rehabilitative exercise equipment is necessary, and exercise postures must be considered during equipment development. This study compared old and young adults using different backrest angles, and found differences between the two groups. However, aside from the backrest angle, various factors, such as the seating angle and the armrest, can affect exercise postures. In this study, we analyzed the backrest angle, as it causes the most significant postural changes. In addition, SG have fewer subjects than YG, due to the problem of experimental progress. So, Additional research is needed for more accurate results. The results of this study may be useful in the development of rehabilitative equipment for the lower and upper limbs of older adults.

References

1. Era P, Lyyra AL, Viitasalo JT, Heikkinen E (1992) Determinants of isometric muscle strength in men of different ages. Eur J Appl Physiol Occup Physiol 64(1):84–91
2. Frontera WR, Hughes VA, Lutz KJ, Evans WJ (1991) A cross-sectional study of muscle strength and mass in 45-to 78-yr-old men and women. J Appl Physiol 71(2):644–650
3. Häkkinen K, Häkkinen A (1991) Muscle cross-sectional area, force production and relaxation characteristics in women at different ages. Eur J Appl Physiol Occup Physiol 62(6):410–414
4. Bottomely JM, Lewis CB (2004) Geriatric rehabilitation: a clinical approach, 2nd edn
5. James B, Parker AW (1989) Active and passive mobility of lower limb joints in elderly men and women. Am J Phys Med Rehabil 68(4):162–167
6. Kramer JF, Vaz MD, Vandervoort AA (1991) Reliability of isometric hip abductor torques during examiner-and belt-resisted tests. J Gerontol 46(2):M47–M51
7. Prud'homme G, Alexander L, Orme S (2018) Management of urinary incontinence in frail elderly women. Obstet Gynaecol Reprod Med 21:281

8. Himann JE, Cunningham DA, Rechnitzer PA, Paterson DH (1988) Age-related changes in speed of walking. Med Sci Sports Exerc 20(2):161–166
9. Corlett EN, Eklund JAE (1984) How does a backrest work? Appl Ergon 15(2):111–114
10. Carcone SM, Keir PJ (2007) Effects of backrest design on biomechanics and comfort during seated work. Appl Ergon 38(6):755–764
11. Shibata N, Maeda S (2010) Determination of backrest inclination based on biodynamic response study for prevention of low back pain. Med Eng Phys 32(6):577–583
12. Shields RK, Cook TM (1988) Effect of seat angle and lumbar support on seated buttock pressure. Phys Ther 68(11):1682–1686
13. Vos GA, Congleton JJ, Moore JS, Amendola AA, Ringer L (2006) Postural versus chair design impacts upon interface pressure. Appl Ergon 37(5):619–628
14. Brattgård SO, Lindström I, Severinsson K, Wihk L (1983) Wheelchair design and quality. Scand J Rehabil Med Suppl 9:15

Effective Musculoskeletal Disorders Prevention

Samson Adaramola(✉)

Ergonomics Society of Nigeria, Complete-Ergo Consultant, 15c Abua Close, Rumuoibekwe Housing Estate, Port Harcourt, Rivers State, Nigeria
totalergo@gmail.com

Abstract. It is important to recognize that Musculoskeletal Disorders (MSD) prevention requires a multi-faceted approach due to the complex nature of MSD. A proactive approach and early intervention strategies are integral to the reduction and elimination of the incidence of work-related MSD. Ergonomics should also be a key consideration in the intervention programs. Most Organizations are supportive of the coordinated approach to MSD prevention being proposed by many regulated and international bodies. An approach of coordinated approach in Nigeria is recommended in this paper has yielded some measure positive results.

Keywords: MSDs · Intervention · Risk factors · Prevention strategies

1 Introduction

Nearly 1 million people each year report taking time away from work to treat and recover from musculoskeletal pain or loss of function due to overexertion or repetitive motion either in the low back or upper extremities as reported by Bureau of Labor Statistics, 1999a. Estimated workers' compensation costs associated with these lost workdays range from $13 to $20 billion annually. However, in order to determine the total economic burden, indirect costs related to such factors as lost wages, lost productivity, and lost tax revenues must be added to the cost of compensation claims, leading to estimates as high as $45 to $54 billion annually for musculoskeletal disorders reported as work-related. These figures are conservative and represent only reported cases. Several studies suggest that many disorders that could be attributed to work are not reported and therefore are not counted in any of the existing databases.

Given the global dimensions of the problem and the diverse positions of interested parties—including medical and public health professionals, behavioral researchers, ergonomists, large and small businesses, labor, and government agencies—on the strength of the evidence regarding causation, prevention of MSDs has been emphasized in multiple jurisdictions.

S. Adaramola—National President.

S. Bagnara et al. (Eds.): IEA 2018, AISC 818, pp. 635–639, 2019.
https://doi.org/10.1007/978-3-319-96098-2_78

2 Global Burden

Despite worldwide attention for more than four decades, musculoskeletal disorders (MSDs) remain a substantial concern at work and result in considerable personal and societal burden. This slow progress is not for want of trying. Prevention of MSDs has been emphasized in multiple jurisdictions. For example, in 2007 the European Agency for Safety and Health at Work organized a major campaign, "Lighten the Load – How to prevent Musculoskeletal Disorders (MSDs)" and the National Institute for Occupational Safety and Health (NIOSH) in the USA specifically identified MSDs as a major focus in their National Occupational Research Agenda. However, the results from surveys, published sick leave, and lost time data indicate we have a way to go in preventing MSDs.

These initiatives mentioned above (and others) have uncovered multiple research questions including: the effects of new forms of work, the interaction of psychosocial and mechanical exposures, changing demographics, risk assessment, identification of best practice, and the implementation of interventions in companies. While all these are clearly relevant questions, we still need to know which of these will drive us forward in the prevention of work-related MSDs.

3 Risk Assessment Approach

In industrialized countries, musculoskeletal disorders and injuries are the most common occupational Work problems. MSDs now account for over 50% of all occupational diseases. The figures are higher in industrially developing nation like Nigeria. They reduce company profitability and add to governments' social costs. Brewer (1996) said despite well-established practices, many commonly recommended workplace interventions have not been examined for their ability to improve health (MSD) outcomes or their ability to change exposures. Recent systematic reviews have not supported the success of widespread interventions.

3.1 Intervention Program in Nigerian Industries

Table 1 illustrates the most common risk factors examined in an intervention program of office workers; workers of agriculture, cleaning workers, workers assemble line,

Table 1. Risk factors for musculoskeletal disorders

Physical Risk Factors (A)	(B) **Psychosocial Risk Factors**	**Individual Risk Factors** (C)
Repetition	Job demands	Age
Force	Job control	Gender
Posture	Social relations at work	Socioeconomic status
Vibration		Pre-existing musculoskeletal disorders

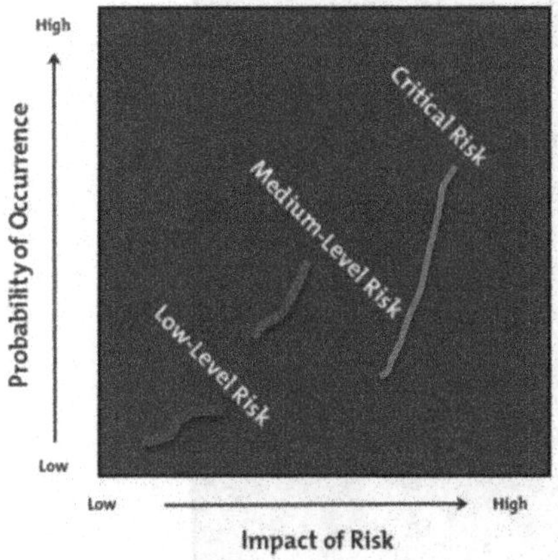

Fig. 1. Office workers

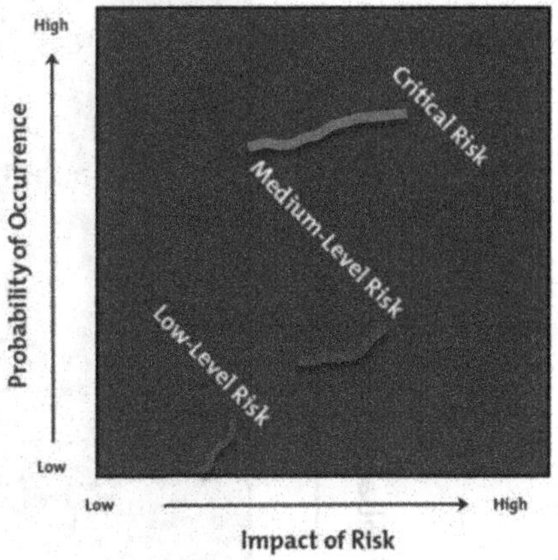

Fig. 2. Agriculture workers

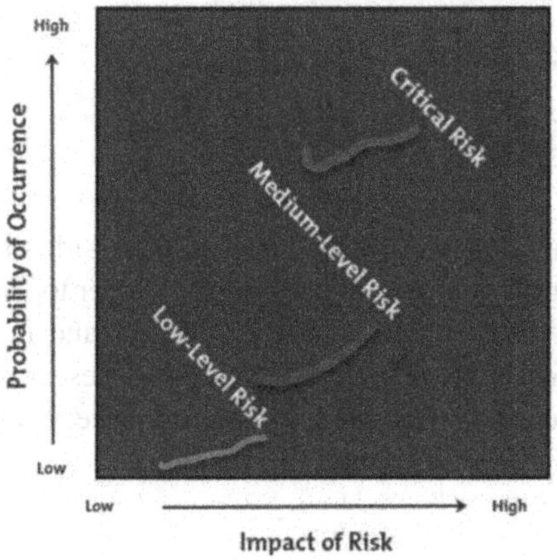

Fig. 3. Cleaning workers

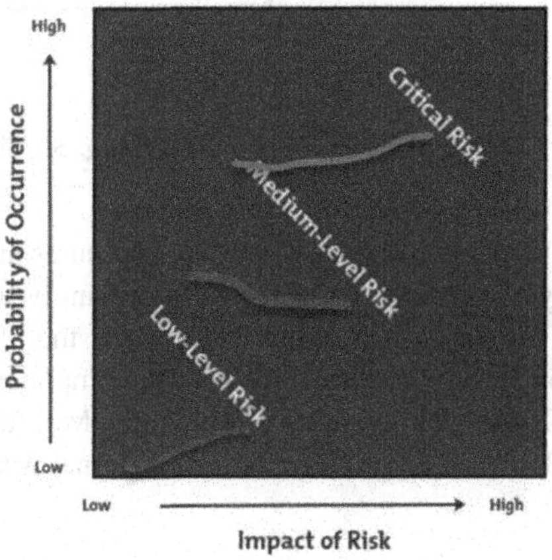

Fig. 4. Assemble workers

workers of fishing industry in Nigeria. Fifty Workmen were selected randomly in each of these industries.

The above risk factors were used as guide to examine the workers and their workplaces. The probability of the risk was determined based on the chances of occurrence and the documented record of the reported cases. The impact of each category of the risks was trended for a period of six month. The results are shown in graphical form in Figs. 1, 2, 3, 4 and 5 below. Among the workers examined, this greatly contributed to creating more and better quality jobs, as has been demonstrated in a number of successful cases in different work sectors as well in Nigeria.

The effectiveness of this risk assessment approach is the combination of physical, psychosocial factors and Individual risk factors in the intervention. However the

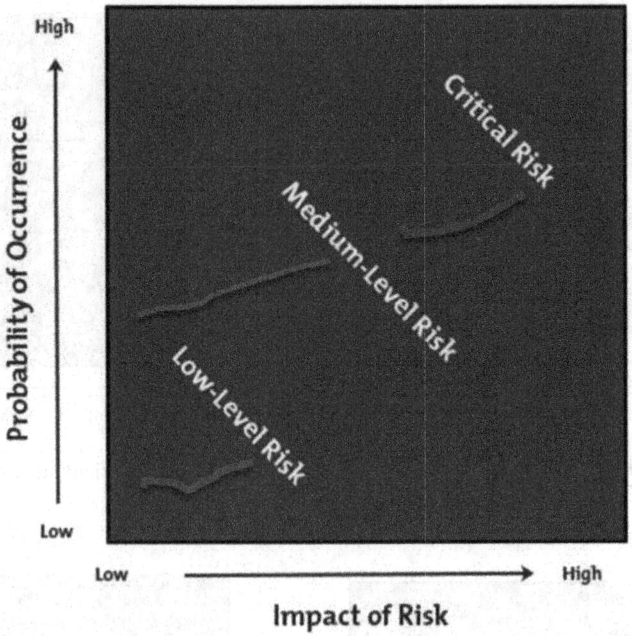

Fig. 5. Fishing workers

challenge is that as the worst exposures are eliminated – the low hanging fruit – identifying further risks becomes increasingly difficult without more training and better tools. The body of case studies supports the effectiveness of ergonomics programs and the business consensus is that multi-component programs are preferable (National Research Council and Institute of Medicine, Musculoskeletal Disorders and the Workplace: Low Back and Upper Extremities, National Academy Press, Washington DC 2001).

4 Conclusion

The prevention of MSDs is on a daily basis in several industries. Yet in order to truly make a difference there is need to fashion and test MSD prevention strategies that are feasible, socially acceptable, affordable, scalable and sustainable – and which have good coverage for substantial parts of the working population. Ideally they should also result in measurable MSD improvements within a reasonable time-frame. Frank et al. (1996) confirms that It seems that some MSDs persist for long periods. Because most people suffer from MSDs at some point in their lives, it has been suggested that to impact MSD burden, prevention activities should take place at the primary, secondary and tertiary levels simultaneously. Yet with limited resources we need to prioritize the issues that are restraining our effectiveness. We need to focus our energies on specific research that will potentially maximize our intervention impact at the societal level while we continue to build the knowledge base around MSDs and improve prevention strategies.

References

National Research Council and Institute of Medicine (2001) Musculoskeletal Disorders and the Workplace: Low Back and Upper Extremities. National Academy Press, Washington DC

Brewer S, Van Eerd D, Amick B III, Irvin E, Daum K, Gerr F, Moore S (2006) Workplace interventions to prevent musculoskeletal and visual symptoms and disorders among computer users: a systematic review. J Occup Rehabil 16:325–358

Frank J, Brooker A-S, DeMaio S, Kerr M, Maetzel A, Shannon H, Sullivan T, Norman R, Wells R (1996) Disability due to occupational low back pain: what do we know about primary prevention? Spine 21(24):2908–2917

Effectiveness of Specific Lifting Techniques and Tools on Workload in a Lifting Situation – A Case Study

Hermien Matthys[1,2(✉)], Willy Bohets[1,3(✉)], and Veerle Hermans[2,4(✉)]

[1] University College Odisee, Brussels, Belgium
hermien.matthys@gmail.com
[2] Department of Ergonomics, Group IDEWE (External Service for Prevention and Protection at Work), Leuven, Belgium
[3] DuPont de Nemours BVBA, Mechelen, Belgium
[4] Department of Experimental and Applied Psychology, Work and Organisational Psychology (WOPS), Faculty of Psychology and Education Sciences, Vrije Universiteit Brussel, Brussels, Belgium

Abstract. A lot of workers complain of backache, muscular pains and working in painful or tiring postures [1]. To reduce workload and consequently the drop out of workers due to these complaints, companies adapt the work place, provide tools and/or give training. The objective of this study is to investigate in a real working environment during picking of boxes, the effect of using good lifting techniques and/or work height adjustments on the physical workload.

Five subjects, with experience in lifting, were asked to pick up 12 boxes of 11 kg from a pallet and place them on a cart. They had to repeat this four times. Two times with the pallet on the ground, using first a bad and then a good technique, and two times with an increased height, also using first a bad and then a good technique. 3D registration of movements (TEA, CAPTIV), was used to measure the postures of back and hips. Due to the small amount of subjects, a descriptive analysis was used.

Results showed that using a good technique and increased working height can help to reduce the time spent in harmful postures of the back during lifting boxes. Further research is needed with larger groups of subjects to evaluate these results.

Keywords: Manual lifting · Lifting technique · Working height

1 Background

According to the European risk observatory report (2010), 24,7% of the European workers complains of backache, 22,8% of muscular pains and 45,5% reports working in painful or tiring postures [1]. To reduce workload and consequently the drop out of workers due to these complaints, companies adapt the work place, provide tools and/or give training. According to Van der Molen et al. [11] optimising the working height has a positive effect to reduce the duration and the frequency of harmful postures. Lavender et al. [6] confirm that the starting height of the lifting movement is an important factor

S. Bagnara et al. (Eds.): IEA 2018, AISC 818, pp. 640–647, 2019.
https://doi.org/10.1007/978-3-319-96098-2_79

to reduce the forces on the spine of employees [7]. But what about a correct lifting technique? Workers are often instructed to lift loads with a 'good' squat technique (bending the knees, keeping the back straight), not with a stoop technique (bending the back with stretched legs). But Straker et al. [8–10] conclude that when the load is positioned between knee and shoulder height, the used lifting technique is negligible. In addition, there is even no agreement which lifting technique is the best. There seems to be insufficient biomechanical evidence to train employees in the squat technique [18] and the lifting technique should be adapted to the lifting situation [3–5, 13].

2 Objective

The objective of this study is to investigate in a real working environment the effect of using good lifting techniques and/or adjusting working height on workload during picking of boxes. It is hypothesized that both using a good technique and adjusting working height, will reduce the percentage of time spent in harmful postures of the back and hips. It is also analysed if the perceived workload will decrease.

3 Methodology

This case study was performed at the medical warehouse of a hospital in Ghent (Belgium). The subjects were 2 male employees working in the medical warehouse and 3 employees responsible for ergonomics training on the work floor (2 women and 1 man). They were all informed about the study and signed the informed consent. The subjects selected lifting boxes without handles that contained liquid bags, as the heaviest lifting task (11 kg). For the case study, the worst-case scenario was used: the bottom layer of 12 boxes on a pallet, with the pallet only reachable from one side.

The five subjects had to repeat the lifting action four times. The first two times with the pallet on the ground, using first a 'bad' and then a 'good' technique. The next two times the height of the pallet was increased so that the boxes could be pushed from the pallet on the cart, also using first a 'bad' and then a 'good' technique. The first 6 boxes were always placed on the cart at 113 cm height. The next 6 boxes were placed on the cart at 72 cm. Based on previous guidelines and literature [3, 5, 13] squat bending with feet beside the box, without rotation in the back was selected as the 'good' lifting technique and shifting without rotation in the back as a good shifting technique. The bad lifting/shifting technique, with the trunk bended forward and a back rotation, was considered as 'bad' lifting technique.

Between each situation there was a 5 min break to exclude the effects of muscle fatigue. The boxes were always placed the same way on the pallet at the start of each situation. All subjects were informed that they could stop at all times when they felt it was a risk for their physical health.

The Tea CAPTIV, a 3D registration movement tool, was used to measure the postures of the back (flexion, extention, rotation and lateral flexion) and both hips (flexion). To measure the postures of back and hips 5 motion sensors were placed: 1 sensor between the shoulder blades on the back, 1 sensor on the right hip and 1 sensor on the outside of each upper leg just above the kneecap. As a reference value to calculate joint angles, the subject was asked to stand up straight with the hands next to the body, before they started with the lifting tasks.

To determine if the postures of the back and hips were "comfortable", "non-comfortable" or "to avoid" the guidelines of TEA Captive were used (Table 1), based on ISO 11226, EN 1005-4 and ISO 11228-3 [2]. For each subject and for each lifting situation, the % time in a 'comfortable', 'non-comfortable' or 'to avoid' posture for the back and hips were calculated.

Table 1. Guidelines joint angles back and hips in TEA CAPTIV

Postures	Comfortable	Non-comfortable	To avoid
Back			
Rotation left	<−15°	−15°–<−30°	>−30°
Rotation right	<15°	15°–<30°	>30°
Flexion	<30°	30°–<45°	>45°
Extention	<−10°	−10°–<−20°	>−20°
Lateral flexion left	<−10°	−10°–<−20°	>−20°
Lateral flexion right	<10°	10°–<20°	>20°
Hips			
Flexion	<70	70°–<100°	>100°

Due to the small number of subjects, a descriptive statistical analysis was used.

4 Results

In the figures below, the percentage of the time in the green (comfortable), orange (non-comfortable) and red (to avoid) zone of each subject in the four conditions is presented for rotation, flexion/extension, and lateral flexion of the back and flexion/extension of the left and right hips.

For all back postures (Figs. 1, 2 and 3), larger differences between subjects during the conditions pallet on the ground (for both situations 1 = bad, 2 = good) are found compared with the condition pallet on height (for both situations 3 = bad, 4 = good). Between the conditions bad and good, it is clear that when the pallet is on a better height, differences between techniques are smaller.

For the hips, large differences are found between subjects in hip flexion of the right or left hips. Also large differences exist between the conditions and the % time in the different zones (Figs. 4 and 5).

In summary, for the back, all the subjects spent a higher percentage of time in an acceptable posture for flexion/extension with the pallet on height than the pallet on the ground. For lateral flexion, a higher percentage of time was found using a bad technique with the pallet on height than using a bad/good technique with a pallet on the ground. Minimum 4 subjects spent a higher percentage of time in an acceptable back posture for forward flexion, rotation and lateral flexion using a good technique with a

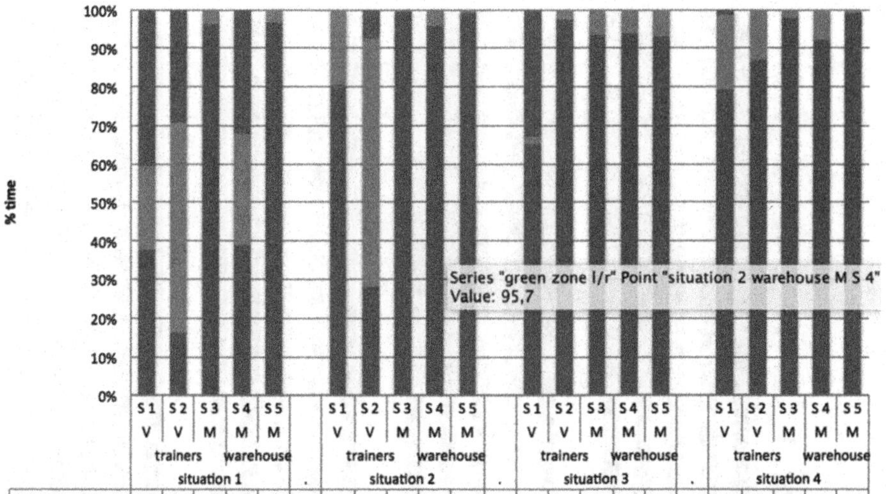

	Sit. 1 S1 V	Sit. 1 S2 V	Sit. 1 S3 M	Sit. 1 S4 M	Sit. 1 S5 M	Sit. 2 S1 V	Sit. 2 S2 V	Sit. 2 S3 M	Sit. 2 S4 M	Sit. 2 S5 M	Sit. 3 S1 V	Sit. 3 S2 V	Sit. 3 S3 M	Sit. 3 S4 M	Sit. 3 S5 M	Sit. 4 S1 V	Sit. 4 S2 V	Sit. 4 S3 M	Sit. 4 S4 M	Sit. 4 S5 M
red zone l/r	40,6	29,2	0	32	0	0	7,6	0	0	0	33	0	0	0	0	1,2	0	0	0	0
orange zone l/r	21,9	54,7	3,8	29,2	3,3	19,5	64,4	0,7	4,2	0,8	1,9	2,7	6,5	6,4	7,1	19,6	13	2,2	7,7	1
green zone l/r	37,5	16,1	96,2	38,8	96,7	80,5	28	99,3	95,7	99,2	65,1	97,3	93,5	93,6	92,9	79,2	87	97,8	92,3	99

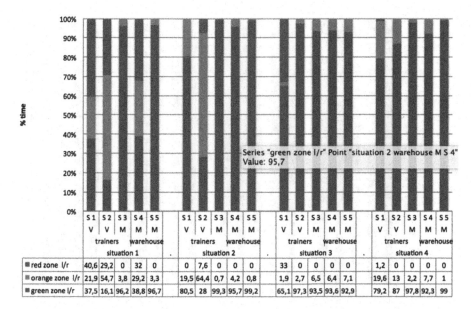

Fig. 1. % time back rotation in green, orange or red zone for all subjects (S1 to S5) in situations pallet on the ground with bad (situation 1) or good technique (situation 2) and in situations pallet at height with bad (situation 3) or good technique (situation 4). (Color figure online)

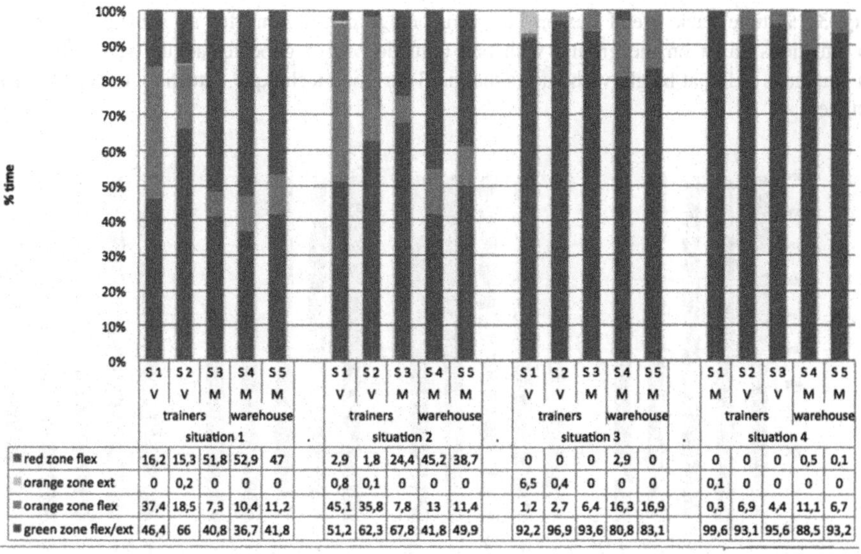

	Sit. 1 S1 V	Sit. 1 S2 V	Sit. 1 S3 M	Sit. 1 S4 M	Sit. 1 S5 M	Sit. 2 S1 V	Sit. 2 S2 V	Sit. 2 S3 M	Sit. 2 S4 M	Sit. 2 S5 M	Sit. 3 S1 V	Sit. 3 S2 V	Sit. 3 S3 M	Sit. 3 S4 M	Sit. 3 S5 M	Sit. 4 S1 M	Sit. 4 S2 V	Sit. 4 S3 V	Sit. 4 S4 M	Sit. 4 S5 M
red zone flex	16,2	15,3	51,8	52,9	47	2,9	1,8	24,4	45,2	38,7	0	0	0	2,9	0	0	0	0	0,5	0,1
orange zone ext	0	0,2	0	0	0	0,8	0,1	0	0	0	6,5	0,4	0	0	0	0,1	0	0	0	0
orange zone flex	37,4	18,5	7,3	10,4	11,2	45,1	35,8	7,8	13	11,4	1,2	2,7	6,4	16,3	16,9	0,3	6,9	4,4	11,1	6,7
green zone flex/ext	46,4	66	40,8	36,7	41,8	51,2	62,3	67,8	41,8	49,9	92,2	96,9	93,6	80,8	83,1	99,6	93,1	95,6	88,5	93,2

Fig. 2. % time back flexion/extension in green, orange or red zone for all subjects (S1 to S5) in situations pallet on the ground with bad (situation 1) or good technique (situation 2) and in situations pallet at height with bad (situation 3) or good technique (situation 4). (Color figure online)

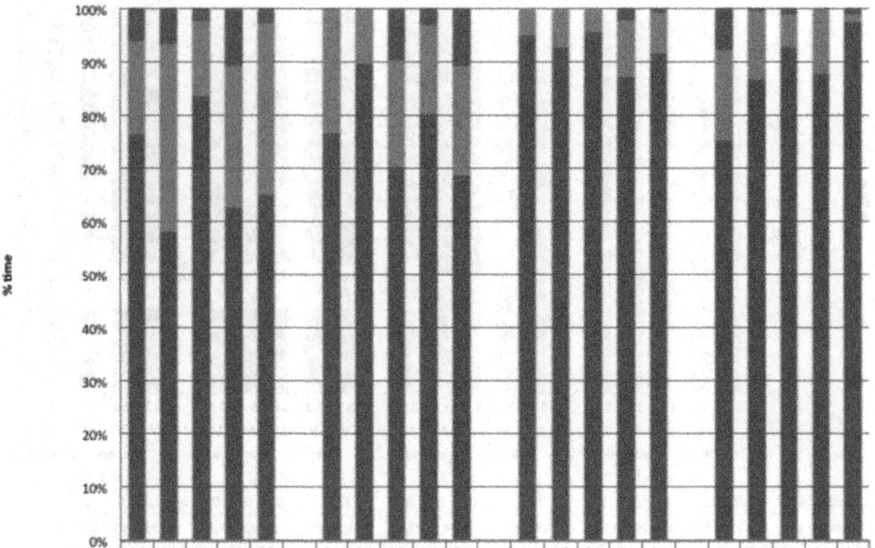

	situation 1					situation 2					situation 3					situation 4				
	trainers			warehouse		trainers			warehouse		trainers			warehouse		trainers			warehouse	
	S1 V	S2 V	S3 M	S4 M	S5 M	S1 V	S2 V	S3 M	S4 M	S5 M	S1 V	S2 V	S3 M	S4 M	S5 M	S1 V	S2 V	S3 M	S4 M	S5 M
red zone l/r	6,2	6,7	2,5	10,9	2,8	0,4	0,2	9,8	3,2	11	0	0	0	2,2	1	7,9	0,6	1,3	0,2	1,1
orange zone l/r	17,6	35,2	14	26,5	32,3	23	10,1	20	16,5	20,3	5,1	7,4	4,5	10,7	7,5	17	12,9	6	12,2	1,4
green zone l/r	76,2	58,2	83,4	62,6	64,9	76,6	89,7	70,1	80,3	68,7	94,9	92,6	95,5	87,1	91,5	75,1	86,4	92,7	87,5	97,5

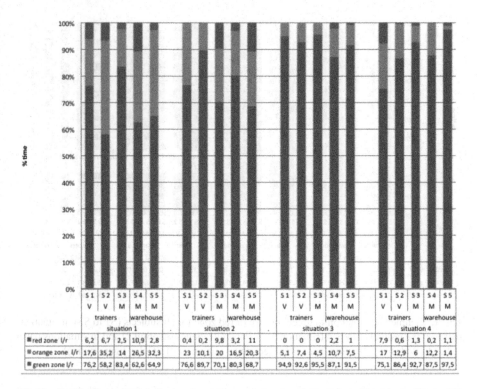

Fig. 3. % time back lateral flexion in green, orange or red zone for all subjects (S1 to S5) in situations pallet on the ground with bad (situation 1) or good technique (situation 2) and in situations pallet at height with bad (situation 3) or good technique (situation 4). (Color figure online)

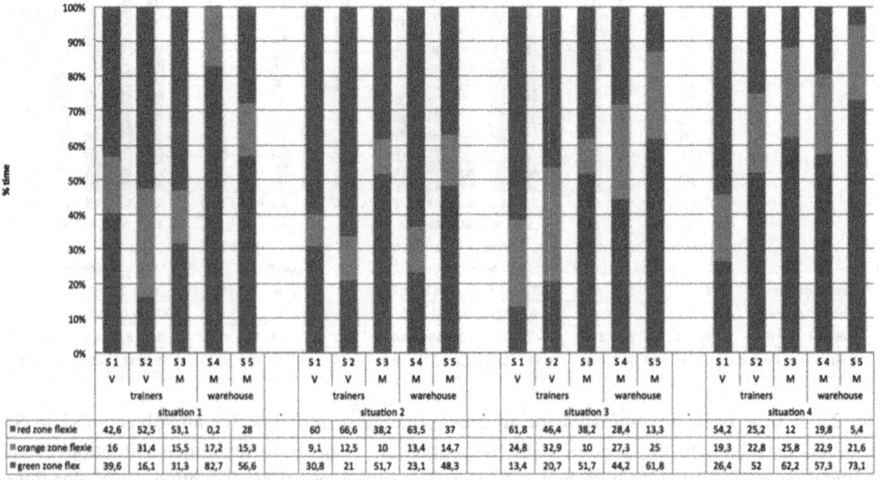

	situation 1					situation 2					situation 3					situation 4				
	trainers			warehouse		trainers			warehouse		trainers			warehouse		trainers			warehouse	
	S1 V	S2 V	S3 M	S4 M	S5 M	S1 V	S2 V	S3 M	S4 M	S5 M	S1 V	S2 V	S3 M	S4 M	S5 M	S1 V	S2 V	S3 M	S4 M	S5 M
red zone flexie	42,6	52,5	53,1	0,2	28	60	66,6	38,2	63,5	37	61,8	46,4	38,2	28,4	13,3	54,2	25,2	12	19,8	5,4
orange zone flexie	16	31,4	15,5	17,2	15,3	9,1	12,5	10	13,4	14,7	24,8	32,9	10	27,3	25	19,3	22,8	25,8	22,9	21,6
green zone flex	39,6	16,1	31,3	82,7	56,6	30,8	21	51,7	23,1	48,3	13,4	20,7	51,7	44,2	61,8	26,4	52	62,2	57,3	73,1

Fig. 4. % time hip flexion left in green, orange or red zone for all subjects (S1 to S5) in situations pallet on the ground with bad (situation 1) or good technique (situation 2) and in situations pallet at height with bad (situation 3) or good technique (situation 4). (Color figure online)

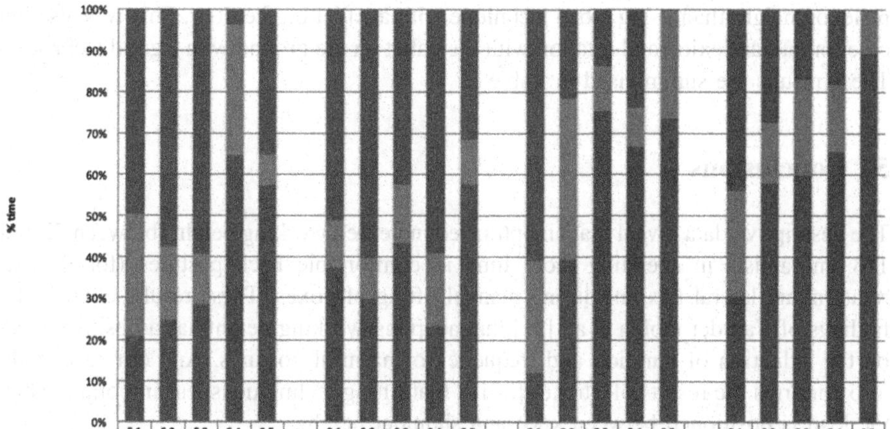

	S1 V	S2 V	S3 M	S4 M	S5 M	.	S1 V	S2 V	S3 M	S4 M	S5 M	.	S1 V	S2 V	S3 M	S4 M	S5 M	.	S1 V	S2 V	S3 M	S4 M	S5 M
	trainers			warehouse			trainers			warehouse			trainers			warehouse			trainers			warehouse	
	situation 1						situation 2						situation 3						situation 4				
red zone flexie	49,5	57,4	59,2	3,7	35,2		51,2	60,2	42,3	59,1	31,6		60,9	21,6	13,7	23,6	0,3		43,9	27,3	17	18,2	0,3
orange zone flexie	29,7	18	12,5	31,8	7,4		11,4	9,9	14,2	17,8	10,9		27,1	38,8	11	9,8	26		25,7	14,7	23,2	16,3	10,5
green zone flex	20,8	24,6	28,3	64,5	57,4		37,5	29,9	43,5	23,1	57,5		12	39,6	75,3	66,6	73,7		30,4	58	59,8	65,5	89,2

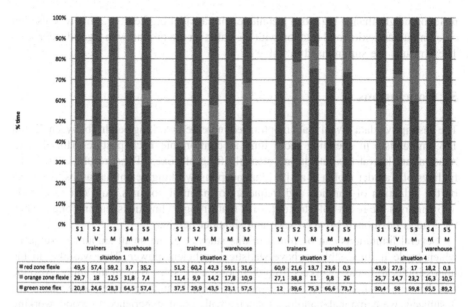

Fig. 5. % time hip flexion right in green, orange or red zone for all subjects (S1 to S5) in situations pallet on the ground with bad (situation 1) or good technique (situation 2) and in situations pallet at height with bad (situation 3) or good technique (situation 4). (Color figure online)

Table 2. % of time spent in a comfortable posture in the back for flexion/extention, lateral flexion and rotation

Pallet	Technique	Subject	Flexion/extention	Lateral flexion	Rotation
Pallet on the ground	Bad technique	S1	46,4	76,2	37,5
		S2	66	58,2	16,1
		S3	40,8	83,4	96,2
		S4	36,7	62,6	38,8
		S5	41,8	64,9	96,7
	Good technique	S1	51,2	76,6	80,5
		S2	62,3	89,7	28
		S3	67,8	70,1	99,3
		S4	41,8	80,3	95,7
		S5	49,9	68,7	99,2
Pallet on working height	Bad technique	S1	92,2	94,9	65,1
		S2	96,9	92,6	97,3
		S3	93,6	95,5	93,5
		S4	80,8	87,1	93,6
		S5	83,1	91,5	92,9
	Good technique	S1	99,6	75,1	79,2
		S2	93,1	86,4	87
		S3	95,6	92,7	97,8
		S4	88,5	87,5	92,3
		S5	93,2	97,5	99

pallet on height than using a bad technique with a pallet on the floor. This was also the case for lateral flexion and rotation with the pallet on the ground with a good technique. These results are summarised in Table 2.

5 Conclusions

The descriptive data reveal that an optimised increased working height (between 72 and 113 cm) assists in spending more time in comfortable back postures (for flexion, rotation and lateral flexion) during manual lifting of boxes. These results confirm the findings of Van der molen et al. [11] that adapting working height has a positive effect on the reduction of duration and frequency of harmful postures. And our case study also confirms the results of Straker [8–10] that lifting technique is less important when the load is positioned between knee and shoulder height.

Several limitations can be found in this study, e.g. the small amount of subjects, which made it impossible to have statistical sufficient power. This was due to the impossibility to release workers from their jobs at the same time of the study. Also the fact that subjects had different experiences in manual handling causes large variations: 3 subjects were manual handling trainers, with good experience in good working techniques, but they had difficulties in performing the bad technique. Furthermore, since the time to release people from work had to be limited, the changing time between conditions had to be as short as possible and no ad random assignment in changing order of conditions between subjects was possible.

References

1. European Agency for Safety and Health and Work (2010). OSH in figures: work-related musculoskeletal disorders in the EU – Facts and figures
2. Kapitaniak B (z.d.) Proposition de zones angulaires. INRS, France
3. Kingma I, Bosch T, Bruins L, van Dieën JH (2004) Foot positioning instruction, initial vertical load position and lifting technique: effects on low back loading. Ergonomics 47 (13):1365–1385
4. Kingma I, Dieën JH, Looze M, Toussaint HM, Dolan P, Baken CTM (1998) Asymmetric low back loading in asymmetric lifting movements is not prevented by pelvic twist. J Biomech 31(6):527–534
5. Kingma I, Faber GS, Bakker AJM, van Dieën JH (2006) Physical therapy 86(8):1091–1105
6. Lavender SA, Andersson GBJ, Schipplein OD, Fuentes HJ (2003) The effects of initial lifting height, load magnitude, and lifting speed on the peak dynamic L5/S1 moments. Int J Ind Ergon 31(1):51–59
7. Schipplein OD, Reinsel TE, Andersson GBJ, Lavender SA (1995) The influence of initial horizontal weight placement on the loads at the lumbar spine while lifting. Spine 20 (17):1895–1898
8. Straker LM (2002) A review of research on techniques for lifting low-lying objects: 1. Criteria for evaluation. Work 19:9–18
9. Straker LM (2003) A review on techniques for lifting low-lying objects: 2. Evidence for a correct technique. Work 20:83–96

10. Straker L (2003) Evidence to support using, squat, semi-squat and stoop techniques to lift low-lying objects. Int J Ind Ergon 31(3):149–160
11. Van der Molen HF, Grouwstra R, Kuijer PPFM, Sluiter J, Frings-Dresen MHW (2004) J Ergon 47(7):772–783
12. Van Dieën JH, Hoozemans MJM, Toussaint HM (1999) Stoop or squat: a review of biomechanical studies on lifting technique. Clin Biomech 14:685–696
13. Wang Z, Wu L, Sun J, He L, Wang S, Yang L (2012) Squat, stoop, or semi-squat: a comparative experiment on lifting technique. J. Huazhong Univ. Sci Technol 32(4):630–636

Safety and Quality of Maternal and Neonatal Pathway: Implementation of the Modified WHO Safe Childbirth Checklist in Two Hospitals of the Tuscany Center Trust, Italy

S. Albolino[1(✉)], G. Dagliana[1], T. Bellandi[1], N. Gargiani[2], F. Ranzani[1], I. Fusco[3], A. Maggiali[4], and L. Ventura[5]

[1] Centre for Clinical Risk Management and Patient Safety, WHO Collaborating Centre in Human Factor and Communication for the Delivery of Safe and Quality Care, Florence, Italy
albolinos@aou-careggi.toscana.it

[2] Human Health Science Department, University of Florence, Florence, Italy

[3] Centre for Clinical Risk Management and Patient Safety, Tuscany Centre Trust, Regional Healthcare of Tuscany, Florence, Italy

[4] Tuscany Center Trust, Regional Healthcare of Tuscany, Florence, Italy

[5] ISPO, Florence, Italy

Abstract. Data monitoring and reporting systems are extremely important for safety and quality of care. The use of tools to support the work of health professionals in managing clinical risk is widespread particularly in high income countries. In 2008 the World Health Organization designed a checklist for the safety during childbirth (safe Childbirth Checklist) initially dedicated to low and middle-income countries, now available also for developed countries. Two studies have been conducted by the Centre for clinical risk management and Patient safety (Florence) in hospitals in Tuscany with the aim of evaluating the impact of a modify version of the WHO checklist on clinical practice, the usability of the tool and the users' compliance through prospective pre- and post-intervention studies based on clinical records review. The effects of the checklist on professionals' adherence to clinical practices and on the standardization of processes have been investigated, in order to design and refine the tool. Both studies show that the presence of correctly compiled partogram tool in the clinical charts is strongly and significantly associated with the checklist implementation ($OR_1 = 14.9$, 95% confidence interval [CI] = 3.5, 63.9 and $OR_2 = 2.51$, 95% confidence interval [CI] = 1.41–4.47) and that the checklist promotes the interdisciplinary work.

Keywords: Checklist · Health care professionals' compliance
Usability

S. Bagnara et al. (Eds.): IEA 2018, AISC 818, pp. 648–659, 2019.
https://doi.org/10.1007/978-3-319-96098-2_80

1 Introduction

Maternal and neonatal mortality and morbidity associated with childbirth are a global health problem of the highest priority. The vast majority of maternal and newborn deaths can be prevented by relatively straightforward effective interventions. Quality of care in delivering these interventions along the continuum of care during pre-pregnancy, antenatal, intrapartum, childbirth and post-natal periods is paramount to ensure progress.

Globally, an estimate of 10.7 million women have died in the 25 years between 1990 and 2015 due to maternal causes (WHO 2015).

According to the most recent UNICEF data updated to 2015, the median number of neonatal death in 2015 was 2,682,438. The number of under-five deaths worldwide has declined from 12.7 million in 1990 to 5.9 million in 2015. Since 1990, the global under-five mortality rate has dropped from 91 deaths per 1000 live births in 1990 to 43 in 2015. Globally, the neonatal mortality rate fell from 36 deaths per 1000 live births in 1990 to 19 in 2015, and the number of neonatal deaths declined from 5.1 million to 2.7 million. However, the decline in neonatal mortality from 1990 to 2015 has been slower than that of post-neonatal under-five mortality: 47% compared with 58% globally (UNICEF et al. 2015). The main causes of children death in 2015, include preterm birth complication (18%), pneumonia (16%), intrapartum- related complications (12%), diarrhea (9%) and sepsis (9%) (WHO 2015). These data, with some minor differences, are confirmed by other recent studies (Liu et al. 2016). UNICEF Report on child mortality (UNICEF et al. 2015) reports that in developed regions the under-five mortality rate decreased from 15 deaths per 1000 live births in 1990 to six in 2015, while neonatal mortality rate fell from eight deaths per 1000 in 1990 to three in 2015. As far for Italy, the under-five mortality rate decrease from 10 deaths per 1000 live births in 1990 to four in 2015, while the neonatal mortality rate decrease from six per 1000 live births in 1990 to two in 2015.

According to estimates generated by the UN Inter-Agency Group for Child Mortality Estimation (IGME) in 2015, in Italy, the number of neonatal deaths decreased from a median number of 3657 in 1990 to a median number of 998 in 2015.

According to a recent study (WHO 2015), the Global Maternal Mortality Rate has fallen globally from the 1990 level of 385 to the 2015 level of 216. This translates to a decrease of over 43% in the estimated annual number of maternal deaths, from 532,000 in 1990 to 303,000 in 2015. In developed regions, the maternal mortality rate (per 100,000 live births) decrease from 23 in 1990 to 12 in 2015, and the number of maternal death from 3500 to 1700. In developing countries, the maternal mortality ratio (per 100,000 live births) was 430 in 1990 and 239 in 2015, while the number of maternal deaths was 539,000 in 1990 and 302,000 in 2015 (WHO 2015). In developed regions, the main direct cause of maternal death is hemorrhage (16,3%), followed by embolism (13,8%), hypertension (12,9%), abortion (7,5%), complication during delivery (5,2%) and sepsis (4,7%). Most of the death due to hemorrhage occurs during post-partum period (8%) followed by antepartum (4,8%) and intrapartum (3,5%) (Say et al. 2014). Effective prevention and management of conditions in late pregnancy, childbirth and the early new-born period are likely to reduce the numbers of maternal deaths, antepartum-and intrapartum-related stillbirths and early neonatal deaths significantly. Therefore, improvement of the quality of preventive and curative care during this critical period could have the greatest impact on maternal, fetal and new-born survival. The following thematic areas are some of those considered as high priorities for evidence-based practices in routine and emergency care: routine care during childbirth, including monitoring of labour and new-born care at birth; management of pre-eclampsia, eclampsia and its complications; management of difficult labour with safe, appropriate medical techniques; management of post-partum hemorrhage and management of maternal and new-born infections (WHO 2016). With regards to **Italy**, WHO reported for 2015 a maternal mortality rate equal to four and a number of maternal death equal to 18 (WHO 2015). For the period 2013–2015, the Italian Institute of Health reported 64 cases of maternal deaths recorded in the eight regions that took place in a maternal mortality surveillance project. Among the direct causes of death, 11 maternal deaths were related to hemorrhage, nine to infection/sepsis, four to hypertensive disorder related to pregnancy, two due to embolism and according to the study every 1000 childbirth, we assist to obstetric near miss, and in particular, every 1000 spontaneous delivery we incur in one near miss and every 1000 caesarean section we incur in three near misses (ISS 2016). According to the report on sentinel events published in 2015 by the Italian Ministry of Health (MoH Italy 2013) and related to the period 2005–2012, the most common sentinel events related to maternal and neonatal care are:

- Any intrapartum (related to the birth process) maternal death (during labour and/or delivery): maternal death or serious disease related to the labour and/or delivery. Delivery has to be intended from the beginning of the labour until three days after discharge;
- Any perinatal death unrelated to a congenital condition in an infant having a birth weight greater than 2500 g that occurs within 48 h after birth.

In the period 2005–2012, the Italian MoH registered 55 sentinel events related to intrapartum and 82 perinatal deaths. For the period 2014–2015, the Italian MoH reports that 8% of the total number of sentinel events are perinatal death (nine perinatal deaths out of 109 sentinel events), while in the same period no intrapartum correlated sentinel events were reported (MoH, Sentinel events reporting system, 2014–2015).

In high-income countries, the use of tools to support the work of health workers with the aim of managing clinical risk and improving patient safety, is quite usual. The adoption of the checklist in clinical practice, as well as the introduction of the surgical checklist has shown a reduction of deaths and complications in intensive care medicine and surgery (Haynes et al. 2009). One of the main objectives of checklist in high income countries is to promote standardized processes of care for which variation in practice may increase patient risk and the chance of medical errors that resulted in increase of complications. The use of checklist contributes also to increase communication among team members, team work and situational awareness.

WHO has always concentrated its efforts on the goal of reducing maternal mortality and morbidity, perinatal and neonatal deaths. In 2008, it developed and designed a pilot checklist for safe childbirth (WHO Safe Childbirth Check List Program) for low- and middle-income countries. The programme and the related pilot checklist were initially tested in Africa and Asia and since 2013, the pilot study has been also available to developed countries (WHO 2013). The WHO Safe Childbirth Checklist was designed as a tool to improve the quality of care provided to women giving birth. The Checklist is an organized list of evidence-based essential birth practices, which targets the major causes of maternal deaths, intrapartum-related stillbirths and neonatal deaths that occur in health care facilities around the world. Each checklist item is a critical action that, if missed, can lead to severe harm for the mother, the newborn, or both. Limited data are available on the impact of Safe Childbirth Checklist-based programs on the adherence to Essential Childbirth-related practices and no published studies have examined the impact on maternal and infant mortality and morbidity (Spector et al. 2012; Patabendige and Senanayake 2015). According to the study carried out by Gawande, the pilot testing of the Safe Childbirth Checklist in Southern India led to a marked increase in delivery of essential childbirth practices linked with improved maternal, fetal and new-born outcomes. Overall, there was an increase in adherence to accepted clinical practices from a mean of 10/29 (34%) to 25/29 (86%). While the pilot study demonstrated that the SCC programme improved health worker performance, the study was not powered to detect a change in maternal and perinatal health outcomes. The new study conducted by Gawande in Uttar Pradesh, India, showed that birth attendants' adherence to essential birth practices was higher in facilities that used the coaching-based WHO Safe Childbirth Checklist program than in those that did not, but maternal and perinatal mortality and maternal morbidity did not differ significantly between the two groups (Semrau et al. 2017).

The study presented in this paper can be defined as an implementation research whose aim is to design and to test a tool for supporting sharp-end healthcare workers coping with critical activities during delivery. The study has been conducted within the framework of the WHO Collaborative. The study is based on quantitative analysis aimed at evaluating: the checklist impact on clinical practice through a prospective pre- and post-intervention study based on clinical records review against process indicators, the usability and feasibility of the tool and the user's compliance by different healthcare operators. The study is an extension of a research conducted in 2015 in nine hospitals of the Tuscany Region, Italy, whose aim was to evaluate the adapted WHO Safe Childbirth Checklist in terms of usability, feasibility and impact on team working, communication among member's team and work organization (Albolino et al. 2015).

1.1 Aim of the Study

The study investigates the effects of the introduction of an operational checklist in terms of work organization and standardization of processes in maternal and child units and whether the WHO Safe Childbirth Checklist, adapted to the context, is able to effectively increase the adherence of the healthcare operators to fundamental clinical practices.

2 Materials and Methods

2.1 The Modified Checklist

In February 2015 the GRC formally adhered to the collaborative promoted by WHO for testing the Safe Childbirth Checklist in hospitals of the Tuscany Region. The pilot started with the adaptation of the checklist adaptation to the local context. This process has included the translation into Italian of each item of the checklist and the definition of the adequacy to the context of the various items. To this aim, a special classification

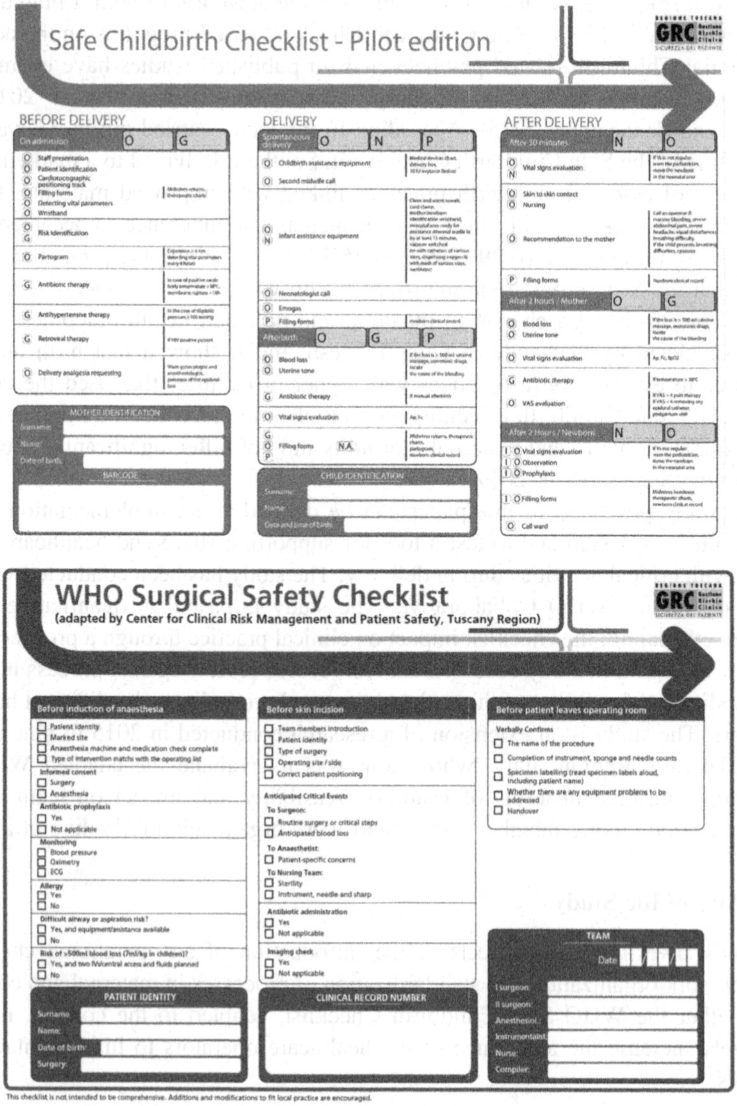

Fig. 1. Adapted WHO safe childbirth checklist

of original items was created, consisting of three categories: “suitable for the context”, “unsuitable”, “difficult to translate” so as to implement a first streamlining of the checklist. The personalization of the tool then required the inclusion of new items according to the local workflow and in line with the results of an ergonomic analysis to evaluate its usability (Albolino et al. 2015). The modified tool was then presented to a multidisciplinary group of specialists consisting of obstetricians, gynaecologists, anaesthesiologists, ergonomists and experts in clinical risk management and patient safety. A final prototype was then realised for the testing (Fig. 1).

The resulting checklist consists of two pages (Albolino et al. 2018). The first page is dedicated to assistance in the delivery room and is divided into four main blocks:

1. Preparation: activities to be verified before delivery, during labor.
2. Intrapartum: activities to be verified during the expulsion period.
3. Postpartum: activities to be verified after childbirth and before discharge.
4. Mother/child identification.

There is also a block dedicated to the identification of health personnel involved in care (obstetricians, gynecologists, anesthesiologists, pediatricians and OSS).

The second page is dedicated to assistance in the event of an emergency caesarean section in labor. It is divided into three main blocks, in turn subdivided into sub-blocks:

1. Room access
2. Time out
3. Room exit

2.2 Phases of the Study

2.2.1 Phase One: Pilot Study

First a pilot study was carried out in the maternity and neonatal unit of the Hospital of Prato, a large size hospital that hosts four delivery rooms and one room for delivery in water; it performs around 2500 deliveries a year, it registered 2603 childbirths in the year before the pilot (2014) and 1435 childbirths during the piloting period (March 2015–October 2015). The checklist was use for 1000 delivery and the enrolment of the women was conducted in accordance with the following eligibility criteria:

- Women who had access to the delivery area for spontaneous delivery.
- Women of any nationality.
- Women of any age.
- Women with any obstetric history and number of deliveries.
- Women with single or twin pregnancy.
- Women with more than 24 weeks of pregnancy.

The exclusion criteria were:

- Women who had access to the delivery area to perform a caesarean section in elective or emergency delivery and who were not in labour
- Women who spent a period in the delivery area for observation for medical reasons.

Table 1. Prepartum, intrapartum and post-partum process measures distribution between before and after checklist implementation (%) (Albolino et al. 2018)

Process measures	Before N = 141	After N = 98	p-Value*
Partogram correctly compiled in medical chart	106 (75.2)	96 (98.0)	<0.001
Hypertensive therapy when blood pressure > 100 mm Hg	1 (50)	1 (50)	1.000
Antibiotic therapy when vaginal swab test +, body heat > 38 °C or premature rupture of membranes	24 (100.0)	31 (100.0)	–
Antibiotic therapy in case of manual removal of the placenta	2 (50)	1 (100)	0.361
Pain evaluation	139 (98.6)	97 (99.0)	0.243
Analgesic treatment when VAS > 4	41 (29.1)	26 (26.5)	0.207
Manual massage when blood loss was over 500 mL	0 (0)	0 (0)	–
Uterotonic drugs when blood loss > 500 mL	7 (70.0)	10 (83.3)	0.457

Table 2. Results of the logistic model for the process measures partogram, analgesic treatment when VAS > 4, and uterotonic drugs when (Albolino et al. 2018)

	Presence of partogram in medical chart		Analgesic treat. when VAS > 4		Uterotonic drugs when blood loss > 500 mL	
	OR	95% CI	OR	95% CI	OR	95% CI
Checklist implementation	14.9	3.5, 63.9	0.8	0.5, 1.5	2.1	0.3, 16.4

Before the beginning of the pilot, the checklist was presented to health professionals through training sessions and job coaching (Albolino et al. 2018).

Researchers conducted a prospective pre and post-intervention study evaluating the effect of introduction of the checklist on some selected process measures. The clinical records were randomly sampled from the list of childbirths in June, July September 2014 and the same period after the checklist introduction, i.e. June, July, September 2015 (Albolino et al. 2018) Then sampling and data collection was carried out enrolling 141 women in the pre-intervention group and 98 women in the post-intervention.

The list, provided by the coordinators of the departments, reported an identification code, the date of delivery, the gestational age at the time of delivery, the mode of delivery and the number of newborns (to identify the twins) but didn't include information connected to the identity of the patients, so as to guarantee anonymity. Results of the data analyses show that the presence of correctly compiled partogram tool is strongly and significantly associated with the checklist implementation (OR = 14.9, 95% confidence interval [CI] = 3.5, 63.9). Compliance to the checklist was high for mid-wives (96%) and very low for obstetricians (3%) (Albolino et al. 2018).

2.2.2 Phase Two: Second Study

The results obtained with the pilot study have opened the way to subsequent studies. The modified version of the WHO Safe Childbirth Checklist was then tested in another hospital of the Tuscany Region, the San Giovanni di Dio Hospital. This is a reference birth center for the entire south-west Florentine area. The garment counted 1760 total shares in 2017 and presents an advanced second level birth point.

Objectives, methods, eligibility and exclusion criteria used were the same as those used in the pilot study (see Sect. 2.2.1). Researchers conducted a prospective pre and post-intervention study on a sample of clinical records randomly selected from the list of childbirths in June, July September 2016 and the same period after the checklist introduction, i.e. June, July, September 2017. Then sampling and data collection was carried out enrolling 100 women in the pre-intervention group and 100 women in the post-intervention.

2.3 Measurement

Any differences in the demographic characteristics of the two groups were evaluated using a t-test for continuous variables and a chi-squared test combined with a Fisher test for categorical variables. The level of education has been codified into three categories: low (less than high school, ≤ 8 years of education), moderate (up to high school or similar, 9–13 years of education), and high (undergraduate or higher, >13 years of education).

To test the various measures of interest in the preparation, in the intrapartum process and in the postpartum, the chi-square test and the Fisher test were used before and after the implementation of the checklist. Furthermore, a logistic model was implemented to study the association of results with the use of the checklist in terms of Odds Ratio (OR).

Indicators of clinical practice were considered regarding health care professionals' compliance on: partogram use, antibiotic treatment administration in case of premature rupture of membranes, temperature > 38°, vagino-rectal swab and manual placenta removal and uterotonic treatment in case of blood loss > 500 ml.

3 Results

Firstly, a comparison between the two groups has been made. As shown in Table 1, no significant differences emerged between the study groups so the two samples are comparable.

The indicators related to the clinical practice were then evaluated and the possible association with the presence of the checklist was investigated. Below, in Table 2, the results of these analyzes.

Table 3. Sociodemographic aspects of the groups (*T-test was used to compare averages for continuous variables. In case of categorical variables, chi-squared test was used)

	Pre N = 100 (%)	Post N = 100 (%)	p-Value*
Mother's average age (DS)	31.7 (7.2)	32.3 (5.8)	0,49
Level of education Mother N (%)			0,82
Low	22 (22.2)	27 (27.0)	
Middle	41 (41.4)	40 (40.0)	
High	33 (33.3)	29 (29.0)	
Other	3 (3.0)	4 (4.0)	
Marital status Mother N (%)			0,77
Maiden	41 (41.4)	46 (46.0)	
Married	54 (54.5)	51 (51.0)	
Divorced/separated	4 (4.0)	3 (3.0)	
Number of parts N (%)			0,29
1	36 (36.0)	46 (46.0)	
2	45 (45.0)	35 (35.0)	
>2	19 (19.0)	19 (19.0)	

Blood pressure was above 100 mm Hg in four women in the pre-intervention group (4%) and in 13 women (13%) in the post-intervention group, while a blood loss > 500 ml was reported in 3 women in the pre-intervention group (3%) and 20 women in the post-intervention group (20%). Since frequency of these events was low, the statistical model of analysis for the evaluation of the association between the implementation of the checklist and the compliance of the operators with regard to these specific welfare measures has not been applied (Tables 3 and 4).

Table 4. Distribution of prepartum, intrapartum and postpartum process measures of the study groups (*Chi-squared test)

	Prima N = 100 (%)	Dopo N = 100 (%)	p-value*	OR	IC 95%
Partogrammi compilati correttamente	31 (31.0)	53 (53.0)	**<0.01**	**2,51**	**1.41–4.47**
Terapia ipertensiva quando PAD > 100 mm Hg	8 (100)	13 (100)	–	–	–
Terapia antibiotica in caso di tampone vaginale +, TC > 38° o PROM	43 (100)	41 (100)	–	–	–
Terapia antibiotica in caso di secondamento manuale	2 (100)	14 (100)	–	–	–
Valutazione dolore	100 (100)	100 (100)	–	–	–
Trattamento antalgico se VAS > 4	100 (100)	100 (100)	–	–	–
Massaggio uterino esterno se PE > 500 ml	NA	NA	–	–	–
Farmaci uterotonici se PE > 500 ml	100 (100)	100 (100)	–	–	–

Antibiotic therapy was also administered in all cases of positive vaginal swab, body temperature > 38 °C and premature rupture of membranes, therefore the statistical model was not applicable.

Regarding the compliance related to the administration of antibiotic therapy in case of manual secondment, it emerged that in all cases (two cases in the group "before implementation" of the checklist and 12 in the group "after implementation of the Checklist") administration of antibiotics was recorded.

The VAS has also been evaluated, and treated if >4, in all cases.

On the contrary, statistically significant differences emerged regarding the presence of the correctly completed partogram in the medical records. In fact, a larger number of correctly completed partograms were found in the post-intervention sample (p value < 0.01). The table shows that the presence of the correctly completed partogram is strongly associated with the use of the checklist (OR = 2.51, 95% [IC] = 1.41–4.47).

4 Discussion and Conclusion

The introduction of the WHO Safe Childbirth Checklist is achieving good results in terms of acceptability and it has received a very positive feedback from end users. Adapting the WHO Safe Delivery Checklist facilitated the acceptance of the tool among health workers and improved its usability. Acceptance by users can be consider one of the most important success factors for the introduction of any new tool or safety intervention. Furthermore, the direct involvement of sharp end operators in the design process has made the process of adaptation and personalization extremely effective and focused, generating the best solution that adapts to their daily work. The multidisciplinary approach adopted by researchers allowed to create a tool consistent with principles and methods of human factors and thus ensured a participatory attitude for the implementation of the checklist.

Easiness of use and economy are factors that have certainly contributed to the acceptability of the instrument. These aspects involve considerable advantages including the speed of compilation, which does not interrupt the delivery of care

The study showed that the safe delivery checklist improves the quality and safety of assistance by encouraging the use of the partogram, an important tool to prevent the risks related to labour. The partogram is in fact a fundamental tool in clinical practice as it is able to provide a graphic representation of the progression of labor and information on the conditions of the mother and the fetus, acting as an early warning system and allowing a quick identification of problems. It therefore supports midwife regarding the timing and the decisions to be made in the management of labor. Failure to use this tool exposes mother and fetus to risks and represents negligence of the assigned obstetric staff. For this reason, the increase in the use of the partogram reported in this study is an important result, on which leverage to promote the implementation of the checklist as a tool that can support doctors and health professionals in avoiding the occurrence of adverse events during labor and delivery.

The strength of this study is certainly the pioneering role it assumes as the first adaptation and evaluations of the WHO Safe Childbirth Checklist in developed countries.

It is intention of researchers to continue with this investigation by involving more delivery units in Tuscany and from other Regions. with the aim of get close to the homogenization of operational tools and assistance in the delivery room. We are confident that the new tool we have adapted can suit a large number of structures and settings and can be a useful ally in improving obstetric care and in preventing clinical risk.

5 Limits of the Study

A first limit concerns the sample size: as described, the sample size was estimated by calculating 9% of the population, corresponding to the average percentage of patients with blood loss > 500 ml (Rogers et al. 1998; Prendiville et al. 2000). However, during the reference period, more than 1000 checklists were used in the post-intervention group, it would therefore be desirable to use a larger sample to have a greater possibility of examining also those items which, due to sample size and prevalence of situations of reference, it was not possible to evaluate, such as administration of antihypertensive therapy when PAD > 100 mm Hg and uterotonic drugs in case of blood loss > 500 ml.

A second limitation is that the study was conducted only in two hospitals, although the pilot study took place in one of the largest birth centers of the region in terms of the number of parts and personnel involved and is therefore the external validity of the study is questionable. We intend to carry out the study at other centers in order to combine data and make the results applicable to all contexts.

Costs. The study used human and technological resources of the hospitals, of the Department of Clinical Risk Management and of the University of Florence. All researchers and operators conducted the study within their working/training schedule.

Disclosure Statement. There are no ethical implications to consider in this type of study. The authors declare the absence of conflict of interest and economic interest related to this study.

References

Albolino S, Dagliana G, Meda M, Ranzani F, Tanzini M (2015) Safety and quality of maternal and neonatal pathway: a pilot study on the childbirth checklist in 9 Italian Hospitals. In: 6th international conference on applied human factors and ergonomics (AHFE 2015) and the affiliated conferences, AHFE 2015, Procedia Manufacturing, vol 3, pp 242–249 (2015)

Albolino S, Dagliana G, Illiano D, Tanzini M, Ranzani F, Bellandi T, Fusco I, Bellini I, Carreras G, Di Tommaso M, Tartaglia R (2018) Safety and quality in maternal and neonatal care: the introduction of the modified WHO Safe Childbirth Checklist. Ergonomics 61 (1):185–193. https://doi.org/10.1080/00140139.2017.1377772

Haynes AB, Weiser TG, Berry WR, Lipsitz SR, Breizat AH, Dellinger EP, Herbosa T, Joseph S, Kibatala PL, Lapitan MCM, Merry AF (2009) A surgical safety checklist to reduce morbidity and mortality in a global population. N Engl J Med 360(5):491–499. https://doi.org/10.1056/NEJMsa0810119

ISS (Istituto Superiore di Sanita) (2016) Sorveglianza della mortalità materna e grave morbosità materna in Italia, Report 2016 [National Health Institute, Surveillance of Maternal Mortality and Severe Morbidity in Italy. Report 2016]

Liu L, Oza S, Hogan D, Chu Y, Perin J, Zhu J, Lawn JE, Cousens S, Mathers C, Black RE (2016) Global, regional, and national causes of under-5 mortality in 2000–15: an updated systematic analysis with implications for the sustainable development goals. Lancet 388:3027–3035

Ministero della Salute (2013) Protocollo di Monitoraggio degli eventi sentinella, IV Rapporto (Settembre 2005–Dicembre 2011) [Ministry of Health, Protocol for Monitoring Sentinel Events, IV Report, (September 2005–December 2011)]

Patabendige M, Senanayake H (2015) Implementation of the WHO safe childbirth checklist program at a tertiary care setting in Sri Lanka: a developing country experience. BMC Pregnancy Childbirth 15:491

Prendiville WJ, Elbourne D, McDonald S (2000) Active versus expectant management in the third stage of labour. Cochrane Database Syst Rev 3(3):CD000007

Rogers J, Wood J, McCandlish R, Ayers S, Truesdale A, Elbourne D (1998) Active versus expectant management of third stage of labour: the hinchingbrooke randomised controlled trial. Lancet 351(9104):693–699

Say LL, Chou D, Gemmill A, Tuncalp O, Moller AB, Daniels J, Gulmezoglu AM, Temmerman M, Alkema L (2014) Global causes of maternal death: a WHO systematic analysis. Lancet Global Health 2(6):e323–e333. https://doi.org/10.1016/S2214-109X(14)70227-X Epub 2014 May 5

Semrau KEA, Hirschhorn LR, Kodkany B, Spector JM, Tuller DE, King G, Lipsitz S, Sharma N, Singh VP, Kumar B, Dhingra-Kumar N, Firestone R, Kumar V, Gawande A (2016) Effectiveness of the WHO safe childbirth checklist program in reducing severe maternal, fetal, and newborn harm in Uttar Pradesh, India: study protocol for a matchedpair, cluster-randomized controlled trial. Trials 17:1609. https://doi.org/10.1186/s13063-016-1673-x

Semrau KEA, Hirschhorn LR, Delaney MM, Singh VP, Saurastri R, Sharma N, Tuller DE, Firestone R, Lipsitz S, Dhingra-Kumar N, Kodkany BS, Kumar V, Gawande AA (2017) Outcomes of a coaching-based WHO safe childbirth checklist program in India. N Engl J Med 377:2313–2324. https://doi.org/10.1056/NEJMoa1701075

Spector JM, Agrawal P, Kodkany B, Lipsitz S, Lashoher A, Dziekan G, Bahl R, Merialdi M, Mathai M, Lemer C (2012) Improving quality of care for maternal and newborn health: prospective pilot study of the WHO safe childbirth checklist program. PLoS ONE 7(5):e35151

UNICEF, WHO, WB, and UN (2015) Levels & trends in child mortality, Report 2015 Estimates Developed by the UN Inter- Agency Group for Child Mortality Estimation United. UNICEF, New York

WHO (2013) Safe childbirth checklist programme: an overview. WHO, Geneva

WHO (2015) Trends in maternal mortality: 1990 to 2015. Estimates by WHO, UNICEF, UNFPA, World Bank Group and the United Nation Population Division. WHO, Geneva

WHO (2016) Standards for Improving Quality of Maternal and Newborn Care in Health Facilities. WHO. ISBN 978 92 4 151121 6

Assessment of Various Extrinsic Risk Factors Causing Pressure Ulceration in People with Spinal Cord Injury

Priyanka Rawal(✉) and Gaur G. Ray

IDC School of Design, Indian Institute of Technology, Mumbai 400076, India
priyanka24rawal@gmail.com, gaurgray@gmail.com

Abstract. Pressure Ulceration (PrU) is a leading cause of insalubrity and even mortality in people with Spinal Cord Injury (SCI). It is a major cause of re-hospitalization and subsequent mental and financial agony among them. Development of PrU in the initial phases of SCI, reduces the chances of recovery as the person becomes bedridden and thus, misses out on important period of physiotherapy and exercise. Moreover, this extended state of inactivity, in turn, mostly results in more PrUs due to prolonged exposure to pressure.

Circumstances and elements leading towards these circumstances, which expose skin to surfeit and prolonged periods of pressure or deplete its pressure bearing ability, are risk factors and risk elements for the development of PrUs.

This study was carried out on civilian SCI patients in India, a developing country, to understand the occurrence and recurrence of PrU in them. This work helped the researcher to assess various risk factors responsible for developing PrUs and conclusively delineate them based on evidence produced through direct data and experience extraction from the patients, medical experts and caregivers through semi-structured interviews and medical reports. The information collected then was analyzed for frequency and severity of PrUs and subsequent re-hospitalization cycles. The intention is to help the SCI medical community to dissipate information at a critical stage, regarding these risk factors to the stakeholders to reduce re- hospitalizations and improve their chances of recovery. Also, this study can assist SCI related product designers and manufacturers to understand product related risk factors so as to improve upon the usability and safety of their products.

Keywords: Spinal cord injury · Pressure ulcer · Re-hospitalization
Patient engagement

1 Introduction

NPUAP defines a Pressure Ulcer as "a localized injury to the skin and/or underlying tissue usually over a bony prominence, as a result of pressure, or pressure in combination with shear. A number of contributing or confounding factors are also associated with pressure ulcers; the significance of these factors is yet to be elucidated" [1].

S. Bagnara et al. (Eds.): IEA 2018, AISC 818, pp. 660–666, 2019.
https://doi.org/10.1007/978-3-319-96098-2_81

Pressure Ulcer (PrU) is a widely known phenomenon amongst the SCI medical fraternity. Pressure has been recognized as the most important extrinsic factor involved in the development of pressure ulcers for many years. Consequently, it features prominently in definitions of pressure ulcers, including the recent definition produced by the National Pressure Ulcer Advisory Panel (NPUAP) and European Pressure Ulcer Advisory Panel (EPUAP) [1, 2]. Also, it is largely known to be an avoidable complication if a few precautions are followed. The frequent recurrence was also observed in this study, as PrUs were found to be the major cause of re-hospitalization among the people with SCI considered in this study [3]. Studies indicate that the cost of treating pressure ulcers is 2.5 times the cost of treating them [4].

A lot of research has already been done in the area of PrU, reasons causing it and its preventions [5–9]. A substantial amount of literature in the domain of its effects on the body is also available, with numerous risk factors enlisted and briefly described [10–12]. But, independent studies are not reported covering the variety of causes and means for developing them, especially studies from developing countries. In most of the studies the broader picture is highlighted but the details regarding the same are either absent or inadequately elaborated. Especially, the risk factors related to the usage of equipment and products meant for people with SCI are not extensively studied and reported.

This research is thus focused on SCI patients from India, a developing country. The subjects chosen are bedridden or are not fully recovered and still have paralysis. Basically, the ones who are more prone to PrUs. Further, efforts will be made to highlight the elements, which we usually miss while providing care and information to the patients and their caregivers, especially after hospitalization apropos to PrU. This study shall take into account only a class of risk factors classified as Extrinsic Risk Factors, which are defined as the risk factors, which are attributed to development of PrU due to the mechanical elements that come in contact with the skin of the patient and also the local temperatures and moistures which affect PrU initiation and development.

2 Backdrop and Methodology

This study was conducted across three SCI rehabilitation centers in the metropolitan city of Mumbai, India. Two of these centers have an experience of more than five decades in handling people with SCI. A semi-structured questionnaire was prepared for the purpose of interviewing patients and rehab team members face to face. Audio recordings were made wherever necessary with due permissions. Care was taken to make the participants feel as comfortable as possible by letting them choose the place and time of their choice for the interviews. Efforts were made to introduce least possible amount of researcher bias in the interview procedure. Each interview was given sufficient time for the participant to get acquainted with the actual purpose of the study and answer accordingly. All the patients were made aware of their right to not answer any question they may feel uncomfortable answering and also of ending the interview whenever they deem fit.

The study included detailed discussions with the patients, caregivers and the rehab team and also cross examination of the information provided by the patients and caregivers with the available medical records. Due approvals were obtained from the ethical committee and consents of the patients were also taken in their native language.

3 Results and Discussion

A few initial questions in the interview were designed to understand the level awareness of patients about PrU. About 60% of the patients were not found to be familiar with the terms Pressure Ulcer/Pressure Sore/Bed sore. On explaining the meaning of the terms and providing description in their native language, in much more layman terms, they understood what was being discussed. Later on, questions regarding the reasons of development of PrU and ways to cure them were also asked. About 74% patients were not found to understand the reasons behind PrU development. Regarding prevention, they could only tell about changing sides every two hours while sleeping. No other reason could be pointed out by them initially (Table 1).

Table 1. The demographics of the participants (N = 33)

Variable	Mean	Range
Age (yrs.)	40 ± 15	64 to 13 yrs.
Duration of SCI (yrs.)	14 ± 13	47 yrs. to 4 months
Variable	Number	Percentage
Gender		
Male	24	73
Female	9	27
SCI Level		
Tetraplegia	6	18
Paraplegia	27	82
SCI Cause		
Traumatic Injury		
Road accident	9	27
Fall from height	17	52
Gun shot	1	3
Sports injury	2	6
Non traumatic injury	4	12

Patients who already had developed PrU had little knowledge about it but had many confusions and doubt about the same. They did not know anything about the pressure ulcer until they develop one. Post that, sole concern they have is to cure it, due to which they are not able to participate in their ongoing rehabilitation process. This leads to more agony to them as its affects the recovery process. Patients with 3rd grade and 4th grade PrU have to put a huge amount of money for operating on it [3]. Even after the patients having such agonizing experiences with PrUs it was observed that they did not have enough knowledge about PrU.

82% patient participants reported developing PrUs when they were in hospitals and still were not educated about them by the hospital staff. It was only in the rehabilitation center that first attempts were made to inform them about PrU. The patient's caregivers, which in most cases are family members, reported thoroughly following the instructions as conveyed by the doctors and nurses in the hospitals but they were informed only about turning sides of the patient every two hours and not about moving everything that comes in contact with them in every two hours like diaper, calipers; which caused pressure sores in many cases. Its reported that the duration for which a surface mattress is being used can also be the reason as the cushion of the mattress is not effective after a certain time as SCI patient spent most of the time in bed and wheelchair their support surfaces condition can play a crucial role in development of the PrU, even they change their position in every two hours [13].

Further, patient PrU history was investigated during the interviews. This then, gradually, yielded many risk factors, which seldom found mention in the available literature. Further detailed discussions then revealed various causes and consequences of these instances. This study is intended to make an attempt to understand them.

Post the interviews, efforts were made to make a logical segregation of the uncovered extrinsic risk factors into broader classes based on their most basic causes, which yielded the following chart.

Extrinsic Risk Factors: Extrinsic Risk Factors are defined as the risk factors, which are attributed to development of PrU due to the mechanical elements that come in contact with the skin of the patient and also the local temperatures and moistures which affect PrU initiation and development (Fig. 1).

A. Pressure: It is the vertical force exerted by the body due to self-weight.

1. Self- Weight: It is the localized pressure on the outer skin due to weight of the bones and tissues above it and a support surface below it.
2. Equipment and Accessories: It is the force exerted by an external assistive equipment on the skin in contact with it.
   a. Size Mismatch: This force can be attributed to the equipment being of wrong size. An equipment mismatched in terms of size can over pressurize the skin accelerating PrU initiation. This factor was observed in 3% of the participants.
   b. Prolonged Use: Longer use of an equipment affects owing to the skin being pressurized for a longer duration than it can take. This was observed to happen to about 21% of the patients under this study. This was observed to happen because most patients and their caregivers were ill informed regarding the duration of use of equipment. To point out an example, a patient contracted a PrU because he let his catheter tube be in contact with his skin for an indefinite duration.
   c. Manual Overtightening of Support Surface: Since the patient has loss of sensation, he/she cannot convey if an assistive equipment is overtightened during use. 6% of people were observed to contract a PrU because of this reason.

B. Shear: It is the distortion of a body by two oppositely directed parallel forces.

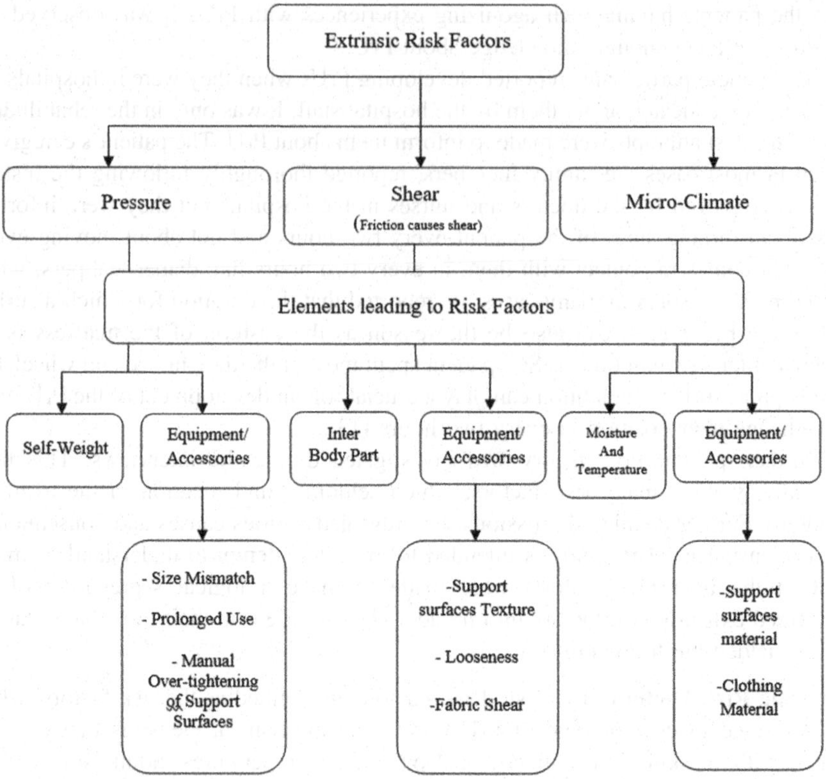

Fig. 1. Extrinsic risk factors and elements causing them.

1. Inter Body Part: When any two parts of a patient's body moving in different directions rub against one another, shear is resulted.
2. Equipment and Accessories: It is the distortion of skin caused by an equipment surface coming in contact with the patient skin.
   a. Support Surface Texture: This factor occurs when a coarse support surface in contact with the skin breaks the skin. 12% of the participants were reported to contract PrU due this factor.
   b. Looseness: Whenever any support surface is lose, its movement against causes breakage of skin. About 9% of participants were observed to contract because of this factor.
   c. Fabric Shear: Whenever a fabric coming in contact with the skin of a patient has wrinkles, it causes the skin to break with continuous rubbing. PrU due to this were observed in about 15% of the participants.

C. Micro-climate: It is the consideration of temperature and humidity effects.

1. Moisture and Temperature: A moist skin softens and then its tendency to break increases. Moisture increases the interface friction coefficients, therefore increasing the possibility of shear induced tissue damage. A higher local temperature acts as a

catalyst to aggravate the initiation of PrUs. 18% of the participants were found to contract PrUs due to this factor.

2. Equipment and Accessories: Assistive equipment when in contact with skin increase the temperature and moisture due to perspiration aiding PrU generation.
   a. Support Surface Materials: About 21% of the participants were observed to suffer PrU due to non-ventilated support surface materials. Prolonged use of assistive equipment also aids this factor. Even diaper rims were observed to cause PrUs in a few cases.
   b. Clothing Material: Ventilation and non-wrinkling of the clothing material is a factor to be considered. 15% of the participants were observed to contract PrU owing to this factor.

4 Results and Discussion

This study was an attempt to find risk factors which can become points of consideration for the designer and manufacturing community for production of future products. Efficient ways of communicating information regarding procedure to use products and their duration of use. Almost all the assistive products used by the people with SCI were found to not bear this information. Simpler steps like printing relevant information on the product covers or products themselves may save patients from mental and financial distress due to re-hospitalizations.

References

1. National Pressure Ulcer Advisory Panel/European Pressure Ulcer Advisory Panel (2009) Pressure Ulcer Prevention & Treatment: Clinical Practice Guidelines. National Pressure Ulcer Advisory Panel, Washington DC
2. European Pressure Ulcer Advisory Panel and National Pressure Ulcer Advisory Panel. Prevention and treatment of pressure ulcers: quick reference guide. National Pressure Ulcer Advisory Panel, Washington DC (2009). www.npuap.org and www.epuap.org. Accessed 23 Nov 2009
3. Pandey VK, Nigam V, Goyal TD, Chhabra HS (2007) Care of post-traumatic spinal cord injury patients in India: an analysis. Indian J Orthop 41(4):295–299
4. Patient Safety and Quality: An Evidence-Based Handbook for Nurses 1, Chap 12, 1–268
5. Berkowitz M (1993) Assessing the socioeconomic impact of improved treatment of head and spinal cord injuries. J Emerg Med 11(Suppl 1):63–67
6. Savic G, Short DJ, Weitzenkamp D, Charlifue S, Gardner BP (2000) Hospital readmissions in people with chronic spinal cord injury. Spinal Cord 38(6):371–377
7. Cardenas DD, Hoffman JM, Kirshblum S, McKinley W (2004) Etiology and incidence of rehospitalization after traumatic spinal cord injury: a multicenter analysis. Arch Phys Med Rehabil 85(11):1757–1763
8. Paker N, Soy D, Kesiktaş N, Nur Bardak A, Erbil M, Ersoy S, Ylmaz H (2006) Reasons for rehospitalization in patients with spinal cord injury: 5 years' experience. Int J Rehabil Res 29 (1):71–76

9. Noreau L, Proulx P, Gagnon L, Drolet M, Laramée MT (2000) Secondary impairments after spinal cord injury: a population-based study. Am J Phys Med Rehabil 79(6):526–535
10. Sving E, Idvall E, Högberg H, Gunningberg L (2014) Factors contributing to evidence-based pressure ulcer prevention. A cross-sectional study. Int J Nur Stud 51:717–725
11. Stevenson R, Collinson M, Henderson V, Wilson L, Dealey C, McGinnis E, Briggs M, Andrea Nelson E, Stubbs N, Coleman S, Nixon J (2013) The prevalence of pressure ulcers in community settings: an observational study. Int J Nurs Stud 50:1550–1557
12. Coleman S, Gorecki C, Andrea Nelson E, José Closs S, Defloor T, Halfens R, Farrin A, Brown J, Schoonhoven L, Nixon J (2013) Patient risk factors for pressure ulcer development: systematic review. Int J Nurs Stud 50:974–1003
13. Rawal P, Ray GG (2017) Risk factors for the development of pressure sores among people with spinal cord injury: results of a case study. In: Chakrabarti A, Chakrabarti D (eds) Research into design for communities, vol 1. ICoRD 2017. Smart innovation, systems and technologies, vol 65. Springer, Singapore

Effects of 'Blue-Regulated' Full Spectrum LED Lighting in Clinician Wellness and Performance, and Patient Safety

Octavio L. Perez[1](✉), Christopher Strother[2], Richard Vincent[3], Barbara Rabin[1], and Harold Kaplan[1]

[1] Department of Population Health Science and Policy, Icahn School of Medicine at Mount Sinai Hospital, New York, NY 10029, USA
octavio@lighting.healthcare

[2] Department of Emergency Medicine, Icahn School of Medicine at Mount Sinai, New York, NY, USA

[3] Icahn School of Medicine at Mount Sinai Hospital, New York, NY, USA

Abstract. Lighting has been recognized in the fields of human factors, ergonomics, and systems engineering, as an environmental factor that can affect wellness and performance, and the occurrence of medical error. Short wavelength ('blue') light is known to influence 'non-visual' effects of light in humans. These effects, that go beyond the pure 'visual' function, can affect human wellness and performance, as has been reported in previous scientific research. The aim and novelty of this research is to study the potentially beneficial 'non-visual' effects of lighting in the clinical environment to advance patient safety, and improve clinician wellness and performance.

The hypothesis of this study was that clinician wellness and performance in the execution of clinical procedures in the emergency department (ED) could be improved through controlled, indirect, 'blue'-regulated, full visible spectrum, tunable, solid state, 'white' lighting.

To conduct our inquiry, we performed a crossover study with current ED clinicians that executed clinical procedures in a high-fidelity, simulated ED setting, under two different lighting conditions. We used the existing fluorescent lighting as the control condition. To provide the appropriate experimental lighting condition, we developed a novel multichannel lighting system for precise control and assessment of light delivery conditions, with specific emphasis in the short wavelength (blue light) spectral area.

The results of this study suggest that it is possible that indirect, 'blue-enriched', full visible spectrum, 'white' lighting, might reduce clinician sleepiness and workload perceptions, might reduce the execution time for clinical procedures, and the occurrence of medical error, while improving clinician wellness.

O. L. Perez—Adjunct Researcher.
C. Strother—Professor.
R. Vincent—MS, FIES, LEED-AP.
B. Rabin—MT, MHA.
H. Kaplan—Professor.

S. Bagnara et al. (Eds.): IEA 2018, AISC 818, pp. 667–682, 2019.
https://doi.org/10.1007/978-3-319-96098-2_82

Future work would expand the scope of our study to advance patient safety in clinical scenarios where prevalence of adverse events has been observed, such as improvement in clinician cognitive recovery from medical error, hand-offs, and teamwork conditions. This study can also be translated to other fields of applications such as 24/7 control centers.

Keywords: Blue-enriched lighting · ipRGC · Randomized control trial
Human factors · Visual ergonomics · Healthcare ergonomics
Emergency department · Patient safety

1 Introduction

1.1 Background

Access to quality healthcare is a major societal concern. The costs associated with providing healthcare have become a tremendous financial burden, rising to 17.5% of the GDP in the USA (average 10% in the rest of the world), approximately three trillion USD in 2014. Aside from, though entwined with costs, healthcare institutions are concerned with issues of patient safety, medical error, and clinician burnout. Though very important, and with much research and effort committed in the last two decades, there has been very little significant improvement in terms of reductions of effects. The yearly costs associated with adverse events (medical error) can come close to 1 trillion USD [2].

The concept of total quality management was introduced in healthcare in the nineties, with a systems-engineering approach vs. the traditional patient-practitioner narrow view from the past. As part of the new vision, ergonomics and human factors engineering concepts started being applied, and the environment (including lighting) recognized as a factor that can affect not only patients' wellbeing, but clinician wellness and performance and consequently patient safety. A major advance in healthcare was the application of human factors engineering (HFE) to improve the performance of clinicians. Factors such as sleepiness, workload, fatigue, and stress can negatively impact the delivery of care, creating an environment more prone to error and burnout. While this issue is being raised, and the 'evidence-based' philosophy is being applied in healthcare design practice, there is very little known about the effect of environmental design on the clinicians giving care, particularly lighting.

Recent discoveries in the fields of light, biology, and lighting show that light has proven biological effects on humans beyond the 'visual' function -to see- and beyond the basic circadian functions. Light can affect human cognition, stress, workload perception, and mood, and ultimately performance. Precise lighting interventions that can reduce sleepiness and workload perception and improve mood and cognition may have the ability to improve clinician performance, reduce medical error and contribute to the advancement of patient safety.

Finally, simulation has become in the last decade a reliable practice for teaching and research in medicine with the advantage of translatability and the avoidance of direct harm to patients. The simulation environment is an appropriate setting to conduct lighting research for improving clinician performance. We will follow this approach in our novel lighting research framework.

1.2 Literature Review

Extensive literature has described the effect of the environment on patients, particularly reduction of mortality, reduction of length of stay, and reduction in pain, anxiety and sleep medication, [3–5, 7, 11, 17, 18, 26, 27]. Most of these studies are related to daylight availability in the patient room.

There is however, a gap in research on clinical lighting design as an intervention to meet clinician lighting needs as 'The Cochrane Collaboration' study points out [24].

Several literature reviews discuss the promotion of healing environments by considering the contribution of lighting [12, 16]; however, none focuses on clinician practices during daytime, except for factors such as medication dispensing [6]. The study of clinicians' wellbeing has been a missing factor in today's healthcare system as a means to improve the quality of care delivery [28], with lighting potentially playing a key role in addressing some of this limitation.

In a very similar intention to our study, the National Academy of Medicine (NAM) launched in 2017 the Action Collaborative on Clinician Well-Being and Resilience, committed to reversing trends in clinician burnout.

Closely related to our research are the multidisciplinary fields of Human Factors, Ergonomics and in particular 'Ergophthalmology', also known as 'Occupational Ophthalmology and the Ergonomics of Vision' [25], or 'Visual Ergonomics' [21]. Our review of different sources converging Human Factors, Ergonomics, and Lighting, and even in the more specific field of 'Visual Ergonomics', found there is a critical gap in recent advances in 'non-visual' effects of light, as considered in this research, that are still not incorporated into the practice of these disciplines. These fields of study hold the general lighting photometric approach (based on the 1924 $V(\lambda)$ function for foveal vision and the visual task with metrics such as Correlated Color Temperature (CCT) based on chromaticity) that does not account for the spectral quality, characterization, and measurement of light, and is not appropriate for the study of 'non-visual' effects, that are modulated by spectral effects, timing and intensity. Because of these limited views of lighting there is a gap in knowledge of what the potential of harnessing this view of lighting can bring to lighting design that will benefit clinicians and clinical practice, as we have seen in the field of Human Factors and Ergonomics [20, 22, 23, 29].

1.3 Aims

The goal of this multidisciplinary research is to provide preliminary evidence in the field of healthcare ergonomics of the beneficial effects of the lighting environment for the healthcare professionals during daytime, overcoming the described limitations in the current body of knowledge. We limit our study to daytime to avoid the complexity of sleep issues related to nighttime and shift-workers, which could be the focus of future studies.

2 Research Methods

The objective of the study is to investigate the effect of indirect, 'blue-enriched', full visible spectrum, solid state, 'white' lighting in clinician wellness and performance during the execution of a single-clinician clinical procedure in the Emergency Department simulation lab. For this purpose, we first define our research questions and hypotheses.

2.1 Research Question

The main research question of this study was to determine if there is a difference in clinician wellness and performance during the execution of clinical procedures in an ED Simulation Center due to specific changes in the lighting environment. Performance was measured in qualitative terms (medical evaluation) and in differences in execution time, together with subjective workload and sleepiness/alertness metrics for wellness evaluation.

This exploratory research is a screening experiment. We are therefore not gathering information at this stage on the optimum spectrum or the most efficacious system. If we obtain satisfactory proof of concept findings, this will lead to other research oriented at optimizing our research question to estimate the best control variables (factors) and obtain the optimum values.

As a secondary question, since 'blue-enriched' lighting associated with high CCT has generally not been liked in previous studies, 'Is our lighting environment with these aspects but with diffused lighting, going to be accepted by our participants?' This would be a critical issue for the translatability of the results.

2.2 Research Hypothesis

We hypothesized that clinician wellness and performance outcomes would be significantly improved with indirect, 'blue-enriched', full visible spectrum, solid state, 'white' lighting (experimental treatment condition) compared to the existing prevalent fluorescent lighting (control treatment condition).

Based on our previous experience and the literature we presumed that 'blue-enriched' lighting would be adequate for our research. It is known that this particular region of the spectrum can be difficult for the human visual system if not delivered properly. We therefore chose to develop an indirect delivery of the lighting conditions, with no direct exposure of the light sources to the participants.

We studied effects in the following dependent variables:

1. Sleepiness/Alertness
2. Workload perception
3. Clinical procedure execution time
4. Clinical procedure performance
5. Occurrence of medical errors

In the data analysis we payed special attention to the direction of the alternative hypothesis. The null hypothesis was based on the difference in the means equal to zero, but the alternate changed depending on the variables. To see an improvement in clinician wellness and performance we tested for lower than zero differences for the first three variables. Under the experimental treatment ('blue-enriched' lighting) we anticipated lower sleepiness perception, workload perception, and execution time of the clinical procedure. For the medical error (iatrogenic events) variable we expected, as with the first three variables, a reduction in the number of iatrogenic events during the execution of the clinical procedures under our experimental treatment. In the case of the qualitative performance we expected an increase in our quantitative markers.

With the secondary research questions, we expected a preference of our experimental treatment ('blue-enriched') compared to control (fluorescent).

2.3 Research Tools

Experimental Design

We conducted a randomized controlled trial (RCT) with '2 × 2 crossover AB/BA' design. This is similar to a 'crossover' clinical trial where, instead of drugs, our treatments are lighting conditions (light might be considered as a drug). Proper documentation of clinical trials has been a matter of discussion due to replication and social implications. There are relevant efforts to properly characterize them and we have adhered to the main recommendations (CONSORT, SPRINT and TIDier) proposed by EQUATOR (Enhancing the QUAlity and Transparency Of health Research). The statistical study includes descriptive and inference statistics, according to the particular characteristics of the 'crossover' methodology (paying special attention to the appropriateness rationale and to carry-over and period effects implications) [20, 22, 23, 29].

Our RCT was conceived, designed and conducted as a proof of concept study. We had two lighting conditions (control and experimental), and we randomized the participants (ED clinicians) into two groups, sequences AB and BA. Group AB performed the test (two sequenced clinical procedures 'airway' + 'chest-tube') first under control lighting condition (fluorescent) and then under experimental ('blue-enriched' LED), and BA group did the reverse order/sequence. Participants had a 'wash-out' period of at least one full week (7 days) between the treatments/conditions (tests). The process flow is shown graphically in Fig. 1.

Equipment

There are no commercial options to provide illumination levels of a room such as the simulation laboratory used in this experimental framework, that would allow full control over the visible spectrum (SPD). For this reason, we built a custom system which provided full SPD controllability with an indirect lighting strategy (Fig. 2).

Research Facilities and Study Population

The research study enrolled emergency department clinicians following the Institutional Review Board (IRB) approved protocol. The study was performed in the Emergency Department Simulation Teaching and Research Center (ED STAR), a teaching simulation center at Mount Sinai Hospital in NYC, NY. Our experimental setup is a permanent installation that will enable further research with the goal of rapid translatability into the real ED clinical environment. For the experimental lighting condition (treatment), we developed our own instruments and systems and used the existing lighting system (fluorescents) as the control.

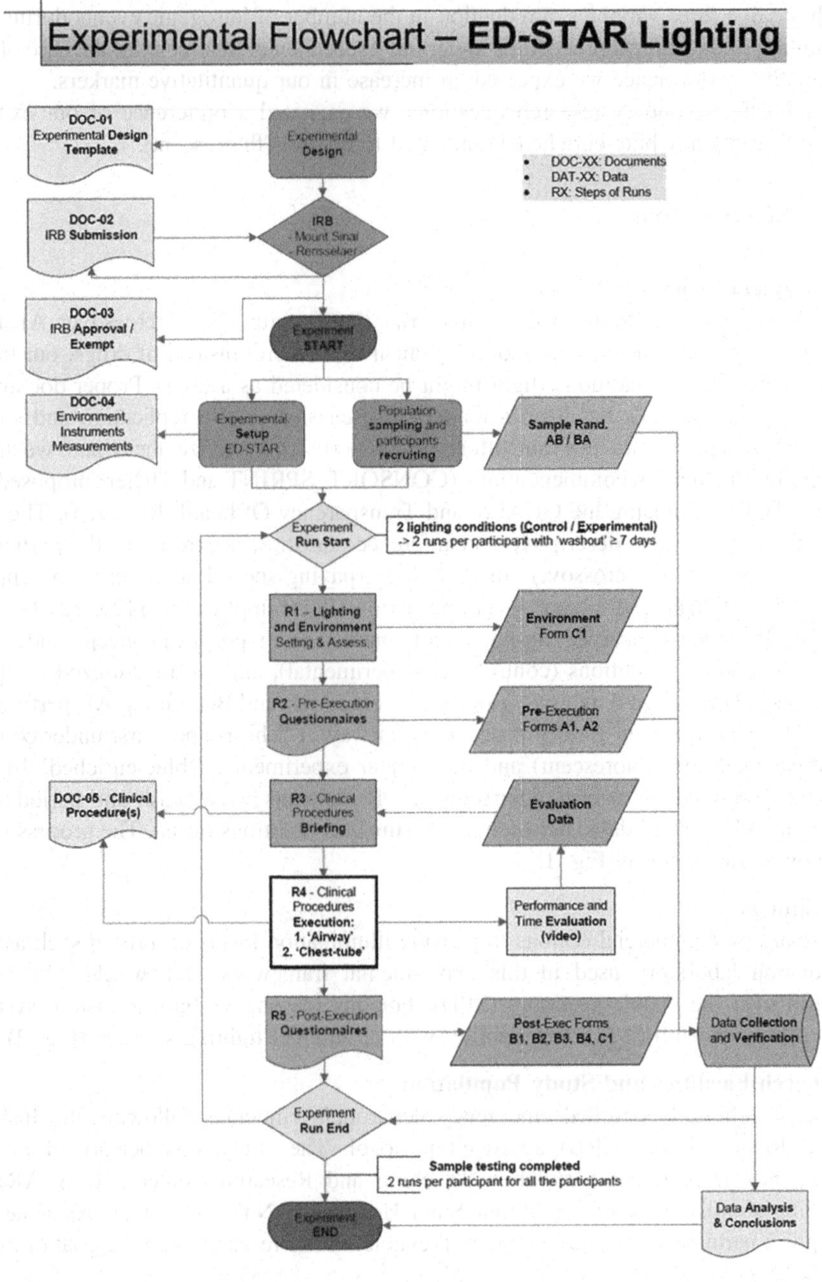

Fig. 1. Experimental flowchart

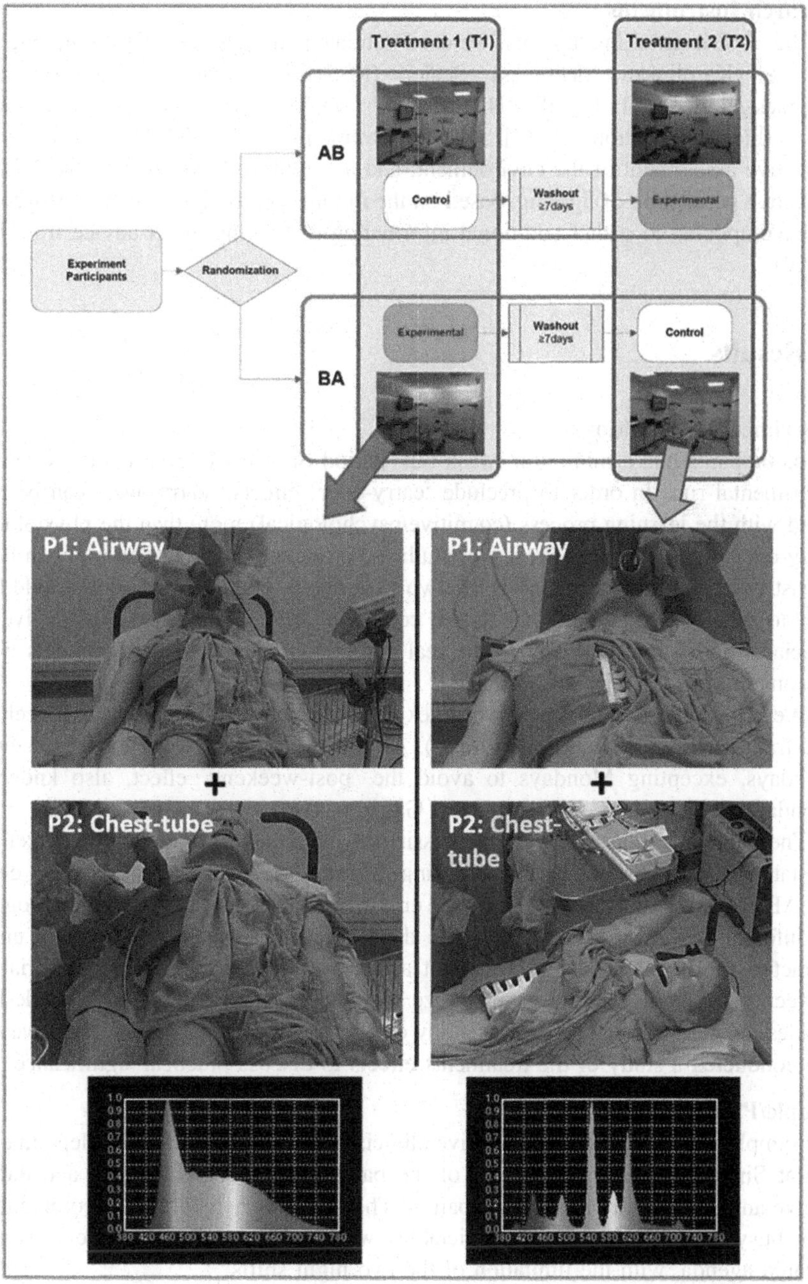

Fig. 2. Crossover Experimental Design (top), experimental conditions appearance and associated spectral power distributions (SPD)

Research Instruments

For the experimental instruments, we used validated surveys and questionnaires, such as: the 'Karolinska Sleepiness Scale' (KSS) for subjective assessment of sleepiness/alertness [1, 15], the 'NASA-TLX' ('TLX': 'Task Load Index) for subjective workload perception [13, 14], a modified version of the ASHRAE IEQ survey' for subjective assessment of the environment, and a 'Lighting Perception Survey' for the subjective perception of lighting, based on the initial work by Prof. Peter R. Boyce that we have updated to gather additional information (following direct advice from Prof. Boyce).

3 Results

Experimental Conditions

All participants had a minimum 'wash-out' period of at least seven (7) days between experimental runs in order to preclude 'carry-over' effects. 'carry-over' can be associated with the learning process (cognitive/psychological) more than the physiological 'carry-over' of typical drug crossover studies. Participants were not on night-shifts for at least two days prior to either of the two treatments. The ideal situation would have been to avoid all night-shifts, but this could not be accomplished with active ED clinicians, and it is also closer to the real conditions of the ED environments where experimental results will be translated.

We also arranged the timing of the experiment in terms of days of the week and time frames within the day (noon-4 pm). All the experiments were conducted during weekdays, excepting Mondays to avoid the 'post-weekend' effect, also known as 'Monday effect' (Butler, Kleinman, and Gardner 2014).

The data collection and initial processing of the data was done in 'MS Excel' and the statistical analysis was performed using 'Minitab v17'. As the experimental design was AB/BA crossover, we have used the crossover statistical analysis methodology for the inferential statistics analysis of the data. The crossover design has the unique characteristic of having each participant as its own control, reducing the variability between participants and the sample size required compared to double-arm designs. Besides the conventional inferential study of p-values, statistical significance, we have also conducted a study of the treatments effects to discuss practical significance.

Sample/Participants

Our sample was composed of ten active clinicians (n = 10) from the ED department at Mount Sinai Hospital in NYC. All of the participants were volunteers and did not receive any compensation for participation. They were very helpful in accommodating their busy, and sometimes unpredictable, working schedule to the experimental research agenda, with the limitation of the two night shifts.

Characterization of the sample:

- size: ten (n = 10) participants (equivalent to 20 on a double arm experiment)
- experimental runs: twenty (20), two per participant, following an AB/BA design
- age: average 35.10 years (SD = 4.95, min = 28, max = 44)

- gender: seven males and three females
- experience level: five residents (3th and 4th year) and five faculty members
- all participants were healthy, and none declared having diabetic conditions, nor color vision problems
- one participant had light-colored eyes

There was no participant drop-out.

Experimental Results and Data Analysis

The experimental lighting environment was positively accepted by the study participants and our indirect, 'blue-enriched', full visible spectrum, solid state, white lighting (experimental condition) was evaluated in our sample as better than the existing and prevalent fluorescent lighting (control condition). There were slight differences in perception of temperature and humidity. The environment was perceived as more humid and warmer under control (fluorescent) conditions. The experimental condition was accepted by all the participants of our study, except one, and had less claims for controllability. Also, the experimental condition was evaluated as better than the control in the 'compared to other workplaces' survey. It is the first time in lighting research that 'blue-enriched' full visible daylight lighting delivery has been reported under experimentation with an equivalent CCT of ≈78,000 K (seventy-eight thousand Kelvin) at the task. Contrary to previous research, this setting was considered better than the conventional lighting (control condition).

We observed in our sample a possible reduction in sleepiness (alertness increase) in the KSS scale (−15.94% observed effect with p-value = 0.022) and a possible reduction in workload perception in the NASA-TLX sum of scales (−21.87 observed effect with p-value = 0.009). We cannot claim a statistically significant reduction in execution times for procedure two and total execution times (−21.04 observed effect with p-value = 0.094). Also, there was no improvement seen in the clinical performance evaluation, even if we can see an observed effect improvement in our study in the scores from control to experimental lighting conditions. Finally, two medical errors occurred under control conditions and none under experimental conditions.

The observations of reduction of workload and sleepiness, factors that may be associated with clinician wellness/stress and medical error causation in the healthcare environment, are of special interest in the emergency setting. In the ED environment, clinicians work under high levels of stress. The potential increased ability of saving lives is a major component of the practical significance of our study (Table 1).

Data Analysis Results Summary

1. **Sleepiness Perception (KSS)**

   In the crossover analysis of KSS, (measured in a scale 9–1), n = 10, we obtained the following results:

   - Treatment effect −0.875, 95% CI [−1.7174, −0.032555]
   - The Mean under Control ($M_C = -0.3$, $SD_C = 1.57$) was higher than the Mean under experimental ($M_E = -1.1$, $SD_E = 1.10$), by 15.94%, DF(8), T-value = −2.3951, p-value = 0.022

Table 1. Summary of inferential statistic tests for numerical responses

Summary of inferential statistic tests						
Control: Existing fluorescent (4,000 K)						
Experimental: Solid State (LED) lighting system (78,000 K)						
H_0 Null hypothesis: Mean (Experimental) – Mean (Control) ≥ 0						
H_1 Alternative hypothesis: Mean (Experimental) – Mean (Control) < 0						
	Values	Effect	Effect %	alpha	P-Value	Result
Karolinska Sleepiness Scale						
KSS	9–1	−0.88	−15.94%	0.05	0.022	Can Reject H_0
NASA-TLX (workload perc.)						
Sum of scales	$1 - 21 \times 6$	−11.50	−21.87%	0.05	0.009	Can Reject H_0
				0.01	0.009	Can Reject H_0
Clinical performance execution time	Seconds					
Procedure 1 (P1): Airway		−6.17	−21.35%	0.05	0.252	*Cannot Reject*
Procedure 2 (P2): Chest-tube		−101.46	−21.02%	0.05	0.090	*Cannot Reject*
Total Execution Time (P1 + P2)		−107.63	−21.04%	0.05	0.094	*Cannot Reject*
H_0 Null hypothesis: Mean (Experimental) – Mean (Control) ≤ 0						
H_1 Alternative hypothesis: Mean (Experimental) – Mean (Control) > 0						
Clinical performance evaluation						
Procedure 1 (P1): Airway						
Checklist (P1C)	1–12	0.21	1.77%	0.05	0.182	*Cannot Reject*
Psychomotor (P1P)	1–7	0.17	1.64%	0.05	0.239	*Cannot Reject*
Novice to Expert (P1N)	1–5	0.25	7.32%	0.05	0.138	*Cannot Reject*
Procedure 2 (P2): Chest-tube						
Checklist (P2C)	1–34	0.63	4.04%	0.05	0.160	*Cannot Reject*
Psychomotor (P2P)	1–7	−0.04	0.00%	0.05	0.534	*Cannot Reject*
Novice to Expert (P2N)	1–5	0.21	5.26%	0.05	0.182	*Cannot Reject*

We observed in our sample that the experimental condition reduces the KSS (sleepiness perception) as hypothesized, in the terms described above.

2. **Workload Perception (NASA-TLX)**
In the crossover analysis of NASA-TLX, (measured 1–126), n = 9 (1 participant's score missing), we obtained the following results:

- Treatment effect −11.5, 95% CI [−20.393, −2.6074]
- The Mean under Control (M_C = 52.33, SD_C = 19.69) was higher than the Mean under experimental (M_E = 40.89, SD_E = 17.07), by 21.87%, DF(7), T-value = −3.0579, p-value = 0.009

We observed in our sample that the experimental condition reduces the NASA-TLX (workload perception) as hypothesized, in the terms described above.

3. **Clinical Procedures Execution Time:**

3.1 **Procedure 1 – 'Airway'**
In the crossover analysis of 'Execution Time (P1, 'airway')' (measured in seconds), n = 10, we obtained the following results:

- Treatment effect –6.1667, 95% CI [−26.480, 14.147]
- The Mean under Control (M_C = 38.40, SD_C = 34.09) was higher than the Mean under experimental (M_E = 30.20, SD_E = 12.42), by 21.35%, DF(8), T-value = −0.70006, p-value = 0.252

3.2 **Procedure 2 – 'Chest-tube'**
In the crossover analysis of 'Execution Time (P2, 'chest-tube')' (measured in seconds), n = 10, we obtained the following results:

- Treatment effect –101.46, 95% CI [−260.56, 57.641]
- The Mean under Control (M_C = 558.90, SD_C = 224.81) was higher than the Mean under experimental (M_E = 441.40, SD_E = 152.92), by 21.02%, DF(8), T-value = −1.4705, p-value = 0.090

3.3 **Procedure 1 + 2 (Total Time)**
In the crossover analysis of 'Total Execution Time (P1 + P2, 'airway' + 'chest-tube')' (measured in seconds), n = 10, we obtained the following results:

- Treatment effect −107.63, 95% CI [−279.88, 64.627]
- The Mean under Control (M_C = 597.30, SD_C = 248.58) was higher than the Mean under experimental (M_E = 471.60, SD_E = 152.19), by 21.04%, DF(8), T-value = −1.4408, p-value = 0.094

4. **Clinical Procedures Performance - Procedure 1 ('Airway'):**

4.1 **Procedure 1. 'Airway':**

4.1.1 **Procedure 1 – 'Airway - Checklist'**
In the crossover analysis of (P1, 'airway - checklist')' (measured 1–12), n = 10, we obtained the following results:

- Treatment effect 0.20833, 95% CI [−0.28977, 0.70644]
- The Mean under Control (M_C = 11.30, SD_C = 0.48) was lower than the Mean under experimental (M_E = 11.50, SD_E = 0.53), by 1.77%, DF(8), T-value = 0.96449, p-value = 0.182

4.1.2 Procedure 1 – 'Airway - Psychomotor'

In the crossover analysis of '(P1, 'airway - psychomotor')' (measured 1–7), n = 10, we obtained the following results:

- Treatment effect 0.16667, 95% CI [−0.34852, 0.68186]
- The Mean under Control (M_C = 6.10, SD_C = 0.99) was lower than the Mean under experimental (M_E = 6.20, SD_E = 1.03), by 1.64%, DF(8), T-value = 0.74600, p-value = 0.239

4.1.3 Procedure 1 – 'Airway - Novice to Expert'

In the crossover analysis of (P1, 'airway – novice to expert) (measured 1–5), n = 10, we obtained the following results:

- Treatment effect 0.25000, 95% CI [−0.24228, 0.74288]
- The Mean under Control (M_C = 4.10, SD_C = 0.57) was lower than the Mean under experimental (M_E = 4.40, SD_E = 0.52), by 7.32%, DF(8), T-value = 1.1711, p-value = 0.138

4.2 Clinical Procedures Performance - Procedure 2 ('Chest-tube'):

4.2.1 Procedure 2 – 'Chest-tube - Checklist'

In the crossover analysis of (P2, 'chest-tube - checklist')' (measured 1–34), n = 10, we obtained the following results:

- Treatment effect 0.62500, 95% CI [−0.73595, 1.9859]
- The Mean under Control (M_C = 22.30, SD_C = 2.75) was lower than the Mean under experimental (M_E = 23.20, SD_E = 2.53), by 4.04%, DF(8), T-value = 1.0590, p-value = 0.160

4.2.2 Procedure 2 – 'Chest-tube - Psychomotor'

In the crossover analysis of '(P2, 'chest-tube - psychomotor')' (measured 1–7), n = 10, we obtained the following results:

- Treatment effect -0.041667, 95% CI [−1.1451, 1.0617]
- The Mean under Control (M_C = 5.50, SD_C = 0.97) was higher than the Mean under experimental (M_E = 5.50, SD_E = 1.08), by 0.00%, DF(8), T-value = −0.087080, p-value = 0.534

4.2.3 Procedure 2 – 'Chest-tube - Novice to Expert'

In the crossover analysis of (P2, 'chest-tube – novice to expert) (measured 1–5), n = 10, we obtained the following results:

- Treatment effect 0.20833, 95% CI [−0.28977, 0.70644]
- The Mean under Control (M_C = 3.80, SD_C = 0.63) was lower than the Mean under experimental (M_E = 4.00, SD_E = 0.47), by 5.26%, DF(8), T-value = 0.96449, p-value = 0.182

5. **Categorical Variables. Effect Size Estimation (Risk Ratio)**

5.1 **Indoor Environmental Quality (IEQ) perception, Overall Rate:**
RR (Very Good) = 6/3 = 2 (greater for experimental condition)

5.2 **Lighting Perception Survey, Comparison to other workplaces:**
RR (comparison to other workplaces) = 5/2 = 2.5 (better for experimental condition)

4 Conclusions and Discussion

We have obtained results from our experimental research that may be supportive to our hypothesis about potential 'non-visual'/'non-image-forming' (NIF) beneficial effects of blue-enriched full visible spectrum white lighting in clinician wellness and performance. The observed reduction in sleepiness (increase in alertness) and the observed reduction in perceived workload, the effect in clinical performance, and the acceptance and preference of the experimental condition (indirect, 'blue-enriched', full visible spectrum, solid state, 'white' lighting) compared to the control condition (fluorescent lighting) are discussed here. These results may contribute to the translatability of the findings from the simulated to the actual clinical environment.

Our experimental lighting condition suggested an aggregate set of beneficial results for clinician wellness and performance that may support our research hypothesis. Using the existing fluorescent lighting as control reinforced the validity of our study, as these are ubiquitous conditions in the current healthcare environment.

The acceptance of the experimental lighting condition is critical for its practical feasibility. Also, our lighting conditions provided general lighting performance, while results from previous research that has been conducted in lightboxes or on 'ad-hoc' laboratory setups are difficult to be translated into real-world environments.

The observed reduction in sleepiness, associated with an increase in alertness, might initially seem inconsistent with the observed reduction in workload perception, as alertness could be erroneously associated with stress and arousal. In our study, the participants interviewed agreed that the experimental lighting provided a 'calming effect' that empowered them to better focus on the execution of the clinical procedures. This 'calming effect', which at the same time improves alertness and focus, requires further research and analysis (explored in fields such as mindfulness). Another potential explanation can be found in the (controversial?) field of 'Syntonic Optometry' where the effect of 'blue' visible radiation is associated with the activation of the parasympathetic nervous system activation (PSNS) (Spitler 1941), (Liberman 1986), (Gottlieb and Wallace 2001).

Taking into account that we performed all of our experiments between noon and 4 pm (trying to have a similar circadian time, 'CT', lapse for the participants), we presume that melatonin suppression, or reduction, was not the cause of the effect measured. Melatonin levels are high in the human body during the night and remain low during the day. In the selection of our timeframe we were also careful to not target the morning peak of cortisol. Considering that our timeframe was coincident with the post-prandial dip, previous research suggests that the orexin/hypocretin neurons that are associated with wakefulness (Lecea and Huerta 2014) (Schwartz and Kilduff 2015) might be inhibited by higher glucose levels associated with high carbohydrate meals (Burdakov et al. 2006). Monk (2005) recognizes the post-meal effect, and also the carbohydrates factor, but also proposes that the decrease in performance in the early afternoon could be driven by a 12-h circadian harmonic that could also be higher in morning-type individuals, and that is not necessarily caused by the post-meal event. Monk also proposes ('bright') light as a coping mechanism for the 'siesta' effect. The orexin wakefulness mechanisms in association with lighting, has been proposed in the doctoral dissertation "Research methods and neurophysiological mechanisms behind the alerting effects of daytime light exposure" (Rautkylä and others 2011). To the best of our knowledge this is the only previous lighting research that considers this mechanism beyond the effects of light in melatonin.

This is clearly an area of future research related to the study of the neurological mechanism underlying the effects of light in human psycho-physiology that would be difficult to conduct in other organisms (such as mice) due to the high level cognitive functions associated with the research, and the fact that mice are nocturnal creatures. Techniques such as fMRI (functional Magnetic Resonance Imaging) have been proposed in the field of neuroscience, even if the statistical methods in neuroimaging have been put recently into question (Eklund, Nichols, and Knutsson 2016). Other neurophysiological metrics, such as heart rate variability (HRV), eye tracking (pupillary response), electroencephalography (EEG), and skin conductivity, look very promising to evaluate sympathetic and parasympathetic nervous system activation and should be explored in future studies.

Limitations of the Study

One limitation of our study is the size of the sample, ten clinicians, due to the size of the population at the ED at Mount Sinai Hospital and their work schedules. Nevertheless, due to the randomized AB/BA crossover design, it is equivalent to a twenty-participant double-arm study. In future experiments, melatonin levels should be measured to fully discard the potential for melatonin having confounding effects and a more precise time-frame, particularly related to the number of hours since wake-up (circadian time) should be accommodated. Cortisol should be considered together with stress assessment. Working in the simulation environment in single-clinician procedures is also a limitation since we do not have the clinician-patient interaction and the team (clinician-clinician) interaction. These factors, even if real in day-by-day clinical practice, would have introduced confounding factors in our research and should be carefully considered for future experiments. Neuro-physiological metrics such as HRV, EEG or eye-tracking, together with quantity and quality of sleep would be also desirable in the future.

Contributions

Griendy? Emergency Department Simulation (ED STAR), Mount Sinai Hospital, Icahn School of Medicine, for subject recruitment, lab setup, coordination. Nicholas Genes, MD, emergency medicine clinician for support with Mount Sinai Innovations. Rensselaer Polytechnic Institute Dean of Graduate School, Dr. Stanley Dunn as the director of the doctoral research of Dr. Octavio L. Perez.

References

1. Akerstedt T, Gillberg M (1990) Subjective and objective sleepiness in the active individual. Int J Neurosci 52(1–2):29–37
2. Andel C, Davidow SL, Hollander M, Moreno DA (2012) The economics of health care quality and medical errors. J Health Care Fin 39(1):39–50
3. Beauchemain KM, Hays P (1996) Sunny hospital rooms expedite recovery from severe and refractory depressions. J Affect Disord 40(1–2):49
4. Beauchemin KM, Hays P (1998) Dying in the dark: sunshine, gender and outcomes in myocardial infarction. J R Soc Med 91(7):352–354
5. Benedetti F, Colombo C, Barbini B, Campori E, Smeraldi E (2001) Morning sunlight reduces length of hospitalization in bipolar depression. J Affect Disord 62(3):221–223. https://doi.org/10.1016/S0165-0327(00)00149-X
6. Buchanan TL, Barker KN, Gibson JT, Jiang BC, Pearson RE (1991) Illumination and errors in dispensing. Am J Health Syst Pharmacy 48(10):2137–2145
7. Canellas F, Mestre L, Belber M, Frontera G, Rey MA, Rial R (2015) Increased daylight availability reduces length of hospitalisation in depressive patients. Eur Archiv Psychiatry Clin Neurosci 266(3):277–280. https://doi.org/10.1007/s00406-015-0601-5
8. Carayon P (2012) Emerging role of human factors and ergonomics in healthcare delivery–a new field of application and influence for the IEA. Work 41(Suppl. 1):5037–5040
9. Carayon P, Xie A, Kianfar S (2013) Human factors and ergonomics. In: Making health care safer II: an updated critical analysis of the evidence for patient safety practices. Evidence Reports/Technology Assessments 211. Agency for Healthcare Research and Quality, Rockville, MD. http://www.ncbi.nlm.nih.gov/books/NBK133393/?report=reader
10. Carayon P, Wetterneck TB, Joy Rivera-Rodriguez A, Schoofs Hundt A, Hoonakker P, Holden R, Gurses AP (2014) Human factors systems approach to healthcare quality and patient safety. Appl Ergon 45(1):14–25. https://doi.org/10.1016/j.apergo.2013.04.023
11. Choi J-H, Beltran LO, Kim H-S (2012) Impacts of indoor daylight environments on patient average length of stay (ALOS) in a healthcare facility. Build Environ 50:65–75. https://doi.org/10.1016/j.buildenv.2011.10.010
12. Dalke H, Little J, Niemann E, Camgoz N, Steadman G, Hill S, Stott L (2006) Colour and lighting in hospital design. Opt Laser Technol 38(4–6):343–365. https://doi.org/10.1016/j.optlastec.2005.06.040
13. Hart SG (2006) NASA-task load index (NASA-TLX); 20 years later. In: Proceedings of the human factors and engineering society (HFES) 50th annual meeting, vol 50, pp 904–908. Sage Publications, San Francisco, California. https://doi.org/10.1177/154193120605000909
14. Hart SG, Staveland LE (1988) Development of NASA-TLX (Task Load Index): results of empirical and theoretical research, vol 52, pp 139–183
15. Hommes V, Giménez MC (2015) A revision of existing Karolinska sleepiness scale responses to light: a melanopic perspective. Chronobiol Int 32(6):750–756. https://doi.org/10.3109/07420528.2015.1043012
16. Iyendo TO, Uwajeh PC, Ikenna ES (2016) The therapeutic impacts of environmental design interventions on wellness in clinical settings: a narrative review. Complement Ther Clin Pract 24:174–188. https://doi.org/10.1016/j.ctcp.2016.06.008
17. Joarder MAR, Price ADF, Mourshed M (2009) Systematic study of the therapeutic impact of daylight associated with clinical recovery. https://dspace.lboro.ac.uk/dspace/handle/2134/6576

18. Joarder A, Price A (2013) Impact of daylight illumination on reducing patient length of stay in hospital after coronary artery bypass graft surgery. Light Res Technol 45(4):435–449. https://doi.org/10.1177/1477153512455940
19. Levin S, France DJ, Hemphill R, Jones I, Chen KY, Rickard D, Makowski R, Aronsky D (2006) Tracking workload in the emergency department. Hum Factors J Hum Factors Ergon Soc 48(3):526–539
20. Li T, Yu T, Hawkins BS, Dickersin K (2015) Design, analysis, and reporting of crossover trials for inclusion in a meta-analysis. PLoS ONE 10(8):e0133023. https://doi.org/10.1371/journal.pone.0133023
21. Long J (2014) What is visual ergonomics? Work 47(3):287–289. https://doi.org/10.3233/WOR-141823
22. Senn S (2002) Cross-over trials in clinical research. Statistics in practice. Second edn. Wiley, Chichester
23. Senn S, Lee S (2004) The analysis of the AB/BA cross-over trial in the medical literature. Pharm Stat 3(2):123–131. https://doi.org/10.1002/pst.106
24. Tanja-Dijkstra K, Pieterse ME (2011) The psychological effects of the physical healthcare environment on healthcare personnel. In: Cochrane database of systematic reviews, Issue 1. The Cochrane Collaboration, London, United Kingdom, 19 January 2011. http://doi.wiley.com/10.1002/14651858.CD006210.pub3
25. Tengroth B (1984) Ergophthalmology. Acta Ophthalmol 62(S161): 13–16
26. Ulrich R (1984) View through a window may influence recovery. Science 224(4647):224–225. https://doi.org/10.1126/science.6143402
27. Walch JM, Rabin BS, Day R, Williams JN, Choi K, Kang JD (2005) The effect of sunlight on postoperative analgesic medication use: a prospective study of patients undergoing spinal surgery. Psychosom Med 67(1):156–163. https://doi.org/10.1097/01.psy.0000149258.42508.70
28. Wallace JE, Lemaire JB, Ghali WA (2009) Physician wellness: a missing quality indicator. Lancet 374(9702):1714–1721
29. Wellek S, Blettner M (2012) On the proper use of the crossover design in clinical. Dtsch Arztebl Int 109:276–281

Contributions of Activity Ergonomics to Design a Virtual Tool for Sharing Mental Health Care

Carolina Maria do Carmo Alonso[1(✉)], Anderson Nogueira de Lima[2], Melissa Ribeiro Teixeira[3], Eliel Prueza Oliveira[1], Emanoela Ferreira de Lima Silva[4], Maria Cristina Ventura do Couto[3], and Francisco José de C. M. Duarte[2]

[1] Department of Occupational Therapy, School of Medicine, UFRJ, Rio de Janeiro, Brazil
carolina.alonso@ufrj.br
[2] Production Engineering Program, COPPE, UFRJ, Rio de Janeiro, Brazil
[3] Research Core on Public Mental Health Policies, Institute of Psychiatry, UFRJ, Rio de Janeiro, Brazil
[4] Department of Speech Therapy, School of Medicine, UFRJ, Rio de Janeiro, Brazil

Abstract. The goal of this paper is to illustrate the contributions generated by Activity Ergonomics (AE) to the participatory design process of an EHR created to support collaborative mental health care of children and youth in the city of Rio de Janeiro. It is qualitative research based on the theoretical framework of AE. In this paper we will present the results of the first step of this research. This preliminary stage used as a form of data collection interviews with professionals from three sectors of the mental health care network of children and youth within Brazil. An analysis of the interviews demonstrated that mental health collaborative care encompasses an intersectoral network of services that are articulated in the construction of therapeutic projects. This type of organization requires an intense exchange of information between experts from distinct fields, like social workers, educators and health care professionals. In fostering the diverse professional realities, there is a difficulty to overcome the challenge of sharing information, due in part to the combination and inundation of the individual accounts. In addition, the difficulties identified in the use of the old device were: fear of wrongly inputting the data into the system, rework, incomprehension of the codes used as landmarks to cases. The useful features identified by the workers for a new tool were: access through smartphones, possibility of choosing to consult a summary of the cases or more detailed information, dynamic communication between the services and generation of alerts for urgent situations.

Keywords: Activity ergonomics · Participatory design
Electronic health records

S. Bagnara et al. (Eds.): IEA 2018, AISC 818, pp. 683–690, 2019.
https://doi.org/10.1007/978-3-319-96098-2_83

1 Background

Since 2014, the Research Core on Public Mental Health Policies (RCPMHP) of the Federal University of Rio de Janeiro has developed a project that aims to certify the mental health care of children and youth. Hence, this project seeks to promote and establish Collaborative Care Management (CCM) strategies among services historically committed to the care of this population [1]. According to Thota *et al.* [2] CCM "is a multicomponent, healthcare system–level intervention that uses case managers to link primary care providers, patients, and mental health specialists".

In this model of care, the integration of services has a prominent role because it allows for the exchange of information and the construction of service flows that are useful in enhancing the use of available resources. Additionally, they favor the qualification of care, through the early detection and management of risk situations of children and youth who demand mental health care [1].

Therefore, within this realm, one of the important consolidating factors for the integration between the services is communication; given that an effective articulation of the work accomplished between the different services and professionals depends, in great part, on the quality and integrity of the information exchanged [3]. Thus, seeking to adhere to the communications demands of the aforementioned project's team, a spreadsheet was developed, available online through a cloud service. Accompanied by a few rules and conventions, this solution was used for a number of months. Subsequently, this solution proved to be inadequate, and its use was terminated.

As a means to substitute this spreadsheet, development for an Electronic Health Record (EHR) project commenced. The choice for the given system derived from the fact that this type of tool directly relates to the improvement of service quality and care coordination [4–6], of which aligns with the goals from the project developed by RCPMHP.

Considering that EHR projects generally focus on the individual use of the system and not integrating the specific needs of the workers to provide support for collaborative activities [7, 8], nevertheless, the project of the platform presented in this paper based itself on the participatory methodologies; which include the analysis of the work of the different users, while seeking to fulfill the collaborative care strategy demands. The goal of this paper is to demonstrate the contributions of Activity Ergonomics (AE) on the participatory designing process of an EHR to support collaborative mental health care of children and youth in the city of Rio de Janeiro.

1.1 Theoretical Framework

The theoretical framework adopted to guide the conception process addressed in this article was Activity-Centered Ergonomics (EA). This method focuses on the development of strategies that lower the gap present between those who plan and those who execute the work. An investigation spearheaded by EA principles allowed for the discovery of that which is effectively done by the workers to correct situations that are prejudicial to health and provide assistance to projects in situations that are more adequate to achieve production objectives and develop workers' competencies [9–12]. In specific, during the conception process of technological devices, EA has primed itself in studying the activities executed with a computer, being able to predict the difficulties relating to work/usability situations of the system; encompassing the following aspects: environment, equipment, interface and dialog, organization, online support, composition, maintenance, etc. [13]. In essence, EA is used in computerized systems projects, while also relying on additional methodological approaches that help incorporate a usage dimension in these processes, of which include: user-centered approach, participatory development and Design Thinking [13].

However, while such approaches include the user as a protagonist in the projects, they are rarely considered a fountain of knowledge when defining the usefulness and usability of said application. In this context, EA contributes by including the user as a co-developer, an agent whose precise objectives must be reached [13, 14]. Additional points which EA approaches technology development processes differently are: usefulness, usability, hazardousness and accessibility. In this article, usefulness and usability will be described to a greater extent [13].

For EA, usefulness applies to significant advantages for the user in any given activity; such advantage can always be fulfilled by their objectives, the available tools, the environment being used and the dependence on additional activities. As such, while in a traditional computerized systems project usefulness is a given fact, in EA, it is relative to the needs of the work/usage situations.

Usability, on the other hand, may be evaluated by its accessibility to learn and memorize, and its level of error. In all, a technological device can have positive usability ratings without the user being able to fulfill their needs. This occurs, mainly, when the user model used in the project is created reductively without considering the aspects connected to usefulness previously listed [13].

The EA procedures in the development processes pertaining to technological systems include: (a) analysis of the work and location where worker/users involved in identifying the demands, operational methods and problems with functionality to be considered in the project definition; (b) analysis of the tools and/or situations considered similar, while anticipating potential problems and solutions already proposed; (c) model and prototype specifications for interfaces to illustrate the visions of the users and designers in a simulation process; (d) satisfaction study; (d) development and implementation of the production proposals for the use of the projected tool.

Based on the results of the actions taken, the designers will materialize solutions in the form of blueprints, models, prototypes and beta versions of the system being produced. This occurs for the purpose of performing simulations that seek to validate the design choices with the users before finalizing the development of the project. In essence, the assessment of the proposed solutions should infiltrate every stage of development. In this sense, two assessment models can be put into motion, one that requires a direct participation of the users, and another that involves analyzing interface characteristics while focusing on specific technological issues. In the wake of the iterative stages of development, assessment and construction of the computer system, the last stage of the project consists of accompanying its use in a real world environment followed by the assessment of its field use, to obtain information about the difficulties encountered by the workers in conditions not anticipated during the testing phase of the project.

This section briefly presented the context and objective of this article, in addition to its theoretical anchor, while discussing the central concepts of EA. Demonstrated in this respect was the role in which EA plays in the development of computerized systems that are operational strictly related to user-centered project methodologies. The following sections will discuss the methods used in this research in addition to displaying the results and discussion generated from the analysis of the data.

2 Methods

This study inserts itself into the qualitative investigation field. Its procedures for collecting and analyzing data are guided by the theoretic methodological reference of EA, that includes the following research stages: 1. demand analysis; 2. organizational, technical and economic, and social environment analysis; 3. analysis of the work activities and situations; 5. rebuilding the results, validating the study and formulating recommendations to improve the work (Guérin et al. 2001).

In this article, the results of the stage analyzing demand will be presented, of which its research procedures include: individual and group interviews and document analysis. The data collection occurred during the period between March and September of 2017.

2.1 Study Setting

This study was executed within three services involving the SM care of children and youth, in a given region in the city of Rio de Janeiro. The services include: a primary care health unit, a mental health specialized unit, and members of the RCPMHP group. This final group is a service connected to the Universidade Federal do Rio de Janeiro (Federal University of Rio de Janeiro) that develops studies that investigate the adequacy of public policies regarding mental health in Brazil, of which required the EHR project addressed in this article.

2.2 Sample and Procedures of Data Collection

The purposeful sampling strategy was used; researchers sought to identify and select workers who used the spreadsheet to share information that were no longer in use, additionally, professionals who had experience in the mental health care of children and youth who needed integration with other services. In this sense, professionals were identified within each service that could provide important information for the problem encountered in the use of the worksheet previously employed, as well as the desired requirements for the design of the new EHR. The participants were recruited through informational meetings, and the anonymity and voluntary participation of the workers in this study were ensured.

We interviewed a physician, a nurse and a community health agent from the primary care unit, two psychologists from the specialized mental health service and a researcher linked to RCPMHP. In addition, the content of the spreadsheet that the professionals previously used to share information regarding cases was analyzed. The results of this analysis served as a basis for eliciting the first data on information sharing between the different services. This information helped construct the first hypotheses of this study.

3 Results and Discussion

The results of our data collection and analysis highlight the following points: the characteristics that mark the different workers' perceptions regarding the usage of a communication tool for the collaborative care; the problems related to the spreadsheet usability; and the desirable attributes that should be considered in the conception phase of a new EHR.

3.1 The Diferente Workers' Perceptions Regarding the Use of a Communication Tool Directed Towards the Collaborative Care

By means of the data analysis, this paper identifies the different perceptions that workers have on the objective and the usage of a shared care tool. This can be seen both between the same service professional as well as those in different service areas. In regard to the different perception of the same service professionals, the first care doctor and nurse wait to access and register the health condition data of a given child; while a Community Health Agent deems dealing with family dynamic aspects important.

In terms of the different service area professionals, this research demonstrates that in the specialized mental health service, psychologists demonstrate an interest to obtain data on school attendance as well as follow up on data of the cases by means of other mechanisms that monitor the children and youth. In addition to this, the professionals within this type of service also deem reporting the activities done within this service important; whereas, the primary care doctor considers the access to information to more concise information more important. The perceptions on the tool usage from the RCPMHP professionals address the related questions on the mental health work prescriptions of children and adolescents more effectively when it comes to public policy; more so than those related to the tool use itself. This scenario shows that a child and youth mental health collaborative care encompasses an intersectoral network of services that are articulated in the construction of therapeutic plan. This type of organization requires an intense exchange of information from experts of distinct areas such as social workers, education and health care professionals. As the workers have different professional worlds [14], the challenge of sharing information is not overcome by means of overlap of narratives.

This being the case, the work done by Madhu *et al.* [7] shows that EHR projects that seek to support collaborative work must go beyond the assessment of the necessity of information. Instead, it must get closer to the multiplicity of formation and perspectives that are at stake. This is so as, according to the above-mentioned study, the differences will have implications on how the information is registered within the system will better favor or not the collaboration. Therefore, the first challenge identified in this research is to build a tool that integrates different values and practices in the children and youth care. Following along this line, it is expected that the EA contributes in a differentiated fashion in this project; this is so as its following research steps will yield a detailed analysis on the different work done by the workers involved in the collaborative care seeking to understand the characteristics of each of these professional worlds [14].

3.2 Problems Related to the Spreadsheet Usage and the Desirable Attributes that Must Be Considered in the Creation of a New EHR

The difficulties in using the spreadsheet used as a communication platform between the different services as identified by the workers within this study are listed on Table 1.

Table 1. Data on the difficulties for spreadsheet usage

Difficulties	ACS	Nurses	Psychologists
Available computer			X
Able to find specific case information		X	X
Lack of data update	x	X	
Software incompatibility		X	
Slowness on responses to requests made through the system	x		
Difficulty in inserting the data		x	X

On the other hand, the requirements that the workers pointed out as desirable for a new tool are listed in Table 2.

Table 2. Desirable Requirements

Atributos desejáveis	Médico	ACS	Nurse	Psicólogo	Psicóloga
Alertas/notificações		x			x
Classificação de informação por serviço					x
Comunicação dinâmica entre serviços		x	x	x	x
Frequência nos serviços				x	x
Acesso por smartphones	x	x		x	x
Função de busca			x		
Histórico resumido	x		x		
Exclusão de informações sigilosas		x			
Compatibilidade com outros softwares			x		

The data regarding the usage problem e desirable requirements for this new tool serve as a base for the understanding of a few matters that will guide the creation of a new EHR. However, this first research stage must be validated by means of a work activity analysis by each of the collaborative care proposal actors.

This is because several studies highlight the need for social technical approach applications for the information technology creation. This has implications on the bottom-up study implementation that are able to cover all of the complexity related to the work, instead of representing it by means of abstract models. Additionally, this is not based on preconceived and formal representations, but based on empirical studies, often times qualitative in nature, to gain a social cultural every day process perspective of such practices as used by EA [16], [17,18].

The relationships between AE and the design of electronic devices have been evidenced by means of the use of user-centered design methodologies and methods for evaluating the usability of interfaces. As such, AE favors the improvement of the project considering that the worker is not only a user of a device, but an operator whose actions have purpose. Therefore, from the perspective of EA's framework, the worker is a co-designer (6). In this sense, early reports from this research indicates the workers' perception on the use of EHR in order to incorporate the work dimension into the design of a new tool. For further analysis, simulations will be conducted using a paper mockup of the software to simulate work situations to confirm or discourage design requirements showed in this paper.

References

1. Teixeira, MR, Couto, MCV, Delgado, PGG (2017) Atenção básica e cuidado colaborativo na atenção psicossocial de crianças e adolescentes: facilitadores e barreiras. Ciência & Saúde Coletiva 22(6):1933–1942. https://dx.doi.org/10.1590/1413-81232017226.06892016
2. Thota AB et al (2012) Collaborative care to improve the management of depressive disorders: a community guide systematic review and meta-analysis. Am J Prevent Med 42 (5):525–538
3. Cifuentes M, Davis M, Fernald D, Gunn R, Dickinson P, Cohen DJ (2015) Electronic health record challenges, workarounds, and solutions observed in practices integrating behavioral health and primary care. J Am Board Fam Med 28(Supplement 1):S63–S72
4. Zlabek JA, Wickus JW, Mathiason MA (2011) Early cost and safety benefits of an inpatient electronic health record. J Am Med Inf Assoc 18(2):169–172
5. Cusack CM (2008) Electronic health records and electronic prescribing: promise and pitfalls. Obstet Gynecol Clin North Am 35(1):63–79
6. Reitz R, Common K, Fifield P, Stiasny E (2012) Collaboration in the presence of an electronic health record. Fam Syst Health 30(1):72
7. Huang ME (2017) IT is from mars and physicians from venus: bridging the gap. PM&R 9 (5):S19–S25
8. Reddy MC, Shabot MM, Bradner E (2008) Evaluating collaborative features of critical care systems: a methodological study of information technology in surgical intensive care units. J Biomed Inform 41(3):479–487
9. Falzon P, Mollo V (2009) Para uma ergonomia construtiva: as condições para um trabalho capacitante.Laboreal 5(1)
10. Guérin F et al (2001) Compreender o Trabalho para Transformá-lo - A prática da Ergonomia. São Paulo: Edgard Blucher
11. Daniellou F, Béguin P (2007) Metodologia da ação ergonômica: abordagens do trabalho real. In: FALZON, PIERRE (Org.). Ergonomia. [S.l.]: Blucher São Paulo, pp 281–301

12. Pizo CA, Menegon NL (2010) Análise ergonômica do trabalho e o reconhecimento científico do conhecimento gerado. Production 20(4):657–668
13. Burkhardt J, Sperandio J (2007) Ergonomia e concepção informática. Ergonomia. Edgard Blücher, São Paulo
14. Bastien C, Scapin D (2007) A Concepção de Programas de Computador Interativos Centradas no Usuário: Etapas e Métodos. Ergonomia. São Paulo, Editora Blücher, pp 383–392
15. Béguin P (2010) Conduite de projet et fabrication collective du travail, une approche développementale, document de synthèse en vue de l'habilitation à diriger des recherches. Université Victor Segalen Bordeaux, 2
16. Berg M, Langenberg C, vd Berg I, Kwakkernaat J (1998) Considerations for sociotechnical design: experiences with an electronic patient record in a clinical context. Int J Med Informatics 52(1–3):243–251

Weak Signals in Healthcare: The Case Study of the Mid-Staffordshire NHS Foundation Trust

Eva-Maria Carman[(✉)], Mike Fray, and Patrick Waterson

Human Factors and Complex Systems Research Group, Loughborough University, Loughborough, UK
{E.Burford,M.J. Fray,P.Waterson}@lboro.ac.uk

Abstract. Most organisational disasters have warning signals prior to the event occurring, which are increasingly appearing in accident reports. In the case of the Mid-Staffordshire Disaster, the disaster was not as a result of component failure or human error but rather an organisation that drifted into failure with precursory warning signals being ignored. It has been estimated that between 400 and 1200 patients died as a result of poor care between 2004 and 2009. The aim of this study was to identify the precursory signals and their rationalizations that occurred during this event. Qualitative document analysis was used to analyse the independent and public inquiry reports. Signals were present on numerous system levels. At a person level, there were cases of staff trying to make management aware of the problems, as well as the campaign "Cure the NHS" started by bereaved relatives. At an organisational level, examples of missed signals included the decrease in the trust's star rating due to failure to meet targets, the NHS care regulator voicing concern regarding the unusually high death rates and auditors' reports highlighting concerns regarding risk management. At an external level, examples included negative peer reviews from various external organizations.

Keywords: Healthcare safety · Systems approach · Weak signals (need 3)

1 Introduction

An organisational disaster can be defined as a low-probability, high impact event with the potential to threaten an organisation's survival [1]. Most organisational disasters have warning signals prior to the event occurring [2], which are increasingly appearing and receiving progressively more attention in accident reports. These warning signals are sensed information regarding emerging events [3], and include indicators or cues from the environment [4] which require interpretation and sensemaking [5]. Many of these warning signals are often also referred to as weak signals, as the information they contain is frequently imprecise and vague in nature [3].

These signals are gaining increasing interest in the research community as they may provide an opportunity to achieve pro-activeness and promote effective risk management, as they provide an opportunity "sooner-rather-than-later" for identifying problems that threaten safety [6]. Furthermore, by using signals, unexpected events may be

S. Bagnara et al. (Eds.): IEA 2018, AISC 818, pp. 691–700, 2019.
https://doi.org/10.1007/978-3-319-96098-2_84

addressed in a more cost-effective and timely manner [7]. In healthcare, this could result in significant benefits particularly with regards to patient health. Furthermore signals may provide an opportunity to render a system more resilient [8] as they provide insight regarding the status of the system and areas of risk [9]. By identifying where these signals originate from and understanding how signals are identified and interpreted, possible changes to work structure and management could be developed to encourage signal identification for promoting patient safety. Despite their potential for improving safety, research exploring signals, especially in healthcare, is limited.

1.1 Case Study Description

This case study focused on the failings surrounding the Mid-Staffordshire NHS Foundation Trust, which resulted in unacceptable standards of care for patients between the years of 2004 and 2009 [10]. During this time, there was a noted rise in patient mortality and complaints relating to clinical care, with an estimate of between 400 and 1200 excess deaths occurring during this time period [10]. As a result of the sub-standard care, many patients were left struggling to care for themselves [11]. The investigation into this organisational disaster found that the system failed to protect patients from unacceptable risk, and in several cases inhumane treatment [12]. The systemic failings were so pronounced, that this event has been described as:

> *"the worst crisis any district general hospital in the NHS can ever have known"* [12, p. 47].

Events contributing to the failings that resulted in unacceptable standards of care to occur at this Trust have been dated back to about 2001 [13]. Initial concerns regarding the Trust were report in August of 2001 and January 2002 to the NHS Executive West Midlands. These were related to staffing levels, leadership and management problems [13]. Over the period from 2004 to 2009, several key events or changes occurred within the organisation, that could be seen as contributing towards the downward spiral the Trust experienced. Some of these included the suspension of reporting patient complaints to the hospital trust board from 2003 to 2006 [14], a financial recovery plan that was put in place in 2005, the modification of the ward structure to include two new units and reconfiguration of the clinical floors and at the end of 2005 the replacement of the Chief Executive [13]. Additionally the director of Nursing was replaced at the end of 2006, the Trust request £1 million for redundancies twice in 2006, and the Trust (the Mid-Staffordshire General Hospital NHS Trust) received foundation trust status beginning of 2008 [12]. Over this period, numerous reviews, audits and reports were conducted, of which many suggested concerns regarding staffing, managerial and other concerns.

In 2008, the Health Care Commission (HCC) were approached based on above average mortality rates to investigate the Trust. The investigation found that the staffing shortages, operational problems and a lack of leadership meant that despite the best efforts of staff, the quality of care was compromised and patient safety was at risk [13]. In response to the HCC investigation, the above-average mortality rates and the persistent complaints made by a group of patients, named "Cure the NHS", an independent inquiry was launched [13].

The independent inquiry, which concluded in 2010 [11] and resulted in the publication of what has become known as the Francis Report, recounted detailed accounts of inadequate care and incidents that identified unexpected risks that patients experienced. These included malnutrition and dehydration [15]. Due to the extensive public outcry, the independent inquiry was followed by a public inquiry which concluded in 2013 [12].

Numerous system factors were identified as contributing to the appalling standards of care patients received [12]. Examples of these included inadequate staffing, negative culture, professional disengagement regarding reporting concerns, poor governance, a lack of focus on standards of service, inadequate risk assessments, and incorrect priorities [16].

The above description of the events surrounding the Mid-Staffordshire NHS Foundation Trust is a brief summary and not an exhaustive description of the events and occurrences. The purpose of this summary is to provide the context for the results included below. Additional events will be mentioned in the results as they directly relate to signals.

2 Method and Analysis Frameworks

Signals and their relation to organizational disasters are gaining increasing interest and traction among the research community. Signals are increasingly mention in accident reports, with the an entire section of the Public Inquiry report [13] of the Mid-Staffordshire NHS Foundation Trust dedicated to them. The aim of this study was to analyse the signals identified in the 2013 report as well as other signals that could be extracted from the event descriptions included in both reports for this example of a healthcare-related organisational disaster using a systems approach. The objectives of this case study included determining the possible sources of signals and the rationalizations that affect action and response to these signals.

An explorative qualitative method was adopted to investigate these due to the fuzzy nature of weak signals. The two reports generated by the independent [11] and public [13] inquiries into this Mid-Staffordshire Disaster were selected for analysis using thematic analysis [17]. The signals and related events leading up to the May 2009 were included in the analysis.

The two models selected for the analysis were the SEIPS 2.0 model [18] and the Weak Signals Framework [19]. The SEIPS 2.0 model was selected as it a systems model that has not only been used to understand healthcare processes better to improve patient safety [20, 21] but also to understand infection outbreaks within acute care settings to identify the larger system contributing factors of the outbreaks [22]. The Weak Signals Framework was selected for the analysis as it provides a structure for the analysis of weak signals in the context of the work, actions and events in the system in which they occur specific for the healthcare context. The framework provided a point of reference for the analysis of related and key information relating to signals. By analysing the signals from this perspective, one identifies the sources and the related rationalizations. This together provides a greater understanding on the context in which these signals occur. The Weak Signals Framework is depicted in Fig. 1.

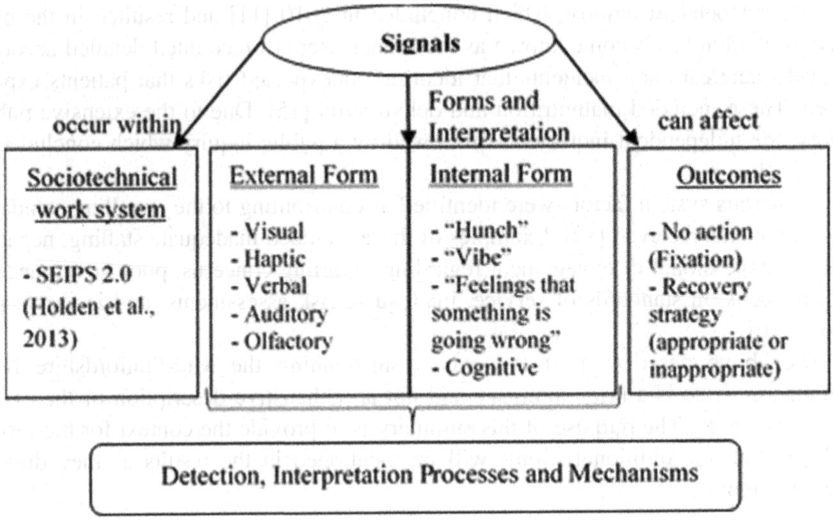

Fig. 1. The Weak Signal Framework for the investigation of weak signals within the healthcare context.

Signals were identified and grouped according to the elements in the sociotechnical work system, as described in the SEIPS 2.0 model, from which they originated, the types, and the associated rationalizations. All data were analysed using NVivo 10 (QSR International, 2014) using thematic analysis [17].

3 Results

From the analysis, an extensive list of signals, categorised according to the system levels from which they originated was created. In addition to the examples of signals collected, corresponding rationalizations were included where available for the signals.

Numerous signals, some stronger than others, were extracted from the official reports. A collection of examples from the different system levels, the year they were presented and the rationalizations for why no action was taken is included in Table 1. As is clear from the table, numerous signals were present over the time period, though many were rationalized away. From the analysis, it is clear that many different factors contributed to this disaster and numerous signals were present, but for brevity, only a few key examples are included here. The majority of the signals were identified from the person, organization and external environment levels of the Work System Element of the SEIPS 2.0 model.

Table 1. A selection of signals, the year they occurred, and their corresponding rationalizations identified in the two reports.

System component	Signal	Examples (Year of occurrence)	Rationalization
Person	Patient complaints	Patients reported inappropriate attitudes of staff (2002)	Patient's feared repercussions
	Staff behaviour and complaints	Staff voiced concerns regarding reconfiguration of wards (2005)	Staff were told their input had been considered
	Whistleblowing	Whistleblowing regarding conduct of senior staff (2005, 2007)	She was advised to "keep her head down"
	Management voiced concerns	The medical director voiced concerns about the surgical department (2007)	
Organization	Loss of star ratings	Loss of star rating due to breaches in waiting times and financial deficit (2004)	The tool was criticized for being a blunt assessment tool
	Financial recovery plan	Despite staffing levels concerns, a plan containing redundancy was implemented (2005)	Staff reductions seemed to be consistent with proposed reduction in beds
	Negative statistics	Mortality statistics were above average	There were concerns with the coding
Internal environment	Outbreaks	Outbreaks of *Clostridium difficile* (2008, 2009)	It was claimed an action plan was underway
External environment	Findings in reviews, audits and surveys	Reviews included Children's Service Review (2003, 2004, 2006, 2008), National Review of Medicines Management (2006), National Cancer Review (2005, 2006), Royal College of Surgeons Review (2007)	Rationalizations included it was thought action was being taken prior to the reports being published, the complaints were limited to one ward and not reflective of wider concerns, and not all data had been submitted for consideration
	Findings in official reports	Reports included Commission of Health Improvement Report (2002), Barry Report (2005), Dr Foster Report (2007)	
	HCC investigations	HCC investigation highlighted concerns regarding basic nursing care and medication (2008)	Other regulation bodies decided to await the outcome of the investigation

Person level signals identified included patient complaints, staff complaints and behaviours, whistleblowing by staff and management staff voicing concerns. Highlighted in both the first and second inquiry reports, was that senior management was not made aware of the concerns of patients and frontline staff [12]. This is highlighted by the following extract:

> *"Incidences of poor care were not formally fed through the system and they were not supplied to commissioners or regulators..."*[13, p. 47]

However, staff did speak up which was visible in several different types of signals. Examples of these include the person level signals relating to staff whistleblowing on senior staff's conduct and staff voicing their complaints regarding the reconfiguration of certain wards. Furthermore, based on the Barry Report, published in 2005, which was a report generated as a result of a whistle blowing complaint, staff were trying make management aware of the current situation, which contradicts the above statement [13, p. 69]. With regards to staff voicing concerns regarding the reconfiguration of wards, this signal was met with the rationalization that the staff had already given input. Another example of a signal related to the concerns of staff, which was identified though an internal staff survey conducted in 2007, which could be seen as an organization level signal, is highlighted by the following extract:

> *"Concern was expressed at the percentage of staff who said they would not want to be treated at the hospital, nor wish a relative to be either."* [13, p. 98]

An example of a signal that was identified where management staff voiced concern was that the medical director reported concerns about the surgical department. Elements of the strength of the signal and the response to signal is indicated in the extract below:

> *"From the time of her appointment as Medical Director, ... had harboured concerns about the Surgical Department. ... She had a number of audits and other reviews undertaken but these came up with no evidence of concern. She approached the National Clinical Assessment Service (NCAS) who agreed with her proposal to invite the Royal College of Surgeons (RCS) to conduct a review..."* [13, p. 111]

This is a prime example of a weak signal, as the staff member was convinced that there was a problem despite evidence contradicting this. With hindsight, it has been confirmed that there were reasons for concern.

Organization level signals identified included the loss of the Trust's star rating; implementation of a financial recovery plan that included redundancies despite staffing level concerns; and negative organizational statistics such as above average mortality statistics. The loss of stars signal was rationalized away in questioning the validity of the assessment tool. The rationalization for the financial recovery plan was that the proposed staff reductions were in-line with the proposed reduction in beds. The signal of the above average mortality rates appears to be one of the strongest signals in this case study. Despite the rationalization of the metric being subject to concerns, especially regarding the coding of the data, this signal, possibly in addition to several weaker ones, initiated the greatest response, namely the HCC investigation and the first inquiry. This signal was identified by several sources, and can be considered as an organization level signal. However, this signal also featured in the external environment signal “Dr Foster report”. The strength of the signal is highlighted by the extract below:

> *“... it has to be concluded that this was a clear alarm bell requiring urgent action to find out whether this result could be explained by a review of the care provided.”* [13, p. 100]

Internal environment level signals identified included two outbreaks of *Clostridium difficile*, one in 2008 and 2009. These signals directly reflect the concerns relating to the key areas identified by the independent inquiry of continence, bladder and bowel care; safety; personal and oral hygiene; and cleanliness and infection control [11]. Furthermore, following the outbreak in 2008, senior staff raised concerns, which should have been a signal to management, as they felt unhappy about the Trust’s reaction to outbreak as there had been an insufficient sense of urgency.

External environment level signals identified included the findings of external and peer reviews, audits and surveys; findings from official reports and the HCC investigation. Several reviews, investigations, audits and surveys were repeated over the time period in question whereby on numerous occasions similar results were found. Numerous concerns highlighted in the report published in 2010 were concerns that had been mentioned in earlier reports, for example the Barry Report in 2005 highlighted inadequate handovers, deficient note keeping, poor standards of care, inappropriate management style and inappropriate behaviour by staff [13, p. 64].

The HCC investigation was an unusual event and should have been considered as a signal to other regulatory bodies that there was a need for concern regarding this Trust’s performance. Other oversight and regulation bodies decided to await the outcome of the investigation, which is a form of rationalization.

4 Discussion and Recommendations

The events surrounding the Mid-Staffordshire NHS Foundation Trust can be qualified as an organizational disaster by the definition provided in the introduction [1], as in its entirety it was a high impact event that threatened the survival of the organization, in this case this specific Trust. As a result of this organizational failure, numerous patients lost their lives, experienced unacceptable standards of care and undignified treatment, and in addition to this, the inquiries into this event have cost approximately £19 million and ultimately have led to the trust being dissolved in 2014 and services being relocated to other centres.

In this case study, the drift into failure through ignoring precursory signals is quite evident. This is visible in that as system components and safety processes were failing, the signals indicating this were rationalised away or the response was too weak to prevent the failures from still occurring. This is visible from the results above, specifically with regards to the numerous signals available over the time period as well as the repetition of various signals.

Through the analysis of signals and their rationalizations in this case study, insights into the characteristics of signals can be extracted. The characteristics included the repetition of signals and the accumulative effect of signals. One would expect the extent and severity of what was occurring, to have been derivable from some individual signals if they were clear and strong enough, for example the higher system level signals such as Commission of Health Improvement Report in 2002 or the loss of the star rating in 2004. But also, a combination of slightly weaker signals, repetition or the accumulation of these signals should have notified management of what was occurring. Examples of repeated signals included the outbreak of *Clostridium difficile* and external peer reviews.

Furthermore the clarity of the information regarding what was occurring at this Trust should have become more apparent with the repetition or accumulation of signals [23]. The findings in official investigations and reports, which provide evidence-based information, and as a result may be considered as relatively strong higher system level signals, can be seen as an accumulation of signals. The reports usually comprised of an accumulation of lower system level signals, including person level signals, for example patient complaints and staff behaviours, task level signals and internal environment level signals. But in this case study, the majority of the signals, both lower and higher system level signals, were rationalized away, and then occurred later in the timeline again.

A common theme among the rationalizations was that the Trust assured external and regulatory bodies that progress had been made in correcting perceived deficiencies at numerous different time points. And there appeared to have been a lack of follow up on the previously generated reports. Additional system factors that prevented signals from being noticed and responded to include a negative culture, professional disengagement regarding reporting concerns, poor governance, a lack of focus on standards of service, inadequate risk assessment of staff reduction, and incorrect priorities [12].

The case study also highlights how rationalizations from one signal may impact the "face-value" of other signals. An example of this was that it was felt that if the severe staff cutting to meet the needs of the cost improvement plan had negative effects, this would have been highlighted in other performance measures. But unfortunately, the Trust's performance measures had been highlighted as unreliable and the systems for collecting and coding for these measures were seen as inadequate.

5 Conclusion

Signals provide an opportunity for insight regarding the status of the system and areas of risk [9]. But unfortunately, these signals are often rationalized away. As with many organisational disasters, this one too highlights the difficulty of recognising and accurately acting upon signals of imminent failure [2]. This case study highlighted examples of numerous different system level signals and their associated rationalizations. The source of the signals was a focus of this case study as incorporating signals in risk management requires being able to identify where these signals originate, namely "knowing where to look for them". The second key focus of this case study was that of the rationalizations of signals. One needs to know the rationalizations for signals to better understand the factors that hinder acting upon them when identified.

It is the hope of this research that by understanding why signals were dismissed previously and the responses taken in the past, a better action plan may be developed for the current unfolding situation. One needs experience or cases to create precedents, which one needs to assist in creating operational policy. This case study is an example of how healthcare can suffer from an organizational failure. It is essential that one learns as much as possible to inform policy and procedure so that safe guards for this kind of failure can be developed and put in place.

References

1. Duncan WJ, Yeager VA, Rucks AC, Ginter PM (2011) Surviving organizational disasters. Bus Horiz 54(2):135–142
2. Wei Choo C (2008) Organizational disasters: why they happen and how they may be prevented. Manag. Decis. 46(1):32–45
3. Ansoff I, Mcdonnell E (1990) Implanting strategic management, 2nd edn. Prentice Hall International, United Kingdom
4. Rasmussen J (1983) Skills, rules, and knowledge; signals, signs, and symbols, and other distinctions in human performance models. IEEE Trans. Syst. Man. Cybern. SMC-13 (3):257–266
5. Weick KE (1995) Sensemaking in organizations. SAGE Publications, London
6. Macrae C (2014) Close calls. Palgrave Macmillan, Basingstoke
7. Vogus TJ, Sutcliffe KM (2007) Organizational resilience: towards a theory and research agenda. In: IEEE international conference on systems, man and cybernetics, pp 3418–3422
8. Hollnagel E (2014) Safety-I and safety-II: the past and future of safety management. Ashgate Publishing Ltd., Farnham
9. Macrae C (2014) Early warnings, weak signals and learning from healthcare disasters. BMJ Qual Saf 23(6):440–445
10. Roberts DJ (2013) The Francis report on the mid-staffordshire NHS foundation trust: putting patients first. Transfus Med 23(2):73–76
11. The Mid Staffordshire NHS Foundation Trust Inquiry (2010) Independent inquiry into care provided by mid Staffordshire NHS foundation trust January 2005–March 2009, vol 1. The Stationary Office Limited, London
12. The Mid Staffordshire NHS Foundation Trust Public Inquiry (2013) Report of the mid Staffordshire NHS foundation trust public inquiry: executive summary. The Stationary Office Limited, London
13. The Mid Staffordshire NHS Foundation Trust Public Inquiry (2013) Report of the mid Staffordshire NHS foundation trust public inquiry. The Stationery Office Limited, London
14. Colin Thomé D (2009) Mid staffordshire NHS foundation trust: a review of lessons learnt for commissioners and performance managers following the healthcare commission investigation
15. Vincent C, Amalberti R (2016) Safer healthcare: strategies for the real world. Springer, New York
16. The Mid Staffordshire NHS Foundation Trust Inquiry (2010) Independent inquiry into care provided by mid Staffordshire NHS foundation trust January 2005–March 2009, vol II. The Stationary Office Limited, London
17. Braun V, Clarke V (2006) Using thematic analysis in psychology. Qual Res Psychol 3(2):77–101
18. Holden RJ, Carayon P, Gurses AP, Hoonakker P, Hundt AS, Ozok AA, Rivera-Rodriguez AJ (2013) SEIPS 2.0: a human factors framework for studying and improving the work of healthcare professionals and patients. Ergonomics 56(11):1669–1686

19. Carman E-M, Fray M, Waterson P (2017) Development of a framework for the analysis of weak signals within a healthcare environment. In: Organizing for high performance: proceedings of the 48th annual conference of the association of Canadian ergonomists and 12th international symposium on human factors in organizational design and management, pp 265–271
20. Rivera AJ, Karsh B-T, Beasley JW et al (2008) Human factors and systems engineering approach to patient safety for radiotherapy. Int J Radiat Oncol Biol Phys 71(1 Suppl): S174–S177
21. Gurses AP, Kim G, Martinez EA, Marsteller J, Bauer L, Lubomski LH, Pronovost PJ, Thompson D (2012) Identifying and categorising patient safety hazards in cardiovascular operating rooms using an interdisciplinary approach: a multisite study. BMJ Qual Saf 21(10):810–818
22. Waterson P (2009) A systems ergonomics analysis of the Maidstone and Tunbridge Wells infection outbreaks. Ergonomics 52(10):1196–1205
23. Hiltunen E (2010) Weak signals in organizational futures learning. Helsinki School of Economics Acta Universitatis Oeconomicae Helsingiensis

An Action Research for Patient Safety and Service Quality in Nursing Homes

G. Lefosse[1], L. Rasero[1], L. Belloni[2], E. Baroni[2], M. Matera[3], E. Beleffi[4], S. Paiva[5], L. Brizzi[6], and T. Bellandi[6(✉)]

[1] School of Nursing Sciences, University of Florence, Florence, Italy
giulia.lefosse@gmail.com
[2] Regional Referral Centre for Relational Criticalities, Regional Health Service of Tuscany, Azienda Ospedaliero Universitaria Careggi, Florence, Italy
[3] Alzheimer Patients' Association (AIMA), Florence, Italy
[4] Centre for Clinical Risk Management and Patient Safety, Regional Health Service of Tuscany, Azienda Ospedaliero Universitaria Careggi, Florence, Italy
[5] Lisbon School of Public Health, Lisbon, Portugal
[6] Tuscany Northwest Trust, Regional Health Service of Tuscany, Lucca, Italy
tommaso.bellandi@uslnordovest.toscana.it

Abstract. The objective of this study is redesigning quality and safety of care in Nursing Homes (NH), through a participatory approach based on human factors and ergonomics principles and techniques. We started with an in-deep qualitative analysis of daily practices in 4 nursing homes, selected on the basis of their performance (good and bad) and location (urban and rural). Then we performed on-site observations and ethnographic interviews with workers and patients. To conduct the explicit, non-participant observations, an original grid was developed, inspired by the definition and dimensions of quality. The crude notes of observations were subsequently re-aggregated on the grid by themes and dimensions of the observation scheme, then subjected to expert reviewers who evaluated weaknesses and strengths. According to experts' review, the most relevant safety related themes are infections and falls prevention, quality of nutrition, behavioural disorders and management of emergencies. On the basis of the elaboration of data, we then designed a multidimensional intervention encompassing interactive training sessions, scenario based redesign of weak care processes, supervision and feedback to front-line workers and management. A follow-up session is scheduled to monitor progress on site and provide feedback, as well as an evaluation of sensitive indicators at the intervention sites compared to the regional benchmark. Pilot study shows feasability of the intervention and active participation of the involved workers.

Keywords: Nursing home · Long-term care · Patient safety
Multidimensional intervention

1 Background

Patient safety in long term care has been under investigated, namely in nursing home settings [1], comparing to hospitals [2–4].

S. Bagnara et al. (Eds.): IEA 2018, AISC 818, pp. 701–714, 2019.
https://doi.org/10.1007/978-3-319-96098-2_85

Although in long term care quantitative studies have been the most used as a research method in the study of patient safety, qualitative analysis have been also refer as an important approach to be developed in this area due to the need of interpretative philosophy and ethnographic studies [3].

Long term care involves different settings focused on various patients groups that require different care supports. These are organized into inpatient units that are divided in convalescence units, rehabilitation, medium and long-term care, palliative care, and outpatient units, for whom is still independent in daily life activities. Nursing homes include both inpatient and outpatient units.

Falls, pressure ulcers, infections (resulting from the care) and medication-related adverse events (AEs) are the most common incidents identified in long term care [2, 5, 6] and are potentially preventable [7, 8]. Masotti, Mccol and Green (2010) refer that line-related adverse events are also very common in long term care context [2]. This kind of events are related with "medical interventions that require the insertion of a line through the skin and other tissue" such us line/catheter occlusion, catheter site infections and catheter-related blood stream infections [9, 10]. Aggressive resident behavior is a reality in the long term care, mostly in psychiatric patients [11].

Long term care focus on improving the quality of health of patient from different age groups. In nursing homes patients are in their majority old people with complex clinical conditions, functional dependence, co-morbidities (chronic diseases) and terminal conditions [5] which lead to higher risk of a patient safety incident [5] and higher risk of hospitalization [12].

Moreover, patients' poor health status involving cognitive and/or functional impairments, presence of depression and low level of health literacy are also identified as risk factors for adverse events occurrence in long term care settings [2, 4, 11, 13]. All of this factors increase the need of complex and prolonged care. For example, the presence of cognitive and behavioral problems can complicate some procedures such as administration of medication which increase the risk of adverse drug events [14, 15].

Adverse events occurrence is also influenced by the organization and system level characteristics.

Patient safety recommendations focus on the importance of having accurate and complete clinical records [11] and on the development of communication skills underlying coordination and collaboration between the management, staff, residents and families for preventing patient safety incidents [2, 11].

The team experience, staff training and education may influence the medication management, unrecognized drug label instructions and inadequate patient monitoring/assessment which can cause harm to the patient [2, 11, 16, 17]. Inappropriate prescription of medication and medical tests were highlighted as important safety failures for patients with comorbidities [4].

Most of long term care facilities have limited financial resources for staff development and education and low access to current research literature [18]. Is therefore important to maximize the resources and create a learning environment where the leaders support the development of the staff skills in the identification of potential hazards and safe resident handling procedures, building a safety culture and support evidence-based practice adoption [17–19]. Also appropriate number of human resources that enable health professionals to have time to attend patients' needs are one of the key factors for the improving of patient safety [19, 20].

Factors that influence safety incidents in nursing homes	
Intrinsic to the Patient	Extrinsic to the Patient
Age Co-morbidities Gender Depression Cognitive impairments Functional status Patient compliance Cognitive and behavioural problems	Team experience Staff training and education Medication management Patient monitoring Patient assessment Reporting Documentation (protocols and guidelines) Communication between different level of operation (staff, management) Staff workload

The literature refers that patient safety improvement requires staff training and education, communication skills as well as increase patient monitoring and reporting [2, 11, 21, 22]. Safety culture has been also highlighted as an important factor for safety incidents prevention such us falls and medication errors [1, 3, 23].

Patient safety underlines leadership involvement from both professionals and managers [24]. It was found that high level of interaction between medical staff and the board improves quality of care. Board meeting spending more than 25% of their time on quality issues and an introduction of a formal quality performance measurement report are some other strategies to improve quality in healthcare and patient safety [24]. For reaching an effective patient safety improvement is therefore necessary to engage all levels of operation in healthcare and prioritize quality of health issues in long term settings and, in particular, in nursing home facilities.

Finally, care of the elderlies is the main public health challenge for the next decades, given the ageing population. Up until now, service quality and patient safety in long term care is not a hot topic in the political and research agenda. The approval of the Italian law 24/2017 for patient safety requires a quick shift, given its emphasis on the integration of risk prevention between health and social-health services.

2 Objectives

To explore quality and safety of care in Nursing Homes (NH), in order to design an improvement multidimensional intervention through a participatory approach based on human factors and ergonomics principles and techniques.

3 Methods

3.1 Setting

This is an action research in a complex environment, based on extensive observations of daily practices, interviews with workers and patients, participatory interactive sessions to interpretate and redesign care processes.

4 nursing homes were selected for the pilot study. The criteria used to selcet the organizations were: scores in the regional evaluation system (2 with a high score and 2 with a low score), location (2 urban and 2 rural), membership health organization (2 in the Central Trust, one in the Northern and one in the Southern Trust). The intervention was conducted by a multidisciplinary team with variable composition of operators from the Regional Centres for Patient Safety and Centre for Human Relations, the Regional Agency of Health, representatives of the Italian Alzheimer's Disease Association, with background in health and social sciences.

3.2 Framework for Data Collection and Elaboration

To conduct the explicit, non-participant observations, an original framework was developed, inspired by the definition and dimensions of quality [1, 2] according to the WHO, the narrative review of literature summarized in the background of this paper and the accreditation criteria for nursing homes. The framework declines for each dimension one or more specific themes within the context of nursing homes, for each of which sensitizing themes and observation units have been identified, according to literature [3] and accreditation requirements (Table 1).

3.3 On-Site Observations

The program of each individual on-site visit was defined to observe the reality of the NH from different points of view: that of the managers and operators, that of the families of the residents and that of the researchers of the working group. The Human Relations staff conducted focus groups and interviews with the operators aimed at gathering information on the quality of relationships and organizational well-being; patient advocates lead group meetings with relatives of residents in the facilities to collect and process life stories of the assisted persons; experts in patient safety conducted structured observations in the two complementary health and social areas to understand the critical points and the strengths in the quality and safety of daily activities.

The method of ethnographic observation is part of qualitative research, it is particularly indicated when the objective of the study is the in-depth knowledge of the object being observed, on which there is no clear and shared definition in the community and in the reference culture [26]. NHs are places on the borders of communities and service organizations, exposed to changes in assisted persons caused by the epidemiological and social transition of the last decades, as well as political and economic pressures due to the limitation of resources for assistance, with a strong performance variability according to available and incomplete data [27, 28]. In addition, the

Table 1. Framework for quality in Nursing Homes

Dimensions of quality [25]	Themes in NH	Units of observation
PATIENT SAFETY delivering health care which minimizes risks and harm to service users	Falls	Handling of assisted persons (presence and use of aids, etc.) Orientation to mobilization (how much is avoided) Promotion of physical activity (even minimal) Safety of internal/external routes (no obstacles in the most frequent routes) Lighting quality Floor maintenance and quality status, timing and alert modes during cleaning Evaluation of the risk of falling Management of the consequences of the fall
	Medications	Drug storage and methods for checking the equipment Therapies registration (completeness and traceability of therapeutic acts) Prescriptions update Allergies annotation Awareness of the characteristics of drugs and the goal in their use (especially for chronic patients with cardiorespiratory problems, diabetes and mental health) Reporting adverse events and reactions
	Infections	Awareness of how to prevent infections (what is done here to prevent infections) Placing baths, soap and hydro-alcoholic gel availability, hand washing frequency and compliance control

(*continued*)

Table 1. (*continued*)

Dimensions of quality [25]	Themes in NH	Units of observation
		Presence of remainders for hands washing Management of antibiotic therapies (something is done for) Management of an infected person (e.g. gastrointestinal virus, influenza, pneumonia, what to do when)
	Pressure ulcers	Presence of aids and handling procedures for patients for the prevention of PUs Evaluation of risk and detection of risk factors Presence of advanced medications' products Nutritional risk
PATIENT CENTREDNESS delivering health care which takes into account the preferences and aspirations of individual service users and the cultures of their communities	Respect for primary needs (hygiene and nutrition)	Methods of preparation, distribution and consumption of meals Cleaning and care of private and community environments both inside and outside the facility Attention to seasonality and variety of foods, to the requests of the assisted persons and the possibility to prepare meals independently and/or with the support of the staff
	Attention to secondary and tertiary needs (self-care, social relationships and fulfillment of desires)	Care of clothing and look of the person Amplitude and quality of common and private spaces Possibility of personalization of the rooms Ability to customize the activities of patients through their involvement Technological equipment to support rehabilitation and play activities

(*continued*)

Table 1. (*continued*)

Dimensions of quality [25]	Themes in NH	Units of observation
		In-house and external recreational and socio-educational program Care for relationships with family members Care for relationships with the surrounding community (district, school, association, etc.)
ACCESSIBILITY delivering health care that is timely, geographically reasonable, and provided in a setting where skills and resources are appropriate to medical need;	Physical	Comfort and quality of private and common environments Routes actually accessible to reach all areas inside and outside the structure Features of the beds, armchairs, chairs and tables used in daily activities Characteristics of the environments dedicated to the cleaning (assisted bathroom) of the person
	Cognitive	Communications to assisted persons that are clear and comprehensible, both verbal and written about the treatments and daily activities Support for participation even for the most vulnerable people or with very limited autonomy Activities aimed at maintaining basic cognitive functions (memory, attention, reasoning)
	Organizational	Clarity of the roles and functions of the operators Knowledge of how to exit/access the facility for assisted persons and guests Relations with the outside

(*continued*)

Table 1. (*continued*)

Dimensions of quality [25]	Themes in NH	Units of observation
EQUITY delivering health care which does not vary in quality because of personal characteristics such as gender, race, ethnicity, geographical location, or socioeconomic status	Respect and enhancement of cultural, religious and social differences	Attitude towards foreign guests or in any case not belonging to the local community Respect for religious practices Prevention of discrimination against the weakest individuals from a social and relational point of view
EFFECTIVENESS delivering health care that is adherent to an evidence base and results in improved health outcomes for individuals and communities, based on need	Adequacy and appropriateness of care with respect to the person's health needs	Management of emergencies in case of deterioration of the clinical condition of the person due to illness or accident Collaboration between internal staff, family doctors and health care specialists Participation in chronic disease management programs Management of transitions in the case of hospital admissions and/or specialist visits and/or return periods at home
EFFICIENCY delivering health care in a manner which maximizes resource use and avoids waste	Ability to use human, technological and organizational resources available in the structure and in the community	Amount of work observed for direct care vs administrative activities Effective use of equipment available for rehabilitation and educational purposes Relations with the community for recreational and cultural activities Training and updating of staff
	Ratio benefits/wastes for caregivers, workers and organization	Report any evident waste (e.g. unused spaces or tools, personnel with overload or underload of work, lack of appreciation of specific skills of the staff, etc.)

concepts of quality and safety of care in NH are also sometimes out of line [29–31], depending on the point of view and the interests at stake, therefore observation is a particularly interesting method to deepen a reality in movement, returning a multidimensional representation that is not always inferable from the results of quantitative surveys.

The unit of analysis was a day of life in NH, divided into two planned observing sessions one day after the other: on the first day of the visit the observers gathered information from lunch to the moment in which the guests they are put to bed for the night; on the second day of the visit the activity started at the beginning of the morning shift until lunch. The choice of the observation period is motivated by the objective of observing and representing the development of daily activities from the point of view of the assisted persons, literally from when they open their eyes and when they close their eyes, and in the perspective of health workers, 'assistance and staff engaged in social and educational activities, both internal and contracted. The observations were provided in an explicit, non-participant manner by two operators for each work session, who followed with the shadowing technique all the professional figures engaged in NH, according to a snowball sampling. Following the unfolding of the activities led by the observation units [32]. In some cases, short ethnographic interviews were conducted with the operators and with the assisted persons, to better understand the meaning of the activities carried out and of the life experiences in NH. The sessions were scheduled in agreement with the managers of the structures and carried out on the same days of the interviews with the operators conducted by the CRCR staff, so as to be able to exchange some hot reflections during the same visit days, adjusting the observations or subsequent interviews on some particularly significant critical issues.

Observational notes were collected in real time through the use of a free application available on smartphones for writing, automatic saving and synchronization of notes.

3.4 Elaboration of Data

The crude notes were subsequently re-aggregated on the grid by themes and units of observation of the framework, then subjected to an expert review, by 6 nurses and 2 social workers who evaluated weaknesses and strengths for each reported episode of care. A color code was used to classify observational notes for each category of observed actors. They were then subjected to a formal verification of data quality and shared with the other members of the working group, for a first assessment of consistency and compatibility with what emerged from the surveys with operators and family members. The revisions of data revealed the lack of a very relevant theme in terms of its impact on patient safety, that is management of behavioural disorders: it was therefore later added to the framework and considered a priority during the intervention phase.

At the end of the review of these data emerged the strengths and weaknesses in terms of safety of care and risk management of the 4 structures.

On the basis of the elaboration of data and substantial confirmation of the validity of the framework, we then designed a multidimensional intervention encompassing interactive training sessions, scenario based redesign of weak care processes, supervision and feedback to front-line workers and management.

3.5 Design of the Intervention

The intervention is dedicated to all staff, in interactive meetings with the participation of representative staff of all the profiles present in NH (nurses, physiotherapists, assistant nurse, social workers, educators, nursing coordinator, core managers). It is delivered in two days, the first with a focus on the critical issues, called "The worst working day", the second with focus on the strengths called "The best working day", so to use examples at the extreme of bad and good care in order to help the participants reflect upon risk and success factors.

The interactive meetings are structured to bring out the critical issues and strengths of the structure through a thematic analysis in real time of care delivery problems and experiences of good practice reported by the operators, without pre-setting the most important thematic areas and connecting the emerging themes to the reference framework. What emerges from the thematic analysis of the case histories reported in the interactive meetings are the working areas themselves, on which to elaborate the improvement actions in a systemic approach, by providing critical scenarios to be resolved with a structured action plan. The research team prepares critical scenarios recombining stories reported during the interactive sessions, representing relevant contributory factors associated with care deliver problems, taking into account the London Protocol [33] as a guiding scheme. For the action plan, a structure is provided to the local teams so to help them to elaborate improvements that are consistent with accreditation requirements for quality and safety management.

Follow-up sessions have been also defined: follow-up A to 2 months from interventions to evaluate the output of the intervention in terms of process indicators (i.e. compliance with standards of safe practice); follow-up B to 6 months from interventions to evaluate impacts on outcomes measures (i.e. prevalence of infections).

4 Results

Eight observational sessions were conducted in 4 nursing homes to validate the framework. Each session took an average of 8 h and 30 min for a total of 134 h of observation. A total of 52 pages (standard A4 format) of observational notes were collected and then classified with the thematic analysis into the grid. The observation scheme resulted appropriate to encompass most of the notes, with the exception of management of behavioural disorders. According to experts' review, the most relevant safety related themes are infections and falls prevention. Quality of nutrition is a crucial theme belonging to the dimension of patient centrality, while the emergent theme of behavioural disorders can be classified within the dimension of effectiveness and paired with management of emergencies, given that a weak management of behavioural disorder can result in an emergency.

The pilot intervention took place in a single NH, selected on the basis of the management's willingness to participate in the pilot and stability of the personnel with respect to the observation period. The structure is located in a rural area and has a low score in the performance evaluation.

80% of the staff of the structure participated in the interactive seminars. A list of priority patient safety issues, consistent with the framework, emerged: healthcare-associated infections, management of behavioral disorders, nutrition and nutritional risk, emergency management.

The groups were asked to perform: analysis of the actions, conventions and competences that they put in place for the management of the issues, critical issues, possible interventions for improvement in terms of safety and quality.

Following the 2 days interactive sessions, a re-elaboration of the work was carried out through an Excel table, which reported the cases, the thematic analysis and the possible interventions enunciated by the working groups. After the analysis, scenarios have been defined on the base of the reported case histories, i.e. a critical condition and an optimal condition for the management of possible cases around the 4 identified themes). A mandate was sent to the 5 working groups of the NH to resolve scenarios, design and implement improvement actions with the support of the research team. On the topic of management of emergencies, two working groups were organized, one dedicated to emergency in low-acuity patients and one to emergency in patients with Alzheimer or other behavioural disorders. This mandate lead to a series of critical reflections starting from the resolutions of scenarios based on case histories, in order to prevent risk factors and leverage on elements of good practice to improve quality and safety of care.

The first follow-up session has been conducted to monitor progress on site and provide feedback.

Starting from the resolution of the scenarios, 4 out of 5 working groups were able to respond to the mandate. The hardest problem resulted the prevention of infection risks associated with the use of nasogastric tube. In particular, the group reported a lack of basic knowledge to identify and manage the risk of misplacement of nasogastric tube. Thanks to this awareness, feedback was provided to reach relevant educational programs on technical skills for assistant nurses and nurses. Of the other groups that resolved their scenarios, just one group was able to formulate an adequate improvement plan including a thorough review of current procedures and practice to manage a clinical emergency that may occur at the low acuity module of the NH. The other group dedicated to emergency reviewed availability of devices and medication to manage emergencies, as well as workers training, but did not produce any improvement action. Group on nutrition made also a review of the current state of risk assessment and management of special diets, especially to prevent ab ingestis, while group on behavioural disorders discussed current practices to address this very important issue, finding out a strong need of better coordination and collaboration between health and social workers to integrate medical and non-medical treatments.

The next follow-up session is scheduled to monitor any possible variation of sensitive indicators at the intervention sites compared to the regional benchmark.

5 Discussion

This paper reports the preliminary results of a complex multidimensional intervention to improve quality and safety in NH. The study demonstrated the validity and feasability of a structured framework designed to analyze quality of care in NH, that was tested in 4 nursing homes with the collection and elaboration of extensive observational data. The following pilot study conducted in one individual NH proved to be feasable, even though at the follow-up just one local working group was able to fully comply with the mandate and start an improvement plan. Limited tutoring and lack of technical and managerial competences at the local level seemed to be the main barriers, the research team recognized after the follow-up session. A closer tutoring and networking with other NH shall be considered to improve the effectiveness of the intervention, as well as a formal inclusion of NH within the quality and safety network of membership health organization, that currently act towards the NH with authorization rules and inspections on structural characteristic, plus accreditation rules and surveys on process requirements and indicators.

References

1. Castle NG, Sonon KE (2006) A culture of patient safety in nursing homes. Qual Saf Health Care 15(6):405–408
2. Masotti P, McColl MA, Green M (2010) Adverse events experienced by homecare patients: a scoping review of the literature. Int J Qual Health Care 22(2):115–125
3. Gartshore E, Waring J, Timmons S (2017) Patient safety culture in care homes for older people: a scoping review. BMC Health Serv Res 17(1):1–11
4. Panagioti M, Blakeman T, Hann M, Bower P (2017) Patient-reported safety incidents in older patients with long-term conditions: a large cross-sectional study. BMJ Open 7(5):1–9
5. Simmons S, Schnelle J, Slagle J, Sathe NA, David Stevenson M, Carlo M et al (2016) Resident safety practices in nursing home settings. Agency for Healthcare Research and Quality, Rockville
6. Rust TB, Wagner LM, Hoffman C, Rowe M, Neumann I (2008) Broadening the patient safety agenda to include safety in long-term care. Healthc Q 11:31–34
7. Ouslander JG, Lamb G, Perloe M, Givens JH, Kluge L, Rutland T et al (2010) Potentially avoidable hospitalizations of nursing home residents: frequency, causes, and costs. J Am Geriatr Soc 58(4):627–635
8. Saliba D, Kington R, Buchanan J, Bell R, Wang M, Lee M et al (2000) Appropriateness of the decision to transfer nursing facility residents to the hospital. J Am Geriatr Soc 48(2): 154–163
9. Waal GHD, Naber T, Schoonhoven L, Persoon A, Sauerwein H, Van Achterberg T (2006) Problems experienced by patients receiving parenteral nutrition at home: results of an open interview study. J Parenter Enteral Nutr 30(3):215–221
10. Shirotani N, Iino T, Numata K, Kameoka S (2006) Complications of central venous catheters in patients on home parenteral nutrition: an analysis of 68 patients over 16 years. Surg Today 36(5):420–424

11. Wagner LM, Rust T (2008) Safety in long-term care settings: broadening the patient safety agenda to include long-term care services. Canadian Patient Safety Institute
12. Wang KN, Bell JS, Chen EYH, Gilmartin-Thomas JFM, Ilomäki J (2018) Medications and prescribing patterns as factors associated with hospitalizations from long-term care facilities: a systematic review. Drugs Aging 26:1–35
13. Lim RHM, Anderson JE, Buckle PW (2016) Work domain analysis for understanding medication safety in care homes in England: an exploratory study. Ergonomics 59(1):15–26
14. Alldred DP, Standage C, Fletcher O, Savage I, Carpenter J, Barber N et al (2011) The influence of formulation and medicine delivery system on medication administration errors in care homes for older people. BMJ Qual Saf 20(5):397–401
15. Tariq A, Georgiou A, Westbrook J (2012) Medication incident reporting in residential aged care facilities: limitations and risks to residents' safety. BMC Geriatr 12(1):67
16. Axelsson J, Elmstahl S (2004) Home care aides in the administration of medication. Int J Qual Health Care 16(3):237–243
17. Silvestre JH, Bowers BJ, Gaard S (2015) Improving the quality of long-term care. J Nurs Regul [Internet] 6(2):52–56. Elsevier Masson SAS
18. Specht JK (2013) Evidence based practice in long term care settings. J Korean Acad Nurs 43(2):145–153
19. Sohn M, Choi M (2017) Factors related to healthcare service quality in long-term care hospitals in South Korea: a mixed-methods study. Osong Public Health Res Perspect 8(5):332–341
20. Shin JH, Hyun TK (2015) Nurse staffing and quality of care of nursing home residents in Korea. J Nurs Scholarsh 47(6):555–564
21. Curatolo N, Gutermann L, Devaquet N, Roy S, Rieutord A (2014) Reducing medication errors at admission: 3 cycles to implement, improve and sustain medication reconciliation. Int J Clin Pharm 37(1):113–120
22. Arnoldo L, Cattani G, Cojutti P, Pea F, Brusaferro S (2016) Monitoring polypharmacy in healthcare systems through a multi-setting survey: should we put more attention on long term care facilities? J Public Health Res 5(3):104–108
23. Institute of Medicine (2001) Crossing the quality chasm: a new health system for the 21th century. National Academy Press, Washington, DC
24. Vaughn T, Koepke M, Kroch E, Lehrman W, Sinha S, Levey S (2006) Engagement of leadership in quality improvement initiatives: executive quality improvement survey results. J Patient Saf 2(1):2–9
25. World Health Organization (2006) Quality of care: a process for making strategic choices in health systems. World Health Organization, Geneva
26. Hutchins E (1995) Cognition in the wild. Massachusetts Institute of Technology Press, Cambridge
27. Nuti S, Rosa A, Trambusti B (2014) La mappatura delle residenze per anziani non autosufficienti in Toscana- Report 2014. Laboratorio Management & Sanità, Scuola Superiore Sant'Anna, Pisa
28. ARS Toscana (2014) Progetto ministeriale CCM: qualità e sicurezza
29. Castle NG, Ferguson, JC (2010) What is nursing home quality and how is it measured? Gerontologist 50(4): 426–442
30. Mor V, Berg K, Angelelli J, Gifford D, Morris J, Moore T (2003) The quality of quality measurement in US nursing homes. Gerontologist 43:S237–S246

31. McHugh M, Shi Y, Ramsay PP, Harvey JB, Casalino LP, Shortell SM, et al (2016) Patient-centered medical home adoption: results from aligning forces for quality. Health Affairs 35(1):141–149
32. Corbetta P (1999) Metodologia e tecniche della ricerca sociale. Il Mulino
33. Vincent C, Taylor-Adams S, Chapman EJ, Hewett D, Prior S, Strange P et al (2000) How to investigate and analyse clinical incidents: clinical risk unit and association of litigation and risk management protocol. BMJ 320:777

Musculoskeletal Disorders Among Physiotherapists Working in a Single Rehabilitation Centre: A Longitudinal Study

Deepak Sharan[1(✉)] and Joshua Samuel Rajkumar[2]

[1] Department of Orthopedics and Rehabilitation, RECOUP Neuromusculoskeletal Rehabilitation Centre, 312, 10th Block, Anjanapura, Bangalore 560108, KA, India
deepak.sharan@recoup.in

[2] Department of Physiotherapy, RECOUP Neuromusculoskeletal Rehabilitation Centre, 312, 10th Block, Anjanapura, Bangalore 560108, KA, India
joshua.samuel@recoup.in

Abstract. Physiotherapists are at higher risk because the major amount of their job task and duration involves physical handling and exerting force. Their tasks are generally complex and involve many physical activities that can lead to acute and chronic Work Related Musculoskeletal Disorders (WRMSDs). A longitudinal survey study was conducted, in which physiotherapists working in a single rehabilitation centre were evaluated through a self-administered questionnaire. Nordic Musculoskeletal Questionnaire (NMSQ) was used to find the prevalence and disability rate of MSD. The data were collected from a total of 250 physiotherapists, who were working in a tertiary level rehabilitation centre in an industrially developing country. On an average the physiotherapists worked for 9.45 h per day for 6 days a week. The highest prevalence of musculoskeletal pain among physiotherapists was in the following anatomical areas: low back (52.55%), upper back (48.35%), wrist/hand (25.25%), and neck (20.6%). A high prevalence of WRMSD was reported among Physiotherapists working in the tertiary level rehabilitation centre.

Keywords: Musculoskeletal disorder · Physiotherapists · Rehabilitation centre

1 Introduction

Health care professionals report a high prevalence of musculoskeletal symptoms due to risk factors like strenuous physical activity, lifting and carrying the patients, assistance in transferring the patients, improper handling of loads, maintaining awkward posture like bending for a prolonged time, repeated twisting and use of different body joint movements, using high frequency vibration machines, overstraining and static body position, and mental and psychological stress. In particular, Physiotherapists are at higher risk because the major amount of their job task and duration involves physical handling and exerting force. Their tasks are generally complex and involve many physical activities that can lead to acute and chronic Work Related Musculoskeletal

S. Bagnara et al. (Eds.): IEA 2018, AISC 818, pp. 715–716, 2019.
https://doi.org/10.1007/978-3-319-96098-2_86

Disorders (WRMSDs). The aim of this study was to evaluate the prevalence and risk factors for WRMSDs among physiotherapists working in a single rehabilitation centre.

2 Methodology

A longitudinal survey study was conducted, in which physiotherapists working in a single rehabilitation centre were evaluated through a self-administered questionnaire. The data were collected from a total of 250 physiotherapists, who were working in a tertiary level rehabilitation centre in an industrially developing country. The participants were a full time physiotherapist, minimum of at least 1 year of experience in the centre, age group 22 to 45 years and both the genders were included, the physiotherapist should not involve in any other part time job and exclusion criteria is that the physiotherapist should not suffer with musculoskeletal symptoms previously or any other pathology, any recent surgeries, medical conditions and post pregnancy, any other systemic health problems. The questionnaire included demographic details like age, gender, height, weight, total work experience, number of working hours in a day, type or department of work, etc. Nordic Musculoskeletal Questionnaire (NMSQ) was used to find the prevalence and disability rate of MSD. The short form of work style questionnaire was used to know about psychological risk factors of adverse work style.

3 Results

The mean age of the participants who participated in our study was 29.45 years. The mean height and weight of the participants was 162.35 cm and 66.75 kg respectively. On an average the physiotherapists worked for 9.45 h per day for 6 days a week. Only 10% of the subjects exercised regularly. The prevalence of musculoskeletal pain among physiotherapists was found to be 80%. The highest prevalence of musculoskeletal pain among physiotherapists was in the following anatomical areas: low back (52.55%), upper back (48.35%), wrist/hand (25.25%), and neck (20.6%).

4 Conclusions

A high prevalence of WRMSD was reported among Physiotherapists working in the tertiary level rehabilitation centre. A risk evaluation tool is therefore required to evaluate the specific tasks carried out by them to help reducing the risk factors for developing WRMSD.

The Burden of Caregiving: Musculoskeletal Disorders in Caregivers of Children with Cerebral Palsy

Deepak Sharan[1(✉)] and Joshua Samuel Rajkumar[2]

[1] Department of Orthopedics and Rehabilitation, RECOUP Neuromusculoskeletal Rehabilitation Centre, 312, 10th Block, Anjanapura, Bangalore 560108, KA, India
deepak.sharan@recoup.in

[2] Department of Physiotherapy, RECOUP Neuromusculoskeletal Rehabilitation Centre, 312, 10th Block, Anjanapura, Bangalore 560108, KA, India
joshua.samuel@recoup.in

Abstract. While caring for the child with cerebral palsy (CP), the caregivers may be exposed to stress and physical load leading to musculoskeletal disorders (MSDs). A longitudinal survey was conducted among 405 (mean age: 37.5 ± 4.4 years) caregivers of children with CP who underwent SEMLS. A self-administered questionnaire was used to collect the relevant data, which included the Borg CR-10 scale and modified Caregivers Strain Index. About 90% of the primary caregivers reported one or more musculoskeletal symptom. Pain was the most common musculoskeletal symptom reported by 94% of the caregivers studied. Over the year after the child underwent SEMLS, nearly 40% of the caregivers reported that their health had worsened further, and this may be due to increased physical stress and strain along with psychosocial energy depletion. Lower back pain was the commonest reported musculoskeletal symptom (65%) followed by knee pain (46%), neck pain (40%) and shoulder pain (36%). Musculoskeletal problems increased the fatigue levels as recorded by Borg CR-10 scale.

Keywords: Caregivers · Musculoskeletal disorder · SEMLS

1 Introduction

Cerebral palsy (CP) causes not only motor disturbances but also sensory, cognitive, social, behaviour, speech and communication, seizure disorder, respiratory illness and other musculoskeletal disorder leading to complex limitations in self-care functions. Nearly two-thirds of children with CP will require a high need for health care services including caretaking. While caring for the child, the caregivers may be exposed to stress and physical load leading to musculoskeletal disorders (MSDs). Research related to the epidemiology of musculoskeletal disorders (MSDs) in caregivers of children with CP is scanty. The aim of this study was to find out the prevalence and risk factors of MSDs among caregivers of children with CP following Single Event Multi Level Surgery (SEMLS).

S. Bagnara et al. (Eds.): IEA 2018, AISC 818, pp. 717–718, 2019.
https://doi.org/10.1007/978-3-319-96098-2_87

2 Methodology

A longitudinal survey was conducted among 405 (mean age: 37.5 ± 4.4 years) caregivers of children with CP who underwent SEMLS. A self-administered questionnaire was used to collect the relevant data, which included the Borg CR-10 scale and modified Caregivers Strain Index. Caregivers who were found to have musculoskeletal symptoms were subsequently assessed by an Orthopaedic Surgeon and Rehabilitation Physician and treated with medication and/or physical therapy.

3 Results

The prevalence of MSDs among caregivers of children with CP post SEMLS varied from caregivers to caregivers depending upon the level of disability of the children. Most of the primary care givers were female and were the children's mothers. About 90% of the primary caregivers reported one or more musculoskeletal symptom. Pain was the most common musculoskeletal symptom reported by 94% of the caregivers studied. Over the year after the child underwent SEMLS, nearly 40% of the caregivers reported that their health had worsened further, and this may be due to increased physical stress and strain along with psychosocial energy depletion. Lower back pain was the commonest reported musculoskeletal symptom (65%) followed by knee pain (46%), neck pain (40%) and shoulder pain (36%). Musculoskeletal problems increased the fatigue levels as recorded by Borg CR-10 scale. In this study, caregivers reported several difficulties in complying with the treatment regimen, including financial difficulties, distance from home to the hospital, lifting and carrying the child during the rehabilitation, lack of family support and lengthy waiting time at the clinics. More than 60% of the caregivers reported a high level of stress with scores of more than 7 in the CSI. Caregivers of children with older age, more severe disabilities (GMFCS IV and V), uncooperativeness and higher body mass index were at higher risk of developing MSD.

4 Conclusions

Caregivers of children with CP undergoing SEMLS are predisposed to various MSDs because of the significant physical and psychological effort required of them while caring for the child.

Biomechanical Overload Evaluation in Manufacturing: A Novel Approach with sEMG and Inertial Motion Capture Integration

Maria Grazia Lourdes Monaco[1(✉)], Lorenzo Fiori[2], Agnese Marchesi[2], Alessandro Greco[3], Lidia Ghibaudo[4], Stefania Spada[4], Francesco Caputo[3], Nadia Miraglia[5], Alessio Silvetti[2], and Francesco Draicchio[2]

[1] University Hospital of Verona, Verona, VR, Italy
mariagrazialourdes.monaco@aovr.veneto.it
[2] INAIL, Monteporzio Catone, RM, Italy
{lo.fiori,al.silvetti,f.draicchio}@inail.it,
agnese.marchesi@gmail.com
[3] Department of Engineering, University of Campania L. Vanvitelli, Aversa, CE, Italy
{alessandro.greco,francesco.caputo}@unicampania.it
[4] FCA EMEA Manufacturing Planning & Control – Ergonomics, Turin, TO, Italy
{lidia.ghibaudo,stefania.spada}@fcagroup.com
[5] Experimental Medicine Department, University of Campania L. Vanvitelli, Naples, NA, Italy
nadia.miraglia@unicampania.it

Abstract. Biomechanical overload represents one of the main risks in the industrial environment and the main possible source of musculoskeletal disorders and diseases. The aim of the this study is to introduce new technologies for quantitative risk assessment of biomechanical overload, by integrating surface electromyography (sEMG) with an innovative motion-capture system based on inertial measurement units (IMU).

The case study was carried out in collaboration with Fiat Chrysler Automobiles Italy S.p.A. and deals with the analysis of the "central tunnel cabinet assembly" activity, performed by two workers of assembly lines during a working task, which lasts about one minute. The electromyography signals were acquired bilaterally, in three different body regions on the right and on the left side of the Erector Spinae, during standard working activities; the progression of trunk postures (flexion-extension, lateral flexion and twisting) was tracked by using an inertial motion-capture system made of wearable inertial sensors, to evaluate the alignment of the major body segments, using a developed algorithm.

Data analysis showed kinematic and muscular activity patterns consistent with the expected ones. In particular, data show that the proposed technologies can be integrated and simultaneously used during workers' real performing activities. Data quality also demonstrates that both types of sensors, EMG

S. Bagnara et al. (Eds.): IEA 2018, AISC 818, pp. 719–726, 2019.
https://doi.org/10.1007/978-3-319-96098-2_88

electrodes and IMU, not influenced each other, neither by electromagnetic noise usually present in an industrial environment. The results of this study show feasibility and usefulness of the integration of kinematic and electromyography technologies for assessing the biomechanical overload in production lines.

Keywords: Surface electromyography · Inertial sensors
Biomechanical overload

1 Introduction

Musculoskeletal disorders (MSDs) account for the most frequently reported health risks at work. Their prevention remains a challenge in the industrial settings, also through a correct evaluation and management of biomechanical overload [1].

Many observational tools have been developed for biomechanical risk assessment; these tools moderately agree with technical measurements [2].

Surface electromyography (sEMG) and kinematic measurements nowadays are the gold standard for biomechanical risk assessment. sEMG measurements give objective information about activation times and intensity of muscle activity, not appreciable otherwise [3, 4].

For industrial applications, motion capture systems are largely used to acquire workers' movements, in order to obtain objective ergonomic risk assessment and to improve working conditions. Motion capture systems composed of inertial measurement units (IMU) represent the best solution for ergonomic application in real work environment. Several researchers have introduced IMU equipment, in order to evaluate workers body motion [5]. However, the above mentioned equipment suffers from possible electromagnetic interference, widely present in industrial environments [6, 7].

The aim of this study is to provide a methodological approach for quantitative risk-assessment of biomechanical overload, based on integration of sEMG with an innovative motion-capture system based on IMU units, immune from electromagnetic interference.

2 Methods

2.1 Study Design, Setting, and Subject

The study was conducted in collaboration with Fiat Chrysler Automobiles Italy S.p.A., in the Melfi plant. Electromyography signals and flexion-extension angles of the trunk were collected from four workers, in order to evaluate muscular activity and posture in a standard work task in the assembly line.

2.2 Working Activity Description

The figures below shows a phase of activity performed by a worker on the right side (RS) (Fig. 1) and left side (LS) (Fig. 2) of the automobile in the reference workstation where the central cabinet is assembled inside the car using: screws, plugs, and cables. The cycle time lasts about one minute (58 s).

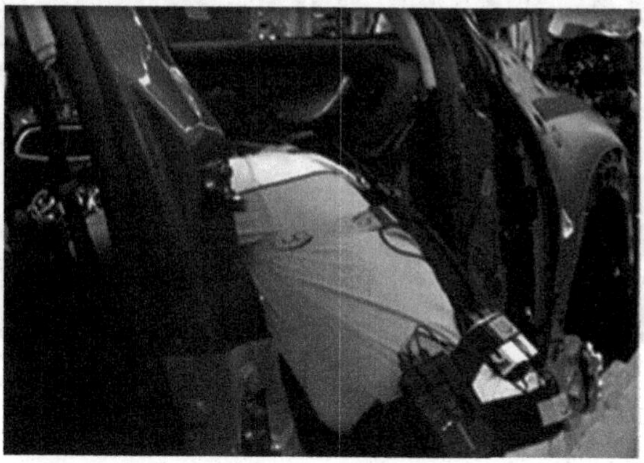

Fig. 1. Working activity on the right side of the car

2.3 Procedure and Data Acquisition

Acquisition and Processing of sEMG Signals. The electrical muscle activity was recorded with an electromyography system (FreeEMG, BTS SpA, Milan, Italy) with eight probes. Each of them disposes of an instrumentation amplifier with a 100 dB CMRR, a Hamming bandpass filter with 10–400 Hz cutting frequencies, an analog-to-digital conversion system with a sampling frequency of 1 kHz and a wireless data transmission system (Wi-Fi). Six of the eight available probes were placed over the muscles concerned by the study, using Ag/AgCl pre-gelled electrodes (H124SG, Kendall ARBO, Donau, Germany), observing the recommendations of the Atlas of the zones of muscular innervation [8]. In particular, the electrical activity was collected bilaterally from paravertebral muscles: Erector Spinae Thoracic (EST), Erector Spinae Lumbar (ESL) and Multifidus (M).

Before starting the work activity, the maximum voluntary contractions (MVC) of the muscles have been acquired, to determine the maximum level of muscle activation, used as reference during the signal-processing phase. Each task of MCV was repeated three times, after a rest period of 3 min between trials [9] and the average value was calculated.

The sEMG signals, thus collected, were after processed with an algorithm, written in the MATLAB computing environment R2017b (verses 9.3.0, MathWorks, Natick, MA, USA). In the algorithm, the electromyography signal is filtered in the band of interest [30–450 Hz] with a Butterworth IIR digital pass filter of the 5th order to reduce movement artifacts (electrode-skin) and other components of the high frequency noise. Afterwards, to extract the muscle activity profile, the signal is rectified and filtered using another Butterworth low pass IIR filter of the 3rd order with a 5 Hz cutoff frequency, thus obtaining the linear envelope. The envelope of the sEMG signal is then

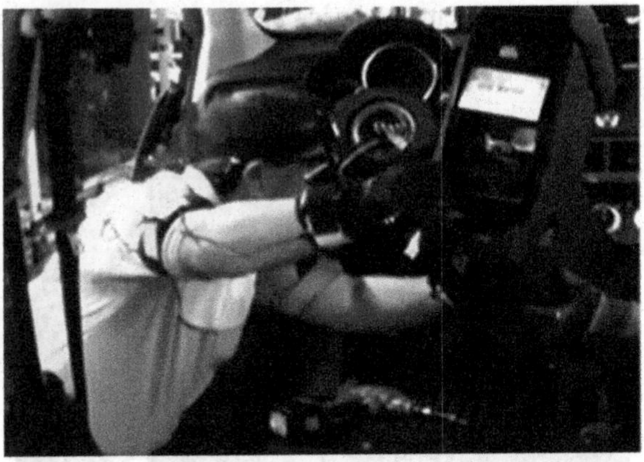

Fig. 2. Working activity on the left side of the car

expressed as the percentage value of MCV, relatively to the own muscle, and the root mean square (RMS) is calculated.

Body Motion Study. A homemade inertial motion capture system has been used in this study. The system, developed by the research teams of Machine Design and Flight Control of the Department of Engineering of the University of Campania Luigi Vanvitelli, is composed of multiple micro inertial measurement units [10, 11].

The system, in upper-body configuration, is composed of two independent modules. Each module is composed of 4 IMUs, positioned on the trunk, the arm, and the forearm. Data are recorded and pre-processed by a raspberry, powered by a battery. A camera positioned on the goggles, and synchronized with the raspberry, records the entire activity from the worker's field of view (Fig. 2).

An algorithm has been developed in order to provide attitude quaternions, attitude Euler angles per each IMU and, combining these data, the posture angle trends over the time considered pelvis rigid rotation; flexion forward, lateral flexion and torsion of the trunk; elevation, lateral flexion and rotation of the arm; flexion and rotation of the forearm.

For this study, only the static posture angles of the trunk were computed, i.e., forward flexion of 20°, sustained at least for 4 s.

3 Results

3.1 Electromyographic Analysis

Figure 3 shows the results of the comparisons between the average values (±SD) of the muscle activity of the left and right paravertebral complex, in the work performance on the right and on the left of the workstation. Left side muscles show a significant statistical difference activity between left and right side of the workstation ($P < 0.05$).

On the right side of the workstation there is an asymmetry of the muscular activity ($P < 0.001$) between the two muscle groups. In particular, right side muscles

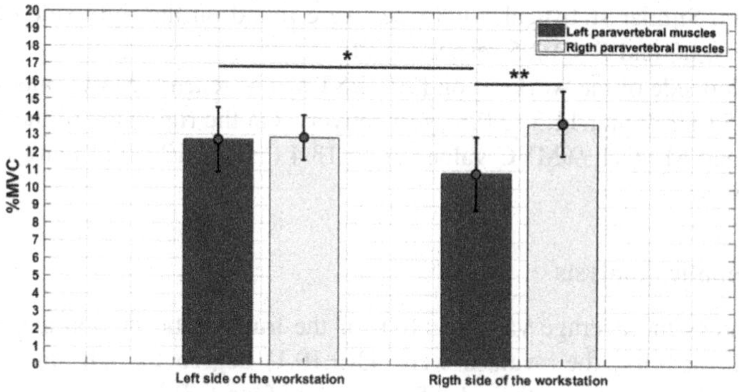

Fig. 3. Activity of the right and left paravertebral muscles.

(13.7 ± 1.8% of MCV) show a higher activity than the left side ones (10.8 ± 2% of MCV). This asymmetry is not present on the left side of the workstation (P > 0.05). Muscles activity recorded from right and left paravertebral muscles is almost the same (12.9 ± 1.3% vs 12.7 ± 1.8% of MCV).

Figure 4 shows the results of the comparisons between the average values (±SD) of muscle activity investigated (EST, ESL and M).

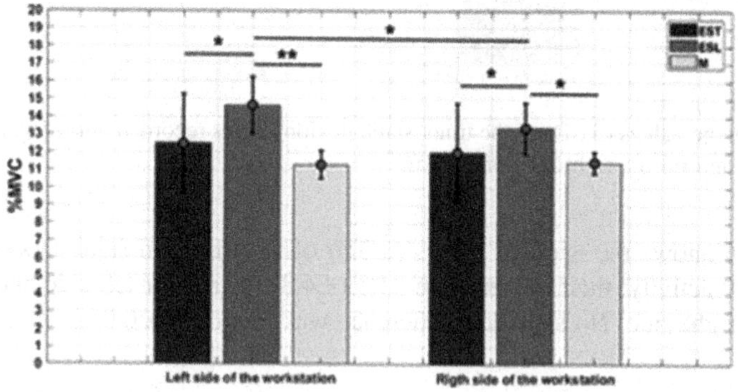

Fig. 4. Activity of the several paravertebral muscles levels (EST = erector spinae thoracic; ESL = erector spinae lumbar region, M = Multifidus).

Results show statistical difference on ESL muscles between right and left side of the workstation (P < 0.05).

There are significant differences on both sides of the workstation (left side P < 0.001 and right P < 0.05) among all muscles investigated. The average activity of the muscles ESL has a higher value if compared to EST muscles and to M muscles.

More in detail, on the left side of the workstation both the comparison between EST vs ESL and EST vs M shows a statistical difference (P < 0.05 and P < 0.001

respectively). Similar statistical differences were found on the right side of the workstation ($P < 0.05$ and $P < 0.05$ respectively).

On the left side of the workstation ESL, EST and M reach %MVC values up to 14.6 (±1.6), 12.5 (±2.8) and 11.3 (±0.8), respectively. On the right side of the workstation ESL, EST and M reach %MVC values up to 13.4 (±1.4), 12 (±2.8) and 11.4 (±0.6), respectively.

3.2 Kinematic Analysis

Figure 5 shows the average values (±SD) of the trunk static flexion angles recorded during work activity; their values were 43(±10,1) degree for left side and 44(±6) degree for right side. No statistical difference was found ($P > 0.05$).

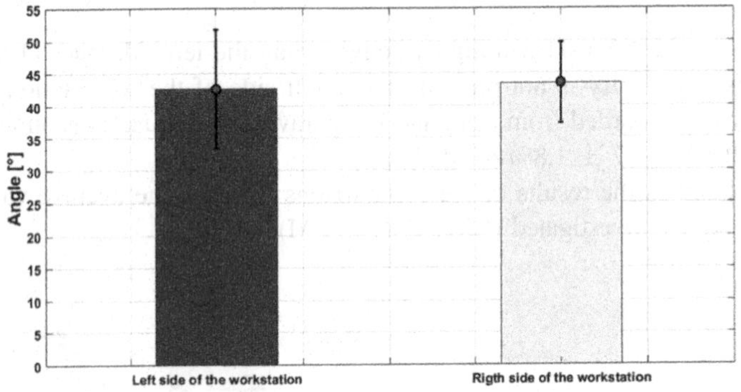

Fig. 5. Average values (±SD) of the trunk static flexion angles recorded during work activity on the left side and on the right side of the car.

Figure 6 shows the average values (±SD) of total time in static flexion recorded during work activity; their values were 21,8 (±4,8) degree for left side and 20 (±3,9) degree for right side. No statistical difference was found ($P > 0.05$).

4 Discussion and Conclusion

The aim of this study is to develop a methodological approach for quantitative risk-assessment of biomechanical overload, based on integration of surface electromyography with an innovative inertial measurement unit.

This study was conducted in a real work environment and the setup was not influenced by potential electromagnetic interference that could affect the signals.

Furthermore, the proposed integration of the sEMG and of the IMU did not interfere with the movement of the worker during his tasks.

The study limits are represented by the small sample size and the lack of the trunk torsion data.

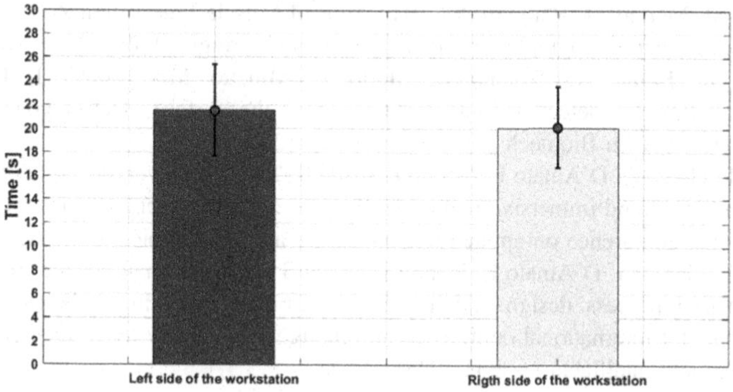

Fig. 6. Average values (±SD) of the time spent in static flexion, recorded during work activity on the left side and on the right side of the car.

Our results suggest a further study of muscle activity, including muscle fatigue of not only trunk muscles, but also of shoulder and upper arm.

sEMG and kinematic data, indeed, can be integrated into a wider evaluation of the biomechanical load due to awkward postures with or without manual load handling. Furthermore sEMG, together with kinematics, can contribute to better analyze both static and dynamic postures.

References

1. Zare M, Sagot JC, Roquelaure Y (2018) Within and between individual variability of exposure to work-related musculoskeletal disorder risk factors. Int J Environ Res Public Health 15(5):E1003
2. Takala E-P, Pehkonen I, Forsman M, Hansson G-Å, Mathiassen SE, Neumann WP, Sjøgaard G, Veiersted KB, Westgaard RH, Winkel J (2010) Systematic evaluation of observational methods assessing biomechanical exposures at work. Scand J Work Environ Health 36(1): 3–24
3. Merletti R, Farina D (2016) Surface Electromyography: physiology, engineering and applications. IEEE Press/J Wiley, USA
4. Peppoloni L, Filippeschi A, Ruffaldi E, Avizzano CA (2015) (WMSDs issue) A novel wearable system for the online assessment of risk for biomechanical load in repetitive efforts Int J Ind Erg 37(6):563–571
5. Filippeschi A, Schmitz N, Miezal M, Bleser G, Ruffaldi E, Stricker D (2017) Survey of motion tracking methods based on inertial sensors: a focus on upper limb human motion sensors (Basel) 1, 17(6)
6. Graham RB, Agnew MJ, Stevenson JM (2009) Effectiveness of an on-body lifting aid at reducing low back physical demands during an automotive assembly task: assessment of EMG response and user acceptability. Appl Ergon 40(5):936–942
7. Buchholz B, Park JS, Gold JE, Punnett L (2008) Subjective ratings of upper extremity exposures: inter-method agreement with direct measurement of exposures. Ergonomics 51(7):1064–1077

8. Barbero M, Merletti R, Rainoldi A (2012) Atlas of Muscle Innervation Zones. Understanding Surface Electromyography and Its Applications. Springer, Milan; New York
9. Merletti R, Botter A, Troiano A, Merlo E, Minetto MA (2009) Technology and instrumentation for detection and conditioning of the surface electromyographic signal: state of the art. Clin Biomech (Bristol, Avon) 24:122–134
10. Caputo F, Greco A, D'Amato E, Notaro I, Spada S (2018) A preventive ergonomic approach based on virtual and immersive reality. In: Advances in intelligent systems and computing, international conference on applied human factors and ergonomics, Los Angeles (2017)
11. Caputo F, Greco A, D'Amato E, Notaro I, Spada S (2018) Human posture tracking system for industrial process design and assessment. In: Advances in intelligent systems and computing. 1st International conference on intelligent human systems integration: integrating people and intelligent systems, IHSI, Dubai

Nurses' Perception and Cognition of Electrocardiogram Monitoring Alarms

Yasuyo Kasahara[1,2(✉)]

[1] Tokyo Metropolitan Institute of Medical Science, 2-1-6, Kamikitazawa, Setagaya Ward, Tokyo, Japan
Kasahara-ys@igakuken.or.jp
[2] Showa University, 1865 Tokaichibamachi, Midori Ward, Yokohama, Kanagawa, Japan

Abstract. In this study, nurses' perception and cognition of electrocardiogram (ECG) monitoring alarms were investigated in relation to their competence and experience of incidents or accidents related to ECGs. Questionnaires were given to 300 nurses who used ECG monitors at a hospital. Relationships between personal attributes and scores on questions about the perception and cognition of alarms were analyzed.

Analysis of variance was performed to determine the tendency of perception and cognition scores among nurses on the clinical ladders I, II and ≧III and pre-acquisition nurses. The group on Ladder I was the most sensitive to the ECG's alarm ($p < .05$). The group ≧III was more observant than the other groups, not only to the sound of the alarm, but also the waveform of the ECG ($p < .01$). The pre-acquisition group thought that if they responded to an alarm, it would affect the work of other groups. Furthermore, they were unable to concentrate on other work if disturbed by the alarm ($p < .05$). Subsequently, t test was conducted to examine the tendency of scores score based on experience of accidents or incidents related to ECG. The results showed that the experienced group had more confidence in alarms than the non-experienced group ($p < .01$). On the other hand, the experienced group believed felt that if alarm frequencies were high, and they could not concentrate on other work and this would have an adverse effect on their performance ($p < .05$).

Keywords: Risk · Perception · Cognition · ECG · Monitor

1 Introduction

In the medical field, at times, some nurses ignore the alarms of the electrocardiogram (ECG) monitor and prioritize other tasks. They cannot afford to check the ECG monitor when they are busy. "These nurses believe that someone will respond to the alarms even if eventually no one responds" [1]. Furthermore, in cardiovascular wards, nurses who are inexperienced may be confused because they are unable to detect immediately which alarm is ringing. Unnecessary alarms are thought to be related to the background outlined above. Furthermore, "After installing an ECG monitor, the alarm setting may not have been changed even if the patient's condition has changed" [2]. Thus, the

S. Bagnara et al. (Eds.): IEA 2018, AISC 818, pp. 727–732, 2019.
https://doi.org/10.1007/978-3-319-96098-2_89

possibility exists that an unnecessary alarm may ring because the setting of the alarm is inappropriate; consequently, when the alarms rings, it may not be viewed abnormally.

Even if the alarm rings, some nurses may believe it is a mistake before checking the monitor or think that the alarm is not alerting them to abnormal changes in the patient's condition. It is also time-consuming to check the waveform of the monitor. Even if nurses check it, they may come to the conclusions that it is not an alarm indicating arrhythmia after all or just as expected, it is a wrong alarm, thus, justifying their assumptions. In this way, nurses' consciousness of alarms is reduced because of the many unreliable alarms. Within this context, it is believed that if the number of unnecessary alarms increases, the alarm will be abandoned. "With so many false alarms, alarm fatigue represents a symptom of a larger problem" [3].

There are various findings about nurses' tendencies to ignore alarms. However, the relationship between the perception and cognition of alarms, and the practical abilities of nurses and incidents or accidents involving ECGs has not been clarified. Accordingly, the purpose of this study was to clarify the perception and cognition of alarms in relation to nurses' practical abilities and experiences of incidents or accidents involving ECGs.

2 Methods

2.1 Subjects

The subjects included 300 nurses who used an ECG monitor in a Japanese hospital.

2.2 Date Collection and Analysis

The researchers designed a questionnaire. The items in the questionnaire are summarized and presented in Table 1. The items that dealt with the perception and cognition about alarms were answered by means of a five-point scale (1: agree〜3: Neither Agree Nor Disagree〜5: Do Not agree).

Table 1. Question items.

Contents	Example
• Clinical department	Respiratory, Cardiovascular, Gastrointestinal, Neurosurgical, Orthopedic, Obstetrics and Gynecology, ICU etc.
• Years of experience as a nurse	—
• Clinical Ladder Level	Ladder I, II, ≧III, Pre-acquisition nurses
• Perception and cognition about alarm	• Do you feel nervous when you hear an alarm? • Do you not notice the alarm? • How dangerous do you consider it to ignore ECG alarms?
• Experience of incidents or accidents related to ECG monitor	Existence and Situation of experience

Each questionnaire was put in an envelope with a request form. The nurses' manager was asked to distribute the questionnaire. The nurses were asked to insert their completed questionnaire in an envelope and place it in a box for collection. Subsequently, the collection box was collected two weeks later.

The relationship between nurses' personal attributes and the total score of the items related to the perception and cognition of alarms was determined by analysis of variance and a t test.

The questionnaire was anonymous. The subjects were assured that the data would only be used for this study. Furthermore, they were told their participation was not compulsory and refusal to participate would not disadvantage them in any way. In addition, the purpose of the study was included in a request letter; it was explained that they would be giving their consent to participate in the study by submitting a questionnaire. This research was approved by the research ethics review committee of belonging facility.

3 Results

Of the 300 nurses to whom questionnaires were distributed, 266 were collected; thus, yielding a response rate of 88.6%. However, only 235 questionnaires could be used for analysis (78.3%). The average number of years of nursing experience was 7.1 years ($SD = 6.1$). Results of the clinical ladder revealed 74 (31.5%), 82 (34.9%) and 53 (22.6%) respondents were classified as being on Ladders I, II and III or more, respectively. Furthermore, 26 (11.1%) were classified as pre-acquisition nurses. Nurses on Ladder I understand the contents of the manual receive support from seniors. Those on Ladder II carry out work independently while those on Ladder ≧III can act as leaders by supporting and managing the staff's work.

3.1 Relationship with Level of Ladder

Variance analysis was performed by establishing the ladder level as the independent variable and the score of the items as the dependent variable. The results revealed there was a significant difference in the five items (Table 2).

In question 1, the average scores of ladders I, II and ≧III were 1.7 ($SD = 0.8$), 2.0 ($SD = 0.9$) and 2.2 ($SD = 1.1$), respectively. The average score for pre-acquisition nurses was 1.8 ($SD = 0.7$). Thus, the main effect was significant ($F\ (3{,}234) = 3.83$, $p = 0.01$). For multiple comparisons, the score was higher for ladder ≧III than I. In question 5, the average scores of ladders I, II, ≧III and the pre-acquisition group were 1.4 ($SD = 0.5$), 1.3 ($SD = 0.5$), 1.2 ($SD = 0.5$) and 1.8 ($SD = 0.7$), respectively. Therefore, the main effect was significant ($F\ (3{,}234) = 7.17$, $p < 0.01$). In multiple comparisons, the score was higher for the pre-acquisition group than the others. In other words, nurses on the higher levels consider the waveform. In question 8, the average score of ladders I, II and ≧III was 2.3 ($SD = 0.9$), and the pre-acquisition group was 3.0 ($SD = 0.9$). Thus, the main effect was significant ($F\ (3{,}234) = 4.52$, $p < 0.01$). In multiple comparisons, the score was higher for the pre-acquisition group than the others. In question 9, the average score of ladders I and II was 2.4 ($SD = 0.9, 1.0$) and

Table 2. Tendency of perception and recognition by ladder level

	Question items	Ladder level	Average	SD	Significant difference
1	Do you feel nervous when you hear an alarm?	Pre-acquisition	1.8	0.7	
			1.7	0.8	
			2.0	0.9	*
			2.2	1.1	
2	Do you not notice the alarm?	Pre-acquisition	2.4	1.0	
			2.2	0.9	
			2.5	1.0	*n.s.*
			2.5	1.2	
3	Do you think that it is safe for the patient to wear anECGs monitor?	Pre-acquisition	2.2	0.6	
			2.1	0.6	*n.s.*
			2.0	0.7	
			2.3	0.8	
4	Do you think that the alarm sounds at a high frequency during work?	Pre-acquisition	2.3	0.8	
			2.3	0.9	*n.s.*
			2.2	0.9	
			2.4	0.9	
5	If you hear an alarm, do you care about the waveform of the ECG?	Pre-acquisition	1.8	0.7	*
			1.4	0.5	**
			1.3	0.5	
			1.2	0.5	
6	Do you trust ECG alarms?	Pre-acquisition	2.3	0.5	
			2.3	0.6	*n.s.*
			2.2	0.7	
			2.4	0.7	
7	Do you check immediately when the alarm sounds?	Pre-acquisition	1.7	0.5	
			1.5	0.6	*n.s.*
			1.5	0.6	
			1.4	0.6	
8	Do you think the alarm would be more likely to interfere with other work?	Pre-acquisition	3.0	0.9	
			2.3	0.8	**
			2.3	0.9	*
			2.3	0.9	
9	Do you think that you will lose concentration if an alarm rings while working?	Pre-acquisition	3.0	1.1	
			2.4	0.9	*
			2.4	1.0	
			2.5	1.1	
10	When you hear the alarm, do you think that other nurses will respond?	Pre-acquisition	2.7	1.0	
			2.6	0.8	*n.s.*
			2.7	0.8	
			2.9	1.0	
11	Do you think that there are patients who do not need an electrocardiogram monitor?	Pre-acquisition	3.4	1.1	*
			2.8	0.9	**
			2.6	1.0	
			2.6	1.0	
12	How dangerous do you consider it to ignore ECG alarms?	Pre-acquisition	1.6	0.6	
			1.4	0.6	*n.s.*
			1.4	0.6	
			1.3	0.5	
13	Do you think an alarm are difficult to notice?	Pre-acquisition	2.7	1.0	
			2.5	0.8	*n.s.*
			2.7	1.1	
			2.9	1.0	

*:$p<.05$, **:$p<.01$

that of ladder ≧III was 2.5 (*SD* = 1.1). The average score of the pre-acquisition group was 3.0 (*SD* = 1.1). Therefore, the main effect was significant (F (3,234) = 2.79, $p < 0.05$). In multiple comparisons, the score was higher for the pre-acquisition group than for those on ladder II. In question 11, the average score of ladder I was 2.8 (*SD* = 0.9) and that of ladders II and ≧III was 2.6 (*SD* = 1.0), and the pre-acquisition group was 3.4 (*SD* = 1.1). Therefore, the main effect was significant (F (3,234) = 5.31, $p < 0.01$). In multiple comparisons, the score was higher for the pre-acquisition group han II and ≧III. In other words, the higher level nurses believed that there was not much necessity.

3.2 Relationship with Existence of Incident or Accident Experience

Of the 235 participants, 45 had experienced an incident or accident with an ECG while 190 participants had not. The existence of an experience was regarded as the independent variable and the score of the items as the dependent variable; the relationship between the two was analyzed by a t test. Results revealed a significant difference in four items.

In question 4, the average score of the group with experience was 1.8 (*SD* = 0.8) while that of the group with no experience was 2.4 (*SD* = 0.9) (t (233) = 4.28, $p < 0.01$). The group with experience heard more frequent alarms than the group with no experience. In question 6, the average score of group with experience was 2.1 (*SD* = 0.5) while the group with no experience was 2.3 (*SD* = 0.7) (t (233) = 1.89, $p < 0.05$). The group with experience had more confidence in alarms than the group with no experience. In question 8, the average scores of the groups with experience and without it was 2.1 (*SD* = 0.8), and 2.4 (*SD* = 0.9) (t (233) = 2.03, $p < 0.05$), respectively. The group with experience believed that the alarm would be more likely to interfere with other work than the group with no experience. In question 9, the average score of group with experience was 2.1 (*SD* = 0.8) while that of the group with no experience was 2.6 (*SD* = 1.1) (t (233) = 2.8, $p < 0.01$). The group with experience was of the view that their concentration would cease when the alarm rang. The group with no experience did not believe this.

4 Discussion

4.1 Tendency of Perception and Recognition by Ladder Level

When an ECG's alarm rang, the nurses on Ladder I were the most sensitive. Furthermore, the group on ladder ≧III was more careful than the other groups; they were alerted to the sound of the alarm and took cognizance of the waveform of the ECG. The pre-acquisition group was the least anxious and had a tendency not to be concerned about the ECG's waveform. As noted previously, those on ladder 1 understand the manual and/or receive support from seniors. Therefore, the highest tension associated with alarms is related to the increase in knowledge about diseases and the recognition of the importance of observation and support for patients. On the other hand, the pre-acquisition group was less tense and unconcerned about the waveform because of their poor understanding of its importance. Those on ladders ≧II were less tense about alarms because they were accustomed to them. However, they considered the waveform and arrived at conclusions after examining it.

The pre-acquisition group believed if they responded to an alarm, it would affect the work of other groups. Furthermore, the pre-acquisition group was unable to concentrate on other work if an alarm rang because they were more unfamiliar with the work than the other groups. Furthermore, they were easily disturbed by alarms because of their lack of technical and non-technical skills. Moreover, their judgment was not advanced.

4.2 Tendency of Perception and Recognition by Experience

The group with experience had more confidence when an alarm rang than the group without experience.

On the contrary, the group with no experience believed that when the alarm frequencies were high, they could not concentrate on other work and this had an adverse effect on their other work. Experience of incidents or accidents related to ECGs enhances reliability and attention to alarms. However, careful attention to alarms may indicate that attending to other tasks is secondary.

5 Conclusions

1. The nurses on Ladder I were the most sensitive to the ECG's alarm.
2. The nurses on ladder ≧III were more attentive than the other groups, not only to the sound of the alarm, but also the waveform of the ECG.
3. The pre-acquisition group thought that if they responded to an alarm, it would affect the work of other groups. Furthermore, they were unable to concentrate on other work when the alarm rang.
4. The group who had experience of accidents had more confidence in alarms than those without experience. On the other hand, the group with experience believed that alarm frequencies were high, and they could not concentrate on other work and this had an adverse effect on their work.

References

1. Matsuda Y, Kobayashi S et al (2012) Changes in nurse's perceptions and behavior by introducing an electrocardiogram monitoring system. In: The 32nd Tokyo Medical University Hospital nursing research acquisition, pp 20–24
2. Yotsuji S, Ooe T, Fuse Y (2013) Change of nurses intention and action to unnecessary alarm. In: The 43rd annual meeting of the Japanese society of nursing, adult nursing, vol I, pp 171–174
3. Gazarian PK, Carrier N, Cohen R et al (2015) A description of nurses' decision-making in managing electrocardiographic monitor alarms. J Clin Nurs 24(1–2):151–159

WHO Safe Childbirth Checklist: The Experience of Kenya According to the WHO African Partnership for Patient Safety

G. Dagliana[1(✉)], B. Tommasini[2], S. Zani[2], S. Esposito[3], M. Akamu[3], F. Chege[3], F. Ranzani[1], M. J. Caldes[4], and S. Albolino[1]

[1] Centre for Clinical Risk Management and Patient Safety, WHO Collaborating Centre in Human Factor and Communication for the Delivery of Safe and Quality Care, Via Dazzi 1, 50141 Florence, Italy
daglianag@aou-careggi.toscana.it

[2] University Hospital of Siena, Strade delle Scotte, 14, 53100 Siena, Italy

[3] World Friends Kenya, Ruaraka Uhai Neema Hospital, Off Thika Highway, P.O. Box 39433, Nairobi 00623, Kenya

[4] Centre for Global Health, Tuscany Region, Viale Gaetano Pieraccini 28, 50141 Florence, Italy

Abstract. The burden of unsafe care is still very high all around the globe. A study conducted in 2012 in African and Middle Eastern Countries reports that in developing countries the incidence of adverse event is 8,2% and of these 83% are preventable. WHO estimates that about 287.000 are maternal deaths, 1 million fetal deaths during intrapartum period and 3 million deaths of infants during the neonatal period. WHO promoted a campaign for adopting the Safe Childbirth Checklist (SCC), that is an organized list of evidence-based essential birth practices, which targets the major causes of maternal deaths, intrapartum-related stillbirths and neonatal deaths that occur in health-care facilities. The objectives of the project are: introducing the WHO SCC in one hospital of Kenya and evaluating the locally adapted tool in terms of impact on safety and quality and its usability and feasibility. The Centre for Clinical Risk Management and Patient Safety, the Centre for Global Health, the University Hospital of Siena the Ruaraka Uhai Neema Hospital undersigned a partnership following the WHO African Partnership for Patient Safety model for implementing safety and quality in the maternal and neonatal area in particular through the use of the WHO SCC. The WHO SCC has been adopted with a positive feedback from midwifes. The childbirth checklist has increased the delivery of some essential childbirth-related care practices and the appropriateness during the administration of antibiotic therapy and antihypertensive treatment. The twinning model proposed by WHO has the potential to go far beyond patient safety issue it can advance efforts towards building resilient health systems.

Keywords: Safe childbirth checklist · Safety · Maternal and neonatal care Kenya

S. Bagnara et al. (Eds.): IEA 2018, AISC 818, pp. 733–740, 2019.
https://doi.org/10.1007/978-3-319-96098-2_90

1 Introduction

The burden of unsafe care is still very high all around the globe. A study conducted in 2012 in African and Middle Eastern Countries reports that in developing countries the incidence of adverse event is 8,2% and of these 83% are preventable e that the main causes of these adverse events could are poor training and difficulties of the health workers to follow hospital procedures and protocols [1] This data reflects the study conducted by the Joint Commission aiming at identifying the main causes of sentinel events in US hospitals between 2004–2014: human factor, communication, teamwork and leadership [2].

In regards to maternal and neonatal care, globally, an estimate of 10.7 million women have died in the 25 years between 1990 and 2015 due to maternal causes [WHO 2015]. According to WHO report published in 2014, 44% percent of stillbirths, 73% of newborn deaths and 61% of maternal deaths occur around the time of labor and birth and in the first week after birth. The main causes of children death in 2015, include: complications of prematurity (18%), pneumonia (16%), intrapartum-related neonatal deaths (including birth asphyxia) (12%) and sepsis (9%) [3].

Most newborn deaths occur in low- and middle-income countries: 39% in South Asia and 38% in Sub-Saharan Africa. 75% of neonatal death could be prevented by providing quality and safe assistance [4]. Effective prevention and management of conditions in late pregnancy, childbirth and the early new-born period are likely to reduce the numbers of maternal deaths, ante-partum- and intrapartum-related stillbirths and early neonatal deaths significantly.

The World Health Organization has long called for the need to widen the focus of interventions in areas with limited resources from the pure guarantee of accessibility to health services to the provision of quality and safety-oriented care for patients and operators [5]. There are various examples of experiences in which accessibility has favored the increase in access to health facilities without, however, corresponding to an improvement in the outcomes. A study conducted in Pakistan tells us that although between 2001 and 2013 the number of women giving birth in hospital or assisted by trained staff increased from 21% to 48%, maternal mortality decreased by less than 1/4 [4].

In the framework of the wider program of reducing maternal and neonatal deaths, in 2008 WHO promoted a campaign for adopting globally the Safe Childbirth Checklist (SCC). The field testing was initially thought to be carried out in Africa and Asia but in 2013, the pilot study has been open also to developed countries. The SCC is an organized list of evidence-based essential birth practices, which targets the major causes of maternal deaths, intrapartum-related stillbirths and neonatal deaths that occur in health-care facilities [6]. Limited data are available on the impact of Safe Childbirth Checklist-based programs on the adherence to Essential Childbirth-related practices and on the impact on maternal and infant mortality and morbidity [7, 8]. According to the study carried out by Gawande, the pilot testing of the Safe Childbirth Checklist in Southern India led to a marked increase in delivery of essential childbirth practices linked with improved maternal, fetal and new-born outcomes. Overall, there was an increase in adherence to accepted clinical practices from a mean of 10/29 (34%) to 25/29 (86%). The study published in 2017 by Gawande and carried out in Uttar

Pradesh, India, showed that birth attendants' adherence to essential birth practices was higher in facilities that used the coaching-based WHO Safe Childbirth Checklist program than in those that did not, but maternal and perinatal mortality and maternal morbidity did not differ significantly between the two groups [9].

2 Objectives

The main objectives of this study is to introduce the WHO Safe Childbirth Checklist in one hospital of Kenya and to evaluate the locally adapted tool in terms of impact on safety and quality, its usability and feasibility.

3 Methods

The methodology used in this project is that of implementation science oriented to improve safety and quality standards in healthcare.

The implementation research can be defined as a method to promote systematic dissemination of research results and evidence-based practices in clinical practice in order to bring improvements in quality and effectiveness of health care [10].

At the ground of the implementation science method is the idea that every solution put in place to face critical situations must be oriented to bring improvement both at the organizational and outcome level in a reasonably rapid time, triggering also virtuous processes towards security that over time become part of the customary approach to assistance [11].

Therefore, improving quality and safety standards does not mean providing only clinical-assistance interventions but also implementing organizational interventions, stimulating the development of the so-called non-technical skills such as culture and attitude to safety, a correct a structured communication between operators, teamwork and the so-called "situational awareness". Improvement projects have indeed the main objective of modifying behaviors, creating virtuous processes in the approach to delivering care, creating a culture and a way of thinking oriented to the safety of patients and operators.

From the methodological point of view, implementation science provides research protocols that combine methods of quantitative analysis and qualitative investigation just to go to identify factors and variables that can influence the process at different levels: patient, caregiver, department, structure, organization, communities and political decision-makers. Clinical research methods, based on the "observation-introduction of a disturbing factor-new observation" model, could be too limited for the evaluation of socio-cultural changes in complex, variable and non-linear contexts such as healthcare.

In this perspective, the effectiveness of an implementation science project relies also on the involvement of expertise and knowledge that are specific of the social, organizational, anthropological and economic sciences as well as those of the ergonomics and human factor science. To this end, both during the design of the intervention and in the evaluation phase it is essential to combine the collection and analyses of epidemiological data with ethnographic investigation and evaluations. The added value of the ethnographic method lies indeed in its ability to intercept and analyze what actually happens in the care settings, to understand how the work is actually done with respect to the work as it is imagined and planned (work as done rather than work as imagined) [12].

Within this described conceptual framework, the Tuscany Region in 2015 promoted a partnership with two hospitals in Kenya with focus on patient safety and quality improvement. The regional Centre for Clinical Risk Management and Patient Safety and the Centre for Global Health, thanks to several years of experience respectively in patient safety and global health, supported the University Hospital of Siena (Italy), the Ruaraka Uhai Neema Hospital and North Kinangop Catholic Hospital to build a partnership for improvement following the APPS approach and focusing on critical areas of intervention for increasing the quality and safety of services delivered both in Kenya and in Italy.

The operative approach promoted for introducing patient safety initiatives in Kenya, combines the WHO African Partnership for patient safety [13] with the Collaborative Breakthrough model promoted by the Institute for Healthcare Improvement [14].

African Partnerships for Patient Safety (APPS) is a WHO Patient Safety Program building sustainable patient safety partnerships between hospitals in countries of the WHO African Region and hospitals in other regions. APPS is concerned with advocating for patient safety as a precondition of health care in the African Region and catalyzing a range of actions that will strengthen health systems, assist in building local capacity and help reduce medical error and patient harm. The program acts as a channel for patient safety improvements that can spread across countries, uniting patient safety efforts.

The approach promoted by APPS emphasizes bidirectional and intercontinental transfer of knowledge, experience and solutions between front line health-care operators and it stimulates structural and behavioral changes at the hospital and policy level through technical cooperation [13].

APPS promotes an approach to partnership for improvement based on 6-step cycle that facilitates the development of the partnership, the identification of patients' safety needs and gaps, the development of an action plan and evaluation of results (Fig. 1).

The APPS model has been integrated by researchers with specific tools to perform a quali-quantitative self-assessment of the state of maturity of a health system from the point of view of safety, both in terms of the c.d. "safety logistics" (equipment, work organization, procedures, etc.) and in terms of safety culture. Tools of ethnographic investigation, such as interviews, observations and questionnaires, have been used in this phase to support the assessment tools in order to have a broader and deeper understanding of the context.

Following then the adapted model of the Collaborative Breakthrough promoted by the IHI, improvement actions have been designed starting from literature, from solutions promoted by international actors (e.g. WHO Campaigns) or from experiences already made in other similar contexts. Each identified intervention has been evaluated by a multidisciplinary working group, adapted to the context, tested and assessed in

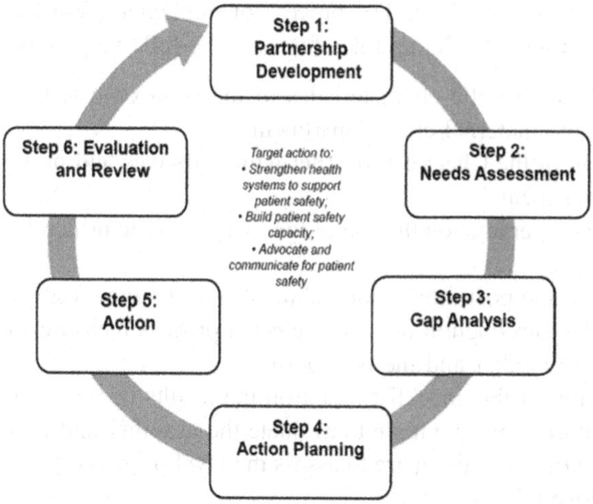

Fig. 1. WHO APPS 6-steps cycle model

terms of usability, feasibility and impact on quality and safety. Quantitative methods have been used in the evaluation phase- such as prospective pre- and post-intervention analyses on the basis of process and outcome indicators- as well as qualitative tools such as questionnaires, interviews and observations (Fig. 2).

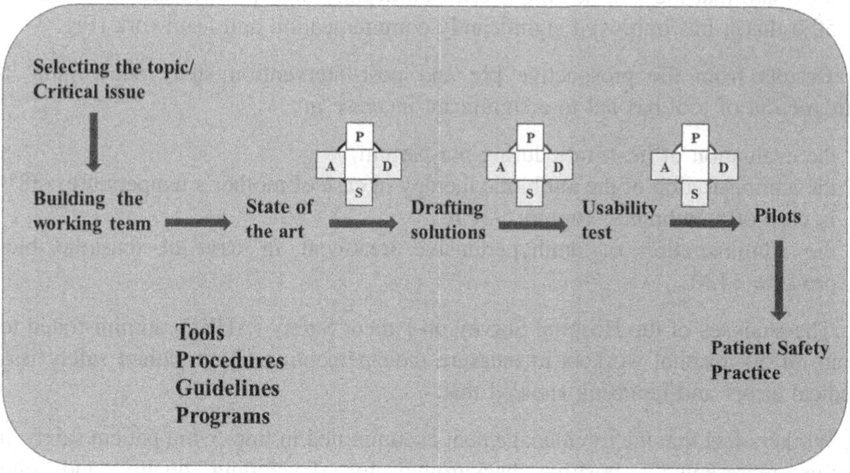

Fig. 2. Collaborative Breakthrough, adapted

According to the results of the qualy-quantitative self-assessment and to evidences emerged by the gap analyses conducted on the ground, the intervention program jointly developed by Italy and Kenya, focused on specific strategies for implementing safety and quality in the maternal and neonatal area and in particular through the use of the

WHO Safe Childbirth Checklist. The process of implementation has follow the Collaborative Breakthrough Model and has foreseen the following steps:

- evaluation of the specific characteristics of the context in terms of resources and work flow in the maternal child department
- personalization of the Checklist together with midwives and nurses of the maternal and child department
- coaching of the operators on the use of the tool and basic principles of patient safety
- 6-month piloting
- prospective pre and post-intervention clinical record review for evaluating the effect of the checklist introduction on some selected process measures related to the care delivered to the mother and the new-born
- re-customization of the checklist based on the results of the pilot test
- administration of a questionnaire to evaluate the usability and feasibility of the tool
- administration of a questionnaire to assess the level of maturity of the safety culture (SOPSTM Hospital Survey).

4 Results

The results of the questionnaire administrated to midwife for the evaluation of usability, efficiency and functionality of the Safe childbirth checklist showed that:

- 70% of the midwifes considers the checklist easy or very easy to use
- 56% said that the tool has improved significantly their practice around childbirth
- 50% that it has improved significantly communication and teamwork (Fig. 3)

Results from the prospective pre and post-intervention study shown that the introduction of tool has led to a significant increase in:

- the evaluation of heart rate during pre-partum
- the administration of the antibiotic therapy in case of mother's temperature >38° or in case of membranes' rupture >24 h
- the administration of antihypertensive treatment in case of diastolic blood pressure >120

The analyses of the Hospital Survey on Patient Safety (AHRQ) administrated to a group of 50 hospital workers to measure their perception about patient safety issue, medical errors and reporting showed that:

- Workers feel that the top management is committed in improving patient safety and this represents a positive background for developing quality and safety interventions
- About 50% of the staff associate the occurrence of an adverse event to potential blaming rather than a learning opportunity
- Most of the headworkers say that there is a limited culture of reporting even related to near misses

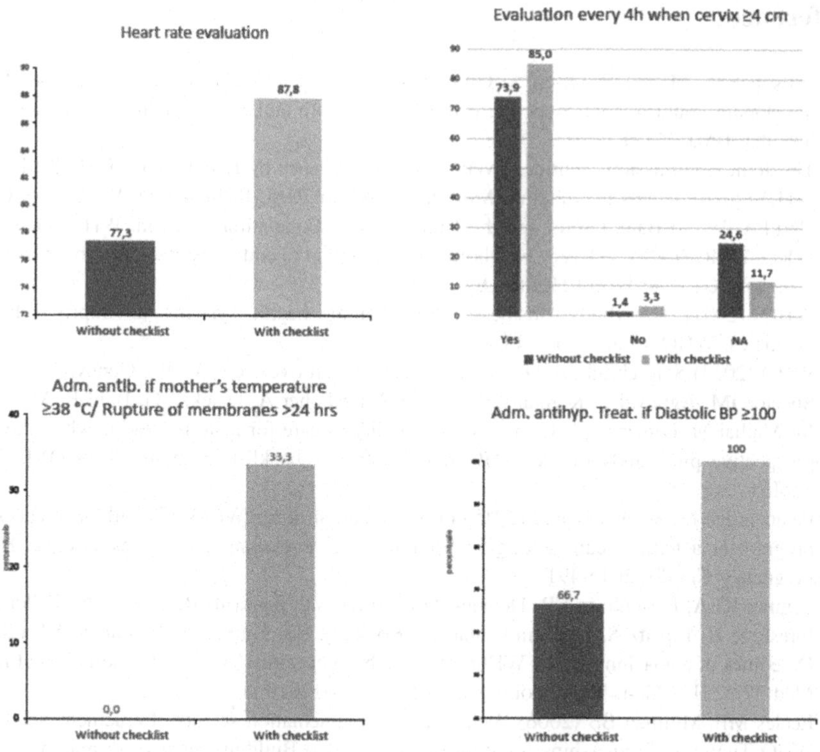

Fig. 3. Results from the prospective pre and post-intervention study

- Even if the adverse events reported and discussed are few they produce positive changed
- Staff feel to be part of a positive environment for teamwork and collaboration also with top management.

5 Conclusions

The WHO SCC has been adopted with a positive feedback from midwifes and it has been judged as a usable and feasible tool. According to midwife perception, the checklist has also increased the so called "non-technical skills" such as communication and teamwork. The childbirth checklist has also had a positive impact on safety and quality promoting a higher adherence to the delivery of some essential childbirth-related care practices and the appropriateness during the administration of antibiotic therapy and antihypertensive treatment.

References

1. Wilson RM, Michel P, Olsen S, Gibberd RW, Vincent C et al (2012) Patient safety in developing countries: retrospective estimation of scale and nature of harm in patient and hospital. BMJ 344:e832
2. The Joint Commission, Sentinel Event Data Root Causes by Event Type 2004–2014
3. WHO (2015) Trends in Maternal Mortality: 1990 to 2015. Estimates by WHO, UNICEF, UNFPA, World Bank Group and the United Nation Population Division. WHO, Geneva
4. WHO, UNICEF (2014) Every Newborn: an action plan to end preventable deaths: executive summary, Geneva: World Health Organization
5. WHO (2007) Everybody business: strengthening health systems to improve health outcomes, WHO's framework for action
6. WHO (2013) Safe childbirth checklist programme: an overview. WHO, Geneva
7. Spector JM, Agrawal P, Kodkany B, Lipsitz S, Lashoher A, Dziekan G, Bahl R, Meri-aldi M, Mathai M, Lemer C (2012) Improving quality of care for maternal and newborn health: prospective pilot study of the WHO safe childbirth checklist program. PLoS ONE 7(5): e35151
8. Patabendige M, Senanayake H (2015) Implementation of the WHO safe childbirth checklist program at a tertiary care setting in Sri Lanka: a developing country experience. BMC Pregnancy Childbirth 15:491
9. Semrau KEA, Hirschhorn LR, Delaney MM, Singh VP, Saurastri R, Sharma N, Tuller DE, Firestone R, Lipsitz S, Dhingra Kumar N, Kodkany BS, Kumar V, Gawande AA (2017) Outcomes of a coaching-based WHO safe childbirth checklist program in India. New Engl J Med 377:2313–2324. https://doi.org/10.1056/nejmoa1701075
10. Eccles MP, Mittman BS (2006) Welcome to implementation science. Implement Sci 1:1
11. WHO Twinning Partnerships for Improvement (2016) Building capacity to reactivate safe essential health services and sustain health service resilience, WHO
12. Hollnagel E, Wears RL, Braithwaite J (2015) From Safety-I to Safety-II: a white paper. The Resilient Health Care Net: published simultaneously by the University of Southern Denmark, University of Florida, USA, and Macquarie University, Australia
13. http://www.who.int/patientsafety/implementation/apps/en/
14. Institute for Healthcare Improvement (2003) The Breakthrough Series: IHI's Collaborative Model for Achieving Breakthrough Improvement. IHI Innovation Series white paper, Boston

Patient Safety in Pediatrics: Ergonomic Solutions for Safer Care of Children

S. Albolino[1(✉)], E. Beleffi[1], F. Ranzani[1], G. Toccafondi[1], A. Savelli[2], K. P. Biermann[2], S. De Masi[2], G. Frangioni[2], F. Festini[2], G. Dagliana[1,2], I. Sforzi[2], G. Merello[2], and S. Guagliardi[2]

[1] Centre for Clinical Risk Management and Patient Safety, WHO Collaborating Centre in Human Factor and Communication for the Delivery of Safe and Quality Care, Via Dazzi 1, 50141 Florence, Italy
albolinos@aou-careggi.toscana.it

[2] Meyer Children's Teaching Hospital, Viale Pieraccini 24, 50139 Florence, Italy

Abstract. Epidemiological data from the literature are few for patient safety in pediatrics and there is a need for comparing experiences and applied solutions in different contexts. A study published in 2012 underlined that the 79% of adverse events in children happened in intensive care unit, the incidence on admissions is of 6,5 and 44,7% of these adverse events are preventable. The promotion of patient safety in pediatrics requires also patient and family programs of information and education to increase awareness about risk factors and behaviors to prevent harms. The aim of project is to design a multidimensional approach to patient safety and to pilot three patient safety practices: preventing children's falls through the use of the Modify Humpty Dumpty Fall Scale validated in Italian; the appropriate transition of care and the early evaluation of patient deterioration through the Pediatric Early Warning Score. All pilots have been realized by adopting a systemic approach with solutions designed by taking into account the principles of ergonomic and human factors and applying them for the definition of cognitive, organizational and physical ergonomic solutions. In order to support the spreading of safety practices, a specific project of designing, implementing and evaluating of four cartoon video-vignettes on pediatric patient safety has been developed.

Keywords: Pews · Pediatrics · Patient safety · Falls · Handover

1 Introduction

Patient safety is a strategic activity in children's care. Epidemiological data and evidences from the literature are few for patient safety in pediatrics and there is a need for comparing experiences and applied solutions in different contexts. At the European level some recommendations have been defined and the need for applying specific solutions and patient safety practices have been underlined. We have now a list of evidence based patient safety practices at the international level. These practices need to be adapted by trials in order to be useful also for the pediatric settings. The patient safety manifesto published by the American Academy of Pediatrics in 2001 underlines

S. Bagnara et al. (Eds.): IEA 2018, AISC 818, pp. 741–751, 2019.
https://doi.org/10.1007/978-3-319-96098-2_91

the importance to evaluate the patient's specific characteristics (such as degree of reached evolution, weight variation, limited capacity to cooperate in delivery of care, high level of dependency and rarity of some pediatric diseases) every time that you are going to design an activity for improving safety for children. It is also important to standardize and make more usable medical devices with the application of a cognitive and organizational ergonomic approach to the development of clinical processes [1]. A study published in 2012, conducted in 14 local healthcare agencies e 8 teaching hospitals in Canada, underlined that the 79% of adverse events in children happened in intensive care unit, the incidence on admissions is of 6,5% (3640 cases in total). 44,7% of these adverse events are preventable [2].

The Tuscany region network of pediatrics hospitals, in collaboration with the Centre for Patient Safety and Clinical Risk Management (Centre GRC) has developed a joint program of activities for promoting safer care for children, based on ergonomics and human factor principles and methodology. The collaboration is aimed at designing and piloting patient safety practices for pediatrics and in particular:

- the Modify Humpty Dumpty Fall Scale validated in Italian for preventing patients' falls
- some structured sheets for performing an appropriate transition of care (Handover)
- the Pediatric Early Warning Score (PEWS) adapted to local workflow for supporting the early evaluation of patient deterioration

2 State of the Art

2.1 State of the Art Regarding the Pediatric Early Warning Score

Patients in general wards can deteriorate, up to the cardiac arrest. These patients can be early detected as the critical conditions which precedes the cardiac arrest are anticipated by a phase where the patient is unstable for a variable period of time. The signs for clinical deterioration can be detected and treated ("track and trigger") starting from the monitoring of the physiological signs [3–8].

The general wards are settings where it is necessary the adoption of a structured way to identify the acute phases of a clinical case and to set up the appropriate care according to the priority criteria for intervention. The surveillance of the clinical deterioration is also appropriate in the sub-intensive areas and in the emergency room for cases under observation for more than 12 h [6–8]. Organizing the surveillance of physiological signs in groups of heterogeneous patients, through scales for rating the patient's state, arises the chance for early detection of clinical deterioration and for an effective evaluation of the degree of instability in order to activate the intensive care unit and the rapid response team.

A rating scale which rates the common physiological parameters and expert clinicians to monitor the patients whose conditions are getting worse are key elements for avoiding adverse events. Together they played an important role to specify the appropriate clinical responses and to support the decision to transfer patients between different health care units [9, 10].

To detect and treat the clinical deterioration, it is therefore necessary to define a structured modality of the intervention, with the assistance of the algorithms and matrixes of responsibility, to correctly manage, with resources already available in the ward, the alert of a patient in deterioration and for the eventual activation of external resources not already located in the ward (rapid response team).

Thus far, however, there are no studies that demonstrate the actual correlation between the use of the scoring system based on criteria of high alert and the reduction of the mortality rate for cardiac arrest. Mainly, there are 3 scoring systems for the premature evaluation of the deterioration of the pediatric patient. Two of these scoring systems demonstrated an adequate capacity to identify the premature deterioration but none of them were evaluated with respect to the reliability. Furthermore, another study was conducted (2009) with respect to the sensibility and specificity of the Pediatric Early Warning Score with positive outcomes. The use of PEWS was also associated with an increase of the number of patients transferred to the PICU (Pediatric Intensive Care Unit). In 2008, a study was conducted on 2,979 patients evaluated with the PEWS system in an arch of the time of 12 months, and it demonstrated that the PEWS system is a valid and reliable instrument to identify the patients at risk for deterioration [9].

2.2 State of the Art Regarding the Humpy Dumpty Falls Scale

It has been calculated that the prevalence of patient falls varies between 3 and 14 for every 1,000 days of recovery [11]. A study conducted in the United Kingdom in 472 hospitals revealed that there were 206,350 hospital falls between September 1, 2005 and August 31, 2006. The same study also exposed that those falls constituted 32,3% of the total accidents verified in the hospital [11]. According to a 2004 study, 30% of the Intensive care accidents had wounds/lesions. Of these, 3–4% represented wounds serious in nature including fractures, subdural hematomas and even death [12].

Even in the pediatric environment, patient falls are a widespread phenomenon. In fact, they constitute the major reason of trauma which cause hospitalization of children under 5 years of age [13]. A study in the United States analyzed the prevalence of hospital falls in patients under than 18 years of age. They calculated that the prevalence of patient falls was 0.84 for every 1,000 patients per day, with 48% of those falls as preventable. 47% of the subjects who had fallen were identified as risks [14].

The individualization of subjects at risk is one of the first duties of all the patient fall prevention programs and this is also valid in the pediatric environment. Over the course of many years, there have been many evaluation instruments for the risks for adults, such as, Downtown Scale (Downtown), the Morse Fall Fall Scale (MFS) (Morse 1988), the St. Thomas Risk Assessment Tool in Falling Elderly Inpatients (STRATIFY) (Oliver 1997), the Tinetti test (Tinetti 1986), the Conley Scale (Coloney 1999), the Hendrich Fall Risk Model (HFRM) (Hendrich 1995) and its last version, the HFRM II (Hendrich 2003). Even if these tools have good predictive features, they are not suitable for pediatric patients.

The Humpty Dumpty Fall Scale (HDFS) is a tool for risk evaluation designed especially for children [15]. It classifies the patients as at high risk or at low risk on the basis of the presence/absence of some specific characteristics: age, sex, diagnosis, presence of cognitive deficits, environmental factors, the reaction to sedation or to anesthesia and the prescribed medication. The final score, that can vary from a minimum of 7 to a maximum of 23, is the result of the sum of the scores collected from each item. The child is at high risk if the total score is >12, while she is at low risk for a score lower than that.

The English version of HDFS has a sensitivity of 0,85 and a specificity of 0,24, while the PPV is of 0,53.

2.3 State of the Art Regarding the Handover

Handover is "the transfer of professional responsibility and accountability for some or all aspects of care for a patient, or group of patients, to another person or professional group on a temporary or permanent basis" [16]. The handover is then a communicative process aiming to achieve effective continuity of clinical care within the patient care pathway, regardless of when the transitions occur. In recent years, the complexity of the health care system have been growing rapidly and health care team are distributed in time and space. From the patient perspective is getting more and more difficult to identify a single healthcare professional in charge for the process of care. Care coordination is not anymore a complementary element but instead a backbone of clinical activity which need to be mindfully designed and maintained. We applied the principles of the collaborative communication model to design a safe communication at care transitions: (1) construction of common ground (2) redundant and complementary information organized in written and oral communication [17]. The handover critical aspects become more sensible in the area of pediatrics where parents and caregivers are deeply involved in the care process. They hold the responsibility of decision making regarding the children's care. As a consequences the handover process need to include the caregiver role in the main categories of handover schema such as "identification of patients" and "contact information" not only with in "contextual and social information". In particular, in this article we are discussing the design of the handover of medical information within the regional-wide pediatric network. The nodes of the network are dislocated in 42 hospitals. Care transitions may occur: between the main hub center, the Meyer pediatric teaching hospital, and secondary hub centers, that is other adult teaching hospitals; between the spoke hospitals, that is emergency department units and pediatric units and the Meyer teaching hospital. The transition "from" and "to" the Meyer teaching hospital may occur in both directions. Distances in between the hospital hub and spoke centers range from 200 km to 5 km and are covered by ambulances and by helicopters.

3 Objectives and Methods

The main objective of the study is to develop customized patient safety practices for pediatric settings through the use of a human factor and ergonomic approach.

Then there are specific objectives related to the different practices we decided to design. Regarding the PEWS, the study aims at customizing, testing and evaluating the tool in terms of usability and impact on clinical practice.

Regarding the Humpty Dumpty Fall Scale, the study aims to a linguistic and cultural validation of a scale for risk assessment and evaluation of the adherence to safety practice as well as to estimate the sensitivity, specificity, PPV, NPV and AUC of the new instrument in Italian language.

The intervention for safe handover aimed at constructing a solid common ground of relevant minimum set of information for managing communication at care transitions between the emergency department of the "hub" Meyer pediatric teaching hospital and the "spoke" hospitals by taking into consideration the role of parents and caregiver. A need assessment questionnaire was distributed within the pediatric network. Afterwards, through a series of 6 focus group with the healthcare workers of the several node the network, the minimum information set was constructed and formalized into two prototypes of handover sheets template for written and verbal communication.

3.1 Early Identification of Patient Deterioration

3.1.1 Design of the Tool

Starting from the international experiences, the PEWS sheet has been adapted to local facilities characteristics and team needs. A multidisciplinary working group, composed by pediatrician, anesthesiologists and nurses, was required to review appropriateness of the items according to clinical practice and organization of clinical activities. A prototype was then designed following criteria of ergonomic and usability by EU certified ergonomists and patient safety experts. The following picture describes the PEWS sheet with the different vital signs to monitor and the scoring system. The sheets are differentiated by age (Fig. 1).

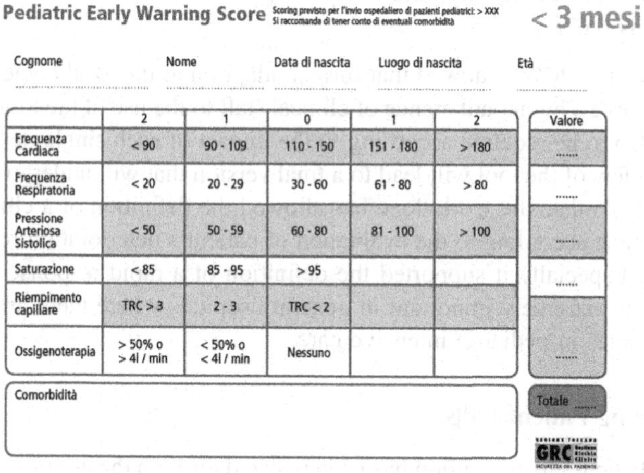

Pediatric Early Warning Score Scoring previsto per l'invio ospedaliero di pazienti pediatrici: > XXX
Si raccomanda di tener conto di eventuali comorbidità

< 3 mesi

Cognome .... Nome .... Data di nascita .... Luogo di nascita .... Età ....

	2	1	0	1	2	Valore
Frequenza Cardiaca	< 90	90 - 109	110 - 150	151 - 180	> 180	....
Frequenza Respiratoria	< 20	20 - 29	30 - 60	61 - 80	> 80	....
Pressione Arteriosa Sistolica	< 50	50 - 59	60 - 80	81 - 100	> 100	....
Saturazione	< 85	85 - 95	> 95			....
Riempimento capillare	TRC > 3	2 - 3	TRC < 2			
Ossigenoterapia	> 50% o > 4l / min	< 50% o < 4l / min	Nessuno			....

Comorbidità

Totale ....

GRC

Fig. 1. An example of PEWS sheet for infants up to 3 months old

3.1.2 Pilot of the Tool

The sheet for scoring the PEWS has been piloted for 6 months in 6 regional hospitals: general pediatrics, short observation unit in the Emergency room, Post surgery unit. The targeted patients were children from 0 to 12 years.

3.1.3 Results

A first analysis of the collected data focused on the assessment of compliance with the use of the tool and the related algorithm. The data relating to the compilation of the PEWS were compared with those present in the discharge letter with respect to the following parameters:

- if patients have been check at least 2 time for each nursing shift with PEWS sheet;
- if the rapid response team have been activated for patients with PEWS score higher than 5 and the patient has been referred to the ICU as predicted by the algorithm

From the point of view of the impact on clinical practice, the statistical analyzes focused on:

- sensitivity and specificity assessment
- predictability assessment

Preliminary results show that:

- Approximately 5.48% of the patients have been assessed twice for each nursing shift
- approximately 97% of patients assessed with PEWS fall into the 0–2 score class
- the higher level of sensitivity and specificity of the score is given for the cutpoint *PEWS score* >2 (Sensitivity 35.6%; Specificity 89%)
- The AUC is 0.6786.

The pilot of the PEWS showed that further adaption to the local context are needed in order to achieve a higher adherence of clinical staff to the use of the tool. The scoring sheet needs also to be redefine according to the impact of each single item on the final score. This review of the tool will lead to a final version that will undergo a new testing on the ground. To date the work done has allowed the definition of a clinical protocol for actions to take according to the evaluation of patient's deterioration realized thanks to the PEWS. Especially it supported the definition of a rapid response team in each hospital. This is extremely important in general hospitals where there are limited staff with competencies in pediatric intensive care.

3.2 Preventing Patient Falls

The pilot for Patient fall prevention has been realized through the use of some key tools:

1. Handout for patient falls evaluation at the admission
2. Handout for patient falls during the hospitalization
3. Evaluation scale for the daily risk
4. Protocol for falls prevention
5. Material for the identification of patients at high risk of falling.

The following picture give an example of the material used for patients' identification in case of high risk for falls. The material has been designed by a certified European ergonomist and promoted in the entire regional network of pediatric units. They are all especially designed for children and very easy to understand (Fig. 2).

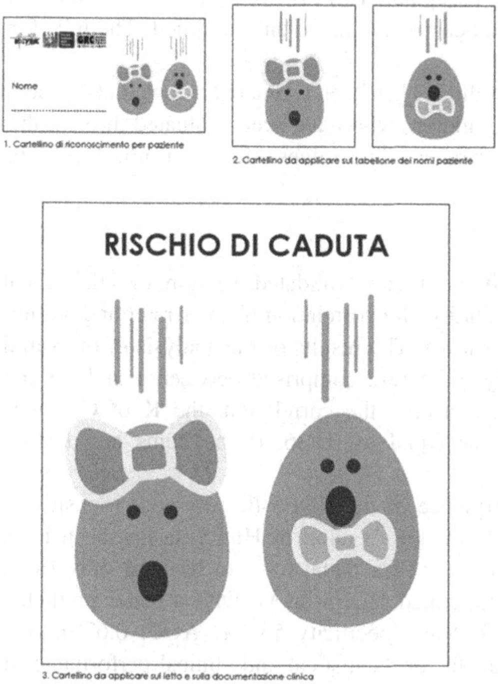

Fig. 2. Material for patient's identification in case of high risk of fall

The key tools have been piloted in 13 regional hospitals in all the wards except for the Intensive Care Unit and the Emergency room. The targeted patients were children who were in the hospital for more than 12 h and the pilot lasted for 12 months.

3.2.1 Translation of HDFS

The original text in English (V0) has been translated from a professional (translator n.1 - T1) and a second translator (T2) realized a forward translation. The two professionals translated in an independent and blinded way and their translations were compared by a third professional (T3). The consensus among the Italian and English version of the scale has been obtained thanks to the joint participation of T1, T2, T3 and the Principal Investigator. This translation represents the first validated Italian version of HDFS (V1). The V1 has been translated from other two translators (T4 e T5). Two different English translations have been realized (V.2 and V3) of V1: the team composed of T1, T2, T3, T4, T5 and the Principal Investigator compared these versions with the original version of the HDFS [16].

3.2.2 Pilot of the Tool

A piloting test on 30 nurses has been realized in order to establish the degree of comprehension of version. Validity and reliability of the scale have been calculated on the same sample. For the validity calculation researchers used the Content Validity Index for both the items (ICVI) and the scale (SCVI). The ICVI value is defined from a group of experts. An acceptable ICVI is >0,78. The items which did not reach the IVCI minimum level have been re-evaluated and modified. The ICV for the scale (SICV) is acceptable if >0,90.

The reliability of the V4 has been estimated by calculating the Inter-rater Reliability on a sample of 100 children, who have been evaluated through the use of the V4 from two independent researchers. The values Alfa of Cronbach >0,7 have been considered acceptable.

3.2.3 Results

The study provided the Italian validated version of HDFS, called HDFS-Ita. It's understandability, validity, the correlation of Inter-rater and the internal consistency of HDFS-Ita has been shown. The results of the analysis of the validity, carried out with the Content Validity Index, are comprised between 1 and 0.8 for ICVI and 0.92 for SCVI. For when it concerns the correlation, the K of Cohen was equal to 0.965. Cronbach's alpha was equal to 0.736, if the items are deleted with low internal consistency.

Predictive performance of the HDFS-Ita was poor (Sensitivity 77.8%, Specificity 36.6%, AUC: 0.593). A new version of HDFS-ita has been identified, consisting of only three items with a cut-off equal to 7, to be used only for subjects between 12 months and 16 years, called HDFS-ita-M, with a better predictive performance than HDFS (Sensitivity 77.8%, Specificity 53.3%, AUC: 0.670). A new scale was thus developed with a satisfactory statistical and clinical performance and predictive value (Sensitivity 88.9%, Specificity 41.9%, AUC: 0.820).

3.3 Developing a Safe Handover

The intervention has been carried out within the pediatric emergency network and in collaboration with AOU Meyer, coordinator of the pediatric network, the emergency departments and pediatric operative units. In order to understand the current perception regarding handover safety gaps and the existing needs, an on-line questionnaire was diffused to healthcare workers. 73 operators (29 nurses and 44 physicians) of the pediatric units and 55 operators (27 nurses and 28 doctors) of the emergency department of the responded to the online survey. The most significant answers in are reported below (Tables 1, 2 and 3).

The high variability in communication methods and perceived risks where taken into account for informing the agenda of 6 structured focus group which allowed further information gathering. The design of the handover sheet templates is shown below. In order to validate the design of the template a training simulation session is being scheduled in order to increase heath worker acceptance and integration with current routine activity. The project allowed the definition of a structured communication among the emergency units of the community hospitals (spoke) with the

Table 1. Verbal handover

Need assessment - verbal handover	Yes (%)	No (%)
Do you use a shared handover template for communicating with hub center at care transition?		
Emergency units	5.5	94.5
Pediatric units	13.7	86.3

Table 2. Written handover

Need assessment	Health care records	Synthetic print-out	Both
What type of written information do you send to the hub center?			
Emergency units	27.4	34.2	3.7
Pediatric units	74.5	21.8	38.4

Table 3. Risk perception

Need assessment	Yes (%)	No (%)
Are you able to recall from you memory a significantly critical care transitions in terms of patient safety?		
Emergency units	83.6	16.4

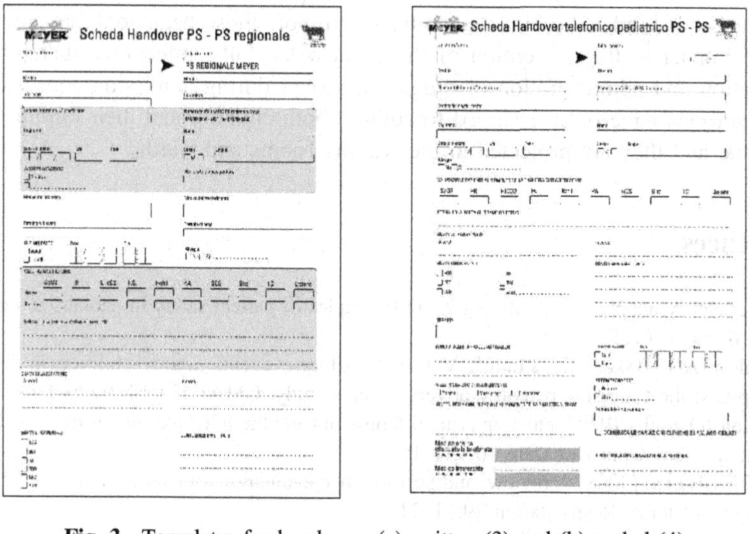

Fig. 3. Templates for handover; (a) written (3) and (b) verbal (4)

emergency room of the pediatric teaching hospital (hub). This new formats for communicating are aimed at improving quality of care and reducing errors in information shared among clinicians in different points of care (Fig. 3).

4 Conclusions

The work done in the Meyer Children's Teaching Hospital in collaboration with the Center for Clinical Risk Management and Patient Safety of the Tuscany Region aimed at defining patient safety practices that can be suitable for children in the hospital but also in the regional network of pediatric units.

All pilots have been realized by adopting a systemic approach to patient safety with solutions designed by taking into account the principles of ergonomic and human factors and applying them for the definition of cognitive, organizational and physical ergonomic solutions. The conducted projects are challenging in terms of management for the large number of operators involved and the variability and dynamicity of the clinical contexts involved. The presence of facilitators trained in ergonomics and human factors as leaders of the projects at the local level together with the realization of training sessions on the job, was a strategic element for the coordination of the different contributions and large adoption of the designed solutions.

Cartoons represented a very important vehicle for sharing information and knowledge with patients a families regarding safety and risk management during hospitalization and at home. The main theme of these multimedia products is the importance of the participation of families and children in the process of care. Their active participation represents indeed an added value to safety and to contributes to take more conscious and shared decision. The messages contained in the cartoons have been projected to be simple and clear so are the graphics of the characters and the environment. In the vignettes hospital setting has been designed to be as much as possible similar to the real environment of the Meyer Hospitals and they try to reflect the real organization of work. The idea was to try to promote those behaviors that can have a positive impact in the prevention of the main risks: falls, infections, therapy errors (preparation and administration of drugs), handover during admission and discharge. These cartoons have been designed to address both children and their families at the same time and they are projected in the waiting rooms and wards,

References

1. American Academy of Pediatrics (2011) Principles of patient safety in pediatrics. Pediatrics 107(6):1473–1475
2. Matlow AG, Baker GR, Flintoft V (2012) Adverse events among children in Canadian hospitals: the Canadian pediatric adverse events study. CMAJ 184(13):E709–E718
3. Jansen JO et al (2010) Detecting critical illness outside the ICU: the role of track and trigger systems. Curr Opin Crit Care 16:184–190
4. Smith GB et al (2008) A review, and performance evaluation, of single parameter "track and trigger" systems. Resuscitation 79:11–21

5. Smith GB, Prytherch DR, Schmidt PE, Featherstone PI (2008) Review and performance evaluation of aggregate weighted "trackand trigger system". Resuscitation 77(1):170–179
6. Duncan H, Hutchison J, Parshuram CS (2006) The Pediatric Early Warning System score: a severity of illness score to predict urgent medical need in hospitalized children. J Crit Care 21(3):271–278
7. Seiger N, Maconichie I, Oostenbrink R et al (2013) Validity of different pediatric early warning scores in the emergency department. Pediatrics 132:e84
8. Way C, Crawford D, Gray J et al (2013) Standards for assessing, measuring and monitoring vital signs in infants, children and young people. Royal College of Nursing
9. Tucker KM, Brewer TL et al (2009) Prospective evaluation of a pediatric inpatient early warning scoring system. J Spec. Pediatr Nurs 14:79–85
10. Hillman KM, Bristow PJ, Chey T et al (2001) Antecedents to hospital deaths. Int Med J 31:343–348
11. Healey F, Scobie S, Oliver D, Pryce A, Thomson R, Glampson B (2008) Falls in English and Welsh hospitals: a national observational study based on retrospective analysis of 12 months of patient safety incident reports. Qual Saf Health Care 17(6):424–430
12. Hitcho EB, Krauss MJ, Birge S, Claiborne Dunagan W, Fischer I, Johnson S, Nast PA, Costantinou E, Fraser VJ (2004) Characteristics and circumstances of falls in a hospital setting: a prospective analysis. J Gen Intern Med 13:732–739
13. Pomerantz WJ, Gittelman MA, Hornung R, Husseinzadeh H (2012) Falls in children birth to 5 years: different mechanisms lead to different injuries. J Trauma Acute Care Surg 73(4 Suppl. 3):S254–S257
14. Jamerson PA, Graf E, Messmer PR et al (2014) Inpatient falls in freestanding children's hospitals. Pediatr Nurs 40(3):127–135
15. Hill-Rodriguez D, Messmer PR, Williams PD, Zeller RA, Williams AR, Wood M, Henry M (2009) The humpty dumpty falls scale: a case-control study. J Spec Pediatr Nurs 14(1):22–32
16. Wong MC, Yee KC, Turner P (2008) Clinical Handover Literature Review. eHealth Services Research Group. University of Tasmania, Australia, pp 111–114
17. Toccafondi G, Alboino S, Tartaglia R et al (2012) The collaborative communication model for patient handover at the interface between high-acuity and low-acuity care. BMJ Qual Saf 21:58–66
18. Sousa V, Rojjanasrirat W (2011) Translation, adaptation and validation of instruments or scales for use in cross-cultural health care research: a clear and user friendly guideline. J Eval Clin Practica 17:268–274

A Novel Intervention to Prevent In-Patient Falls

G. Terranova[1(✉)], I. Razzolini[1], S. Giovannini[2], S. Vallini[3], and T. Bellandi[1]

[1] Patient Safety Unit, Tuscany Northwest Trust - Regional Health Service, Florence, Italy
giuseppina.terranova@uslnordovest.toscana.it
[2] Medical Unit, Lotti Hospital, Tuscany Northwest Trust - Regional Health Service, Florence, Italy
[3] Emergency Department, Lotti Hospital, Tuscany Northwest Trust - Regional Health Service, Florence, Italy

Abstract. Falls in healthcare facilities are a frequent adverse event and a critical patient safety problem. Since 2008 we joined the Falls Prevention Campaign of the Center for Clinical Risk Management and Patient Safety of the Tuscany Region and sustained a continuous improvement process involving the governing body, medical staff, nurses, patients and their families. Despite the effort to reduce falls, 3 sentinel events resulting in patient deaths occurred in 2016, therefore the local patient safety team decided to revise and update the preventive interventions. The aim of the study is to evaluate the effects of an improvement program to prevent in-patient falls, on the basis of systemic analysis of reported incidents in one of the province that belongs to Tuscany Northwest Trust.

Keywords: Falls · Prevention · Program

1 Background

Falls in healthcare facilities are a frequent adverse event and a critical patient safety problem. National statistics about inpatients falls are not available; according to a regional survey [1] we can estimate that every year in Italy approximately 120 thousands patients (1% of the admissions) are victims of falls in healthcare facilities, with 30% to 50% resulting in injury requiring additional treatment, prolonged hospital stay, permanent functional impairment and death. Falls with serious injury/death are the first (n. 471, 24,6%) sentinel event type reported to the Ministry of Health within SIMES (Information System for Monitoring Errors in healthcare) from September 2005 to December 2012 [2]. Research shows that close to one third of falls can be prevented, so falls rates reduction is a strategic patient safety goal [3]. In 2006 a pilot, multicentric study (including Pisa healthcare trust) was conducted by the Center for Clinical Risk Management and Patient Safety of the Tuscany Region in the framework of a Patient Safety Campaign, in order to validate a measurement system and solutions. The project resulted in a Patient Safety Practice approved in 2008 and a prevention toolkit

S. Bagnara et al. (Eds.): IEA 2018, AISC 818, pp. 752–758, 2019.
https://doi.org/10.1007/978-3-319-96098-2_92

including an implementation Vademecum, a fall risk assessment scale (ReTos), a checklist to assess environmental risk factors, training materials for nurses and doctors, charts and procedure to promote local audit on falls. Since 2008 we joined the Falls Prevention Campaign and sustained a continuous improvement process involving the governing body, medical staff, nurses, patients and their families. Despite the effort to reduce falls, 3 sentinel events resulting in patient deaths occurred in 2016, therefore the local patient safety team decided to revise and update the preventive interventions.

The aim of the study is to evaluate the effects of an improvement program to prevent in-patient falls, on the basis of systemic analysis of reported incidents in one of the province that belongs to Tuscany Northwest Trust.

2 Standard Falls Prevention Program

Falls prevention, according to the Center for Clinical Risk Management and Patient Safety of the Tuscany Region, is based on the detection and control of environmental factors through the Environmental Safety Data Sheet (SAP Checklist, i.e. *Checklist Sicurezza Ambienti e Presidi*), on the fall risk assessment by means of the ReTos scale, on the implementation of multifactorial interventions, finally on the detection, analysis and monitoring of falls.

2.1 Environmental Risk Assessment

The SAP Checklist is a tool for the evaluation of fall risks concerning environmental factors in hospitals. The Nurse Coordinator, possibly with the collaboration of an occupational health professional, is responsible for the implementation of the SAP checklist. It is recommended the application of the checklist on an annual basis, or in the event of significant changes in the environment or equipments in use in the Department. The first section focuses on the evaluation of floors, corridors, stairwells, access areas to departments. The second section contains items related to equipments used for handling non-ambulatory patients. The third section explores risks related to hospital environments where patients spend most of their time. Based on the results of the assessment, the Nurse Coordinator in agreement with the Medical Director is in charge of defining an improvement plan, with the goal of prioritazing and decreasing the number of risk factors.

2.2 Individual Fall Risk Assessment

ReTos scale is used for patients aged greater than or equal to 65 years, in state of consciousness, for admissions lasting more than 24 h; it explores 11 items divided into three sets:

(1) questions to ask your patient (falls, dizziness, dizziness in the past six months, low vision/hearing, difficulty in performing everyday activities)
(2) conditions of patients' health: agitation, lack of sense of danger, impaired gait, mobility issues or need for help on transportation

(3) clinical documentation: cerebral, cardiovascular, neurological, psychiatric diseases, medications (diuretics, antihistamines, antihypertensive, vasodilator, painkillers, psychotropic drugs)

The total score resulting from ReTos scale varies between 0 and 21 and directs the judgment of doctors and nurses about the real risk of fall and the actions to be taken.

You have not defined any threshold score in order to keep the attention of the healthcare staff on the meaning of each risk factor and to allow each unit to define, possibly over time, cut-offs tailored to its target of patients, structural and organisational conditions.

2.3 Multifactorial Interventions

For the purpose of patient information a special flyer and posters were developed by the Center for Clinical Risk Management and Patient Safety of the Tuscany Region. Clinical and nursing recommendations have been adopted concerning: early mobilization, which in all cases reduces the risk of falling apart from injury from pressure; for the higher scores, increase in the frequency of nursing observations to monitor any relevant change in the conditions of the patient requiring reassessment with ReTos or immediate care interventions; patient assistance during personal hygiene, since most of the falls occur when the patient gets off the bed to go to the bathroom, in the path from the bed to the bathroom or in the bathroom; the review of drug therapies by the doctor in order to reduce any non-essential medication that increases the risk of falling.

For patients with higher scores the eventual intervention of the physiotherapist was considered, to maintain mobility and self-sufficiency. Other measures to be considered in second instance: managing vision or hearing deficits and cognitive-behavioral disorders with the involvement of specialists in counselling.

With regard to equipments, as most of the falls is from the bed, it was recommended to monitor over time the maintenance condition of beds and in particular the proper functioning of wheels and brakes. Bedrails must be considered with caution; the use of bedrails is best when the patient is cooperative and able to understand the utility to reduce the risk of slipping or falling out of bed, while it is not recommended if the patient has an altered mental state. In the latter case it is appropriate to reduce to a minimum the level of the bed, to increase nursing observations and/or informal caregiver assistance. The use of bedrails is especially recommended in the post-operative period and at the time of the partial recovery of mobility as a result of a prolonged period of hospitalization.

2.4 Detection, Analysis and Monitoring

Reporting the event to the clinical risk Facilitator triggers a process aimed at the analysis of contributing factors and fall risk prevention. The Facilitator receives the alert, collects informations, evaluates the case and decides whether archiving or perform in-depth analysis through a clinical audit. As a result of the audit the Facilitator draws up an alert report containing a summary of the analysis of the main problems encountered, a detailed description of the recommendations and the action plan to implement them. The essential content of the report must be shared among the participants during the discussion. Actions that can be managed at the unit level are implemented and monitored over time under the supervision of the Facilitator, the facility Manager and the Clinical Risk Manager. Actions that may require support from the Healthcare Trust management are included in a plan of priorities for patient safety.

3 Methods

Since 2008 we joined the regional recommendations and the prevention toolkit of the Patient Safety Practice. In this phase we formed the healthcare professionals, introduced the environmental risk assessment, on an annual basis, and the individual risk assessment by means of the ReTos scale for eligibile patients; we implemented systemic analysis and monitoring of reported incidents and multifactorial interventions suggested by the Fall Prevention Campaign (see par. 2).

In the second phase we revised interventions to prevent falls at the microsystem level as a result of data analysis and local audit.

The Regional Incident Reporting Database (Si-GRC) was retrospectively searched for reports of falls occurring from January 2016 through December 2017. Inhospital falls rates were standardised as number of falls per 1000 inpatients days. We analysed patient informations, associated factors, specific circumstances and dynamics of each incident and patient outcome.

By acknowledging preventive measures already in place, the patient safety team with front-line professionals explored additional interventions or revision of the existing tools to anticipate or control preventable risk factors emerged during incidents analysis. As a result, a new educational intervention was introduced to educate patients and their families with personal consultation on admission and orientation to the wards facilities.

A simple flyer with instructions to be observed during the hospital stay and recommendations about footwear and clothes suitable for the purpose of falls prevention (nonslip, comfortable, well-fitting) was delivered and discussed with patients and caregivers.

Moreover, the use of bedrails was further minimized thanks to a shared evaluation of patients' autonomy involving doctors, nurses and patients or their relatives.

To collaborating patients with impaired mobility was proposed a one-side bedrail, applied by the side from which the patient doesn't get off the bed; the one-side bed-rail proved to be a useful and welcome support to delimit space and for postural changes.

Regular rounds were performed by the nursing staff, particularly before sleep, as a proactive measure to offer help using the toilet, offer hydration, help the patient get into a comfortable position or turn immobile patients to maintain skin integrity.

In the Emergency Department the old stretchers were all replaced with those adjustable in height and safety belts made of soft material were introduced in selected cases and during patient transportation. Flyers containing information on the preventive measures available in the unit and on the possible use of safety belts have been distributed in the waiting room, coining the slogan "When you get on board fasten your seatbelts".

4 Results

Over a two-year period (2016–2017), we observed a 21% reduction in the total number of falls recorded in the Si-GRC system. In the hospitalized subpopulation falls reduction was 15% with falls rates of 1,25 (2016) vs 1,06 (2017) per 1,000 inpatient days (see Fig. 1). Sentinel events dropped from 3 in 2016 to 0 in 2017.

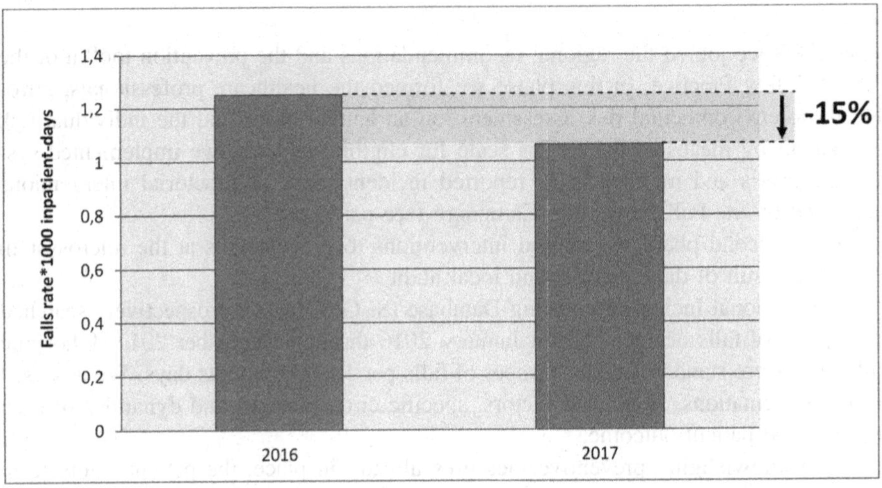

Fig. 1. Falls reduction over a two-year period (2016–2017)

5 Discussion

From 2008 to the present, research and literature on falls prevention have reconsidered the role of systematic risk score assessment tools and have, instead, exploited an individualized approach in all patients aged over-65 or over-50 years when suffering from morbid conditions which predispose to falls.

The SIGN guidelines [4], for example, consider these two categories of patients as those who, by definition, are at risk of falling and recommend a multidimensional evaluation, multi-factorial interventions to manage individual factors, in adjunct to identification and address of environmental risk factors.

The multidimensional assessment on the admission, carried out systematically by the nursing staff, explores the overall functional condition of the patient: mobility, autonomy, elimination, cognitive status and communication, nutritional conditions, risk of pressure ulcers, social frailty, sensory deficits.

The doctor's anamnestic and clinical evaluation complements the nursing assessment with information on previous illnesses, medication intake, clinical conditions that may predispose to falls or aggravate the outcome (osteoporosis).

In Italian hospitals, however, it is unusual that patients are asked systematically if he or she has suffered falls in the last year and what the consequences have been, a fact that places the patient in the risk class regardless of whether or not he/she is over-65; moreover, a systematic evaluation of clothing and footwear in patients at risk, as prescribed by the same SIGN guidelines, is not carried out.

Delirium, particularly hyperactive, is a major risk for falls [5] and there is good evidence that delirium prevention interventions reduce fall risk; nevertheless this is a condition that is often under-diagnosed and untreated.

At the current state of the evidence, it is therefore necessary, for the future, to experiment with a new approach to risk assessment, reconsidering the systematic application of screening scales and scores and focusing on multidimensional evaluation, and to develop systematic and integrated strategies for the prevention, at the same time and possibly with a single, comprehensive and simple method of risk detection, of Delirium and falls in frail, elderly patients with mild cognitive impairment.

Our experience shows that few well-designed interventions can achieve good results at the microsystem level but we are sure that much more can still be done, especially in the prevention of falls linked to Delirium.

Strategic elements to achieve value improvement in healthcare are a social and collective approach in the design phase and the involvement of the right people in the right way, as Øvretveit teached us [6]: trained professionals and experts who derive their own perception and gratification from their daily practice will support value improvement only if they are actively involved by leaders who can drive change and convince them that this will lead to better assistance.

This report is part of a safety improvement program, not designed as a research, with all the limitations of such an approach for the effectiveness of the interventions [7]. Despite its limitations, this project suggests how improvement can be considered within a well structured system for patient safety management, by looking at novel interventions which after this experience could be tested in a well designed research project.

References

1. Centre for Clinical Risk Management and Patient safety (2011) Falls prevention at hospital. 1st edn. ETS, Pisa. http://www.regione.toscana.it/documents/10180/603668/La+prevenzione+delle+cadute+in+ospedale.pdf/96a548bc-23e5-4eea-b7ea-665f30bfb2fe. Accessed 18 May 2018
2. Italian Ministry of Health (2015) Fifth Monitoring Report on Sentinel Events. http://www.salute.gov.it/imgs/C_17_pubblicazioni_2353_allegato.pdf. Accessed 18 May 2018
3. Cameron ID, Murray GR, Gillespie LD et al (2010) Interventions for preventing falls in older people in nursing care facilities and hospitals. Cochrane Database Syst Rev 2010(1), Art. No CD005465
4. Nice Clinical Guideline Falls in older people: assessing risk and prevention, 12 June 2013. https://www.nice.org.uk/guidance/cg161. Accessed 18 May 2018
5. Pendlebury ST, Lovett NG, Smith SC et al (2015) Observational, longitudinal study of delirium in consecutive unselected acute medical admissions: agespecific rates and associated factors, mortality and readmission. BMJ Open 5:e007808. https://doi.org/10.1136/bmjopen-2015-007808

6. Øvretveit J (2014) Il miglioramento del valore nei servizi sanitari. Un metodo basato sulle evidenze. 1st italian edn. Il Mulino, Pisa
7. Burke RE, Shojania KG (2018) Rigorous evaluations of evolving interventions: can we have our cake and eat it too? BMJ Qual Saf. https://doi.org/10.1136/bmjqs-2017-007554. 09 Feb 2018

An Action Research to Study and Support the Transition to a Comprehensive Electronic Patient Record in Acute Care

T. Bellandi[1(✉)], G. Luchini[1], A. Reale[1], M. Micalizzi[2], and M. Mangione[2]

[1] Tuscany Northwest Trust, Regional Health Service, via A. Cocchi, 7/9, 56124 Pisa, Italy
tommasobellandi@yahoo.it
[2] Fondazione Toscana Gabriele Monasterio, Regional Health Service of Tuscany, via G. Moruzzi 1, 56124 Pisa, Italy

1 Background

In the last twenty years, health organizations have started the gradual shift from paper-based information systems to digital systems for recording internal and external communication [1–3]. This is an epochal change, which is going through all the productive systems, yet in the health systems this passage is particularly prolonged and difficult [4], mainly due to the complexity of the clinical and care activities [5], the variety of organizations that produce services [6] and a hyper-regulated institutional context [7]. These three factors typical of the health system have generated a tension on the clinical level between standardization and personalization of care, on the organizational level between autonomy and integration of health providers, at the institutional level between centralization and location of planning and control functions. This tension is reflected in projects and programs for the development and implementation of digital healthcare, which have sometimes failed miserably despite the investments and commitment of key stakeholders [4], due to inconsistencies generated on the clinical, organizational and institutional levels, or open clashes between professionals, managers and policy makers for decision-making power on technological innovation [8].

On the other hand, the intrinsic quality of IT products for health care is sometimes lacking, both in terms of functionality, and above all of ergonomics and systems integration [8]. Functionality means here the ability of the tool to meet the objectives for which it was designed, the ergonomics of the product can be summarized in the ease of use and in the ability of the tool to support the production process, integration is the possibility and fluidity of communication with other applications involved in the same or other related production process.

A well-known systematic review of electronic medical records conducted in the United States has promoted only a very limited number of products, however, highlighting how the few co-developed products between computer scientists and clinicians within healthcare facilities were superior to commercial products [9]. An extensive evaluation program of the digitization process of health information systems conducted

S. Bagnara et al. (Eds.): IEA 2018, AISC 818, pp. 759–766, 2019.
https://doi.org/10.1007/978-3-319-96098-2_93

in the UK has shown all the limitations of existing products and even the new risks that may result in clinical activities (e.g. errors of prescription of drugs induced by automatisms in the data entry) and managerial (e.g. prolonged waits for exam results due to difficulty in finding relevant information), as well as the ways that operators and managers use to overcome software limitations, through routine violations of procedures and good working practices (e.g. transcription of operational notes on paper notes to prepare the letter of resignation, use of commercial applications for internal communications such as informal consultations or handover) [4]. Also worthy of mention is the limited evidence regarding the efficiency of digital systems in comparison with paper ones [10], the attitude to multitasking and the risk of disruptive interruptions [11], the intrusion of electronic devices in human relationships between colleagues [12] and between operators and patients [13].

Beyond the easy enthusiasm towards digital technology and in the prejudice that these can magically solve the aforementioned tensions [14], the transition between paper and bits is initiated and at this point requires a structured reflection to guide its developments [15].

In Italy, both the national addresses included in the so-called "digital agenda", as well as the recent decisions taken at the regional level in Tuscany, impose the complete digitalization of production processes on the healthcare companies.

Specifically, given the evidence regarding the risks of a non-controlled and partially unconscious digitization of the medical record, it is therefore desirable to monitor the implementation in order to support further developments based on the user experience and the effects on the processes of care, both from the point of view of patient safety and the health of the operators.

2 Objectives

To describe the critical issues and the elements of good practice in the current process of implementation of the electronic patient record (EPR), to support the working groups responsible for the digitization processes to integrate in the developments the return of experience from the operators, so to prevent the main risks associated with the transition from paper to bits in 3 urban hospitals at the Tuscany Northwest Trust.

3 Methods

Being a descriptive and intervention study, it involves the joint use of field observations, semi-structured interviews, cognitive walkthrough sessions, analysis of official documents, participation in some meetings of the working groups that have responsibility for the development of EPR and in some sessions of staff training. Settings are medical and surgical wards at three urban hospitals. An expert ergonomist collects field data with shadowing technique of doctors and nurses, sampling one morning shift one week before and 2 weeks after the implementation of EPR at each ward. EPR product

"C7" is being developed in-house on the basis of a 15 years program carried out at the FTGM hospitals.

The observations at hospital 1 were conducted following nurses and doctors in the two areas (internal medicine and general surgery) for a shift of work a week before and two weeks after the implementation of C7, from October to December 2017, with the technique of shadowing, explicit and non-participating. The activities covered by the observation were the preparation and administration of medications, patients' round and the preparation of the discharge letter. During the field visits, some ethnographic interviews were conducted with the medical and nursing staff on duty, with the nursing coordinators and the managers of the operative unit. A training session dedicated to nurses in the surgical area was also observed. The notes were collected with a simple application that allows to take notes on smartphones and record the times of the activities observed. The collected notes were then subjected to thematic analysis, which was associated with an analysis of times and methods of sampled activities. Indicators used are time for completion of each medication, time per patient in the round, number of care delivery problems, elements of good practice and workarounds (= informal ways of solving interaction problems). The contents of the notes have been reported on a matrix scheme with the aforementioned observed activities and themes related to software ergonomics, with an adaptation of the classical criteria of Green and Jordan [16] for the analysis of the interactions between operator-software-process. The selected criteria are safety, functionality, efficiency, usability and pleasantness. Finally, interruptions of the workflow were also highlighted, counting them and identifying their cause and duration.

4 Results

Four field observation sessions were conducted, for a total of 28 h of observation and ethnographic interviews, half of which before and half after the implementation of C7, to which 4 h of observation of the training course were added.

Table 1 shows the processing of data relating to the times of the activities observed, the critical points, the strengths and workarounds that emerged before and after the implementation of C7.

After the introduction of C7, a reduction of the time for single administration of 10.8% and of the times for the visit to a patient of 6.2% was observed. Care delivery problems were reduced by 53% and 8 elements of good practice emerged in the interactions mediated by the EPR, of which previously no trace was found.

The most numerous and potentially serious problems with the use of paper documentation concern the inconsistencies between the same data reported in different forms or the inability to find information relevant to the planning and management of care:

[Hospital 1 Med pre - @med adminstration #safety] prescription of oxygen therapy with incongruity to be resolved, in prescription order is indicated 2 l per minute, while in nurses' notes 1 l per min

Table 1. Summary of results at hospital 1

	Internal medicine	Surgery	Total
Before			
Time for medication	2 min 7 s	2 min 11 s	2 min 9 s (average)
Time for patient	7 min 48 s	7 min 13 s	7 min 30 s (average)
Care delivery problems	13	6	19
Elements of good practice	0	0	0
Workarounds	1	5	6
Interruptions	10	9	19
After			
Time for medication	1 min 59 s	1 min 40 s	1 min 55 s (average)
Time for patient	7 min 49 s	6 min 14 s	7 min 2 s (average)
Care delivery problems	5	4	9
Elements of good practice	2	6	8
Workarounds	1	2	3
Interruptions	16	9	25

[Hospital 1 Med pre - @med adminstration #safety] Insulin prescriptions inconsistent and written unclearly [...]. Prescribed Humalog if blood sugar greater than 150? 160? The patient has 158.

[Hospital 1 Med pre - @patients round #functionality] an incongruity between anamnesis on home-based NIV and the patient's report, as well as a problem of lack of a drug in the ward

[Hospital 1 Surg pre - @patients round #safety] The first doctor complains about the difficulty of finding the trend of the radiodiagnostics tests and finally decides to request an additional chest x-ray control

[Hospital 1 Surg pre - @patients round #usability] Reviewing the A&E report, the doctor notices the indication of a CT scan with contrast material, of which the report is not present

With C7 the observed criticalities appear less serious in terms of possible consequences for patients, while they deserve attention to the negative effects they may have on clinical and care processes. The first three examples that follow are due to difficulties in interacting with the new working methods introduced by C7, while the last two concern problems connected to the network and the software structure:

[Hospital 1 Med post - @med administration #safety] The nurses notice a discrepancy between what is prescribed and a non-administration of a potassium infusion, represented in red, verifying the problem from the profile of the doctor, it is canceled due to exceeding 50% of the time set for the infusion

[Hospital 1 Med post - @med administration #usability] The nurse realizes that the patient has parenteral nutrition with the full bag, even on C7 it appears in progress.

[Hospital 1 Chir post - @med preparation #safety] The nurse observes that 2 drugs are indicated for 11 am and one for 10 am, she interprets this problem as time errors committed by the prescriber from the operating room

[Hospital 1 Med post - @registration of vital signs #functionality] They spend 15 min to solve a problem with the software: first it must be restarted three times by re-entering the login credentials, then it does not register the MEWS giving a date and time error message that are not reported automatically

[Hospital 1 Surg post - @med administration #functionality] in the summary screen of the alarms, the list of therapies not administered by 9 o'clock remains, telling me that at other times these alarms have been active for two hours, making them useless

Even if to a limited extent, from 6 in the pre- to 3 in the post-implementation stage of C7, there remain some workarounds, some of which are explicitly adopted to meet some limits of digitization, such as the persistence of informal paper supports, whose functions also according to operators could be integrated into C7:

[Hospital 1 Surg post - @patients round #safety] for the diet they use an old flying sheet written in pencil by the nurse carrying the medication trolley, from which they then transcribe onto another flying sheet that is sent by fax to order for the catering service

[Hospital 1 Med post - @patients round #functionality] The expert daily nurse shows me a simple table with column names, in-line schedules and blood glucose levels in the cells. She tells me that in C7 there is no possibility of grouping of diabetic patients for a synthetic view of the glycemia to be done (for example an icon could be inserted in the right part of the screen with the list of patients)

Only in one case is an informal practice connected with usability problems of the interface, in particular related to the criterion of visibility of contents:

[Hospital 1 Surg post - @patients round #usability] Nurse writes notes in the diary in capital letters, as she tells me that some colleagues have requested her to read better

Regarding the preparation of the discharge letter, only two sessions were observed in the phase prior to the introduction of C7 and short ethnographic interviews were collected in the post phase.

In the medical area, before the move to C7, the letter was written using the application of administrative management of patients, with serious problems in finding the information in the folders, usability and functionality of the instrument. In the observed session, lasting one hour, three doctors (two senior consultants and one resident) were involved in the provision of a transfer and a discharge and were unable to complete either of them, due to interaction problems (3 criticalities) and to due to the numerous and long lasting interruptions (6 in total) provided by colleagues.

In the surgical area, the discharge letter was written to a computer on a preset word module, also in this case with difficulty in finding and transcribing information from the paper folder, which made the process inefficient and a source of possible transcription errors.

In both departments, the doctors interviewed in the phase following the implementation of C7 reported to appreciate the new modality of management of the discharge letter, above all for the possibility to select the diary texts, the therapies and the relevant reports to be included in discharge. However, a partial awareness of the functionality of C7 emerged, attributable to limited training and practice in the field with the tool. Basic computer skills are a factor that can facilitate the use of the tool, not to be taken for granted.

The strengths in the interactions between operators-software-process emerged with the use of C7 mainly concern efficiency (6 out of 8), in addition to an observed improvement in terms of safety (just in time nursing assessment) and functionality (access with two clicks to all the reports of radiodiagnostics and laboratory):

[Hospital 1 Med post - @patients round #efficiency] An operator asks for documentation to the nurse who answers by saying that everything is digital, if she wants, she can print the therapy; then they realize that patient is being transferred to Massa hospital, where they can access the electronic folder so there is no need to print anything

[Hospital 1 Med post - @patients round #efficiency] takes care of a patient transferred from another setting that already sees from C7 and to which the bed is assigned by the nurse of the sending ward

[Hospital 1 Surg post - @administration #efficiency] the nurse on ward calls the doctor asking to suspend the prescription, which he does directly from the operating room.

[Hospital 1 Surg post - @patients round #efficiency] as they move the trolleys to go to the next patient, they already open C7 and the nurse describes patient's condition

[Hospital 1 Surg post - @patients round #efficiency] The doctor checks with nurses the availability of one medication requested by the patient, not finding it they decide jointly for an alternative present in the pharmacy's list visible from the prescription form

Regarding interruptions, a relatively high frequency of these was observed both before and after the introduction of C7, with a relative increase of 24%. The frequency of interruptions does not appear to be correlated with the use of paper or electronic documentation, as the cause of *interruptions* was in most cases a verbal communication by a colleague (9 out of 19 in the pre and 18 out of 26 in the post). To consider interruptions connected with technical problems due to the lack of a drug or a device in the trolley (4 in the pre and 7 in the post) that could partly be resolved with better information on the process when preparing the trolleys, to which follow those less preventable due to communications from family members and patients (6 in the pre and 3 in the post).

5 Discussion

The transition to EPR has improved the efficiency in the conduct of the departmental assistance activities subject to observations, thanks above all to a better organization of the information in the EPR, the ease of access and sharing of data between various actors involved in the process. Safety also improved, thanks to the elimination of manual transcriptions, the clarity of writing and reading on screen and the possibility of solving any inconsistencies in real time.

There remain some critical issues related to consistency between what is shown on interface and what actually performed on patients, as well as the partial awareness of how to effectively use the tool, to be clarified through the preparation of a business manual and the reinforcement of training.

The extent of the network to the servers in which the database is located is a problem to be solved quickly, to avoid the disservices observed and reported by the operators and to maintain a constructive attitude in the phase of change. The full integration with the other applications in use and, possibly, with the remaining paper documentation is another area to work on, to bring patients' management in an integrated information environment.

Interruptions are a problem to be prevented regardless of the use of the electronic record, especially in the preparation phase of the discharge letter, with actions aimed at identifying interruption free zones and demanding that the rest of the staff does not disturb the doctors in the time dedicated to discharge. During the activities of the department, it emerged that EPR facilitates recovery of the interrupted task, as the doctor or nurse easily find on the interface of C7 the progress of the activity that they had to abandon due to an interruption.

The present analysis and reflections are to be considered as preliminary, within the boundaries of a data collection limited to only two departments and conducted by a single ergonomic specialist both in the collection and processing of the data, subject to possible selection and anchor biases and based on low numbers of observed occurrences. Currently the research is ongoing in 2 more hospitals, following the implementation plan in medical and surgical departments.

References

1. Jha AK, Doolan D, Grandt D, Scott T, Bates DW (2008) The use of health information technology in seven nations. Int J Med Inform 77(12):848–854
2. Jha AK, DesRoches CM, Kralovec PD, Joshi MS (2010) A progress report on electronic health records in US hospitals. Health Aff 29(10):1951–1957
3. Nguyen L, Bellucci E, Nguyen LT (2014) Electronic health records implementation: an evaluation of information system impact and contingency factors. Int J Med Inform 83 (11):779–796
4. Robertson A, Bates DW, Sheikh A (2011) The rise and fall of England's national programme for IT. J R Soc Med 2011(104):434–435
5. Westbrook JI, Ampt A, Kearney L, Rob MI (2008) All in a day's work: an observational study to quantify how and with whom doctors on hospital wards spend their time. Med J Australia
6. Cresswell KM, Mozaffar H, Lee L, Williams R, Sheikh A (2016) Safety risks associated with the lack of integration and interfacing of hospital health information technologies: a qualitative study of hospital electronic prescribing systems in England. BMJ Qual Saf. bmjqs-2015
7. Black AD, Car J, Pagliari C, Anandan C, Cresswell K, Bokun T, Sheikh A (2011) The impact of eHealth on the quality and safety of health care: a systematic overview. PLoS Med 8(1):e1000387
8. Middleton B, Bloomrosen M, Dente MA, Hashmat B, Koppel R, Overhage JM, Zhang J (2013) Enhancing patient safety and quality of care by improving the usability of electronic health record systems: recommendations from AMIA. J Am Med Inform Assoc 20(e1):e2–e8
9. Chaudhry B, Wang J, Wu S, Maglione M, Mojica W, Roth E, Shekelle PG (2006) Systematic review: impact of health information technology on quality, efficiency, and costs of medical care. Ann Int Med 144(10):742–752

10. Poissant L, Pereira J, Tamblyn R, Kawasumi Y (2005) The impact of electronic health records on time efficiency of physicians and nurses: a systematic review. J Am Med Inform Assoc 12(5):505–516
11. Westbrook JI, Reckmann M, Li L, Runciman WB, Burke R, Lo C, Day RO (2012) Effects of two commercial electronic prescribing systems on prescribing error rates in hospital in-patients: a before and after study. PLoS Med 9(1):e1001164
12. Callen JL, Braithwaite J, Westbrook JI (2007) Cultures in hospitals and their influence on attitudes to, and satisfaction with, the use of clinical information systems. Soc Sci Med 65 (3):635–639
13. Shachak A, Reis S (2009) The impact of electronic medical records on patient–doctor communication during consultation: a narrative literature review. J Eval Clin Pract 15 (4):641–649
14. Shojania KG, Duncan BW, McDonald KM, Wachter RM, Markowitz AJ (2001) Making health care safer: a critical analysis of patient safety practices. Evid Rep Technol Assess (Summ) 43(1):668
15. Carayon P (ed) (2016) Handbook of human factors and ergonomics in health care and patient safety. CRC Press, Boca Raton
16. Green WS, Jordan PW (eds) (2003) Pleasure with products: beyond usability. CRC Press, Boca Raton

Effectiveness Evaluation of the Conceptual Framework for the International Classification Patient Safety (ICPS) in the Reporting and Learning System of the Tuscany Region

M. Tanzini[1(✉)], F. Ranzani[1], G. Falsini[2], C. Sestini[2], I. Fusco[3], F. Venneri[3], G. Terranuova[4], and R. Tartaglia[1]

[1] Clinical Risk Management and Patient Safety Centre Tuscany Region - WHO Collaborating Centre in Human Factors and Communication for the Delivery of Safe and Quality Care, Villa La Quiete alle Montalve - via Pietro Dazzi 1, 50141 Florence, Italy
tanzinim@aou-careggi.toscana.it
[2] Tuscany South East Trust, Regional Health Service of Tuscany, Florence, Italy
[3] Tuscany Centre Trust, Regional Health Service of Tuscany, Florence, Italy
[4] Tuscany North West Trust, Regional Health Service of Tuscany, Florence, Italy

Abstract. RLS is a best practice and represent an important surveillance and awareness-raising tool that assess how health systems behave. This research focuses on two key elements related to reporting systems: the learning component and dissemination strategies regarding the lessons learned to physicians and the accuracy of event analysis.

Two different online surveys has been carried out: the first one, to detect clinicians' perception of the usefulness, when and how they use the framework, and the taxonomy's criticalities; the second one, to verify the accuracy of the description, the adequacy of the type of accident classification, the attribution of the main identified contributing and mitigating factors useful to provide information on the context useful for identifying future preventive organizational solutions. This research highlighted some areas needing considerable improvement, including better definition of terms, more emphasis on assessing reliability of coding and great opportunities related to the classification results analysis focusing on facilitation and improvement strategies dissemination connected to learned lessons.

Keywords: Adverse events · Taxonomy · Causal factors · Reliability

1 Introduction

RLS represents a cornerstone of safe health care practice and they represent an important surveillance and awareness-raising tool which assess how health systems are performing [8]. Reporting is cumbersome and time-consuming: a good internal reporting system ensures that all responsible parties are aware of major hazards. Identification of adverse events is critical for improving patient safety, yet medical

S. Bagnara et al. (Eds.): IEA 2018, AISC 818, pp. 767–773, 2019.
https://doi.org/10.1007/978-3-319-96098-2_94

errors and adverse events can be challenging to measure [4]. Key elements determining success, as a response to the healthcare staff's commitment in adverse events reporting and analysis, are the dissemination of lessons learned and the sharing of "best practices" implemented. One of the prerequisites in facilitating the analysis and reflection on contributing and mitigating factors linked to the adverse event is strictly related to the taxonomy used for its classification. The WHO has recognized the need for such standard terminology in the patient safety areas, developing a conceptual framework for patient safety [3], which has been adapted and implemented in the Tuscany Regional Clinical Risk management Reporting System – named Si-GRC – as tool to support clinicians in classifying the adverse events, in order to led them identify common elements related to organizational and clinical aspects related to the adverse event. This research represent a first step in trying to validating the framework thanks to the Tuscany experience in the field.

2 Purpose

This research focuses on two key elements related to the reporting systems: the learning component and the dissemination strategies concerning the learned lessons to clinicians and the event analysis accuracy. To be precise, the focus of the first part of the study concerns the detection of usefulness, ease of use and any critical issues related to the functionality to classify adverse events based on the International Classification Patient Safety (ICPS) framework. The second part, on the other hand, aims to verify the effectiveness of this framework for the classification of the types of incidents, the contributing and mitigating factors of all sentinel events that occurred during the year 2016.

3 Methods and Materials

The conceptual framework for the ICPS was designed to provide a much needed method of organizing patient safety data and information so that it can be aggregated and analyzed to compare patient safety data across disciplines, between organizations, and across time and borders and to examine the roles of system and human factors in patient safety [2]. In order to understand the framework effectiveness, two different online surveys has been carried out using Google Forms: the first one, to detect clinicians' perception of the usefulness, when and how they use the framework, and the taxonomy's criticalities; the second one, to verify the accuracy of the description, the adequacy of the type of accident classification (verifying the assigned priorities), the attribution of the main identified contributing and mitigating factors useful to provide information on the context useful for identifying future preventive organizational solutions. In the first survey, both the clinical risk manager and patient safety manager network have been involved, as well as some nurses (identified by the Healthcare Trust) who act as facilitators (i.e. they receive the near miss reports and take charge of them for a more deep analysis). To discuss the analysis and classification applied to sentinel events in 2016, a working group consisting of clinical risk managers from the different Tuscan Healthcare Trust has been set up.

The first survey, ICPS framework, has been divided into 8 sessions for a total of 16 questions, included both open questions to express more detailed opinions and evaluations on some aspects of the taxonomy, and 7-point Likert scales to detect clinicians perception regarding the functionality, but also the degree of satisfactions related to the categories of the framework: type of incident, contributing factors, mitigating factors, organizational outcomes, ameliorating actions and risk reduction actions. An excel file has been also distributed to the survey recipients which reported all the taxonomy leaves: we expected to detect comments and opinions in terms of clarity, ambiguity, not pertinence related to their experience of use during audit. The survey has also been used to detect the usefulness perception of the ICPS as a method of aggregating data, trying to identify its strengths and weaknesses and the improvement opportunities in order to plan upgrade to the regional reporting system. The second survey, Sentinel Event Analysis Quality Assessment and ICPS application concordance, divided into 7 sessions for a total of 22 questions, included both open questions and 7-point Likert scales in order to detect the opinion of the experts on the coherence of the sentinel event narrative part with the classification related to the main categories.

4 Results

The first survey has been addressed to 16 regional Clinical Risk Manager and 30 facilitators: 24 healthcare provider responded (33.3% Clinical Risk Manager; 33.3% nurses; 8.3% Patient Safety manager; 8.3% doctors; 8.4% ergonomists, 4.2% obstetrician and 4.2% physiotherapist). 45.8% of respondents have over 10 years of experience in audit management A total of 14 clinicians have expressed a positive opinion regarding the ICPS usefulness to describe the adverse event (adding all the users answers who placed themselves in the points going from 5 to 7 of the Likert scale). 8 users - 33.3% of the interviewees have placed on the value 3 of the scale (Fig. 1).

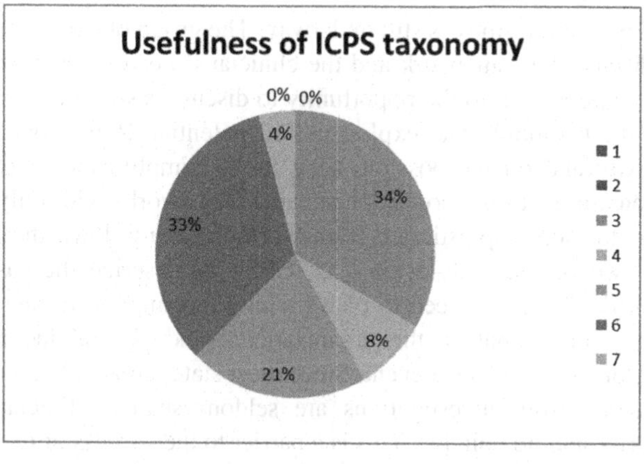

Fig. 1. Survey results – ICPS usefulness to describe the adverse event

75% of clinicians have used the framework to classify the adverse event after the audit have been took place. The reasons associated with a subsequent use are mainly linked to these factors: the lack of IT tools during the audit; the priority of managing relationships with clinicians during the analysis and the perception that taxonomy can be more useful after having already reflected and analyzed the event, taking advantage of the classification to give coherence to the descriptive and structured part of the report. Those have used the taxonomy during the audit (only 6 operators - 25% of the respondents), exploited it as a tool to support the analysis, encouraging the discussion to let the audit become a way of training. A total of 17 clinicians have expressed a positive opinion regarding the ease of learning of taxonomy (adding all the users answers who placed themselves in the points going from 4 to 6 of the Likert scale) (Fig. 2).

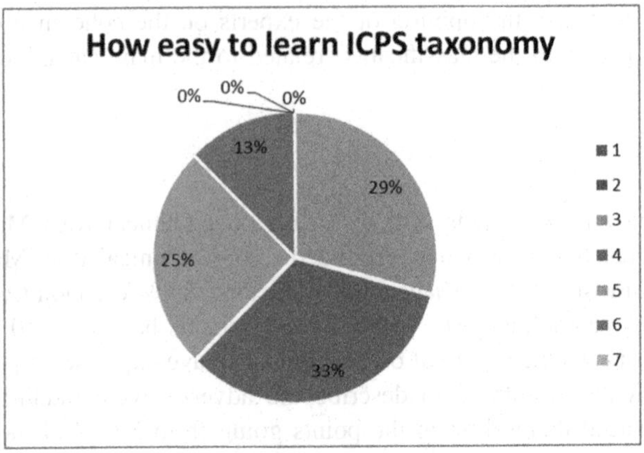

Fig. 2. Survey results – how easy to learn to use ICPS taxonomy

The second part of the study is still underway. The group of experts has already met and began to discuss the framework and the clinicians' need to share the best way of applying ICPS, supposing also the opportunity to discuss case studies in order to learn a better use of the taxonomy and exploit its full potential. Reflections about the taxonomy complexity and on the possible need for its simplification have been made. Single-group categories, to be more sophisticated frameworks with only single broad-level categories, to more sophisticated frameworks breaking down these single broad categories into two or four sub-levels - as ICPS - to describe the casual factors of adverse events [6]. There is a need to plan training meetings to further improve clinicians capacity to deep analyze the organizational and clinical factors as possible contributing factors to the adverse event. Since most state reports elicit no response and the lessons learned from investigations are seldom shared, clinicians often view reporting as all risk and no gain [7]. This is a barrier to the success of reporting systems that must be overcome, by investing resources and time in data processing and in carrying out case studies to be disseminated and shared. First elaborations have been

Table 1. Sentinel event type distribution (2016)

Type	N.
1 - Surgery or other invasive procedure performed on the wrong patient	1
10 - Patient suicide, attempted suicide, or self-harm resulting in serious disability, while being cared for in a health care facility	6
12 - Violence on healthcare staff	7
13 - Death or serious damage resulting from a malfunction of the transport system (intra-hospital, extra-hospital)	1
15 - Death or severe unforeseen damage resulting from surgery	3
16 - Any other adverse event that causes death or serious harm to the patient	13
2 - Wrong surgical or other invasive procedure performed on a patient	1
4 - Unintended retention of a foreign object in a patient after surgery or other procedure	3
6 - Death, coma or severe harm resulting from errors in drug therapy	1
7 - Maternal death or serious injury associated with labor or delivery	1
8 - Death or permanent disability in healthy newborn weighing > 2500 grams unrelated to congenital disease	2
9 - Patient death or serious injury associated with a fall while being cared for in a health care setting	19

Table 2. Death or severe unforeseen damage resulting from surgery

Year	Code	Factors
2015	ES104-15-00001	Patient characteristics - > Patho-Physiologic/Disease Related Factors
	ES902-15-00027	Patient characteristics - > Patho-Physiologic/Disease Related Factors
2016	ES902-16-00034	Patient characteristics - > Patho-Physiologic/Disease Related Factors
	ES902-16-00035	External Factors - > Products, Technology & Infrastructure
	ES903-16-00063	Other, Organizational/Service Factors - > Protocols/Policies/Procedures/Processes
2017	ES104-17-00105	Patient characteristics - > Patho-Physiologic/Disease Related Factors Organizational/Service Factors - > Protocols/Policies/Procedures/Processes Human Factors - > Cognitive Factors
	ES111-17-00154	Patient characteristics - > Cognitive Factors Human Factors - > Communication Factors
	ES901-17-00113	Other
	ES903-17-00106	Patient characteristics - > Social Factors
	ES903-17-00109	Patient characteristics - > Patho-Physiologic/Disease Related Factors

made on the use of the framework on the 58 sentinel events reported in 2016 (Table 1). Analysis of many events at different sites could lead to the identification of common contributing factors. Successful experiences could be widely shared [8].

By comparing the contributing factors of the same type of sentinel event, we can identify common organizational or clinical factors on which to implement preventive strategies and interventions, without forgetting the context topicalities' in which they will be implemented and then adapted. Table 2 shows an example of analysis that illustrates that in the 10 sentinel events occurred between 2015 and 2017 following surgical procedures, common contributing factors have been identified.

5 Discussion

The study, still underway, is proceeding with the compilation of the second survey focusing on the experts second level analysis regarding the consistency and correctness of the ICPS framework application compared to the narrative and structured part describing the sentinel event. We are also discussing strategies and methods for disseminating and sharing case studies. This research highlighted some areas needing considerable improvement, including better definition of terms, more emphasis on assessing reliability of coding and great opportunities related to the classification results analysis focusing on facilitation and improvement strategies dissemination connected to learned lessons. While comprehensiveness is an essential attribute of a patient safety classification system, to be useful, the classification and coding must also be reliable. The coding of factors must be consistent and reproducible between coders and between occasions of coding [9]. Reliability could be improved by ensuring that casual factor categories are mutually exclusive, by developing a data dictionary that contains clear definition of each casual factor, along with examples of casual factor classifications. Improvements can also be made by conducting exhaustive pilot testing of the framework and by providing training opportunities for staff in using the framework, and regular discussion of difficult event classifications [10].

References

1. Lawton R, McEachan RR, Giles SJ, Sirriyeh R, Watt IS, Wright J (2012) Development of an evidence-based framework of factors contributing to patient safety incidents in hospital settings: a systematic review. BMJ Qual Saf 21(5):369–380
2. Murff HJ, Patel VL, Hripcsak G, Bates DW (2003) Detecting adverse events for patient safety research: a review of current methodologies. J Biomed Inform 36(1–2):131–143
3. The conceptual framework for the international classification for patient safety (ICPS) (2009) World Health Organization, WHO Patient Safety
4. Gandhi TK, Seger DL, Bates DW (2000) Identifying drug safety issues: from research to practice. Int J Qual Health Care
5. Davis KSSC, Collins KS, Tenney K, Hughes DL, Audet AJ (2002) Room for improvement: patients reports on the quality of their health care. The Commonwealth Fund

6. Mitchell RJ, Williamson AM, Molesworth B, Chung AZ (2014) A review of the use of human factors classification frameworks that identify causal factors for adverse events in the hospital setting. Ergonomics 57(10):1443–1472
7. Rosenthal J, Booth M, Flowers L, Riley T (2001) Current state programs addressing medical errors: an analysis of mandatory reporting and other initiatives. National Academy for State Health Policy, Portland
8. Leape LL (2002) Reporting of adverse events. N Engl J Med 347(20):1633–1638
9. Stemler, S (2004) A comparison of consensus, consistency and measurement approaches to estimating interrater reliability. Practical Assessment, Research and Evaluation (2004)
10. Gliklich R, Dreyer N (2014) Registries for evaluating patient outcomes: a user's guide. Agency for Healthcare Research and Quality, Rockville

Safe Transitions of Care: A Participatory Human Factors Approach for Improving Safety in the Communication of Healthcare Organizations

G. Toccafondi[1(✉)], S. Albolino[1], T. Bellandi[4], A. Savelli[2], G. Frangioni[2], O. Elisei[4], M. Baroni[5], and A. Molisso[3]

[1] Clinical Risk Management and Patient Safety Centre Tuscany Region - WHO Collaborating Centre in Human Factors and Communication for the Delivery of Safe and Quality Care, Villa La Quiete Alle Montalve - via Pietro Dazzi 1, 50141 Florence, Italy
toccafondig@aou-careggi.toscana.it
[2] AOU Meyer Pediatric Hospital, Regional Health Service of Tuscany, Florence, Italy
[3] Tuscany Centre Trust, Regional Health Service of Tuscany, Florence, Italy
[4] Tuscany North West Trust, Regional Health Service of Tuscany, Lucca, Italy
[5] FTGM Hospital, Regional Health Service of Tuscany, Massa, Italy

Abstract. Care transitions are critical moments which may expose patients to adverse events and generate organizational failures. Ineffective care transition processes lead to higher hospital readmission rates and costs and patients can be harmed when the many moving parts of their care process are not effectively coordinated.

The human factor and patient safety approach can provide effective methodologies for the design of tools to improve the ability of health care workers to make available key information at the right time, ensuring patient safety and continuity of the clinical pathway.

In order to unveil what promotes or hinders effective communications at care transitions we involved health care workers of 10 dyads of inpatient care units (250 operators accounting for 1500 care transitions) in an action research process. The aim was to endow the participants with the skills necessary for evaluating the organizational context in which the handovers occur and give them support in prompting the interventions for constructing an organizational context underpinning safer communications at care transitions.

In particular through the application of the FMEA technique the highest priority of interventions have been assigned to 7 pitfalls which need to be taken into account in order to amplify the capability of organizations to implement the handover patient safety practice and fruitfully maintain it. Communication at care transitions is a fundamental testbed for the resilience of complex healthcare organizations. We attempt to increase the safety of communication during care transitions in order to allow healthcare organization to sustain required operations, in the presence of continuous stress. To achieve that we tried to endow the healthcare workers with the methodological tools for analyzing the current situations and adapt it in order to embrace the handover patient safety practice.

S. Bagnara et al. (Eds.): IEA 2018, AISC 818, pp. 774–780, 2019.
https://doi.org/10.1007/978-3-319-96098-2_95

Keywords: Handover · Care transitions · Human factors

1 Introduction

Handover is "the transfer of professional responsibility and accountability for some or all aspects of care for a patient, or group of patients, to another person or professional group on a temporary or permanent basis" [1]. The handover is then a communicative process aiming to achieve effective continuity of clinical care within the patient care pathway, regardless of when the transitions occur. In recent years the complexity of the health care system have been growing rapidly and health care team are distributed in time and space. From the patient perspective is getting more and more difficult to identify a single healthcare professional in charge for the process of care. Care coordination is not anymore a complementary element but instead a backbone of clinical activity which need to be mindfully designed and maintained.

In general, care transitions are critical moments which may expose patients to adverse events and generate organizational failures. Ineffective care transition processes lead to higher hospital readmission rates and costs and patients can be harmed when the many moving parts of their care process are not effectively coordinated [3, 8].

2 Purpose

The contemporary distributed structure of health care system are making necessary the construction of an ecosystem that enables the sharing of information [4]. The World Health Organization (WHO) has identified the handover as a key process for patient safety and the Alliance for Patient Safety released an important document "Communication During patient Handover" [13], in which is explicitly stated that the transfer of information about a patient from one healthcare practitioner to another, between medical teams and from health care practitioners to the patient and to the caregiver, should ensure patient safety and continuity of care.

Communication and cognition are imbricated activities. Human mind evolved for communicative purposes and the human cognition is an eminently social process deeply superimposed on our communication skills [12]. Consequently human communication can be hardly studied without considering the social context in which communication occurs. In the healthcare sector human factors and ergonomics propose a set of methods which can be used to study the context in which activities occurs. The main focus of HFE is the interactions between humans and the elements of a systems. HFE is fundamentally endowed with a contextual approach which turns to be useful for accounting of communicative events and to underpin the redesign of workflows. In order to scan the critical elements of care transitions [6] the analysis of handover documentation alone does not account for important contextual elements which needs to be considered in order to improve the safety of care transitions. The direct observation of handovers is then particularly important to make explicit contextual elements which may be otherwise taken for granted. Such insight regarding the contextual nature of handover has been confirmed by the research performed in the last years [2]. There are two important aspects that emerged as far as safety in care transitions. Handover, as every human communicative endeavor, is characterized by interdependence; that is the joint participation of the social actors taking part in the communication activities. Therefore the analysis should include the way in which the dyads of actors interacts. Moreover it should consider the various type of media used to support such interactions (written and verbal communication mediated by ICT, paper, face-to-face). In this regards the aim of the pilot study which originated the handover patient safety practice (PSP) of Tuscany region [11] was to evaluate the communicative ability enabled by the social context during cross-unit handover between high and medium intensity of care. The study was intended to investigate if the sharing of a common conceptual ground could reduce the potential threats to patient safety. The study traced the transfers of information and responsibility for 11 patients between two chosen settings (an intensive care units as a source and a sub-intensive therapy as a receiving unit), for a total of 22 observations in 2 hospitals, with the aim of assessing the level of information sharing between the two units involved and the communication tool adopted. The overall analysis of data and the findings from the focus groups detected discontinuities in the information transferred between health care practitioners which in some cases were connected with a lack of common ground in communication. It is suggested that handovers are most of the time unidirectional, with the sender's vision as a primary scaffold of the communication, rather than being bidirectional shared endeavors. An important aspect highlighted by this study was the different way in which doctors and nurses are participating to the handover process. In fact, while doctors participate to direct handover (face-to-face or on the phone) and have the opportunities to lay the foundations for the creation of common ground, nurses remain outside of these dynamics. The second important aspect regards organizational ergonomics. The handover is a communication process located in a context that involves many actors, teams and micro-organizational systems. Despite the attention given to the handoff moments - the "here and now" of communication among healthcare workers - the overarching workflows, which encompasses the singles handoffs moments and the interactions with other processes, needs to be carefully assessed in order to amplify the benefits of the standardized communication of relevant cues for patient care.

In particular, patient transitions among different settings, such as the step down from critical care to medical care or the transfer from emergency department to the medical wards, constitute critical touchpoints whose safety and effectiveness relies both on well-structured handover and on the way in which the wider workflows unfolds. We argue that the latent factors crystallized into the history of the organization needs to be elicited and considered in order to maximize the introduction of handover patient safety practice given the profound link between communication and the organizational context in which communication occurs.

Under this perspective the recent work carried out [9] outlined the correlations between safety and the way in which the handover was integrated in the whole health care organization. It was investigated the correlation between the effectiveness of handover and the occurrence of adverse events. The study aimed to determine whether the introduction of a handover bundle improved quality and safety. In particular the bundle was composed of the following elements: standard tool for verbal Handover, constitution of an integrated team (nurses and physicians), regular supervision, introduction of a computerized system, reducing interruptions during the Handover. In particular, the study tried to correlate the introduction of the bundle with the reduction of medical errors and preventable adverse events, with a reduction of the loss of information in the written documentation and improving the verbal handover. The results shown that after the implementation of the bundle errors decreased from 33.8% to 18.3%. preventable adverse events decreased from 3.3% to 1.5%. Moreover It was detected a decrease in the losses of key information in written handover.

3 Methods and Materials

The human factor and patient safety can provide effective methodologies for the design of tools to improve the ability of health care workers to make available key information at the right time, ensuring patient safety and continuity of the clinical pathway.

In order to unveil what promotes or hinders effective communications at care transitions we involved health care workers of 10 dyads of inpatient care units (250 operators accounting for 1500 care transitions) in an action research process. The aim was to endow the participants with the skills necessary for evaluating the organizational context in which the handovers occur and give them support in prompting the interventions for constructing an organizational context underpinning safer communications at care transitions. The project started with construction of an handover team consisting of a doctor and a nurse for each dyad involved (sender and receiving unit). In each setting a clinical risk management team trained in human factors had the role of detecting the base-line of handover through a wizard-based data collection of process and outcome indicators about continuity of treatments at care transitions. The practitioner's team supervised by clinical risk manager was responsible for the definition of a minimum set of handover information shared between the dyads, that suited both written and verbal handover, and which integrated the medical and nursing staff.

The selected handover local groups of health care workers have been trained to use a toolbox of instruments – the FMEA (Fault Modalities and Effect Analysis), the value stream map and the flow matrix – to evaluate how the actual organizational context could respond to the implementation of the handover patient safety practice [14]. The intervention was articulated in two steps. Initially the local groups addressed the organizational latent factors preventing the implementation of a seamless and effective handover during patient care transitions. Secondly the group worked on the implementation of the requirements of the patient safety practice: adopting a standardized format (e.g. SBAR - Situation Background Assessment Recommendation), using both oral and written channels in order to create redundant handover systems, involving disciplines and professions in the handover process, enabling the participation of patient and the family caregiver before the transitions occur and the definition of training sessions for the all the health care workers [10].

The implementation of the requirements and the analysis of the wider workflows are part of the interventions carried out by the local handover groups. The latent factors have been addressed and considered in order maximize the capacity of the organization to integrate standardized handover into existing workflows by means of tailored handovers adaptations to local needs. Moreover the impact on the intervention was be measured before and after using a three pronged approach consisting in (a) the compliance to the handover patient safety practice requirements (b) the extent to which continuity of care is guaranteed throughout the transactions between the dyads of healthcare unit (c) the perception of the health care workers regarding the handover process before and after the intervention.

4 Results

Before and after the intervention the clinical risk managers gathered the data related to the process and outcome indicators of handover by means of the structured review of medical records the observation of transitions of patients and a questionnaires to the professionals involved. For the review of medical records it was used a dedicated tool that considered the data in the health care documentation until 72 h after the transition occurred. The planned sample size is of 200 patients transferred to each participating unit before the interventions and 200 patients transferred to 3 months after the start of the intervention. Direct observation of the transitions was conducted on 30 patients before the intervention and at a distance of 30 patients 3 months after the intervention.

The questionnaire for measuring the satisfaction of healthcare workers regarding handover was given to doctors and nurses in the source and receiver units before the interventions and again at 3 months after the intervention.

The planned statistical analyzes are descriptive, with an examination of the significance of changes in the endpoint before and after the local intervention, using both linear regression analysis of covariance techniques to evaluate the possible impact of the individual instruments of the handover patient safety practice.

In this section we will show the outcome of the analysis of the latent factors carried out by the local handover groups by means of the methodological toolbox. In particular through the application of the FMEA technique the highest priority of interventions have been assigned to 7 pitfalls which need to be taken into account in order to amplify the capability of organizations to implement the handover patient safety practice and fruitfully maintain it (Table 1).

Table 1. Pitfalls of handover implementation

Priorities of intervention
Continuity of information – communication network
Patient participation and patient involvement
Integrated workflow
Team working and common ground construction
Tailoring of handover PSP to the local context
Management of beds and human resource management
Support from general management

Many handover groups identified the access to a stable network of communication lines as enabling factors allowing for timely communication and exchange of information among health care workers. The presence of a reliable communications system is considered as relevant as electronic health care records for accessing to time-sensitive patient information.

The bed management, especially in the transitions from emergency department to medical wards, may absorb many of the time healthcare workers spend in communicating and may compress the time allocated for handover. The lack of an inter-professional workflow supporting the shared responsibilities of physicians and nurses in the care transitions is considered crucial for the success of handover process.

Communication at care transitions is a fundamental testbed for the resilience of complex healthcare organizations [4, 7]. We attempt to increase the safety of communication during care transitions in order to allow healthcare organization to sustain required operations, in the presence of continuous stress. To achieve that we tried to endow the healthcare workers with the methodological tools for analyzing the current situations and adapt it in order to embrace the handover patient safety practice.

The absence of designed communication strategies at care transitions may lead to an increase in the duration of patients' stays, claims, and costs. Standardized ways for constructing patient handover has revealed effective in reducing the numbers of errors and potential adverse event [3]. Moreover communication research has demonstrated the profound symbiotic relationship between communication and the social context in which communications occurs [6]. Introducing from above constraints on communication performance disregarding the wider organizational context may lead to scarce implementation results. We focused on the way in which the handover practices are integrated and supported by the pre-existing workflows in order to enhance the capacity of the organization to retain organizational changes and improve communication safety.

References

1. Wong MC, Yee KC, Turner P (2008) Clinical Handover Literature Review. eHealth Services Research Group University of Tasmania, Australia, pp 1–114
2. Abraham J, Kannampallil T, Patel VL (2014) A systematic review of the literature on the evaluation of handoff tools: implications for research and practice. J Am Med Inform Assoc 21(1):154–162
3. Arora V, Johnson J, Lovinger D et al (2005) Communication failures in patient sign-out and suggestions for improvement: a critical incident analysis. Qual Saf Healthcare 14:401–407
4. Bergman AA, Flanagan ME, Ebright PR et al (2015) Mr Smith's been our problem child today…: anticipatory management communication (AMC) in VA end-of-shift medicine and nursing handoffs. BMJ Qual Saf 25(2):84–91
5. Bomba D, Prakash R (2005) A description of Handover processes in Australian public hospital. Aust Health Rev 29:68–79
6. Catchpole KR, De Leval MR, Mc Ewan A et al (2007) Patient handover from surgery to intensive care: using Formula 1 pit-stop and aviation models to improve safety and quality. Paediatr Anaest 17:470–478
7. Hollnagel E, Leonhardt J, Licu T et al (2013) From safety-I to safety-II: a white paper Eurocontrol–European Organisation for the Safety of Air Navigation

8. Hesselink G, Flink M, Olsson M et al (2012) Are patients discharged with care? A qualitative study of perceptions and experiences of patients, family members and care providers. Bmj Qual Saf Suppl 1:i39–i49
9. Starmer AJ, Spector ND, Srivastava R et al (2014) Changes in medical errors after implementation of a handoff program. N Engl J Med 371:1803–1812
10. Jeffcott SA, Ibrahim JE, Cameron PA (2009) Resilience in healthcare and clinical handover. Qual Saf Health Care 18:256–260
11. Toccafondi G, Albolino S, Tartaglia R et al (2012) The collaborative communication model for patient handover at the interface between high-acuity and low-acuity care. BMJ Qual Saf 21:58–66
12. Tomasello M (2008) The origins of human communication. MIT Press, Cambridge
13. WHO Collaborating Centre for Patient Safety, (2007) Patient Safety Solutions, vol 1, solution 3
14. Potts H et al (2014) Assessing the validity of prospective hazard analysis methods: a comparison of two techniques. BMC Health Serv Res 14:41

Author Index

S. Bagnara et al. (Eds.): IEA 2018, AISC 818, pp. 781–785, 2019.
https://doi.org/10.1007/978-3-319-96098-2

Printed in the United States
By Bookmasters